MW00716021

Sets

$\{\ \}$	(to denote a set)	1
$\ldots$	and so on	2
$\mathbf{N}$	set of all positive integers	2
$\mathbf{Z}$	set of all integers	2
$\mathbf{Q}$	set of all rational numbers	2
$\mathbf{R}$	set of all real numbers	2
$\in$	is an element of	3
$\notin$	is not an element of	4
$\subseteq\ \not\subseteq$	is [is not] a subset of	4
$A = B$	equality between sets A, B	5
$\sqcup$	shoebox	10
$\cup$	union	10
$\cap$	intersection	11
$\bigcup, \bigcap$	union operator, intersection operator	12
$A - B, A \backslash B$	relative complement of B in A	16
$\overline{A}$	complement of set A	17
$A \times B$	Cartesian product of sets A and B	19
(a, b)	ordered pair	19
$\mathcal{P}(X)$	the power set of X	21
$(a_1, \ldots, a_n)$	ordered n-tuple	26
$A \Delta B$	symmetric difference of sets, A, B	29

Propositional Logic

$\sim$	*not*	34
$\wedge$	*and*	34
$\vee$	*or*	34
$\longrightarrow$	*if ... then ...*	34
xor	*exclusive or*	36
$\longleftrightarrow$	*if and only if*	37
iff	*if and only if*	37
$\Longrightarrow$	logical implication for prop. formulas	41
$\Longleftrightarrow$	logical equivalence for prop. formulas	43
$T(f)$	truth-set of f	45–47
$F(f)$	falsity-set of f	45–47
$\downarrow$	Peirce arrow (*nor*)	64
$\mid$	Sheffer stroke (*nand*)	64

Predicate Logic

$P(x)$	predicate	70–71
D_x	domain of x	70
$\{x;\ x \in D_x,\ P(x)\}$	set-builder notation	72–73
$T_p = T(P(x))$	truth set of $P(x)$	73
$\Longrightarrow$	logical implication for predicates	77

Symbol	Meaning	Page
$\Leftrightarrow$	logical equivalence for predicates	77
$\forall$	for all	78
$\exists$	for some	79
$\bigwedge$	big *and*	80
$\bigvee$	big *or*	80

Functions

Symbol	Meaning	Page
$f : X \longrightarrow Y$	function from X to Y	161
$\widehat{f}$	function from $\mathcal{P}(X)$ to $\mathcal{P}(Y)$	164
max		27, 173
min		27, 173
$\widehat{f}^{-1}$	an inverse of $\widehat{f}$	176
f^{-1}	an inverse of f	350
$g \circ f$	composition of f and g	179
i_X	identity function on X	181
f^{-1}	inverse of bijective f	183
$f\|_E$	f restricted to E	184
$(1, 2, 3)(4, 5)$	disjoint cycle notation for permutations	191
f_A	characteristic function of A	192
$\sim$	is asymptotic to	203
$f = O(g)$	f is big oh of g	204
$f = \Theta(g)$	$f = O(g)$ and $g = O(f)$	206
$n!$	n factorial	126, 203
$\|t\|$	absolute value of t	217
$\lfloor t \rfloor$	floor of t	217
$\lceil t \rceil$	ceiling of t	217
$\gcd(a, b)$	greatest common divisor of a and b	227–228
$\operatorname{lcm}(a, b)$	least common multiple of a and b	242
φ	Euler phi function	283
$\binom{n}{k}$	binomial coefficient "n choose k"	304
$\|X\|$	number of elements in set X	309
Pr	probability function	373
	probability distribution	377
Ef	expected value of f	378
$\operatorname{Var} f$	variance of f	379
$D_b(\ ) = D_b(n, p;\ )$	binomial distribution	389

Relations on **Z** *and Associated Sets*

Symbol	Meaning	Page
$b \mid a$	b divides a	216
$b \nmid a$	b does not divide a	245, 471
$\equiv$	congruence	252
$[a]_m$	congruence class of a mod m	253
$\mathbf{Z}_m$	the set of all congruence classes mod m	253

(Continued on inside of back cover)

Discrete Mathematics with Applications

Discrete Mathematics with Applications

H. F. Mattson, Jr.

John Wiley & Sons, Inc.
New York · Chichester · Brisbane · Toronto · Singapore

Acquisitions Editor Barbara Holland
Marketing Manager Susan Elbe
Copyediting Supervisor Richard Blander
Production Manager Lucille Buonocore
Senior Production Supervisor Savoula Amanatidis
Text Designer Sheila Granda
Cover Design Madelyn Lesure
Cover Photo Kit D. Walling/Index Stock International, Inc.
Illustration Coordinator Sigmund Malinowski
Manufacturing Manager Susan Stetzer

This book was set in Palatino by Publication Services, Inc.

Recognizing the importance of preserving what has been written, it is a policy of John Wiley & Sons, Inc. to have books of enduring value published in the United States printed on acid-free paper, and we exert our best efforts to that end.

Library of Congress Cataloging in Publication Data:

Mattson, H. F. (Harold F.). 1930–
Discrete mathematics with applications / by H.F. Mattson, Jr.
p. cm.
Includes index.

1. Computer science–Mathematics. I. Title.
QA76.9.M35M48 1933 92-33772
004'.01'51–dc20 CIP

Printed in Singapore.

10 9 8 7 6 5 4 3 2 1

To Jeanette, David, and Jennifer

To the Student

What does it mean to think mathematically? In this text I've tried to interest you in doing that. And I've tried to show you how.

The purpose of this book is to introduce you to the language and techniques of discrete mathematics. If the book succeeds you'll be well started in mathematics or computer science—both use the same language and much of the same kind of thinking. A good portion of it is useful in electrical or computer engineering as well.

Discrete

What does *discrete* mean?

Answer: The opposite of *continuous*. The integers are *discrete*. The real numbers are continuous. The thing considered is discrete if it consists of a finite number of irreducible parts. Certain infinite sets, like the integers, are also considered discrete. But we don't really try to define *discrete* as a general idea. We simply present the topics here—graphs, for example—and remark that they are a part of discrete mathematics.

Discrete mathematics has grown enormously in recent decades. Part of the reason is that digital computers work *discretely*. A computer does not represent all real numbers, but a large finite set of real numbers. It adds, compares, divides, and so forth, within that set. Careful study of what a computer does requires discrete math. (You'll see such uses of discrete math in this book.) Discrete math also has applications in business, social sciences, electrical and computer engineering, and cryptography, among other areas.

If you are majoring in mathematics or computer science you'll need to learn most of the topics in this book. If you're in electrical or computer engineering you'll find some topics useful; the rest will broaden you.

Some topics are not solely discrete. *Set* and *function*, for example, are ideas used throughout mathematics and computer science; so are *equivalence relation* and *partial order*. The principles of logical inference govern in all areas. But the rest of the topics are discrete; and we need sets, functions, relations, and—certainly—logic to discuss them.

Working Problems

You can have some fun working the problems in this book. Usually you'll need to read the text first. If you already know a topic, just skim it to make sure I haven't sneaked in any curve balls. Working problems will cement your knowledge of chapters. It's all very well to nod "yes" at an idea; the way to understand it thoroughly is to do some problems involving it.

There's a secret to working a tough problem: Try it repeatedly between periods of other activity, like sleeping. Go as far as you can, quit, and see what happens the next morning. Often you'll have made progress, "by osmosis."†

So my advice is to work a little bit every day on this subject rather than a lot just before a deadline. You may spend the same total amount of time either way; but you learn far more, or find better solutions to problems, if you spread out your efforts. Why? Because you'll have more periods of osmosis. You'll also get more out of class.

Special Notations

I usually use the notation $:=$ to define things, or sometimes to assign a value to a variable. Thus "$x := a$" means "x is defined to be a," or, depending on the context, it may mean "x is set to the value a." I may even use it backwards. Thus "$a =: x$" means "$x := a$." Sometimes I use $=$ instead of $:=$.

Three dots making the vertices of a triangle $\therefore$ stand for therefore.

CAUTION

A caution sign in the margin means "Be careful—this part can be tricky."

The symbol "$\doteq$" means "is approximately equal to." For example $\pi \doteq 3.14$. We also indicate roundoff of decimals by, for example, "$\pi = 3.14+$," or "$\pi = 3.1416-$."

When a term is first defined it is **boldfaced**.

Mathematical symbols defined in the text are listed inside the front cover.

The end of the statement of a Theorem, Lemma, Remark, etc., is signaled either by the word ***Proof*** in the next line, or by the symbol ■.

Numbering System

Sections within chapters are numbered consecutively. Inside each section there is just one sequence of numbers for theorems, displayed formulas, figures, examples, practice problems, even some paragraphs. But figures carry an F after their number, and sometimes that number is the same as the number on something else nearby.

The numbers look like this: $(s.x)$, or for figures, $(s.xF)$, where s is the number of the section, and x is usually a positive integer. (Once in a while x is broken down into, for instance, 6.1, as in the Dewey decimal system, when it refers to something strongly subsidiary to what the "6" refers to.)

For example, in Chapter 3, Section 16, there are five formulas numbered (16.1) to (16.5), followed by a paragraph headed (16.6), a figure (16.6*F*), a displayed statement (16.7), two figures (16.7*F*) and (16.8*F*), two more formulas (16.9) and (16.10), and a practice problem (16.11P).

One exception: the problems at the ends of the chapters are numbered separately. Except in Chapter 3, problems are grouped according to the section they pertain to and are numbered roughly in order of increasing difficulty, for example, 1.1, 1.2, 1.3,... for Section (1).

†You could ask your psychology professor why this approach works. I think the reason is that, as my colleague Dr. Gustave Solomon once said, the unconscious is faster than the conscious.

An asterisk before a problem number, for instance, *3.7, means that problem is more difficult than those not so marked. Some chapters have *General* problems, numbered G1, G2, and so on. Some have test questions, numbered T1, T2, and so forth, from in-class tests. They are not ordered according to difficulty.

#: This symbol before a problem number means the result of the problem is used in the text.

Ans: 3.7^{Ans} for instance means that an answer to problem 3.7 appears at the end of the book under Answers to Selected Problems.

Note: A problem that mentions say, (11.2), is referring to (11.2) of the text, not to a problem **11.2**, which is boldfaced and carries no parentheses.

Practice Problems

Larded into the text of each chapter you'll find several practice problems. The number of a practice problem carries a "P," as "(11.7P)." They vary in difficulty, but most should be within your grasp. I've put the answer to almost every one of them at the end of the chapter in which it appears. You will learn more if you try hard on these problems, even waiting until the next day before looking up answers to the tough ones. But I hope you will read the answers eventually, because some of them expand on the topic of the problem.

Abbreviations and Their Translations

ca.	approximately the same date as (*circa*)
et al.	and others
ff	and following
i.e.	that is
QED	that which was to be proved (*quod erat demonstrandum*)

Greek letters are listed with their English names under "alphabet" in dictionaries.

To the Teacher

This book is aimed at second- and third-year students in colleges and universities. Most of them will be majoring in computer science or mathematics, some perhaps in electrical or computer engineering. I suggest that a semester of calculus is a reasonable prerequisite, if only for the elusive mathematical maturity that it gives the student. A trifle of differentiation appears in Chapter 8 ff., and some integration in Chapters 11 and 13. This book grew out of a two-semester course I've taught for about 10 years at Syracuse University. It is required for computer science students. The first semester is required for math majors.

Topics

I chose the topics by consulting with my colleagues at Syracuse University and with guidance from the 1986 report of the Committee on Discrete Mathematics in the First Two Years of the Mathematical Association of America. The topics and themes constitute most of those recommended by that committee. The committee unanimously agreed that a course in discrete structures should be a mathematics course and that it should be a one-year course.

Almost every assertion in the book receives a careful, though usually informal, proof. To the extent that you wish, you may emphasize the idea of proof, for example, by discussing the proofs in the text and by assigning problems calling for proof. The many other problems, ranging from the routine to those that challenge your students' mathematical insight, allow you considerable leeway in designing the course.

Note: Some problems with more than one part can be long. You may wish to assign only some of the parts for those.

After the idea of proof, induction is the most recurrent theme in the book. Following its first appearance in Chapter 3, every chapter except Chapter 4 appeals to it, at least in the problem section. Recursion and algorithms appear in several places, but algorithmic thinking is less emphasized than more traditional mathematical approaches.

I've tried to give a substantial introduction to the topics in this book without writing a treatise on any of them. Formal languages, for example, appear briefly in Chapter 14, as an application of graphs. Students of mathematics don't need to know more of that topic than is here, and those in computer science may elect a whole course in it.

What to Include

My experience is that students at this level in the U.S. need to be taught almost all the material in Chapters 1–5. After that the teacher has a great deal of choice. I like to include at least the beginnings of Chapters 6 and 7. They are good topics for students of computer science, and necessary ones for those math majors who won't have any more number theory. If I answer in class lots of questions about homework problems (which I assign for each of the three days a week that the class meets), I may end the first semester with Chapter 5.

In the second semester I've always taught Chapters 8, 9, and 12 (binomial theorem, counting, and matrices and order relations). Then I choose any other chapters (usually two) as time permits. Chapters 10–14 are independent of each other.

I mention my choices simply to give some concrete examples of possibilities. I've taught each chapter once (Chapters 10, 11, 14), or many times (the others). Several times I've even taught two nonexistent chapters (coding and groups).

Realistically, a year course would complete 10 chapters, maybe as many as 12, maybe only 9. It depends entirely on your aims and on how quickly you think your students can learn the subject to your satisfaction. For help in choosing, please see Figure (0.1*F*) below.

Dependence of Chapters

Chapters 1–5 are basic to the rest of the book. Figure (0.1*F*) shows the dependence relation of the chapters:

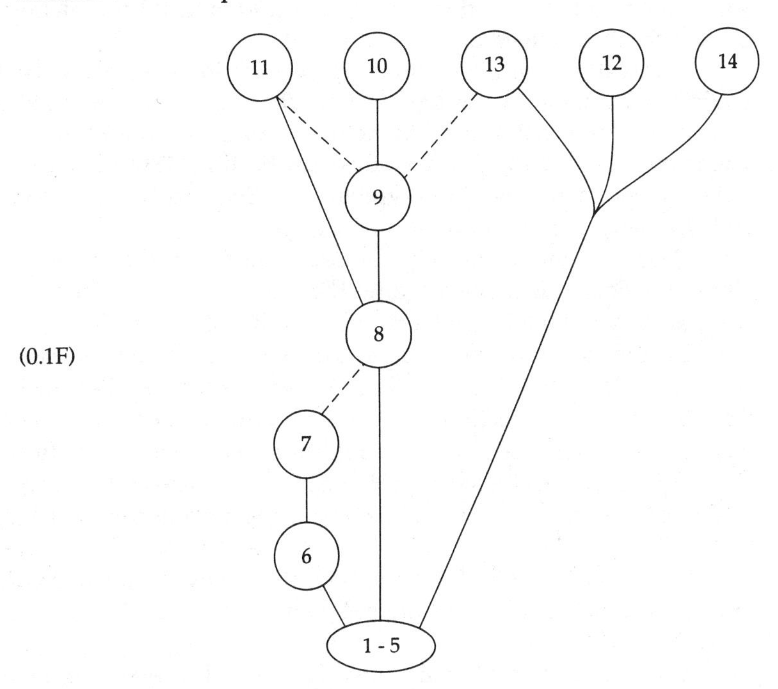

(0.1F)

A dashed line means light dependence, as in only one section, or a few problems.

What to Omit

Within most chapters it's possible to omit parts and still be able to read any later chapter. Here are suggestions.

I usually omit all the material between blue triangles (▸ ◂), except for disjunctive normal form in Chapter 2. I also omit the proof of the substitution theorem in Chapter 2, Part I.

I think all of Chapter 3, on induction, is necessary. I sometimes omit refinements from Chapter 4. It's possible to omit Propositions Left and Right from Chapter 5, and permutations, too. You could also omit Sections (19)–(21), but (19) is needed for Chapter 10.

Minimal coverage of Chapter 6 would stop with Section (2), the division algorithm. It would be nice to go through Section (4), however. Fuller coverage would go through Section (8), the Euclidean algorithm, including algorithm EE or not, as taste and time dictate. Although the pouring problems of (10) are fun, I omit them for lack of time. You could even omit Section (11), on primes; if you include it, the sections following it are optional.

Minimal coverage of Chapter 7 goes through Section (4); the examples there are so simple that you could easily include them. You might omit Section (2), but math students ought to know that computers use mod m adders, and computer science students need to see the cleanness of a mathematical description of 2's-complement, and of the simple motivation for it. (You could omit 1's-complement if you wish.) For this much, all you need of Chapter 6 is (2).

The next stopping point in Chapter 7 would be the end of Section (6), for which you'd need Chapter 6 through Section (9). After that you could stop at the end of Section (8). After that you could include or omit Section (9), and Section (10) as well. Then you could present Fermat's theorem, Section (11) and, if you like, continue with Euler's theorem and the RSA cryptosystem, Sections (12) and (13). Sections (11), (12), and (13) depend on Section (11) of Chapter 6.

Chapter 8 goes fairly fast. I present it all except that a few times I have omitted zipper merge; I almost always omit the next application. There is not much point in stopping short of the binomial theorem itself, in Section (7).

Chapter 9, on counting, is one of the more difficult because, when trying to solve a problem, students often don't know which counting principle to use. I usually teach all of it except Section (8). A minimal treatment would go through Section (4), inclusion-exclusion. Section (5) would not take much more effort, especially if you omit surjections. The birthday problem is nice to present; I've never failed to find two students in the class with the same birthday even when the number present was less than 23.

In Section (7), sets with repetition, you'd need to present only one of the three proofs. In sum, though I recommend going through Section (7), you could stop after (4) or (5).

Chapter 10, on probability, goes fairly quickly. A subminimal treatment would stop after Section (4), but I suggest going at least through Section (10). A next stopping place is Section (14). You could omit the queueing of Sections (15) and (16) if you wish.

Chapter 11, on recurrence relations, lends itself to a blitzkrieg approach: omit the proofs. Or you could give the general idea of the proofs without the details. This chapter depends somewhat on Chapter 8.

Chapter 12, on matrices and order relations, is more elementary than several prior chapters. I find that students at this level need to be told how to multiply matrices. I often include all sections, even Section (15). You could omit Sections (19)–(22), but I've never done so.

Chapter 13, on rooted, ordered trees, is the closest chapter to computer science. It moves rapidly. You might omit Section (7), or the analysis in that section. You could even omit Sections (8)–(10). I have always enjoyed counting binary trees in Section (11), however. Often it's the last lecture of the course. That section depends a bit on Chapter 9, and is a slight substitute for Chapter 11 if you've omitted it.

Chapter 14, on graphs, is not difficult. You could omit Section (9), on interval graphs, and Section (12). Although Section (13), a simple introduction to formal languages and the pumping lemma, is omissible, I recommend including it for its stretch of students' logical skills. Section (14), showing how graph theory was used to derive a workable algorithm for computerized tomography, is of course omissible, but students might like to see it.

A **one-semester course** could consist of Chapters 1–5 done briskly, and selections from Chapters 6, 8, 12, 13, or 14. Although Chapter 12 might be the most essential choice, you could actually select parts of Chapters 6 and 7, or of Chapters 8 and 9, even of Chapter 11—anything but Chapter 10, since it depends too heavily on Chapters 8 and 9.

If your class is advanced, you could assign Chapters 1–5 as reading and go through the entire book in one semester. Such an approach might fit a class of first-year graduate students in computer science who needed to solidify their discrete mathematical background.

Acknowledgments

My first thanks go to my wife, Jeanette, for cheerfully putting up with my long hours and periods of abstraction. I am also grateful to the many students who have used earlier versions of this book at Syracuse. Their comments have led to several improvements. I particularly thank Denise Moriarty for suggesting that I include answers to the Practice Problems. I also thank the many graduate students who have assisted me in the course.

I am grateful to the many people who have helped me with this book. My greatest thanks go to Lockwood Morris, whose ideas and precision have influenced much of this book. He has been the reliable source to whom I turned repeatedly for guidance throughout the years of work on this book. He has also read the entire manuscript with great care, suggesting many corrections and improvements, particularly to the Problems and Answers.

Colleagues who taught from earlier versions and gave me valuable comments are A. Deleanu, V. Fatica, J. Kwiatkowski, L. G. Lewis, Jr. K. Mehrotra, and M. E. Watkins. My warm thanks go to them.

Other colleagues suggested topics, served as expert advisers, or read chapters and suggested changes. They are E. F. Assmus, Jr., R. M. Blumenthal, N. J. McCracken, K. Jabbour, H. Janwa, T. E. Patton, V. Pitchumani, E. F. Storm, D. E. Troeger. I thank them for their excellent help. I also thank Jeff Pepper of Osborne McGraw-Hill for kindly commenting on one of the chapters.

The different perspectives of the several reviewers for John Wiley & Sons, Inc., have helped to shape this text. Their specific comments enhanceed its clarity and enabled me to avoid many errors. They are Professor Charles Marshall, SUNY-Potsdam, whom I want to thank particularly; Professor Alfred Patrick, Adirondack Community College; Professor Stoyanka Zalateva, Boston University; Professor Sung Y. Song, Iowa State University; Professor Richard Sot, University of Nevada-Reno; Professor Hudson V. Kronk, SUNY-Binghamton; Professor Ken Plochinski, University of Michigan; and Professor Dorothy Blum, Millersville University.

I am also grateful to David and Jane Otte and to Dianne Borawski for valuable advice and encouragement.

Syracuse University has provided support. I particularly thank W. Semon, E. E. Sibert, G. Frieder, and C. R. P. Hartmann for that support and for their valuable suggestions on content.

I thank Elaine Weinman for applying her superb skills at TeX to this manuscript over so many years. I am also grateful to D. Steinberg and R. Gonda for early typesetting help. K. Purdy provided excellent editorial assistance, and Ecole Nationale Supérieure des Télécommunications hospitality and support, in the end game.

It has been a pleasure to work with Bob Pirtle, Valerie Hunter, Bob Macek, and Barbara Holland at John Wiley & Sons, and with Marlo Welshons and Shelley Clubb at Publication Services.

I welcome your comments on this book. If you find typographical or more serious errors, and especially places that are hard to understand, I'd like to hear from you.

Finally, I tell you, on advice of counsel, that some but not all of the quotes heading the chapters are bogus.

H. F. Mattson, Jr.
School of Computer and Information Science
4-116 Center for Science & Technology
Syracuse, New York 13244-4100

Contents

CHAPTER 1. *Sets*

CHAPTER 2. *Logic*

CHAPTER 4. *Equivalence Relations and Partitions*

CHAPTER 5. *Functions*

CHAPTER 7. *Congruences*

CHAPTER 8. *The Binomial Theorem*

CHAPTER 9. *Counting*

CHAPTER 10. *Probability*

CHAPTER

1

Sets

A set is known by its elements.
A. M. Gleason

1. Beginnings

Set is the most basic term in mathematics and computer science. Hardly any discussion in either subject can proceed without *set* or some synonym such as *class* or *collection*.

The term *set* is sometimes used in everyday speech in much the same sense as in mathematics and computer science: The library has a set of the *Encyclopaedia Britannica*; Arnold Palmer has several sets of golf clubs.

But what *is* a set in mathematics or computer science? Strictly speaking, it is an undefined term! Since we don't want to be *too* strict, let's explore the term *set* intuitively. We'll start with (1.1), an admittedly vague statement. In the next few pages we'll extend it, clarify it, and show how to work with sets.

(1.1) A **set** is a collection of objects of a mathematical nature that is defined precisely enough so that we can in principle determine whether any given mathematical object is or is not an object in that set.

DEFINITION

The objects of a set are called the **elements**, or the **points**, or the **members** of the set.

(1.2) FIRST NOTATION FOR SETS: the **list** notation. We sometimes denote a set by listing its elements, separated by commas, between braces (curly brackets). Different orderings of the elements in list notation, or repetitions of some of the elements (which may occur in different descriptions of the set) do not alter the set.

Examples (1.3) Here are some sets defined by the first notation for sets.

$$A_1 := \{1,2,3,4,5\}$$
$$A_2 := \{5,2,1,4,3\}$$
$$A_3 := \{1,1,1,2,3,4,4,5\}$$
$$B_1 := \{1,3,5\}$$
$$B_2 := \{1,5,3\}$$
$$B_3 := \{1,5,1,3,1,5,5,5\}$$

Because of the statement in (1.2) that different orderings or repetitions of the elements of a set don't alter the set, $A_1 = A_2 = A_3$ and $B_1 = B_2 = B_3$. In (4) we'll explain more fully why they are equal.

(1.4) Next we illustrate one of the most useful of all mathematical notations, the triple dot. We sometimes replace part of the list by three periods—when the omitted part is, or is supposed to be, obvious from the context. Thus the set

$$\{1,2,3,4,5,6,7,8,9,10\}$$

could be denoted as

(1.5) $$\{1,2,\ldots,10\}.$$

In using the list notation we *must* rely on the triple dot if the set is infinite. Thus the set **N** of all positive integers might be denoted

(1.6) $$\mathbf{N} = \{1,2,3,\ldots\},$$

and the set **Z** of all integers might be denoted

(1.7) $$\mathbf{Z} = \{0,\pm 1,\pm 2,\ldots\}.$$

We'll use the following notation throughout this book.

N:= the set of all positive integers

Z:= the set of all integers

Q:= the set of all rational numbers

R:= the set of all real numbers

Denoting **Q** and denoting **R** with the triple dot notation are, respectively, difficult and impossible. We assume you know enough about these sets so we may speak of them. We will not define them from scratch.

(1.8) We've slipped inexorably into the second notation for sets. Although we discuss it much more fully in Chapter 2, (19)ff, we express it here briefly and informally. We define a set using the second, or **set-builder**, notation by calling it the set of all objects having a specified property.

set - builder p.72

Examples (1.9) A_1 in (1.3) could be defined as the set of all integers between 1 and 5, inclusive. The "property" is: being an integer between 1 and 5 inclusive. A_1 is the set of *all* objects having that property.

Similarly, B_1 in (1.3) is the set of all odd integers between 1 and 5, inclusive. The property here is: being an odd integer between 1 and 5, inclusive. B_1 is the set of *all* objects with this property.

In (1.6) the property is this: being a positive integer. Thus **N** is the set of all objects with that property. In (1.7) the property is simply: being an integer. **Z** is the set of all objects with that property.

Examples (1.10) If health inspectors find a hepatitis carrier working at a restaurant, they would search for all people having the property that they ate at that restaurant during some specified period.

The set S of outputs for all possible inputs of a program that computes x^2 for the input x on a given computer is finite but too large to list. Questions of the smallest and largest numbers representable in the computer and of round-off errors make S hard to describe explicitly (i.e., hard to list), but S is perfectly well defined implicitly (i.e., by the set-builder notation). A number y is in S if and only if there is a number x admissible as an input to the computer such that the given squaring program produces y as output for the input x. (S should also have some words like *overflow* or *error* in it to allow for inputs that are too large or small or not in correct form.)

2. Membership

Now we introduce *membership*, a relation between objects and sets.

(2.1) If a set is defined via the list notation (1.2), let's say as

$$\{a_1, a_2, \ldots\},$$

then we say that each object or item in the list **is a member of** that set. Thus a_1 is a member of $\{a_1, a_2, \ldots\}$, a_2 is a member of $\{a_1, a_2, \ldots\}$, and so on for each item a_i in the list $a_1, a_2, \ldots$

$$\text{NOTATION: } a_1 \in \{a_1, a_2, \ldots\}$$
$$a_2 \in \{a_1, a_2, \ldots\}.$$

In general, if A is the name we give to the set $\{a_1, a_2, \ldots\}$, then for every item x in the list $a_1, a_2, \ldots$ we may write

$$x \in A.$$

Examples (2.2)

$$1 \in \{1,3,5\}$$
$$3 \in \{1,3,5\}$$
$$5 \in \{1,3,5\}$$
$$7 \in \{1,2,\ldots,10\}$$
$$-100 \in \{0, \pm 1, \pm 2, \ldots\}.$$

(2.3) If a set B is defined via the second notation (1.8), then we would say that the object x is a member of the set B, and write $x \in B$, if and only if x had the property used to define B.

Examples Suppose B is defined as the set of all even integers between 1 and 1001. If someone gave us an integer x, we could test x for evenness and for being between 1 and 1001. If it passed both tests, we'd say $x \in B$. For example, $x := 732 \in B$; $x := -12$ is not in B because -12 is not between 1 and 1001.

(2.4) *Nonmembership.* To express that an object y is not a member of the set A, we write

$$y \notin A.$$

Examples (2.5)

$$2 \notin \{1,3,5\}$$
$$-1 \notin \{1,3,5\}$$
$$-12 \notin \{1,2,\ldots,10\}$$
$$11 \notin \{1,2,\ldots,10\}$$
$$0 \notin \{1,2,\ldots,10\}$$

The symbol $\in$, reserved exclusively for membership, is a stylized epsilon, ϵ.

Let us agree to give names to sets and to confuse freely the name of a set with the set itself. We will often use capital letters for sets, as $A := \{1,2,3\}$, for example. Then we can say $1 \in A$, $2 \in A$, $3 \in A$, and $4 \notin A$.

3. Inclusion

We define *inclusion*, a relation between two sets.

DEFINITION (3.1) For all sets A and B, we say that A is **included** in B if and only if every member of A is a member of B.
NOTATION: $A \subseteq B$

We may also say "B includes A" and write $B \supseteq A$ for $A \subseteq B$.

In terms of the list notation (1.2), $A \subseteq B$ if and only if every item in the list used to denote A is an item in the list used to denote B. "Inclusion doesn't care" where in the lists the item may appear.

Example (3.2) $\{1,2,3,1,1\} \subseteq \{4,3,2,1\}$ since each of the five items in the list $1,2,3,1,1$ is an item in the list $4,3,2,1$.

The negation of the inclusion relation is denoted with the usual slash: "$A \not\subseteq B$" means "A is not included in B." That is, some element of A is not an element of B. If the sets are denoted as lists in braces, it means that some item in the list for A is not an item in the list for B.

More examples (3.3) $\{1,2\} \subseteq \{1,2,3\}$ because every item in the list 1, 2 is an item in the list 1, 2, 3.

$$\{1,2,3\} \not\subseteq \{1,2\} \quad \text{because} \quad 3 \in \{1,2,3\} \quad \text{but} \quad 3 \notin \{1,2\}.$$

$$\{1,3,1,2,2,1\} \subseteq \{1,2,3\}$$

because each item in the list 1, 3, 1, 2, 2, 1 is an item in the list 1, 2, 3. It is also true that

$$\{1,2,3\} \subseteq \{1,3,1,2,2,1\}$$

because every item in the list 1, 2, 3 is an item in the list 1, 3, 1, 2, 2, 1. Anticipating the next definition, we say

$$\{1,2,3\} = \{1,3,1,2,2,1\}.$$

In the list notation (1.2) checking whether each item in one list is an item in the other list is purely a mechanical matter.

$$\{3,1,4,1,5,9,2,6,5,3,5,8,9,7,9,3\} \quad \subseteq \quad \{1,2,3,4,5,6,7,8,9\}.$$

The procedure becomes more interesting and challenging with the set-builder notation for sets (1.8). We postpone examples to Chapter 2, (24).

Practice (3.4P) Prove that every set A satisfies $A \subseteq A$.

Practice (3.5P) Explain which of the following sets A, B, C, D includes $X := \{1,2,2,1,1,3\}$.

$$A := \{1,2,3\}, \quad B := \{1,1,1,3,4,5,6\},$$
$$C := \{3,1,4,1,5,9,2\}, \quad D := \{0,1,2,\ldots,10\}.$$

4. Equality of Sets

Finally we define *equality* between sets.

DEFINITION (4.1) For all sets A and B, we say that

CAUTION

A equals B if and only if
$A \subseteq B$ and $B \subseteq A$.

NOTATION: $A = B$

To take definition (4.1) back to basics, i.e., membership (2), we say "$A = B$ if and only if every member of A is a member of B, and vice versa." In turn this statement means that, in the list notation (1.2),

> every item in the list used to denote A is an item in the list used to denote B, and every item in the list used to denote B is an item in the list used to denote A.

Examples (4.2) $\{1,2,3\} = \{3,1,2\}$ and $\{1,2,3\} = \{2,3,1\}$. Also, $\{1,2,3\} = \{1,2,1,3,2,1\}$ and $\{\{1,2,1\},1,2\} = \{1,2,\{1,2\}\}$.

These examples illustrate this important principle:

(4.3) The ORDER in which the items appear in a list does not influence the *set* made from that list.

Equally important,

(4.4) The REPETITION[†] of items in a list does not influence the *set* made from the list.

Examples $\{1,1,1,1,1\} = \{1,1\}$, and $\{1,1\} = \{1\}$. Also,

$$\{1,2,1,3,1,4\} = \{4,3,2,1\} \neq \{1,2,3\}.$$

The definition of equality in (4.1) and the remarks (really, theorems) in (4.3) and (4.4) explain the last sentence of (1.2). Any change of order of the items in a list used to denote a set A or any repetition of the items produces a set that we've defined in (4.1) to be equal to A. The definition of set equality "doesn't care" how many times an item may be repeated nor where it appears in the list.

In terms of (1.2) there can be many different lists, all of which produce the same set when enclosed in braces. (Examples appear in (4.2).) Though the lists are different, the set is the same. "A set is known by its elements."

(4.5) *Properties of equality between sets.* For all sets A, B, and C

(*i*) $A = A$

(*ii*) If $A = B$ then $B = A$

(*iii*) If $A = B$ and $B = C$, then $A = C$.

These claims require proof, which is so easy that we leave it for the problems. Note that (4.5) is another theorem. It states properties we certainly would want to be true of any relation called "equality."

Practice (4.6P) Prove the assertion of equality between sets here:

$$\{3,1,4,1,5,9,6,2,5,3,5\} = \{9,6,3,3,1,2,4,5\}.$$

5. Terminology

There are several other ways to say "x is a member of the set A." We may say

x	is an **element** of	A
x	is a **point** of	A
x	is **in**	A
x	**belongs** to	A
x	is **contained** in	A
A	**contains**	x

[†]In the programming language APL there is an "idiom" (i.e., a one-line program) that strips repetitions from a list Y. Called the **nub**, it is $((\iota\rho Y) = Y\iota Y)/Y$.

Some other terms related to inclusion are *subset* and *superset*. For sets A and B we say

$$\begin{array}{lll} A & \text{is a } \textbf{subset} \text{ of} & B \\ B & \text{is a } \textbf{superset} \text{ of} & A \end{array}$$

if and only if $A \subseteq B$.

Practice. (5.1P) List all the supersets of the set $\{1,2\}$ that are subsets of the set $\{1,2,3,4\}$.

6. Membership vs. Inclusion

We return to membership and inclusion. Here are some examples:

$$\begin{gathered} \{1,2,3\} \subseteq \{1,2,3,4,5\} \\ \{1,3\} \subseteq \{3,2,1,0\} \\ \{1\} \subseteq \{1,3,6\} \\ 1 \not\subseteq \{1,2\} \\ 1 \in \{1,2\} \\ \{1\} \notin \{1,2\} \\ \{1\} \subseteq \{1,2\} \end{gathered} \tag{6.1}$$

Note carefully these last four examples; they illustrate the distinction between membership and inclusion. The point 1 is not included in the set $\{1,2\}$, because it is not true that every element of the set 1 (whatever that means!) is an element of $\{1,2\}$. Since (3.1) is not satisfied, $1 \not\subseteq \{1,2\}$.† On the other hand, 1 appears in the list of elements of $\{1,2\}$; thus 1 is an element of $\{1,2\}$, and we may write "$1 \in \{1,2\}$".

In the next example the question is whether $\{1\}$ is an element of $\{1,2\}$. The answer is no, because $\{1\}$ does not appear in the list of $\{1,2\}$. Those elements are 1 and 2, and $1 \neq \{1\}$, because "1" is the element 1, and "$\{1\}$" is the set consisting of the element 1. Another way to see that these are not equal is to remark, as we just did, that $1 \not\subseteq \{1\}$; since (4.1) is not satisfied, $1 \neq \{1\}$. In other words, the set $\{1\}$ has 1 as an element, but 1 does not have 1 as an element, so they are not equal.

But every element of $\{1\}$, namely 1, appears in the list of elements of $\{1,2\}$, so $\{1\} \subseteq \{1,2\}$.

(6.2) *A mechanical procedure for checking membership.* To check whether $x \in A$, if A is in list notation, strip the outermost pair of braces $\{\ \}$ from A and see whether x is equal to some item in the resulting list.

Examples

This procedure is laughably easy for "simple" sets like $\{1,2,3\}$. But consider

$$x := \{1,2,2,1\} \in \{1,2,\{1,2\},3\} =: A. \tag{6.3}$$

†We might say instead that if 1 is not a set, then "$1 \subseteq \{1,2\}$" and "$1 \not\subseteq \{1,2\}$" are syntax errors (ungrammatical). Similarly, "$1 \neq \{1\}$." But see Chapter 5, (25).

If we strip the braces from A we get the list $1,2,\{1,2\},3$. (Note: we are *not* instructed in (6.2) to strip braces from x.) Yes, x is equal to an item in that list, because $\{1,2,2,1\} = \{1,2\}$. We said that equality of objects is identity; but if the object, for example, $\{1,2,2,1\}$, is a set, we must use the definition of set equality (not identity of objects) to determine whether x equals an item in the list for A.

(6.4) *A mechanical procedure for checking inclusion.* To check whether $A \subseteq B$, when both sets are in list notation, strip the outermost pair of braces from A, do the same for B, and then see whether each item in the list resulting from A is an item in that from B.

Examples (6.5) We check that $\{7,8,7,8,9\} \subseteq \{1,2,\ldots,10\}$ by stripping the outer braces from each set, getting the lists 7,8,7,8,9 on the left and $1,2,3,\ldots,10$ on the right. Then we check each item in the left-hand list: 7 is in the right-hand list. So is 8. So is 7. So is 8. So is 9. That's obvious.

(6.6) Now consider whether

$$A := \{\{1\},\{1,2\},\{\{1\}\},1,2,\{1,2,2,1\}\} \subseteq \{1,\{1\},\{\{1\}\},2,\{2\},\{2,1\}\} =: B.$$

We are to strip only one pair, the outer braces, from each of A and B. We get these lists:

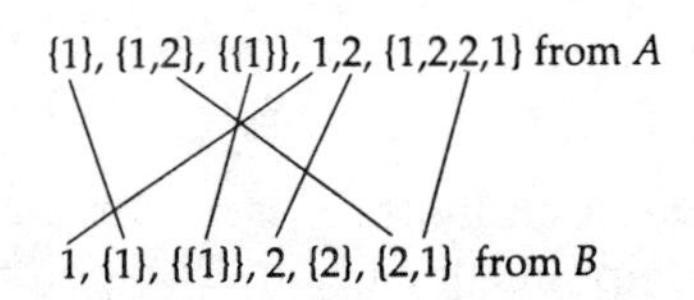

We've drawn lines to show equality of each item in the list for A with an item in that for B. Notice that we have used the fact that $\{1,2,2,1\} = \{2,1\} = \{1,2\}$, which follows from our definition of equality between sets. Therefore $A \subseteq B$.

Notice also that $B \not\subseteq A$ because $\{2\} \in B$ but $\{2\} \notin A$.

7. The Empty Set

One special set has no elements; we call it the **empty set**, and denote it $\varnothing$. (The symbol $\varnothing$ is a letter of the Scandinavian alphabet. It is not the Greek letter ϕ (phi).)

The empty set $\varnothing$ is a subset of every set: $\varnothing \subseteq A$ for every set A. This follows from (3.1), because every element of $\varnothing$ is indeed an element of A. (Since $\varnothing$ has no elements, all the elements of $\varnothing$ have any property whatever! More on this logical tactic in Chapter 2, (25).) It follows that the empty set is unique. (Why?) In particular

(7.1) $$\varnothing \subseteq \varnothing \quad \text{but} \quad \varnothing \notin \varnothing.$$

Why have an empty set? It's a natural idea. If Jim has no money, then the set of coins in his pocket is empty. If we want to prove that there is nothing with some property, we then want to prove that the set of elements having that property is empty. For example, the set of integers n such that n is even and n is odd is empty. Another example is the famous last "theorem" of Fermat, which states that for each integer $n \geq 3$, the set S_n of all triples of positive integers x, y, z satisfying $x^n + y^n = z^n$ is empty.† More generally, in a discussion a set might vary. We want to be able to consider it even if it should be empty.

Example (7.2) Suppose we use the set-builder idea (1.8) to define R as the set of all real roots of the polynomial $ax^2 + bx + c$, allowing a, b, c to vary over **Z**. Then R varies; if $b^2 - 4ac < 0$, then $R = \varnothing$.

Practice (7.3P) List all subsets of the set $\{1,2\}$.

8. Singleton, Proper Subset, Universal Set

A set consisting of a single element is called a **singleton**. Thus $\{3\}$ is a singleton; so is $\{\varnothing\}$.

We say that A is a **proper subset** of B if $A \subseteq B$ and $A \neq B$. That is, A is not the whole set. Thus $\varnothing, \{1\}$, and $\{2,3\}$ are proper subsets of $\{1,2,3\}$. We sometimes denote $A \subseteq B$ *and* $A \neq B$ by $A \underset{\neq}{\subset} B$.

Practice (8.1P) Prove that if A, B, and C are sets such that A is a subset of B and B is a proper subset of C, then A is a proper subset of C.

Practice (8.2P) List the supersets of $\{1,2,3\}$ that are proper subsets of $\{1,2,3,4,5\}$.

Practice (8.3P) What are the nonsingleton proper subsets of $\{1,2,3\}$?

(8.4) In some mathematical discussions there is a **universal set**. Often it is implicit. Its (vague) general definition is, "a set including all the sets under discussion." Usually it is a "standard" set like **N**, **Z**, or **R**.

(8.5) For example, if we were discussing finite sets of positive integers it would be natural to consider **N** as the universal set. If we were discussing three sets A, B, C, we could take $V := A \cup B \cup C$ as the universal set, or a superset of V.

Often we have no need to mention a universal set. Its main use for us is to allow complementation (see (12)).

9. Shoeboxes

In discussing such sets as $\{1\}$ and $\{\{1\}\}$ it is sometimes helpful to think of the braces $\{\ \}$ surrounding the list of elements as a shoebox, with the elements placed within the

†This claim (ca. 1650) of Fermat has so far been proved only for certain values of n. A stunning result proved in 1983 by Faltings implies that for each $n \geq 3$, S'_n is a finite set, where S'_n is the set of all triples x, y, z as above but with no common factor ≥ 2. The claim made in late 1987 of a solution to Fermat's problem proved to be premature.

box. Two sets are then equal if and only if the contents of the shoebox of one are the same as those of the other. Thus $\{1,2\}$ is represented in this way as $\lfloor 1,2 \rfloor$. The empty set is

$$\varnothing = \lfloor\ \rfloor,$$

the empty shoebox. The singleton set of the empty set is nonempty; it is

$$\{\varnothing\} = \lfloor \lfloor\ \rfloor \rfloor,$$

a shoebox with one empty shoebox in it. More examples:

$$\{1\} = \lfloor 1 \rfloor$$
$$\{\{1\}\} = \lfloor \lfloor 1 \rfloor \rfloor$$
$$\{1,\{1\}\} = \lfloor 1, \lfloor 1 \rfloor \rfloor.$$

You can see that our shoebox is just a more visually compelling pair of braces. If we join the braces by a line below and then straighten the curls, we get our shoebox,

$$\{\quad\} \rightarrow \{\underline{\quad\quad}\} \rightarrow \lfloor\underline{\quad\quad}\rfloor.$$

The bottom line tells you which left and right braces belong together.

10. Union and Intersection

We now describe two fundamental ways of combining sets. We must use the set-builder idea (1.8) to define the new terms.

(10.1) *Union.* Let A and B be two sets. The **union**, denoted $A \cup B$, of A and B is defined to be the set of all points x such that x is in A or x is in B. We emphasize that if a point x belongs to either A, or B, or to both A and B, then x belongs to $A \cup B$. (Also see Chapter 2, (2).)

Example (10.2) If $A = \{1,2,3,4,5\}$ and $B = \{1,3,5,7,9\}$, then

(10.3) $$A \cup B = \{1,2,3,4,5,7,9\}.$$

Strictly speaking, it would be correct to represent $A \cup B$ as the list obtained by juxtaposing the list of elements of A with that for B, as

(10.4) $$A \cup B = \{1,2,3,4,5,1,3,5,7,9\}$$

for this example, since by Definition (4.1) it equals $A \cup B$. But we avoid repetitions of elements in specifying a set unless it is inconvenient or cumbersome to eliminate the duplicates. Most teachers would prefer you to write (10.3), because if you wrote (10.4) they'd wonder if you knew that (10.3) is true.

(10.5) *Intersection.* If A and B are sets, we define the **intersection** of A and B, denoted $A \cap B$, as the set of all points y such that y is in A and y is in B. Thus $A \cap B$ is the subset of A and of B consisting of all the points common to both A and B. In example (10.2)

$$A \cap B = \{1,3,5\}.$$

We may define union and intersection by tables, as follows:

$x \in A$	$x \in B$	$x \in A \cup B$	$x \in A \cap B$
T	T	T	T
T	F	T	F
F	T	T	F
F	F	F	F

The "T" stands for "true," the "F" for "false." For example, row 2 means that "$x \in A$" is true and "$x \in B$" is false, and that in that case "$x \in A \cup B$" is true and "$x \in A \cap B$" is false. This table anticipates the *truth-tables* of Chapter 2.

If A and B have no elements in common, we call them **disjoint** and we may write

$$A \cap B = \varnothing.$$

This situation furnishes another motivation for the definition of the empty set, for with it we can consider the intersection of every two sets, even if disjoint.

If $A \cap B \neq \varnothing$, we may say that A **meets** B.

As an example of the use of Definition (3.1), we prove that for all sets A and B,

(10.6) $$A \cap B \subseteq A \subseteq A \cup B.$$

There are two (or three) inclusions claimed here. One is $A \cap B \subseteq A$, another is $A \subseteq A \cup B$, and, if you like, the third is $A \cap B \subseteq A \cup B$. We prove the first two. To see that $A \cap B \subseteq A$, we must prove that every point of $A \cap B$ is a point of A. We use the definition of intersection: $A \cap B$ is the set of all points common to A and B. Therefore, every point of $A \cap B$ is, in particular, a point of A. Thus we have proved the first inclusion.

The second inclusion, $A \subseteq A \cup B$, is equally easy to prove. We need to show, by Definition (3.1), that every point of A is a point of $A \cup B$. But by the definition of union, $A \cup B$ consists of all points x such that $x \in A$ or $x \in B$. In particular, every point of A is a point of $A \cup B$.

The third inclusion is a special case of the following result:

(10.7) For all sets A, B, C, if $A \subseteq B$ and $B \subseteq C$, then $A \subseteq C$.

You should test your understanding by trying to prove (10.7) on your own. Go ahead—it's easy.

Both union and intersection are commutative and associative, that is,

(10.8) $$A \cup B = B \cup A, \quad A \cap B = B \cap A$$

and

(10.9) $$(A \cup B) \cup C = A \cup (B \cup C), \quad (A \cap B) \cap C = A \cap (B \cap C)$$

for all sets A, B, C.

The equations (10.8) and (10.9) are claims of equality between sets. The only way open to us now to prove that two sets, say X and Y, are equal is to prove $X \subseteq Y$ and $Y \subseteq X$, the two parts of the definition (4.1) of equality of sets. The four equalities here are easy to prove in this fashion. We do just one of the eight inclusions as an example.

(10.10) To prove that $A \cup B \subseteq B \cup A$, we must show that the definition (3.1) of set inclusion is satisfied. That is, we must show for all points x of $A \cup B$ that x is a point of $B \cup A$. So let $x \in A \cup B$. Then, by definition (10.1) of union, $x \in A$ or $x \in B$. For x to be an element of the right-hand side $B \cup A$, x must satisfy the condition

$$x \in B \quad \text{or} \quad x \in A.$$

It has come down to this: We know $x \in A$ or $x \in B$. Is it true that $x \in B$ or $x \in A$? Of course it's true; that's how we understand the word *or*. Therefore $A \cup B \subseteq B \cup A$. The other seven inclusions all rest on similarly obvious properties of *or* and *and*.

Using (10.8) and (10.9) we define for each finite set $S = \{A_1, \ldots, A_n\}$ of sets

(10.11)
$$\bigcup_{1 \leq i \leq n} A_i := A_1 \cup \ldots \cup A_n$$
$$\bigcap_{1 \leq i \leq n} A_i := A_1 \cap \ldots \cap A_n,$$

if $n \geq 1$.

Because of (10.9) no parentheses are needed, and by (10.8) the sets defined in (10.11) are the same for all listings of S. The sets in (10.11) are sometimes denoted

(10.12) $$\bigcup_{C \in S} C \quad \text{and} \quad \bigcap_{C \in S} C,$$

respectively. The following convention extends the definition to the case when S is empty. Let V be a universal set. Then

(10.13) $$\bigcup_{C \in \varnothing} C = \varnothing; \quad \bigcap_{C \in \varnothing} C = V.$$

The second equation of (10.13) is a highly arbitrary convention. We need (10.13) because when (10.12) is in play, S may vary and become empty.

There is an analogy for sums and products:

(10.14) $$\sum_{i \in \varnothing} = 0; \quad \prod_{i \in \varnothing} a_i = 1.$$

Finally, here are two simple properties of union and intersection: For all sets A,

(10.15) $$A \cup \varnothing = A, \quad \text{and} \quad A \cap \varnothing = \varnothing.$$

11. The Distributive Laws

The following relations between union and intersection come up frequently, and proving them calls for the use of definition (4.1) for equality between sets. Suppose that A, B and C are sets. Then

(11.1) $$A \cup (B \cap C) = (A \cup B) \cap (A \cup C)$$

and

(11.2) $$A \cap (B \cup C) = (A \cap B) \cup (A \cap C).$$

Equations (11.1) and (11.2) are called the *distributive laws for union and intersection;* in (11.1), we say union distributes over intersection. In (11.2) intersection distributes over union. These equations also are theorems.

Let us prove (11.1). Using (4.1) we must show that

(11.3) $$A \cup (B \cap C) \subseteq (A \cup B) \cap (A \cup C)$$

and also the reverse of this inclusion ((11.6) following). First let us prove (11.3). According to (3.1), the definition of inclusion, we must prove that every element of $A \cup (B \cap C)$ is an element of $(A \cup B) \cap (A \cup C)$. Thus we let x be any element of $A \cup (B \cap C)$. By definition (10.1) of union ($\cup$), there are two possibilities: x is in A or x is in $B \cap C$ (or is in both).

> (11.4) In the first case, that is, if $x \in A$, then $x \in A \cup B$ and $x \in A \cup C$, again by definition of union; now by definition (10.5) of intersection ($\cap$), $x \in (A \cup B) \cap (A \cup C)$.
>
> In the second case, that is, if $x \in B \cap C$, then $x \in B$ and $x \in C$, by definition (10.5) of intersection. But if $x \in B$, then $x \in A \cup B$; and if $x \in C$, then $x \in A \cup C$. In this second case, x is in $A \cup B$ and in $A \cup C$, and hence is in their intersection (by definition of that operation).

Thus we have proved (11.3).

Practice (11.5P) Why is there no need to prove (11.3) in a third case, that is, when x is in both A and $B \cap C$?

Now let us prove the reverse inclusion,

(11.6) $$A \cup (B \cap C) \supseteq (A \cup B) \cap (A \cup C).$$

Let y be any element of the right-hand side $(A \cup B) \cap (A \cup C)$. Notice how often we referred to the definitions of union and intersection in proving (11.3). We'll do so here, too, but let's abbreviate the references to "DU" and "DI," respectively. Thus y is in both $A \cup B$ and $A \cup C$ (DI). Thus y is in A or B (DU applied to "$y \in A \cup B$") and y is in A or C (DU applied to "$y \in A \cup C$"). There are two cases: y is in A, or y is not in A.

> (11.7) In the first case, if y is in A, then y is in $A \cup X$ for all sets X (DU), so y is certainly in $A \cup (B \cap C)$. Thus (11.6) holds in the first case.
>
> In the second case, if y is not in A, then y is in both B and C, because we showed that y is in A or B, and also y is in A or C. Therefore, y is in $B \cap C$ (DI), so y is in $A \cup (B \cap C)$ (DU) in the second case as well.

This completes the proof of (11.1).

(11.8) Here's a proof of (11.3) with less talk. For explanations it uses the abbreviations DU and DI just mentioned.

To prove: $A \cup (B \cap C) \subseteq (A \cup B) \cap (A \cap C)$

Proof Let $x \in A \cup (B \cap C)$. Then

$$x \in A \quad \text{or} \quad x \in B \cap C \qquad \text{(DU)}$$
$$x \in A \quad \text{or} \quad (x \in B \quad \text{and} \quad x \in C) \qquad \text{(DI)}$$

Case 1: $x \in A$. Then

$$x \in A \cup B \quad \text{and} \quad x \in A \cup C \qquad \text{(DU)}$$
$$\therefore x \in (A \cup B) \cap (A \cup C) \qquad \text{(DI)}$$
$\therefore$ (11.3) holds in Case 1.

Case 2: $x \in B$ and $x \in C$. Then

$$x \in A \cup B \quad \text{and} \quad x \in A \cup C \qquad \text{(DU)}$$
$$\therefore x \in (A \cup B) \cap (A \cup C) \qquad \text{(DI)}$$
$\therefore$ (11.3) holds in Case 2.

Since Cases 1 and 2 exhaust all possible homes for x, (11.3) is true. QED

CAUTION

(11.9) I call the argument used in the proof just concluded an **epsilon-argument** because it uses (3.1) explicitly. An epsilon-argument proves inclusion of one set in another by explicit use of the membership relation $\in$ of (2.1) or (2.3). By contrast, there are proofs of inclusion that don't need explicit epsilon-arguments because they use known results (which themselves may have been proved by epsilon-arguments).

Example (11.10) Assume $A \cap B = \varnothing$ and $A \subseteq B \cup C$. Prove $A \subseteq C$.

Our target is to prove $A \subseteq C$. By the definition of inclusion, we are to let x be any element of A and then show x must be an element of C.

So, let $x \in A$. Since we are given $A \subseteq B \cup C$, by definition of union we know

(★) $$x \in B \quad \text{or} \quad x \in C.$$

We are also given $A \cap B = \varnothing$. By definition of intersection, this implies

$$x \notin B,$$

since we know $x \in A$. (That is, if x were also in B, then $A \cap B$ would not be empty.) But (★) is true, so $x \in C$ must be true. QED

Notice that except for the statements of (4.5) and (11.1) we have done nothing in Chapter 1 but define terms and symbols, except for our remark that repetition and order are irrelevant in the list of elements of a set. Therefore, we could appeal only to those definitions in seeing whether the connections claimed in (11.1) and (11.2) between union and intersection of sets were true. We now preach a sermon on that point.

Strategy. The preceding proofs of (11.3) and (11.6) illustrate the only strategy so far available to you for making proofs. That strategy is:

(11.11) Go to the definitions.

We did just that. We were trying to prove (11.1), a claim of equality between certain sets. So the first thing we did was use the definition (4.1) of equality between sets. Doing so broke our problem into two parts, the proofs of (11.3) and (11.6). Since the latter were claims of inclusion between certain sets, we went to the definition (3.1) of inclusion. We also had to use the definitions of union (10.1) and intersection (10.5), because those operations appeared in (11.3) and (11.6).

The order in which we call in these various definitions is imposed on us by the problem. The last assertion or operation used to form the statement (11.1) is the assertion of equality, the equal sign in (11.1). So that is the definition (4.1) we go to first, getting (11.3) and (11.6). Since the proofs of (11.3) and (11.6) are similar, we discuss only the former now.

In (11.3) the last assertion or operation is the inclusion $\subseteq$. So we call in that definition (3.1). *It* tells us to focus first on the left-hand side of (11.3), $A \cup (B \cap C)$ and to consider all points x of that set.

In the formation of $A \cup (B \cap C)$, the last operation is the union $\cup$. So now we invoke that definition (10.1). It tells us that x is in A or x is in $B \cap C$. The *or* forces two cases on us, as you see in (11.4).

In the second case we have $x \in B \cap C$, so we must now call in the definition of intersection (10.5).

In working on (11.3) we peeled the formula like an onion, especially in our work on the left-hand side of it.

In the concluding parts of the two cases under (11.4) we reversed the procedure. That is, when we focused on the right-hand side, we first established that x was an element of each "piece," $A \cup B$ and $A \cup C$, and then concluded that x was in the intersection of those two "pieces." We could equally well have proceeded instead by the same order that we used on the left. We'd set ourselves the target to show

$$x \in (A \cup B) \cap (A \cup C).$$

Then we'd say: By (10.5), which is (DI), we must show

$$x \in A \cup B \quad \text{and} \quad x \in A \cup C;$$

and then we could show that these are both true as we did in (11.4).

You will eventually find the little sermon just concluded so obvious that you'll never need to look at it again, but it may help you understand epsilon-arguments. Many find them baffling at first (hence the warning in the margin), but if peeling the onion doesn't bring tears to your eyes, you'll grasp the idea sooner or later.

We will say more about proofs in Chapter 2, (11), (12), (13), (18), (21), (26).

Example (11.12) Here is another example of an epsilon-argument. Let A, B, C be sets. We will prove

(11.13) $$\text{If} \quad A \subseteq B, \quad \text{then} \quad A \cap C \subseteq B \cap C.$$

Our target is the statement that $A \cap C \subseteq B \cap C$. We restate it using the definition (3.1) of inclusion:

(11.14) $$\text{For all } x \text{ in } A \cap C,\ x \text{ is in } B \cap C.$$

Now we peel off one layer by saying

(11.15) $$\text{If} \quad x \in A \cap C, \quad \text{then} \quad x \in A \quad \text{and} \quad x \in C$$

from the definition (10.5) of intersection. Now we can use the "given" (i.e., the hypothesis $A \subseteq B$), which we restate in terms of (3.1) as

(11.16) $$\text{Every } y \text{ satisfying} \quad y \in A \quad \text{satisfies} \quad y \in B.$$

We apply this "given" to x. That is, we know $x \in A$ is true. From (11.16) we conclude that $x \in B$ is true.

Now go back to (11.15). We have $x \in A$ and $x \in C$. But since we just showed $x \in B$ if $x \in A$, we have

$$x \in B \quad \text{and} \quad x \in C.$$

By definition (10.5) we know $x \in B \cap C$, a subsidiary target. Thus we've proved (11.14), which is our target (in $\in$-form).

Notice how we went to the definitions and identified our targets.

12. The Complement

(12.1) NOTATIONS: (older) $A - B$ (newer) $A \setminus B$

$$A - B := A \setminus B := \text{relative complement of } B \text{ in } A.$$

In this book I use the older notation. The **relative complement** of B in A is defined as the set of all elements of A that are not elements of B. Other terms for $A - B$ are the **complement of** B **in** A, or the **set-theoretic difference**.

Examples (12.2) Letting $A = \{1,2,3\}$ throughout, we have

B	$A - B$
$\{1,2\}$	$\{3\}$
$\{3,4\}$	$\{1,2\}$
$\{1,2,3,4\}$	$\varnothing$
$\{3,4,5,6,7,8\}$	$\{1,2\}$
$\varnothing$	A
$\{7,8\}$	$\{1,2,3\}$

If we have agreed on a universal set V (8.4), we call the relative complement of A in V the **complement** of A, and denote it $\overline{A}$. Thus

$$\overline{A} := V \setminus A = V - A.$$

In this situation $A - B = A \cap \overline{B}$.

A list of identities satisfied by sets involving union, intersection and complement appears in Chapter 2, (13.25). Two of the most important are the **De Morgan laws**, which we leave you to prove in the problems. These laws are: for all sets A, B

(12.3) $$\overline{A \cup B} = \overline{A} \cap \overline{B} \quad \text{and} \quad \overline{A \cap B} = \overline{A} \cup \overline{B}.$$

Practice (12.4P) For all sets A and B, prove the following:

(*i*) $A - B \subseteq A$;

(*ii*) If $A \cap B = \varnothing$, then $A - B = A$;

(*iii*) $\overline{\overline{A}} = A$.

13. Venn Diagrams

We sometimes represent a set schematically as the interior of a circle or rectangle. For example, we could consider sets A and B as

(13.1*F*)

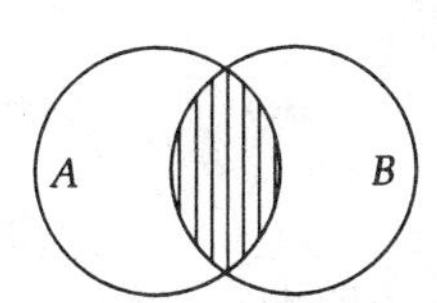

The shaded portion represents $A \cap B$, and the part inside the circles $A \cup B$.

We often represent the universal set V by a rectangle. Thus in this diagram,

(13.2*F*)

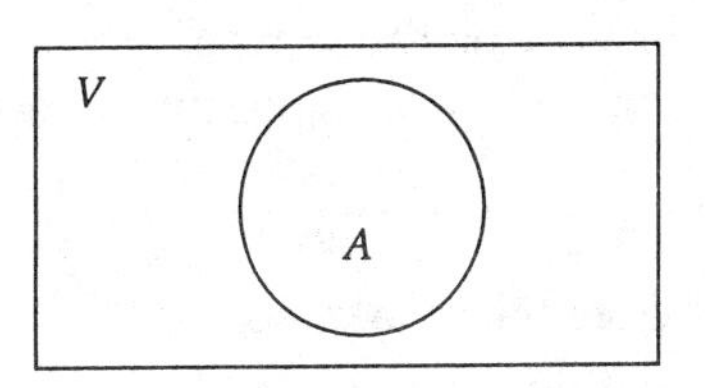

the exterior of A inside the rectangle is $\overline{A}$.

Venn diagrams are handy for convincing oneself of identities involving three or fewer sets (plus a universal set), the operations union, intersection, complementation, and the relations equality and inclusion. For example, let us prove[†] the distributive law (11.2) this way. It says

$$A \cap (B \cup C) = (A \cap B) \cup (A \cap C).$$

[†]See Chapter 2, (29). You *can* prove things with Venn diagrams.

We shade the left-hand side $A \cap (B \cup C)$ in one Venn diagram, and the right-hand side in another. Thus

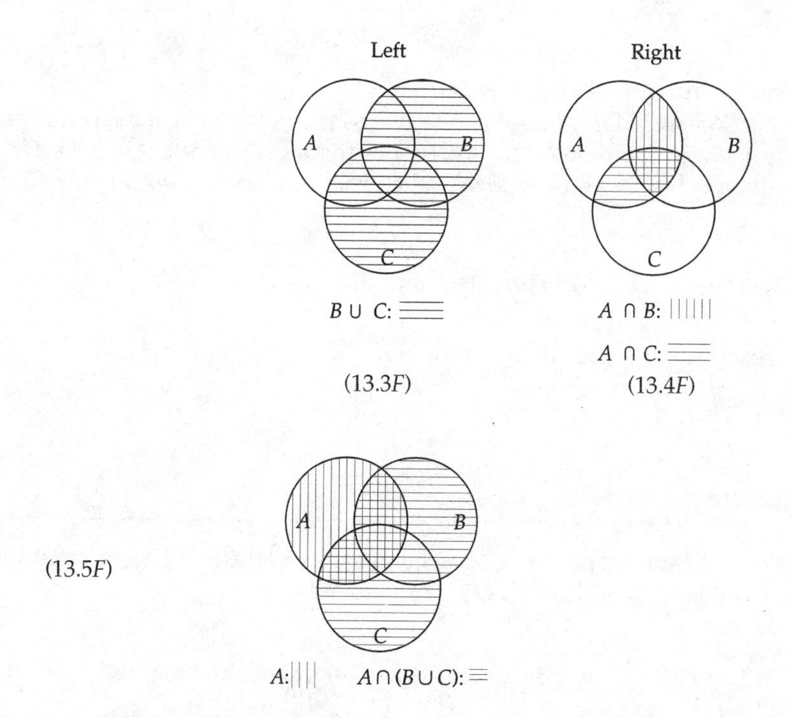

(13.3*F*) (13.4*F*)

(13.5*F*)

On the left, $A \cap (B \cup C)$ is the crosshatched portion of (13.5*F*). On the right $(A \cap B) \cup (A \cap C)$ is the marked portion of (13.4*F*). The two portions are the same subset, so the equality is established.

Writing out a proof by Venn diagrams so as to convince someone who is to read it later can be tedious, but showing someone such a proof one step at a time allows you to overwrite on the same diagram. Thus you could do the above proof with only one left diagram. Venn diagrams are better acted out than recorded.

Practice (13.6P) Prove $(A - B) \cup (B - C) \cup (C - A) = (A \cup B \cup C) - (A \cap B \cap C)$.

It is essential to take account of all possible intersections in a Venn diagram. Do not draw one as

(13.7*F*)

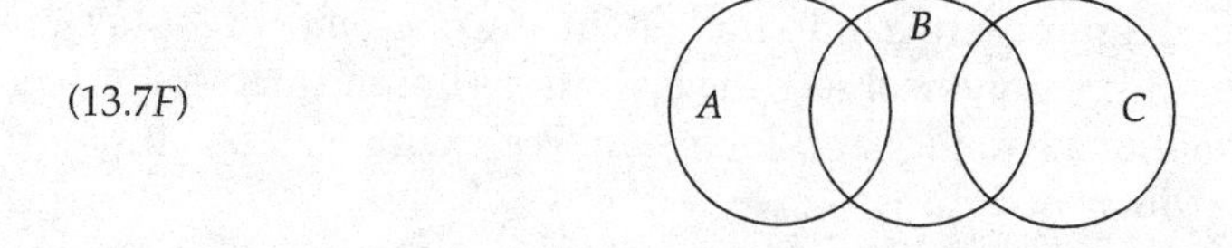

unless you know that A and C are disjoint. (Venn himself showed how to draw correct Venn diagrams for four, and even five, sets. See, for example, page 175 of Lewis Carroll, *Symbolic Logic* [Chapter 2, (31.2)]).

But if you know that special conditions hold for your problem, by all means incorporate them in your Venn diagram. For example, suppose you are to prove that

$$\text{if } A \subseteq B, \quad \text{then} \quad \overline{B} \subseteq \overline{A}.$$

The proof is easy with Venn diagrams if you draw yours to show the hypothesis $A \subseteq B$. (A and B don't *have* to be circles.) You don't even have to draw any more if you are just talking to yourself, because $\overline{B}$ is the outside of the oval, and $\overline{A}$ is obviously $\overline{B} \cup (B - A)$, a superset of $\overline{B}$.

(13.8F)

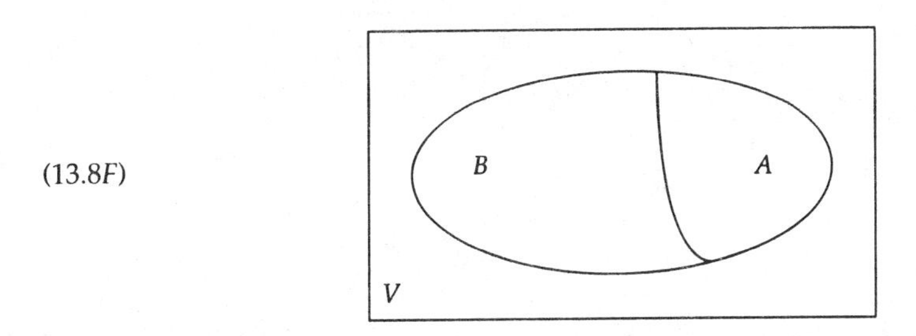

To summarize, Venn diagrams are perfectly good methods of proof for three or fewer sets, but they don't "like" to be written.

Practice (13.9P) Suppose there are three sets A, B, C such that $A \cap B \subseteq C$. Draw a Venn diagram showing this condition but not restricting the sets in any other way.

14. The Cartesian Product

Suppose A and B are two sets. The **Cartesian product** of A and B is defined to be the set of all *ordered pairs* (a, b) where a is in A and b is in B. The notation is $A \times B$.

For example, let $A = \{1, 2\}$ and $B = \{2, 3, 4\}$. Then

(14.1) $$A \times B = \{(1,2), (1,3), (1,4), (2,2), (2,3), (2,4)\}.$$

For each element a of A we include in $A \times B$ the subset consisting of the elements (a, b), for all b in B. We could equally well say that for each element b in B we include the subset consisting of the elements (a, b), for all a in A.

(14.2) We define an **ordered pair** as a list of two items enclosed in parentheses. (The more traditional definition in terms of sets is in (17).) It is always important to say when two newly defined mathematical objects are equal, so we say that two ordered pairs are equal if and only if their underlying lists are equal. Lists are equal if and only if they are identical. More formally,

(14.3) The ordered pair (a, b) equals the ordered pair (x, y) if and only if $a = x$ and $b = y$.

Although we ignore ordering and repetition in sets, they are all-important in ordered pairs.

Example The ordered pair (2,3) is *not* equal to the ordered pair (3,2). The *set* $\{2,3\}$ equals the set $\{3,2\}$. For that reason it is important that you express sets with braces $\{\ \}$ and not with parentheses. It is a syntax error to write "the set (2,3)."

The plane in analytic geometry is represented as the set of all ordered pairs of real numbers, that is, as the Cartesian product $\boldsymbol{R} \times \boldsymbol{R}$, where $\boldsymbol{R}$ is the set of all real numbers. By analogy we sometimes schematically represent $A \times B$ as the rectangle in the following figure,

(14.4F)

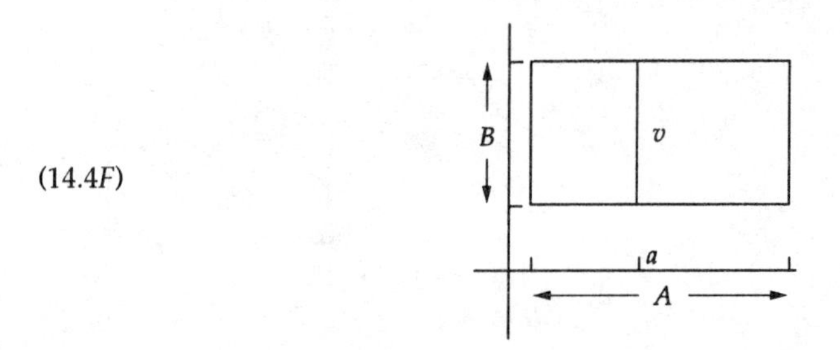

where the indicated intervals stand for A and B, even when A and B are not subsets of $\boldsymbol{R}$. The subset of all points (a,b) for fixed a in A and for all b in B is then represented by the vertical line v in the figure.

(14.5) Notice that $A \times B$ is not the same set as $B \times A$ unless $A = B$ or unless A or B is empty. (You would expect this result intuitively because ordering matters in ordered pairs.) Let us prove this claim as another example of the use of our definitions of set inclusion and equality, (3.1) and (4.1).

If $A = B$, then there is nothing to prove. If either A or B is the empty set then $A \times B = \varnothing = B \times A$.

Practice (14.6P) Prove this claim: $A \times \varnothing = \varnothing = \varnothing \times B$.

So let us assume that neither A nor B is empty, and that $A \neq B$; then either $A \not\subseteq B$ or $B \not\subseteq A$, by (4.1). Suppose first that $A \not\subseteq B$. Then there is a point a in A that is not in B. Let b be some point of B (b exists because B is not empty). Then (a,b) is a point of $A \times B$; and since a is not in B, (a,b) cannot be a point of $B \times A$. Therefore, $A \times B$ is not a subset of $B \times A$.

The other possibility, that $B \not\subseteq A$, is handled by the same argument with "B" and "A" interchanged. This procedure works because the conclusion, $A \times B \neq B \times A$, remains the same when "A" and "B" are interchanged. Therefore, we need not write any more of the proof; it is finished. We usually summarize such dismissal of the second part of a proof with the words "By symmetry of argument, ..."

The Cartesian product has the following properties (which we allow you to prove as problems). For all sets A, B, C, D

(14.7) $$A \times (B \cup C) = (A \times B) \cup (A \times C)$$

(14.8) $$A \times (B \cap C) = (A \times B) \cap (A \times C).$$

(14.9) $$\text{If } A \subseteq B, \text{ then } A \times C \subseteq B \times C, \text{ and } C \times A \subseteq C \times B.$$

(14.10) $$\text{If } A \times A = B \times B, \text{ then } A = B.$$

(14.11) $$(A \times B) \cap (C \times D) = (A \cap C) \times (B \cap D).$$

(14.12) If A has exactly m elements and B exactly n elements, then $A \times B$ has exactly mn elements (see Chapter 9).

(14.13) In the Cartesian product of a set A with itself, the **diagonal** is defined as the set of all ordered pairs (a,a) as a runs over A. The diagonal of $A \times A$ is a subset of $A \times A$. The name comes from the Venn diagram (14.4F) for $B = A$, in which the diagonal (subset) is the diagonal line.

(14.13F)

A

A×A

A

Example The diagonal of $\{1,2,3\} \times \{1,2,3\}$ is $\{(1,1), (2,2), (3,3)\}$. Of course, there are six other pairs in this Cartesian product.

Practice 14.14P List five points of $C \times D$, where $C := \{1,2,3,4\}$ and $D := \{5,6,7,8\}$.

15. The Power Set

(15.1) For each set X we may consider the set $\mathcal{P}(X)$ of all subsets of X. $\mathcal{P}(X)$ is called the **power set** of X. A rare alternate notation for $\mathcal{P}(X)$ is 2^X.

Examples (15.2)

$$\mathcal{P}(\varnothing) = \{\varnothing\}$$

$$\mathcal{P}(\{1\}) = \{\varnothing, \{1\}\}$$

$$\mathcal{P}(\{1,2\}) = \{\varnothing, \{1\}, \{2\}, \{1,2\}\};$$

if $X = \{1,2,3\}$, then

$$\mathcal{P}(X) = \{\varnothing, \{1\}, \{2\}, \{3\}, \{1,2\}, \{1,3\}, \{2,3\}, X\}.$$

Notice that $\varnothing$ is a member of $\mathcal{P}(X)$ for every set X, because the empty set is a subset of every set. Also, $X \in \mathcal{P}(X)$. If X is a finite set of exactly n elements, then $\mathcal{P}(X)$ consists of 2^n elements. (We prove this claim in Chapter 3, (11.1).)

Some properties of the power set follow: For all sets A, B

(15.3)
$$(i) \quad \mathcal{P}(A) \cap \mathcal{P}(B) = \mathcal{P}(A \cap B);$$
$$(ii) \quad \mathcal{P}(A) \cup \mathcal{P}(B) \subseteq \mathcal{P}(A \cup B),$$

with inequality holding in some cases of the latter.

Practice. (15.4P) Let A and V be sets such that $A \subseteq V$. Show that

$$\mathcal{P}(A) \subseteq \mathcal{P}(V) \quad \text{and} \quad \mathcal{P}(A) \cup \mathcal{P}(V - A) \subseteq \mathcal{P}(V).$$

NOTE: The elements of the power set are *sets*. ■

16. Postscript

The mathematical idea of set may not be commonly understood in everyday thought and speech. As expressed in (1.1) and (1,2), a set is a collection of things in which order and repetition are irrelevant. In ordinary language, however, we often think of collections as ordered: the encyclopedia has a first volume, a second, and so on; golf clubs are often thought of as ordered according to the numbers on them. A baseball team is not just the set of its nine players but carries with it, for example, the batting order. Every year someone wins the singles championship at the tennis club, which posts the winners for each year. At many clubs the same person wins more than once. For example, it might be that the winners for the past five years, in order, were A, B, A, C, A. Mathematically the *set* of winners is $\{A, B, C\}$, but most members have the list A, B, A, C, A in mind when they think of the recent winners.

Because of examples like these I have informally made the more intuitive idea of *list* the primitive term in explaining what a set is in the first notation (1.2).

One more example: you may know the Gilbert and Sullivan operetta *The Mikado*. In current parlance its character Pooh-Bah wears many hats. Thus an organization chart for the royal court would include, in our terms, several entries like

First Lord of the Treasury	{ Pooh-Bah }
Lord Chief Justice	{ Pooh-Bah }
Commander in Chief	{ Pooh-Bah }
Groom of the Back Stairs	{ Pooh-Bah }

and many more titles. Each office has listed opposite it the singleton set of people who hold that office. If the set of officeholders were all that mattered, we would then have to say that all the offices were the same, because the same set would define each office. But such crudity on our part would ruin the story. As Pooh-Bah himself points out:

> ...as First Lord of the Treasury I could propose a special vote that would cover all expenses, if it were not that, as Leader of the Opposition, it would be my duty to resist it, tooth and nail. Or, as Paymaster-General, I could so cook the accounts that, as Lord High Auditor, I should never discover the fraud. But then, as Archbishop of Titipu, it would be my duty to denounce my dishonesty and give myself into my own custody as First Commissioner of Police.

The mathematical idea of set strips away all these features: order of the elements, repetition of the elements, and any added information about the elements. (Which hat is Pooh-Bah wearing? Which club is Palmer's favorite? How did Bob Lemon pitch to left-handed batters on Sundays?) You are left with just the elements. "A set is known by its elements."

All right, then, what strictly *is* a set? It is a *collection* of *things*, a *class* of *objects*, an *ensemble* of *elements*. And what do these words mean? Everything and nothing! "Everything" because logicians and mathematicians can build all of mathematics from only those words, and "nothing" because they are undefined! If we defined them in terms of other words, we would have to agree that the other words were undefined, or else define *them* using still other terms, and so on. We would be forced to stop somewhere with undefined terms. Otherwise we'd have an *infinite regress* of definitions. The common agreement is to stop with *set* and *element* (and the relation $\in$).

In this text I use *list* as another undefined term. That helps to explain finite sets and gives you some intuition about sets in general.

We have mentioned the second way (1.8) to define sets, in terms of properties. We'll do a lot more with that idea in Chapter 2, (19)ff.

▶ 17. More on Ordered Pairs[†]

I trusted your intuition and experience to supply you with an understanding of what an ordered pair is. The ordered pair (a,b) is not the set $\{a,b\}$, because $\{a,b\} = \{b,a\}$ but $(a,b) \neq (b,a)$ unless $a = b$. That is, the definition of equality between ordered pairs is (we repeat (14.3))

(17.1) $$(a,b) = (x,y) \text{ if and only if } a = x \text{ and } b = y.$$

Yet ordered pairs can be defined strictly in terms of sets (an example of how all of mathematics rests on the term *set*):

DEFINITION (17.2) The ordered pair (a,b) is the set $\{\{a\},\{a,b\}\}$.

The condition (17.1) now is a consequence of the definition (17.2), strictly as an instance of equality between sets. Let us prove this claim.

Proof: *Part 1.* "If $a = x$ and $b = y$, then $\{\{a\},\{a,b\}\} = \{\{x\},\{x,y\}\}$." This part is self-evident.

Part 2. "If

(17.3) $$\{\{a\},\{a,b\}\} = \{\{x\},\{x,y\}\},$$

then $a = x$ and $b = y$."

From (17.3), we break this proof into two cases.

Case 1. $a = b$.

Then $\{a,b\} = \{a,a\} = \{a\}$, so $\{\{a\},\{a,b\}\} = \{\{a\}\}$. In this case (17.3) becomes $\{\{a\}\} = \{\{x\},\{x,y\}\}$.

But this expression tells us that $x = y$; otherwise the right-hand side would be a set of two elements $\{x\}$ and $\{x,y\}$. It could not then be equal to the singleton set $\{\{a\}\}$, because at least one of its elements would not be equal to $\{a\}$, an element of $\{\{a\}\}$.

Thus (17.3) becomes $\{\{a\}\} = \{\{x\}\}$, which forces $\{a\} = \{x\}$, and in turn, $a = x$. Since $b = a$ and $x = y$, we have shown that $a = x$ and $b = y$, as we set out to do.

[†] Sections included between triangles (▶) may be omitted.

Case 2. $a \neq b$.
Now $\{a\} \neq \{a,b\}$, so $\{\{a\},\{a,b\}\}$ is a set of two elements, and so, therefore, is $\{\{x\},\{x,y\}\}$. For (17.3) to hold, $\{a\}$ must be one of $\{x\}$, $\{x,y\}$; and $\{a,b\}$ is then the *other* one of these two sets, for otherwise $\{a\}$ would equal $\{a,b\}$ and b would equal a. So we consider the possibilities.

If $\{a\} = \{x\}$, then $a = x$. Also $\{a,b\} = \{x,y\}$ and $\{a,b\} = \{x,b\}$. Thus $\{x,b\} = \{x,y\}$; and from (4.1), b is either x or y. But b cannot be x, for then b would equal a. Therefore, $b = y$. Again we have proved what we set out to prove.

If $\{a\}$ were not equal to $\{x\}$, then $\{a\}$ would equal $\{x,y\}$. But by Definition (1.2) this means $\{x,y\} \subseteq \{a\}$, so $x = a$ and $y = a$. But then $x = y = a$ and $\{x,y\} = \{a\} = \{x\}$, a contradiction.

Thus we have proved Part 2 and so we have proved that condition (17.1) is a logical consequence of the definition (17.2). (A colleague of mine calls this proof "an ugly piece of Fortran code.")

Practice (17.4P) Express the ordered pair (3, 3) as a set.

As we said, if we take *list* as primitive concept, then there is no problem at all: an ordered pair is simply a list of two items enclosed in parentheses.

18. Summary

We have presented sets intuitively with emphasis on

- the distinction between membership and inclusion
- the use of (4.1) in proving two sets equal.

We have introduced Cartesian products and power sets. A detailed list of new terms and symbols with their place of definition follows.

Symbol	*Term*	*Where*
	Set	(1.1)
$\{1,2,3\}$	Set as list in braces	(1.2)
$\in$	Member, element	(2)
$\subseteq$	Inclusion	(3.1)
	Equality between sets	(4.1)
	Point, belongs, contains	(5)
	Subset ($\subseteq$), superset ($\supseteq$)	(5)
$\varnothing$	Empty set	(7)
$\sqcup$	Shoebox	(9)
$\cup, \cap$	Union, intersection	(10)
$\cup, \cap$	... of several sets	(10.10)ff.
	Epsilon-argument	(11.9)
$\overline{A}$	Complement (of A)	(12)
	Venn diagrams	(13)
$A \times B$	Cartesian product	(14)
(a, b)	Ordered pair	(14.2)
$\mathcal{P}(X)$	Power set	(15)

There is a hierarchy in our definitions. Equality of sets is defined in terms of inclusion between sets:

$$A = B \quad \text{if and only if} \quad A \subseteq B \quad \text{and} \quad B \subseteq A.$$

Inclusion is defined in terms of membership:

$$A \subseteq B \text{ if and only if every element of } A \text{ is an element of } B$$

Membership is the most basic idea. An epsilon-argument reduces questions of inclusion or equality to questions of membership.

19. More about Sets

We comment here on selected points from the chapter. We also define *ordered* n*-tuples*, *strings*, and min and max.

(19.1) *On the "definition" of a set.* We said in (1.1) and (1.2) that a set was a collection of things in which order and repetition are irrelevant. That statement fails to be a definition because *collection* is not defined. Trying to define *collection* would start the infinite regress mentioned in (16).

We have not defined in (1) the Platonic idea of set (that is, what a set "really" is), but we've shown how to define some specific sets (and we'll show more in Chapter 2).

(19.2) *Precise definition of sets.* The definition of a set may be precise even if we can not determine its elements.

Example $X := \{a\}$, where a is the millionth decimal digit of π. Some Japanese mathematicians computed π to 16 million decimal places in 1983. This result made U. S. newspapers under the headline "16 million decimal places but still just a piece of pi" (see *The Mathematical Intelligencer*, vol. 7, no. 3, pp. 65–67, 1985). They know X, but we don't. If we defined $X' = \{b\}$, with b the digit in position 10^{20}, presumably no one would know X'; but it's still defined precisely. (There is a fascinating human-interest story on mathematicians, mathematics, and a homemade supercomputer. It only seems to be about calculating π to more than 10^9 digits. I recommend "The Mountains of Pi," by Richard Preston, in *The New Yorker*, March 22, 1992, pp. 36–67.)

(19.3) *Set equality.* Any proof that two sets are equal rests ultimately on Definition (4.1). In particular, you must prove two inclusions. We'll discuss at length in Chapter 2, (19), (20), and (24), the interesting case of proving sets equal when they are defined by properties as in (1.8).

(19.4) *The triple dot notation, (1.4).* This usage for sets is a hybrid of the list and set-builder notations. You are to infer the property defining the set from the given fragment of the list of its elements.

(19.5) *The Cartesian product of more than two sets.* Another way to state (14.5) is to say that the Cartesian product is not commutative. Strictly speaking it is not associative either, that is,

$$(A \times B) \times C \neq A \times (B \times C) \tag{19.6}$$

because, e.g., if $A = B = C = \{1,2,3\}$ then $(1,2) \in A \times B$ and $((1,2),3) \in (A \times B) \times C$. But the corresponding element $(1,(2,3))$ of $A \times (B \times C)$ is not equal:

$$((1,2),3) \neq (1,(2,3))$$

because $(1,2) \neq$ (equally well because $3 \neq (2,3)$).

Nevertheless we define an **ordered n-tuple** for integral $n \geq 2$ as a *vector of n coordinates*, that is, as

$$a := (a_1, a_2, \ldots, a_n),$$

where the a_i are chosen from specified sets. The ordered n-tuple a (above) and $b := (b_1, \ldots, b_n)$ are defined to be equal if and only if

(19.7) $$a_1 = b_1,\ a_2 = b_2, \ldots, a_n = b_n,$$

that is, if and only if they are identical.

Thus an ordered 2-tuple is an ordered pair. We may call an ordered 3-tuple an *ordered triple*.

(19.8) With this generalization of the ordered pair we define the Cartesian product of the sets $A_1, A_2, \ldots, A_n$ as $A_1 \times A_2 \times \ldots \times A_n :=$ the set of *all* ordered n-tuples $(a_1, \ldots, a_n)$ such that $a_1 \in A_1, a_2 \in A_2, \ldots,$ and $a_n \in A_n$.

In other words, we defined ourselves out of trouble.

Examples Three-dimensional space is $\mathbf{R} \times \mathbf{R} \times \mathbf{R}$. Einstein's four-dimensional space-time continuum is simply $\mathbf{R} \times \mathbf{R} \times \mathbf{R} \times \mathbf{R}$, in which the fourth factor records time, not geometrical distance.

Suppose we wanted to record in great detail the change carried by each student in a class. We could ask each one to count his or her pennies, nickels, and so on. We could record the counts, for example, in a table.

	Pennies	*Nickels*	*Dimes*	*Quarters*	*Half-dollars*	*Silver dollars*
Bob	3	2	4	3	0	0
Alicia	7	1	3	5	1	0
Angela	1	4	11	6	0	0
Jim	3	3	3	3	0	0

Each row of the table can be viewed as a 6-tuple; Bob's 6-tuple is (3,2,4,3,0,0). The first coordinate is the number of pennies, the second is the number of nickels, and so on.

(19.9) *Strings*. We define a **string** in the symbols of a set A (a **string over** A) as an ordered n-tuple in $A \times A \times \cdots \times A$ (with n "factors" all equal to A) for some integer $n \geq 0$. But we write a string without parentheses or commas. This notation makes juxtaposition of strings a natural operation, called **catenation**, or concatenation. Thus if x and y are strings over A, so is xy. If $n = 0$, there is only the empty string, denoted λ.

Example Take $A := \{0, 1\}$. Then 011, 1000, 000, and 11 are strings over A. 1000 is the catenation of 10 and 00, also of 1 and 000, and of 100 and 0.

Equality of strings is identity. For example, $101 \neq 110$.

We could give a more "primitive" definition for *string* by saying that a string over A is a finite list of items all from A but written without commas.

You will need to know this simple concept because it comes up often.

(19.10) To generalize (14.12), if the set A_i has exactly q_i elements for each $i = 1, \ldots, n$, then the Cartesian product $A_1 \times A_2 \times \cdots \times A_n$ has exactly $q_1 q_2 \cdots q_n$ elements. In the special case when $A_1 = A_2 = \cdots = A_n = A$ with q elements, the Cartesian product of n copies of A, denoted A^n, has q^n elements (see problem 30 of Chapter 3).

(19.11) It will be useful to have these definitions: Let $A := \{a_1, \ldots, a_n\}$ be a nonempty finite subset of **R**. Then $\min\{a_1, \ldots, a_n\}$ denotes the algebraically smallest element of A, and $\max\{a_1, \ldots, a_n\}$ the largest.

Example If $A := \{-3, 0, 1\}$, then $\min A := \min\{-3, 0, 1\} = -3$ and $\max A := \max\{-3, 0, 1\} = 1$.

(19.12) *Numbers as sets.* Finally, consider the question: If everything in mathematics can be defined in terms of sets, what are the numbers $0, 1, 2, \ldots$? They have elegant definitions in terms of sets alone; see Chapter 5, Section (25), for a sketch of how to construct them.

20. Further Reading

(20.1) Andrew M. Gleason, *Fundamentals of Abstract Analysis*, Addison-Wesley, Reading, 1966.

(20.2) Paul R. Halmos, *Naive Set Theory*, Van Nostrand, Princeton, 1960, and Springer-Verlag, New York, 1974.

(20.3) L. E. Sigler, *Exercises in Set Theory*, Van Nostrand, Princeton, 1966.

21. Problems for Chapter 1

Problems for Section 2

2.1 Explain why each of the following is or is not an element of **Z**, the set of all integers: $\{3\}, \{1, 2\}, 7, \sqrt{2}, a :=$ the 3000*th* decimal digit in the base-10 expression for π.

— · —

The next two problems concern the sets

$A := \{1, 2, 3, 4, 5, 6\}$ $\quad D := \{4, 1, 3\}$
$B := \{1, 3, 5, 3, 1\}$ $\quad E := \{\{1\}, 1, 2, \{2\}, 3, \{3\}, 4, \{4\}\}$
$C := \{4, \{3\}, 2, 1\}$ $\quad F := \{\{1, 2, 3, 4\}\}$

2.2 For each of $a_1, \ldots, a_5$ below, say which of the sets $A, \ldots, F$ above it is a member of. Present your answer in a table:

	A	B	$\cdots$	F
a_1				
$\vdots$				
a_5				

where $a_1 := 1, a_2 := \{2\}, a_3 := 2, a_4 := \{2,1\}, a_5 := \{3\}$. Use an "$\in$" to stand for membership and a blank to stand for nonmembership. Explain each $\in$ and blank for a_2.

Problems for Section 3

3.1 For each of $a_1, \ldots, a_5$ in Problem 2.2, say which of the sets $A, \ldots, F$ it is a subset of. Use another table like that in Problem 2.2 for your answer, and explain each $\subseteq$ and blank for a_4.

3.2 Define $U := \{2,1,2\}$ and $G := \{1,2,\{1\},\{2\},\{1,2\}\}$. Is $U \in G$? Is $U \subseteq G$? Explain.

3.3 For each of the following sets or elements U, V, W, X, Y, and Z, say which of the sets it is in the relation $\in$ with; also, is in the relation $\subseteq$ with.

$U := \{1,2,3,4\}$ $\qquad V := \{1,4,5\}$
$W := 2$ $\qquad X := \{3, \varnothing\}$
$Y := \varnothing$ (the empty set) $\qquad Z := \{\{1\}, 2, \{3\}, 4\}$.

Present your answer in a table, as in Problem 2.2, with entries $\in$, $\subseteq$, or blank. Explain your answers, including the blanks, for row Y and for column Z.

Problems for Section 4

4.1 What relations ($\subseteq$ or $=$), if any, are there between the sets $A, \ldots, F$ of Problem 2.2?

4.2 Define S as the subset of the integers $\{0,1,\ldots,9\}$ that are used in the expression of π to the fifth decimal place (base 10). Define T as the set of all integers $x \in \{0,1,\ldots,9\}$ such that x is a positive square or x is a prime not dividing 14. Prove that $S = T$. [An integer x *is a square* if for some integer $y, x = y^2$. A *prime* is an integer ≥ 2 not divisible by any positive integer except itself and 1. The integer *a divides* the integer *b* if there is an integer c such that $b = ac$.]

4.3 Let A be the set of digits in the base-10 expression of the rational number $\frac{41}{333}$. Let B be the same for $\frac{44}{333}$. Prove that $A = B$.

4.4 Prove or disprove: the set C of digits in the base-10 expression of

$$\frac{40363\ 63637}{3\ 33000\ 00000}$$

equals the set of A the previous problem.

4.5 Prove (4.5) (*i*), (*ii*), and (*iii*), properties of equality between sets.

Problems for Section 5

5.1 List all the subsets of $\{1,2,3\}$.

5.2 (*i*) Write all subsets of the set $\{2,4,6\}$ once each.
(*ii*) What are all the subsets of $\{2,4,6,4,2\}$?

5.3 (*i*) Write all the subsets, once each, of the set of letters in the word *radar*.
(*ii*) Same question for *queue*.

Problems for Section 6

6.1 Define sets A, B, C, D, and E as follows: $A := \{1,2,3,4\}$. $B := \{1,2,\{2\},\{\{4\}\}\}$. $C := \{1,\{1,2\}, -\{\{1,2,3\}\}\}$. $D := \{1,2,2,1\}$. $E := \{1,1,1,2\}$. For each set W, X, Y, Z defined below, say whether it is an element or a subset of each of the sets A, B, C, D.

$$W := \{1,3,5\} \quad X := \{1,2,3\} \quad Y := \{4\} \quad Z := \{2\}$$

Problems for Section 7

7.1 Prove that the empty set is unique.

Problems for Section 8

8.1 Describe all sets that have no proper subsets.

8.2 Describe all sets having exactly one proper subset. Do the same for two proper subsets. Do the same for three proper subsets.

8.3 Prove or disprove the following: for all sets A, B, C, if A is a subset of B and B is a proper subset of C, then A is a proper subset of C.

Problems for Section 10

10.1 (*i*) Define A, B, C, and D as in problem 6.1.
(*ii*) Calculate the following sets; write them in the most economical way:

$$A \cap C \qquad B \cap C \qquad A \cup E$$
$$\star D \cap E \qquad \star A \cup (B \cap C) \qquad A \cap E.$$

Explain your answers for those marked $\star$.

10.2 [Recommended] Let A and B be any sets. From the definition of union it follows that

(*i*) If $x \in A \cup B$, then $x \in A$ or $x \in B$.

Consider now the analogue of (*i*) for inclusion:

(ii) For any set X, if $X \subseteq A \cup B$, then $X \subseteq A$ or $X \subseteq B$. Prove or disprove *(ii)*.

10.3 State the analogues for intersection of *(i)* and *(ii)* of the preceding problem and prove or disprove each. [Note: change $\cup$ to $\cap$ and *or* to *and*.]

10.4 Use (4.1) to prove that

$$\text{if } A \cup B \subseteq A \cap B, \text{ then } A = B.$$

10.5 Let $A_i := \{1, \ldots, i\}$ for $i = 1, 2, 3, 4$. Let $S := \{A_1, \ldots, A_4\}$. Calculate

$$\bigcup_{C \in S} C \text{ and } \bigcap_{C \in S} C.$$

Problems for Section 11

11.1 Give a careful proof of (11.2) using an epsilon-argument: if A, B, C are sets, then

$$A \cap (B \cup C) = (A \cap B) \cup (A \cap C).$$

Problems for Section 12

12.1 Let $A := \{1, 2, \ldots, 10\}$ and $B := \{4, 6, 8, 9, 10, 12, 14\}$.

(i) Find $A - B$.
(ii) Find $B - A$. (No explanation needed.)

12.2[Ans] Let X and Y be subsets of the set V. Prove that $X = Y$ if and only if

$$(\star) \quad (X \cap \overline{Y}) \cup (\overline{X} \cap Y) = \varnothing.$$

Notation: $\overline{X} := V - X$; $\overline{Y} := V - Y$. [There are two results to prove here. First, the "only if" part: assume $X = Y$ and prove $(\star)$. Second, the "if" part: assume $(\star)$ and prove $X = Y$.]

12.3 Prove the De Morgan laws (12.3) using *(i)* Venn diagrams (do both laws) *(ii)* An epsilon-argument (do just one law).

12.4 Define the symmetric difference of sets A, B as

$$A \Delta B := (A - B) \cup (B - A).$$

(i) Prove by an epsilon-argument that

$$A \Delta B = (A \cup B) - (A \cap B).$$

(ii) Show that if $A \Delta B = \varnothing$, then $A = B$.

12.5 Prove that symmetric difference is associative and commutative. Prove also that for all sets A

$$A \Delta A = \varnothing.$$

Problems for Section 13

13.1 Let A, B, C be subsets of the set V. Use a Venn diagram to simplify the expression $\overline{A \cap (B - C) \cup A}$.

13.2 For each of the problems *(i)*–*(iv)*, copy this Venn diagram below and then shade the indicated subset:

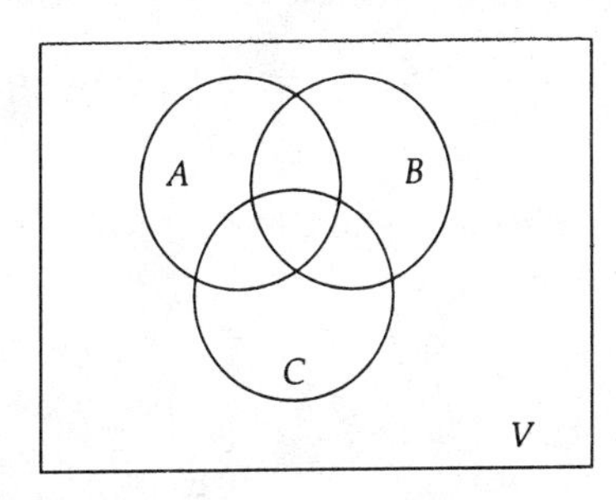

(i) $(A \cap \overline{B}) \cup (\overline{A} \cap B)$
(ii) $A \cap (B - C)$
(iii) $A \cup (C - B)$
(iv) $\overline{A} \cap \overline{B} \cap C$.

13.3 *(i)* In a large-enough Venn diagram, label each region with the number next to the set it depicts:

1	$A \cap B \cap C$	5	$\overline{A} \cap B \cap C$
2	$A \cap B \cap \overline{C}$	6	$\overline{A} \cap B \cap \overline{C}$
3	$A \cap \overline{B} \cap C$	7	$\overline{A} \cap \overline{B} \cap C$
4	$A \cap \overline{B} \cap \overline{C}$	8	$\overline{A} \cap \overline{B} \cap \overline{C}$

(ii) Let $V = \{1, 2, \ldots, 8\}$. Find subsets A, B, C of V such that each of the eight intersections in *(i)* is equal to $\{j\}$, where j is the number next to the set. For example, $A \cap B \cap C$ should equal $\{1\}$ and $\overline{A} \cap B \cap C$ should equal $\{5\}$.

13.4 Use Venn diagrams to show that $A \subseteq B$ (for sets A, B) if and only if $A \cap B = A$. [See comment in Problem 12.2.]

13.5 Prove the following identities for sets A, B, C:

(i) $A - B = A \cap \overline{B}$.
(ii)[Ans] $A \subseteq B$ if and only if $A - B = \varnothing$.
(iii) $A - (A - B) = A \cap B$.

(iv) $A \cap B \subseteq (A \cap C) \cup (B \cap \overline{C})$.
(v) $(A \cup C) \cap (B \cup \overline{C}) \subseteq A \cup B$.
(vi) $A \cap B = \emptyset$ if and only if $A \subseteq \overline{B}$ if and only if $B \subseteq \overline{A}$.
(vii) $A \subseteq B$ if and only if $\overline{B} \subseteq \overline{A}$.

13.6 Use Venn diagrams to prove or disprove the following for all sets A, B, C

(i) $\overline{(\overline{A} \cup B)} = A \cap \overline{B}$.
(ii) $\overline{A} \cup \overline{B} = \overline{(A \cup B)}$.
(iii) $(A \cup B) \cap C = (A \cup C) \cap B$.

13.7 Use Venn diagrams to show that for all sets X, E, F

$$X \cap (E - F) = (X \cap E) - F.$$

Can you find a shorter proof?

13.8 On a Venn diagram, shade each of the following areas:

$$(\overline{X} \cap Y) \cup (X \cap \overline{Z}),\ \overline{(X \cup Y)} \cap Z,\ (X \cup \overline{Z}) \cup \overline{Y}.$$

13.9 Let A, B, C be any sets.

(i) Prove $(A \cup B) - (A \cup C) \subseteq B - C$.
*(ii) Find the most general condition on A, B, C making equality hold in (i).

13.10 Use Venn diagrams to prove or disprove that for all sets A, B, if $A \cup B \subseteq A \cap B$ then $A = B$.

13.11 Use Venn diagrams to prove

(i) For all sets A, B, C, $\overline{(A \cap \overline{B}) \cup C} = (\overline{A} \cap \overline{C}) \cup (B \cap \overline{C})$.
(ii) For some sets A, B, C, $A \cup (B - C) \neq (A \cup B) - (A \cup C)$
*(iii) For which sets A, B, C does equality hold in (ii)?

13.12 Prove or disprove the following: For all sets A, B, C, $A - (B \cup C) = (A - B) \cup (A - C)$.

13.13 Let A, B, and C be sets.

(i) Use Venn diagrams to prove $A \cap B = \emptyset$ if and only if $A \subseteq \overline{B}$.
(ii) Prove using Venn diagrams or otherwise: If $A \subseteq C$, then $A \cap (B \cup \overline{C}) = A \cap B$.

13.14 (i) Prove or disprove the following: for all sets A, B, C,

$$(A \cup B) - (A \cup C) = B - C.$$

If true, prove by an epsilon-argument. If false, give a counterexample. Do the same for

(ii) $(A - B) \cup B = A$, and
(iii) $(A \cup B) - B = A$.

13.15 Prove or disprove: for all sets A, B, C,

(i) $A \cap (B - C) = (A \cap B) - (A \cap C)$.
(ii) $A \cap (B - C) = (A \cap B) - C$.
(iii) $A - (B \cap C) = (A - B) \cup (A - C)$.

If true, use an epsilon-argument. If false, give a counterexample.

13.16[Ans] Can you conclude at a glance that one of (i) and (ii) is false in the preceding problem? In other words, explain whether the following cancellation law is valid for sets X, Y, Z:

$$\text{If } X - Y = X - Z, \text{then } Y = Z \text{ (?)}$$

13.17 (i) Prove, using Venn diagrams, that for some sets A, B, C

$$A \cup \overline{B} \cup C \neq \overline{(\overline{A} \cap B)}.$$

*(ii) Find the most general condition on these sets making equality hold in (i).

13.18 Let A, B, C be any sets. Prove

(i) $(A \cup B) - C = (A - C) \cup (B - C)$
(ii) $A - (B - C) = (A - B) \cup (A \cap C)$.

13.19 Let A, B, C be any sets. Prove that $B - C \subseteq \overline{A}$ if $A \cap B \subseteq C$.

(i) Prove by using Venn diagrams.
(ii) Prove by using set algebra; see sections (11) and (12).
(iii) Prove by an epsilon-argument.

Problems for Section 14

14.1 Let $A := \{1,2\}$ and $B := \{1,3\}$. List the elements of $A \times B$.

14.2 Consider $A := \{1,\ldots,5\}$. Let $B := \{2,3\}$. Draw a schematic rectangle for $A \times A$ as in (14.4F).

(i) Show the two subsets $A \times B$ and $B \times A$.
(ii) Show $(A \times B) \cup (B \times A)$
(iii) Show and list $(A \times B) \cap (B \times A)$.

14.3 Let A, B, U, and V be any sets such that $A \subseteq U$ and $B \subseteq V$. Then is $A \times B \subseteq U \times V$ true? Explain.

14.4 Prove:

(i) (14.7)
(ii) (14.8)

(*iii*) (14.9)
(*iv*) (14.10)
(*v*) (14.11).
(NOTE: these are statements in the text, not problems.)

14.5 Prove that for all sets A, B,

(*i*) $(A \times B) \cap (B \times A) = (A \cap B) \times (A \cap B)$.
(*ii*) $(A \times B) \cup (B \times A) \subseteq (A \cup B) \times (A \cup B)$.

14.6 Let $C := A \cup B$ for sets A, B. Prove by using a result in the text that

$$A \times B \subseteq A \times C \subseteq C \times C.$$

That is, do not use an epsilon-argument (except as a last resort).

14.7[Ans] Let A and B be sets and let $R \subseteq A \times A$ and $S \subseteq B \times B$. Prove that

$$R \cap S \subseteq (A \cap B) \times (A \cap B).$$

14.8 Consider arbitrary sets A, B with $B \subseteq A$. Prove (*i*) and (*ii*) by ϵ-arguments.

(*i*) $(A \times B) \cap (B \times A) = B \times B$.
(*ii*) (diagonal of $A \times A) \cap (A \times B) =$ diagonal of $B \times B$.

Problems for Section 15

15.1 Let $X = \{1,2\}$ and $C = \{X - \{0\}, X - \{1\}, X - \{2\}, \{3\} - X\}$. List most economically the elements of $C \cap \mathcal{P}(X)$.

15.2 Let X be the set $\{1,2,3,4,5,6,\{1\}\}$. Let $\mathcal{P}(X)$ be the power set of X. Find

$$Y := X \cup (X \cap \mathcal{P}(X)).$$

Caution: think first.

15.3 List the elements of the following:

(*i*) $\varnothing \times \mathcal{P}(\varnothing)$
(*ii*) $\{\varnothing\} \times \mathcal{P}(\varnothing)$
(*iii*) $\mathcal{P}(\mathcal{P}(\{3\}))$
(*iv*) $A \times \mathcal{P}(A)$, where $A = \{1,2\}$.

15.4 Prove or disprove: for all sets A, B, $\mathcal{P}(A \cup B) = \mathcal{P}(A) \cup \mathcal{P}(B)$.

15.5 Prove for all sets A, B that $\mathcal{P}(A \cap B) = \mathcal{P}(A) \cap \mathcal{P}(B)$.

15.6 Let A and B be sets. Explain which, if either, of the following two sets is a subset of the other.

$$\mathcal{P}(A) \times \mathcal{P}(B), \quad \text{and} \quad \mathcal{P}(A \times B).$$

15.7 Let $X = \{\varnothing, 1, \{1\}\}$. Find $X \cap \mathcal{P}(X)$.

15.8 Let A and V be sets. What is $\mathcal{P}(A) \cap \mathcal{P}(V - A)$? Prove your answer.

22. Answers to Practice Problems

(3.4P) Let x be any element of A. We must show that $x \in A$. But that is what we are given. $\therefore A \subseteq A$.

(3.5P) The elements of X are only 1, 2, and 3. Each appears in the lists of elements of A, C, and D; but 2 is not an element of B. $\therefore X \subseteq A, X \subseteq C$, and $X \subseteq D$.

(4.6P) On the left we have the list L of elements 3, 1, 4, 1, 5, 9, 6, 2, 5, 3, 5. Each of those elements is an element in the list R on the right: 9, 6, 3, 3, 1, 2, 4, 5. For example, 3 is the third element in the list R; 1 is the fifth. Conversely, each element in the list R is an element in the list L: 9 is the sixth element of L, and so on. Therefore, since each set is a subset of the other, they are equal.

(5.1P) $\{1,2\}, \{1,2,3\}, \{1,2,4\}, \{1,2,3,4\}$.

(7.3P) The subsets of $\{1,2\}$ are $\varnothing, \{1\}, \{2\}, \{1,2\}$.

(8.1P) We are given $A \subseteq B$ and, from the definition of proper subset, $B \subsetneq C$. To prove; $A \subsetneq C$.

We know $A \subseteq B$ and $B \subseteq C$. Therefore, $A \subseteq C$, because every point of A is a point of B, and every point of B is a point of C.

Since $B \subseteq C$ and $B \neq C$, we know from (4.1) that $C \not\subseteq B$. Therefore there is a point of C not in B. Since $A \subseteq B$ that point is not in A. Therefore $C \not\subseteq A$, so $A \neq C$. Therefore $A \subseteq C$ and $A \neq C$: A is a proper subset of C.

(8.2P) If $\{1,2,3\} \subseteq A \subsetneq \{1,2,3,4,5\}$, then A is one of the sets $\{1,2,3\}, \{1,2,3,4\}, \{1,2,3,5\}$.

(8.3P) For this problem the answer is $\varnothing$ and all subsets of two elements, namely

$$\{1,2\}, \{1,3\}, \text{ and } \{2,3\}.$$

(11.5P) When x is in both A and $B \cap C$, it has been dealt with twice already—first in the case "$x \in A$," second in the case "$x \in B \cap C$." Nothing in the argument in either case ruled x out of $A \cap (B \cap C)$.

(12.4P) (*i*) By definition, every point of $A-B$ is a point of A. Thus $A-B \subseteq A$.

(*ii*) We are given $A \cap B = \varnothing$. Target: $A - B = A$. By (*i*) we know $A - B \subseteq A$ for any two sets A, B. By definition (4.1) of set equality, we must prove the opposite inclusion

$$A \subseteq A - B.$$

So, let x be any point of A. Since $A \cap B = \varnothing$, x is not a point of B. (Draw a Venn diagram for A, B with $A \cap B = \varnothing$.) By definition of $A - B$, x is in $A - B$, since x is a point of A not in B. $\therefore A \subseteq A - B$. Since we proved each is a subset of the other, we proved $A = A - B$ when $A \cap B = \varnothing$.

(13.6P)

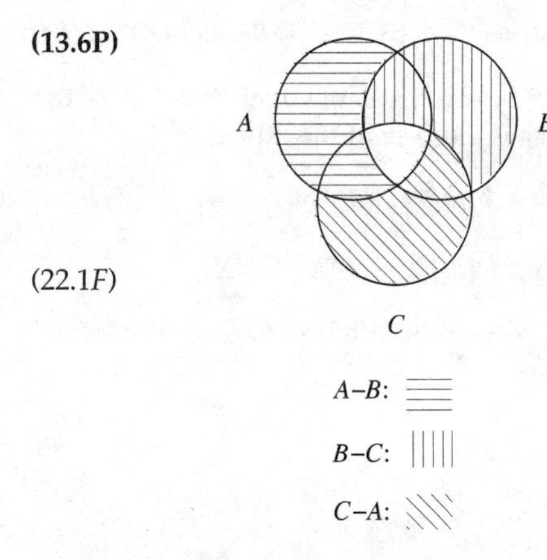

(22.1*F*)

The only unmarked part of $A \cup B \cup C$ is $A \cap B \cap C$.

(13.9P) Here is a Venn diagram for sets A, B, C, showing $A \cap B \subseteq C$ at the left.

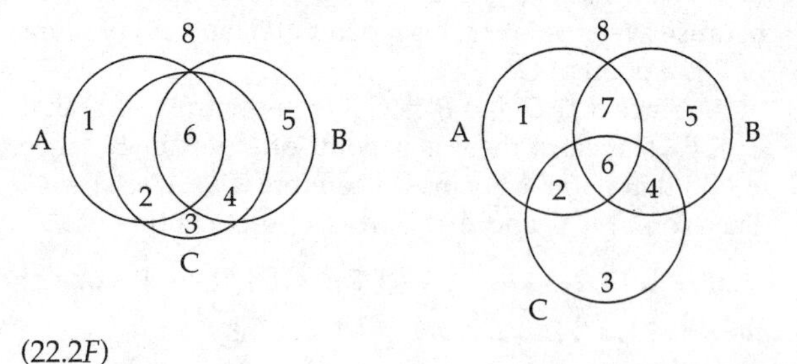

(22.2*F*)

I have numbered the eight regions of the most general three-set Venn diagram at the right. The difference is that region 7 has disappeared in our answer because it is supposed to be empty; "7" is the part of $A \cap B$ that is outside of C.

(14.6P) $A \times \varnothing = \varnothing$ because if $(x, y) \in A \times \varnothing$, then $x \in A$ and $y \in \varnothing$. Since $y \in \varnothing$ is false for all points y, (x, y) is not in $A \times \varnothing$. Therefore, $A \times \varnothing$ is empty.

Similarly for $\varnothing \times B = \varnothing$.

(14.14P) There are 16 points in $C \times D$, namely all ordered pairs (x, y) with $x \in C$ and $y \in D$. So one list of five of these points is

$$(1,5),\ (1,7),\ (2,5),\ (3,8),\ (4,5).$$

(15.4P) Since $A \subseteq V$, every subset of A is a subset of V—that follows immediately from "Every point of A is a point of V."

$$\therefore \mathcal{P}(A) \subseteq \mathcal{P}(V).$$

Since also $V - A \subseteq V$, we see that for the same reason

$$\mathcal{P}(V - A) \subseteq \mathcal{P}(V).$$

$$\therefore \mathcal{P}(A) \cup \mathcal{P}(V - A) \subseteq \mathcal{P}(V).$$

The idea at the end is simply this: $X \subseteq Z$ and $Y \subseteq Z$;

$$\therefore X \cup Y \subseteq Z,$$

which follows immediately from the definitions of union and inclusion.

(17.4P) $(3,3) := \{\{3\},\ \{3,3\}\} = \{\{3\},\ \{3\}\} = \{\{3\}\}$.

CHAPTER

2

Logic

If [Westway] is the road to progress,
then I am the retired Gaekwar of Baroda.
Robert Moses

Logic is a description of how we reason—when we are at our clearest and most convincing, that is. It is not some mysterious, otherworldly set of rules, despite its presentation that way in some advanced treatments.

This chapter is in two parts: Propositional Logic and Predicate Logic.

I. PROPOSITIONAL LOGIC

1. Propositions

(1.1) At first, we confine ourselves to **propositions:** statements of mathematical content, the truth or falsity of which is unambiguous.

Examples *of propositions:*

$$1 = 1 \qquad 2 \in \{1,2,3\}$$
$$2 < 3 \qquad \varnothing \subseteq \{1,2,3\}$$
$$7 > 11 \qquad 1 \in \varnothing$$
$$2 = 1 + 1 \qquad 8 = 2^n \text{ for some } n \in \mathbf{N}$$

2 is an odd integer

For all sets, $A, B, C, A \cap (B \cup C) = (A \cap B) \cup (A \cap C)$

The millionth decimal digit of π is 3.

(1.2) Some of the propositions in the preceding examples are true, some are false. Even the last is definitely true or false. We therefore say that it too is a proposition, even though we don't know which—true or false—applies to it.

Here is an example of a statement with mathematical content that is not a proposition: $x = 1$. Since x might change in value during the discussion, we can't say that the statement is true; neither can we say it is false. Some would say that $x = 1$ is not a statement but a form of statement, something that becomes a statement when an integer, say, is put in place of x.

Another way to speak of the truth or falsity of a proposition is with the term **truth-value.** A true proposition has the truth-value true; a false one, false. The term truth-value is less cumbersome than it may at first seem: consider that we may restate the point of (1.2) as, "A statement unambiguously true or false is a proposition even if we don't know its truth-value."

Let us reserve the letters T and F to stand for the truth-values *true* and *false*, respectively. (When we use T and F with other meanings we shall say so.)

A great deal, maybe all, of mathematics and computer science is made up of propositions, such as "The area of a triangle equals half the product of the base and the altitude," and "All differentiable functions are continuous."

We shall denote propositions by lowercase letters. Thus if p is (our name for) a proposition, then p is either true (T) or false (F).

Practice (1.3P) For each of the following expressions, state its truth-value if it is a proposition. If it isn't, explain why.

(i) $2 + 2 = 7$

(ii) $x + 2 = 5$

(iii) $(7^2 + 1) \div 5$

(iv) $3 \in \mathbf{N}$

(v) If it rains tomorrow, then we will get wet.

(vi) Let $x \in \mathbf{Z}$. If $x^2 - 3x + 2 = 0$, then $x = 2$ or 3.

2. Logical Connectives

We can make new propositions out of one or more propositions by means of the **logical words** or **logical connectives:** *not*, *and*, *or*, and *if ..., then ....* Thus let p and q stand for any propositions. We make the following definitions.

DEFINITION (2.1) **Not-p**, denoted $\sim p$, is the proposition that is true if p is false and false if p is true.

DEFINITION (2.2) **p and q**, denoted $p \wedge q$, is the proposition that is true if both p and q are true, and false if p is false or if q is false (or both).

DEFINITION (2.3) **p or q**, denoted $p \vee q$, is true if p is true or if q is true or if both p and q are true; it is false if both p and q are false.

DEFINITION (2.4) **If p then q**, denoted $p \longrightarrow q$, is true if p is true and q is true; it is true if p is false, irrespective of whether q is true or false; it is false if p is true and q is false.

Thus if p and q are propositions, then each of "$\sim p$," "$p \wedge q$," "$p \vee q$," and "$p \longrightarrow q$" is also a proposition; the rules (2.1)–(2.4) tell us how to determine the truth-value from those of p and q.

Examples (2.5) $\sim (1 + 1 = 2)$ is the proposition $1 + 1 \neq 2$; since "$1 + 1 = 2$" is true, $\sim (1 + 1 = 2)$ is false.

$\sim (7 > 3)$ is the proposition $7 \leq 3$;

$\sim (1 \in \{3,2\})$ is the proposition $1 \notin \{3,2\}$.

$(1 + 1 = 2) \wedge (3 > 7)$ is a false proposition, but

$(1 + 1 = 2) \vee (3 > 7)$ is a true proposition.

$(1 + 1 = 2) \longrightarrow (3 > 7)$ is false, but

$(3 > 7) \longrightarrow (1 + 1 = 2)$ is true. (Why?)

We can make more complicated **compound propositions** than propositions like $p \wedge q$ by using more than one logical connective. For example,

$$(p \vee q) \longrightarrow ((\sim p) \wedge (q \vee p)) \tag{2.6}$$

is a proposition if p and q are propositions, because we can determine its truth-value unambiguously from the definitions (2.1)–(2.4) and the truth-values of p and q. For example, if p is true and q is false, then (2.6) is

$$T \longrightarrow (F \wedge (\cdots)),$$

which is $T \longrightarrow F$, which is false by (2.4). (We ignored the truth-value of $q \vee p$ because it occurred in $F \wedge (q \vee p)$, which is F in any case, by (2.2).)

The compound propositions must conform to the rules of grammar implicit in (2.1)–(2.4). Thus $p \sim q$ is ungrammatical, and so are $(p \longrightarrow q) \longrightarrow$, $p \wedge q \vee r$, and $p \longrightarrow q \longrightarrow r$. The first two are obvious syntax errors, because "$\sim$" applies to only one proposition, so $p \sim q$ is the meaningless juxtaposition of p with $\sim q$; if..., then... requires two propositions, the second of which is missing for the second arrow.

The second two are errors because they are ambiguous. They need parentheses to give them a definite truth-value. That is, for some truth-values of p, q, and r, the two propositions

$$(p \wedge q) \vee r \qquad \text{and} \qquad p \wedge (q \vee r) \tag{2.7}$$

have different truth-values from each other,[†] and so do

$$(p \longrightarrow q) \longrightarrow r \qquad \text{and} \qquad p \longrightarrow (q \longrightarrow r). \tag{2.8}$$

(2.9) **Parentheses.** Having shown how important parentheses are, we now explain how to get rid of some—though not those in (2.7) and (2.8). We agree that

(2.10)
$\sim$ is the "strongest" connective
$\wedge$ and $\vee$ are equal in strength to each other and "next strongest"
$\longrightarrow$ is the "weakest" connective.

[†] You should verify this claim.

These statements mean that we agree that such expressions as

$$p \wedge q \longrightarrow \sim q \vee p \tag{2.11}$$

are bona fide compound propositions. This one, (2.11), means

$$(p \wedge q) \longrightarrow ((\sim q) \vee p).$$

Why? The $\wedge$ and $\vee$ are stronger than $\longrightarrow$, so they don't allow the arrow to "grab" the q and $\sim q$ as $\ldots \wedge (q \longrightarrow \sim q) \vee \ldots$. The $\sim$ applies, by this agreement, to only the first symbol after it, not to the later ones. Thus $\sim q \vee p$ is not $\sim (q \vee p)$ but is $(\sim q) \vee p$.

3. Truth-Tables

The definitions (2.1) through (2.4) can be summarized in *truth-tables*, which will also turn out to be useful devices when we discuss the more abstract propositional formulas. Thus we have

(3.1)

p	(2.1) $\sim p$
T	F
F	T

(3.2)

p	q	(2.2) $p \wedge q$	(2.3) $p \vee q$	(2.4) $p \longrightarrow q$	(3.3) p **xor** q
T	T	T	T	T	F
T	F	F	T	F	T
F	T	F	T	T	T
F	F	F	F	T	F

In the truth-table (3.1) we have listed in the left-hand column all the possible truth-values of the proposition p, which is the only constituent proposition of $\sim p$; in the column (2.1) we list the truth-value of $\sim p$ corresponding to the truth-value of p in the same row. In (3.2), which is really four truth-tables in one, we list all the four possible combinations of truth-values of p and q in the two columns at the left. Then in the columns on the right we put the truth-values of $p \wedge q$ and so on that belong with the truth-values of p and q in the two columns at the left, according to the definitions (2.2), (2.3), and (2.4). We have added a definition of the **exclusive or** in column (3.3). It is like the other *or* in (2.3) except that it is defined to be false if both propositions are true: T *xor* T is F.

4. Terminology

Stuffier names for the logical connectives are

(4.1)

$\sim p$	the **negation** of p
$p \wedge q$	the **conjunction** of p and q
$p \vee q$	the **disjunction** of p and q
$p \longrightarrow q$	**material implication (**p **materially implies** q**)**

Still more names for *if ..., then ...* are *only if,* and *conditional.* Thus, "if p, then q" is the same as both of

p **only if** q,

the **conditional of** p **and** q.

(4.2) The conditional of p and q is *not* the same as the conditional of q and p; the latter is "if q, then p." See Section (11).

More terminology: The propositions in a conditional have names of their own. In "$p \longrightarrow q$"

(4.3)
p is called the **antecedent;**
q is called the **consequent.**

Notice from column (2.4) in (3.2) that a true proposition is materially implied by every proposition, and that a false proposition materially implies every proposition.

One final logical connective, **material equivalence,** denoted $\longleftrightarrow$, is defined in terms of the other connectives: $p \longleftrightarrow q$ means $(p \longrightarrow q) \wedge (q \longrightarrow p)$. Its truth-table is

(4.4)

p	q	$p \longleftrightarrow q$
T	T	T
T	F	F
F	T	F
F	F	T

Two propositions are materially equivalent if and only if they have the same truth-value. Another term for material equivalence is *if and only if.* Thus the proposition p **if and only if** q means

p if q (that is, $q \longrightarrow p$)

and

p only if q (that is, $p \longrightarrow q$).

You will often see "if and only if" abbreviated as **iff.**

Example "1 + 1 = 2" is materially equivalent to "The area of a triangle is half the product of the base and the altitude." Why? Because both propositions are true.

"1 + 1 = 2" is not materially equivalent to "2 × 2 = 3," because the first is true and the second is false.

Practice (4.5P) Let p, q, and r be any propositions. Prove that $p \longleftrightarrow p$; that if $p \longleftrightarrow q$, then $q \longleftrightarrow p$; and that if $p \longleftrightarrow q$ and $q \longleftrightarrow r$, then $p \longleftrightarrow r$.

5. The Common Meaning of the "Logical Words"

The definitions (2.1) through (2.4) correspond closely to ordinary English usage of the terms *not, and, or,* and *if ..., then ....* Indeed, we have defined them in terms of the ordinary English words *not, or,* and *and!* One point about *or* is that we have defined

$p \vee q$ to be true when both p and q are true. Sometimes an *exclusive or* is desired; that has the truth-table given before in (3.3).

(5.1) The definition of *if..., then...* may seem strange, but as the quote from Robert Moses at the head of this chapter shows, even the line "$F \longrightarrow F$ is T" of the truth-table for material implication corresponds to our usage in everyday speech. Another example (adapted from (31.8), p. 198) of this line of the truth-table occurred when Bertrand Russell was challenged to give the actual reasoning to show how one false proposition could imply another. He was asked to assume that $1 = 2$ and prove that he was the Pope. After a moment's reflection, he said "If I am not the Pope, then the Pope and I are two. But two equals one, so the Pope and I are one." (We analyze this argument in (10).)

The line "$F \longrightarrow T$ is T" of the truth-table for material implication comes up, too, from time to time in ordinary speech. For example, we would consider both the following statements true: "If professors discuss *Alice in Wonderland* on the Quad, then they might get rained on," and "If professors don't discuss *Alice in Wonderland* on the Quad, then they might get rained on." One of these is an instance of "$F \longrightarrow T$."

Practice (5.2P) Explain the last sentence.

Examples

(i) From a bill for payment on a student loan: "Payments sent to any address other than the one noted on the ... bill will result in a delay in crediting." We know from experience, however, that payments sent to the correct address will also be credited late.

(ii) (Dave Barry) If I fail to water, fertilize, and mow my lawn, it will die. If I water, fertilize, and mow my lawn, it will die.

6. Propositional Formulas; Truth-Tables

If lions do math, then tigers fly jets.
Lions do math. Therefore tigers fly jets.

That's good logic, but it's nonsense. Logic depends on form, not content.

To analyze the form of compound propositions, logicians introduced an abstraction called *propositional formula.* You'll find it takes getting used to, but you will understand more once you master the idea. It totally ignores the meaning of propositions and focuses only on their truth-values.

CAUTION

(6.1) A **propositional variable** is a symbol $a, b, c, \ldots$ to which we may give the truth-value T or F as we please, independently of the truth-value that we may give to any other propositional variable. Sometimes they are called **atomic propositions,** or **atoms.** More generally, *we may substitute any proposition* for a propositional variable.

When we join propositional variables by means of our logical symbols $\vee, \wedge, \sim, \longrightarrow$ we get what we call a **propositional formula.** For example,

$$(a \vee b) \longrightarrow (\sim a \wedge c)$$

is a propositional formula in the three atoms a, b, c. The use of the logical symbols must be grammatical, of course; "$(a \vee b) \wedge \longrightarrow c$" is not a propositional formula. The general definition is this:

DEFINITION (6.2) "T" and "F" are propositional formulas; each atom is a propositional formula; for all propositional formulas f_1 and f_2, the following are also propositional formulas:

$$\sim (f_1)$$
$$(f_1) \vee (f_2)$$
$$(f_1) \wedge (f_2)$$
$$(f_1) \longrightarrow (f_2).$$

Every propositional formula must be formed from these rules.

(6.3) Definition (6.2) allows us to build up propositional formulas of any desired complexity. We start with atoms:

$$a, b, c, \ldots \text{ are propositional formulas.}$$

Next we may construct, for example (omitting parentheses),

$$\sim a, \qquad a \vee b, \qquad a \wedge b, \qquad c \longrightarrow a.$$

The next level contains such propositional formulas as

$$(\sim a) \vee b, \qquad \sim (a \vee b), \qquad a \longrightarrow (a \wedge b), \qquad (a \vee b) \longrightarrow (a \wedge b).$$

And we may keep on with this, forever. In fact, the definition encompasses all expressions obtained like this, no matter how complex. See Chapter 3, (12.4).

The only equality that we recognize between propositional formulas is that of identity. For example, we don't even regard $f_1 = a \vee b$ and $f_2 = b \vee a$ as equal, although we will soon see, in (7.9), that they are "logically equivalent."

(6.4) In each propositional formula, we are allowed to substitute any propositions (not just T and F) for the atoms.

A propositional variable is not a proposition. It is really a placeholder, into which we may put any proposition. Thus '$a \longrightarrow a \vee b$' is '$\square \longrightarrow \square \vee \bigcirc$'. We might substitute "$1 + 1 = 2$" for a, and "$7 < 3$" for b. We'd get $(1 + 1 = 2) \longrightarrow (1 + 1 = 2) \vee (7 < 3)$. (See also the examples in (7.4).)

A propositional formula is analogous to an algebraic formula, like $(x + y)/(x^2 + 1)$, for example. In the latter we may substitute any real numbers for the variables x and y and evaluate the result as some real number. Thus $x := -2$ and $y := 3$ yields $\frac{1}{5}$. In a propositional formula we may substitute any *propositions* for the variables and use the truth-values of those propositions to **evaluate** the resulting expression (that is, to find the truth-value of the resulting compound proposition). We now detail such evaluations.

(6.5) *Evaluating propositional formulas; truth-tables.* We *evaluate* a propositional formula, as *true* or *false*, when we substitute T or F for each variable in the formula and then find the truth-value of the resulting proposition, using the definitions (2.1), (2.2), (2.3), (2.4), of *or*, *and*, *not*, and *only if*.

The substitution must, of course, be consistent; for example, if a is made T in one of its occurrences, it must be made T in each of its occurrences in the formula.

(6.6) If a formula has exactly n propositional variables in it, then a total of 2^n different substitutions of T or F can be made for its variables.† A table of 2^n rows, listing all these substitutions together with the truth-value of the resulting expression for each substitution, is called the **truth-table of the propositional formula**.

For example, the zero-variable formulas are T and F; since there are no substitutions to be made, they each have a truth-table of one ($= 2^0$) line.

Truth-table for T
T

Truth-table for F
F

Consider a one-variable example, say $(\sim a) \longrightarrow a$.

Truth-table for $(\sim a) \longrightarrow a$

(6.7)

a	$\sim a$	a	$(\sim a) \longrightarrow a$
T	F	T	T
F	T	F	F

We introduced the columns for $\sim a$ and a to make the calculation easier. By the definition, (6.2), they are not part of the truth-table; but in practice we include such auxiliary columns to try to avoid errors.

Here is a two-variable example: Let the propositional formula be $(a \vee b) \longrightarrow (a \wedge b)$. It has the truth-table

(6.8)

a	b	$a \vee b$	$a \wedge b$	$(a \vee b) \longrightarrow (a \wedge b)$
T	T	T	T	T
T	F	T	F	F
F	T	T	F	F
F	F	F	F	T

The truth-tables (2.1)–(2.4) are the *definitions* of the truth-tables for the basic connectives $\sim$, $\vee$, $\wedge$, and $\longrightarrow$ if we regard p and q as propositional variables instead of as the propositions they were when (2.1)–(2.4) were set down.

There are more examples of truth-tables in the next section.

7. Logical Implication and Equivalence

Logical implication is a relation between propositional formulas, say f_1 and f_2. We say

(7.1)
f_1 **logically implies** f_2
if and only if

†We shall prove this claim in Chapter 3, Theorem (11.1).

every substitution for the atoms (of f_1 and f_2) that makes f_1 true also makes f_2 true.

$$\text{NOTATION:} \qquad f_1 \Longrightarrow f_2.$$

(7.2) Another way to say (7.1) is this: Consider the joint truth-table of f_1 and f_2. Then on every line where f_1 is true, f_2 is also true. Thus in (6.8) we see that $a \vee b$ does not logically imply $a \wedge b$, because there is a line where $a \vee b$ is T but $a \wedge b$ is F. (In fact, there are two such lines, but only one is needed to show that (7.2) does not hold here.)

(7.3) A consequence of definition (7.1) is that *every* substitution of propositions for the propositional variables in $f_1 \Longrightarrow f_2$ results in a true material implication. That is, if we substitute *propositions* $p, q, \ldots,$ for the *atoms* $a, b, \ldots$ (as $a := p,\ b := q, \ldots$) we get *propositions* $f_1(p,q,\ldots)$ and $f_2(p,q,\ldots)$; and if $f_1 \Longrightarrow f_2$, then the proposition

$$f_1(p,q,\ldots) \longrightarrow f_2(p,q,\ldots)$$

is true.

Examples (7.4) (justified using (7.2)).

(*i*) a logically implies $a \vee b$ because of the truth-tables for a and $a \vee b$,

a	b	a	$a \vee b$
T	T	T	T
T	F	T	T
F	T	F	T
F	F	F	F.

On every line where a is true, $a \vee b$ is also true, so we may write $a \Longrightarrow a \vee b$.

(*ii*) We show that $f := (a \longrightarrow b) \longrightarrow c \Longrightarrow a \longrightarrow (b \longrightarrow c) =: g$. We need a truth-table of eight lines. Just this once we show another way to work out truth-tables. You may prefer it. If so, fine. The numbers show the order in which we filled in the columns. Column 1 holds the truth-values of $a \longrightarrow b$; we copied c into column 2, and between them we then wrote column 3, the truth-values of $(a \longrightarrow b) \longrightarrow c$.

			1	3	2	4	6	5
a	b	c	$(a \longrightarrow b)$	$\longrightarrow$	c	a	$\longrightarrow$	$(b \longrightarrow c)$
T	T	T	T	T	T	T	T	T
T	T	F	T	F	F	T	F	F
T	F	T	F	T	T	T	T	T
T	F	F	F	T	F	T	T	T
F	T	T	T	T	T	F	T	T
F	T	F	T	F	F	F	T	F
F	F	T	T	T	T	F	T	T
F	F	F	T	F	F	F	T	T

On every line where f is true (lines 1, 3, 4, 5, 7) g is also true. Thus $f \Longrightarrow g$. Also note that g does not logically imply f, because there is a line on which g is T but f is F, for example, line 8.

Let us use the truth-table in (7.4*i*) to work up an example of the substitution idea in (7.3). (The truth-table records the substitutions $a := T$ and $b := T, a := T$ and $b := F$, and so on.) Here (7.3) says that *in each line* of the truth-table the truth-value of

$f_1 := a$ materially implies the truth-value of $f_2 := a \bigvee b$. In the first line a is T and $a \bigvee b$ is T, and $T \longrightarrow T$ is T. In the third line, a is F and $a \bigvee b$ is T, and $F \longrightarrow T$ is T. And so on.

To take (7.3) to the ultimate, we may substitute any propositions for a and b. For example, in (7.4*i*) again,

> If John is tall then John is tall or fat,
>
> If down is up, then down is up or soft,
>
> If apples have stems, then apples have stems or Robert Moses is the retired Gaekwar of Baroda.

The result is a *true* "if..., then..." proposition.

Let us go back for a moment to (7.2). We claimed it was another way to state the definition (7.1) of logical implication. Such a claim is really a *theorem*, i.e., a statement of mathematical fact that can be proved from the definitions of the terms in it. Although (7.2) is pretty obvious, let's go through a proof so you can see *(i)* the logical structure of an "if and only if" proof, and *(ii)* how a proof proceeds from definitions of terms to the conclusion. First we'll state our little theorem clearly:

Theorem (7.5) Let f_1 and f_2 be propositional formulas. Then $f_1 \Longrightarrow f_2$ if and only if the set of lines of the truth-table where f_1 is true is a subset of the set of lines where f_2 is true.

Proof To prove the *only if* part, we first assume $f_1 \Longrightarrow f_2$. Then we use the definition (7.1) of $f_1 \Longrightarrow f_2$ to see that on each line (of the joint truth-table of f_1 and f_2) where f_1 is true, f_2 is also true. Why? Because the truth-tables for f_1 and f_2 arise from substitutions for the atoms of f_1 and f_2. Therefore, the *only if* part of (7.5) is true.

To prove the *if* part (the converse of the *only if* part) of (7.5), we assume that the set of lines of the truth-table where f_1 is true is a subset of the set of lines where f_2 is true. We then try to prove $f_1 \Longrightarrow f_2$. To achieve the proof we notice that

> (7.6) *Every* substitution for the atoms of f_1 and f_2 is mirrored by a line in their joint truth-table, because the latter contains all possible substitutions of T and F for the atoms. The truth-value of such a substitution depends only on the truth-values of the propositions substituted for the atoms.

For example, if f_1 and f_2 depend on the two atoms a and b, then the evaluations of f_1 and f_2 when we substitute $1 + 1 = 2$ for a and $1 + 2 = 4$ for b have the same truth-value as appears on the line for $a := T$ and $b := F$ of the joint truth-table of f_1 and f_2.

Thus under our assumption, no substitution for the variables makes f_1 true and f_2 false.

QED

(7.7) Notice that (7.5) reduces a general case (all substitutions for the atoms) to a particular case (the 2^n lines of the truth table). We aren't always this lucky.

CAUTION

NOTE If f_1 and f_2 are propositional formulas, then $f_1 \Longrightarrow f_2$ is *not* a propositional formula. (After all, since "$\Longrightarrow$" is not found in (6.2), "$f_1 \Longrightarrow f_2$" doesn't satisfy the formation rules for propositional formulas.) It is a particular statement *about* the propositional formulas f_1 and f_2. In fact, "$f_1 \Longrightarrow f_2$" is a proposition.

Practice (7.8P) Explain the last sentence.

Logical equivalence is defined in terms of logical implication:

(7.9) Two propositional formulas f_1 and f_2 are called **logically equivalent** if and only if

$$f_1 \Longrightarrow f_2 \qquad \text{and} \qquad f_2 \Longrightarrow f_1.$$

NOTATION: $f_1 \Longleftrightarrow f_2$.

It follows immediately from (7.5) that

(7.10) $f_1 \Longleftrightarrow f_2$ if and only if f_1 and f_2 have the same truth-table.

The statement (7.10) will be our working definition of logical equivalence. It is concrete and practical as opposed to (7.9), a two-way version of (7.1).

Example (7.11) $a \Longleftrightarrow (a \wedge b) \vee (a \wedge \sim b)$. We verify this claim by showing that the truth-tables are the same.

a	b	$a \wedge b$	$a \wedge \sim b$	$(a \wedge b) \vee (a \wedge \sim b)$
T	T	T	F	T
T	F	F	T	T
F	T	F	F	F
F	F	F	F	F

The truth-table for a is the first column; it is the same as that for the other formula, in the last column.

Two more examples:

$$a \vee (b \longrightarrow c) \not\Longleftrightarrow a \vee (\sim b)$$
$$a \vee (b \wedge c) \Longleftrightarrow (a \vee b) \wedge (a \vee c).$$

(A slash through a symbol negates that symbol.) Their truth-tables have $8 = 2^3$ lines because they depend on three atoms a, b, c:

			(i)	*(ii)*	*(iii)*	*(iv)*	*(v)*	*(vi)*	*(vii)*	*(viii)*	*(ix)*
a	b	c	$b \longrightarrow c$	$a \vee (b \longrightarrow c)$	$\sim b$	$a \vee (\sim b)$	$b \wedge c$	$a \vee (b \wedge c)$	$a \wedge b$	$a \wedge c$	$(a \wedge b) \vee (a \wedge c)$
T	T	T	T	T	F	T	T	T	T	T	T
T	T	F	F	T	F	T	F	T	T	T	T
T	F	T	T	T	T	T	F	T	T	T	T
T	F	F	T	T	T	T	F	T	T	T	T
F	T	T	T	T	F	F	T	T	T	T	T
F	T	F	F	F	F	F	F	F	T	F	F
F	F	T	T	T	T	T	F	F	F	T	F
F	F	F	T	T	T	T	F	F	F	F	F

We have combined the two examples in one array, which is really nine truth-tables. We see that *(ii)*, $a \vee (b \longrightarrow c)$, is not logically equivalent to *(iv)*, $a \vee (\sim b)$, because they don't have the same truth-table: they differ on line 5. That is, the former is true, and the latter is false, when a is false and b and c are true. In the second example the two propositional formulas *(vi)* and *(ix)* have the same truth-table and are therefore logically equivalent.

The reason we included the five extra columns,—(i), (iii), (v), (vii), and ($viii$)—in (7.12) was purely practical: we needed them to calculate columns (ii), (iv), (vi), and (ix) without error. (Even so, I made an error my first time through.) I like to enter T and F in separate subcolumns to help me distinguish these two similar-looking letters. It is still easy to make mistakes. Constructing truth-tables is tedious, especially if you want to get them right.

(7.13) Logical equivalence has the same three properties as equality of sets (Chapter 1, (4.5)): For all propositional formulas f, g, h,

$$f \Leftrightarrow f$$
$$\text{If } f \Leftrightarrow g, \text{ then } g \Leftrightarrow f$$
$$\text{If } f \Leftrightarrow g \text{ and } g \Leftrightarrow h, \text{ then } f \Leftrightarrow h.$$

These properties follow immediately from (7.10).

Practice (7.14P) Explain how (7.13) follows from (7.10).

(7.15) What is the difference between $\Leftrightarrow$ and $\longleftrightarrow$? Logical equivalence ($\Leftrightarrow$) is a relation between propositional formulas. To say $f \Leftrightarrow g$ means that f and g have the same truth-*table*.

Material equivalence ($\longleftrightarrow$) is a relation between propositions. To say $p \longleftrightarrow q$ means that p and q have the same truth-*value*.

8. Tautologies and Contradictions

(8.1) A **tautology** is defined as a propositional formula that is true on every line of its truth-table.

With the idea of tautology one can define logical implication more succinctly: Let f_1 and f_2 be propositional formulas. Then

$$f_1 \Longrightarrow f_2 \text{ if and only if } (f_1) \longrightarrow (f_2) \text{ is a tautology.} \tag{8.2}$$

(Strictly speaking, (8.2) is, like (7.5), a theorem that follows immediately from the definitions.)

And, of course,

$$f_1 \Leftrightarrow f_2 \text{ if and only if } (f_1) \longleftrightarrow (f_2) \text{ is a tautology.} \tag{8.3}$$

Examples *of tautologies*

(8.4)

T, the true formula made up of 0 atoms

$(a \vee \sim a)$

$(a \vee b) \vee ((\sim a) \wedge (\sim b))$

$(a \longrightarrow b) \longleftrightarrow ((\sim a) \vee b)$

You can easily verify these examples. The fourth one is important because it expresses *only if* in terms of *negation* and *or*. The fact that the truth-tables for *only if* and *or* each consist of three T's and one F might lead you to discover this fourth example on your own.

(8.5) A **contradiction** is defined to be a propositional formula that is false on every line of its truth-table. Thus a contradiction might equally well be defined as the negation of a tautology. Example: $a \wedge \sim a$.

Sometimes tautologies are expressed instead as logical equivalences, using (8.3) implicitly. Thus **De Morgan's laws** are

$$\begin{aligned} \sim (a \vee b) &\Longleftrightarrow (\sim a) \wedge (\sim b) \\ \sim (a \wedge b) &\Longleftrightarrow (\sim a) \vee (\sim b). \end{aligned} \tag{8.6}$$

You can verify (8.6) using truth-tables. But let us demonstrate in the next section a method that sometimes is faster and less error-prone.

9. A Shortcut

To prove that $\sim (a \vee b) \Longleftrightarrow \sim a \wedge \sim b$ we observe that $a \vee b$ is F on only one line (of its truth-table), that where a and b are both F. Thus $\sim (a \vee b)$ is T on only that line. But "$\sim a$ and $\sim b$" is T on only the line where both $\sim a$ and $\sim b$ are T; that is, the line where both a and b are F. Therefore, the two expressions have the same truth-tables, because we showed that each is true if and only if a and b are each false.

Here is the general procedure: Let *table* mean the truth-table of the propositional formula f. Then define **T**(f) to be the set of all lines of the table where f is true, and define **F**(f) to be the set of all lines of the table where f is false. Thus $T(f) \cup F(f)$ is the set of all lines of the table. In the joint table of two formulas f and g, if $T(f) = T(g)$, then necessarily $F(f) = F(g)$, and conversely, so f and g have the same truth-table and are thus logically equivalent.

Examples (9.1) *Illustrations of T(f) and F(f)*: Please look at the truth-table in (7.11). Let's number the lines 1, 2, 3, 4. Then

$$\begin{aligned} T(a) &= \{1,2\}, & T(b) &= \{1,3\}, \\ T(a \wedge b) &= \{1\}, & F(a \wedge \sim b) &= \{1,3,4\}. \end{aligned}$$

We return to (8.6). We prove its second part in the same way. Since $a \wedge b$ is T on only the line $a = b = T$, $f := \sim (a \wedge b)$ is F on only that line. The propositional formula $g := (\sim a) \vee (\sim b)$ is an *or* and thus is F if and only if $\sim a$ and $\sim b$ are both F; that occurs exactly on the line $a = b = T$. Since $F(f) = F(g)$ we conclude that $\sim (a \wedge b) \Longleftrightarrow \sim a \vee \sim b$.

(9.2) We'll now use this method to disprove the equivalence of $a \vee (b \longrightarrow c)$ and $a \vee \sim b$ more economically and convincingly than we did by use of truth-tables in (7.12). We argue that $f := a \vee (b \longrightarrow c)$ is false only when a is F and $b \longrightarrow c$ is F. But $b \longrightarrow c$ is F only when b is T and c is F. Therefore, $F(f)$ consists of only $a = c = F, b = T$. But $g := a \vee \sim b$ is F not only when $a = c = F$ and $b = T$ but also when $a = F$ and $c = b = T$. Therefore, $F(f) \neq F(g)$, so $f \not\Longleftrightarrow g$. Quine (31.7) calls this approach the "fell swoop" and working out the whole truth-table the "full sweep."

If we look back at the truth-table (7.12) we see that $F(a \vee (b \longrightarrow c)) = \{\text{line } 6\}$ and $F(a \vee \sim b) = \{\text{lines } 5,6\}$. We just found these facts more quickly by our shortcut.

Since $F(f) \subseteq F(g)$ in this example, it follows that $T(g) = \overline{F(g)} \subseteq \overline{F(f)} = T(f)$; so $g \Longrightarrow f$.

Although these examples are probably enough to teach the method, we give a brief general discussion. The shortcut is a method to find the truth-table of a propositional formula g. It is based on the following observations:

(9.3)
$$\begin{aligned} a \wedge b &\text{ is } T \text{ only when } a = T \text{ and } b = T \\ a \vee b &\text{ is } F \text{ only when } a = F \text{ and } b = F \\ a \longrightarrow b &\text{ is } F \text{ only when } a = T \text{ and } b = F. \end{aligned}$$

To find $T(g)$ or $F(g)$ we dissect g so as to find the *last operation*, the logical connective that would be evaluated last if we were to work out the truth-table of g. If it is $\wedge$ we usually determine $T(g)$; if it is $\vee$ or $\longrightarrow$ we usually find $F(g)$.

If $T(g)$ or $F(g)$ is small (well fewer than half the lines of the table), then the shortcut is easier than working out the whole table.

(9.4) *Naming the points of* $T(f)$. We've called them lines of the truth-table of f, but we'd like to bypass the truth-table. It's more accurate to say that the points of $T(f)$ are ordered pairs, triples, or whatever in T and F, once we fix an ordering of the atoms of f.

Examples (9.5)

(i) In (9.2) we found that $F(f) = \{(F,T,F)\}$ if the order of the atoms is a,b,c. We found that

$$F(g) = \{(F,T,F), (F,T,T)\}.$$

(ii) In (7.12) we found via the full sweep that

$$F(a \vee (b \longrightarrow c)) = \{(F,T,F)\}.$$

The fell swoop makes it obvious.

(iii) We show $f := (a \longrightarrow b) \wedge (a \longrightarrow c) \Longleftrightarrow a \longrightarrow (b \wedge c) =: g$ using the shortcut method.

$$F(g) = \{(T,T,F), (T,F,T), (T,F,F)\};$$

i.e., for g to be F we must have a be T and $b \wedge c$ be F.

$$F(f) = \{(T,F,*), (T,*,F)\},$$

where "$*$" temporarily means the variable takes both values. That is, f is F exactly when a is T and b is F and c has either value ("$*$"), or when a is T, b has either value, and c is F. Thus

$$F(f) = \{(T,F,T), (T,F,F), (T,T,F), (T,F,F)\},$$

which equals $F(g)$.

Here was a case where f had the form $f_1 \wedge f_2$, but we calculated $F(f)$, not $T(f)$ as suggested after (9.3).

Practice (9.6P) Use the shortcut to give a fast proof of (7.4*ii*).

The idea of $T(f)$, the **truth-set** of the propositional formula f, and $F(f)$, the **falsity-set** of f, is useful because it allows concise, workable expressions of the conditions for logical implication and equivalence:

(9.7) $$f \Longrightarrow g \text{ iff } T(f) \subseteq T(g)$$

(9.8) $$f \Longleftrightarrow g \text{ iff } T(f) = T(g).$$

Statement (9.7) is a restatement of Theorem (7.5) in this new language; (9.8) follows immediately from (9.7) and from the definition (7.9) of logical equivalence.

Practice (9.9P) Verify the last sentence.

The concepts $T(f)$ and $F(f)$ will also come up prominently in Part II of this chapter.

10. More Tautologies; Rules of Inference

Rules of inference are tautologies (or logical implications or equivalences) used to draw conclusions in logical reasoning. They are useful in part because every substitution of propositions for the variables in a tautology results in a true proposition.

Some common rules of inference:

(10.1)	CONTRAPOSITIVE	$(a \longrightarrow b) \Longleftrightarrow (\sim b \longrightarrow \sim a)$
(10.2)	*MODUS PONENS*	$(a \wedge (a \longrightarrow b)) \Longrightarrow b$
(10.3)	*MODUS TOLLENS*	$(a \wedge (\sim b \longrightarrow \sim a)) \Longrightarrow b$
(10.4)		$((\sim a) \longrightarrow a) \Longrightarrow a$
(10.5)		$a \vee \sim a$
(10.6)		$(a \vee b) \wedge (\sim b) \Longrightarrow a$
(10.7)		$a \wedge b \Longrightarrow a$
(10.8)		$a \wedge \sim a \Longleftrightarrow \sim (a \vee \sim a)$
(10.9)		$\sim (\sim a) \Longleftrightarrow a$

These are all easy to verify. The first one expresses the implication $a \longrightarrow b$ in its **contrapositive** form $\sim b \longrightarrow \sim a$. We can verify it in several ways. For example, from (8.4) we know that $(a \longrightarrow b) \Longleftrightarrow \sim a \vee b$. We use (8.4) also on the right-hand side of (10.1):

$$\sim b \longrightarrow \sim a \Longleftrightarrow (\sim (\sim b)) \vee \sim a \Longleftrightarrow b \vee \sim a \Longleftrightarrow \sim a \vee b.$$

The result is that each side of (10.1) is logically equivalent to $\sim a \vee b$, so (7.13) tells us that (10.1) is true. We could also prove it with truth-tables or by the shortcut method.

The Latin name **modus ponens** is given to the famous rule of inference (10.2): $a \wedge (a \longrightarrow b) \Longrightarrow b$. It assures us that if p is true, and if p implies q, then q is true. This is the rule of inference that we most often use to bring a known mathematical theorem to bear in an argument. The theorem usually has the form $p \longrightarrow q$; if in some argument we can show that p is true, then we assert (correctly) that therefore q is true.

For example, in the Pythagorean theorem p is the statement: A right triangle has sides of lengths u, v, and w, of which w is the largest; and q is the statement: $u^2 + v^2 = w^2$. If in the course of some other investigation we prove that a certain triangle is a right triangle with sides of lengths x, y, and z, then we may justifiably conclude that $x^2 + y^2 = z^2$ for that triangle (where z is the maximum of the three lengths).

This step of logic is so natural that we make it automatically, as easily as we walk (an activity the psychologists classify as "learned automatic movement"). But we lay bare the structure of the reasoning in *modus ponens*.

The third rule of inference is sometimes called **modus tollens**; it is just the combination of the first two.

Example We can use *modus tollens* to analyze the quote from Robert Moses at the head of the chapter. Let p be the statement "Westway is not the road to progress" and q the statement "I [Robert Moses] am not the retired Gaekwar of Baroda." Then the quote becomes $\sim p \longrightarrow \sim q$. Since we are meant to know that Robert Moses is not the retired Gaekwar of Baroda (a famous prince in India of the nineteenth century), we may assert q. So, assuming the quote is true, we have

$$q \wedge (\sim p \longrightarrow \sim q).$$

From *modus tollens*, setting $a := q$ and $b := p$, we get p : Westway is not the road to progress. That is the conclusion Moses meant us to draw. Or we can just say that the quote is an instance of "$F \longrightarrow F$ is T."

The fourth one is charming. Let us work out the truth-table of each side.

a	$\sim a \longrightarrow a$	$(\sim a \longrightarrow a) \longrightarrow a$
T	T	T
F	F	T

So $\sim a \longrightarrow a \Longrightarrow a$. (It's even a logical equivalence, but there's no interest in the converse.) We can make the argument in words instead as follows: p is a proposition that we wish to show is true. We show somehow that $\sim p \longrightarrow p$ is true. Proposition p is either true or false. If p is false, then $\sim p \longrightarrow p$ is $T \longrightarrow F$, which is false. That statement contradicts that $\sim p \longrightarrow p$ is true. Therefore, p must be true.

The Bertrand Russell argument, in (5.1), is an example of this last rule of inference. Let p be the statement "Bertrand Russell is the Pope" and q the statement "$1 = 2$." Russell argued that in a little universe in which q is true, $\sim p \longrightarrow p$. He then counted on us to know and use the fourth rule of inference (10.4) to conclude p.

The fifth rule, $a \vee \sim a$, is called the **law of the excluded middle**. We effectively adopted it in our definition (1.1) of "proposition." Did you notice that we tacitly used it to justify the fourth rule?

The sixth rule is another variant of *modus ponens*, via the equivalence $(\sim b \longrightarrow a) \Longleftrightarrow a \vee b$. We could use it to prove *modus ponens* since it is easy to prove directly.

(10.10) **Substitution.** Suppose we wish to simplify a propositional formula g. Then g has one of the forms (6.2). Consider the case $g = f_1 \vee f_2$. Suppose we know that $f_1 \Longrightarrow h_1$ for some simpler propositional formula $h_1 (f_1 \Longrightarrow h_1$ could be one of our rules of inference, for example). We will show that we may substitute h_1 for f_1 in $f_1 \vee f_2$. Similarly we can do the same for $\wedge$. In sum,

(10.11) Let $f_1 \Longrightarrow h_1$. Then $f_1 \vee f_2 \Longrightarrow h_1 \vee f_2$; and $f_1 \wedge f_2 \Longrightarrow h_1 \wedge f_2$.

Let us prove (10.11) using Theorem (7.5) as restated in (9.7): $f_1 \Longrightarrow h_1$ if and only if $T(f_1) \subseteq T(h_1)$. Then

$$T(f_1 \vee f_2) = T(f_1) \cup T(f_2).$$
$$\text{Since } f_1 \Longrightarrow h_1,\ T(f_1) \subseteq T(h_1).$$
$$\text{Therefore, } T(f_1) \cup T(f_2) \subseteq T(h_1) \cup T(f_2);$$
$$T(f_1 \vee f_2) \subseteq T(h_1 \vee f_2);$$
$$f_1 \vee f_2 \Longrightarrow h_1 \vee f_2.$$

QED

This proves the first part of (10.11). You prove the second part the same way, just replacing $\vee$ by $\wedge$ and $\cup$ by $\cap$.

Practice (10.12P) Prove the second part of (10.11).

In the other two forms in (6.2) negation complicates the picture. We state the result for pure negation:

(10.13) If $f_1 \Longrightarrow h_1$, then $\sim h_1 \Longrightarrow \sim f_1$.

This is a contrapositive for logical implication.

Proof of (10.13) We are given that $T(f_1) \subseteq T(h_1)$. It follows (from Chapter 1, (13)) that $F(f_1) = \overline{T(f_1)} \supseteq \overline{T(h_1)} = F(h_1)$. But $F(h_1) = T(\sim h_1)$ and $F(f_1) = T(\sim f_1)$. Thus

$$T(\sim h_1) \subseteq T(\sim f_1)$$

so $\sim h_1 \Longrightarrow \sim f_1$.

Finally, for material implication:

(10.14)
(*i*) If $f_1 \Longrightarrow h_1$, then $h_1 \longrightarrow f_2 \Longrightarrow f_1 \longrightarrow f_2$
(*ii*) If $f_2 \Longrightarrow h_2$, then $f_1 \longrightarrow f_2 \Longrightarrow f_1 \longrightarrow h_2$.

QED

Proof of (10.14) It's easiest to remark that for all propositional formulas f and g

$$T(f \longrightarrow g) = \overline{T(f)} \cup T(g).$$

This equation follows from (8.4) and (8.3): $f \longrightarrow g \Longleftrightarrow \sim f \vee g$, and logically equivalent formulas have the same truth-table—that is (7.10).

We prove (*i*) by noting that $T(f_1) \subseteq T(h_1)$ is our hypothesis. Using it, we see that

$$T(h_1 \longrightarrow f_2) = \overline{T(h_1)} \cup T(f_2) \subseteq \overline{T(f_1)} \cup T(f_2) = T(f_1 \longrightarrow f_2).$$

We leave the proof of (*ii*) to you.

QED

An important use of (10.11), (10.13), and (10.14) is to prove the "obvious" result that if $f \Leftrightarrow g$, then we may substitute g for f in any propositional formula (with f in it) and obtain a logically equivalent formula. We state this as follows:

Theorem (10.15) **The Substitution Theorem**. Let f, g, and h be propositional formulas such that $f \Leftrightarrow h$. Then

(i) $\sim f \Leftrightarrow \sim h$
(ii) $f \vee g \Leftrightarrow h \vee g$
(iii) $f \wedge g \Leftrightarrow h \wedge g$
(iv) $f \longrightarrow g \Leftrightarrow h \longrightarrow g$
(v) If in addition $g \Leftrightarrow h'$, then $f \longrightarrow g \Leftrightarrow h \longrightarrow h'$.

Proof We shall use the definition (7.9) of $\Leftrightarrow$. It tells us that

(10.16)
$$f \Longrightarrow h \quad \text{and} \quad h \Longrightarrow f.$$

To prove *(i)* we apply (10.13) to each part of (10.16). We get

$$\sim h \Longrightarrow \sim f \quad \text{and} \quad \sim f \Longrightarrow \sim h.$$

Therefore, from (7.9), $\sim f \Leftrightarrow \sim h$. Parts (ii)–(v) are equally easy to prove if we use (10.11), (10.13), and (10.14), so we leave them for your amusement. QED

The point of this section was to prove the substitution theorem (10.15)—and to convince you that it needed to be proved. It is not enough to say "Equivalent things can always be substituted for each other," because at first the term *logical equivalence* is only a word (well, two words). Just because the name and notation are similar to "equality" and "=" doesn't automatically make it share the substitution property with equality.

Examples (10.17) From (10.9), $\sim (\sim a) \Leftrightarrow a$, we know that we can replace $\sim (\sim f)$ with f for any formula f in any larger formula. Thus to prove one De Morgan law (8.6) from another, we start with

$$\sim (a \vee b) \Leftrightarrow \sim a \wedge \sim b.$$

Next we use (10.15*i*), which allows us to negate each side and preserve the logical equivalence. We get

$$\sim (\sim (a \vee b)) \Leftrightarrow \sim (\sim a \wedge \sim b);$$

now on the left use (10.9) and then (7.13) to get

(10.18)
$$a \vee b \Leftrightarrow \sim (\sim a \wedge \sim b).$$

Now introduce new variables

$$c := \sim a \quad \text{and} \quad d := \sim b.$$

Then $\sim c = \sim (\sim a) \Leftrightarrow a$ and $\sim d = \sim (\sim b) \Leftrightarrow b$. The left side $a \vee b$ of (10.18) now satisfies $a \vee b \Leftrightarrow \sim c \vee \sim d$, by (10.15). Thus using (7.13) again we see that

$$\sim c \vee \sim d \Leftrightarrow \sim (c \wedge d),$$

which is the other De Morgan law. (Note that c and d are propositional variables; they have the values T and F as a and b are F and T.) As you gain experience you will learn to substitute like this quite rapidly.

Substituting h for f in propositional formulas when $f \Longleftrightarrow h$ is like the use of algebraic identities to substitute in algebraic formulas. If we have, say, $(x^2-y^2)/(x-y)$ we may use the identity $x^2 - y^2 = (x - y)(x + y)$. We get

$$\frac{x^2 - y^2}{x - y} = \frac{(x - y)(x + y)}{x - y} = x + y \quad (\text{if } x \neq y).$$

***Practice** (10.19P) Suppose $f \Longrightarrow g$ is true for propositional formulas f and g. Justify the implication after substitution of propositional formulas for the atoms of f and g.

For example, we know that $a \wedge (a \longrightarrow b) \Longrightarrow b$. (It's *modus ponens*.) Justify asserting that $f_1 \wedge (f_1 \longrightarrow f_2) \Longrightarrow f_2$ for all propositional formulas f_1 and f_2. Then justify in the general case. ◀

11. Some Wrong Inferences

We saw in (10.1) that $a \longrightarrow b$ is logically equivalent to $\sim b \longrightarrow \sim a$, and that each of these conditionals is called the contrapositive of the other. But $a \longrightarrow b$ is not logically equivalent to any other variant $x \longrightarrow y$, where x and y are in $\{a, \sim a, b, \sim b\}$ (aside from $a \longrightarrow b$ itself, of course). In particular, $a \longrightarrow b$ is not logically equivalent to $b \longrightarrow a$; in fact $a \longrightarrow b$ does not even logically imply $b \longrightarrow a$. (If it did, they would be logically equivalent. Why?) We prove this claim by showing the following:

(11.1) $(a \longrightarrow b) \longrightarrow (b \longrightarrow a)$ is not a tautology.

We must choose truth-values that make $b \longrightarrow a$ false; thus we set $b = T$ and $a = F$. Then $a \longrightarrow b$ is T, and so (11.1) is F for $a = F$ and $b = T$. Notice also that (11.1) is not a contradiction.

When we say in some specific situation "$p \longrightarrow q$, and conversely," we mean "$p \longrightarrow q$ and $q \longrightarrow p$," i.e., $p \longleftrightarrow q$. For example, a triangle is isosceles if two of its angles are equal, and conversely. Equivalence between propositions is a common occurrence in mathematics and computer science; but the assertion of equivalence is the assertion of two conditionals, neither of which logically forces the other to be true. That the two conditionals are both true is a result of the meaning of the statements p and q.

Here is another new term:

(11.2) The **converse** of $p \longrightarrow q$ is $q \longrightarrow p$ if p and q are propositions.

If f and g are propositional formulas, the **converse** of

- $(f) \longrightarrow (g)$ is $(g) \longrightarrow (f)$,
- $f \Longrightarrow g$ is $g \Longrightarrow f$.

Thus the converse of the proposition "If 1 + 1 = 2, then 2 + 2 = 3" is "If 2 + 2 = 3, then 1 + 1 = 2." (The first is false, the second true.)

It follows that $a \longrightarrow b$ is the converse of $b \longrightarrow a$, and from (11.1) that *there is no relation of logical implication between a conditional and its converse.*

Confusing an implication with its converse is a common pitfall, even for the experienced.

You know that $p \longrightarrow q$; in an argument you find that you have established the truth of q; you get mixed up about $p \longrightarrow q$ and think you know $q \longrightarrow p$ instead, so you use *modus ponens* and wrongly assert p. Another way to make this mistake is this: Say you know $p \longrightarrow q$, and you then establish $\sim p$. You then conclude $\sim q$, by error. Since $\sim a \longrightarrow \sim b \Leftrightarrow b \longrightarrow a$, you've made the same mistake.

12. Theorems and Proofs

(12.1) The main points of mathematical texts are almost always expressed as theorems. To show their importance, and to help you find them, authors set them apart from the rest of the text. If they were embedded in ordinary paragraphs, you might be less able to focus on the sometimes tricky ideas involved—on exactly what is assumed, or what is to be proved.

Authors could present other subjects "mathematically," in a theorem-proof format. For example, the important causes of the American revolution could be stated separately, as

> "The colonists were heavily taxed"—followed by evidence (the "proof" of the "theorem").
>
> Another: "The colonists had little voice in their own governance" (evidence...).

But history is more fun in narrative form. Literature in mathematical form would be deadly. Imagine the first lines of *Moby Dick* as "Let my name be Ishmael, let the captain's name be Ahab, let the boat's name be Pequod, and let the whale's name be as given in the title."† Mathematics presented as narrative, however, would be too hard to follow.

What is a theorem? It is simply a statement of mathematical fact. It can be obvious, like $1 + 1 = 2$, or less so, like the Pythagorean theorem, or even both at once, like the Jordan curve theorem, which says that a simple closed curve†† in the plane divides the plane into two regions, an inside and an outside, an "obvious" result infamous for the difficulty of the various proofs given for it. Here's an example from computer science: the number of comparisons needed to sort n things is at least $cn \log n$ for some constant c and for all large n.

Theorems may have other names, such as *lemma* and *corollary*. A **lemma** is a theorem that is not so interesting in itself but is used to prove other theorems. A **corollary** is a theorem, maybe even the main result, that is an immediate consequence of the result just preceding it. They could both be called *theorems*, but this hierarchy of names will quicken the reader's grasp of the roles of the results. (You will also see *Proposition* used, usually for a result between *Lemma* and *Theorem* in importance. I'm avoiding that usage for several chapters, since *proposition* is a new technical term for us.)

†B. A. Cipra, *Math. Intell.* 10 (1988), p. 27.

††One that begins and ends at the same point and does not intersect itself. Example: a circle.

A theorem often has the form $p \longrightarrow q$, where p and q are propositions. The **hypothesis** is p, the **conclusion** q. Both p and q may be compound propositions, of course.

What does the theorem $p \longrightarrow q$ say? It says that $p \longrightarrow q$ is true. (Recall that whenever we assert a proposition r, we are saying that r is true.)

(12.2) *The sixty-four dollar question*: How does one prove a theorem? If the theorem is $p \longrightarrow q$, we must show that $p \longrightarrow q$ is true.

Consider the truth-table for $p \longrightarrow q$:

p	q	$p \longrightarrow q$
T	T	T
F	T	T
F	F	T
T	F	F

We have changed the usual order of rows for emphasis. The propositions p and q are real-life statements, like "A is a right triangle," "n is a prime number," "f is a differentiable function," and so on. To show $p \longrightarrow q$ all we have to do is assume that p is true, and then show by some convincing argument that q is necessarily true.

If p is false, then $p \longrightarrow q$ is true whatever the truth-value of q, so on lines 2 and 3 there is no work for us to do. Another way to put it is this: We must show that if p is true, then q cannot be false. The bottom line is that we must show the bottom line is impossible.

There is no set procedure for finding a proof. There are, however, two steps to take if you want to improve your ability to construct proofs. One is to study examples of proofs; the other is to practice making your own proofs. You've seen several simple examples in this book already:

Chapter 1	(10.8)	$A \cup B = B \cup A$
	(10.1)	$A \cup (B \cap C) = (A \cup B) \cap (A \cup C)$
	(12.5F)	$A \cap (B \cup C) = (A \cap B) \cup (A \cap C)$
	(12.8F)	If $A \subseteq B$, then $\overline{B} \subseteq \overline{A}$
	(16.1)	$(a,b) = (x,y)$ iff $a = x$ and $b = y$
Chapter 2	(7.3)	Substitution in a logical implication gives a material implication
	(7.5)	$f_1 \Longrightarrow f_2$ iff $T(f_1) \subseteq T(f_2)$.
	(7.13)	Properties of logical equivalence
	(8.6)	De Morgan's laws
	(10.15)	The substitution theorem

Several problems ask you to prove things. They should provide appropriate practice at this stage.

Each problem to prove something has a *given*, something you are to assume is true. You then must find some way to connect that *given* to the *target*, the assertion to be proved. Often you are to prove $p \longrightarrow q$ for certain propositions p and q. Then p is the given and q is the target. Finding the connection between p and q can be obvious, easy, tedious, tricky, difficult, or impossible. It all depends on the problem and on the person trying to solve it.

So far we have found the connection by starting from the definitions of the given terms. Since our targets were close to our givens we didn't have to work too hard. As things get more advanced, we will use already-proved theorems about our givens to help us hit the target. So far we've done that only in proving the substitution theorem (10.15), where we used the auxiliary theorem (7.5) that $f \Longrightarrow g$ iff $T(f) \subseteq T(g)$. That result is itself close to the definition of $f \Longrightarrow g$.

Some books that may help you find the missing link between given and target are the classic by Polya (*How To Solve It* (31.6)) and the text by Solow (*How To Read and Do Proofs* (31.11)).

(12.3) *Proof by contrapositive.* One technique of proof you should know is the contrapositive. To prove $p \longrightarrow q$ we sometimes instead prove $\sim q \longrightarrow \sim p$. Since $a \longrightarrow b \Longleftrightarrow \sim b \longrightarrow \sim a$, if we can show that $\sim q \longrightarrow \sim p$ is true, then we've shown that $p \longrightarrow q$ is true. We use this technique because sometimes it's easier to make the connection from $\sim q$ to $\sim p$ than from p to q. We'll point out our uses of this technique as we come to them. For now, consider the following:

Example

Suppose p, q, and r are propositions such that both $(i) \sim p \vee q \longrightarrow r$ and $(ii)\ p \wedge \sim r \longrightarrow q$ are true. Prove that r is necessarily true. We prove it by the contrapositive method. Let r be false. Then from (i) and the definition of material implication, $\sim p \vee q$ is F; thus p is T and q is F. When r is F, (ii) becomes $p \wedge T \longrightarrow q$, which is $p \longrightarrow q$. But we said p is T and q is F. Thus $p \not\rightarrow q$. So if r is F, (ii) cannot hold.

The structure of the proof is this: We tried to prove $s_1 := (i) \wedge (ii) \longrightarrow r$. We showed $(i) \wedge \sim r \longrightarrow \sim (ii)$, by confining ourselves to the case where both $s_2 := (i)$ and $\sim r$ are true. The propositions s_1 and s_2 are materially equivalent because of the following logical equivalence:

$$f := a \wedge b \longrightarrow c \Longleftrightarrow a \wedge \sim c \longrightarrow \sim b =: g.$$

How to prove this? Easy: $F(f) = \{(T,T,F)\}$ and $F(g) = \{(T,T,F)\}$. A way to make this logical equivalence look more like the contrapositive is to state f as "When a is true, $b \longrightarrow c$." Then state g as "When a is true, $\sim c \longrightarrow \sim b$."

(12.4) *Proof by contradiction.* Another technique of proof, related to contrapositive, in fact, is proof by contradiction. The classic example is the proof (dating to the Pythagoreans in the sixth century B.C.) that $\sqrt{2}$ is irrational. We shall prove it by assuming its negation and then deriving a contradiction. Thus define the proposition q as

$$q: \sqrt{2} \text{ is irrational.}$$

Then $\sim q$ is: $\sqrt{2}$ is rational. That means that $\sqrt{2} = a/b$ for some integers a, b with $b \neq 0$. Assuming a little number theory (which you'll find in Chapter 6), we may cancel any factor m common to a and b. The result is that $\sqrt{2} = c/d$ where $a = mc$ and $b = md$, and c and d have no common factor except 1. We take the last as a proposition r:

$$r\text{: } c \text{ and } d \text{ have no common factor.}$$

Thus the assumption $\sim q$ implies that there are certain integers c and d such that r is true. That is, we've shown that $\sim q \longrightarrow r$.

Now we square both sides of $\sqrt{2} = c/d$, getting $2 = c^2/d^2$, or

$$2d^2 = c^2.$$

Assuming a little more of Chapter 6, we see that since 2 is a factor of c^2 and 2 is prime, then 2 must be a factor of c. Thus $c = 2e$ for some integer e. Now

$$2d^2 = c^2 = 4e^2.$$

So $d^2 = 2e^2$. By the same reasoning 2 is a factor of d, making c and d have the common factor 2. Thus we've shown that $\sim q \longrightarrow \sim r$.

Since we know already that $\sim q \longrightarrow r$, we've shown $\sim q \longrightarrow r \wedge \sim r$. The proposition $r \wedge \sim r$ is false no matter what the truth-value of r. The only way $\sim q \longrightarrow F$ can be true (and we just showed it is true) is for $\sim q$ to be false. Thus q is true, as we wanted to prove.

(12.5) *If-and-only-if proofs*. To prove $p \longleftrightarrow q$ for propositions p and q you need to prove both

$$p \longrightarrow q \qquad \text{and} \qquad q \longrightarrow p.$$

Why? Because material equivalence ($\longleftrightarrow$) is *defined* that way, in (4.4). You may substitute the contrapositive for either or both of the implications $p \longrightarrow q$ and $q \longrightarrow p$. In particular you've proved p iff q if you've proved

$$p \longrightarrow q \qquad \text{and} \qquad \sim p \longrightarrow \sim q.$$

13. Conjunctive and Disjunctive Normal Forms

Another name for *and* is *conjunction* and for *or* is *disjunction*. Thus we may say "$a \vee b$ is the disjunction of a and b." Suppose we consider a truth-table in the two propositional variables a and b, for example, exclusive-or:

a	b	f
T	T	F
T	F	T
F	T	T
F	F	F

f is a xor b

The truth-value of the propositional formula f depends on that of a and of b, as indicated by the table. Both $(a \vee b) \wedge (\sim a \vee \sim b)$ and $(\sim a \wedge b) \vee (a \wedge \sim b)$ have the same truth-table as f.

Practice (13.1P) Prove these claims.

They are called, respectively, the *conjunctive* and *disjunctive normal forms* for f. The general definition (for two variables a, b) follows:

DEFINITION (13.2) If the propositional formula f is not a tautology or a contradiction, then the **disjunctive normal form (DNF)** of f is the disjunction (logically equivalent to f) of one, two or three distinct terms of the form $x \wedge y$, where x is a or $\sim a$ and y is b or $\sim b$;†

(13.3) The **conjunctive normal form (CNF)** of f is the conjunction (logically equivalent to f) of one, two, or three distinct terms of the form $x \vee y$, with x and y as already defined.†

If f is a tautology, its normal form of either type is defined to be T; if a contradiction, F.

Note that we have not yet shown that every propositional formula has a DNF and a CNF. We do so in a moment.

From (8.4) we know that $a \longrightarrow b \Longleftrightarrow \sim a \vee b$. Therefore $\sim a \vee b$ is the conjunctive normal form of $a \longrightarrow b$. What is the disjunctive normal form of $a \longrightarrow b$? To answer this question it is best to develop a general approach. Consider the four possible terms of disjunctive normal form. They are $a \wedge b$, $\sim a \wedge b$, $a \wedge \sim b$, $\sim a \wedge \sim b$. Let's call them the **building blocks** for DNF in two variables. Their truth-tables (and that of $a \longrightarrow b$ as well) follow.

(13.4)

a	b	$a \wedge b$	$a \wedge \sim b$	$\sim a \wedge b$	$\sim a \wedge \sim b$	$a \longrightarrow b$
T	T	T	F	F	F	T
T	F	F	T	F	F	F
F	T	F	F	T	F	T
F	F	F	F	F	T	T

You can see that each of these building blocks $a \wedge b$ and so on takes the truth-value T on only one line of the table. If we join two of these terms with *or*, we get a formula true on the same two lines on which the T's of our two terms appear, because $T \vee F = T$, and $F \vee F = F$. For example, $(a \wedge b) \vee (a \wedge \sim b)$, the disjunction of the first two of these terms, has the truth-table with entries T, T, F, F in rows 1 through 4 respectively. Thus for the DNF of $a \longrightarrow b$ we have

(13.5) $$(a \longrightarrow b) \Longleftrightarrow (a \wedge b) \vee (\sim a \wedge b) \vee (\sim a \wedge \sim b).$$

We have formed the disjunctive normal form of $a \longrightarrow b$ by picking for each T in the truth-table of $a \longrightarrow b$ the building block which has its lone T-value in the same row.

(13.6) *The first way to find the DNF or CNF.* We may mechanize our approach to finding the DNF with the truth-table (13.4): We start with a T-value in the column for the *target* formula ($a \longrightarrow b$ in this example). We follow its row to the only T-value in it (under the building blocks). We then go to the head of that column. Do this for all T-values in the target column, and combine all the building blocks chosen this way with *or*. It's simple, as the marked truth-table (13.4), repeated here, shows.

†Since $h \vee g \Longleftrightarrow g \vee h$ for all formulas g, h, the order of the terms in the DNF is immaterial. We therefore think of the two DNFs as the same if they differ only in the order of their terms. The same is true for CNF.

(13.7)

a	b	$a \wedge b$	$a \wedge \sim b$	$\sim a \wedge b$	$\sim a \wedge \sim b$	$a \longrightarrow b$
T	T	Ⓣ ←	F	F	F	Ⓣ
T	F	F	T	F	F	F
F	T	F	F	Ⓣ ←	F	Ⓣ
F	F	F	F	F	Ⓣ ←	Ⓣ

This procedure obviously works for any compound proposition in a and b; it also shows us that the disjunction of all four terms produces a tautology. We could extend the definition of disjunction to say that a contradiction is the disjunction of zero terms of this type. Then we could restate the definition by saying that the DNF of a propositional formula depending only on a and b is the disjunction of zero or more terms $x \wedge y$ as specified in (13.2). No special statement would be needed for tautology or contradiction.

(13.8) The analogue of (13.6) for CNF is to find all the F-values in the target column, trace back to the unique F-value in that row, and go to the building block (for CNF now) heading that column. You then join all such building blocks with *and*.

(13.9) *More than two variables.* The number of building blocks doubles each time we add a new variable. For example, the building blocks for DNF in the three atoms a, b, c are

$$a \wedge b \wedge c, \quad a \wedge b \wedge \sim c, \quad a \wedge \sim b \wedge c, \quad a \wedge \sim b \wedge \sim c,$$
$$\sim a \wedge b \wedge c, \quad \sim a \wedge b \wedge \sim c, \quad \sim a \wedge \sim b \wedge c, \quad \sim a \wedge \sim b \wedge \sim c,$$

There are eight; there were four for just a, b. The procedures for finding DNF and CNF are the same for more variables as the procedures in (13.6) and (13.8) except that the truth-tables are bigger. The DNF of a formula f becomes more complicated if we increase the number of variables. Thus a has DNF a in the setting of the one variable a. In the setting of two-variables a and b, a has as DNF the first two-term formula in (13.13). In the setting of three variables a, b, and c the DNF of a has four building blocks, those on the first row displayed just above.

(13.10) *The second way to find DNF or CNF.* Another way to find the DNF or CNF of a propositional formula f in the n atoms $x_1, \ldots, x_n$ is this:

For DNF, find $T(f)$ by the shortcut method. For every n-tuple $(t_1, \ldots, t_n)$ in $T(f)$, the DNF has in it the building block $y_1 \wedge y_2 \wedge \cdots \wedge y_n$, where

$$(13.11) \qquad y_i := \begin{cases} x_i & \text{if } t_i = T \\ \sim x_i & \text{if } t_i = F \end{cases} \qquad \text{for } i = 1, \ldots, n.$$

This claim follows immediately from (13.6).

Example

Suppose $n = 3$ and $T(f) = \{(T, F, T), (F, F, T)\}$. Then the DNF of f is $(x_1 \wedge \sim x_2 \wedge x_3) \vee (\sim x_1 \wedge \sim x_2 \wedge x_3)$. These building blocks are the ones that are true on the points of $T(f)$.

For CNF, find $F(f)$. Then for each n-tuple $(t_1, \ldots, t_n)$ in $F(f)$, the CNF of f has the building-block $y_1 \vee \cdots \vee y_n$, where

$$(13.12) \qquad y_i := \begin{cases} \sim x_i & \text{if } t_i = T \\ x_i & \text{if } t_i = F \end{cases} \qquad \text{for } i = 1, \ldots, n.$$

This claim follows from (13.8).

Example If $F(f) = \{ (F,T,F),(T,F,T) \}$, then the CNF of f is $(x_1 \vee \sim x_2 \vee x_3) \wedge (\sim x_1 \vee x_2 \vee \sim x_3)$. These building blocks are precisely the ones which are false on the points of $F(f)$.

The DNF and CNF are not always the simplest expressions for a given formula. For example, consider the proposition a. Nothing could be simpler, but notice:

$$a \Leftrightarrow (a \wedge b) \vee (a \wedge \sim b) \Leftrightarrow (a \vee b) \wedge (a \vee \sim b). \tag{13.13}$$

These are the DNF and CNF, respectively, for a in the setting of the two atoms a, b.

We can now state and prove the result that CNF and DNF exist and are unique.

Theorem (13.14) Each formula f in n propositional variables is logically equivalent to exactly one DNF and one CNF in those n variables.

Proof We construct the truth table for f and choose the building blocks in accordance with procedure (13.6) or (13.8). Clearly no other choice is possible. The *set* B of these building blocks is thus unique; since *or* and *and* are commutative and associative the disjunction (or conjunction) of the building blocks in B produces formulas logically equivalent to f no matter what order we use to combine them. QED

(13.15) *Applications*. One use of CNF and DNF is to turn truth-tables into propositional formulas. Another is to check whether two formulas are logically equivalent. The latter works this way (sometimes): You use the distributive laws and other tautologies to express each formula in, say, DNF. Then just pair off the terms to see whether or not the two formulas are equivalent.

Example Consider the two formulas

$$(a \wedge b) \longrightarrow c, \quad \text{and} \quad a \longrightarrow (b \longrightarrow c).$$

We work on these first by getting rid of the $\longrightarrow$, using $a \longrightarrow b \Leftrightarrow \sim a \vee b$.

$$\begin{aligned} (a \wedge b) \longrightarrow c &\Leftrightarrow \sim (a \wedge b) \vee c \\ &\Leftrightarrow (\sim a \vee \sim b) \vee c \end{aligned}$$

The last is a single building-block for CNF; therefore, it is in CNF. So we turn the other into its CNF:

$$\begin{aligned} a \longrightarrow (b \longrightarrow c) &\Leftrightarrow \sim a \vee (b \longrightarrow c) \\ &\Leftrightarrow \sim a \vee (\sim b \vee c). \end{aligned}$$

Therefore, since "$\vee$" is associative, we've shown that $(a \wedge b) \longrightarrow c$ is logically equivalent to $a \longrightarrow (b \longrightarrow c)$.

This method for finding DNF and CNF is sometimes more convenient than truth-tables.

Another application of the DNF is to the realization of Boolean functions (: = propositional formulas) in computer hardware. *Programmable logic arrays* (PLAs) are the standard means for this; they require that the function be first expressed in DNF. See (13.16) and Chapter 5, (21).

Terminology. Some texts call our DNF the *sum-of-products* form, and our CNF the *product-of-sums* form. Still others call them the *minterm* form and *maxterm* form, respectively. Such texts use DNF to mean the *simplest* logically equivalent expression made with DNF building blocks in fewer variables if possible. Thus $a \longrightarrow b$ would have DNF $\sim a \vee b$, since the latter is the disjunction of two one-variable building blocks for DNF. It's all semantics.

(13.16) *Karnaugh maps.* Even though propositional formulas in DNF can be easily set up in a PLA, designers want to simplify the formula as much as reasonably possible to save space on the chip. The Karnaugh map is a way to simplify a propositional formula. It rests on a two-part foundation, namely, a distributive law and the law of the excluded middle:

$$a \vee \sim a \Leftrightarrow T \tag{13.17}$$

$$(a \wedge c) \vee (b \wedge c) \Leftrightarrow (a \vee b) \wedge c. \tag{13.18}$$

For example, set $b := \sim a$ in (13.18). Then

$$\begin{aligned}(a \wedge c) \vee (\sim a \wedge c) &\Leftrightarrow (a \vee \sim a) \wedge c \\ &\Leftrightarrow T \wedge c \text{ by (13.17)} \\ &\Leftrightarrow c.\end{aligned} \tag{13.19}$$

The left-hand side simplifies to c. (Compare (13.13).) The result in (13.19) seems trivial, but when there are many variables or terms it can be hard to recognize the possibility of such simplification. That is where the Karnaugh map comes in. For example, assume three variables a, b, c and consider the formula

$$f = (a' \wedge b' \wedge c') \vee (a' \wedge b \wedge c') \vee (a \wedge b' \wedge c) \vee (a \wedge b \wedge c),$$

where $a' := \sim a$, and so on. We represent f in a **Karnaugh map** as follows:

(13.20F)

c \ ab	00	01	11	10
0	1	1		
1			1	1

where 00 stands for $a' \wedge b'$, 01 for $a' \wedge b$, and so forth. The entry 1 in the position where $c = 0$ and $ab = 00$ represents the term $a' \wedge b' \wedge c'$, and so on for the others. The 0 stands for the primed variable, the 1 for the unprimed.

Now we search the Karnaugh map for adjacent 1's. In the first row we find two such 1's; they stand for the terms

$$\begin{aligned}(a' \wedge b' \wedge c') \vee (a' \wedge b \wedge c') &\Leftrightarrow (a' \wedge c') \wedge (b' \vee b) \\ &\Leftrightarrow (a' \wedge c') \wedge T \\ &\Leftrightarrow a' \wedge c'.\end{aligned} \tag{13.21}$$

So that part of f can be simplified.

The reasoning of (13.21) can be made automatic. If we look at the 1's in the map we see that their terms 000 and 010, respectively, differ in only one place, that for b. The argument of (13.21) shows that the disjunction of the two terms is equivalent to $a' \wedge c'$, which is 00 (for a and c). In other words, whenever you find two terms differing in only one variable you simply delete that variable and keep the remaining as a single term of one fewer variable.

Thus we may conclude that the other two terms, 111 and 101, come down to 11 (for ac), and so

$$f \Leftrightarrow (a' \wedge c') \vee (a \wedge c).$$

The Karnaugh map makes it easier to find pairs of terms differing in only one variable, because it is set up so that such pairs (almost) always show up as adjacent 1's in the map. The order of the column headers

00 01 11 10

is chosen so that adjacent headers differ in only one coordinate.

Here is an example where simplification is possible but the 1's are not adjacent:

(13.22*F*)

c \ ab	00	01	11	10
0				
1	1			1

This stands for

$$g := (a' \wedge b' \wedge c) \vee (a \wedge b' \wedge c).$$

You see that the terms are 001 and 101, differing in only the first (a-) coordinate. They simplify to

$$g \Leftrightarrow b' \wedge c.$$

The column-headers are 00 and 10 for the 1's in the map, and they differ only in the a-coordinate. That is, we should regard the map as drawn on a cylinder with the columns wrapped around so they touch. Then (13.22*F*) would be simply another instance of a map with adjacent 1's; we could call it a Karnaugh cycle.

One more example: Suppose our Karnaugh map were

(13.23*F*)

c \ ab	00	01	11	10
0				1
1	1			1

We note that $a \Leftrightarrow a \vee a$ (another piece of the Karnaugh-map foundation), and so we use the 101 term twice. Wrapping around as in (13.22*F*) gives $b' \wedge c$ as before. Put it back and use the two vertical adjacent 1's to get 10 (for ab), that is, $a \wedge b'$. Thus (13.23*F*) represents $(b' \wedge c) \vee (a \wedge b')$.

It gets more complicated with more variables. ◀

(13.24) *Analogy with sets.* There is a perfect analogy—to be explored later, in (29)—between sets combined by ∪, ∩, and $\bar{\ }$ (complement), and propositional formulas in DNF and CNF. For now, we merely show the analogy in the following table (13.25).

(13.25) Basic Facts on

Propositional Formulas		*Sets*
$a \vee a \Leftrightarrow a \wedge a \Leftrightarrow a$	Idempotence	$A \cup A = A \cap A = A$
$(a \vee b) \vee c \Leftrightarrow a \vee (b \vee c)$	Associativity	$(A \cup B) \cup C = A \cup (B \cup C)$
$(a \wedge b) \wedge c \Leftrightarrow a \wedge (b \wedge c)$		$(A \cap B) \cap C = A \cap (B \cap C)$
$a \vee b \Leftrightarrow b \vee a$	Commutativity	$A \cup B = B \cup A$
$a \wedge b \Leftrightarrow b \wedge a$		$A \cap B = B \cap A$
$a \wedge (b \vee c) \Leftrightarrow (a \wedge b) \vee (a \wedge c)$	Distributivity	$A \cap (B \cup C) = (A \cap B) \cup (A \cap C)$
$a \vee (b \wedge c) \Leftrightarrow (a \vee b) \wedge (a \vee c)$		$A \cup (B \cap C) = (A \cup B) \cap (A \cup C)$
$\sim (\sim a) \Leftrightarrow a$	Involution	$\overline{(\overline{A})} = A$
$\sim (a \vee b) \Leftrightarrow (\sim a) \wedge (\sim b)$	De Morgan Laws	$\overline{(A \cup B)} = \overline{A} \cap \overline{B}$
$\sim (a \wedge b) \Leftrightarrow (\sim a) \vee (\sim b)$		$\overline{(A \cap B)} = \overline{A} \cup \overline{B}$
$a \vee F \Leftrightarrow a \qquad a \wedge F \Leftrightarrow F$	Identity	$A \cup \emptyset = A \qquad A \cap \emptyset = \emptyset$
$a \vee T \Leftrightarrow T \qquad a \wedge T \Leftrightarrow a$		$A \cup V = V \qquad A \cap V = A$ (V = Universal Set)
$a \vee (a \wedge b) \Leftrightarrow a$	Absorption	$A \cup (A \cap B) = A$
$a \wedge (a \vee b) \Leftrightarrow a$		$A \cap (A \cup B) = A$

14. Logical Puzzles

(14.1) Here is a famous old puzzle: The king's son is up for marriage to the first of three finalists who solves the following problem. The king tells them he will place a green or red hat on the head of each of the women in a darkened room. He does so. He instructs them to hold up their hand when they see a red hat, and to announce the color of their own hat (which they will not be able to see) as soon as they know it. Then he turns on the lights. Through the one-way glass we see that all three hats are red. Each woman raises her hand, and after a few minutes one announces that her hat is red. How did she know?

You probably remember the solution: If we call the women A, B, and C, A can reason that if her hat were green, then one of B and C, say B, would be able to deduce that if B's hat were green, then C could not see a red hat. So C's hand would not be up. Thus B could know her hat is red. Since C's hand *is* up, and neither B nor C announces her hat color, A concludes that hers is red.

A has to wait to give B and C a chance to do the reasoning; and since this is the finals, A can safely assume that B and C are capable of this reasoning.

The situation is symmetric so each contestant has an equal chance. It is a matter of which one thinks it through first.

Practice (14.2P) Try it on your friends. Sometimes use two reds and one green hat, or one red and two greens.

Let's analyze this solution with propositional logic. Define the propositions

$$A_1: \text{ } A\text{'s hat is red}$$
$$A_2: \text{ } A\text{'s hand is up}$$
$$A_3: \text{ } A \text{ has announced her hat color.}$$

and similarly for B and C. Remember—each woman knows the truth-value of eight of these nine propositions. A does not know A_1, B does not know B_1, and C does not know C_1.

When the lights go up we have

$$A_2 \wedge B_2 \wedge C_2 \quad \text{and} \quad \sim A_3 \wedge \sim B_3 \wedge \sim C_3.$$

Also, A knows $B_1 \wedge C_1$, B knows $A_1 \wedge C_1$, and C knows $A_1 \wedge B_1$.

From the king's rules

(14.3) $$C_2 \longrightarrow A_1 \vee B_1$$

We now look at the situation from A's point of view. A argues by contradiction. She assumes $\sim A_1$. From (14.3) she concludes

(14.4) $$\sim A_1 \wedge C_2 \longrightarrow B_1.$$

But $\sim A_1$, if true, is known to B, and C_2 also is clear to B. Therefore, B could conclude B_1 from (14.4) and would announce (that is, B_3 would become true).

Since $\sim B_3$ is still true, we have the contradiction $B_3 \wedge \sim B_3$. Therefore, $\sim A_1$ is false; so A_1 is true. (A's reasoning is strengthened by the symmetry between B and C. $B_2 \longrightarrow A_1 \vee C_1$ holds, so C could reason as B would if $\sim A_1$.)

For a nicer solution, solve problem 54 of Chapter 3.

Practice (14.5P) Derive (14.4) from (14.3). Do it by showing

$$c \longrightarrow a \vee b \Longleftrightarrow \sim a \wedge c \longrightarrow b.$$

(14.6) Here is another old one: A tourist in a land where some people always tell the truth and the rest always lie wants to ask directions of a native. He needs first to see whether the fellow is a truth-teller. What *one* yes-or-no question can he ask the native to determine his truth-value, so to speak?

Answer: "What would you say if I ask, 'Are you a truth-teller?' "

Practice (14.7P) Work out why this is an answer.

Smullyan's book (31.9) is full of problems like (14.6). We have included a few of them in (16).

15. Summary of Propositional Logic

Proposition: Statement unambiguously true or false. $p, q, r \ldots$ (1.1)

Propositional variable, atom: Placeholder that can be replaced by any proposition including T, F. $a, b, c, \ldots$ (6.1)

Propositional formula: T, F, any atom, or if f and g are propositional formulas then $\sim f, (f) \wedge (g), (f) \vee (g), (f) \longrightarrow (g)$ are propositional formulas. A propositional formula does *not* have a truth-*value*. It has a truth-*table*. (6.2)

$T(f) :=$ the set of lines of the truth-table of f where f is T. (9)

Logical implication ($\Longrightarrow$) is a *relation* between propositional formulas. (7.5)

$$f \Longrightarrow g \text{ iff } T(f) \subseteq T(g). \quad (9.7)$$

$$f \Longrightarrow g \text{ is a proposition.} \quad (7.7)$$

A propositional formula is a *tautology* iff it is T on every line of its truth-table. (8.1)

Theorem. $f \Longrightarrow g$ iff $(f) \longrightarrow (g)$ is a tautology. (8.2)

Logical equivalence ($\Longleftrightarrow$) : $f \Longleftrightarrow g$ iff $f \Longrightarrow g$ and $g \Longrightarrow f$. (7.9)

$$f \Longleftrightarrow g \text{ iff } T(f) = T(g). \quad (9.8)$$

Logical equivalence ($\Longleftrightarrow$) is a relation between propositional formulas. Material equivalence ($\longleftrightarrow$) is a relation between propositions. (7.15)

16. Problems on Propositional Logic

Problems for Section 1

1.1 A, B, and C are sets, defined as $A := \{1, \ldots, 10\}$, $B := \{3, 7, 11, 12\}$, and $C := \{0, 1, \ldots, 20\}$. Which of the following are propositions? Explain.

(*i*) $1 + 1 = 3$
(*ii*) $A \cap B$
(*iii*) $7 \in A$
(*vi*) $(8 + 22)^3 / 10^2$
(*v*) $A \cup B \subseteq C$
(*iv*) $9 \in B$
(*vii*) $B \cap C \in 9$
(*viii*) C is an infinite set.

1.2 What are the truth-values of the propositions in problem 1.1?

Problems for Section 2

2.1 What is the truth-value of

(*i*) $2 \cdot 2 = 4$ or $3 \cdot 3 = 6$?
(*ii*) if $2 + 2 = 5$, then $3 + 3 = 7$?

2.2 What is the truth-value of

(*i*) $(1 + 2 = 3) \wedge (3 + 3 = 7)$? Explain.
(*ii*) if $2 + 3 = 6$, then $3 + 4 = 9$? Explain.

2.3 Let p, q, r, and s be propositions such that q is T and r is F. Determine the truth-value of the compound proposition $(p \longrightarrow q) \wedge (r \longrightarrow s)$. Naturally, you should explain your answer.

2.4 Suppose $p \rightarrow q$ is true. What can you say about the converse $q \longrightarrow p$? Explain. Consider the same question, but assume instead that $p \rightarrow q$ is false.

2.5 (*i*) Find truth values for p, q, and r that make $p \longrightarrow (q \wedge r)$ have a truth-value different from that of $(p \longrightarrow q) \wedge r$. Explain.
(*ii*) Which proposition in (*i*) is understood to be the definition of $p \longrightarrow q \wedge r$? Why?

2.6 Express the proposition "p unless q" in terms of the propositions p, q, and the logical symbols $\wedge, \vee, \sim, \longrightarrow$. [Some dictionaries mumble about "unless." Mine says it means "if not."]

2.7 Let $A := \{1, 2, 3, 4, 5\}$. Write the truth-value of each of the following statements (*i*) through (*vi*):

(*i*) $(3, 4) \in \mathcal{P}(A)$, the power set of A.
(*ii*) $\{(1, 3), (4, 2)\} \subseteq A \times A$.
(*iii*) $\{(2, 4), (5, 2)\} \in A \times A$.
(*iv*) $((1, 5) \in A \times A)$ or $(\{(2, 2)\} \in A \times A) \longrightarrow (1, 3) \in A \times A$.
(*v*) $\{2\} \in A$.
(*vi*) $\{3, \{3\}\} \subseteq \mathcal{P}(A)$.

Explain your answer to (*vi*).

Problems for Section 3

3.1 Let p, q, and r be propositions such that $p \rightarrow q$, $(\sim p) \rightarrow r$, and $r \rightarrow (p \vee q)$ are all true. Show that q is true.

3.2 If $p \rightarrow q$ is false, what can be said about the truth-value of $((\sim p) \wedge q) \longleftrightarrow (p \vee q)$?

3.3 Say whether the information given about the following proposition is enough to determine its truth-value: q is F and r is T. Explain.

$$(p \rightarrow q) \wedge (r \rightarrow s).$$

3.4 Let p and q be propositions. If $(\sim p) \wedge q$ is false, what can you prove about the truth-value of

$$[p \vee (\sim q) \vee (p \vee q)] \wedge [(\sim (p \vee q)) \vee ((\sim p) \wedge q)]?$$

Problems for Section 5

5.1[Ans] Express the following in terms of propositional logic "If wishes were horses beggars would ride." Explain why it is considered to be true.

Problems for Section 6

6.1 [Recommended] (from Gleason (31.1)) Let X be a set, x and y points in X, and A and B subsets of X. Assume that the following two propositions are true.

If $x \in A$ and $y \in B$, then $y \in A$

If $x \notin A$ or $y \in A$, then $y \notin B$.

What conclusion can you draw, and why?
[Hint: let p be the proposition $x \in A$, and so on.]

6.2 Which propositional formula has the truth-table (i)? Answer the same question for (ii).

a	b	(i)	(ii)
T	T	F	T
T	F	T	T
F	T	T	F
F	F	T	T

6.3 Find truth-values in each case below so that the propositional formula is false—*if* such truth-values exist.

(i)[Ans] $((a \rightarrow b) \wedge b) \rightarrow a$
(ii) $((a \rightarrow b) \wedge a) \rightarrow b$
(iii)[Ans] $a \vee \sim a \rightarrow b$
(iv) $(a \vee b) \rightarrow (a \rightarrow b)$.

6.4 Write the truth-tables for

(i) $b \vee (a \longrightarrow b)$,
(ii) $a \longrightarrow (b \longrightarrow a)$,
(iii) $(a \longrightarrow b) \wedge (\sim b \vee c)$.

6.5 Write the truth-table for

(i) $b \vee (a \rightarrow b)$,
(ii) $a \rightarrow (b \rightarrow a)$.

6.6 Write the truth-table for

(i) $a \wedge (b \longrightarrow a)$,
(ii) $(a \vee b) \longrightarrow a$.

6.7 Find the truth-table of

(i) $(a \longrightarrow T) \wedge (F \longrightarrow b)$,
(ii) $(F \vee a) \longrightarrow (b \wedge F)$,
(iii) $(a \vee b) \wedge (a \vee \sim b)$.

Problems for Section 7

7.1 Prove that $a \wedge b \Longrightarrow a \vee b$.

7.2 (i) Prove or disprove the following logical implications between propositional formulas:

$$(a \vee b) \wedge (\sim a \vee b) \Longrightarrow b$$
$$a \wedge (b \vee c) \Longrightarrow (a \longrightarrow b) \longrightarrow c.$$

(ii) Consider the same question for the logical converses. That is, substitute $\Longleftarrow$ for $\Longrightarrow$.

7.3 Prove $\sim (a \longrightarrow b) \Longleftrightarrow a \wedge \sim b$.

7.4 Show that $a \wedge b \longrightarrow c \Longleftrightarrow (a \longrightarrow c) \vee (b \longrightarrow c)$.

7.5 Show that $a \vee b \longrightarrow c \Longrightarrow a \wedge b \longrightarrow c$ but not conversely (that is, not $\Longleftarrow$).

7.6[Ans] (i) Prove that $a \longleftrightarrow b \Longleftrightarrow \sim a \longleftrightarrow \sim b$.
(ii) Prove that $a \longrightarrow b \not\Longleftrightarrow \sim a \longrightarrow \sim b$.

7.7 The *Peirce arrow* $\downarrow$ and Sheffer stroke $|$ are the logical connectives defined by the truth-tables

a	b	$a \downarrow b$	$a \mid b$
T	T	F	F
T	F	F	T
F	T	F	T
F	F	T	T

(i) Prove that $\sim a$, $a \vee b$, and $a \wedge b$ can all be expressed in terms of the Peirce arrow alone. Explain how you found your answer.

(*ii*) Answer the same question for the Sheffer stroke. [Note. The Peirce arrow is also called *nor* and the Sheffer stroke *nand*.]

Problems for Section 8

8.1 Which of the following are tautologies? Prove your answers.

(*i*) $(a \longrightarrow b) \wedge (a \longrightarrow c) \longleftrightarrow (a \longrightarrow (b \wedge c))$
(*ii*) $(a \longrightarrow c) \wedge (b \longrightarrow c) \longleftrightarrow ((a \vee b) \longrightarrow c)$
(*iii*) $\sim (a \longleftrightarrow b) \longrightarrow (\sim a \longleftrightarrow b)$
(*iv*) $(a \longrightarrow (b \longrightarrow c)) \longrightarrow ((a \longrightarrow b) \longrightarrow c)$

8.2[Ans] Negate the following propositional formulas. Express for each negation a logically equivalent formula in which no compound formula is negated, only some atoms.

(*i*) $a \longrightarrow (\sim b)$, (*ii*) $a \longrightarrow (b \longrightarrow c)$, (*iii*) $(\sim a) \longrightarrow b$.

8.3 Prove that (*i*) is not, and that (*ii*) is, a tautology:

(*i*) $(a \longleftrightarrow b) \longrightarrow (a \wedge b)$,
(*ii*) $(a \longleftrightarrow b) \longleftrightarrow (a \wedge b) \vee (\sim a \wedge \sim b)$,

8.4 Work out the truth-tables for *modus ponens* (10.2) to show that it is a logical implication but not an equivalence.

8.5 Recall that a xor $b \Longleftrightarrow (a \vee b) \wedge \sim (a \wedge b)$. Prove (*i*) that a xor b is logically equivalent to $a \longleftrightarrow (\sim b)$. (*ii*) What relation is there between xor and $\longleftrightarrow$?

8.6 (*i*) Prove that $((a \longrightarrow b) \longrightarrow a) \longrightarrow a$ is a tautology.
(*ii*) Is $(a \longrightarrow b) \longrightarrow a$ a tautology? Explain.

8.7 Prove (8.2) and (8.3). (These are assertions in the text.)

8.8 Prove $a \wedge b \longrightarrow c \Longleftrightarrow b \wedge \sim c \longrightarrow \sim a$.

8.9 Negate each of the following propositional formulas f by finding a formula logically equivalent to $\sim f$ in which the negation symbol $\sim$ applies only to individual atoms.

(*i*) $a \vee \sim b$
(*ii*) $a \longrightarrow \sim b$
(*iii*) $a \longrightarrow (b \longrightarrow c)$
(*iv*) $\sim a \longrightarrow b$
(*v*) $(a \wedge b) \vee c$
(*vi*) $a \wedge (b \longrightarrow c)$

Problems for Section 9

9.1 Find all values of a, b in $\{T, F\}$ such that $(a \longrightarrow b) \longrightarrow (b \longrightarrow a)$ is F (false). Prove your answer.

9.2 Use a shortcut to find $T(g)$ or $F(g)$ for $g := (a \longleftrightarrow b) \longrightarrow (a \vee b)$. Explain.

9.3 Use a shortcut to find $T(g)$ or $F(g)$ for the formula $g := (a \vee b \longrightarrow \sim c) \wedge (\sim a \vee \sim b)$.

9.4 Use the shortcut method to find the set of truth-values for a, b, c such that $a \longrightarrow (\sim b \wedge c) \vee (a \wedge \sim c)$ is F. Explain your procedure.

9.5 Find a simple propositional formula logically equivalent to that in problem 9.4. Explain.

9.6[Ans] For propositional formulas g_1 and g_2, prove

$$T(g_1) \cap T(g_2) = T(g_1 \wedge g_2)$$
$$T(g_1) \cup T(g_2) = T(g_1 \vee g_2)$$
$$F(g_1) \cap F(g_2) = F(g_1 \vee g_2)$$
$$F(g_1) \cup F(g_2) = F(g_1 \wedge g_2).$$

9.7 How can you quickly determine from the previous problem what $T(f \longrightarrow g)$ is?

9.8 Let a, b, and c be atomic propositions. Prove

(*i*) $a \longrightarrow b \Longleftrightarrow (a \wedge \sim b) \longrightarrow b$,
(*ii*) $a \longrightarrow b \Longleftrightarrow (a \wedge \sim b) \longrightarrow (c \wedge \sim c)$.

9.9 Prove $(a \longrightarrow b) \longrightarrow c \Longrightarrow b \longrightarrow (a \longrightarrow c)$.

9.10 State and prove conditions like (9.7) and (9.8) in terms of the falsity-sets $F(f)$ and $F(g)$.

9.11[Ans] Find $T(f)$ or $F(f)$ for the following formulas. Use the shortcut method.

(*i*) $(a \longrightarrow b) \wedge b$
(*ii*) $a \longrightarrow (b \longrightarrow a)$
(*iii*) $((a \wedge b) \longrightarrow c) \longrightarrow (\sim c \longrightarrow a)$.

9.12 Use a shortcut to find all truth-values of a, b, c such that $(a \wedge b) \vee c \longrightarrow (b \wedge c)$ is F (false). Explain.

Problems for Section 11 and 12

The problems of Section 15 are also suitable for this Section.

11.1 From: Frank likes Cobol or John does;
If Frank likes Cobol so does Nancy;

If Nancy likes Cobol so does John;
If John likes Cobol so does Bourbaki.
Deduce: Bourbaki likes Cobol.

11.2[Ans] (*i*) Assume (*h*1) and (*h*2):

(*h*1) If roses are red, then violets are blue.

(*h*2) Violets are not blue.

Question: Can you deduce that roses are not red? Explain.

(*ii*) Assume (*h*3) and (*h*4):

(*h*3) If problems are hard, then we study long.

(*h*4) Problems are not hard.

Question: Can you deduce that we do not study long? Explain.

11.3 A U.S. Attorney General was quoted as saying, "If a person is innocent of a crime, then he's not a suspect." [Take *guilty* and *innocent* as negations of each other.] What is the contrapositive of the quoted statement (QS)? Explain your answer by representing QS symbolically.

11.4 Determine whether the conclusion below follows logically from the premises. Explain by representing the statements symbolically and using rules of inference (state which one or ones). Premises: If Claghorn has wide support, then he'll be asked to run for the senate. If Claghorn yells "Eureka" in Iowa, he will not be asked to run for the senate. Claghorn yells "Eureka" in Iowa. Conclusion: Claghorn does not have wide support.

11.5 (Source: (31.12).) Write the symbolic counterpart of this sentence: "If both the police and the courts are corrupt, then not only does crime diminish if and only if the citizens are alert, but also crime does not diminish if the citizens are easily intimidated." Now determine whether the foregoing sentence is true under the assumption that: the police are corrupt, the courts are not corrupt, crime diminishes, and the citizens are neither alert nor easily intimidated. Naturally, you should explain.

11.6 [Recommended] (Source: (31.12).) First, assume items (*i*)–(*iii*). Is inference (*iv*) sound? Explain why or why not.

(*i*) If Jones is transferred to North Africa, then Smith is going to a new post.
(*ii*) Brown will not be appointed undersecretary if Jones is transferred to North Africa or Smith is going to a new post.
(*iii*) Smith is not going to a new post.
(*iv*) Therefore, Brown will be appointed undersecretary.

11.7 Answer "Yes," "No," or "Not enough information" to the following question Q. Assume $P1$ and $P2$ are true and use only logic to find your answer. Then explain your answer.

($P1$.) If it is not raining, then we hold the party.
($P2$.) We hold the party.
(Q.) Is it raining?

11.8 Prove the rest of the substitution theorem (10.15): parts (*ii*)–(*v*).

Problems for Section 14

14.1 Express each of the following formulas in DNF and contrast them with each other:

$$a \rightarrow b, \quad (\sim a) \rightarrow (\sim b), \quad b \rightarrow a.$$

14.2 What is the DNF of $a \vee \sim a$?

14.3 What is the CNF of

(*i*) $a \longrightarrow \sim b$? Explain.
(*ii*) $(a \wedge b) \vee c$? Explain.

14.4[Ans] Express in CNF

$$(a \longleftrightarrow b) \rightarrow (a \vee b).$$

14.5 What is the DNF of $(a \vee b) \wedge (a \vee \sim b)$ in the setting of two atoms a, b? Explain your procedure for finding the answer.

14.6 Answer the same question for

$$(a \vee b) \wedge (a \vee \sim b) \wedge (\sim a \vee b).$$

14.7 In the setting of only the two propositional variables a, b, what is the CNF of $a \wedge b$?

14.8 Find the DNF of $b \wedge c$ in the realm of the three atoms a, b, c. Explain.

14.9 What is the DNF of $(a \longrightarrow b) \wedge (\sim a \longrightarrow \sim b)$? Explain.

14.10 Find the disjunctive normal form (DNF) of $(a \longrightarrow b) \wedge (a \longrightarrow \sim b)$. Explain how you get your answer.

14.11 Express $a \wedge b \longrightarrow c$ in conjunctive normal form.

14.12 Let a, b, c be propositional variables. Find the disjunctive normal form (DNF) of the propositional formula $a \wedge (b \longleftrightarrow c)$. Hint: Use a shortcut.

Problems for Section 15

These problems are from Smullyan (31.9). In the first group the knights always tell the truth and knaves always lie. Everyone is either a knight or a knave.

15.1 There are two people, A and B. A says, "At least one of us is a knave." What are A and B?

15.2 Suppose A says, "I am a knave or B is a knight." What are A and B?

15.3 Now we have three people, A, B, C. A and B make the following statements:

A says: All of us are knaves.
B says: Exactly one of us is a knight.

What are A, B, C?

15.4 Again three people A, B, and C. A says, "B and C are of the same type." Someone then asks C, "Are A and B of the same type?" What does C answer?

In the second group of problems we are at the Transylvania station, where the workers are of four types: (1) sane humans; (2) insane humans; (3) sane vampires; (4) insane vampires. Whatever a sane human says is true; whatever an insane human says is false; whatever a sane vampire says is false; and whatever an insane vampire says is true.

15.5 I once met a Transylvanian who said, "I am human or I am sane." Exactly what type was he?

15.6 Another worker said, "I am not a sane human." What type was he?

15.7 Another worker said, "I am an insane human." Is he of the same type as the last worker?

15.8 I once met a worker and asked him, "Are you an insane vampire?" He answered "Yes" or "No," and I knew what he was. What was he?

Problems for Section 16

16.1 What propositional formula g in a, b, and c has falsity-set $F(g) = \{(T,T,F), (T,F,F), (F,T,F)\}$? Explain how you found g.

16.2 Jack was killed on a lonely road two miles from Trenton at 3:30 A.M. on March 17. A week later Shorty, Hank, Tony, Joey, and Red were picked up by the police and questioned about Jack's murder. Each of these men made four statements, of which three were true and one was false. Did one of these five men kill Jack? If so, which one? Explain.

Shorty: I was in Chicago when Jack was murdered. I never killed anyone. Red is the guilty one. Joey and I were pals.

Hank: I did not kill Jack. I never owned a revolver in my life. Red knows me. I was in Philadelphia the night of March 17.

Tony: Hank lied when he said he never owned a revolver. The murder was committed on March 17. Shorty was in Chicago at the time of the murder. One of us is guilty.

Joey: I did not kill Jack. Red has never been in Trenton. I never saw Shorty before. Hank was in Philadelphia with me the night of March 17.

Red: I did not kill Jack. I have never been in Trenton. I never saw Hank before. Shorty lied when he said I am guilty.

Some Test Questions

T1. Does either of the statements (i) and (ii) imply the other? Explain.

(i) If the exam is over, then we can go home.
(ii) The exam is over or we cannot go home.

T2. Answer "Yes," "No," or "Not enough information" to questions Q1 and Q2. Explain.

(i) Assume: If the program runs, then the computer works. The computer works.

Q1. Does the program run?

(ii) Assume: If wishes are horses, then beggars will ride. Wishes are not horses.

Q2. Will beggars ride?

T3. Consider the following statements (i) and (ii):

(i) If the computer makes a mistake, then we are flooded with bits.

(ii) The computer makes no mistake or we are not flooded with bits.

First, express (i) and (ii) symbolically as compound propositions. Second, determine whether (i) and (ii) are materially equivalent. Explain.

T4. Explain how $\sim a \bigvee \sim b$ could be considered the converse of $a \bigvee b$?

17. Answers to Practice Problems

(1.3P) (i) $2 + 2 = 7$ is a false proposition.

(ii) $x + 2 = 5$ is not a proposition because we do not know what x stands for. We cannot determine a truth-value for $x + 2 = 5$.

(iii) $(7^2 + 1) \div 5$ has no verb. It is just a formula representing the number 10. It is not a proposition.

(iv) $3 \in \mathbf{N}$ is a true proposition.

(v) We have to say that the lack of mathematical content makes this not a proposition. Even if we could tell unambiguously if it rains, and if we get wet, we can not know that we will get wet in the future, nor that we will not. So we cannot give it a truth-value.

(vi) $x^2 - 3x + 2 = (x - 2)(x - 1)$. If this expression equals 0, then $x = 1$ or $x = 2$. If $x = 1$, then "$x = 2$ or 3" is false; so the given statement is a false proposition, since x might be 1. (The reasoning here is officially intuitive at the moment. It is explained in (2) and later.)

(4.5P) $p \longleftrightarrow p$ because the truth-values are the same on each side of the $\longleftrightarrow$.

If $p \longleftrightarrow q$ then $q \longleftrightarrow p$, because if p and q have the same truth-value, then q and p have the same truth-value. (Our understanding of *and* makes this true.)

If $p \longleftrightarrow q$ and $q \longleftrightarrow r$, then $p \longleftrightarrow r$, because if p and q have the same truth-value, and also q and r have the same truth-value, then p and r have the same truth-value.

(5.2P) Here we relax our restriction that propositions have mathematical content. Assume that p and q are propositions, where

p: professors discuss *Alice in Wonderland* on the Quad.

q: they might get rained on.

We not only assume q is a proposition but also that it is true. Now we have two propositions; one is

$$p \longrightarrow q.$$

The other is $\sim p \longrightarrow q$. In spoken language we accept both as true. Since p is a proposition, one of p and $\sim p$ is false. Since q is true, one of $p \longrightarrow q, \sim p \longrightarrow q$ is $F \longrightarrow T$.

(7.8P) $f_1 \Longrightarrow f_2$ is a proposition, because it has an unambiguous truth-value. If $T(f_1) \subseteq T(f_2)$ (which is also a proposition), then $f_1 \Longrightarrow f_2$ is true. Otherwise it is false.

(7.14P) How does (7.13) follow from (7.10)? The latter is

$f_1 \Longleftrightarrow f_2$ if and only if f_1 and f_2 have the same truth-table.

The former is

$$f \Longleftrightarrow f,$$
$$\text{If } f \Longleftrightarrow g, \text{ then } g \Longleftrightarrow f.$$
$$\text{If } f \Longleftrightarrow g \text{ and } g \Longleftrightarrow h, \text{ then } f \Longleftrightarrow h.$$

The proof is word-for-word the same as that for the same properties of "$\longleftrightarrow$" (see (4.5P) preceding) if you substitute *truth-table* for *truth-value*. (The definition of $p \longleftrightarrow q$ is that p and q have the same truth-*value*.)

(9.6P) $(7.4ii)$ is $f \Longrightarrow g$, where

$$f := (a \longrightarrow b) \longrightarrow c \quad \text{and} \quad g := a \longrightarrow (b \longrightarrow c).$$

The shortcut method shows $T(f) \subseteq T(g)$ by showing $F(f) \supseteq F(g)$. We focus on the falsity-sets because f and g are implications.

So: $F(g)$ must have $a = T$ and $(b \longrightarrow c) = F$. Thus it has just one triple in it, namely, $(a, b, c) = (T, T, F)$. For these values f becomes

$$f(T, T, F) = (T \longrightarrow T) \longrightarrow F = T \longrightarrow F = F.$$

$$\therefore F(g) \subseteq F(f). \quad \therefore f \Longrightarrow g.$$

(9.9P) (9.7) is

$$f \Longrightarrow g \text{ iff } T(f) \subseteq T(g),$$

and (7.5) is the theorem about logical implication: $f_1 \Longrightarrow f_2$ iff the set of lines of the truth-table where f_1 is true is a subset of that where f_2 is true. (9.7) is indeed a restatement of this theorem.

(9.8) is the statement

$$f \Longleftrightarrow g \text{ iff } T(f) = T(g),$$

and the definition (7.9) of $f \Longleftrightarrow g$ is

$$f \Longrightarrow g \quad \text{and} \quad g \Longrightarrow f.$$

If we apply (9.7) to (7.9), we get

$$f \Longleftrightarrow g \text{ iff } T(f) \subseteq T(g) \quad \text{and} \quad T(g) \subseteq T(f).$$

From the definition of set equality this becomes

$$f \Longleftrightarrow g \text{ iff } T(f) = T(g),$$

which is (9.8).

(10.12P) We are to prove

$$\text{If } f_1 \Longrightarrow h_1, \quad \text{then} \quad f_1 \wedge f_2 \Longrightarrow h_1 \wedge f_2.$$

We use (9.7). Thus we are given, in (10.11),

$$T(f_1) \subseteq T(h_1)$$

and we want to prove

$$T(f_1 \wedge f_2) \subseteq T(h_1 \wedge f_2).$$

But $T(f_1 \wedge f_2) = T(f_1) \cap T(f_2)$ (prove this as a subpractice problem), and $T(f_1) \subseteq T(h_1)$, so

$$T(f_1 \wedge f_2) \subseteq T(h_1) \cap T(f_2) = T(h_1 \wedge f_2).$$

$$\therefore f_1 \wedge f_2 \Longrightarrow h_1 \wedge f_2.$$

***(10.19P)** f and g are propositional formulas such that $f \Longrightarrow g$. We are to substitute propositional formulas for the atoms of f and g and prove that the resulting ("big") propositional formulas, call them f^* and g^*, satisfy $f^* \Longrightarrow g^*$.

Suppose f and g are expressed in n atoms $a_1, \ldots, a_n$. We substitute for a_1 a propositional formula in the new atoms $b_1, \ldots, b_t$, and similarly for $a_2, a_3, \ldots, a_n$.

Consider $T(f)$, which is some set of n-tuples $(a_1, a_2, \ldots, a_n) = \text{element of} \{T, F\} \times \{T, F\} \times \cdots \times \{T, F\}$ with n factors. $T(g)$ is some superset of $T(f)$.

If we look at f^* we see that every assignment of truth-values to $b_1, \ldots, b_t$ produces a truth-value for a_1, for $a_2, \ldots$, for a_n. The only way that a t-tuple of T's and F's can make f^* true is to turn $(a_1, \ldots, a_n)$ into an n-tuple in $T(f)$. But such a choice of truth-values for $b_1, \ldots, b_t$ forces g^* to be true also, because every n-tuple in $T(f)$ is also in $T(g)$.

(13.1P) We are to prove that both $g := (a \vee b) \wedge (\sim a \vee \sim b)$ and $h := (\sim a \wedge b) \vee (a \wedge \sim b)$ have the same truth-table as the f of the text, where f is a xor b. Thus

$$T(f) = \{(T, F), (F, T)\}.$$

We will use a shortcut.

We compute $F(g)$. It is clear that $(F, F) \in F(g)$, since it is the only pair making $a \vee b$ false. Similarly $(T, T) \in F(g)$ is the only choice making $\sim a \vee \sim b$ false. So

$$F(g) = \{(T, T), (F, F)\};$$

therefore, $f \Longleftrightarrow g$.

We compute $T(h)$. The only pair making $\sim a \wedge b$ true is (F, T). The only pair making $a \wedge \sim b$ true is (T, F). Thus

$$T(h) = \{(F, T), (T, F)\},$$

so $h \Longleftrightarrow f$ also.

(14.5P) Let us use the shortcut method. $F(c \longrightarrow a \vee b) = \{(F, F, T)\}$, since $(a, b) = (F, F)$ is the only way to make $a \vee b$ false. For the other formula, $\sim a \wedge c \longrightarrow b$, its falsity set has $b = F$; and both $\sim a$ and c must be true. So it is the same. Therefore, the formulas are logically equivalent.

(14.7P) If you asked a truth-teller, "Are you a truth-teller?", the answer would be "Yes." So if you asked him, "What would you say if I ask 'Are you a truth-teller?'" he would truthfully describe his answer as "Yes."

If you asked a liar, "Are you a truth-teller?" he would say "Yes" (because the truth is "No"). But if you ask the "double level" question, "What would you say if I ask (etc.)?" he would have to describe his "Yes" answer as "No," since he's a liar. So a truth-teller answers "Yes," and a liar "No."

II. PREDICATE LOGIC

18. Predicates

Propositional logic is fine as far as it goes, but it is not rich enough to model many of the statements and arguments that we need in mathematics and computer science. For example, consider this famous syllogism:

(18.1)

All men are mortal.
Socrates is a man.
Therefore Socrates is mortal.

Each of these statements is a proposition (if we relax for a moment our requirement that propositions have mathematical content), but although the argument seems perfectly correct, we cannot find anything in our rules of inference for propositional logic to justify it.

To analyze such statements, logicians introduced the idea of *predicate* (or *propositional scheme*).

DEFINITION (18.2) A **predicate** is a statement involving zero or more variables $x, y, \ldots$, each variable with an associated set called its **domain** (D_x the domain of x, D_y that of $y, \ldots$), such that for any replacement of each occurrence of each variable by a point of its domain the statement becomes a proposition.

This definition is less complicated than it seems.

Examples (18.3) Here are some instances of predicates:

$$\begin{array}{rl} x < 3 & D_x = \mathbf{N} \\ y \text{ is even} & D_y = \mathbf{N} \\ x^2 + y^2 = 9 & D_x = D_y = \mathbf{Z} \\ 1 + 1 = 2 & \end{array}$$

These predicates have, respectively, 1, 1, 2, and 0 variables. (We see that a predicate of 0 variables is a proposition.) Consider the predicate $x < 3$. For every choice of positive integer a, the statement $a < 3$ resulting from the substitution of a for x is a proposition (true for $a = 1$ or 2, false for $a = 3, 4, \ldots$). If we substitute integers for x and y in $x^2 + y^2 = 9$, we get a proposition, such as, for example, $1^2 + 2^2 = 9$, or $0^2 + (-3)^2 = 9$.

The role of the variable in a predicate is to be a *placeholder*. It tells you all the places where you must make the substitution of a point from its domain. (When you substitute for the x in a predicate, you do *not* substitute for the x in D_x. D_x does not change. It is the fixed name for the set over which x runs, the set from which you pick the points that you may put into the predicate in place of x.)

The most basic predicate of all is $x \in A$, where A is some defined set and D_x is some given domain.

(18.4) *Compound predicates.* Predicates may be combined with the logical symbols $\vee, \wedge, \sim, \longrightarrow, \longleftrightarrow$ in the obvious way. For example, if $P(x)$ and $Q(x)$ are predicates, then $P(x) \vee Q(x)$ is the predicate that becomes the proposition $P(a) \vee Q(a)$ on the substitution of a for x, where $a \in D_x$. That is, we define $P(x) \vee Q(x)$ by saying, for each $a \in D_x$, what proposition results from substitution of a for x. And we do the same for the rest: $P(x) \longrightarrow Q(x)$ is *by definition* the predicate that predicate that becomes the proposition $P(a) \longrightarrow Q(a)$ when we set $x := a$. (All it takes to define a predicate over D_x is to say for each point of D_X what proposition the predicate must turn itself into on substitution.)

Examples (18.5) (*i*) If $P(x)$ and $Q(x)$ are the predicates $x > 1$ and x is odd, respectively, then $(x > 1) \vee (x \text{ is odd})$ is a predicate of this form. If $D_x = \mathbf{Z}$, and if we substitute $x := -2$, we get the (false) proposition

$$(-2 > 1) \vee (-2 \text{ is odd}).$$

If we substitute -3 or 2 for x we get a true proposition.

(*ii*) $$(x < 2) \wedge (x \text{ is odd}), \quad D_x = \mathbf{N}$$

(*iii*) $$(x < 2) \longrightarrow (x^2 < 4), \quad D_x = \mathbf{R}$$

(*iv*) $$\sim (x < 2) \longleftrightarrow (x \text{ is even}), \quad D_x = \mathbf{Z}$$

(18.6) *Notation for predicates.* If a predicate has just one variable x, we could denote it $P(x)$ (or $Q(x)$, or $C(x)$, or whatever(x); the letter P is not sacred). If it has two variables x, y, then we could write $P(x, y)$. This notation is a name for the actual predicate. In (18.3) we might write

(18.7) $$P(x) : x < 3 \qquad (D_x = \mathbf{N})$$

to express that $P(x)$ is our name for the predicate $x < 3$ (and that $D_x = \mathbf{N}$). The proposition resulting from the substitution of a for x in the predicate is then denoted $P(a)$. Thus $P(1)$ is the proposition $1 < 3$.

It is a poor idea to use "=" in place of the colon in (18.7) because, for one thing, an assertion of equality is often the main verb of a predicate. (Yes, a predicate must have a verb.) The expression $Q(x) = x = 3$ would be awkward at best; at worst it could lead to the outright syntax error $Q(x) = 3$. Let's use the colon instead, and translate it as *is the predicate.*

(18.8) $Q(x) : x = 3$ means $Q(x)$ is the predicate "$x = 3$."

Even $P(x) = x < 3$ could lead to the syntax error $P(x) < 3$.

Practice (18.9P) Explain why $Q(x) = 3$; $Q(x) = x$; and $P(x) < 3$ are syntax errors here.

When the predicate is as short as $x < 3$ it may be unnecessary to give it a name like $P(x)$, but if it is long and complicated a name for it can be handy.

In general if a predicate depends on n variables $x_1, \ldots, x_n$, then we show all the variables in naming it, as $P(x_1, \ldots, x_n)$.

19. Set-Builder: The Second Notation for Sets

(19.1) We spoke informally in Chapter 1 of defining sets by properties. Instead of listing the elements, we said we could define a set as the set of all things having a specified property, like the set of all integers between 1 and 10, inclusive $\{1, \ldots, 10\}$, or the set of all integral powers of 2

(19.2) $$\{\ldots, 2^{-3} = \tfrac{1}{8}, 2^{-2} = \tfrac{1}{4}, 2^{-1} = \tfrac{1}{2}, 2^0 = 1, 2, 4, \ldots\}.$$

We've just *listed* these sets, thanks to the triple dot, as well as defined them by properties. The property in the first set is that of being an integer between 1 and 10, inclusive. In the second it is that of being an integral power of 2.

(19.3) Let's express the property of being an integer between 1 and 10 in mathematical symbols. We could say x is an integer and $1 \le x \le 10$. We can compress it further with our membership notation:

(19.4) $$x \in \mathbf{Z} \quad \text{and} \quad 1 \le x \le 10.$$

Since we are trying to express the set of *all* x with the property (19.4), we must either say exactly that

(19.5) $$\text{the set of all } x \text{ such that } x \in \mathbf{Z} \quad \text{and} \quad 1 \le x \le 10,$$

or set up a notation that will compress the words out of it. Mathematicians have agreed on such a notation. It is

(19.6) $$\{x;\ x \in \mathbf{Z},\ 1 \le x \le 10\};$$

it means exactly what is stated in (19.5). What we have is the predicate

(19.7) $$P(x) : 1 \le x \le 10 \quad (D_x := \mathbf{Z}).$$

The set in (19.6) is defined as the set of all points of the domain for which the predicate $P(x)$ is true. The predicate and its domain express in mathematical symbols the property we began with: that of being an integer between 1 and 10.

For the second example, consider the predicate

(19.8) $$Q(x) : \text{There is an } m \in \mathbf{Z} \text{ such that } x = 2^m \quad (D_x = \mathbf{R}).$$

The set of all real numbers b (all $b \in D_x$) such that $Q(b)$ is true is exactly the set in (19.2). $P(x)$ and $Q(x)$ express in mathematical notation the properties we introduced in (19.1).

The general idea is simple: the predicate $P(x)$ determines the subset of the domain D_x of all points $u \in D_x$ for which $P(u)$ is true. We call this set the **truth-set** of the predicate $P(x)$; it is denoted

(19.9) $$\{x;\ x \in D_x,\ P(x)\},$$

or sometimes

$$\{x \in D_x;\ P(x)\} \quad \text{or} \quad \{x;\ P(x)\}.$$

The latter is used if the domain is understood from the context. More often one sees a vertical bar or colon in place of the semicolon in this notation for sets, but I prefer the semicolon because the bar and colon have other mathematical uses. Let us summarize:

DEFINITION (19.10) The **truth-set** T_P of the predicate $P(x)$ with domain D_x is

$$T_P := \{x;\ x \in D_x,\ P(x)\}.$$

CAUTION T_P is the set of *all* points $x \in D_x$ such that $P(x)$ is true. (When $P(x)$ is complicated we write $T(P(x))$ instead of T_P, as in (19.13).)

The "all" in this definition is *always* to be understood. It is the reason for the warning in the margin. The comma has the logical role of *and*. You read $x \in D_x, P(x)$ as "x is in D_x *and* $P(x)$ is true." Again—you must take *all* such x to form the truth-set. Another way to say (19.10) is

$$\text{For all points } a \text{ of } D_x,\ a \in \{x;\ x \in D_x,\ P(x)\} \text{ iff } P(a).$$

Notice also in (19.10) that we have let the variable or placeholder x stand for an actual point in the set D_x; this usage is, strictly speaking, a syntax error, but it is one that all the experts commit. Introducing a new symbol to stand for a point of the domain D_x can be cumbersome, maybe even confusing. From now on we will almost always use the same symbol for the variable of a predicate as for a point of its domain. But, strictly, x is a placeholder; it is not a point in the set D_x.

Examples (19.11) Here are some sets defined by predicates:

$$S_0 := \{x;\ x \in \mathbf{Z}, -2 \le x \le 3\}$$
$$S_1 := \{x;\ x \in \mathbf{N}, x^3 - 12x^2 + 44x - 48 = 0\}$$
$$S_2 := \{n;\ n \in \mathbf{N}, d_n = 1, \text{ where } d_n \text{ is the } n\text{th decimal digit of } \pi\}$$
$$S_3 := \{x;\ x \in \{1,2,3\}\}$$
$$S_4 := \{x;\ x \in \mathcal{P}(S_3), \text{ and } x \text{ consists of exactly 2 elements}\}$$
$$S_5 := \{n;\ n \in \mathbf{N},\ n \ge 3 \text{ and for some } x,y,z \in \mathbf{N},\ x^n + y^n = z^n\}.$$

Thus S_1 is the set of all positive integers that are roots of the polynomial $x^3 - 12x^2 + 44x - 48 = m(x)$. Since $m(x) = (x-2)(x-4)(x-6)$, $S_1 = \{2,4,6\}$. If we defined S_6 as $S_6 := \{x;\ x \in \mathbf{N},\ (x-1)x(x+2) = 0\}$, then S_6 would be just $\{1\}$, even though the polynomial $(x-1)x(x+2)$ has the other roots 0 and -2. The latter integers do not satisfy the condition $x \in \mathbf{N}$ because they are not positive. Therefore they are not in S_6.

Consider S_2. It is the truth-set of $Q(n)$: the nth decimal digit d_n of π is 1, where $D_n = \mathbf{N}$. Since $\pi = 3.141592\ldots$ we see that $1 \in S_2$ and $3 \in S_2$ (if we start counting at the decimal point). That is, $d_1 = 1, d_2 = 4$, $d_3 = 1$, $d_4 = 5, d_5 = 9$, so $Q(1)$ and $Q(3)$ are true, but $Q(2)$, $Q(4)$, and $Q(5)$ are false. Thus $2 \notin S_2, 4 \notin S_2$, and $5 \notin S_2$.

Practice (19.12P)

(*i*) List S_3 and S_4. Explain.

(*ii*) Find a third point of S_2. Explain.

(*iii*) What can you say about S_5?

(19.13) *Truth-sets of compound predicates.* These have natural expressions in terms of the truth-sets of the individual predicates. For example, letting, $T(P(x) \vee Q(x))$ stand for the truth-set of $P(x) \vee Q(x)$, we get

(*i*) $T(P(x) \vee Q(x)) = T_P \cup T_Q$

(*ii*) $T(P(x) \wedge Q(x)) = T_P \cap T_Q$

(*iii*) $T(\sim P(x)) = \overline{T}_P$

(*iv*) $T(P(x) \longrightarrow Q(x)) = \overline{T}_P \cup T_Q$.

Leaving the others for you, we prove (*i*):

$$
\begin{aligned}
T(P(x) \vee Q(x)) :&= \{x;\ x \in D_x,\ P(x) \vee Q(x)\} \quad \text{from Definition (19.10)} \\
&= \{x;\ x \in D_x,\ P(x)\} \cup \{x;\ x \in D_x,\ Q(x)\} \\
&\quad \text{from definition of union, Chapter 1, (10.1)} \\
&= T_P \cup T_Q \quad \text{from (19.10).}
\end{aligned}
$$

QED

In words instead, $T(P(x) \vee Q(x))$ is the set of all points x in D_x such that $P(x) \vee Q(x)$ is true. But by (18.4)

$$P(x) \vee Q(x) \text{ is true iff } P(x) \text{ is true or } Q(x) \text{ is true.}$$

The set of all x for which $P(x)$ is true is T_P, and similarly for T_Q. Therefore, $T(P(x) \vee Q(x)) = T_P \cup T_Q$, by definition of union, Chapter 1, (10.1).

We now discuss domain and truth-set for predicates with more than one variable.

DEFINITION (19.14) If the predicate has n variables $x_1, \ldots, x_n$ with domains $D_1, \ldots, D_n$, respectively, then its truth-set is the set of all points in $D := D_1 \times \cdots \times D_n$ for which the predicate is true. We define D to be **the domain of the predicate**.

Let's restate (19.14) in the special case $n = 2$. If $P(x,y)$ is a predicate of two variables, then its truth-set is the subset of all points of the Cartesian product $D_x \times D_y$ for which $P(x,y)$ is true. That is, the *domain* of $P(x,y)$ is the Cartesian product $D_x \times D_y$.

(19.15) The truth-set of the predicate $C(x,y) : x^2 + y^2 = 1$ (if $D_x = D_y = \mathbf{R}$) is $T_C = \{(x,y);\ x, y \in \mathbf{R},\ x^2 + y^2 = 1\}$. In geometric terms T_C is the circle of radius 1 with center at the origin.

(19.16) If $D_x = D_y = \{r;\ r \in \mathbf{R}, 0 \leq r \leq 1\}$, and $P(x,y)$ is $x \geq y$, then the truth-set of $P(x,y)$ is the shaded triangle shown in the figure.

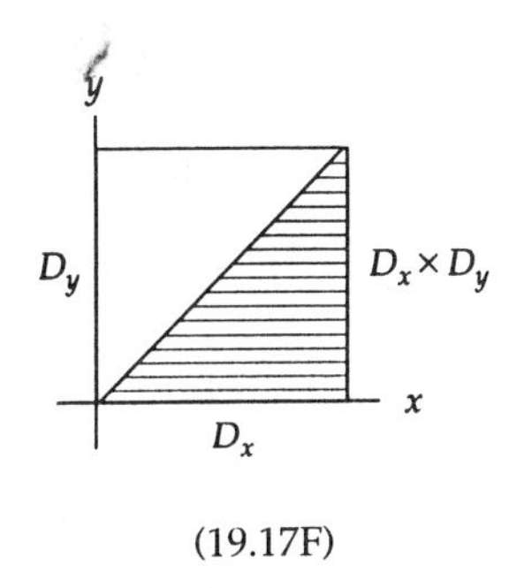

(19.17F)

Practice (19.18P)

(*i*) Find the truth-set of $Q(x,y)$: $x - y$ is even, $D_x = D_y = \{1,2,\ldots,5\}$.

(*ii*) Suppose you are rolling one red die and one green die. You win if the result is 7, that is, if the sum of the two numbers on top is 7. Express winning as a predicate. What is the domain? the truth-set?

Example A physician doing a blood test to measure five properties might specify "normal" as a certain range of values for each property. We could express normality as the predicate

(19.19) $$P(x_1,\ldots,x_5): (a_1 \leq x_1 \leq b_1) \wedge (a_2 \leq x_2 \leq b_2) \wedge \cdots \wedge (a_5 \leq x_5 \leq b_5),$$

where x_1 is, say, the red cell count, x_2 the white cell count, x_3 the cholesterol level, and so on. The a_i's and b_i's are constants, representing the lower and upper limits for normality. If each variable has the domain **R**, then $D = \mathbf{R} \times \cdots \times \mathbf{R}$ (with five factors) $=: \mathbf{R}^5$ is the domain of $P(x_1,\ldots,x_5)$. The truth-set T_P is the five-dimensional "box" specified by the bounds in (19.19), that is, by the predicate. Thus

(19.20) $$T_P = \{(x_1,\ldots,x_5);\ x_i \in \mathbf{R},\ a_i \leq x_i \leq b_i,\ i = 1,\ldots,5\}.$$

The expression $i = 1,\ldots,5$ is a common way of writing the predicate $i \in \{1,\ldots,5\}$.

The "i" is introduced in (19.20) to save writing. The expression to the right of the semicolon in (19.20),

$$x_i \in \mathbf{R}, \qquad a_i \leq x_i \leq b_i, \qquad i = 1,\ldots,5,$$

means exactly the predicate (19.19). In general $Q(i)$, $i = 1,\cdots,5$, for example, means $Q(1) \wedge Q(2) \wedge \cdots \wedge Q(5)$.

You probably read (19.20) properly without batting an eyelash. I put in this explanation just in case.

Practice (19.21P) Suppose you were rolling 10 dice and could win only if each die came up with an even number. Express this game as a predicate with the truth-set being the set of winning outcomes. What is the domain? The truth-set? Do not list these sets!

Think of a predicate as a *property*, like redness, or being an integer, or being $1 + n^2$ for some integer n. A predicate determines its truth-set, which is the subset of *all* points of the domain *with that property*. A predicate can provide a compact description of its truth-set, and using that description is sometimes the only way to prove anything about the truth-set.

Example (19.22) Let $P(x)$ be the predicate "x can be spelled with some letters from 'albatross.' No letter may appear in x more often than it appears in 'albatross.'" Let D_x be the set of all English words. Call V the truth-set of $P(x)$. Suppose you are asked to determine which ones of a small list L of English words are elements of V. It would be far easier to test each word w of L to see if w could be spelled from the letters of 'albatross' than it would be to list all the elements of V and then check L against that list. Thus if $L = \{$ balk, labor, toss, trot $\}$, then you see immediately that

$$\text{labor} \in V, \text{toss} \in V,$$
$$\text{balk} \notin V, \text{trot} \notin V.$$

In doing this you use the predicate to learn about the truth-set. You do not list the whole truth-set. You use the second rather than the first notation for sets.

Look at this situation from another point of view. Suppose someone handed you a listing of V but gave no other information. It would not be easy to find the simple unifying principle of that list, namely, that we used $P(x)$ to define it. But to some people defining V as the truth-set of $P(x)$ is more satisfactory than defining it as a list of words. It is the big picture, or *gestalt*, versus the gritty details, the forest versus the trees.

Actually it might not be too hard to figure out how V was constructed if we allowed the nine-letter word "albatross" itself to be in V. So suppose we redefine $P(x)$ to add the condition (predicate!) that x has at most eight letters. Then it might be difficult to find $P(x)$ from its truth-set.

The first and second notations for sets are sometimes called the **extrinsic** and **intrinsic** notations, respectively. The second notation is also called the **set-builder** notation. This use of predicates to define sets is basic and pervasive. Predicates allow us to define sets that are impossible to list.

Practice (19.23P) If set A is defined by the first (or list) method, then by what method does the statement $B := A \times A$ define B?

(19.24) *Dummy variables.* There is another comment to make about the set-builder notation: The variable is a dummy. You can see that if you define the set in words.

Example The unit interval $I := \{t;\ t \in \mathbf{R},\ 0 \le t \le 1\}$ is the set of all real numbers between 0 and 1 inclusive. The variable t has nothing to do with the set. Any other variable in its place would define the same set. Thus

$$I = \{r;\ r \in \mathbf{R},\ 0 \le r \le 1\} = \{s;\ s \in \mathbf{R},\ 0 \le s \le 1\}.$$

In the present context the symbols r, s, and t stand for variables. None is an element of I.

20. Logical Implication between Predicates

A propositional formula is a special case of a predicate. Consider, for example, the formula $a \longrightarrow b$. Define the predicate $P(a,b)$ as $a \longrightarrow b$, and define $D_a = D_b = \{T, F\} = \{$ true, false $\}$. The truth-set of $P(a,b)$ is (see (19.14))

$$\{(T,T),\ (F,T),\ (F,F)\}.$$

Let $x_1,\ldots,x_n$ be variables with domains $D_1,\ldots,D_n$, respectively. As we said in (19.14), the domain of predicates in $x_1,\ldots,x_n$ is $D := D_1 \times \cdots \times D_n$, and their truth-sets are subsets of D.

DEFINITION (20.1) We say that the predicate $P(x_1,\ldots,x_n)$ **logically implies** the predicate $Q(x_1,\ldots,x_n)$

if and only if

the truth-set T_P of $P(x_1,\ldots,x_n)$ is a subset of the truth-set T_Q of $Q(x_1,\ldots,x_n)$:

$$T_P \subseteq T_Q.$$

NOTATION: $P(x_1,\ldots,x_n) \Longrightarrow Q(x_1,\ldots,x_n)$

Using the same notation '$\Longrightarrow$' for this concept as for that between propositional formulas is reasonable, because this definition generalizes the former one (why?).

Now comes, naturally, a definition of logical equivalence between predicates.

DEFINITION (20.2) Two predicates (as above) are called **logically equivalent** if and only if each one logically implies the other.

NOTATION: $P(x_1,\ldots,x_n) \Longleftrightarrow Q(x_1,\ldots,x_n)$.

It follows immediately from the definitions (20.1) and (20.2) that

(20.3) $$P(x_1,\ldots,x_n) \Longleftrightarrow Q(x_1,\ldots,x_n) \text{ iff } T_P = T_Q.$$

Practice (20.4P) Prove (20.3).

The definition (20.1) and the characterization (20.3), which could be taken as the definition of logical equivalence, are natural outgrowths of the shortcut idea of Section 9, where we were first motivated to define the truth-set.

Examples (20.5) Let $D_x = \{1,2,\ldots,10\}$, and define

$$P(x) : (x-1)(x-3)(x-5) = 0.$$

Then the truth-set of $P(x)$ is $T_P = \{1,3,5\}$. If $Q(x)$ is the predicate "x is odd," then

$$P(x) \Longrightarrow Q(x),$$

because the truth-set of $Q(x), T_Q = \{1,3,5,7,9\}$, includes that of $P(x)$. If we define $R(x)$ to be the predicate

$$(x-1)(x-3)(x-5)(x-7)(x-9) = 0,$$

then $Q(x) \Longleftrightarrow R(x)$, because their truth-sets are equal: $T_Q = T_R$.

(20.6) Since a predicate is something that can be true of an object, to prove two predicates logically equivalent is to find two characterizations of their truth-set. Thus the predicate $Q(x)$ in example (20.5) is the property of oddness over the domain D_x, and the predicate $R(x)$ is the property of being a root of a certain polynomial. These are two characterizations of $\{1,3,5,7,9\}$, the truth-set of both predicates: it is the subset of all odd integers in D_x, and it is the set of all roots in D_x of the polynomial $(x-1)(x-3)(x-5)(x-7)(x-9)$.

21. Quantification

For a moment let us for simplicity consider a predicate $P(x)$ of one variable. We know that substitution of a point a of D_x for x turns $P(x)$ into a proposition $P(a)$. There are two procedures related to substitution that also turn $P(x)$ into a proposition, usually a more complicated one than $P(a)$. They are universal quantification and existential quantification. The definitions follow.

DEFINITION (21.1) The **universal quantifier** $\forall$ is defined as follows:

$$\forall x \in D_x \; P(x)$$

is the proposition that *for every a* (or, *for all a*, or, *for each a*) in the domain D_x of x, the proposition $P(a)$ is true.

DEFINITION (21.2) The **existential quantifier** $\exists$ is defined as follows:

$$\exists x \in D_x \; P(x)$$

is the proposition that *there is* (or, *there exists*) an a in the domain D_x of x for which the proposition $P(a)$ is true. We may equally well say that *for some* $a \in D_x$ the proposition $P(a)$ is true.

Examples (21.3)

(*i*) Suppose $P(x)$ is the predicate $x^2 < 5$, and $D_x = \{1,2,3\}$. Then $\forall x \in D_x \; P(x)$ is the proposition that for each integer a in $\{1,2,3\}$, a^2 is less than 5. We see that $\forall x \in D_x \; P(x)$ is false, because $P(3)$, which is $3^2 < 5$, is false.

It is clear that $\forall x \in D_x \; P(x)$ is the proposition

$$P(1) \wedge P(2) \wedge P(3)$$

for this domain D_x for any predicate $P(x)$.

If we take for $Q(x)$ the predicate $x^2 < 10$ with the same domain, then $\forall x \in D_x \; Q(x)$ is true.

(*ii*) Let us take $P(y)$ to be the predicate $y^2 > 1$ but define D_y to be the set of all positive integers. Now $\forall y \in D_y \; P(y)$ is the statement that the square of every positive integer is bigger than 1, which is clearly false. D_y is an infinite set, but if $P(a)$ is false for one (or more) element(s) of D_y, then $\forall y \in D_y \; P(y)$ is false.

(*iii*) In both the preceding examples $\exists x \in D_x\, P(x)$ and $\exists y \in D_y\, P(y)$ are true, because if there is one (or more) element(s) of the domain for which the predicate is true, then the existentially quantified predicate is true. It is clear that $\exists x \in D_x\, P(x)$ is the proposition $P(1) \vee P(2) \vee P(3)$ for any predicate $P(x)$ if $D_x = \{1,2,3\}$.

(*iv*) Let $Q(z)$ be the predicate $z > -1$, and set $D_z = \mathbf{N}$, the set of all positive integers. Then $\forall z \in D_z\; Q(z)$ is the proposition that every positive integer is greater than -1, clearly true. The proposition $\exists z \in D_z\; Q(z)$ is the statement that some positive integer is greater than 1—also true.

(21.4) In terms of truth-sets,

$\exists x\; P(x)$ says that the truth-set of $P(x)$ is nonempty:

$$\exists x\; P(x) \longleftrightarrow T_P \neq \varnothing.$$

$\forall x\; P(x)$ says that the truth-set of $P(x)$ is the whole domain of x:

$$\forall x\; P(x) \longleftrightarrow T_P = D_x.$$

In particular, $\exists x\; P(x)$ says that there is *at least one* point a in the domain of x such that $P(a)$ is true.

Examples

We used the universal quantifier in words in Chapter 1. Definition (3.1) of set inclusion says $A \subseteq B$ iff every member of A is a member of B. The "every" is the universal quantifier. In symbols we could say it this way (for subsets A, B of D_x):

(21.5)
$$A \subseteq B \text{ iff } \forall x \in D_x,\; x \in A \longrightarrow x \in B.$$

We also used the universal quantifier in the definition (19.10) of the truth-set T_P of a predicate $P(x)$. To make the role of $\forall$ clearer we may restate (19.10) in if and only if form as

(21.6)
$$\forall x \in D_x,\; x \in T_P \longleftrightarrow P(x).$$

In trying to find T_P in a particular problem you may at first miss some of its points. That's why I made such a to-do about the "all" in the definition (19.10).

We used the existential quantifier in words in defining S_5 in Example (19.11). We also use it implicitly in defining the set (19.2), and we expressed the predicate explicitly, using words for $\exists$, in (19.3).

(21.7) *Proof and disproof.* To prove $\exists x\;\; P(x)$ you need find only one $a \in D_x$ such that $P(a)$ is true. That proves the truth-set $T_P \subseteq D_x$ of $P(x)$ is nonempty. To disprove $\exists x\; P(x)$ you must prove that $T_P = \varnothing$. That is the same as proving its negation $\forall x\;\; \sim P(x)$. (See Section 22.)

To prove $\forall x\;\; P(x)$ you must prove that $T_P = D_x$. For some thoughts about that job, see Chapter 3, (11). To disprove $\forall x\; P(x)$ you prove $\exists x\;\; \sim P(x)$, that is, you find one $a \in D_x$ such that $P(a)$ is false.

(21.8) *Generalized* and *and* or. What we saw under examples (21.3*i*) and (21.3*iii*) holds more generally: Whenever D_x is a finite set $\{a_1, \ldots, a_n\}$

(21.9)
$$\forall x \in D_x\; P(x) \text{ is } P(a_1) \wedge P(a_2) \wedge \ldots \wedge P(a_n) =: \bigwedge_{1 \le i \le n} P(a_i)$$

and

(21.10) $$\exists x \in D_x\ P(x) \text{ is } P(a_1) \vee P(a_2) \vee \ldots \vee P(a_n) =: \bigvee_{1 \le i \le n} P(a_i).$$

These assertions follow immediately from the definitions of the quantifiers and from the definitions, commutativity, and associativity of *and* and *or*. When the domain is an infinite set you may thus think of these quantifiers $\forall$ and $\exists$ as replacements for *and* and *or* over the whole domain.

(We have introduced two new notations here, *big and* and *big or*, to the right of the equality signs in (21.9) and (21.10).)

(21.11) *Converses.* Here we extend the term *converse*, defined for propositions in (11.2). Let $P(x)$ and $Q(x)$ be predicates, and consider the two propositions

$$\forall x,\ P(x) \longrightarrow Q(x) \quad \text{and} \quad \forall x,\ Q(x) \longrightarrow P(x).$$

Strictly speaking, we must admit that these propositions are not implications. Nevertheless we say that each one is the converse of the other.

22. Negation of Quantified Predicates

By now we have little trouble negating propositions like $\forall x\ P(x)$ or $\exists x\ P(x)$. The results are

(22.1)
$$\sim (\forall x \in D_x\ P(x)) \longleftrightarrow \exists x \in D_x \sim P(x).$$
$$\sim (\exists x \in D_x\ P(x)) \longleftrightarrow \forall x \in D_x \sim P(x).$$

Thus, symbolically,

(22.2) $$\sim \forall \longleftrightarrow \exists \sim \quad \text{and} \quad \sim \exists \longleftrightarrow \forall \sim .$$

Let us justify these in words. To say, "It is not true that for all x in D_x $P(x)$ is true", amounts to saying, "There is some x in D_x for which $P(x)$ is not true." To say, "It is not true that for some x in D_x $P(x)$ is true", amounts to saying, "For every x in D_x $P(x)$ is false."

Another justification: The identities in (22.1) are just generalizations of De Morgan's laws (8.4). You can verify this by choosing $D_x = \{1, 2\}$ and using (21.9) and (21.10).

A third justification (Who's kidding whom? It's a proof.) in terms of truth-sets: let's omit $\in D_x$ for brevity. We know from (21.4) that

$$\forall x\ P(x) \longleftrightarrow T_P = D_x.$$

Therefore, negating both sides (an operation which in *propositional* logic preserves the biconditional) we find that

$$\sim (\forall x\ P(x)) \longleftrightarrow T_P \neq D_x.$$

Now T_P is a subset of D_x, by definition (19.10). Thus $T_P \subseteq D_x$. For $T_P \neq D_x$ to hold, therefore, it must be true that

(22.3) $$D_x - T_P \neq \varnothing.$$

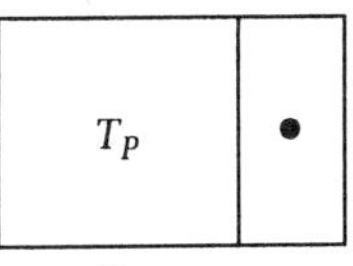

See the Venn diagram, in which the dot indicates the region containing it is nonempty. But $D_x - T_P = \overline{T}_P = T_{\sim P}$. The complement of the truth-set is the falsity-set: $P(a)$ is false iff $\sim P(a)$ is true. From (22.3) we see that $T_{\sim P} \neq \varnothing$. By (21.4) this expression tells us that $\exists x \sim P(x)$.

Practice (22.4P) Give such a proof for the other relation in (22.1).

Notice that if we replace $P(x)$ by $\sim P(x)$ in (22.1) and leave out $\in D_x$ for brevity we get

(22.5)
$$\frac{\sim \forall x \sim P(x) \longleftrightarrow \exists x\, P(x)}{\sim \exists x \sim P(x) \longleftrightarrow \forall x\, P(x).}$$

Thus we may say, symbolically,

$$\sim \forall \sim \longleftrightarrow \exists$$
$$\sim \exists \sim \longleftrightarrow \forall\,.$$

If we now finally negate both sides of (22.5), we see that the two assertions of (22.1) have been interchanged with each other. Thus each one implies the other, in a simple manner. (See also problem 22.3, and compare example (10.17).)

(22.6) In spoken language there is a common error that we should pause here to look at. Consider "Everyone doesn't go to Florida for the winter." We take this to mean

Some people do not go to Florida for the winter,

which is the same as

Not everyone goes to Florida for the winter.

But what the first statement really says is that *no one* goes to Florida for the winter. It reads $\forall x \sim G(x)$, where $G(x)$ is: x goes to Florida for the winter. The error is to negate the predicate without changing the quantifier.

To sum up, the common statement "All is not . . ." should instead be "Not all is . . ." (unless, of course, you really mean "Nothing is. . .").

Other instances: "All that glitters is not gold" and "All is not lost." Sometimes people get it right: "All cognac is brandy, but not all brandy is cognac" (Frank Prial). Some people enjoy mixing up the quantifiers, for example, Benny Hill: "People think because I'm in show business I've had special opportunities with the beautiful women there. I've worked with dancers since I was 16, and I never put a hand on one of them. [Rolling eyes upward] I think her name was Susie. . . ."

23. Quantifying Predicates in More than One Variable

Suppose $P(x,y)$ is a predicate in two variables x and y. What is the meaning of

(23.1)
$$\exists x \in D_x\, P(x,y)?$$

It is a predicate of *one* variable y. For example, if $D_x = \{1,2,3\}$, then by (21.10) $\exists x \in D_x\ P(x,y)$ is the predicate

$$P(1,y) \vee P(2,y) \vee P(3,y).$$

You see that the variable x is gone. In the general case $\exists x \in D_x\ P(x,y)$ is the predicate "For some a in D_x, $P(a,y)$."

Example Let $P(x,y)$ be $x < y$ and take $D_x = D_y = \mathbf{N}$. Then $\exists x \in D_x\ P(x,y)$ is the predicate "For some positive integer a, $a < y$." This statement becomes a proposition whenever we substitute an element of D_y for y. Thus if we set $y = 1$ we get "For some positive integer a, $a < 1$"—a false proposition. If we substitute any other positive integer b for y we get a true proposition, because the value $a = 1$ then satisfies $1 < b$.

Example (23.2) Consider the predicate $P(x,y):\ 2x + 5y \leq 15$ with domains $D_x := D_y := \{1,2,\ldots,5\}$. Then the domain D of $P(x,y)$ is $D_x \times D_y$ by (19.14). Let's work out the truth-set of $P(x,y)$. If we graph the line $2x + 5y = 15$ we see that T_P is the set of all points of D below or on the line. (D consists of the 25 points of intersection of the horizontal and vertical lines.) T_P consists of seven points, marked • in (23.2*F*):

$$T_P = \{(x,y);\ x \in D_x, y \in D_y,\ y = 1\} \cup \{(1,2), (2,2)\}$$

(23.2*F*)

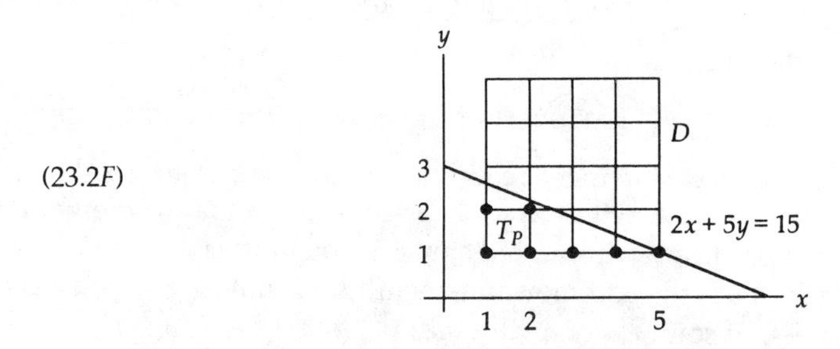

Now let us find the truth-set of $\exists x \in D_x\ P(x,y)$. From the definition (21.10), $\exists x\ P(x,y)$ is $P(1,y) \vee P(2,y) \vee P(3,y) \vee P(4,y) \vee P(5,y)$. Thus $\exists x\ P(x,y)$ is a predicate in y alone, and by (19.13) its truth-set is $T_1 \cup \cdots \cup T_5$, where $T_i :=$ the truth-set of $P(i,y)$ for $i = 1,\ldots,5$. Of course, $T_i \subseteq D_y$. We can see from (23.2*F*) that

(23.3)
$$T_1 = T_2 = \{1,2\};$$
$$T_3 = T_4 = T_5 = \{1\}.$$

Therefore, the truth-set of $\exists x P(x,y)$ is $\{1,2\} \cup \{1\} = \{1,2\}$ in this example.

Now we find the truth-set of $\exists y \in D_y\ P(x,y)$. This predicate is

$$P(x,1) \vee P(x,2) \vee P(x,3) \vee P(x,4) \vee P(x,5),$$

and its truth-set is $S_1 \cup \cdots \cup S_5$, where $S_j :=$ the truth-set of $P(x,j)$ for $j = 1,\ldots,5$. From (23.2F) we see that

$$S_1 = \{1,2,3,4,5\} = D_x \quad S_2 = \{1,2\} \quad S_3 = S_4 = S_5 = \varnothing. \tag{23.4}$$

Therefore the truth-set of $\exists y\ P(x,y)$ is $D_x \cup \{1,2\} \cup \varnothing = D_x$ in this example.

We made good use of (19.13*i*) here. We even extended it from two to five predicates.

You see that $\exists x \in D_x\ P(x,y)$ is a predicate depending only on the variable y. The x has been "quantified out" by the application of $\exists x \in D_x$. (Logicians say x is a *bound* variable, *bound* by the quantifier $\exists x$.)

Suppose we had $\forall x \in D_x\ P(x,y)$. We can see what it means in the same way. First suppose $D_x = \{1,2,3\}$. Then $\forall x \in D_x\ P(x,y)$ is the predicate

$$P(1,y) \wedge P(2,y) \wedge P(3,y). \tag{23.5}$$

Examples (23.6) If $P(x,y)$ is $x < y$ and $D_x = D_y = \mathbf{N}$, then $\forall x\ P(x,y)$ is the predicate in the one variable y "For all positive integers a, $a < y$." For any substitution $y := b \in \mathbf{N}$, this predicate becomes a false proposition, because it says in particular that $b < b$. That is, it asserts in particular $a < b$ for $a = b$.

As a second example, consider again the predicate $P(x,y): 2x + 5y \leq 15$ of (23.2). The predicate $\forall x \in D_x\ P(x,y)$ is, by (21.9),

$$P(1,y) \wedge P(2,y) \wedge P(3,y) \wedge P(4,y) \wedge P(5,y).$$

Its truth-set is, by (19.13*ii*),

$$T_1 \cap T_2 \cap T_3 \cap T_4 \cap T_5 \tag{23.7}$$

where T_i, as in (23.2), is the truth-set of $P(i,y)$. By (23.3) we see that (23.7) is $T(\forall x P(x,y)) = \{1,2\} \cap \{1\} = \{1\}$.

For $\forall y \in D_y\ P(x,y)$ we have

$$P(x,1) \wedge P(x,2) \wedge P(x,3) \wedge P(x,4) \wedge P(x,5).$$

Its truth-set is $S_1 \cap \cdots \cap S_5$ with $S_i :=$ truth-set of $P(x,i)$; and this is $\varnothing$ since $S_5 = \varnothing$ by (23.4).

Again you see that $\forall x \in D_x,\ P(x,y)$ is a predicate depending only on the variable y.

Now consider $P(x,y)$ quantified on both variables, for example

$$\exists x \in D_x\ \exists y \in D_y,\ P(x,y). \tag{23.8}$$

Question: How do we read (23.8)? Answer: from left to right.

It says, "There is an x in D_x and there is a y in D_y such that $P(x,y)$ is true." If you want to be fancy you can say that it is the proposition

$$q := \bigvee_{(u,v)\in D_x\times D_y} P(u,v),$$

which means this: "or" together the propositions $P(u,v)$ for all the ordered pairs (u,v) in the Cartesian product $D_x \times D_y$ (as we said in 21.10). Because the order of the terms in q is unimportant, the order of the two quantifiers $\exists x$ and $\exists y$ is irrelevant. Thus $\exists x\ \exists y\ P(x,y)$ is the same proposition as $\exists y\ \exists x\ P(x,y)$ (we have omitted the domains for the moment to save writing). Indeed, q is simply the proposition

For some point (u,v) in $D_x \times D_y$, $P(u,v)$.

In terms of truth-sets, q says $T_P \neq \varnothing$ (recall (19.14) and (21.4)).

Similar comments apply to the double $\forall$. $\forall x\ \forall y\ P(x,y)$ is the same proposition as $\forall y\ \forall x\ P(x,y)$. In fact it is the proposition, "For every point (u,v) in $D_x \times D_y$, $P(u,v)$." In terms of truth-sets, $T_P = D_x \times D_y$.

(23.9) When both variables are quantified by the same quantifier we usually write the quantifier only once, as $\exists x,y\ P(x,y)$ or $\forall x,y\ P(x,y)$, which mean, respectively, $\exists x\ \exists y\ P(x,y)$ and $\forall x\ \forall y\ P(x,y)$.

Practice (23.10P) Work out the truth-value of the proposition in (23.8) in terms of the truth-values of $P(x,y)$ at each point of its domain when $D_x = \{1,2\}$ and $D_y = \{a,b\}$.

(23.11) When different quantifiers are applied to the variables the order makes a difference: $\forall x\ \exists y\ P(x,y)$ is not the same proposition as $\exists y\ \forall x\ P(x,y)$. To see this, consider the following example.

Example (23.12) Let the predicate $P(x,y)$ be: $x = y$, with $D_x = D_y = \{1,2\}$. Here is a diagram of its domain and truth-set (•).

(23.12F)

2	○	•
1	•	○
y/x	1	2

It is obvious that $\forall x\ \exists y\ P(x,y)$ is true, because it says, "For every x there is a y such that $P(x,y)$ is true." For $x = 1$, y is 1. For $x = 2$, y is 2.

The other proposition, $\exists y\ \forall x\ P(x,y)$, is, however, false. It reads, "There is a y such that for all x $P(x,y)$ is true." But there is no such y.

Notice in the first case, "$\forall x\ \exists y$," that to make $P(x,y)$ true, the value of y varies as x varies.

Practice (23.13P) For a general predicate $P(x,y)$ with $D_x = \{1,2\}$ and $D_y = \{a,b\}$, express in terms of the truth-values of $P(x,y)$ at the points of its domain the truth value of (*i*) $\forall x\ \exists y\ P(x,y)$; (*ii*) $\exists y\ \forall x\ P(x,y)$.

(23.14) Although it is more complicated, we can express doubly quantified predicates in terms of truth-sets. To be specific let us say that D_x is the interval of all real numbers from 1 to 2, and D_y the same from a to b, for some real numbers $a < b$. Then $D_x \times D_y$ can be sketched as the rectangle in figure (23.15*F*).

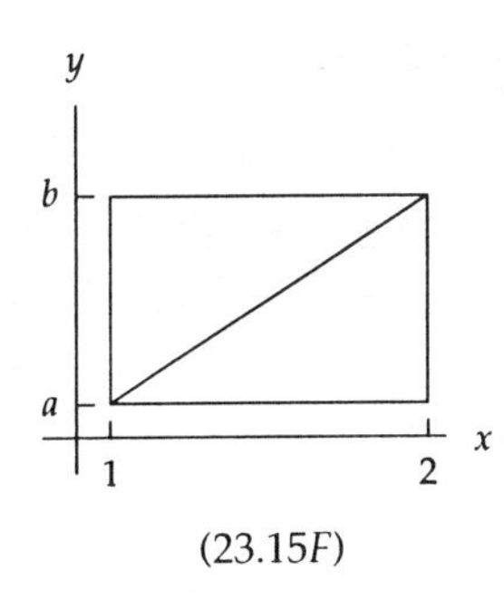

(23.15*F*)

When we say $P(x,y)$ is true at a point (u,v) we mean that $P(u,v)$ is true. Now the proposition $\forall x\ \exists y\ P(x,y)$ means, "On every vertical line in the rectangle there is a point at which $P(x,y)$ is true." It says:

$$\text{For every vertical line } L \text{ in } D_x \times D_y,\ T_P \cap L \neq \varnothing.$$

On the other hand, the proposition $\exists y\ \forall x\ P(x,y)$ means "There is a horizontal line at every point of which $P(x,y)$ is true." This one says:

$$\text{For some horizontal line } H \text{ in } D_x \times D_y,\ H \subseteq T_P.$$

Thus the second proposition implies the first, but not conversely. For example, $P(x,y)$ might be true only at the points of the diagonal line shown in (23.15*F*). (For example, $a = 1, b = 2$ and $P(x,y)$ is $x = y$.) Then the first proposition $\forall x\ \exists y\ P(x,y)$ would be true, but $\exists y\ \forall x\ P(x,y)$ would be false.

Example (23.16) Consider (23.2) again. A glance at (23.2*F*) is enough to show that

$$\text{For every vertical line } L \text{ in } D_x \times D_y,\ T_P \cap L \neq \varnothing.$$

Therefore, $\forall x\ \exists y\ P(x,y)$ is a true proposition in this example. And you can also see that it's true simply by reading it in words. It says "For all x there is a y such that $(x,y) \in T_P$." That's obvious from a glance at (23.2*F*).

Now consider the same quantifiers in the other order: $\exists y\ \forall x\ P(x,y)$. Since there is a horizontal line (in $D_x \times D_y$) at every point of which $P(x,y)$ is true, this too is true. (The line is $y = 1$.)

Practice (23.17P) Determine the truth-values of $\exists x\ \forall y\ P(x,y)$ and $\forall y\ \exists x\ P(x,y)$ for $P(x,y)$ of (23.2). Explain.

Practice (23.18P) Work out criteria analogous to those in (23.14) for $\exists x\ \forall y\ P(x,y)$ and $\forall y\ \exists x\ P(x,y)$.

There are altogether six ways to quantify both variables in $P(x,y)$, namely,

$$\forall x,y \qquad \exists x,y \qquad \forall x\ \exists y \qquad \exists y\ \forall x \qquad \exists x\ \forall y \qquad \forall y\ \exists x,$$

of which we have discussed the first four. The last two are the middle two after an interchange of variables.

To deal with quantifications of three or more variables is just a matter of extending the preceding ideas. The model of the Cartesian product of the dòmains is helpful.

Example (23.19) The definition of continuity that you saw in calculus is a good example of how these procedures come up in real life. You recall that a function f of one real variable is called *continuous* (at the point 0) if and only if for every $\varepsilon > 0$ there is a $\delta > 0$ such that for all x satisfying $|x| < \delta$, the function satisfies $|f(x) - f(0)| < \varepsilon$. To express this definition symbolically, we write

$$\forall \varepsilon\ \exists \delta\ \forall x\ (\varepsilon > 0) \wedge (|x| < \delta) \longrightarrow |f(x) - f(0)| < \varepsilon.$$

The variables of this predicate are ε, δ, and x; we take all three to have domain equal to the set $\mathbf{R}$ of all real numbers.

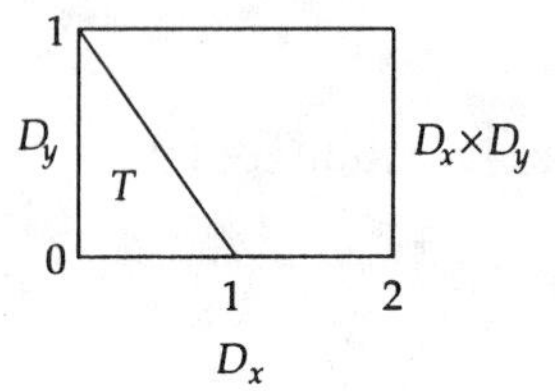

Practice (23.20P) Consider the truth-sets T, U, V of $P(x,y)$, $\exists y P(x,y)$, and $\forall y P(x,y)$, respectively, where $P(x,y)$ is $x + y \leq 1$ and

$$D_x = \{t;\ t \in \mathbf{R},\ 0 \leq t \leq 2\}$$
$$D_y = \{t;\ t \in \mathbf{R},\ 0 \leq t \leq 1\}.$$

(23.20F)

$D_x \times D_y$ and its subset T appear in the sketch.

Now U and V are subsets of D_x. Verify that $U = \{t;\ t \in \mathbf{R},\ 0 \leq t \leq 1\}$ and $V = \{0\}$. (See problem 23.13.)

(23.21) *Negation.* To negate doubly quantified predicates is just a matter of using negation of one-variable predicates (22.1) with the explanations in this section. Thus, leaving out the domains for brevity, we have

$$\begin{aligned} \sim (\forall x,y\ P(x,y)) &\longleftrightarrow\ \sim (\forall x\ (\forall y\ P(x,y))) \\ &\longleftrightarrow \exists x\ (\sim \forall y P(x,y)) \\ &\longleftrightarrow \exists x\ \exists y\ \sim P(x,y) \\ &\longleftrightarrow \exists x,y\ \sim P(x,y) \end{aligned}$$

Similarly, if we have any number of quantifiers alternately $\forall$ and $\exists$ on the predicate $P(x_1, \ldots, x_n)$, the negation is obtained by replacing all $\forall$'s by $\exists$'s and all $\exists$'s by $\forall$'s and finally negating the predicate. (Use (22.2) to see this.)

Example The statement that $f(x)$ is not continuous at 0 (see (23.19)) is

$$\exists \varepsilon\, \forall \delta\, \exists x (\varepsilon > 0) \wedge (|x| < \delta) \wedge |f(x) - f(0)| \geq \varepsilon.$$

This is the negation of the implication in (23.19).

Examples (23.22)

(*i*) Suppose you wanted to express the proposition "The sum of real numbers is real" as a quantified predicate. How to do it? First you need to make it more precise. Let us confine it to two real numbers. It is true for any real numbers, so we can restate it in words as "Any two real numbers have a real sum." Now it becomes as easy as falling off a log: The predicate "x is a real number" can be expressed more briefly as "$x \in \mathbf{R}$"; the *any* tells us to use the universal quantifier. Our answer is

$$\forall x,y \in \mathbf{R},\ x + y \in \mathbf{R}.$$

(*ii*) Let us express in terms of predicates the simple result that each nonzero real number is positive or negative. We start with the predicates $x > 0$ and $x < 0$.

First answer: $\forall x \in \mathbf{R}, (x \neq 0) \longrightarrow (x > 0) \vee (x < 0)$. Second answer: We replace $x < 0$ with "$-x > 0$," which is equivalent. Third answer: $\forall x \in \mathbf{R},\ x \neq 0 \longrightarrow \exists y \in \{1, -1\}, xy > 0$. The third answer replaces the disjunction $(x > 0) \vee (-x > 0)$ of the second answer by the existentially quantified two-variable predicate $xy > 0$. From (21.10) we know that $\exists$ corresponds to $\vee$.

We also see from the placement of $\exists y$ after $x \neq 0 \longrightarrow$ that the choice of y depends on x. We would usually write the third answer as

$$\forall x \in \mathbf{R},\ \exists y \in \{1, -1\}\ x \neq 0 \longrightarrow xy > 0,$$

but perhaps that way hides the dependence of y on x.

(*iii*) Here is a simple example using the quantifiers in the other order. "There is an irrational number" can be represented simply as $\exists x \in \mathbf{R} - \mathbf{Q}$, where $\mathbf{Q}$ denotes the set of all rational numbers. We can involve the universal quantifier by using it to define $\mathbf{Q}$. We get

$$\exists x \in \mathbf{R},\ \forall a \in \mathbf{Z},\ \forall b \in \mathbf{Z} - \{0\},\ x \neq a/b. \tag{23.23}$$

Practice (23.24P) Write a smooth translation of (23.23) into standard English.

(*iv*) Quoting a result from Chapter 6, let's express with predicates the statement "Every integer bigger than 1 is prime or is the product of primes." Define $P(x) : x$ is prime. Then

$$\begin{aligned} &\forall n \in \mathbf{N},\ n > 1 \longrightarrow P(n) \bigvee (\exists r \in \mathbf{N}\ r > 1, \\ &\exists p_1, \ldots, p_r \in \mathbf{N}\ P(p_1) \bigwedge \ldots \bigwedge P(p_r) \bigwedge \text{ (finally) } n = p_1 \ldots p_r). \end{aligned} \tag{23.25}$$

is clearly it, except for the editorial comment. You see that sometimes words are better than logical symbols.

(*v*) Suppose we express the predicate of (*iv*), $P(n) : n$ is prime, in terms of more basic predicates. A prime is defined as an integer $n \geq 2$ such that n is not the product of any two positive integers except n and 1. That tells us how to write the predicate; we just "symbolize" the defining sentence. We get

$$P(n) : (n \geq 2) \bigwedge (\forall a,\ b \in \mathbf{N}\ n = ab \longrightarrow \{a, b\} = \{1, n\}). \tag{23.26}$$

Notice that "$\{a, b\} = \{1, n\}$" allows us not to worry which of a, b is 1 and which is n. The implication in (23.26) is the contrapositive of the one lurking in the definition of prime. We could express it in words as "n is the product of two positive integers only if they are 1 and n."

Our main use of predicates will be to define sets. The mathematical symbolism in the predicate makes it easier to prove things about its truth-set than if the predicate were expressed in words.

24. Epsilon-Arguments with Predicates

Suppose we want to prove that one set A is a subset of another set B, that is, that $A \subseteq B$. We know from the definition in Chapter 1, (3.1), that we must show that every point of A is a point of B. Here we discuss how to show this inclusion when A and B are defined by predicates.

So let A and B be the truth-sets of predicates $P(x)$ and $Q(x)$ defined over some domain D_x. Thus

$$\begin{aligned} A &= T_P = \{x;\ x \in D_x, P(x)\} \\ B &= T_Q = \{x;\ x \in D_x, Q(x)\}. \end{aligned}$$

To show $A \subseteq B$ is, by (20.1), to show $P(x) \Longrightarrow Q(x)$, of course, but the point is that we must prove

$$\forall x \in D_x, \text{ if } P(x) \text{ is true, then } Q(x) \text{ is true:} \qquad (24.1)$$
$$\text{i.e., } \forall x \in D_x \; P(x) \longrightarrow Q(x).$$

(In fact, (24.1) is materially equivalent to $P(x) \Longrightarrow Q(x)$. It is analogous to (8.2), which says of propositional formulas f and g that $f \Longrightarrow g$ iff $f \longrightarrow g$ is a tautology.)

Practice (24.2P) Explain the analogy.

To carry out the proof of (24.1) we must let x be any point of D_x, assume that $P(x)$ is true—that puts x in A—and then prove that $Q(x)$ is necessarily true—that puts x in B. We work with the predicates instead of lists of the elements of A and B.

Example (24.3) Let $P(x)$ and $Q(x)$ be defined as the predicates

$$P(x) : x \text{ can be spelled from the letters of "table,"}$$
$$Q(x) : x \text{ can be spelled from the letters of "alphabet,"}$$

where $D_x :=$ the set of all English words. We agree that these spelling predicates are like that in Example (19.22): No letter may be used more often than it appears in the word defining the predicate.

We can easily show that $T_P \subseteq T_Q$ without listing either set. Here is how:

Step 1. Let x be *any* element of T_P.

Step 2. Thus x can be spelled with the letters t, a, b, l, e—by the definition of T_P.

Step 3. Each of these five letters appears in "alphabet." (This is the connection between the *given* and the *target*.)

Step 4. By Step 3, we see that x is spelled from the letters of "alphabet."

Step 5. Therefore, $x \in T_Q$—by the definition of T_Q.

Step 6. Therefore, $T_P \subseteq T_Q$.

These steps were all simple, but we spelled them out to illustrate the structure of such a proof. (Perhaps you would prefer to combine Steps 3 and 4. If so, fine.)

We can illustrate two-variable predicates with these ideas. Define the predicate

$$S(x, y) : x \text{ can be spelled from the letters of } y. \qquad (24.4)$$
$$D_x = D_y = \text{ the set of all English words.}$$

Then $P(x) = S(x, \text{table})$, and $Q(x) = S(x, \text{alphabet})$. Substitution for one variable of a two-variable predicate turns it into a one-variable predicate, as we said in (23).

Practice (24.5P) Let a and b stand for particular English words. Prove that

$$S(x, a) \Longrightarrow S(x, b)$$

if and only if $S(a, b)$.

Example (24.6) Let $P(x,y)$ be the predicate $x + y \geq 1$ ($D_x = D_y = \{t;\ t \in \mathbf{R},\ 0 \leq t \leq 1\}$). Let $Q(x,y)$ be $(x + 2y \geq 1) \wedge (2x + y \geq 1)$ with the same domains. We may prove that $T_P \subseteq T_Q$ as follows:

Step 1. Let $(x,y) \in T_P$.

Step 2. By definition of truth-set,

$$x + y \geq 1.$$

[Target: prove that both $x + 2y \geq 1$ and $2x + y \geq 1$.]

Step 3. Add y to both sides of the inequality in Step 2 to get

$$x + 2y \geq 1 + y.$$

Step 4. Since $y \geq 0$, $1 + y \geq 1$. Therefore, $x + 2y \geq 1$.

Step 5. Repeat Steps 3 and 4 with x in place of y to get

$$2x + y \geq 1.$$

Step 6. Therefore, $(x + 2y \geq 1) \wedge (2x + y \geq 1)$, that is, $Q(x,y)$ is true.

Step 7. (x,y) was *any* point of T_P because we imposed no condition on (x,y) except that it satisfy $P(x,y)$. Therefore, Steps 1 through 6 showed

$$\forall (x,y) \in D_x \times D_y \ \ P(x,y) \longrightarrow Q(x,y),$$

that is, that all points (x,y) making $P(x,y)$ true also make $Q(x,y)$ true. Therefore, $T_P \subseteq T_Q$.

Practice (24.7P) Draw the domain $D_x \times D_y$ and T_P and T_Q, using analytic geometry.

REMARKS The identification of the target between Steps 2 and 3 is not strictly part of the epsilon-argument, but it is helpful practice. Step 3 shows where we make the connection between the given and the target. It is the key to the proof.

Example (24.8) Every differentiable function is continuous. This statement, in terms of predicates, is (for $X :=$ the set of all real functions)

$$T_d := \{f;\ f \in X,\ f \text{ is differentiable}\} \subseteq \{f;\ f \in X,\ f \text{ is continuous}\} =: T_c.$$

The predicate version is clumsy compared to that in words. But to prove it we use the definition of differentiability at a point $x := a$ (say), namely,

$$\text{The limit } \frac{f(x) - f(a)}{x - a} \text{ exists as } x \longrightarrow a.$$

Now the denominator approaches 0; therefore, the numerator must approach 0, both as $x \longrightarrow a$. Otherwise the limit would not exist. But $f(x) - f(a)$ approaching 0 is the definition that f is continuous at a. Thus if $f \in T_d$, then $f \in T_c$.

25. Quantification over the Empty Set

Logic is neither science nor art but a dodge.
Benjamin Jowett (ca. 1870)

We said in Chapter 1, (7) that because the empty set has no elements, its elements have any property whatever. This observation in symbols is

(25.1) $$\text{"}\forall x \in \varnothing\, P(x)\text{" is true}$$

for any predicate $P(x)$. If (25.1) seems unwarranted, notice that $\forall x \in D_x\; P(x)$ is equivalent to

(25.2) $$\forall x \in V\; (x \in D_x) \longrightarrow P(x)$$

where V is a superset of D_x. If now $D_x = \varnothing$, then (25.2) is an instance of $\forall x \in V$, $F \longrightarrow P(x)$, which is $T \wedge T \wedge \ldots \wedge T$ (since $F \longrightarrow P(a)$ is T), with one T for each point a of V.

We gave in Example (21.3*iv*) a predicate $Q(z)$ such that $\forall z \in \mathbf{N}\; Q(z)$. And we remarked that, of course, $\exists z \in \mathbf{N}\; Q(z)$ was also true. The only thing that prevents us from saying in general that $\forall x \in D_x\; P(x) \longrightarrow \exists x \in D_x\; P(x)$ is (25.1). Why? Because if $D_x = \varnothing$ this becomes $T \longrightarrow F$, which is false, of course. If the domain D_x is nonempty it is true.

26. Analysis of Syllogisms: Truth-Sets and Venn Diagrams

Logic is only the art of going wrong with confidence.
Joseph Wood Krutch, *The Modern Temper* (1914)

Let us go back to the syllogism (18.1). We can express the first statement (the *major premise*), "All men are mortal," as

(26.1) $$\forall x \in D_x\; P(x) \longrightarrow M(x),$$

where D_x is the set of all living things, $P(x)$ and $M(x)$ are the predicates "x is a man" and "x is mortal," respectively. Then the second statement (the *minor premise*), "Socrates is a man," is P(Socrates). We can now see easily how our rules of inference for propositions, specifically *modus ponens*, allow us to justify the logic of this syllogism, which reads

$$\forall x \in D_x\; P(x) \longrightarrow M(x)$$
$$P(\text{Socrates})$$
$$\text{Therefore, } M(\text{Socrates}).$$

Using the major premise, we may assert $P(\text{Socrates}) \longrightarrow M(\text{Socrates})$, because $P(a) \longrightarrow M(a)$ holds for every a in D_x. (It goes without saying that Socrates is an element of D_x.) The second premise tells us that P(Socrates). By *modus ponens* we conclude M(Socrates).†

You can see how the concept of predicate allows us to take apart a proposition like "All men are mortal." Then we can apply the rules of inference of propositional logic.

Now consider another way to check the validity of a logical argument containing predicates. It is probably easier, too. What we do is draw the Venn diagram of the truth-sets. Thus $\forall x\ P(x) \longrightarrow M(x)$ (all men are mortal) becomes

(26.2F) T_M T_P

where T_P and T_M are the truth-sets of $P(x)$ and $M(x)$, respectively. The point representing Socrates is clearly in T_M, so the syllogism is valid.

Examples (from p. 57 of Lewis Carroll (31.2)): Consider

(26.3)
All cats read French.
Some chickens are cats.
Therefore, some chickens read French.

We diagram this one with sets RF (beings that read French), P (chickens), and C (cats). It is

(26.3F)

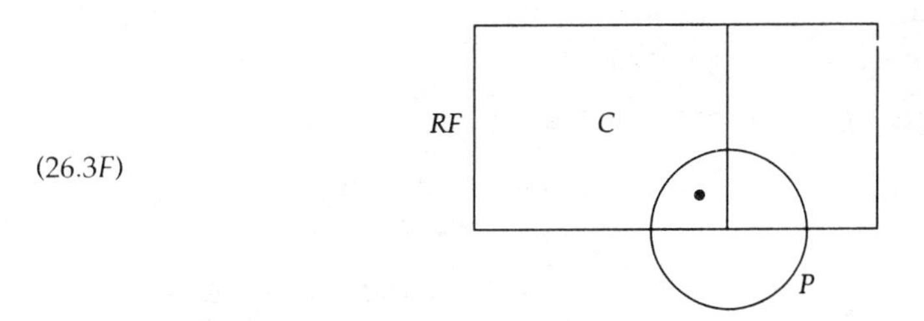

Why? $C \subseteq RF$ is the first premise, $P \cap C \neq \varnothing$ the second. We should show all possible intersections in our diagram, so we are careful to have P intersect both C and $RF - C$. We put the dot in $C \cap P$ to indicate diagrammatically that it is nonempty. Now since it *is* nonempty, we know that the conclusion follows logically from the premises.

The fact that all the statements are false and ridiculous has nothing to do with the logical correctness of the argument.

This syllogism is like the Socrates one except that the second premise involves a set of possibly more than one element.

†This syllogism has a historical point. It is an ironic counter assertion to Socrates's cherished belief that he, and all men, were immortal.

Example Consider now the Carroll syllogism (31.2), p.68:

(26.4)

All soldiers can march.
Some babies are not soldiers.
Therefore, some babies cannot march.

With sets S (soldiers), M (those who can march), and B (babies) we diagram it as

(26.4F)

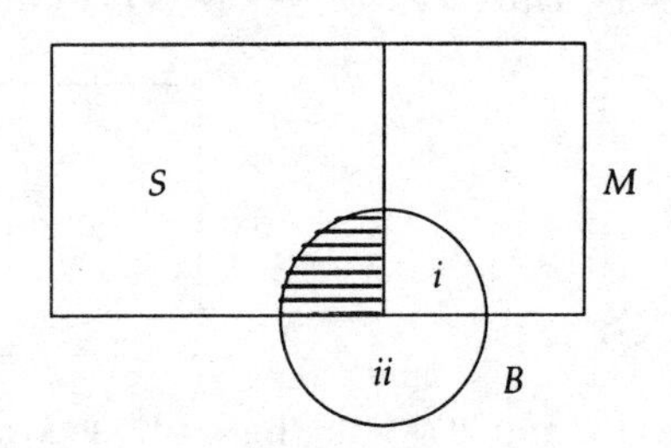

just as before. But now there's a difference, because the second premise tells us only that $B - S \neq \varnothing$, i.e., that there is a baby outside the shaded part $B \cap S$. The baby could be in either (i) or (ii), where

$$(i) \text{ is } (B \cap M) - S$$
$$(ii) \text{ is } B - M.$$

The conclusion says that some baby b is in $B - M$, but the premises do not force that on us; they allow b to be in (i). Therefore, the logic is bad; the conclusion does not logically follow from the premises. These statements are all true, or at least plausible, but the logic is faulty.

Example To check the logic of the syllogism

(26.5)

No pigs have wings;
That bee has wings;
Therefore, that bee is not a pig,

it is easiest to bypass the predicates and go directly to the truth-sets. Thus let P and W be the sets of all pigs and things with wings, respectively, and let b be "that bee." Then the Venn diagram of this syllogism is

(26.5F)

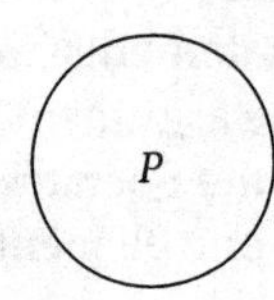

showing that the conclusion follows logically from the premises.

The only tricky part is to diagram "No pigs have wings." It means $P \cap W = \varnothing$, hence the sets P and W are drawn as disjoint. To see why, it helps to look at the alternatives:

$$(26.6) \quad \left.\begin{matrix}\text{No}\\ \text{Some}\\ \text{All}\end{matrix}\right\} \quad \text{pigs have wings} \quad \text{means} \quad \begin{cases} P \cap W = \varnothing \\ P \cap W \neq \varnothing \\ P \subseteq W \end{cases}$$

The first is the negation of the second, not of the third. "No pigs have wings" means "There is no pig that has wings." It is the negation of

$$(26.7) \qquad \exists x \in D \quad P(x) \wedge W(x)$$

where $P(x)$ and $W(x)$ are the predicates "x is a pig" and "x has wings," respectively (and D is the set of all animals, let's say). The proposition (26.7) is $P \cap W \neq \varnothing$, so the major premise is $P \cap W = \varnothing$.

This last example shows how to model premises as quantified predicates. Perhaps beginners should not bypass the predicates after all.

Practice (26.8P) What is the negation of "All pigs have wings"?

Example (26.9) Let's test the validity of the syllogism

$P1$. All computer programs have variables.

$P2$. This document has variables.

C. Therefore this document is a computer program.

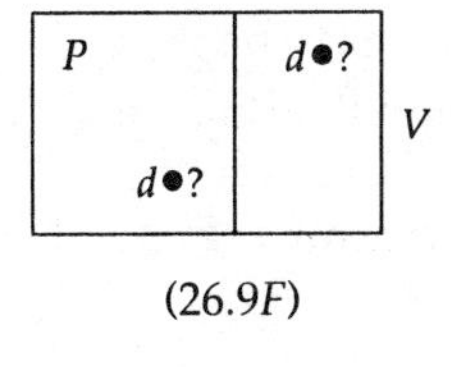

(26.9F)

Let P and V be, respectively, the sets of all computer programs and things with variables. The Venn diagram of the major premise $P1$ is figure (26.9F) without the two "d • ?'s, for it shows $P \subseteq V$. $P2$ says $d \in V$, where d is "this document." But we cannot tell from $P1$ and $P2$ where d may be inside V. It might be in P, and it might be in $V - P$.

Therefore, $P1$ and $P2$ do not allow us to draw the indicated conclusion. The syllogism is invalid.

27. The Unwritten ∀

Often, especially in mathematics, we see statements with unquantified variables in them. Strictly speaking, such things are predicates and not statements, but the understanding is that

All unquantified variables in definitions and theorems have the universal quantifier applied to them.

Often this omission does not inconvenience the reader; for example, one might see in a discussion of complex numbers the assertion that $e^{ix} = \cos x + i \sin x$. That this claim is being made for all real x is probably clear from the context. Or, the statement of the Pythagorean theorem as "In a right triangle with sides a, b, c, where c is the largest, $a^2 + b^2 = c^2$" would be clear enough. No one would miss the quantifiers in a "correct" version:

$\forall\, a, b, c \in \mathbf{R}$ if a, b, c are the lengths of the sides of a right triangle, with c the largest, then $a^2 + b^2 = c^2$.

28. Proofs

A proof is a convincing argument establishing the truth of some proposition. Fine. What does *convincing* mean, and what is an *argument*?

First of all, before answering those two questions, we remark that the hypothesis of the theorem is the given part of it, for example, "A is an isosceles triangle," "n is an odd integer," and so on. If the theorem has the form "$p \longrightarrow q$," then the hypothesis is p.

(28.1) An argument (or proof) is then a series of sentences (propositions), each of which is true as

(*i*) a definition

(*ii*) part or all of the hypothesis

(*iii*) a previously proved result

(*iv*) a consequence of the preceding statements of the proof as a result of rules of inference (or tautologies).

The last sentence of the series is the conclusion of the theorem, what was to be proved.

We have set down a few proofs already in this book. For example, in Chapter 1, (10.6), we proved that for any sets A, B it is true that $A \cap B \subseteq A$. Our proof used the definition of inclusion and then that of intersection. Let's restate our argument, in symbols for brevity. We said

(definition) $\forall x, (x \in A \cap B \longleftrightarrow (x \in A) \wedge (x \in B))$

(inference) $\forall x, (x \in A \cap B \longrightarrow x \in A)$.

Therefore $A \cap B \subseteq A$.

We did not bother to justify or explain in Chapter 1 the rule of inference that we used *because it is so obvious*: $a \wedge b \longrightarrow a$ is a tautology. We apply this rule of inference to each point x of $A \cap B$ in turn and conclude that every such x is in A.

We labored the point about the rule of inference for a good practical reason. You must understand that some things are left unsaid in proofs. If every point were stated, proofs would become forbiddingly long. The (almost) formal definition of proof that appears in (28.1) is never followed to the letter.

The key in writing a proof (assuming you understand why the result being proved is true) is to know what to include and what to leave out. And that depends on your audience. If you are a beginner, then you are writing your proof for an audience of one—the person grading the homework or exam problem. That grader wants to be convinced you know why the assertions made in the proof are true. So you should err, if at all, on the side of more inclusion.

But do not drown the grader with an elaboration like that for the inference drawn in the proof that $A \cap B \subseteq A$. That's too obvious to need explanation. Similarly, if you establish for certain predicates that $\forall x\ P(x) \longrightarrow M(x)$ and also for a certain point a that $P(a)$, then the juxtaposition of those two propositions followed by the conclusion $M(a)$ requires no explanation.

You need not write your proofs symbolically; you may put them in words as we did in Chapter 1, (10.6) if you prefer. As you gain experience you may come to prefer the brevity of mathematical symbolism.

Experienced mathematicians or computer scientists, writing a proof for their peers, would leave out a great deal more than the beginning student could get away with doing. Experts can bank on the experience of their peers; the more experience they have, the more is obvious. You will gradually be able to leave out more and more, but at first you had better put almost all of it in.

What is left out or put in determines how convincing the proof is. And that decision depends on the audience. What might convince an expert could baffle the novice.

The strategy of a proof is propositional: direct, contrapositive, contradiction. We discussed these in (12). The tactics of a proof may involve predicates, usually quantified into propositions. The truth-set of the predicate is all-important in your proof, because it determines the truth or falsity of the quantified predicate. We discussed truth-sets in most of Part II of this chapter.

Example

In the course of a proof you might need to show that $\forall x \in D_x\ P(x)$ is false. From (21.4) that's equivalent to showing $T_P \neq D_x$, i.e., that there is a point in D_x for which $P(x)$ is false. (In other words, prove the negation is true, and by (22.1) the negation is $\exists x \in D_x \sim P(x)$.)

Finally, and to shift gears, how does one understand why the result being proved is true? That is the unanswerable question. If $p \longrightarrow q$ is the result, this understanding calls for seeing a connection between the content of p and that of q strong enough to force q to be true if p is true. It is a proof that you, who thought it up, understand, but that may need filling in and clarifying before others can understand it. No one we know of has proved Fermat's last "theorem" yet, although Fermat said he had done so. You will probably be able to solve most of the problems in this book by just thinking about them long enough, but you may want to refer for inspiration to the book by Polya (31.6).

29. A Correspondence between Sets and Propositional Formulas

Consider a universal set V with subsets A, B, C in *general position*. This means that each of the eight regions $A \cap B \cap C$, $\overline{A} \cap B \cap C$, and so on, in the Venn diagram is nonempty.

(29.1F)

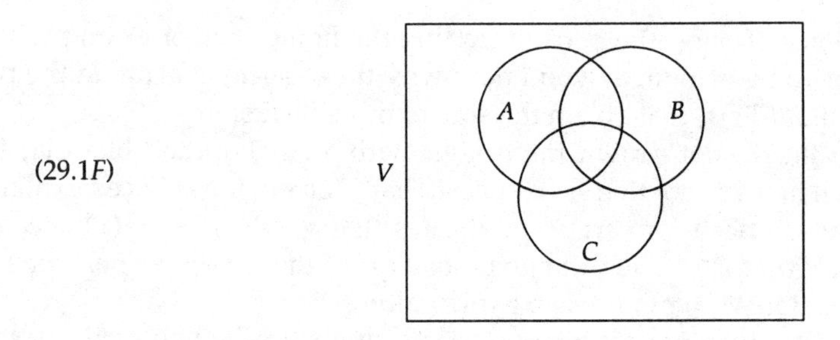

For each subset E of V we consider the predicate $x \in E$. We define $D_x := V$. It is easy to verify that such identities as the following hold:

(29.2)
$$\begin{aligned} \sim (x \in A) &\Leftrightarrow x \in \overline{A}, \\ x \in A \cup B &\Leftrightarrow (x \in A) \vee (x \in B), \\ x \in A \cap B &\Leftrightarrow (x \in A) \wedge (x \in B), \end{aligned}$$

where $\Leftrightarrow$ between predicates is defined, you recall, to mean that the predicate on the left of $\Leftrightarrow$ has the same truth-set as that on the right. Indeed, the verifications are *immediate* consequences of the definitions of $\cup, \cap, \vee, \wedge$, and the like. Using such identities we can, for example, go from the De Morgan law for sets

(29.3)
$$\overline{A \cap B} = \overline{A} \cup \overline{B}$$

to the corresponding De Morgan law for propositional formulas

$$\sim (a \wedge b) \Leftrightarrow (\sim a) \vee (\sim b).$$

That is, from (29.3) we see that

(29.4)
$$x \in \overline{A \cap B} \Leftrightarrow (x \in \overline{A}) \vee (x \in \overline{B}).$$

But the left-hand side is (equivalent to)

$$\sim (x \in A \cap B) \Leftrightarrow \sim ((x \in A) \wedge (x \in B))$$

and the right-hand side of (29.4) is, by (29.2),

(29.5)
$$(x \in \overline{A}) \vee (x \in \overline{B}) \Leftrightarrow \sim (x \in A) \vee \sim (x \in B).$$

Now consider $x \in A$ as x runs over the points of V. The predicate $x \in A$ becomes a proposition whenever we substitute a point of D_x for x. Thus $x \in A$ is true whenever x is in A, and false otherwise. Let us call $x \in A$ the propositional variable a; by appropriate choice of $x \in V$ we may make a T or make it F.

We do the same for $x \in B$ and $x \in C$, calling them b and c, respectively. Because the sets A, B, C are in general position, we can make the triple (a, b, c) take any of the eight combinations of truth-values if we choose a value of x properly.

Example $(a,b,c) = (T,F,T)$ if $x \in A \cap \overline{B} \cap C$.

Under this definition of a b, c, the right-hand side of (29.5) becomes $\sim a \vee \sim b$. The left-hand side of (29.4), $x \in \overline{A \cap B}$, is

$$\sim (x \in A \cap B) \Leftrightarrow \sim ((x \in A) \wedge (x \in B)),$$

which is $\sim (a \wedge b)$. So we turned a De Morgan law for sets into one for propositional variables. You can see now why there is such a tight correspondence between the equivalences of Table (13.25) and the set-theoretic identities opposite them.

The same correspondence sheds light on the concepts of disjunctive and conjunctive normal forms for propositional formulas. For simplicity, consider the case of two variables a, b. The building blocks for DNF are

(29.6) $$a \wedge b, \ \sim a \wedge b, \ a \wedge \sim b, \text{ and } \sim a \wedge \sim b.$$

These correspond to the subsets

(29.7) $$A \cap B, \ \overline{A} \cap B, \ A \cap \overline{B}, \text{ and } \overline{A} \cap \overline{B}$$

shown in the Venn diagram below.

(29.8F)

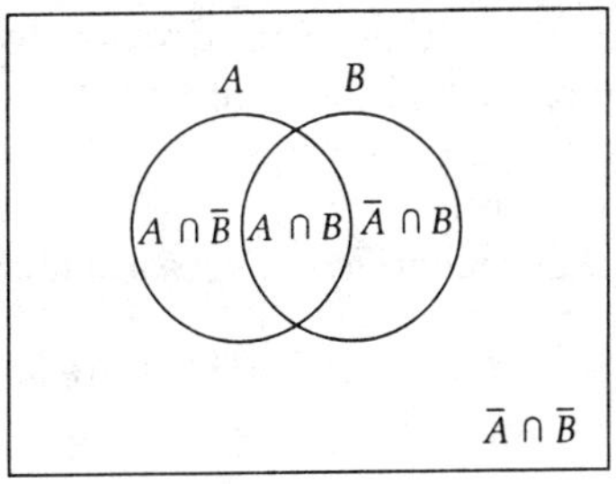

V

A B

$A \cap \overline{B}$ $A \cap B$ $\overline{A} \cap B$

$\overline{A} \cap \overline{B}$

We can uniquely represent any subset of V expressible with the sets A, B, and operators $\cup$, $\cap$, $\overline{}$ as a union of some of the four sets in (29.7). For example,

(29.9) $$A \cup \overline{B} = (A \cap \overline{B}) \cup (A \cap B) \cup (\overline{A} \cap \overline{B}).$$

Under the foregoing correspondence (also denoted $\longleftrightarrow$) between propositional formulas and our predicates, we see that

$$x \in A \cup \overline{B} \Leftrightarrow (x \in A) \vee (x \in \overline{B}) \longleftrightarrow a \vee \sim b.$$

When we similarly translate the right-hand side of (29.9) into a propositional formula, we get exactly the DNF of $a \vee \sim b$.

Verifying set identities by Venn diagrams thus corresponds to verifying logical equivalence of propositional formulas by reducing the formulas to DNF.

Dually we can represent such subsets of V as intersections of some of

$$A \cup B, \quad A \cup \overline{B}, \quad \overline{A} \cup B, \quad \overline{A} \cup \overline{B}.$$

For example,

$$B = (A \cup B) \cap (\overline{A} \cup B). \tag{29.10}$$

The corresponding propositional formulas are

$$b \Leftrightarrow (a \vee b) \wedge (\sim a \vee b), \tag{29.11}$$

the CNF of b.

30. Postscript

Logic is nothing more than a knowledge of words.
—Charles Lamb (1801)

You may have noticed that we used lots of Chapter 2 in Chapter 1! We discussed sets using quantifiers (expressed in words), *if and only if*, *or*, *and*, and *if..., then ....* Why not? After all, logic is just a description of how we think. You knew what these logical terms meant before you picked up this book. Still, it would have been nice to have a cleaner start to our introduction to discrete mathematics, but if we had discussed logic before sets, how could we have handled truth-sets? There just has to be some tugging at the bootstraps. Now that we know sets and logic, we are equipped to go to the real stuff.

(30.1) *On quantifiers.* There are many possible quantifiers for predicates, such as

(30.2)
(*i*) There are at most two $x \in D_x$ such that
(*ii*) There are at least two $x \in D_x$ such that
(*iii*) For all but a finite number of $n \in \mathbf{N}$....

The last is useful in discussing convergence of sequences, for example. These examples can be expressed in terms of the existential and universal quantifiers. We leave (*i*) and (*ii*) to the Problem section (32); here is (*iii*):

$$\exists M \in \mathbf{N}\ \forall n \in \mathbf{N}\ \ n > M \longrightarrow P(n) \tag{30.3}$$

It says for some positive integer M, $P(n)$ is true for all $n > M$. (It may also be true for some values of $n \leq M$, but that's irrelevant.)

Example Define $P(n)$ as "$f(n) := (n-5)(n-10)(n-15) > 0\ \ (D_n = \mathbf{N})$." Then

$$T_P = \{6, 7, 8, 9\} \cup \{m;\ m \in \mathbf{N},\ m > 15\}.$$

Thus in particular $f(n) > 0$ for all $n > 15$, so we could take M to be 15 *or any larger integer* and thereby satisfy (30.3) for this $P(n)$.

Here is another quantifier: There is exactly one x in D_x such that .... This one is (rarely) written as $\exists!$ Although it too can be expressed in terms of the two standard quantifiers, we usually just state it and others like those in (30.2) in words when we use them.

(30.4) *Paradoxes*. The army captain orders his company barber to shave all members of the company provided they do not shave themselves. The barber is so busy at first that his own beard begins to be unsightly. Just as he lathers up, the impossibility of his position strikes him: If he shaves himself, he disobeys the captain's order. (How?) If he does not shave himself, then by the captain's order he is supposed to shave himself.

Paradoxes such as the barber's came to light in about 1900 when Bertrand Russell posed this question:

Practice (30.5P) (This question is called Russell's paradox.) Define the set A by $A := \{X;\ X \text{ is a set}, X \notin X\}$. Is $A \in A$? Explain.

The common element in these two paradoxes is self-reference. Logicians resolved them by changing the rules. Take the barber out of the company; then the captain's order does not apply to him. In general, to avoid paradoxes like Russell's, logicians set up rules of set formation that disallow too much self-reference. "The set of all sets that are not members of themselves" flashes red lights at any mathematician, computer scientist, or logician.

The ultimate mystery is this theorem from advanced logic: In any mathematical system rich enough to be interesting (that is, including the integers) it is *impossible* to prove that there are no paradoxes. Gödel proved this in 1931. For more on this topic, see *Gödel, Escher, Bach* (31.5). Reference (31.1) has a clear introduction to Gödel's ideas, and a fuller account appears in Smullyan's new book (31.10).

(30.6) *Summary*. After *propositions*—statements that are true or false—we studied *propositional formulas*, which are made of variables called atoms, put together into "molecules" by the logical connectives. [Please let me use molecule to mean propositional formula.] We allowed substitution of any proposition, including T and F, for the atoms. The array of all possible substitutions of T and F for the atoms of a molecule, together with its truth-value for each substitution, we called the *truth-table of the molecule*.

A shortcut for finding the truth-table led to the idea of the *truth-set of a molecule*, the set of lines of the table (strictly, the set of tuples of truth-values of the atoms) where the molecule is true.

We then introduced the idea of *predicate*, which allows us to take apart some propositions. Viewed from the other end, predicates can be used to make propositions. The molecule is a special predicate where the variables of the predicate are the atoms of the molecule and the domains of the atoms are sets of propositions. The general predicate allows its variables to have any sets as domains and it uses any grammatical "glue," not just the logical connectives, to put these variables into sentences that become propositions when substitutions are made for the variables. We may think of a predicate as a compact description of this whole set of propositions, one for each point of the domain. Each of these propositions has the same form.

For us, the most important use of a predicate is to define a set, its *truth-set*. It is the subset of the domain of the predicate where the predicate is true. This idea vastly augments our ability to define sets.

Quantifiers are operators that turn predicates into propositions (or into predicates in fewer variables). The two standard quantifiers may be defined in terms of the truth-set T_P. For a predicate of one variable x, say $P(x)$ with domain D_x,

$$\forall x \in D_x \ P(x) \longleftrightarrow T_P = D_x;$$
$$\exists x \in D_x \ P(x) \longleftrightarrow T_P \neq \emptyset.$$

Clearly the statements $T_P = D_x$ and $T_P \neq \emptyset$ are propositions—they are either true or false. *By definition* these propositions have the same truth-value as $\forall x \ P(x)$ and $\exists x \ P(x)$, respectively. Therefore, $\forall x \ P(x)$ is a proposition. You could say if you like that it is the proposition $T_P = D_x$. Similarly you could say that $\exists x P(x)$ is the proposition $T_P \neq \emptyset$.

We asked how to check for logical validity a series of quantified predicates that supposedly lead to a conclusion. Instead of using formal rules of deduction, we drew the Venn diagram of the truth-sets of the predicates involved.

Logic is logic. That's all I say.
Oliver Wendell Holmes (1858)

31. Further Reading

(31.1) Bryan H. Bunch, *Mathematical Fallacies and Paradoxes*, van Nostrand Reinhold, New York, 1982.

(31.2) Lewis Carroll, *Symbolic Logic*, Dover, New York, 1958.

(31.3) Joseph Warren Dauben, *Georg Cantor, His Mathematics and Philosophy of the Infinite*, Harvard Univ. Press, Cambridge, 1979; Princeton Univ. Press, Princeton, 1990.

(31.4) Andrew M. Gleason, [cited in Chapter 1, (20.1)].

(31.5) Douglas R. Hofstadter, *Gödel, Escher, Bach: An Eternal Golden Braid*, Basic Books, New York, 1979.

(31.6) George Polya, *How To Solve It*, Princeton University Press, Princeton, 1945.

(31.7) Willard van Orman Quine, *Methods of Logic*, 4th ed., Harvard University Press, Cambridge, 1982.

(31.8) Paul Rosenbloom, *Elements of Mathematical Logic*, Dover, New York, 1950.

(31.9) Raymond M. Smullyan, *What Is the Name of This Book?: The Riddle of Dracula and Other Logical Puzzles*, Prentice-Hall, Englewood Cliffs, 1978; Simon and Schuster, New York, 1986.

(31.10) Raymond M. Smullyan, *Gödel's Incompleteness Theorems*, Oxford University Press, New York, 1992.

(31.11) Daniel Solow, *How To Read and Do Proofs*, John Wiley & Sons, New York, 1982.

(31.12) Mathematics Staff, The College of the University of Chicago, Chapters 1 and 2 of *Numbers Statements and Connectives, Concepts and Structures of Mathematics*, Univ. of Chicago Press, 1956.

32. Problems on Predicate Logic

Problems on Section 18

18.1 Let $D_x := \mathbf{R}$. Which of the following are predicates? Explain.

(*i*) $x^2 + 1 < 0$.
(*ii*) x is odd.
(*iii*) $(x^2 - 1)/(x + 1)$.
(*iv*) $1 + 2 = 3$.
(*v*) $x \in \mathbf{N}$.
(*vi*) $\sin^2 x + \cos^2 x$.

18.2 [Recommended] (M. Watkins) Express the predicate "x is a multiple of 4" in terms of the predicate $E(x)$: x is even. Take $D_x := \mathbf{N}$. Explain why your answer is right.

Problems on Section 19

19.1 Show that for every set A there is a predicate $P_A(x)$ such that the truth-set of $P_A(x)$ is A.

19.2 What is the truth-set of the predicate $S(a,b)$: $a \longrightarrow \sim b$ if the domains $D_a = D_b$ are the following set of propositions:

$$p: \quad 1 + 1 = 2$$
$$q: \quad 5^2 + 12^2 = 13^2$$
$$r: \quad 1 = 2.$$

19.3[Ans] Prove that sets C and D are equal, where

$$C := \{ v;\ v \in \mathbf{Z},\ \exists a \in \mathbf{Z},\ v \cdot a = 10,\ v > 0 \},$$
$$D := \{ x;\ x \in \mathbf{Z},\ \exists y \in \mathbf{Z},\ x = 1 + y^2 \text{ and } |y| < 4 \}.$$

19.4 [Recommended] Let $A := \{ x;\ x \in \mathbf{N},\ x \leq 10 \}$. Let $B := \{ y;\ y \in \mathbf{N},\ y \leq 15,\ y \text{ is even} \}$.

(*i*) List $A - B$ and list $B - A$ without explanation.
(*ii*) Write predicates of which $A - B$ and $B - A$ are the truth-sets.

19.5 Find the truth-set T_P of the predicate $P(x,y)$: $2x + y > 5$, where $D_x := \{1,2,3\} =: D_y$. Draw a diagram of T_P or list T_P, as you prefer. Give a brief explanation why two of the points of T_P are in T_P.

19.6 Find the truth-sets of the following predicates. Prove your answer. Let $D_x = \mathbf{Z}$.

$P(x)$ is "$1 \leq 2^x \leq 20$."
$Q(x)$ is "$1 + x + x^2 \leq 20$."

19.7 Find the truth-sets of the following predicates $P(x)$ and $Q(x)$, where $D_x = \mathbf{Z}$, the set of all integers.

$P(x)$ is "$-2 \leq 1 + x + 2^x \leq 30$"
$Q(x)$ is "$(x - 3)/2 \in \mathbf{Z}$."

19.8 List the set $\{ x;\ x \in \mathbf{N},\ x^2 + 7 \text{ is an integral power of } 2, x < 40 \}$. Explain how you found the elements of this set.

19.9 List the set $\{ y;\ y \in \mathbf{N} \times \mathbf{N},\ y = (a,b),\ 1 \leq a \leq 3,\ 5 \leq b \leq 7 \}$. Explain.

19.10 List the elements of the set $\{ (i,j);\ 1 \leq i \leq 5,\ 1 \leq j \leq 5, i \text{ is odd}, j \text{ is even} \}$. Write a predicate of which this set is the truth-set. Check your answer. Be sure to say what the domains are. [We usually write "$1 \leq i,j \leq 5$" to mean that both i and j satisfy the inequalities.]

19.11 Let $P(x)$ and $Q(x)$ be any two predicates over the domain D_x. Prove that the truth-set of the predicate $P(x) \wedge Q(x)$ is $T_P \cap T_Q$, where T_P is the truth-set of $P(x)$ and T_Q is that of $Q(x)$. See (19.13).

19.12 Express the truth-set of $P(x) \longrightarrow Q(x)$ in terms of T_P, T_Q, and D. Draw the Venn diagram. Prove your answer. (See (19.13).)

19.13 Let $P(x)$ and $Q(x)$ be predicates each with domain $D_x = D$. Let T_P and T_Q denote the truth-sets of $P(x)$ and $Q(x)$, respectively.

(*i*) Express the truth-set V of $P(x) \longrightarrow \sim Q(x)$ in terms of T_P, T_Q, and D.
(*ii*) Draw the Venn diagram of the sets in (*i*), showing V.

Problems on Section 20

20.1 [A test question.] Write your social security number (SSN). Define $D_x := \{0\} \cup \mathbf{N}$ and consider the predicates $P(x)$: "x is a digit of your SSN" and $Q(x)$: "$x - 1$ or $x + 1$ is a digit of your SSN." Is it true that $P(x) \Longrightarrow Q(x)$? Is it true that $Q(x) \Longrightarrow P(x)$? Explain.

20.2 Explain carefully how the concept of logical implication for predicates generalizes that for propositional formulas, as claimed under Definition (20.1).

20.3 What are the most general conditions on the truth-sets T_P of $P(x)$ and T_Q of $Q(x)$ making (*i*) true?... making (*ii*) true?

(*i*) $(P(x) \longrightarrow Q(x)) \Longrightarrow P(x)$.
(*ii*) $(P(x) \longrightarrow Q(x)) \Longrightarrow Q(x)$.

20.4 What are the most general conditions on the truth-sets so that

$(i)^{\text{Ans}}$ $P(x) \vee Q(x) \Longrightarrow P(x) \wedge Q(x)$? Explain.
(ii) $P(x) \longrightarrow Q(x) \Longrightarrow P(x) \vee Q(x)$? Explain.

Problems on Section 21

21.0 Find three instances of the use of quantifiers in Chapter 1 (other than those mentioned in Chapter 2).

21.1 Which of the following are propositions? Explain. Assume that $P(x)$, $Q(x)$, and $S(x)$ are predicates.

(i) $(\forall x(P(x) \wedge Q(x))) \wedge (\exists x S(x))$
(ii) $(\forall x(P(x) \wedge Q(x))) \wedge S(x)$.

21.2 [A test question.] Let $P(x)$ and $Q(x)$ be predicates over the domain D. Consider

(i) $P(x) \wedge Q(x)$
(ii) $P(x) \Longrightarrow Q(x)$
(iii) $\forall x \in D\, P(x)$
(iv) $P(x) \wedge (\exists x \in D\, Q(x))$

Assume T_P and T_Q are proper subsets of D.

I. Write the most accurate term for each of $(i), \ldots, (iv)$ from among: "proposition, predicate, both of the foregoing, none of the preceding."
II. If it's a predicate, determine its truth-set in terms of T_P, T_Q, and D. If it's a proposition, determine its truth-value if you can.

21.3 Suppose the domain D for all predicates is $\{a, b, c\}$. Express the following propositions without using quantifiers.

$$\forall x\; P(x).$$
$$(\forall x\; R(x)) \wedge (\exists x\; S(x)).$$

21.4 Express as quantified predicates. Define your symbols.

(i) Every cloud has a silver lining.
$^*(ii)$ None but the brave deserve the fair.

21.5 Find the truth-value of

(i) $\forall x \in \mathbf{N}, x^2 + x < 1{,}000{,}000$.
(ii) $\exists x \in \mathbf{N}, (x + 1 = 7) \vee (x^2 = 10)$. Explain.

21.6 Represent the following quotations as quantified predicates, using the indicated predicates. Define your domains (and apologize to Shakespeare).

(i) Something is rotten in the state of Denmark. [$R(x)$: x is rotten.]
(ii) There is more in heaven and earth than is dreamt of in your philosophy. [$M(x)$: x is dreamt of in your philosophy.]

21.7 Prove for any predicates $P(x)$ and $Q(x)$ over any domain $D_x := D$ that

$$[\exists x \in D\, P(x) \wedge Q(x)] \longrightarrow [\exists x \in D\, P(x)] \wedge [\exists x \in D\, Q(x)].$$

21.8 Find a domain $D_x := D$ and predicates $P(x)$ and $Q(x)$ making the following proposition false:

$$[\forall x \in D\;\; P(x) \vee Q(x)] \longrightarrow [(\forall x \in D\;\; P(x)) \vee (\forall x \in D\;\; Q(x))].$$

[Comment: This problem says the universal quantifier does not distribute over *or*.] If you wish, you may specify the truth-sets of $P(x)$ and $Q(x)$ without saying explicitly what $P(x)$ and $Q(x)$ are. Contrast this result with that of the previous problem.

21.9 Let the domain D_x be $\mathbf{N}$. Determine logical implication or equivalence, if either holds, between the following predicates:

$P(x) : x$ is odd.
$Q(x) : \exists t \in \mathbf{N},\ x = 4t + 1.$
$R(x) : x = 1$, or $(\exists t \in \mathbf{N},\ x = 4t + 1$ or $x = 4t - 1)$.

21.10 [Recommended] Express the following with some of $\forall$, $\exists$, and the logical symbols:
There is at most one x in D_x for which $P(x)$ is true.

21.11 (i) Express in terms of quantifiers the statement that there are exactly two x in D_x for which $P(x)$ is true.
$(ii)^{\text{Ans}}$ Do the same for "at most two x."

21.12 Express as a quantified predicate, "There's more than one way to skin a cat." Use $S(x)$: x is a way to skin a cat. Let $D_x :=$ the set of all "ways."

Problems on Section 22

22.1 Consider "No pigs have wings," which is the same as "There is no pig which has wings." Write this proposition as a quantified predicate. (Careful!)

22.2 Express the statement, "There's no jack in the pulpit" (Peter de Vries), using the universal quantifier. Explain your answer. [Use the predicates $J(x)$: x is jack; and $P(x)$: x is "in," or can be derived from, the pulpit. For D_x take any set containing all instances of "jack."] Can you explain the quote?

22.3 Let $P(x)$ be any predicate, and let V be the truth-set of $P(x)$. Prove the second law in (22.1) [of Section (22) of the text] by arguing on V.

Problems on Section 23

23.1 Express the negation of $\forall x \exists y\, P(x,y)$ without simply writing "$\sim$" in front.

23.2 Represent the following proverbs as quantified predicates, using the indicated predicates. Define your domain(s). Explain your steps.

(i)[Ans] There is no fool like an old fool. [$F(x)$: x is a fool; $A(x)$: x is old; $L(x,y)$: x is like y.]

(ii) It's an ill wind that blows nobody good. [$W(x)$: x is a wind; $I(x)$: x is ill; $G(x,y)$: x blows good to y].

23.3 Express the negation of the proposition p:

$$\forall n \in \mathbf{N}\ \exists y \in \mathbf{N} \text{ such that } y > n$$

using neither the negation symbol ($\sim$) nor the word not. Which of p, $\sim p$ is true?

23.4[Ans] Express the following statements as predicates, quantified as necessary. The variable n has domain $\mathbf{N}$ and should not be quantified.

(i) n is an integral multiple of 2 and of 7

(ii) n is not the square of an integer

23.5 Express the following proposition as a quantified predicate. Define your symbols, and specify the domains of the variable(s).

(i) For all integers $n > 2$ there is an odd prime dividing n. [Use $P(x)$: x is prime.]

(ii) Express the negation of (i). Which is true—(i) or (ii)?

23.6 Express the following in terms of quantifiers, variable(s), and the inequality symbols $<$ or $>$. Be careful to define the domain(s) of your variable(s).

(i) For every integer n bigger than 1, there is a prime strictly between n and $2n$.

(ii) Express the negation of the proposition in (i) without use of $\sim$ in your final answer.

Note: (i) is true. It is called Bertrand's postulate.

23.7 Find an example of a predicate $P(x,y)$ and a domain $D_x = D_y = D$ such that

$$\forall x\ \exists y\ P(x,y) \text{ is true and } \exists y\ \forall x\ P(x,y) \text{ is false.}$$

23.8 Let $D = \{1,2\}$ and let x and y be variables with domain D. Give an example of a predicate $P(x,y)$ such that

$$\forall x\ \exists y\ P(x,y) \text{ is true and}$$
$$\exists x\ \forall y \sim P(x,y) \text{ is false.}$$

23.9 Give an example of a predicate $P(x,y)$ with $D_x = D_y \neq \varnothing$ for which the proposition $\exists x\ \forall y\ P(x,y)$ is true.

23.10 Diagram the truth-set of the predicate

$$Q(x,y) : (x + y \geq 1) \wedge (x + 2y \leq 2),$$

where $D_x := D_y := \{\, t;\ t \in \mathbf{R},\ 0 \leq t \leq 1 \,\}$.

23.11 Let $D_x = D_y = \{1,2,3,4,5\}$. Define the predicate $P(x,y)$ as "$(y \geq x)$ or $(x + y > 6)$." Find the truth-sets of the following predicates:

(i) $P(x,y)$. Use notation T_P for truth-set.

(ii) $\exists x \in D_x\ P(x,y)$. Use notation U for truth-set.

(iii) $\exists y \in D_y\ P(x,y)$. Use notation V for truth-set.

(iv) $\forall x \in D_x\ P(x,y)$. Use notation W for truth-set.

(v) $\forall y \in D_y\ P(x,y)$. Use notation X for truth-set. Explain your answers.

23.12 Consider the predicate $Q(x,y) : 3x + y \leq 15$, where $D_x = \{1,2,3,4\}$ and $D_y = \{5,6,7,8\}$. Explaining your procedure, find the truth-set of each of the following:

(i) $\exists x \in D_x, Q(x,y)$

(ii) $\forall x \in D_x, Q(x,y)$

(iii) $\exists y \in D_y, Q(x,y)$

(iv) $\forall y \in D_y, Q(x,y)$.

***23.13** Let V be the truth-set of $P(x,y)$. Thus $V \subseteq D_x \times D_y$.

(i) Prove that the truth-set of $\exists x \in D_x\ P(x,y)$ is the set of all second coordinates of ordered pairs in V.

(ii) Prove that the truth-set of $\forall x \in D_x\ P(x,y)$ is

$$\{\, b;\ b \in D_y,\ D_x \times \{b\} \subseteq V \,\}.$$

23.14 [Recommended] Consider the predicate $Q(x,y)$: $-1 \leq x - y \leq 1$ for $D_x := D_y := \{1,2,3,4,5\}$.

(i) Diagram $D_x \times D_y$ and show the truth-set T_Q.

Explaining your answers, find the truth-set of

(ii) $\exists x\ Q(x,y)$.

(iii) $\forall x\ Q(x,y)$.

Explaining your answers, find the truth-value of
(*iv*) $\exists x,y\, Q(x,y)$.
(*v*) $\exists x \forall y\, Q(x,y)$.
(*vi*) $\forall x \exists y\, Q(x,y)$.

Problems on Section 24

24.1 Prove that $T_P \subseteq T_Q$, where $P(x)$ is the predicate $x/6 \in \mathbf{Z}$ and $Q(x)$ is the predicate $x/3 \in \mathbf{Z}$, and $D_x = \mathbf{Z}$.

24.2 Prove that $T_P \subseteq T_Q$, where $P(x,y)$ is the predicate $(x + y \leq 1) \wedge (y - x \leq 1)$, and $Q(x,y)$ is the predicate $x^2 + y^2 \leq 1$. Set $D_x := \{t;\ t \in \mathbf{R},\ -1 \leq t \leq 1\}$ and $D_y := \{r;\ r \in \mathbf{R},\ 0 \leq r \leq 1\}$.

Problems on Section 26

26.1[Ans] Does the conclusion C follow logically from the premises $P1$ and $P2$? Explain.

$P1$. No Cobol program is free of "do" loops.
$P2$. All programs with "do" loops are well written.
C. Therefore all Cobol programs are well written.

26.2 Assume that every computer has a memory and that this device has a memory. Does it follow logically that this device is a computer? Explain.

26.3 See whether the following arguments (from Lewis Carroll) are logically correct. You may use any correct method of solution as long as you explain it.

(*i*) All who are anxious to learn work hard.
Some of these boys work hard.
Therefore, some of these boys are anxious to learn.

(*ii*) There are men who are soldiers.
All soldiers are strong.
All soldiers are brave.
Therefore, some strong men are brave.

26.4 [Recommended] Suppose $P(x)$ and $Q(x)$ are predicates and that we take such statements as "All P are Q" to mean that everything with the property P has the property Q. Express the following statements in set notation, in Venn diagrams, and as quantified predicates for the truth-sets U of $P(x)$ and V of $Q(x)$.

(*i*) All P are Q
(*ii*) Some P are Q
(*iii*) No P are Q
(*iv*) All P are not Q (Careful!)
(*v*) Not all P are Q
(*vi*) Some P are not Q
(*vii*) No P are not Q
(*viii*) Not all P are not Q

26.5 (Lewis Carroll) Take $P1$ through $P7$ as premises. See what conclusion you can logically derive. Explain.

$P1$: All the policemen on this beat sup with our cook.
$P2$: No man with long hair can fail to be a poet.
$P3$: Amos Judd has never been in prison.
$P4$: Our cook's cousins all love cold mutton.
$P5$: None but policemen on this beat are poets.
$P6$: None but her cousins ever sup with the cook.
$P7$: Men with short hair have all been in prison.

26.6 (Lewis Carroll) Determine whether the conclusion follows logically from the premises. Explain.

Premises { All soldiers can march.
Some babies are not soldiers.

Conclusion: Some babies cannot march.

26.7 (Lewis Carroll) Does the conclusion logically follow from the premises P and P'?

P : No misers are unselfish.
P' : None but misers save eggshells.

Conclusion: No unselfish people save eggshells.

26.8 What conclusion can you draw from the following three premises $P1$, $P2$, and $P3$? Explain your reasoning in detail.

$P1$: Everyone who is sane can do logic.
$P2$: No insane persons are fit to serve on a jury.
$P3$: None of your sons can do logic. (Lewis Carroll)

26.9 (From Professor Joe Kwiatkowski) What conclusion can you draw from P1 ... P5?

P1: When I work a homework problem without grumbling, you may be sure it is one that I can understand.
P2: This problem is not stated clearly.
P3: No easy homework problem ever makes my head ache.
P4: I can't understand problems that are not stated clearly.
P5: I never grumble at a problem unless it gives me a headache.

26.10 [Recommended](From Lewis Carroll, (31.2).) Explain why or why not the conclusion C follows from the premises P and P':

P : No one who means to go by train and cannot get a taxi and has not enough time to walk to the station can do without running.

P' : This party of tourists mean to go by train and cannot get a taxi, but they have plenty of time to walk to the station.

C : This party of tourists need not run.

Problems on Section 27

27.1 Express using predicates: the sum of continuous functions is continuous. Use the predicate $P(f) : f$ is continuous. Do not write a predicate expressing continuity in more basic terms.

33. Answers to Practice Problems

(18.9P) The correct statement is "$Q(x) : x = 3$." "$Q(x) = 3$" looks like the statement that some function $Q(x)$ is equal to 3. If "$Q(x)$" stands for the predicate "$x = 3$," then $Q(x)$ is a name for that predicate. The same comments apply to the other two errors.

(19.12P) (*i*)

$$S_3 = \{1,2,3\}.$$
$$S_4 = \{\{1,2\}, \{1,3\}, \{2,3\}\}.$$

S_3 is defined as $\{x; x \in \{1,2,3\}\}$, which means by (19.10) that S_3 is the set of all x satisfying the predicate "$x \in \{1,2,3\}$." That set is $\{1,2,3\}$.
S_4 is defined as $\{x; x \in \mathcal{P}(S_3)$, and x consists of exactly two elements$\}$.
Thus S_4 is the set of all x satisfying the predicate "$x \in \mathcal{P}(S_3)$ and x consists of exactly two elements." We can easily list all subsets of S_3 if need be and choose those of two elements. The result is S_4 listed above.

(*ii*) We need to find a long-enough decimal expansion of π. From Knuth (cited in Chapter 3, (17.2)), page 613, we find that the next place where "1" appears in the decimal expression for π is the 37th. Thus

$$S_2 = \{1,3,37,\dots\},$$

and I have no idea what the triple dots mean here except that there may be more elements in S_2. You could read "The Mountains of Pi," cited in Chapter 1, (19.2).

(*iii*) The famous "Fermat's last 'theorem' " is the statement that S_5 is empty. From books on number theory (e.g., Starke, cited in Chapter 7, (16.2), pages 145–146) we find that many integers n have been proved *not* to belong to S_5, but for the remaining values of n no one knows. Certainly to date there are no known elements of S_5. Fermat himself proved $4 \notin S_5$. Euler gave a complicated proof that $3 \notin S_5$.

(19.18P) (*i*) $Q(x,y) : x - y$ is even with $D_x = D_y = \{1,\dots,5\}$ has truth-set

$$T_Q = \{(1,1),(1,3),\dots,(2,2),\dots\}$$
$$= \{(x,y);\ x \in D_x, y \in D_y, x \text{ and } y \text{ are both odd, or } x \text{ and } y \text{ are both even}\}.$$

(*ii*) Let r be the number on the top of the red die and g that for the green. The predicate for winning is

$$W(r,g) : r + g = 7$$

and $D_r = D_g = \{1,\dots,6\}$. The domain of W is $D_r \times D_g$, the Cartesian product. It is the set of all ordered pairs (r,g) with $r \in D_r$ and $g \in D_g$.
The truth set of W is

$$T_W = \{(1,6), (2,5), (3,4), (4,3), (5,2), (6,1)\}.$$

These are *all* the ordered pairs (r,g) (points of the domain of W) that make $W(r,g)$ true.

(19.21P) Here the predicate is $G(x_1,x_2,\ldots,x_{10})$: (x_1 is even) $\wedge$ (x_2 is even) $\wedge \cdots \wedge$ (x_{10} is even). For each i, $D_{x_i} := D = \{1,\ldots,6\}$. The domain of G is the Cartesian product

$$D_{x_1} \times D_{x_2} \times \cdots \times D_{x_{10}}.$$

The truth set of G is

$$T_G = E \times E \times \cdots \times E \quad \text{(with 10 factors)}$$

where $E := \{2,4,6\}$. In words, T_G is the set of all 10-tuples of integers from $\{2,4,6\}$.

(19.23P) We could say that B is defined by the "property" method, since the elements of B have the property of being ordered pairs from some set A. In saying "$B := A \times A$" we are certainly not listing the elements of B, no matter how A may be defined.

(20.4P) By (20.1) we have

$$P(x) \Longrightarrow Q(x) \text{ iff } T_P \subseteq T_Q.$$

By (20.2) $P(x) \Longleftrightarrow Q(x)$ iff $P(x) \Longrightarrow Q(x)$ and $Q(x) \Longrightarrow P(x)$. Thus $P(x) \Longleftrightarrow Q(x)$ iff $T_P \subseteq T_Q$ and $T_Q \subseteq T_P$, i.e., iff $T_P = T_Q$.

(22.4P) We show $\sim (\exists x\ P(x)) \longleftrightarrow \forall x \sim P(x)$.
By (21.4) $\exists x\ P(x) \longleftrightarrow T_P \neq \varnothing$.
Negating it thus gives

$$(\star) \qquad \sim (\exists x\ P(x)) \longleftrightarrow T_P = \varnothing.$$

Now $\forall x\ P(x) \longleftrightarrow T_P = D_x$. This is true for any predicate $P(x)$, in particular for $\sim P(x)$. Thus

$$\forall x \sim P(x) \longleftrightarrow T_{\sim P} = D_x.$$

But $T_{\sim P} = D_x - T_P$ by definition of truth-set (19.10), and of complement. Thus $T_{\sim P} = D_x \longleftrightarrow T_P = \varnothing$. From $(\star)$ we get our desired conclusion.
We could more simply go back to $(\star)$ and say: $T_P = \varnothing$ means $P(x)$ is false at all points of D_x; thus $\forall x \sim P(x)$.

(23.10P) First of all, (23.8) is the same as $\exists x \in D_x\ (\exists y \in D_y\ P(x,y))$. We have first "quantified out" the y with $\exists y$ and then quantified the resulting predicate of the one variable x.

Let us see what this double quantifier does in a finite example. Take $D_x = \{1,2\}$ and $D_y = \{a,b\}$. Then $\exists x \in D_x\ \exists y \in D_y\ P(x,y)$ is the proposition

$$\exists x \in D_x \quad (P(x,a) \vee P(x,b)),$$

which is the same as

$$(P(1,a) \vee P(1,b)) \vee (P(2,a) \vee P(2,b)).$$

It can be expressed without the outer parentheses because of the associativity of *or*, and with the terms in any order because of the commutativity of *or*.

(23.13P) (i) First we work out the "inner" predicate.

$$\exists\, y\ P(x,y) \Longleftrightarrow P(x,a) \vee P(x,b)$$

$$(\star) \qquad \forall x\ \exists y\ P(x,y) \longleftrightarrow [P(1,a) \vee P(1,b)] \wedge [P(2,a) \vee P(2,b)].$$

$$\forall x\ P(x,y) \Longleftrightarrow P(1,y) \wedge P(2,y)$$

(ii)

$$(\star\star) \qquad \exists y\ \forall x\ P(x,y) \longleftrightarrow [P(1,a) \wedge P(2,a)] \vee [P(1,b) \wedge P(2,b)].$$

To read such a doubly quantified expression, just go from left to right as if it were an ordinary English sentence. The left side of $(\star)$ reads: For every x there is a y [which may change as x changes] such that $P(x,y)$. Since $(\star)$ is an equivalence of propositions, you can verify this reading by looking at the right side of $(\star)$. The right side is true if and only if both $P(1,a) \vee P(1,b)$ and $P(2,a) \vee P(2,b)$ are true. The first of these is true if and only if for $x = 1$ there is a y such that $P(1,y)$; and similarly with $x = 2$ for the second.

To read the left side of $(\star\star)$, do the same: It says "There is a y [fixed once it is chosen] such that for every x, $P(x,y)$." The right side of $(\star\star)$ provides justification for this reading.

(23.17P) $P(x,y) : 2x + 5y \leq 15$; $D_x = D_y = \{1,\ldots,5\}$. "$\exists x\ \forall y\ P(x,y)$" says

$$\text{For some } x, \quad 2x + 5y \leq 15 \quad \text{for all } y.$$

That is, there is an $x \in D_x$ such that for all y in D_y

$$5y \leq 15 - 2x.$$

This holds iff (under the same quantifiers)

$$\exists x\ \forall y\ \ y \leq 3 - \tfrac{2}{5}x.$$

Even the smallest choice for x says

$$\forall y,\ y \leq 2$$

since y must be an integer. Thus $\exists x\ \forall y\ P(x,y)$ is false.

The other is $\forall y\ \exists x\ P(x,y)$. The quantifiers are reversed. This says

For each y there is an x (which may depend on y) such that $2x + 5y \leq 15$.

But if we choose $y = 3$ we see that it claims

$$\exists x \in D_x \text{ such that } 2x \leq 0.$$

Since $D_x = \{1, \ldots, 5\}$, it too is false.

(23.18P) Consider a general predicate $P(x,y)$ and general domains D_x and D_y. Draw the Venn diagram schematically as a rectangle as in (23.15*F*). Then

$$\exists x \in D_x\ \forall y \in D_y\ P(x,y)$$

is true iff there is an entire vertical line from top to bottom of $D := D_x \times D_y$ in the truth-set T_P. In the example $2x + 5y \leq 15$ we see from (23.2*F*) that no such line exists, confirming our conclusion of the prior problem.

$$\forall y \in D_y\ \exists x \in D_x\ P(x,y)$$

is true iff T_P, the truth-set of $P(x,y)$, when projected horizontally onto the y-axis, hits every point of D_y. "For each $y \in D_y$ there must be a point x of D_x such that $P(x,y)$ is true" is a way of reading it.

(23.20P) $\exists y\ P(x,y)$ is a predicate depending on x. It is true for all $x \leq 1$ in D_x because $P(x,y)$ is "$x + y \leq 1$." Since $0 \in D_y$ we may set $y := 0$ to make $x + y := x + 0 \leq 1$ whenever $x \leq 1$. If $x > 1$ no value of y in D_y works.

$\forall y\ P(x,y)$ depends on x also, but is true only for $x = 0$. Why? If $x > 0$, then $(\forall y \in D_y)\ x + y > y$. Since $1 \in D_y$, we must consider the choice $y := 1$. But $x > 0, y = 1$ makes $P(x,y)$ false. Thus if $x > 0$, then $P(x,y)$ is false for at least one value of y; that shows $\forall y\ P(x,y)$ is false for any such x.

(23.24P) Some real number is irrational. That is about what we started with. We could say

Some real number is not expressible as a quotient of integers.

(24.2P) The analogy between logical implication for propositional formulas f and g and logical implication for predicates $P(x)$ and $Q(x)$ is this:

$$f \longrightarrow g \text{ is a tautology,}$$

which means that $f \longrightarrow g$ is true on every line of its truth-table.

$$P(x) \Longrightarrow Q(x)$$

means that $P(x) \longrightarrow Q(x)$ is true at every point of its domain D_x. The "domain" of $f \longrightarrow g$ is the set of ordered n-tuples of truth-values of the n atoms involved in f and g; there is one such n-tuple for each line of the truth-table.

In fact, propositional formulas are special cases of predicates, and the definition of "$\Longrightarrow$" for predicates becomes that for propositional formulas in that special case.

(24.5P) $S(x,a) \Longrightarrow S(x,b)$ iff for all x making $S(x,a)$ true, we find that $S(x,b)$ is true.

First, assume $S(x,a) \Longrightarrow S(x,b)$. We prove $S(a,b)$.

$S(x,a)$ is defined to be "x can be spelled from the letters of a" (where x and a are English words). Therefore, $S(a,a)$ is true, since a can be spelled from its own letters. By definition of $\Longrightarrow$ we see that $S(a,b)$ is true.

Conversely, suppose $S(a,b)$ is true. Then a can be spelled from the letters of b. Therefore any word x that can be spelled from the letters of a can be spelled from the letters of b. Thus

$$S(x,a) \Longrightarrow S(x,b).$$

(24.7P)

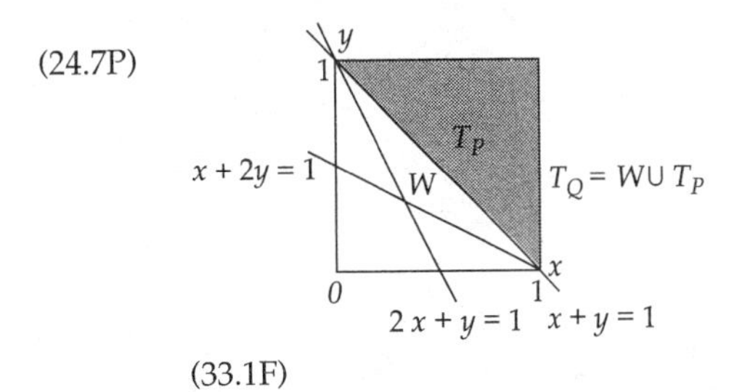

(33.1F)

(26.8P) $\forall x \in D\ \ x \in P \longrightarrow x \in W$ is one way to express "All pigs have wings." Its negation is, by (22.1)

$$\exists x \in D\ \ \sim (x \in P \longrightarrow x \in W)$$

which is, by (8.4),

$$\exists x \in D\ \ \sim (\sim (x \in P) \vee (x \in W)).$$

The De Morgan laws tell us now that this is (materially) equivalent to

$$\exists x \in D \ \ x \in P \wedge \sim (x \in W).$$

In words, "Some pigs do not have wings." But we could tell that right from the start without all the logic. "Not all pigs have wings" is the same as "Some pigs do not have wings."

(30.5P) There are two possibilities, namely, $A \in A$ or $A \notin A$. If $A \in A$, then $A \notin A$, since by the definition A consists precisely of those sets that are not members of themselves. By our rule of inference (10.4) we see that $A \notin A$; i.e., the second case must hold after all. But wait—if $A \notin A$, then by the definition of A, $A \in A$. Thus $A \in A$ iff $A \notin A$.

It's an elegant example, isn't it? With it, in 1903, Bertrand Russell devastated Frege's then-reigning work on logical foundations of mathematics. See (31.3) for a discussion.

CHAPTER

3

Mathematical Induction

Je suis descendu au vrai coeur des mathématiques.

Pierre de Fermat

1. Introduction

Mathematical induction is the most powerful technique for verifying assertions in all of mathematics. The reason for this claim is that mathematical induction is the key property of the integers, and that without the integers in all their glory, mathematics would be a feeble and pale enterprise, not worth anyone's time. We'll not see just why this claim should be true—that's a matter for more advanced courses—but it is true nonetheless. We shall, however, look back at this astounding claim from a practitioner's point of view later on, in order to put it in perspective.

We begin by making a key assumption about the integers. We assume that we know not only the basic properties of addition, multiplication, subtraction, and division, but also the following property, called the **well-ordering** principle:

(1.1) Every nonempty subset of the positive integers has a least element.

This principle is intuitively obvious: If A is a nonempty set of positive integers, then we may imagine checking for the least element of A by looking at the positive integers in turn, first 1, then 2, then 3, and so on, until we first come to an element of A. We must come to an element of A this way, because A is nonempty. The first element of A we come to will be the least element of A, say m. Therefore the process would succeed in finding the least element after a finite number, m, of steps. In fact, the well-ordering principle may be taken as one of the defining axioms for the integers in a treatment that develops them from scratch, but we have agreed not to start that far back. The intuitive procedure just mentioned is really induction; in fact induction and well ordering are equivalent, as slightly more advanced treatments than this one show. At any rate, we now pave the way for our main result by making a new definition.

This chapter is written without heavy reliance on the language of predicate logic from the previous chapter. There are a few places, however, where we use such language to summarize the ideas. If you haven't studied predicate logic, just skip over such places. I do assume knowledge of propositional calculus but want to make induction accessible to those who've not studied predicates.

Example (Gelfand and others) If the first person in a line is a woman, and if the person just behind each woman is a woman, then everyone in line is a woman.

2. Sequences of Propositions

> **A sequence of propositions** is a rule S that selects, for each positive integer n, a proposition, denoted $S(n)$.

A proposition is defined as a statement that is unambiguously true or false, even if we don't know which. For example, we might consider the following sequences S_1 and S_2:

$$
\begin{array}{ll}
S_1(1)\colon\ 1 = 1 & S_2(1)\colon\ 1 < 10 \\
S_1(2)\colon\ 1 + 2 = 2 \cdot 3/2 & S_2(2)\colon\ 2 < 10 \\
S_1(3)\colon\ 1 + 2 + 3 = 3 \cdot 4/2 & S_2(3)\colon\ 3 < 10 \\
S_1(4)\colon\ 1 + 2 + 3 + 4 = 4 \cdot 5/2 & S_2(4)\colon\ 4 < 10 \\
\quad\vdots & \quad\vdots \\
S_1(n)\colon\ 1 + 2 + \cdots + n = n(n+1)/2 & S_2(n)\colon\ n < 10 \\
\quad\vdots & \quad\vdots
\end{array}
$$

The preceding notations mean that our sequence S_1 of statements is defined for each positive integer n to be the statement that the sum of the first n positive integers is the product of n and $n+1$ divided by 2. For each positive integer n, the nth statement for S_2 is "$n < 10$." As to the truth of these statements, in the second sequence, $S_2(n)$ is obviously true for $n = 1, 2, \ldots, 9$ and false for $n = 10, 11, \ldots$. The fact that every statement in the first sequence is true we shall prove shortly. The point for now is to understand what a sequence of statements is.

Here is another example: Let $S_3(n)$ be the statement that the last two decimal digits of n are equal. You can easily work out the values of n for which $S_3(n)$ is true.

Strictly speaking, $S(n)$ is the name of the proposition assigned to the integer n, but for convenience we say "the proposition $S(n)$" instead of "the proposition named $S(n)$." Notice also that we write, for example, $S_2(n) : n < 10$. The colon means *is the statement* or *is the statement that*. Thus $S_2(n) : n < 10$ is our shorthand for $S_2(n)$ *is the statement that* $n < 10$. To write instead $S_2(n) = n < 10$ would be an error. Why? Because $S_2(n)$ is (the name of) the statement "$n < 10$;" $S_2(n)$ is not the integral variable n.

If you've read Part II of Chapter 2, you'll recognize that a sequence of statements, at least when defined by a single formula such as $S_2(n) : n < 10$ or the formula for $S_1(n)$ that we used in our first example, is a predicate of one variable with domain equal to the set of all positive integers. But it is perfectly feasible to discuss these ideas without using the term *predicate*, so we shall do without it for a while longer.

3. Mathematical Induction

We use mathematical induction to prove statements of the form

(3.1) For all positive integers n, $S(n)$ is true,

where S is some sequence of statements as already described. Of course, if $S(n)$ is false for some n then we can't prove (3.1) by any method.

Example $S_2(10)$ is false, since $10 < 10$ is false. Therefore "For all $n \geq 1$, $S_2(n)$" is false. That's obvious. But suppose the definition of $S_2(n)$ were somehow masked from us, so that we could find out only the truth-value of $S_2(1)$, $S_2(2)$, and so on, by asking an oracle for them one at a time. We might ask for $S_2(1)$ through $S_2(4)$, say, and, finding them all true, conclude erroneously that $S_2(n)$ is true for all n. That would be scientific induction as opposed to mathematical induction.

As we'll see, $S_1(n)$ *is* true for all $n \geq 1$. The proof will be our first example of the use of the theorem of mathematical induction.

We now state and prove our first main result.

Theorem (3.2) *(First form of mathematical induction).* Let $S(1), S(2), S(3), \ldots$ be a sequence of propositions having the following two properties:

(*i*) $S(1)$ is true, and

(*ii*) for every integer $n \geq 1$, if $S(n)$ is true, then $S(n+1)$ is true.

Then for every positive integer n, $S(n)$ is true. ∎

For an intuitive proof, look at (*ii*) with $n = 1$. It says: If $S(1)$ is true, then $S(2)$ is true. But by (*i*), $S(1)$ *is* true. Therefore, $S(2)$ must be true. Now (*ii*) with $n = 2$ says: "If $S(2)$ then $S(3)$," so $S(3)$ must be true, by the same kind of reasoning. And so on. Therefore, $S(n)$ is true for every positive integer n. Induction is an infinite *modus ponens*. The more formal proof below uses the well-ordering principle to avoid the "And so on."

Proof We'll prove Theorem (3.2) by proving its contrapositive. That is, we'll assume that the desired conclusion is false and from that assumption show that the hypothesis is false. Thus

(3.3) we assume our conclusion is false; then there is some positive integer n such that $S(n)$ is false.

In other words, the set G of all integers t such that $S(t)$ is false is nonempty. Now we bring in the well-ordering principle to tell us that the set G has a least element. Call it z.

There are two possibilities for z: $z = 1$ or $z > 1$.

Case 1. If $z = 1$, then $S(1)$ is false; this negates our first hypothesis (*i*).

Case 2. If $z > 1$, then $z - 1 \geq 1$. Now z is the *least* element for which $S(z)$ is false; therefore, $S(z-1)$ is true. Thus

$$S(z-1) \text{ is true and } S(z) \text{ is false.}$$

This negates our second hypothesis (*ii*).

The assumption (3.3) that the conclusion of Theorem (3.2) is false thus implies that one or the other of the hypotheses (*i*) and (*ii*) of (3.2) is false. Therefore (3.3) must be false, and we've proved the theorem. QED

(3.4) The logical structure of the proof just finished is simple. Let's name the hypotheses (*i*) and (*ii*) of (3.2) as p and q, respectively. And call the conclusion r. Theorem (3.2) says

$$p \wedge q \longrightarrow r. \tag{3.5}$$

Here's how our proof went. We assumed $\sim r$ (not-r). We then showed that

(3.6)
$$\sim r \longrightarrow \sim p \text{ in Case 1;}$$
$$\sim r \longrightarrow \sim q \text{ in Case 2.}$$

Since Case 1 *or* Case 2 must hold, we therefore showed

$$\sim r \longrightarrow \sim p \vee \sim q,$$

the contrapositive of (3.5). That proves (3.5).

Comment. It's tempting to say in describing (3.6) that $\sim r \longrightarrow \sim p$ contradicts the hypothesis p and then to say that the proof is by contradiction. But we really proved the contrapositive of the theorem.

Another way to summarize the logic of the proof is this: We used well ordering and the two hypotheses of the theorem to show that $G \neq \emptyset$ implies $G = \emptyset$. Now the rule of inference, (not-$r \longrightarrow r) \Longrightarrow r$, tells us that $G = \emptyset$.

(3.7) *How to use Theorem (3.2)*. Suppose you want to prove that each of the propositions $S(1), S(2), \ldots$ is true. One way is to try induction, as Theorem (3.2) is called for short. To succeed, all you need do (!) is show that the two hypotheses of the theorem hold for your sequence. That is, you carry out the following steps:

> *The basis step*. Prove that $S(1)$ is true.
>
> *The inductive step*. Prove for every integer $n \geq 1$ that if $S(n)$ is true, then $S(n+1)$ is true. To do this let n be a fixed but arbitrary integer, $n \geq 1$, and assume that $S(n)$ is true. Under that assumption prove that $S(n+1)$ is true. The assumption that $S(n)$ is true is called the **inductive**, or **induction, assumption**.

Notice that in carrying out the basis and inductive steps we are verifying that hypotheses (*i*) and (*ii*) of Theorem (3.2) hold for the sequence $S(n)$. We are then justified in drawing the conclusion of Theorem (3.2), that for all $n \geq 1$, $S(n)$ is true. That is, Theorem (3.2) says $p \wedge q \longrightarrow r$, in the notation of (3.4). If we establish that p and q are both true, we may use *modus ponens* to conclude that r is true.

Example (3.8) For each positive integer n, let $S(n)$ be the statement that the sum of the first n integers is $n(n+1)/2$.

We use Theorem (3.2).

The basis step: We verify that $S(1)$ is true by noting that it simply says: $1 = 1 \cdot 2/2$, which is true.

The more difficult part is to show that for every positive integer n, if $S(n)$ is true, then $S(n+1)$ is true. That proof is the inductive step. To do this, we let n be a fixed but arbitrary positive integer, and we assume that $S(n)$ is true. Thus we assume that for this n,

(3.9) $$1 + \cdots + n = n(n+1)/2.$$

Our goal is to prove that

$$1 + \cdots + n + (n+1) = (n+1)(n+2)/2$$

must be true, for this latter equation is the statement $S(n+1)$. All we need do is add $n+1$ to each side of the equation (3.9), which then becomes

$$1 + \cdots + n + (n+1) = n(n+1)/2 + (n+1).$$

The left-hand side is that of $S(n+1)$; it's a matter of simple algebra to show that the right-hand side is the same as that of $S(n+1)$ also: It is $(n+1)(n/2+1) = (n+1)(n+2)/2$. Therefore we've shown that $S(n+1)$ is true if $S(n)$ is true. Having shown that both hypotheses, (*i*) and (*ii*), of Theorem (3.2) hold for this sequence of statements, we can use the conclusion of Theorem (3.2) to see that each statement in the sequence is true.

(3.10) *A comment on the preceding proof.* To show that for all positive integers n, if $S(n)$ is true, then $S(n+1)$ is true, we argued on only one fixed but arbitrary integer n. It is important in proofs by induction that this step be correct. That is, it must indeed be a proof that is correct for every such n. Here we manipulated $S(n)$ to turn it into $S(n+1)$. There were no hidden assumptions about the value of n in our procedures—all we used were the distributive law and the fact that adding the same number to both sides of an equation preserves the equality. Each of these procedures is valid for every value of n.

4. Standard Induction Problems

(4.1) All texts on induction—and this is no exception—have many problems for the student just like the preceding example, in that the identity to be proved has the form

$$\sum_{1 \le i \le n} a(i) := a(1) + a(2) + \cdots + a(n) = c(n),$$

where $a(i)$ is usually some simple formula in i, such as, for example, i or $i(i-1)$, and $c(n)$ is some other formula.

We approach each of these problems in the same way: if we call the sum on the left-hand side $L(n)$, then we are to prove that $L(n) = c(n)$ for all $n \ge 1$. The basis step is to show that $L(1) := a(1) = c(1)$. Once we do that we let n be a fixed but arbitrary positive integer, assume $L(n) = c(n)$ (the induction assumption), and try to prove that

$L(n+1) = c(n+1)$. The key is that $L(n+1) = L(n) + a(n+1)$, so all that's needed is to show that $c(n) + a(n+1) = c(n+1)$. Here we have used the induction assumption. The correctness of the implication $S(n) \longrightarrow S(n+1)$ for each value of n follows from the algebraic rules used to show that $c(n) + a(n+1) = c(n+1)$—if, of course, it's true. We now know that $S(n)$ is true for all n by invoking Theorem (3.2), because we have proved that both hypotheses of Theorem (3.2) hold for the sequence of propositions in question. (It seems we did this in such generality that we didn't prove anything; instead we indicated how a proof would go in any particular case of this problem. But see problem 21(*i*).)

Examples (4.2) (*i*) Prove that for all $n \geq 1$

$$\sum_{0 \leq k \leq n-1} 2 \cdot 3^k := 2 + 6 + 18 + \cdots + 2 \cdot 3^{n-1} = 3^n - 1.$$

In this problem $a(k) = 2 \cdot 3^k$ for all k, and $c(n) = 3^n - 1$. If we assumed the analysis of (4.1), all we'd have to say is that

$$c(n) + a(n+1) := 3^n - 1 + 2 \cdot 3^n = (1+2)3^n - 1 = 3^{n+1} - 1 =: c(n+1).$$

The rest of the proof by induction is included in (4.1).

I recommend against this approach until you thoroughly understand the form of proofs by induction.

(*ii*) Problems 35 and 37 are examples of summation problems to which this method does *not* apply.

Practice (4.3) Redo the proof of Example (3.8) by this method.

(4.4) A common error made by students in solving problems of the type just mentioned is to write

$$S(n) = \sum a(i) = c(n),$$

which confuses the name $S(n)$ of the assertion with either $L(n)$ or $c(n)$—or with both in some papers. One can avoid this syntax error by replacing the leftmost equality sign by a colon (:) to stand for *is the statement that*.

$$S(n) : \sum a(i) = c(n).$$

This error, also mentioned in Chapter 2, (18.6), is common and glaring.

5. The Second Form of Induction

There is another theorem on induction more useful than Theorem (3.2) because it allows us to assume more in the inductive step. Here is the theorem.

Theorem (5.1) (*Second form of mathematical induction*).

Let $S(1), S(2), \ldots$ be a sequence of statements with the following two properties:

(*i*) $S(1)$ is true, and

(*ii*) For all positive integers n, if $S(1)$ and $S(2)$ and $\ldots$ and $S(n)$ are all true, then $S(n+1)$ is true.

Then for all positive integers n, $S(n)$ is true. ■

Although this theorem appears weaker than the first form of induction, because of the need to assume more in the inductive step to get the same conclusion, it is in truth equivalent to the first form, a result we leave as a problem.

Proof (The proof is almost the same as that of Theorem (3.2), so some details are omitted.) If the conclusion is false, then the set of all integers t for which $S(t)$ is false has a least element z, greater than 1 by hypothesis (i). Thus the set of integers $\{1, \ldots, z-1\}$ is nonempty and $S(i)$ is true for each integer i in that set. By hypothesis (ii), then, $S(z)$ is true, a contradiction. QED

(5.2) *An alternate statement of Theorem (5.1).* If we recall the result of Chapter 2, (25) that any predicate universally quantified over the empty subset of its domain is true, then we can state Theorem (5.1) more compactly as follows: Suppose that the sequence of propositions $S(n)$ has the following property:

(5.3) For all positive integers n, if $S(i)$ is true for every positive integer i less than n, then $S(n)$ is true.

Then for all positive integers n, $S(n)$ is true.

The new condition (5.3) for $n = 1$ is just the statement that $S(1)$ is true—since there are no positive integers less than 1. For $n > 1$ condition (5.3) is the second (inductive) hypothesis of Theorem (5.1).

Example (5.4) The following problem provides a nice example of the need for the second form of induction. Suppose we try to prove that every integer greater than 1 is a prime or a product of primes. (A *prime* is defined as an integer greater than 1 that is not the product of any two positive integers except itself and 1. The first few primes are 2, 3, 5, 7, 11, 13, 17, 19, 23. The first few nonprimes are 4, 6, 8, 9, 10, 12. Notice that, by definition, every nonprime has a nontrivial factorization: $4 = 2 \cdot 2, 6 = 2 \cdot 3, 9 = 3 \cdot 3, 12 = 3 \cdot 2 \cdot 2$, and so on. We'll discuss primes more fully in Chapter 6.)

We express our n^{th} statement as $S(n)$: n is prime or is the product of primes. We wish to prove that for every $n > 1$, $S(n)$ is true.

For our basis step we observe that $S(2)$ is true, since 2 is prime.

For the inductive step we let n be a fixed but arbitrary integer greater than 2 and make the inductive assumption: All the statements $S(2), \ldots, S(n-1)$ are true. In trying to prove that $S(n)$ is true, i. e., that n is prime or is the product of primes, we separate our argument into two cases:

Case 1. If n is prime, we are finished.

Case 2. If n is not prime, then by definition, n is the product of two integers different from 1 and n. Let us say that $n = ab$ in this case. We know that $1 < a < n$, and $1 < b < n$. Therefore, our inductive assumption applies to both a and b; that is, $S(a)$ is true and so is $S(b)$. Therefore, a and b are each prime or products of primes, so their product n must be a product of primes.

Since we have correctly carried out both the basis step and the inductive step for this sequence of statements, we are justified in drawing the conclusion of Theorem (5.1), that $S(n)$ is true for every $n \geq 2$. That is, every such n is prime or is the product of primes. QED

6. An Extension

(6.1) Perhaps the two small changes from our formal procedure that we made in carrying out the proof just concluded will not have caused problems for you. The first of these changes is that our domain is now the set of all integers greater than 1 instead of the usual set of all positive integers. It should be obvious by now that both forms of induction hold for a sequence of statements defined over any set of integers $b, b+1, b+2, \ldots$, whatever the value of b, provided only that the two hypotheses of the theorem hold. That is, the basis step, now $S(b)$, must be true; and the inductive step, which in the first form of induction now reads

$$\text{for all } n \geq b, \text{ if } S(n) \text{ then } S(n+1),$$

must also be true. The inductive step would be correspondingly changed in the second form of induction. In other words, all we have changed is the starting point, and since the integers starting at 1 and going upward look like the integers starting at b and going upward—no matter where you are in either set, you get to the next element by adding 1—there is no logical difference between the two situations.

(6.2) The second change is that we reformulated the inductive step to read

$$\text{for all } n > b,\ S(b) \textit{ and } \ldots \textit{ and } S(n-1) \longrightarrow S(n).$$

But if we replace n by $n-1$ in the second form of induction we get just this new formulation. In other words, the inductive step consists of the same set of statements in each formulation. (This set of statements is displayed in (7.4) below.)

7. Wrong Proofs by Induction

Consider now the following two "proofs" and try to find the errors before reading the explanations of them.

Example (7.1) We shall prove by induction that all nonnegative integers are equal to 0. For $S(n)$ we take the statement $n = 0$. Obviously, $S(0)$ is true, so let n be a fixed but arbitrary integer, $n \geq 0$. Using the second form of induction, assume that $S(0)$ and ... and $S(n)$ are true. To prove $S(n+1)$, note that $n+1$ is the sum of 1 and n, both of which are 0 by the inductive assumption. Thus $n + 1 = 0 + 0 = 0$.

Having shown that both the hypotheses of Theorem (5.1) hold for this sequence $S(n)$, we may invoke the conclusion of that theorem, that $S(n)$ is true for every $n \geq 0$. That is, we have proved that every integer is 0. (Or have we?)

Example (7.2) For each integer $n \geq 0$, let $S(n)$ be the statement

$$0 + 1 + \cdots + n = n(2n^2 - 5n + 5)/2.$$

$S(0)$ is true, for it reads $0 = 0$. $S(1)$ is true, for it reads $1 = 2/2$. And $S(2)$ is true, for it reads $3 = 2(8 - 10 + 5)/2$. These extra verifications beyond the basis step clearly show that we could do the inductive step easily if we wanted to do so. Thus we have proved that $S(n)$ is true for all n. Or we might just say that since $S(0)$, $S(1)$, and $S(2)$ are true, $S(n)$ is true for all n—all too often students do just that.

(7.3) *Explanations of these examples.* The error in Example (7.1) is that $S(0)$ does not imply $S(1)$—obviously, since $S(0)$ is true and $S(1)$ is false. Where the reasoning breaks down is in the claim that $1 = 0$ in the inductive step. We implicitly assumed that n was positive—true enough if $n + 1$ is 2 or more, but crashingly false when $n + 1$ is 1. The argument given to prove the inductive step was fine—except when $n = 0$. In this example it would be necessary to prove the inductive step for all $n \geq 0$, but our proof really worked only for all $n \geq 1$. The argument used to prove the inductive step must be convincing for all the values of n for which the statements are defined; it must not exclude any of those values by hidden or careless assumptions—or by any other means.

Let us dwell a moment longer on the last point. In verifying the inductive step, we must prove a whole infinite sequence of implications, namely (for the case like the Example (7.1) in which our sequence starts at $n = 0$),

$$\begin{array}{rr} & S(0) \longrightarrow S(1) \\ & S(0) \;\; and \;\; S(1) \longrightarrow S(2) \\ & S(0) \;\; and \;\; S(1) \;\; and \;\; S(2) \longrightarrow S(3) \\ (7.4) & \vdots \\ & S(0) \;\; and \;\; \ldots \;\; and \;\; S(n) \longrightarrow S(n+1) \\ & \vdots \end{array}$$

In making the proof, we say, "Let n be a fixed but arbitrary integer greater than or equal to 0, and assume $S(0)$ *and* ... *and* $S(n)$." We then try to prove $S(n+1)$. Whatever proof we come up with must be good for *all* these values of n; it must not fail for even one value of n. In Example (7.1) our "proof" failed for $n = 0$, because we did not prove $S(0) \longrightarrow S(1)$.

Example (7.2) is more transparent. We have already proved in (3.8) that the sum of the first n integers is $n(n + 1)/2$; since this example claims to show that this same quantity is $n(2n^2 - 5n + 5)/2$, we must be at least suspicious. The error was to substitute a meager version of scientific induction for the inductive step of the theorem of mathematical induction. It's not enough to verify $S(n)$ for the first few, or even the first several, values of n. In this case $S(3)$ is false, for it reads $6 = 12$. And, in fact, $S(n)$ is false for all $n > 2$, for

$$n(2n^2 - 5n + 5)/2 = n(n + 1)/2 + n(n - 1)(n - 2). \tag{7.5}$$

That is, (7.5) is the correct value for the sum of the first n integers plus a polynomial in n that is 0 for $n = 0, 1$, or 2 but not 0 for any other value of n. We could produce an example that would be correct for as many initial values of n as we like but false for all the rest; for example, we could add $n(n - 1)\cdots(n - 105)$ instead of $n(n - 1)(n - 2)$ and make $S(0), S(1), \ldots, S(105)$ all true, but $S(106)$ and all later statements false.

(7.6) PROOFS IN GENERAL: The cautions made in the explanations of the errors in Examples (7.1) and (7.2) apply to the proof of *any* statement of the form

$$\text{For all } x \text{ in } D, P(x),$$

where $P(x)$ is a predicate.

8. The Domino Analogy

Imagine an infinite set of dominoes labeled $1, 2, 3, \ldots$, one for each member in some given sequence $S(1), S(2), \ldots$ of propositions. We can mimic a proof by induction that all the propositions are true by the following game: We set up the dominoes on end along an infinite half-line as in Figure (8.1F).

(8.1F)

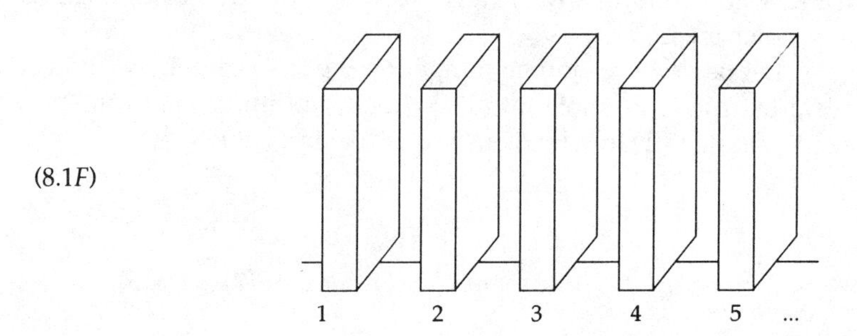

The rule governing this setup is that if $S(n)$ implies $S(n+1)$, then we set up the nth domino close enough to domino $n+1$ so that the former would knock over the latter if it fell down toward the right. If, however, $S(n)$ does not imply $S(n+1)$, then we must set up these two dominoes far enough apart so that the nth could not knock down the $(n+1)$st if it fell. We imagine setting up all the dominoes according to this rule.

The only other rule of this game is that we are to knock over (toward the second domino) the first domino if $S(1)$ is true. What then is the status of the dominoes? If all the propositions are true, then all the dominoes are knocked over. But if one of the propositions is false, then all dominoes numbered z and higher remain standing, where z is the least integer for which $S(z)$ is false. (Why?)

Figure (8.1F) shows the setup of the dominoes for a sequence in which every proposition is true. Figure (8.2F) shows the setup for Example (7.2), in which $z = 3$. (Here $n = 0$ is our starting point instead of $n = 1$.)

(8.2F)

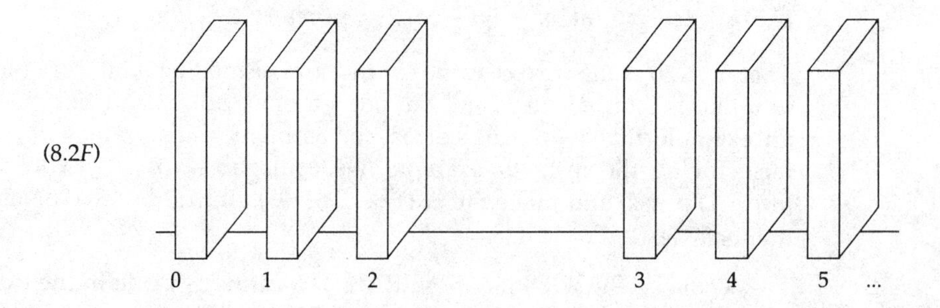

Notice that dominoes $3, 4, 5, \ldots$ are correctly placed with respect to their neighbors.

The point of this imaginary game of dominoes is to give you a concrete image of a correct proof by induction—and the same for an incorrect proof, or for a sequence of propositions in which some are false.

9. How to Solve Problems by Induction

(9.1) To recapitulate, when we use induction to prove that each of a sequence of statements is true, we go through the following procedure:

Basis step. We prove that $S(1)$ is true.

Inductive step. We let n be a fixed but arbitrary integer $n \geq 1$, and assume that $S(1), \ldots, S(n)$ are all true. Under this assumption we prove that $S(n+1)$ is true.

Having done these steps correctly, we use the theorem of the second form of induction to conclude that $S(n)$ is true for all positive n.

In this how-to manual we take our propositions to begin with $n = 1$ for definiteness. You can easily make the necessary changes to accommodate a different starting point. We also use the second form of induction, because it has a stronger inductive hypothesis than the first form. That is, if you can prove the result using the first form, you can certainly do so using the second. (Of course, as you may be asked to show in a problem, anything provable by either form is provable by the other.)

The catch in the above how-to manual is that sometimes it is difficult to prove $S(n+1)$. We shall now try to help you over that hump.

The first point to make is that there is no magic formula. If there were one, mathematics would be an easy game, one where all the problems were solvable—hence there would be no problems! Proving $S(n+1)$ calls for your creative ingenuity.

So what do you do? You try to find a logical or mathematical connection between the statements $S(1), \ldots, S(n)$ on the one hand, and $S(n+1)$ on the other. You try to transform the assumptions into the target. Sometimes this connection is easy to find, as in Example (3.8), where we added $n+1$ to both sides of the equation $S(n)$ to change the left-hand side of $S(n)$ into that of $S(n+1)$. And, as we pointed out in Section (4), the connection between $S(n)$ and $S(n+1)$ for that class of problems is that addition of $a(n+1)$ to each side of equation $S(n)$ turns its left-hand side into that of $S(n+1)$. (Here we refer to the sequence

$$S(n) : \sum_{1 \leq i \leq n} a(i) = c(n).)$$

But the connection can be elusive. Look at the examples given so far and focus on that connection. In most problems that you meet you will be able to do something to either $S(n)$ or $S(n+1)$ to make it look like the other. And perhaps the examples in Section (12) will help you to make your own connections in future problems.

Please understand that finding this connection is a general mathematical problem: Given A, prove B. The only flavor that induction imparts to it is that A and B are statements of the same form, $S(n)$ and $S(n+1)$. The simplest cases, such as the "standard" problems of (4), are routine. In more difficult cases you have your work cut out for you. For help with those, you might like to read Polya's classic *How to Solve It* (17.4). This book can help you find new approaches to mathematical problems in general. And it has a nice treatment of induction. Another book that may help is by Solow, cited in Chapter 2, (31.10).

(9.2) *Induction proofs step by step.* Except for the difficult part just discussed, we can tell you how to write any induction proof. There are several customary steps. For the first form, Theorem (3.2), they are

1. Identify $S(n)$. Do not quantify n.
2. Do the basis step, say for $n = 1$.
3. Write "Let n be a fixed but arbitrary integer ≥ 1, assume $S(n)$, and try to prove $S(n+1)$."
4. Express the target $S(n+1)$.
5. Prove $S(n+1)$—see above remarks, in (9.1). This proof may take several steps.
6. Write "Therefore the inductive step is proved, and by the theorem of mathematical induction, $S(n)$ is true for all positive n."

We may call the third and sixth steps the "boilerplate" of induction proofs.

10. In Logical Terms

Let us summarize part of this chapter by recasting Theorem (3.2), the first form of mathematical induction, in the language of the predicate calculus. The statement becomes more compact; whether you prefer one version or the other will depend on your experience and taste. There is no intrinsic superiority in either one.

Theorem (10.1) Let $S(n)$ be a predicate with domain **N**. Then

$$(S(1) \wedge (\forall n \in \mathbf{N},\ S(n) \longrightarrow S(n+1))) \longrightarrow \forall n \in \mathbf{N},\ S(n).$$

We now state (5.2), the alternate form of Theorem (5.1), in this language. Remember, the variables i and n both have domain **N**, the set of all positive integers. Our "fancy" version of the second form of induction now reads

$$(\forall n\ (\forall i < n\ S(i)) \longrightarrow S(n)) \longrightarrow \forall n\ S(n).$$

Some people like these logical expressions.

11. Subsets and Strings

We now illustrate the use of induction by proving a useful result on the number of subsets of a set.

Theorem (11.1) The total number of subsets of a set having exactly n elements is 2^n. ■

We take the assertion in (11.1) as our statement $S(n)$, and we also supply a domain for n: Since n is the number of elements in a set, n is nonnegative. So let us prove this assertion for all $n \geq 0$. The statement of this theorem is a typical instance of the unwritten $\forall$ (see Chapter 2, (27)).

Proof *The basis step.* At $n = 0$ our assertion is that the empty set has exactly one subset. Since the empty set has only itself as a subset, $S(0)$ is true.

The inductive step. Let n be a fixed but arbitrary integer, $n \geq 0$. Using the first form of induction, (3.2), we assume every set of exactly n elements has exactly 2^n subsets. Now we must try to prove that every set of $n+1$ points has 2^{n+1} subsets (the "exactly" being now understood). So we let T be any set of $n+1$ points, say

$$T = \{ y, x_1, \ldots, x_n \}.$$

We can divide the subsets of T into two classes: those that have y in them, and the rest, those that don't have y in them. The latter class is the collection of all subsets of $\{ x_1, \ldots, x_n \}$; by the induction assumption, it consists of 2^n subsets. But the former class is in one-to-one correspondence with this latter class, because a subset of T with y in it is a subset of $\{ x_1, \ldots, x_n \}$ to which the point y has been "added." Thus the total number of subsets of T is $2^n + 2^n = 2^{n+1}$. QED

Example We illustrate the one-to-one correspondence for the case $n = 2$. $T = \{ y, x_1, x_2 \}$ and the subsets of $\{ x_1, x_2 \}$ are on the left in (11.2), the corresponding ones on the right.

$$\text{(11.2)} \qquad \begin{array}{cc} \varnothing & \{ y \} \\ \{ x_1 \} & \{ y, x_1 \} \\ \{ x_2 \} & \{ y, x_2 \} \\ \{ x_1, x_2 \} & \{ y, x_1, x_2 \} \end{array}$$

We have $2^2 = 4$ subsets of $\{ x_1, x_2 \}$ and $4 + 4 = 8$ subsets of T.

At the end of the proof we omitted our bow to Theorem (3.2). Such omission is customary for experienced "provers," but beginners should include it to show that they understand proofs by induction.

We have proved that there are exactly 2^n elements in the power set of a set with exactly n elements.

(11.3) The result of Theorem (11.1) is important. We look at it from another standpoint, that of representing subsets as "strings" of 0's and 1's. We defined strings in Chapter 1, (19.9). Consider first a concrete example. We may represent the subsets $\varnothing$, $\{ 1 \}$, and $\{ 1,3 \}$ of $\{ 1,2,3 \}$ as rows of the diagram

	1	2	3
$\varnothing$	0	0	0
$\{ 1 \}$	1	0	0
$\{ 1,3 \}$	1	0	1

where a 1 below the header i ($i = 1, 2,$ or 3) indicates that i belongs to the subset named to the left of that row, and a 0 below i indicates that i is not in that subset.

We may do the same for any subset of $\{ 1,2,3 \}$, making it correspond to a row of three symbols, each 0 or 1. Thus $\{ 2,3 \}$ corresponds to the row 0 1 1. The result is that each subset of $\{ 1,2,3 \}$ corresponds to just one *string* of three symbols, each 0 or 1. Conversely, each such string corresponds to a unique subset of $\{ 1,2,3 \}$. For example, the string 0 1 0 corresponds to the singleton $\{ 2 \}$.

(11.4) The general result is that, for all $n \geq 0$, and for any set X of exactly n elements, there is a one-to-one correspondence between the set of all subsets of X and the set of all strings of n symbols, each symbol being 0 or 1.

By now, this result (11.4) is probably obvious; but if not, you will understand all after mastering Chapter 5, on functions. The string corresponding to a subset B of X is closely related to the so-called characteristic function of B.

(11.5) By (11.4), the total number of such strings is 2^n, since we proved in Theorem (11.1) that 2^n is the total number of subsets of X. Such strings allow a natural representation of subsets of X in computers.

12. Examples

Here we give more examples of proofs by induction.

Example (12.1) *An algebraic identity*. Let n be a positive integer, and prove that

(12.2)
$$\sum_{0 \le k \le n-1} 1/(n+k)(n+k+1) = 1/2n.$$

Comment. Before proving this result, let us note an omission in the statement of the problem. Some students see the words "Let n be a positive integer" and think that it's enough to choose one value, like 3 or 7, for n and verify the result for that value. The problem would be more precisely introduced with "Prove that for all positive integers n ...." But we have deliberately stated it in its given form because that form is so commonly used. You should always understand such a statement to mean that you must prove the result for all n in the indicated set, here **N** (See also Chapter 2, (27)).

Now the proof. For the basis step our assertion is that $\frac{1}{2} = \frac{1}{2}$, a true statement. For the inductive step, let n be a fixed but arbitrary positive integer, and assume that (12.2) holds for this value of n. We must try to prove

(12.3)
$$\sum_{0 \le k \le n} 1/(n+k+1)(n+k+2) = 1/(2n+2)$$

Equation (12.2) is $S(n)$, and equation (12.3) is $S(n+1)$. Notice how we had to replace n by $n+1$ in every occurrence of n to pass from (12.2) to (12.3). In particular we had to change the upper limit of summation. How do we find a connection between these two equations? Perhaps the best way is to write out for both equations the first few and the last few terms. We get

$$S(n): \frac{1}{(n)(n+1)} + \frac{1}{(n+1)(n+2)} + \cdots + \frac{1}{(2n-1)(2n)} = \frac{1}{2n}$$

$$S(n+1): \frac{1}{(n+1)(n+2)} + \cdots + \frac{1}{(2n-1)(2n)} + \frac{1}{(2n)(2n+1)}$$
$$+ \frac{1}{(2n+1)(2n+2)} = \frac{1}{(2n+2)}.$$

Now we see that the left-hand sides of these two equations are not so different after all: letting $L(n)$ and $L(n+1)$ stand for the respective left-hand sides, we get

$$L(n+1) = L(n) - \frac{1}{n(n+1)} + \frac{1}{(2n)(2n+1)} + \frac{1}{(2n+1)(2n+2)}$$

Now we use the induction assumption to set $L(n) = \frac{1}{2n}$. It then becomes a matter of straightforward algebra to show that $L(n+1) = 1/(2n+2)$, a task we leave to you.

Example (12.4) *The sportswriter's angle.* There are n teams, each of which plays one match against each other team (this arrangement is called a *round-robin* tournament). A sportswriter then notices that one of the teams, say a_1, beat a_2, that a_2 beat a_3, and so on. Finally, a_{n-1} beat a_n. He proclaims that a_1 is the best team.

Let's call such an ordering of the teams a *victory chain*. If n is small, it's pretty easy to see that a victory chain always exists.

(12.5) For example, for $n = 3$, let the teams be A, B, C. If one team beats both others, say B beats A and C, then the ranking is forced to be

$$B, \quad A, \quad C$$

(if A beat C; otherwise it's B, C, A).

The only other possibility is that each team wins one and loses one match. (Why?) In this case, say that A beats B. Then necessarily B beats C, and also C beats A. So now three rankings are possible:

$$A, B, C; \quad B, C, A; \quad C, A, B.$$

It is a nice proof by induction to show that, for every $n \geq 1$,

(12.6) At least one victory chain exists.

Before proving (12.6) we introduce some helpful terminology:

(12.7) The **complete graph** on n vertices, K_n, is a set V of n points called **vertices,** together with a set of *edges*, defined as follows: An **edge** is a line or curve joining two distinct vertices. A complete graph has an edge between every two distinct vertices (and has no other edges).

Think of the drawings of K_n, as in (12.9*F*).

Examples (12.8)

(12.9*F*)

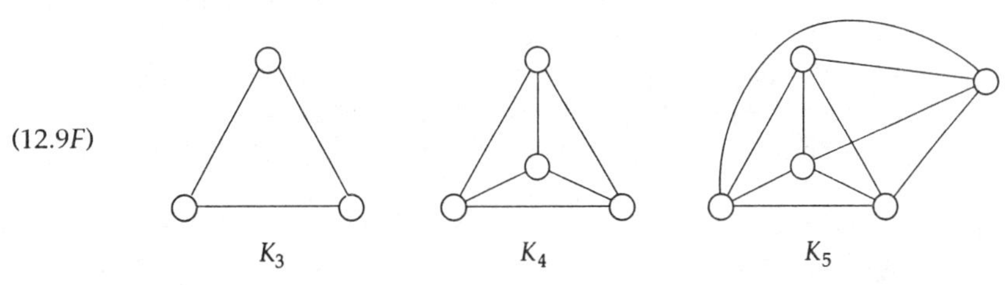

For us, each vertex represents a team, and each edge the match played by the teams at the endpoints. It remains to record which team won each match. We can do that by putting an arrowhead on each edge as follows: Suppose the edge is between teams a and b. Then

if a beat b, we draw the arrow from a to b.

Example (12.10) The two cases of (12.5) are drawn as

(12.10*F*)

The general problem is now this: Prove that for all $n \in \mathbf{N}$,

(12.11) If on each edge of K_n we assign an arrow arbitrarily, then we can always find an ordering

$$a_1, a_2, \ldots, a_n$$

of the vertices of K_n, so that for each successive pair a_i, a_{i+1} (for $i = 1, 2, \ldots, n-1$) the direction of the arrow is from a_i to a_{i+1}.

We take (12.11) as $S(n)$. Since $S(1)$ is vacuously true, the basis case is done. To prove the inductive step, let $n \geq 1$ be any integer. Using the second form of induction, assume $S(m)$ is true for all $m \leq n$. Now consider K_{n+1} with any assignment of arrows. Choose any vertex v of K_{n+1}.

Divide the remaining n vertices into two subsets B_{in} and B_{out}, defined as

B_{in} is the set of all vertices u of K_{n+1} such that the arrow goes from u to v. B_{out} is the same, except that the arrows go from v to the vertices in B_{out}. If B_{in} or B_{out} is empty, we ignore it.

(12.12*F*)

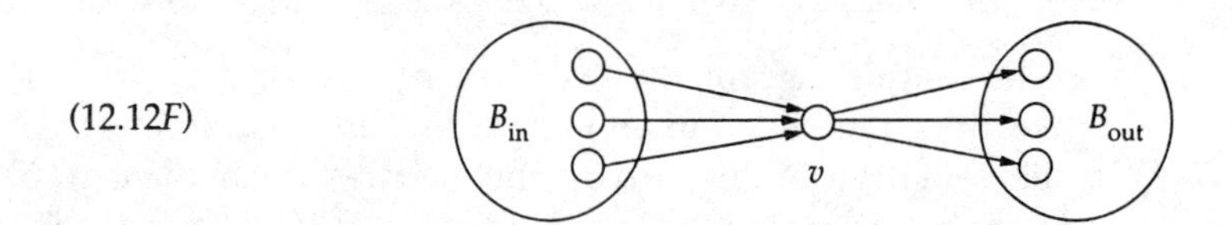

There are arrows between B_{in} and B_{out}, but we ignore them.

The set B_{in} has $m \leq n$ points, and there is an edge with an arrow between each pair of distinct points of B_{in}. Therefore B_{in} is a K_m with arrows. By the inductive hypothesis there is a victory chain for B_{in}. Similarly there is one for B_{out}. We solve the problem by inserting v between the two chains. QED

Induction makes a beautifully simple proof of a nonintuitive result, doesn't it?

Example (12.13) *Maximizing a function.* Here is a final example of the use of induction. Let m be a positive integer. Consider the set V_m of all vectors v satisfying for some $h \geq 0$

$$v = (n_0, n_1, \ldots, n_h)$$

(12.14)
$$\forall i \quad n_i \in \mathbf{N}$$
$$m = n_0 + \cdots + n_h = \sum_i n_i.$$

Thus V_m is the set of all vectors (of whatever length) of positive integers with sum m. For each such vector define the sum

(12.15) $$s(v) := \sum_{0 \le i \le h} i n_i .$$

Example $m = 3$.

v	n_0	n_1	n_2	h	$s(v)$
v_1	1	1	1	2	3
v_2	2	1		1	1
v_3	1	2		1	2
v_4	3			0	0

$V_3 = \{ v_1, \ldots, v_4 \}$.

Practice (12.16) Find V_4 for $m = 4$ and calculate $s(v)$ for each v in V_4.

(12.17) Notice that if $v = (1, \ldots, 1)$, i.e., if $n_i = 1$ for each i, then $h = m - 1$ (and conversely) and $s(v) = \sum_{0 \le i < m} i = (m-1)m/2$, by (3.8). We now prove by induction that this value

(12.18) $(m-1)m/2$ is the maximum value attained by $s(v)$ as v runs over all vectors satisfying (12.14), and $v = (1, 1, \ldots, 1)$ is the only vector for which $s(v)$ is maximal.

Here is the proof. For the predicate $S(m)$ we take (12.18). The basis step is trivial: For $v := (1)$, $V_1 = \{ v \}$, and $s(v) = 0 = 1(1-1)/2$. Thus $S(1)$ is true.

Inductive step. Letting m be a fixed but arbitrary positive integer, we assume $S(m)$ and try to prove $S(m+1)$. Thus let

$$m + 1 = \sum_{0 \le i \le h} n_i$$

and suppose that the vector $(n_0, \ldots, n_h)$ is *not* $(1, 1, \ldots, 1)$. That is, for some i, say for $i = j, n_j > 1$. Let $\sum$ stand for $\sum_{0 \le i \le h}$. We'll show that

$$\sum i n_i < m(m+1)/2.$$

Once we do so, we'll have proved $S(m+1)$.

We define a vector $v' = (n'_0, \ldots, n'_h)$ as follows:

(12.19) $$n'_i = \begin{cases} n_i & \text{if } i \neq j \\ n_j - 1 & \text{if } i = j. \end{cases}$$

Notice that v' is a vector of positive integers summing to m. By our induction assumption, $s(v') \le (m-1)m/2$. But

$$s(v') := \sum i n'_i = \sum i n_i - j$$

by (12.19). Remember j is fixed and summation is over i. Therefore

$$s(v') = \sum in_i - j \leq (m-1)(m/2),$$

which gives us

$$\sum in_i \leq (m-1)(m/2) + j.$$

Now by (12.17) with $m := m+1$, h is less than m, since not all the n_i are 1; therefore $j \leq h < m$. Therefore

$$\sum in_i \leq (m-1)(m/2) + j < (m-1)(m/2) + m = (m+1)(m/2).$$

We have proved that $S(m)$ implies $S(m+1)$. Therefore by Theorem (3.2) on mathematical induction, $S(m)$ is true for all $m \geq 1$. QED

13. Recursion

Recursion is induction put to use. We define or construct certain things by induction and call the process *recursion*. For example, consider the following definition of a sequence of integers $f(0), f(1), \ldots$:

DEFINITION (13.1) $f(0) := 1$, and for all $n \geq 1$ $f(n) := n \cdot f(n-1)$

Thus $f(1) = 1 \cdot f(0) = 1 \cdot 1 = 1$; $f(2) = 2 \cdot f(1) = 2 \cdot 1 = 2$, and so on. Most students readily accept (13.1) as a clear definition and even recognize this sequence as the factorials: $f(n) = n!$. But forget your prior knowledge for a moment and look at (13.1) with fresh eyes. Doesn't it seem to define f in terms of itself?

In fact (13.1) defines $f(n)$ in terms of a *previously* defined value, here $f(n-1)$. This characteristic holds in all recursive definitions: The thing being defined is specified in terms of smaller or prior instances of itself.

To be formal, we use induction to prove that (13.1) really defines a sequence, and that it is unique.

Theorem (13.2) There is one and only one sequence of integers $f(0), f(1), \ldots$ satisfying (13.1).

Proof For $S(n)$ we take "There is one and only one value of $f(n)$ satisfying (13.1)." $S(0)$ is true—immediately from (13.1). For any $n \geq 1$, $S(n-1) \longrightarrow S(n)$. Why? Because $S(n-1)$ tells us that $f(n-1)$ is uniquely defined. Since (13.1) defines $f(n)$ as $f(n) := n \cdot f(n-1)$, we see that now $f(n)$ is uniquely defined. Therefore $S(n)$ is true if $S(n-1)$ is true. We proved the basis and inductive steps for this $S(n)$. Therefore Theorem (3.2) tells us that for all $n \geq 0$, $S(n)$ is true. QED

Besides the factorial function there are other familiar functions defined by recursion. Exponentiation is one:

(13.3) Let a be any real number not 0. Define $a^0 := 1$ and $\forall n \geq 0 \quad a^{n+1} := (a^n)a$.

(13.4) We sneaked recursion into Chapter 2, (6.2) in defining propositional formulas. We said that T, F, and the atoms were propositional formulas—that's the basis step. The inductive step was the next part of definition (6.2): For all propositional formulas f and g, the following are also propositional formulas: $\sim (f)$, $(f) \bigvee (g)$, $(f) \bigwedge (g)$, $(f) \longrightarrow (g)$. Finally we said every propositional formula must be formed in this way. So, we could think of the set PF of all propositional formulas as being defined by induction on the *height* of the formulas. For the meaning of *height* consider first these examples:

$$\begin{aligned}
&\text{Height 0: } T, F, a, b, c, \ldots \\
&\text{Height 1: } a \bigvee b, \sim a, \ldots \\
&\text{Height 2: } (a \bigvee b) \longrightarrow c, \ (a \bigvee b) \longrightarrow \sim a, \ldots \\
&\qquad\vdots
\end{aligned}$$

For $n \in \mathbf{N}$, we say a formula has **height** n if it is $\sim (f)$ and f has height $n-1$, or if it is $(f) \bigvee (g)$, $(f) \bigwedge (g)$, or $(f) \longrightarrow (g)$ for any two formulas f and g of height less than n if f or g has height $n-1$.

Thus we may think of PF as being constructed one level at a time, by induction. First we are given the "foundation," the elements of height 0. We combine those with one connective to get height 1. Then we may combine anything of height 1 and anything of height 0 or 1 to get height 2. In general, to get height n we combine any formula of height $n-1$ with any formula of height less than n (or we negate any formula of height $n-1$).

The foregoing elaboration should make plain the inductive nature of definition (6.2) of Chapter 2. To make the definition look more recursive, consider it in set-builder form: For convenience, we first define B as the set consisting of T, F, and all atoms. Then

$$\begin{aligned}
PF := \{x;\ x \in B, \quad &\text{or} \quad \exists f, g \in PF, \\
&x \in \{\sim (f), (f) \bigvee (g), (f) \bigwedge (g), (f) \longrightarrow (g)\}\}.
\end{aligned} \tag{13.5}$$

Is this definition valid? Or is it like Russell's paradox (Chapter 2, (30)) in its self-reference? It is valid because it presents us with the propositional formulas already formed and asks us to verify that they are in PF by taking them apart, connective by connective. If we did so, we would eventually get to atoms (for a grammatical formula) and thus conclude our formula was in PF. That is the intuitive justification of the definition, that it refers to *smaller* versions of itself and leads ultimately to an explicitly defined basis, here B.

Notice also that we defined height recursively.

14. Applications

Induction is everywhere in mathematics—in logic, number theory, algebra, analysis, what have you. In computer science it is just as widespread, and it is essential in the assertion method of proving the correctness of programs (see, for example, Knuth (17.2, pages 10–18) and Chapter 6 of this text). And recursion, so widely used in computer science, *is* induction, as we just saw in (13). Induction was born in the seventeenth century as Fermat's method of descent, which actually looks more like recursion at first glance than it does like induction.

15. Fermat's Method of Infinite Descent

Fermat stated the inductive step $S(n) \longrightarrow S(n+1)$ in contrapositive form:

$$\sim S(n+1) \longrightarrow \sim S(n).$$

For the second form of induction we would have

$$\sim S(n+1) \longrightarrow \sim S(n) \vee \sim S(n-1) \vee \ldots \vee \sim S(1).$$

Descent can look different from induction in actual situations. It is most often used in Diophantine problems, those involving integral variables. For example, that there are no integers $x, y, k > 1$ such that

$$3x^2 + y^2 = 2^{2k} \qquad \text{with } x \text{ and } y \text{ odd}$$

is a Diophantine result. We shall prove it by Fermat's method of descent at the end of Chapter 6, and by a shortcut at the end of Chapter 7. Other examples of Fermat's method of descent appear in (17.3, pages 199–203) and (17.5, pages 155–161).

16. Postscript

Now that you have learned all about induction, I am loath to tell you this, but in all honesty I must do so: There are problems solvable by induction for which easier, more instructive methods of solution exist. For example, the familiar method for finding the sum of an arithmetic progression would allow you to get the formula for the sum of the first n integers even if you didn't know it in advance—you recall that we were given the answer in advance before we proved it correct by induction. Thus we let A be the sum of the first n integers, and writing its terms out twice in opposite orders, we see that

$$A := 1 + 2 + \cdots + (n-1) + n$$
$$A = n + (n-1) + \cdots + 2 + 1.$$

So if we add these two equations we get $2A = n(n+1)$, from which our result is immediate.

Example One day in 1787 the 10-year-old Gauss was asked, along with the rest of his schoolmates, to find the sum of the first 100 integers. As the other pupils began to add and erase numbers one by one on their slates, Gauss paused for a moment and then wrote "5050" on his slate.

Similarly, we can find the sum of the terms of a geometric progression: let $a \neq 1$, and set

$$T := 1 + a + a^2 + \ldots + a^n = \sum_{0 \le i \le n} a^i .$$

If we multiply T by $(1 - a)$ we get

$$(1 - a)T = 1 - a^{n+1}. \tag{16.1}$$

Thus we see that T is easily determined, since dividing (16.1) by $(1 - a)$ yields the familiar closed-form expression for the sum of terms of a geometric progression. We derived this result more easily than if we had used induction to prove

$$T = \frac{1 - a^{n+1}}{1 - a}, \tag{16.2}$$

although we could of course do that if we wished. (And the purist might insist that we had used a disguised or tacit induction to prove (16.1). That's true, but how much easier it is to prove (16.1) by induction than (16.2).)

Here is another example along these lines: Find the value of

$$\sum_{1 \le k \le n} \frac{1}{k(k+1)}, \tag{16.3}$$

where n is a positive integer. We could calculate the first several values of this sum and then guess at a general formula for it; that process would be called scientific induction. We would then try to prove our guess correct by mathematical induction. Often such an approach is necessary; as you continue with mathematics or computer science you will do lots of such guessing and proving. But with this problem, at least, there's a much easier way, which starts with the observation that

$$\frac{1}{k(k+1)} = \frac{1}{k} - \frac{1}{k+1}. \tag{16.4}$$

Now we see that the sum equals $1 - 1/(n+1)$, because the individual terms "telescope," which means here that the negative term for each value of k cancels the positive term for the next higher value of k:

$$\sum_{1 \le k \le n} \frac{1}{k(k+1)} = \left(\frac{1}{1} - \frac{1}{2}\right) + \left(\frac{1}{2} - \frac{1}{3}\right) + \cdots + \left(\frac{1}{n-1} - \frac{1}{n}\right) + \left(\frac{1}{n} - \frac{1}{n+1}\right). \tag{16.5}$$

Only the first and last terms of (16.5) are untouched by this cancellation process, so their sum is the whole sum. Perhaps you agree that this proof is better than a proof by induction of the same result. This possibility of shortcutting the induction proof is what I had in mind when I mentioned getting a practitioner's viewpoint on induction at the beginning of this chapter.

Example (16.6) Here is a more complicated example of shortcutting induction. We rework an example from Reference (17.6). The problem deals with n lines in the plane in *general position*, which means that no two are parallel, and no three meet in the same point. We are asked to find how many regions of the plane are formed by such a set of n lines. To get a feel for the problem, let's calculate this number for the first few values of n (that is, let's do a bit of scientific induction). For $n = 0$ there is one region, the whole plane; for $n = 1$, there are two regions; for $n = 2$, four regions, for $n = 3$, seven regions—see (16.6F). Presumably we could calculate more values, but at this point an important question arises, namely, how do we know that the number of regions is the same for every set of n lines in general position? We can argue that seven is the right value for $n = 3$ by saying that the three lines determine a triangle (since they are in general position), the interior of which is one region. The other six are opposite the three sides and the three internal angles, respectively. But must we make such an argument, ever more complicated, for each higher value of n? Is the number truly independent of the placement of the n lines? No and yes, respectively.

(16.6F)

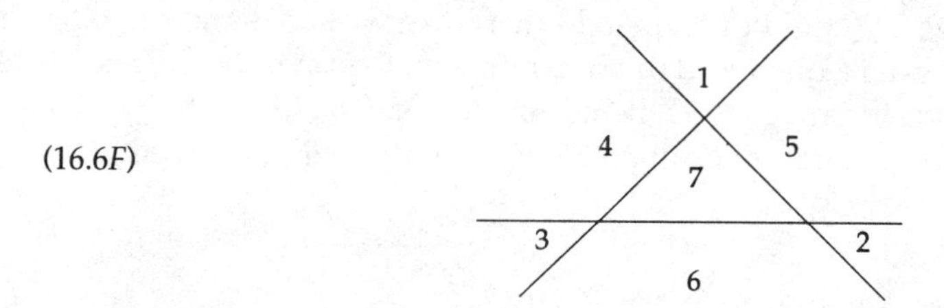

Let us define $A(L_1, L_2, \ldots, L_n)$ to be the number of regions determined by the n lines $L_1, \ldots, L_n$. We'll prove that this function depends only on n, not on the choice of the n lines $L_1, \ldots, L_n$:

(16.7) Every set of n lines in the plane in general position determines the same number of regions.

At the same time we'll determine that number. We let n be any positive integer and consider any set of $n + 1$ lines $L, L_1, L_2, \ldots, L_n$ in general position. We'll compare the number of regions made by the n lines $L_1, \ldots, L_n$ with the number made by the $n + 1$ lines $L, L_1, \ldots, L_n$. Call them the *old* and *new* regions, respectively.

Because the lines are in general position, the line L intersects each of the lines $L_1, \ldots, L_n$ in a different point.

Call these n points $p_1, p_2, \ldots, p_n$; and suppose the subscripts are chosen so that as we travel along L from one infinite end toward the other, we encounter these points in the order $p_1, p_2, \ldots, p_n$. These n points divide L into $n + 1$ segments

(16.7F)

$L \quad p_1 \; p_2 \quad \cdots \quad p_{n-1} \quad p_n$

Each segment cuts one *old* region into two *new* regions. That is the key observation. See (16.8*F*).

(16.8*F*)

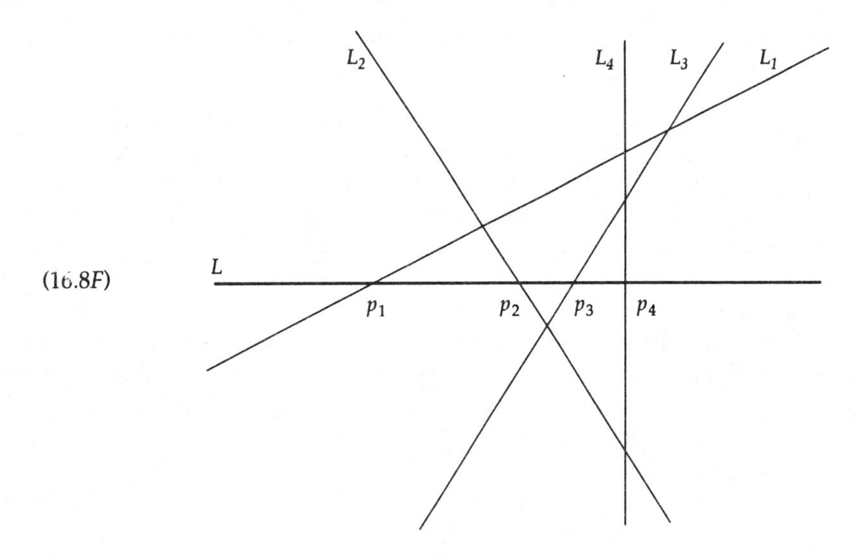

So introducing L into the plane increases the total number of regions by $n + 1$. Thus

$$A(L, L_1, \ldots, L_n) - A(L_1, \ldots, L_n) = n + 1 \tag{16.9}$$

for any set of $n + 1$ lines in general position and for any positive integer n. It is even true for $n = 0$:

$$A(L) - A(\varnothing) = 2 - 1 = 1.$$

If we now apply (16.9) to the set of n lines $L_1, \ldots, L_n$ in general position, and then to the $n - 1$ lines $L_1, \ldots, L_{n-1}$, and so on down, we get the n equations

$$\begin{aligned} A(L_1, \ldots, L_n) - A(L_1, \ldots, L_{n-1}) &= n \\ A(L_1, \ldots, L_{n-1}) - A(L_1, \ldots, L_{n-2}) &= n - 1 \\ &\cdots \\ A(L_1, L_2) - A(L_1) &= 2 \\ A(L_1) - A(\varnothing) &= 1 \end{aligned}$$

We now add all these equations and find that

$$A(L_1, \ldots, L_n) - A(\varnothing) = 1 + 2 + \cdots + (n - 1) + n$$

since the terms telescope; the term $-A(L_1, \ldots, L_i)$ in the ith equation (from the bottom) cancels the term $A(L_1, \ldots, L_i)$ in the equation below it. Therefore

$$A(L_1, \ldots, L_n) = 1 + \sum_{1 \le k \le n} k = 1 + \frac{n(n+1)}{2}, \tag{16.10}$$

from Example (3.8). Since the right-hand side of (16.10) depends only on n, we see that, for all integers $n \geq 0$, the number of regions formed by n lines in general position is the same for any set of n such lines.

Practice (16.11) Let E be any set of lines in the plane in general position. Prove that any subset of E has the same property.

Although there are problems, such as (5.4), to prove that every integer greater than 1 is a product of primes, where induction seems to be essential to the proof, there are others where some different approach is more economical, yields more insight into the problem, and in some cases produces an answer that would have to be guessed at before induction could begin.

These "induction-undercutters" seem to be concentrated in the algebraic or analytic areas, however. Even there, some famous problems have so far required the use of induction for their solution. I am thinking of van der Waerden's theorem on arithmetic progressions: in simplest form, the result is that no matter how the set $\mathbf{N}$ of positive integers is expressed as the union of two disjoint subsets X and Y, an arbitrarily long arithmetic progression exists in either X or Y. The interested reader may find this problem fully explained in (17.1).

Students of computer science have not been "had," however. They will use induction in so many areas, for example, trees, and especially formal languages, that all the ideas of this chapter will become second nature to them. In fact, I hope for that result by the time you finish this book. There will be many uses of induction in later chapters.

17. Further Reading

(17.1) A. Khinchin, *Three Pearls of Number Theory*, Graylock Press, Rochester, 1952.

(17.2) Donald E. Knuth, *The Art of Computer Programming, Volume 1: Fundamental Algorithms*, 2nd ed., Addison-Wesley, Reading, 1973. (See section 1.2.1.)

(17.3) Oystein Ore, *Number Theory and Its History*, McGraw-Hill, New York, 1948. (Also Dover, 1988.)

(17.4) George Polya, *How To Solve It*, Princeton University Press, Princeton, 1945.

(17.5) Harold M. Stark, *An Introduction to Number Theory*, The MIT Press, Cambridge, 1978.

(17.6) I. S. Sominskii, *The Method of Mathematical Induction*, Blaisdell, Waltham, 1961.

(17.7) B. K. Yousse, *Mathematical Induction*, Prentice-Hall, Inc., Englewood Cliffs, 1964.

18. Problems for Chapter 3

In the absence of explicit instructions, you are to prove the assertions made. The notation is uniform: all unspecified sums are over the variable k, which runs over all values satisfying $1 \leq k \leq n$. The domain of the variable n is the set of all nonnegative integers (careful!) unless stated otherwise.

1. $\sum (2k)^2 = \frac{2}{3}n(n+1)(2n+1)$.

2. Prove that 6 divides $n(n+1)(2n+1)$. (Definition: we say that the integer a divides the integer b iff there is an integer x such that $b = ax$.)

3. Let $A_1, \ldots, A_n$ be sets. Prove De Morgan's law by induction:

$$\overline{\bigcup_{1\le i\le n} A_i} := \overline{A_1 \cup A_2 \cup \cdots \cup A_n}$$

$$= \bigcap_{1\le i\le n} \overline{A_i} := \overline{A_1} \cap \cdots \cap \overline{A_n}.$$

4.[Ans] $\sum_{1\le k\le n}(3k-2) := 1 + 4 + 7 + \cdots + (3n-2) = n(3n-1)/2$.

5. $\sum_{1\le k\le n}(3k^2 + k + 3) := n(n^2 + 2n + 4)$.

6. Prove that $\sum k^2 = \frac{1}{6}n(n+1)(2n+1)$.

7. $\sum k(k+1) = \frac{1}{3}n(n+1)(n+2)$.

8. Use Example (3.8) and the result of the previous problem to derive a formula for $\sum k^2$.

9. $\sum k(k+1)(k+2) = \frac{1}{4}n(n+1)(n+2)(n+3)$.

10. Prove that $\sum k(k-1) = \frac{1}{3}(n+1)n(n-1)$.

11. Prove that $\sum k(k-1)(k-2) = \frac{1}{4}(n+1)n(n-1)(n-2)$.

12. Use the previous two problems to derive a formula for $\sum k^3$. Now prove that formula by induction.

13. $\sum k^2(-1)^k = \frac{1}{2}(-1)^n n(n+1)$.

14. If n is any positive odd integer, then $1 + 3^n$ is divisible by 4.

15. Prove that for some positive integer k all integers n greater than k satisfy $2^n > n^2$.

16. Prove that

$$\sum 2/(k+2)k = \tfrac{3}{2} - (2n+3)/(n+1)(n+2).$$

17. Prove that $\sum 1/(3k-2)(3k+1) = n/(3n+1)$.

18. Prove that $\sum 1/(2k-1)(2k+1) = n/(2n+1)$.

19.[Ans] Prove a generalization of the last two examples.

20. Prove by induction or otherwise that for all positive odd n

$$\sum_{0\le k\le n} (-2)^k = (1/3)(1 - 2^{n+1}).$$

21. (*i*) Contrary to the statement in the text, section (4.1) did prove something. State formally as a theorem what was proved there.

(*ii*) Consider Theorems (3.2) and (5.1). Prove that any sequence of statements provable true by either theorem is provable by the other. [A sequence S(n) is "provable true by Theorem (3.2)," for example, if it satisfies the hypotheses of Theorem (3.2). So you must show that if $S(n)$ satisfies the hypotheses of Theorem (3.2), then $S(n)$ satisfies the hypotheses of Theorem (5.1), and conversely.]

22. Frank recently proved by induction that all people are of the same height. He took for $S(n)$: "In every set of exactly n people, all have the same height." He said, "It's obvious that $S(1)$ is true; and for fixed but arbitrary $n \ge 1$ we assume $S(n)$ is true and consider any set G of exactly $n+1$ people. Let one of them be Helen. Remove Helen from G, making a set of n people, all of whom have the same height as each other by the induction hypothesis. Now put Helen back in G and remove Elinor, making another set of n people, again of the same height by the induction hypothesis. But another person, say Betty, was in G all along, and we saw that Betty has the same height, first, as Elinor, and then as Helen. So all $n+1$ of the people in G have the same height. Thus $S(n)$ implies $S(n+1)$ for all $n \ge 1$, so by the theorem of mathematical induction, $S(n)$ is true for all $n \ge 1$. QED."

Frankly, I didn't believe him, but I couldn't find the fallacy. Can you?

23. Guess and prove a formula for $\sum (k-2)^2$, good for all positive n. Do by induction or otherwise. Explain how you made your guess.

24. Prove by induction for all $n > 1$: $\sum 1/\sqrt{k} > \sqrt{n}$.

25. Prove that $2^{2n+1} + 3^{2n+1}$ is a multiple of 5 for all integers $n \ge 0$.

26. Prove that $4^{2n+1} + 3^{n+2}$ is a multiple of 13 for all $n \ge 0$.

27. Prove that $2^{n+2} + 3^{2n+1}$ is a multiple of 7 for all integers $n \ge 0$.

28. $\sum 1/(4+k)(5+k) = \frac{1}{5}n/(n+5)$.

29. [Recommended] Define a sequence of numbers $b(n)$ as follows: After the first two values, $b(n+2) + 2b(n+1) + b(n) = 0$ for all $n \ge 0$. The sequence is thus determined by the recursion just given and the first two values of the sequence.

(*i*) Define $b(0) = b(1) = 1$ and prove $b(n) = (1 - 2n)(-1)^n$ for all $n \ge 0$.

(*ii*) Define $b(0) = 1$, $b(1) = -3$ and prove $b(n) = (1 + 2n)(-1)^n$ for all $n \geq 0$.

#30. Let q be a fixed positive integer. From q distinct symbols $a, b, c, \ldots$ one can form exactly q^n strings of $n \geq 0$ symbols each (with repetition allowed). Prove this assertion by induction on n. [*String* is defined in Chapter 1, (19.9).]

31. We continue with strings. [Any nonempty string in the symbols 0,1 can be uniquely represented as $0^i 1^j 0^k 1^l \ldots$, where $i \geq 0$, and if any later superscript is 0 then all those after it are 0 too. This notation means that the string starts with i 0's, then has j 1's, then k 0's, l 1's, and so on. Thus 00111011 is denoted $0^2 1^3 0^1 1^2$, and 1100001 is denoted $1^2 0^4 1^1$.] Prove that the set S of all strings in 0,1 in which no pair of consecutive 1's is to the right of any 0 equals the set V of all strings of the form $(1^k)s$, where $k \geq 0$ and s is any string in 0, b, where b is 01.

***32.** (Hopcroft and Ullman) Let's denote the empty string by E. Let A be any finite nonempty set. A *palindrome* over A can be defined as a string that reads the same forward as backward. We may also define a set P as follows:

(*i*) $E \in P$

(*ii*) $\forall a \in A,\ a \in P$

(*iii*) $\forall a \in A\ \forall x \in P,\ axa \in P$.

(*iv*) Nothing is in P unless it follows from (*i*), (*ii*), (*iii*).

Prove by induction that P equals the set of all palindromes over A. [The only trick here is to find the predicate $S(n)$.]

***33.** (Hopcroft and Ullman) $E :=$ empty string. Consider the following two definitions of strings of balanced parentheses. A is the set consisting of the two elements "(" and ")".

Definition. 1. A string w over A is in B_1 iff

(*i*) w has an equal number of ('s and)'s; and

(*ii*) Any prefix of w has at least as many ('s as)'s. [A *prefix* of w is w or the string remaining after the erasure of any symbol a (from A) in w and of all symbols to the right of a in w. For example, 1011010 has E, 1, 10, 101, 1011, and so on, as prefixes.]

Definition 2. The set B_2 is defined as:

(*iii*) $E \in B_2$

(*iv*) $\forall w \in B_2,\ (w) \in B_2$

(*v*) $\forall x, w \in B_2,\ xw \in B_2$

(*vi*) Anything in B_2 must come from (*iii*), (*iv*), (*v*).

Prove by induction on the length of a string that $B_1 = B_2$.

34.[Ans] Suppose that in setting up the dominoes according to the rules in the text for a sequence of propositions $S(n)$ you find that the only gap occurs between the third and fourth dominoes. What are the possible distributions of truth-values of $S(1), S(2), S(3), \ldots$?

35. Guess and then prove a formula for the sum of the first n odd integers.

36. Find the value of

$$\sum_{0 \leq k \leq n-1} \frac{1}{(n-k)(n-k+1)}.$$

#37. Prove that the total number of subsets having exactly two elements in a set of n elements is $n(n-1)/2$.

38. Prove by induction or otherwise that for all positive odd integers n

$$\sum_{0 \leq k \leq n-1} \frac{1}{(n-2k)(n-2k+2)} = \frac{n}{4-n^2}.$$

39. Using (13.3), let a, b be any real numbers not 0, and define for all $n \geq 1$, $a^{-n} := (1/a)^n$. Prove that for all i, j in $\mathbf{Z}$

$$(a^i)(a^j) = a^{i+j}$$
$$(a^i)^j = a^{ij}$$
$$(ab)^i = (a^i)(b^i).$$

40. Prove by induction that for all $n \in \mathbf{Z}$

$$(-1)^n = \begin{cases} 1 & \text{if } n \text{ is even} \\ -1 & \text{if } n \text{ is odd.} \end{cases}$$

41. Prove that the alternating sum of the first n odd integers is $n(-1)^{n-1}$. That is,

$$\sum (2k-1)(-1)^{k-1} = n(-1)^{n-1}.$$

42. Prove that $\sum (4k-2) = 2n^2$.

43. Let $c(n)$ be a sequence of integers such that $c(0) = c(1) = 1, c(2) = 3$, and for all $n \geq 1, c(n+2) = 3c(n+1) - 3c(n) + c(n-1)$. Prove that for all $n \geq 0, c(n) = n^2 - n + 1$.

44. Guess and prove a formula for $1 - 2 + 3 - 4 + \cdots + n(-1)^{n-1}$. Explain how you made your guess.

45. Prove: for all $n \geq 0$, $\sum 2^l = 2^n - 1$. Here the sum is over all l satisfying $0 \leq l \leq n-1$. Careful! Recall that any summation over the empty set is 0.

46. (From (17.6)). Prove that n planes in three-dimensional space all going through a common point but no three of which go through the same line divide space into $B(n) = n(n-1) + 2$ regions. Supply a correct domain for n.

47. Find a shortcut method for proving (12.2).

48.[Ans] Ans Prove by a noncalculus shortcut: For some integer k, all integers $n > k$ satisfy $n^2 > 5n - 2$. (A proof by induction is worth half credit.)

49. Prove that for all positive odd integers n, 8 divides $n^2 - 1$.

50. Prove by induction or otherwise that

$$\sum \frac{k^2}{(2k-1)(2k+1)} = \frac{n(n+1)}{4n+2}$$

51. Prove by induction that for all integers $n \geq 2$ the set of all points of intersection of n distinct lines in the plane has no more than $n(n-1)/2$ elements. Give examples showing exactly that many, and also fewer.

***52.** (Chartrand and Liu) The host and hostess, who are married to each other, greet some of the n married couples coming to their party. After various people shake hands the host asks each one how many hands he or she has shaken. They give him $2n + 1$ different answers. No spouses shake hands with each other. How many hands did the hostess shake? (No two people shake hands twice with each other.) Prove your answer by induction. [Hint: do it by trial-and-error for $n = 1$ and for $n = 2$; then guess a general answer and prove it.]

53. *Euler's formula*. Consider a *connected planar graph*, i.e., a finite set of points, called *vertices*, in the plane, some pairs of which are joined by *edges*, which are curves, none of which intersects any others. There are only a finite number of edges. So far we've defined a planar graph; it is *connected* iff for any two vertices $a \neq b$, there is a sequence of edges leading from a to b.

Two points of the plane are said to be in the same *face* of a planar graph iff there is a continuous curve in the plane joining the two points and not touching any edge or vertex of the graph. In particular, the set of all points "outside" the graph make up the *infinite* face.

(*i*) Draw examples of connected, also of disconnected, planar graphs. Count the numbers of vertices, edges, and faces in each example.

(*ii*) Draw an example of a connected planar graph with only one face.

(*iii*) Draw a connected planar graph that becomes disconnected on the removal of one of its edges. Label the edge d. Now do the same for one that remains connected.

(*iv*) Draw an example of a connected planar graph that remains connected after removal of one of its edges and one vertex at an endpoint of that edge. Label the edge d' and the vertex a.

*(*v*) For any nonempty connected planar graph, let v denote the number of vertices, e the number of edges, and f the number of faces. Prove *Euler's formula:*

$$v - e + f = 2.$$

[Hint: Use induction on e, but be careful to keep your answers to (*i*)–(*iv*) in mind. Start with $e = 0$.]

54. A polygon of n sides is a set of n distinct points in the plane together with n distinct line-segments joining pairs of these points so that each point is joined to exactly two other points. It is called *convex* if, for any two points inside it, the line segment joining those two points is entirely inside the polygon. Prove that the sum of the interior angles of a convex polygon of n sides is $(n-2)180°$, for all $n \geq 3$. [Hint: Break the polygon into two smaller convex polygons; use the second form of induction.]

***55.** (Martin Gardner) Let $n \geq 2$. There are n women, and n red hats, and n green hats. The women are blindfolded, a red hat is put on each one, and the blindfolds are removed. Each woman sees $n - 1$ red hats but not the one on her own head. Anyone seeing a red hat is to hold up her hand. Anyone who knows her hat color is to announce it. After a while one of the women (cleverer than the others) says, "My hat is red." How did she know? [Compare Chapter 2, (15).]

***56.** People on the island of St. Martin's have either blue or brown eyes. The peculiar law of St. Martin's is that anyone who discovers that he or she has blue eyes must leave at midnight of the day of discovery, never to return. No one ever leaves for any other reason. The catch is that no islander is allowed to discuss, or communicate in any way at all, anything about anyone's eye color. Furthermore, there are no mirrors, cameras, or other devices by which one could find out the color of one's eyes.

They all know each other and they are all excellent at reasoning. After generations of isolation, one day a mainland visitor appears among them. She says to an assemblage of all the islanders, "I see blue eyes here." She speaks the truth, and everyone believes her.
What happens? Explain your answer.

19. Answers to Practice Problems

(4.3P) Example (3.8) gave a proof by induction that $\forall n \in \mathbf{N}\ \Sigma k = n(n+1)/2$.

The equation to be proved does fit Section (4).

$$L(n) := \sum_{1 \le k \le n} k$$

$$a(k) := k$$

$$c(n) := \frac{n(n+1)}{2}$$

By (4) we need prove only that

$L(1) = c(1)$, the basis step—which we did in (3.8)

and that

For all $n \in \mathbf{N}, c(n) + a(n+1) = c(n+1)$,

the inductive step. To carry out the latter:

$$\frac{n(n+1)}{2} + n + 1 = (n+1)\left(\frac{n}{2} + 1\right)$$
$$= (n+1)\,\frac{(n+2)}{2}$$
$$= c(n+1).$$

(12.16P) V_4 is the set of all vectors of positive integers with sum 4. Thus V_4 is

n_0	n_1	n_2	n_3	$s(v)$	
1	1	1	1	6	max = 1
2	1	1		3	max = 2
1	2	1		4	
1	1	2		5	
2	2			2	
3	1			1	max = 3
1	3			3	
4				0	max = 4

and $s(v) := 0 \cdot n_0 + 1 \cdot n_1 + 2 \cdot n_2 + 3 \cdot n_3$.

(16.11P) E is a set of lines in the plane such that no two are parallel and no three meet in one point. Any subset B of E has the same property, for if two lines are in B they are in E, hence cannot be parallel, and similarly for the rest.

Since E is defined with the universal quantifier the result holds also when $B = \varnothing$, by Chapter 2, (25).

CHAPTER

4

Equivalence Relations and Partitions

Now we take up an idea fundamental for both computer science and mathematics. Indeed, you have seen equivalence relations from the beginning of your study of mathematics and computer science. (You may be less thrilled to know that than Molière's *bourgeois gentilhomme* was to learn he had been speaking prose all his life, but it is true nonetheless.)

1. Relations, Graphs, and Matrices

First we define *relation*, and then quickly specialize to *equivalence relation*, leaving other relations to Chapter 12.

DEFINITION (1.1) A **binary relation** on a set A is a subset of the Cartesian product[†] $A \times A$.

As you can see, the concept of relation is very general. Let us look at different ways to define a relation. We defined it as a set of ordered pairs. It can also be viewed as a *directed graph* or as a *relation matrix*.

(1.2) The **directed graph** of the relation is the set of points of A together with the ordered pairs of the given relation; these two sets are often represented in a sketch as a set of dots, one dot for each point of A, and a set of arrows joining the dots, one arrow for each ordered pair (a, b) in the given relation, and drawn as an arrow from a to b: $a\bullet \longrightarrow \bullet b$. (The "dot" may be instead a small circle.)

[†] Defined in Chapter 1, (14).

Example (1.3) The relation $\{(1,2),(1,3),(1,4),(2,3),(4,4),(4,2)\}$ on the set $A = \{1,2,3,4,5\}$ has the graph

(1.3*F*)

The arrow from "1" to "2" stands for the ordered pair (1,2).

Sometimes such a sketch is helpful in thinking about relations.

(1.4) The **relation-matrix** is a square array of 0's and 1's, with rows and columns labeled by the elements of A, one row and one column for each such element. For each a, b in A, the entry in row a and column b of the matrix is 1 if (a,b) is one of the ordered pairs of the relation, and 0 if it is not.

Example (1.5) The relation-matrix of example (1.3) is

	1	2	3	4	5
1	0	1	1	1	0
2	0	0	1	0	0
3	0	0	0	0	0
4	0	1	0	1	0
5	0	0	0	0	0

The "1" in row 1, column 2 stands for the ordered pair (1,2).

When setting up a relation-matrix, we list the rows and columns in the same order. The relation-matrix is often useful for computation.

Example (1.6) An example of a relation R defined by the set-builder notation (Chapter 2, (19.10)) is

$$R := \{(a,b);\ a,b \in \mathbf{N},\ b = a+1\}.$$

The graph of R is

1 2 3 4

○→○→○→○→ ···

Its relation-matrix is an infinite matrix that in its upper left corner is

	1	2	3	4	···
1	0	1	0	0	
2	0	0	1	0	···
3	0	0	0	1	
⋮		⋮			

It is apparent that we could take either the directed graph or relation-matrix as our definition of relation instead of the ordered-pair definition, because each of the three immediately gives rise to the other two. Logically it would be enough to discuss any problem on relations in terms of just one (any one) of the three viewpoints. I shall, however, take all three viewpoints on the examples below so as to illustrate them all.

▶To convince you how general the idea of relation is, we now observe that our five-point set A of Example (1.3) has a total of more than 32,000,000 different relations on it. Precisely, there are 2^{25} of them, because every different subset of the 25-point set $A \times A$ is a different relation; and as we saw in Chapter 3, (11.1), any n-point set has 2^n different subsets. So we shall narrow our inquiry to a much smaller class of relations, the *equivalence* relations. ◀

2. Equivalence Relations

We start right off with the definition:

DEFINITION (2.1) A relation on a set A is called an **equivalence relation** if and only if it is reflexive, symmetric, and transitive.

What do these new terms mean?

DEFINITION (2.2) A relation R is **reflexive** if and only if for all $a \in A, (a,a)$ is in the relation R.

Thus in graphical terms, there must be a loop at every point of A, like the loop at "4" in example (1.3). In the relation-matrix, the **main diagonal** (by convention, the one from upper left corner to lower right) must be all 1's.

Examples (2.3) The prior examples are not reflexive. The complete relation $R := A \times A$ on A *is* reflexive, of course. So is $R' := \{ (x,y);\ x,y \in \mathbf{N},\ x \leq y \}$, for another set-builder example.

Practice (2.4P) Prove that R' is reflexive.

▶We can easily count the number of reflexive relations on our five-point set A. There are 2^{20}, about 1,000,000, of them. This is so because a reflexive relation R may be any subset of $A \times A$ that includes $D = \{ (x,x);\ x \in A \}$, a set of five points: $D \subseteq R \subseteq A \times A$. Our answer is the total number of subsets of $(A \times A) - D$. This set has $25 - 5 = 20$ points and, therefore, 2^{20} subsets. So to impose reflexivity cuts down our set of relations drastically, by a factor of 32 in this example. ◀

DEFINITION (2.5) A relation R is **symmetric** if and only if for every ordered pair (a, b) in R, the *reversed* ordered pair (b, a) is also in R.

Thus whenever we have one arrow in the graph, there must be another arrow "going backward" (except for loops, of course). The relation-matrix must be symmetric about the main diagonal (if, as usual, we list rows and columns in the same order).

Examples (2.6) Consider $R := A \times A$ again, the *complete* relation. It is symmetric, of course. Its relation-matrix consists entirely of 1's.

Define now R'' on $\mathbf{N}$ as $R'' := \{ (a, b);\ a, b \in \mathbf{N},\ a + b \text{ is odd} \}$. Thus two integers are related by R'' iff one is even and the other is odd. (Why?) R'' is symmetric. The upper left corner of the relation-matrix of R'' is

	1	2	3	4	$\cdots$
1	0	1	0	1	
2	1	0	1	0	$\cdots$
3	0	1	0	1	
$\vdots$		$\vdots$			

You can see in various ways that R'' is not reflexive.

The graph and matrix of R are shown in Example (2.10.4) on page 143 for $A = \{ 1, 2, 3, 4, 5 \}$. The graph of R'' is, in part

(2.7F)

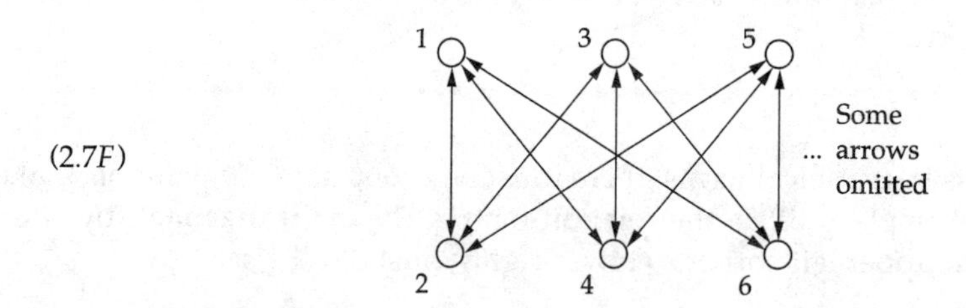

There are too many arrows to draw even here. They exist in both directions between each odd and each even integer.

Let us continue our counting game. How many reflexive and symmetric relations are there on our five-point set A? Here the counting is easiest if we look at the relation-matrix. It has five 1's on the main diagonal (because it is reflexive), and it must be symmetric about the main diagonal. Thus if we fill in the 10 entries in the matrix above the main diagonal, symmetry determines all the entries below. The 10 comes from $1 + 2 + 3 + 4 = 10$ or from $(5^2 - 5)/2 = 10$. Thus there are only $2^{10} = 1{,}024$ relations on our five-point set A that are both reflexive and symmetric.

DEFINITION (2.8) A relation R is **transitive** if and only if, for all a, b, c in A, $[(a,b) \in R$ and $(b,c) \in R]$ implies $(a,c) \in R$.

In a graph of R this property tells of certain shortcuts: whenever we can go from a to c in two steps, we can go there in one step. (We imagine a "step" as a traversal along one arrow.)

(2.9F)

In the relation-matrix, transitivity is harder to describe. We could work up a description in terms of triangles (the three 1's corresponding to (a,b), (b,c) and (a,c) always make the vertices of a right triangle in the matrix), but it may be too complicated to be worth the bother. (See problem 2.16.) (We discuss tests for transitivity in Chapter 12.)

▶The number of relations on a five-point set that are reflexive, symmetric, and transitive is only 52.† ◀

(2.10) Here are a few examples of relations with (or without) these properties, all on the set $A = \{1,2,3,4,5\}$. We denote the relation by R; thus R is a set of ordered pairs from A. The graph we call G, and the matrix M, with rows and columns all labeled as in the first example.

Example (2.10.1) A reflexive but not symmetric and not transitive relation, $R_1 := \{(1,1),(2,2),(3,3),(4,4),(5,5),(1,3),(3,4)\}$.

(2.10.1F)

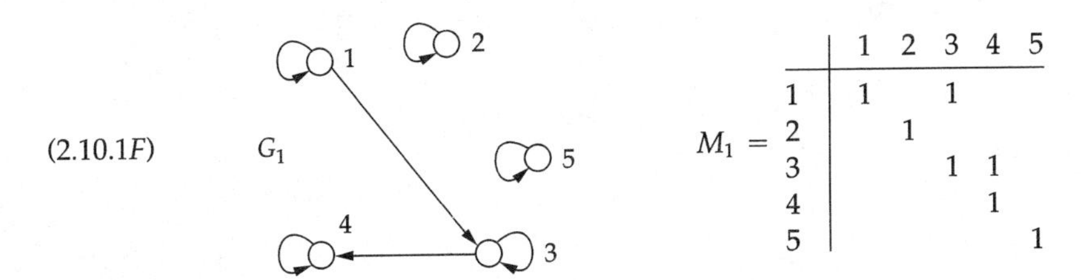

	1	2	3	4	5
1	1		1		
2		1			
3			1	1	
4				1	
5					1

In the matrix we have left blanks instead of writing the 0's. The presence of all the ordered pairs of the form (a,a) in R_1 is the requirement of the definition of reflexivity. We can also see that R_1 is reflexive by noting that there is a loop at every point of the graph G_1, or that the main diagonal of M_1 consists entirely of 1's. R_1 is not symmetric because $(1,3) \in R_1$ but $(3,1) \notin R_1$. In G_1 there is no backward arrow to correspond to the arrow from 1 to 3; the matrix M_1 is not symmetric about the main diagonal. Since $(1,3)$ and $(3,4)$ are in R_1 but $(1,4)$ is not in R_1, it is not transitive. (To repeat: we

†The number e_n of equivalence relations on an n-point set is much harder to derive than the preceding counts. A so-called Bell number (after E. T. Bell), e_n is the coefficient of $x^n/n!$ in the power-series expansion of $\exp(e^x - 1)$. (See Chapter 9, reference (10.1).)

explain here the first two properties in terms of all three viewpoints—(1.1), (1.2), and (1.4)—although any one of the explanations is sufficient.)

Example (2.10.2) Let $R_2 = \{(1,1),(1,2),(1,5),(2,1),(2,2),(2,5),(3,5),(4,4),(5,1),(5,2),(5,3),(5,5)\}$. Then R_2 is symmetric but is neither reflexive nor transitive.

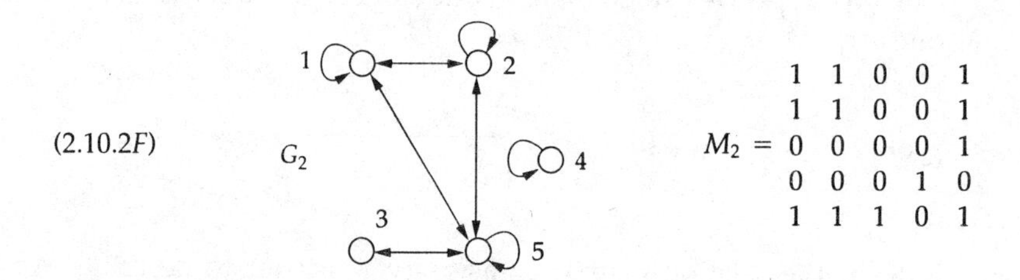

In this example the symmetry is obvious from all three points of view. R_2 is not reflexive because $(3,3)$ is not in R_2; there is a 0 on the main diagonal. The nontransitivity is most apparent from the graph: there is a path from 3 to 5 to 1 but there is no shortcut from 3 to 1. Similarly for 3 and 2, and for the paths in the reverse directions. But there is a still more subtle failure of the condition for transitivity: If G_2 were transitive there would be a loop at 3. That is, $(3,5)$ and $(5,3)$ are in R_2, so if R_2 were transitive ($a = 3,\ b = 5,\ c = 3$) we would have to have $(3,3)$ in R_2.

Example (2.10.3) Take

$$R_3 = \{(1,2),(2,3),(1,3),(3,3),(4,4),(5,5),(2,2),(3,2)\}.$$

Then R_3 is transitive but neither reflexive nor symmetric.

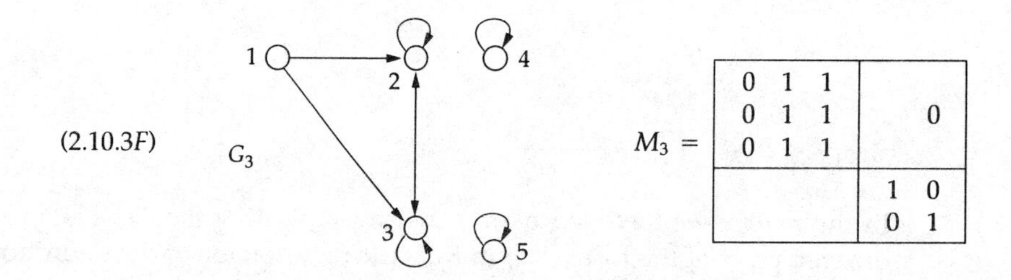

A blank submatrix box, or one with a lone 0, is understood to have all its entries 0. Thus in M_3 the upper right and lower left corners are all 0. Reflexivity fails because the point $(1,1)$ is not in R_3; the loop at 1 does not exist in G_3; there is a 0 on the main diagonal of M_3. Symmetry fails because the back arrow is absent from 2 to 1—for *every* arrow there must be a back arrow, or else the graph (relation) is not symmetric. Or you can simply notice that the matrix is not symmetric about the main diagonal.

Now we discuss equivalence relations, those that are reflexive *and* symmetric *and* transitive. First we look at some examples:

Example (2.10.4) $R_4 = A \times A$. For R_4 we have taken the set of all possible ordered pairs from the set A. Obviously R_4 is an equivalence relation. We show the graph G_4 and the matrix M_4 in the case that $A = \{1, 2, 3, 4, 5\}$.

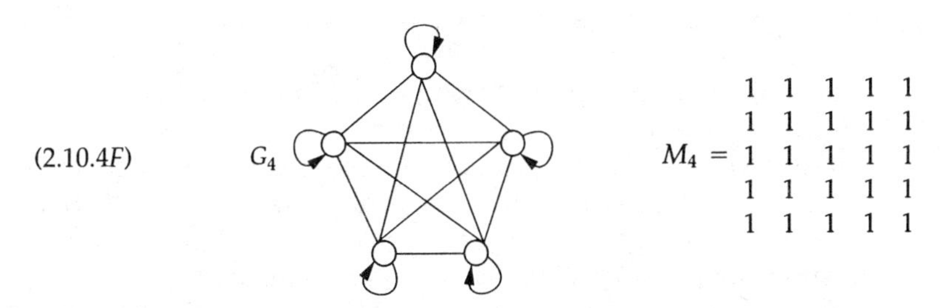

In order to clarify the picture, we have drawn lines with no arrowheads to stand for two arrows, one in each direction.

Example (2.10.5) Let $R_5 = \{(1,1), (2,2), (3,3), (4,4), (5,5)\}$. The graph and matrix are

(2.10.5F) G_5

$$M_5 = \begin{matrix} 1 & & & & 0 \\ & 1 & & & \\ & & 1 & & \\ & & & 1 & \\ 0 & & & & 1 \end{matrix}$$

This equivalence relation is the other extreme from R_4. It is the smallest subset of $A \times A$ that is reflexive and symmetric and transitive. (The transitivity is trivial, in that, for example, $(2,2)$ and $(2,2) \in R_5 \longrightarrow (2,2) \in R_5$. That is, $a = b = c = 2$ in the definition of transitivity.) R_5 is called the **diagonal** of $A \times A$; every equivalence relation, considered as a set of ordered pairs on the set A, must include $\{(a,a); a \in A\}$, the diagonal of $A \times A$.

Example (2.10.6) Take

$$R_6 = R_5 \cup \{(1,2), (2,1), (1,3), (3,1), (2,3), (3,2), (4,5), (5,4)\}.$$

The graph and matrix are

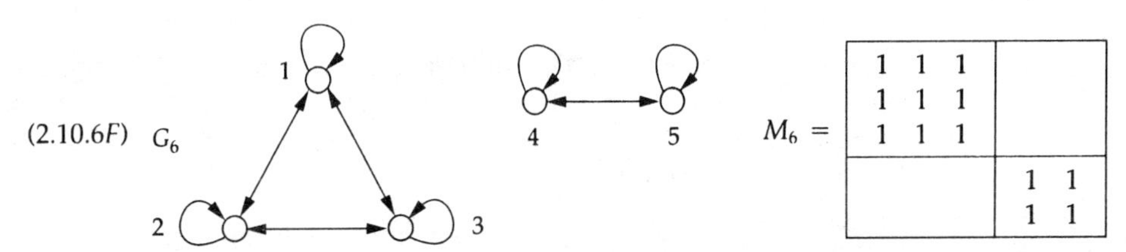

The blank boxes in M_6 are 0. Notice that the blank box in the upper right means there are no arrows from any point of $\{1, 2, 3\}$ to any point of $\{4, 5\}$. Similarly, the blank box in the lower left means that there are no arrows from 4 or 5 to any of 1, 2, 3.

Notice also that $R_6 = (A' \times A') \cup (A'' \times A'')$, where $A' = \{1, 2, 3\}$ and $A'' = \{4, 5\}$.

To generalize Example (2.10.4) we make a definition and a remark:

DEFINITION (2.11) The relation $A \times A$ is called the **complete relation** on A.

For any set A, the complete relation on A is an equivalence relation on A.

Why is R_6 in Example (2.10.6) an equivalence relation? We can best see the reflexivity and symmetry by looking at the matrix M_6, but these characteristics are also easy to see from the graph or the list of points of R_6. The transitivity is most obvious from the graph G_6. We can also see it in R_6, since it is the union of complete relations on mutually disjoint subsets of A, here A' and A''. Since we have already made in (2.10.4) the easy observation that $A \times A$ is an equivalence relation, it is transitive, hence $A' \times A'$ and $A'' \times A''$ are transitive. Since $A' \cap A'' = \varnothing$, the union of $A' \times A'$ and $A'' \times A''$ is transitive. That is, from the hypothesis of the condition defining transitivity, if (a, b) and (b, c) are in R_6, then *both* are in the same one of $A' \times A'$ or $A'' \times A''$. Therefore no "mixing" is possible, and the union is transitive if (and only if) each "piece" is transitive.

(2.12) Look again at Example (2.10.2). We can easily generalize it: Suppose for some a, b in A, both (a, b) and (b, a) are in the *transitive* relation R. Then both (a, a) and (b, b) are in R.

In a figure,

(2.13*F*) Transitivity and 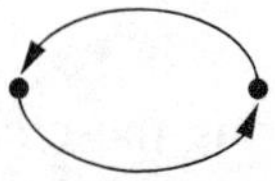 imply 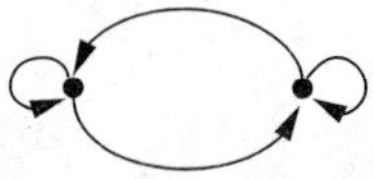

But: Symmetry and transitivity do not imply reflexivity (see problem 2.7).

The first three examples—(2.10.1), (2.10.2), and (2.10.3)—show that no one of these three properties (reflexivity, symmetry, transitivity) implies any of the others. It is also true that no two of them imply the third; you are asked to construct such counterexamples in the problems.

3. The Main Theorem

Example (2.10.6) is typical of all equivalence relations. The idea is simple, and we'll state it as Theorem (3.3). But first we define a term we'll need in the proof. If R is an equivalence relation on the set A, for each x in A we define the subset $E(x)$ of A, called the **equivalence class of** x, as

(3.1) $$E(x) = \{ a;\ a \in A, (x, a) \in R \}.$$

(In the graph, $E(x)$ is the set of all points a such that an arrow goes from x to a. In the matrix it corresponds to the set of 1's in row x.)

Practice (3.2P) List $E(x)$ for various x in the examples of (2.10), even if not equivalence relations.

You should understand that for different points a and b in A, it may happen that $E(a) = E(b)$. Indeed, we shall show that $E(a) = E(b)$ if and only if $a \in E(b)$ or $b \in E(a)$. Going back to Example (2.10.6), we see that

$$E(1) = E(2) = E(3) = \{1,2,3\}$$
$$E(4) = E(5) = \{4,5\}.$$

Theorem (3.3) Any equivalence relation R on A consists of the union of complete relations on each subset of a collection P of subsets of A. This collection P has the following three properties:

(*i*) $$\forall A' \in P, \quad A' \neq \varnothing$$

(*ii*) $$\forall A', \ A'' \in P, \ A' \neq A'' \longrightarrow A' \cap A'' = \varnothing$$

(*iii*) $$\bigcup_{A' \in P} A' = A.$$

The converse is also true. ∎

In words, property (*i*) says that every set in the collection P is nonempty. Property (*ii*) says that any two sets in P are disjoint or equal. Finally, property (*iii*) says the union of all the sets in P is A.[†] In symbols, the words of Theorem (3.3) say

$$R = \bigcup_{A' \in P} A' \times A'.$$

In Example (2.10.6), $P = \{\{1,2,3\},\{4,5\}\}$. We give two proofs of (3.3), first in terms of ordered pairs, second in terms of graphs in (3.12).

First Proof of (3.3) We assume R is an equivalence relation on A. We'll show that the equivalence classes $E(x)$, for various x, are the mutually disjoint subsets $A', A'', \ldots$. To see this, we'll first prove that for all x,y in A,

(3.4) $$E(x) \cap E(y) \neq \varnothing \longrightarrow E(x) = E(y).$$

((3.4) is the contrapositive of condition (*ii*).) To prove (3.4), let $a \in E(x) \cap E(y)$. Then by definition of $E(x)$ the ordered pair (x,a) is in R. Similarly, the ordered pair (y,a) is in R. By symmetry (a,y) is in R. Now by transitivity, (x,y) is in R, since (x,a) and (a,y) are in R. But now $E(y) \subseteq E(x)$, because for all b in $E(y)$, (y,b) is in R, hence by transitivity (x,b) is in R, and b is in $E(x)$. By *symmetry of argument* $E(x) \subseteq E(y)$. Thus $E(x) = E(y)$. Therefore we have shown the validity of (3.4), that any two equivalence classes are disjoint or equal.

Practice (3.5P) Explain why the "symmetry of argument" part is valid.

Another observation is this: every point a in A belongs to one equivalence class, namely, to $E(a)$, by reflexivity. Therefore, (*iii*) holds.

Finally, we define $P := \{E(a);\ a \in A\}$ and observe that we have already proved (*i*), (*ii*), and (*iii*) for this set P. What remains is to show that

$$R = \bigcup_{a \in A} E(a) \times E(a).$$

[†] We use here the notation of Chapter 1, (10.12).

First, for any a in A we show that $R \supseteq E(a) \times E(a)$. Let b, c be any points of $E(a)$. Then both (a,b) and (a,c) are in R. By symmetry $(b,a) \in R$. Now (b,a) and $(a,c) \in R$; by transitivity we see that (b,c) is in R. Since b and c were arbitrary points of $E(a)$, we've shown $E(a) \times E(a) \subseteq R$. Second, for the reverse inclusion, let (x,y) be any point of R. Then $y \in E(x)$, and of course $x \in E(x)$, so $(x,y) \in E(x) \times E(x)$.

Proof of the converse The relation R on A consists of the union of complete relations on the subsets in a collection P with properties (i), (ii), and (iii) stated in the theorem (3.3). We are to prove that R is an equivalence relation. It's obvious. QED

Remarks. (3.6) (i) The last sentence before the proof of the converse elegantly rules out the possibility of "mixing" between equivalence classes: it says for any $(x,y) \in R$, x and y both belong to the same equivalence class.

(ii) When we defined $P = \{E(a);\ a \in A\}$ we made good use of the fact that repetitions of elements in a listing of a set do not affect the set. For Example (2.10.6)

$$P = \{E(1), E(2), E(3), E(4), E(5)\} = \{E(1), E(4)\}. \blacksquare$$

Now let us focus for a moment on the graph and matrix of an equivalence relation. We see that in Example (2.10.6) the graph is in two parts, namely, the two graphs on the subsets $\{1,2,3\}$ and $\{4,5\}$, respectively. Within each part all possible arrows are drawn, but no arrows go from any part to any other part.

(3.7) The matrix consists of square submatrices of all 1's centered on the main diagonal, provided the rows and columns are ordered appropriately, i.e., with all the points of one equivalence class set down first, and all those of another equivalence class set down next, and so on. We showed this feature in (2.10.6F).

Example If we ordered the rows and columns of the matrix in Example (2.10.6) as 1,4,2,5,3 the matrix would be

	1	4	2	5	3
1	1		1		1
4		1		1	
2	1		1		1
5		1		1	
3	1		1		1

which makes the fact that R_6 is an equivalence relation harder to see.

Example (3.8) Let us look at some examples of equivalence relations in real life. Consider the set A of all triangles in a plane, and define R as the relation of *similarity*. That is, triangles T and T' are similar if and only if the three angles in T are the equal, one by one, to the three angles in T'. Thus we would say that R consists of all ordered pairs (T, T') in which T and T' are similar to each other. One of the equivalence classes, for example, consists of all equilateral triangles. Another consists of all "$30°, 60°, 90°$" triangles. There are infinitely many other equivalence classes. It is easy to check that this relation is indeed an equivalence relation.

Example (3.9) Here is another example: equality between the points of a set A. This is the smallest equivalence relation on the set A; each singleton is an equivalence class. Example (2.10.5) is a case of equality. And in Chapter 1, (4.5) we stated that equality between sets is an equivalence relation. Congruence is another equivalence relation on the set of all triangles in a plane. Congruent triangles are similar, but similar triangles are not necessarily congruent. (Two triangles are called *congruent* if the three sides of one have the same lengths, one by one, as those of the other.)

(3.10) We can define two computer programs as equivalent if and only if for each input the output of one equals that of the other. Clearly this relation is reflexive, symmetric, and transitive.

(3.11) Logical equivalence of propositional formulas is an equivalence relation (Problem 2.8). There are many other equivalence relations in mathematics and computer science. You will see several more in later chapters. They are somewhat technical so we won't preview them here.

Second Proof of (3.3) (3.12) Let's prove (3.3) in terms of graphs. The theorem says: G_R consists of complete graphs on subsets of A; these subsets are nonempty (3.3*i*), mutually disjoint (3.3*ii*), and they cover A (3.3*iii*).

The set $E(x)$ is the set of points a of A for which there is an arrow from x in G_R. Thus $E(x)$ and the arrows from x to the points of $E(x)$ might be as shown in (3.13*F*).

(3.13*F*)

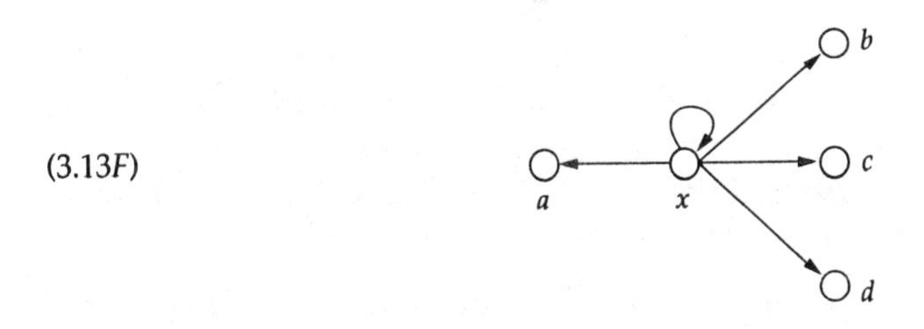

(Here you see a limitation of proof by pictures: they can force you to be too specific.)

To prove (3.3*i*) and (3.3*iii*) using G_R, we note that there is a loop at *every* point of A, so $x \in E(x)$ for all $x \in A$. Thus $E(x) \neq \emptyset$, and the union of the $E(x)$'s is A.

To prove the contrapositive of (3.3*ii*), we let $E(x)$ and $E(y)$ have a common point z:

(3.14*F*)

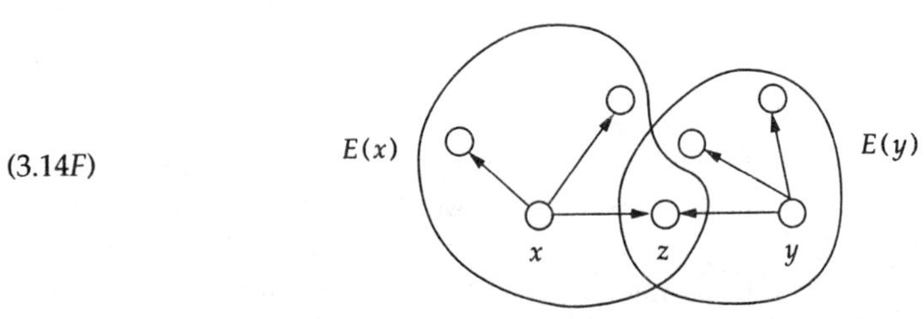

$E(x)$ and arrows from x $\quad$ $E(y)$ and arrows from y

We now sketch the points x, y, z and two arbitrary points, a in $E(x)$ and b in $E(y)$,

(3.15F)

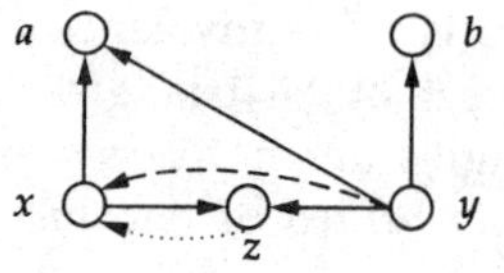

together with arrows implied first by symmetry (dotted) and second by transitivity (dashed). The arrow from y to a comes third, by transitivity again. It shows $a \in E(y)$, and, therefore, $E(x) \subseteq E(y)$. The symmetric picture would show $b \in E(x)$, hence $E(y) \subseteq E(x)$. Thus (3.3*ii*) is proved.

It remains to show that G_R consists of a complete graph on each of the $E(x)$'s. Let a, b be any points in $E(x)$. Thus the graph is, in part,

(3.16F)

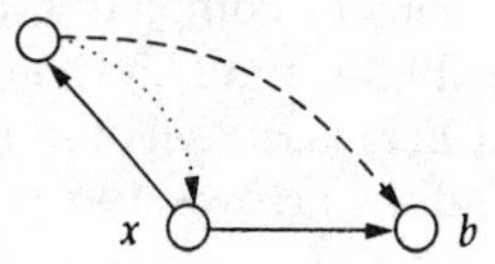

where symmetry yields the dotted arrow and transitivity the dashed. Therefore an arrow from a to b exists in G_R. Since a and b are arbitrary, G_R includes the complete graph on $E(x)$. Since x is arbitrary, G_R includes the complete graph for each of the $E(x)$'s. Conversely, if there is an arrow from c to d in G_R, then d is in $E(c)$; since $c \in E(c)$, that arrow is in the complete graph on $E(c)$. QED

(3.17) As an application of Theorem (3.3) we prove that the relation R on $A := \{1, \ldots, 6\}$ given by the relation-matrix M is not transitive.

$M =$

	1	2	3	4	5	6
1	1	0	1	0	1	0
2	0	1	0	1	0	0
3	1	0	1	0	1	0
4	0	1	0	1	0	1
5	1	0	1	0	1	0
6	0	0	0	1	0	1

Clearly R is reflexive and symmetric. If R were also transitive it would be an equivalence relation. But from row 1 of M the equivalence class of 1 would be $\{1,3,5\}$. From row 2, $E(2)$ would be $\{2,4\}$. But from row 4, $E(4)$ would be $\{2,4,6\}$. Since $E(2)$ must equal $E(4)$ if both 2 and 4 are in the same equivalence class, this contradicts the "fact" that R is an equivalence relation. Hence R is not transitive.

Another way to see that R is not transitive is to rearrange the rows and columns of M to try to achieve a form like that of M_6 in (2.10.6). We have

	1	3	5	2	4	6
1	1	1	1	0	0	0
3	1	1	1	0	0	0
5	1	1	1	0	0	0
2	0	0	0	1	1	0
4	0	0	0	1	1	1
6	0	0	0	0	1	1

It's in the wrong form for the subset $\{2,4,6\}$ and no further shuffling will fix it.

4. The Infix Notation

The **infix notation** for relations is an often-used convenience. Instead of writing $(a,b) \in R$, we write simply

$$a\,R\,b.$$

For example, $a = b$, or $a < b$, or $a \mid b$. These are, respectively, for the *equality* relation $\{(x,x);\ x \in A\}$ on some set A; for the *less than* relation on, say, a subset of $\mathbf{N}$; and the *divisibility* relation on $\mathbf{N}$: $a \mid b$ means "a divides b," i.e., for some integer c, $b = ac$. How cumbersome it would be to have to write $(x,x) \in Eq$ instead of $x = x$, or $(x,y) \in Div$ for $x \mid y$.

Logical implication is a relation on any set of propositional formulas. Its infix notation is $\Longrightarrow$.

Our notations $\in$ and $\subseteq$ are the infix notations for the membership and inclusion relations, respectively. We spoke of these as *relations* from the beginning; it was impossible not to do so.

5. Partitions

The idea of partition is closely related to that of equivalence relation, and we have already seen how; but we shall proceed from the beginning with partitions and make the connection later.

DEFINITION (5.1) Let A be a set. A **partition** of A is a collection of mutually disjoint nonempty subsets of A that have A as their union. Thus if P is the partition, then

$$P = \{A_1, \ldots\} = \{A_i; i \in I\}$$

(*i*) $$\forall i \in I,\ A_i \neq \varnothing$$

(*ii*) $$A_i \neq A_j \quad \longrightarrow \quad A_i \cap A_j = \varnothing$$

(*iii*) $$\bigcup_{i \in I} A_i = A$$

The sets A_i of P are called the **cells** of the partition P.

Thus every point of A is an element of exactly one of the cells of a partition. (Here I is an *index* set. If P has n cells, we may take $I = \{1, \ldots, n\}$.)

If A is empty, then its only partition is the *empty* set of cells.

It is easy to draw a partition schematically. Consider the sketch

(5.1F)

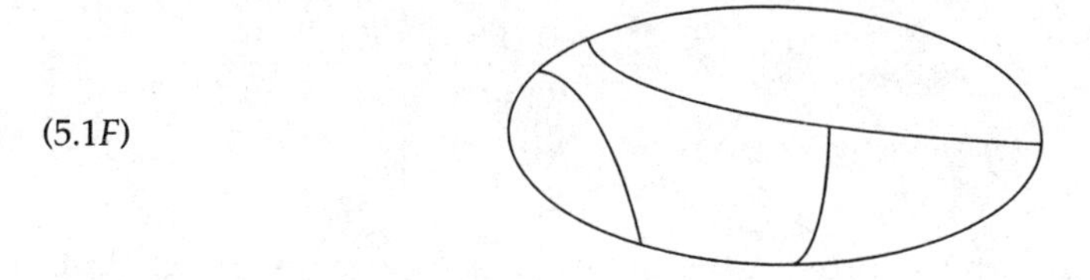

It represents a partition with four cells.

Examples (5.2)

(i) We partition the set $\mathbf{Z}$ of all integers into two cells. A_1 is the set of all odd integers, and A_2 is the set of all even integers. Clearly $A_1 \cap A_2 = \emptyset$ and $A_1 \cup A_2 = \mathbf{Z}$, so $P_1 := \{A_1, A_2\}$ is a partition of $\mathbf{Z}$. (Note that *partition* is both a noun and a verb. As a noun it represents the entire set of cells and is not at all used in the everyday sense of the dividing wall between two rooms. It is more like the set of the rooms in a house than like anything to do with the walls.)

(ii) Here is another partition of $\mathbf{Z}$:

$$P_2 := \{\mathbf{Z}\}.$$

(iii) And here is another: $P_3 = \{\{x\};\ x \in \mathbf{Z}\}$ is the partition of $\mathbf{Z}$ in which each cell is a singleton.

Any nonempty set has partitions like P_2 and P_3 for $\mathbf{Z}$: the whole set as one cell, and each point in a singleton cell.

In general a set has many partitions unless it has only very few elements.

Practice (5.3P) Work out how many partitions there are for a set of exactly n elements for the cases $n = 0, 1, 2, 3$.

(5.4) *Notation.* A convenient notation for partitions is to list the elements of cells between vertical bars.[†]

Thus, instead of the cumbersome

$$\{\{1,2,3,4\},\{5,6\},\{7\},\{8\}\}$$

we would simply write

$$1,2,3,4 \mid 5,6 \mid 7 \mid 8.$$

We use this notation freely in the rest of the book.

[†] We also use the vertical bar for divisibility, but the context will make clear which sense is meant.

6. The Connection between Partitions and Equivalence Relations

(6.1) As you may have already noticed, there is a simple one-to-one correspondence between the class of all partitions on a given set A and the class of all equivalence relations on A. In one direction, to each partition $Q = \{ A_i;\ i \in I \}$ of A we associate the equivalence relation E_Q consisting of all ordered pairs (x,y), chosen subject only to the restriction that x and y must belong to the same cell of the partition Q. In symbols

(6.2)
$$E_Q = \{ (x,y); \exists i \in I, x,y \in A_i \} = \bigcup_{i \in I} A_i \times A_i .$$

The graph of E_Q consists of all possible arrows drawn between the pairs of points of the same cell with no arrows between points in different cells. The matrix of E_Q may be represented as

(6.3F)
$$\begin{array}{c|cccc} A_1 & \boxed{1} & & & 0 \\ & A_2 & \boxed{1} & & \\ & & A_3 & \boxed{1} & \\ & 0 & & & \ddots \end{array}$$

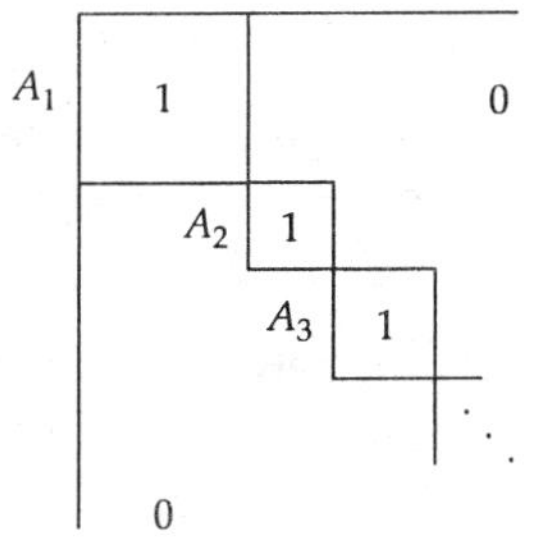

where the box marked A_i with the 1 in it stands for the square matrix of all 1's with rows and columns indexed by the points of A_i.

(6.4) In other words, E_Q is the equivalence relation on A with equivalence classes equal to the cells of the partition Q.

(6.5) In the other direction, let's observe that if we start with an equivalence relation R on A, then the set of all equivalence classes of R is a partition P_R of A; we proved that fact in proving (i), (ii), and (iii) of Theorem (3.3).

(6.6) To summarize: We assigned to each partition Q of A an equivalence relation E_Q on A, and to each equivalence relation R on A a partition P_R. To wrap up, notice that the equivalence relation we would get from the partition P_R by this assignment is just R, and the partition assigned to the equivalence relation E_Q is Q. In symbols

(6.7)
$$E_{(P_R)} = R \quad \text{and} \quad P_{(E_Q)} = Q.$$

Once you see this correspondence—and it is truly simple—you will have no trouble with equivalence relations, because you can think of them in terms of their corresponding partitions.

Example (6.8) Start with the partition

$$Q = 1,2,3 \,|\, 4,5$$

that came up in Example (2.10.6). Set $A_1 = \{ 1,2,3 \}$ and $A_2 = \{ 4,5 \}$. Then E_Q is, by definition (6.2),

$$E_Q = \{ (x,y);\ x,y \in A_1 \quad \text{or} \quad x,y \in A_2 \}.$$

In other words, $E_Q = (A_1 \times A_1) \cup (A_2 \times A_2)$, exactly the equivalence relation R_6 of Example (2.10.6).

To go in the other direction, start with $R_6 = R$. We saw in Example (2.10.6) that R_6 gives rise to the partition $1,2,3 \mid 4,5$, which is the partition P_R with the equivalence classes as its cells. We saw in (6.6) that $P_R = Q$ and that, therefore, $E_{(P_R)} = E_Q$. This equals R, by (6.7).

Since $R = E_Q$, we get $P_R = P_{E_Q}$; from (6.7) this equals Q. In sum, $E_Q = R$ and $P_R = Q$.

7. Refinements

Now we introduce a refinement of the idea of partition. Suppose P and Q are partitions of the same set A. We define

DEFINITION (7.1) The partition P is a **refinement** of the partition Q if and only if for every cell of P there is a cell of Q including it.

Put a little less precisely, (7.1) says every cell of P is a subset of some cell of Q.

In symbols, let $P = \{ A_i;\ i \in I \}$ and let $Q = \{ B_j;\ j \in J \}$; then P is a refinement of Q if and only if

(7.2)
$$\forall i \in I\ \exists j \in J \quad A_i \subseteq B_j.$$

Example (7.3) Let $Q = 2,4,6,8,10 \mid 1,3,5,7,9$. Q is a partition of $\{1,\ldots,10\} = A$. The following partitions of A are refinements of Q:

$$P_1 = 1,3,5,7,9 \mid 2,6,10 \mid 4,8$$
$$P_2 = 1,3,5 \mid 7 \mid 9 \mid 2,4 \mid 6,8,10.$$

The partition P_3 of A is *not* a refinement of Q:

$$P_3 = 1,2,3,4,5 \mid 6,7,8,9,10.$$

We may sketch a refinement schematically as follows:

(7.4F)

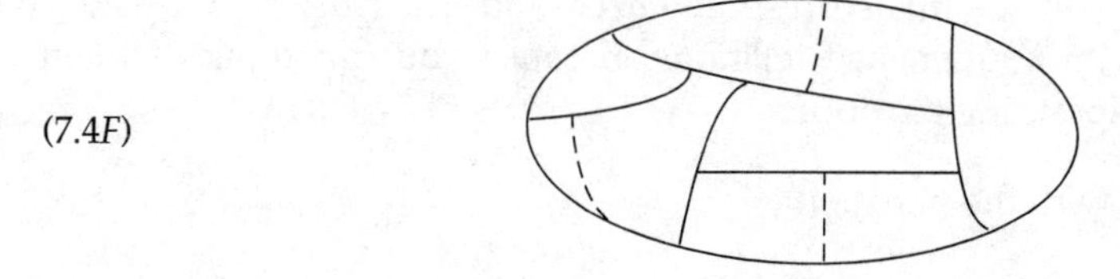

Here the original partition is represented by the solid lines, and the refinement by the dashed lines *together with* the solid lines.

Example (7.5) $\{A\}$ is a partition of the nonempty set A. Every partition of a set is a refinement of itself. Consider the following partitions of the set $A = \{1,2,3\}$:

$$P_4 = 1,2,3 \text{ (i.e., } P_4 = \{A\})$$
$$P_5 = 1,2 \mid 3$$
$$P_6 = 1 \mid 2 \mid 3.$$

Here P_5 is a refinement of P_4, and P_6 is a refinement of P_4 and of P_5. But

$$P_7 = 1,3 \mid 2$$

is not a refinement of P_5, and P_5 is not a refinement of P_6.

(7.6) Another way to think of a refinement of a partition Q is this: If we *partition each cell* of Q, the set of all the resulting *subcells* is a refinement of Q. Conversely, if P is a refinement of Q, then for every cell C of Q some set of cells from P constitutes a partition of C. See problem 7.5.

Example (7.7) Here are some further examples of refinements.

(*i*) P = odds | evens is a partition of $\mathbf{N}$. If we further partition the cell evens, say via evens $= \{4x;\ x \in \mathbf{N}\} \cup \{4x - 2;\ x \in \mathbf{N}\}$, we get a refinement Q of P:

$$Q = \text{odds } \mid 4,8,12,16,20,\ldots \mid 2,6,10,14,\ldots$$

(*ii*) In plane geometry, congruence is a refinement of similarity, say on the set of all triangles. (Explain how.)

(*iii*) Let A be the set of all propositional formulas in the two atoms a, b. Then logical equivalence (Ch. 2, (7)) is an equivalence relation $\Leftrightarrow$ on A (see problem 2.8) which partitions A into cells. We further partition each cell C by dividing it into two subsets:

C_1 is the set of all formulas in C using only one variable.
C_2 is the set of all formulas in C using both a and b.
If $C_1 = \varnothing$ then we partition C merely as $\{C\}$.
Thus for the cell C containing $a \vee \sim a$, the cell of all tautologies,

$$C_1 = \{a \vee \sim a,\ b \vee \sim b,\ (a \longrightarrow \sim a) \vee (\sim a \longrightarrow a),\ (\sim b \longrightarrow b) \longrightarrow b, \ldots\},$$
$$C_2 = \{a \wedge (a \longrightarrow b) \longrightarrow b,\ (a \longrightarrow b) \vee (b \longrightarrow a), \ldots\};$$

but in the cell D containing $a \longrightarrow b$ there are no one-atom formulas. That is, $D_1 = \varnothing$. Thus D is untouched in this refinement.

Practice (7.8P)

(*i*) Why are there no one-atom formulas in D?

(*ii*) How many cells are there for $\Leftrightarrow$?

*(*iii*) How many cells contain one-atom formulas?

*(*iv*) How many cells contain one-atom formulas in a and also some in b?

8. A Relation on Partitions

(8.1) If P and Q are two partitions of a set A, then P is, or is not, a refinement of Q. Thus *being a refinement of* is a relation on the class of all partitions of a given set. As we remarked in (7.5), this relation is reflexive: Every partition is a refinement of itself. We also showed by example in (7.5) that it is not symmetric. We now point out that it is a transitive relation: If P, P', and P'' are partitions of a set A, and if P is a refinement of P' and P' is a refinement of P'', then P is a refinement of P''. The proof of transitivity follows immediately from the definition of refinement: If every cell of P is a subset of a cell of P', and every cell of P' is a subset of a cell of P'', then merely from the transitivity (Chapter 1, (10.7)) of the set-inclusion relation every cell of P is a subset of a cell of P''. Therefore P is a refinement of P''. We shall discuss this situation further in Chapter 12, on order-relations.

Example (8.2) You see in (7.5) that P_6 is a refinement of P_5 and that P_5 is a refinement of P_4. And P_6 is a refinement of P_4.

The Venn diagram of partitions makes the transitivity obvious. Let all lines mark the partition P, the dashed and solid lines P', and the solid lines P''.

(8.3F)

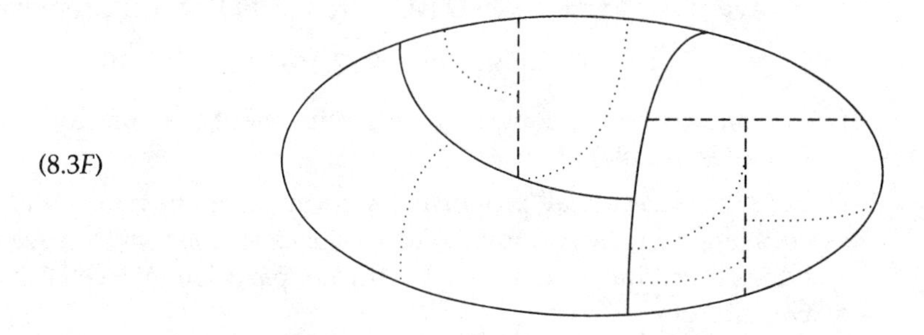

By now you can see that *refinement* is an appropriate name, because a refinement cuts the cells of a partition into finer pieces.

9. Postscript

Let $P(x,y)$ be a predicate with $D_x = D_y = A$. The truth-set T_P of $P(x,y)$ is defined in Chapter 2, (19.10)ff. According to our definition (1.1), T_P is a relation on A. Thus relations may be thought of as truth-sets of two-variable predicates:

$$\forall a,b \in A \quad (a,b) \in R \quad \text{iff} \quad P(a,b).$$

Practice (9.1P) Express reflexivity, symmetry, and transitivity in terms of the predicate $P(x,y)$.

Example Problems 7.6 and T1 define relations in terms of predicates.

10. Further Reading

(10.1.) Andrew M. Gleason (cited in Chapter 1, (20.1)).

(10.2.) Paul R. Halmos (cited in Chapter 1, (20.2)).

Both these books are more advanced than the present text.

11. Problems for Chapter 4

Problems for Section 1

1.1 Draw the graphs of the following relations on the set $\{1,2,3,4,5\}$:

(i) $\{(1,2),(1,4),(1,5),(2,5),(3,3),(3,5),(5,4),(5,5)\}$.
(ii) $\{(3,3),(2,3),(2,5),(3,1),(2,1),(1,3)\}$.

1.2 Write the relation matrices of the relations in the previous problem.

1.3 The inclusion relation $\subseteq$ is a relation on the power set $\mathcal{P}(X)$ of a set X. For $X = \{1,2\}$, write the relation matrix for $\subseteq$ on $\mathcal{P}(\{1,2\})$.

Problems for Section 2

2.1[Ans] For each of the following relations determine whether the relation is reflexive. Is it symmetric? Is it transitive? Explain.

(i) A is the set of all lines in the plane. R_1 is the relation of perpendicularity:

For $L, L' \in A (L,L') \in R_1$ iff L is perpendicular to L'.

(ii) Same A. Definition of R_2:

$(L,L') \in R_2$ iff L is perpendicular or parallel to L'.

(iii) Let $\mathbf{N}$ be the set of all positive integers. Define R_3 on $\mathbf{N}$:

For all $a,b \in \mathbf{N}, (a,b) \in R_3$ iff $a \neq b$.

(iv) Same $\mathbf{N}$. Definition of R_4:

$\forall a,b \in \mathbf{N}, (a,b) \in R_4$ iff $a/b = 2^i$ for some integer $i \geq 0$.

2.2 Define R_4 as in Problem 2.1, (iv). Find all integers b such that $(23,b) \in R_4$. Do the same for $(24,b) \in R_4$.

2.3 Consider the relation matrix on $\{a,b,c,d,e,f\}$.

	a	b	c	d	e	f
a	1	1	0	0	1	1
b	1	0	0	1	0	1
c	0	0	1	1	1	0
d	0	0	1	0	1	0
e	1	0	1	1	1	0
f	1	1	0	0	0	1

Explain why, or why not, the relation it defines is reflexive; symmetric; transitive.

2.4 Define the relation X on $\mathbf{N}$ as $\forall c,d \in \mathbf{N}\ (c,d) \in X$ iff $c+d$ is even.

(i) Prove that X is an equivalence relation.
(ii) How many equivalence classes does X have? Explain.

2.5 Define the relation S on the set B of all students at a school: For all x,y in B, $x\ S\ y$ iff x and y have a class together. Determine whether S is an equivalence relation.

2.6 Define the relation R_5 on $\{1,2,\ldots,10\}$ as follows: For all a,b in that set, $a\ R_5\ b$ if and only if $-2 \leq a-b \leq 2$. Determine whether R_5 is an equivalence relation.

2.7 Prove (by supplying counterexamples) that no two of reflexivity, symmetry, and transitivity imply the third.

2.8 Prove that logical equivalence is an equivalence relation on the set of all propositional formulas in a fixed set of atoms.

2.9 Let R be a reflexive and transitive relation on a set C. Define a relation Y on C as follows:

$$\forall a,b \in C, (a,b) \in Y \text{ iff } (a,b) \in R \text{ and } (b,a) \in R.$$

Prove that Y is an equivalence relation on C.

2.10 What is wrong with the following "proof" that every symmetric and transitive relation R is reflexive? If $(a,b) \in R$ then $(b,a) \in R$ by symmetry. By transitivity $(a,a) \in R$ (just as in (2.12)). Therefore, R is reflexive.

2.11 Let R and R' be equivalence relations on a set A. Prove that $R \cap R'$ is an equivalence relation on A.

2.12 Define an *isolated point* for a relation R on a set A as a point of A that is not the first or second coordinate of any ordered pair in R. Prove that a symmetric, transitive relation on A with no isolated points is an equivalence relation.

2.13 Let A be a set having a total of n elements, where $n \geq 1$. Let R be any equivalence relation on A. Prove that the total number of ordered pairs in R is odd if n is odd, even if n is even. [Hint: Look at the relation matrix.]

2.14 Sometimes 0 and the negative integers are defined via the following equivalence relation on $\mathbf{N} \times \mathbf{N}$:

$$(a,b)\ R\ (x,y) \quad \text{iff} \quad a + y = b + x.$$

Prove that R is an equivalence relation.

***2.15** In the notation of the previous problem, 0 is defined as the equivalence class of $(1,1)$, and

$$-1 := \lfloor (2,1) \rfloor$$
$$-2 := \lfloor (3,1) \rfloor$$

and for all n in $\mathbf{N}$,

$$-n := \lfloor (n+1, 1) \rfloor.$$

The idea is that -1 is the solution of the equation $x + 2 = 1$, and so on. Then we say that n corresponds to $\lfloor (1, n+1) \rfloor$ (the solution of $x + 1 = n + 1$), and we define addition, $+$, on the set of equivalence classes by

$$\lfloor (a,b) \rfloor + \lfloor (c,d) \rfloor = \lfloor (a+c, b+d) \rfloor$$

(*i*) Prove that this definition of addition is independent of the choice of elements from the equivalence classes.

(*ii*) Prove that $0 + \lfloor (a,b) \rfloor = \lfloor (a,b) \rfloor$ for all equivalence classes.

(*iii*) Prove that for all a, b, c, d in $\mathbf{N}$ the equation

$$x + \lfloor (a,b) \rfloor = \lfloor (c,d) \rfloor$$

has a solution $x =$ some equivalence class. Express x in terms of a, b, c, d. Thus $\mathbf{Z}$ can be defined, together with addition $+$ on $\mathbf{Z}$, with only the positive integers, addition of positive integers, and the concept of equivalence relation.

***2.16** Maybe it's not so complicated, after all, to find a "geometric" check for transitivity on the relation matrix. Find and explain it.

Problems for Section 3

3.1 Let $A = \{1,2,3,4\}$. For each of the following three relations on A, prove or disprove that it is an equivalence relation and, if it is one, write down its equivalence classes.

$$R = \{(1,1),(2,2),(3,4),(3,3),(4,4)\}$$
$$R' = \{(1,1),(2,2),(3,4),(4,4),(1,2),(2,1),(3,3),(4,3),(1,3),(1,4),(3,1),(4,1)\}$$
$$R'' = \{(1,1),(2,2),(3,4),(3,3),(4,4),(1,2),(2,1),(4,2),(2,3)\}.$$

3.2 Determine whether the relation with matrix M' below is transitive on the set $\{1,\ldots,8\}$.

	1	2	3	4	5	6	7	8
1	1	1		1		1		1
2	1	1			1			1
3			1	1			1	
4	1		1	1		1		1
5		1			1	1	1	
6	1			1	1	1	1	
7			1		1	1	1	1
8	1	1		1			1	1

$=: M'$.

3.3 (*i*) Draw the graphs of all the equivalence relations on a set of four points, except draw only one of each "shape." (Two such relations have the same "shape" if they are the same except for a relabeling of the points.)

(*ii*) Use your answer to (*i*) to count the total number of equivalence relations on a set of four points.

Problems for Section 5

5.1 List all the partitions of $\{a,b,c\}$. Use notation (5.4).

5.2 What is wrong with the statement "$\{\varnothing\}$ is a partition of $\varnothing$"? After all, if A is a nonempty set, then $\{A\}$ is a partition of A.

5.3[Ans] Prove that the empty set (of cells) is the only partition of $\varnothing$.

5.4 Let A, B, C be any three sets. Consider the three sets

$$A - (B \cup C), \quad (A - C) \cap B, \quad \text{and} \quad A \cap C.$$

Prove that if each of these last three sets is nonempty, then they are the cells of a partition of A. Use an epsilon-argument, but first draw a Venn diagram to help yourself see what's going on.

5.5 Let P and Q be partitions of the set A. Prove that if $P \subseteq Q$ then $P = Q$.

Problems for Section 6

6.1 The relation R is defined on $\mathbf{N}$ as follows: $\forall a, b \in \mathbf{N}, (a, b) \in R$ iff $a/b = 2^m$ for some integer $m \in \mathbf{Z}$. Note: $\mathbf{Z} = \{0, \pm 1, \pm 2, \ldots\}$ and m can be negative. First, show R is an equivalence relation. Second, what are the cells of the partition associated with R?

6.2 How many equivalence relations are there on $A = \{1, 2, 3\}$? List them by merely listing the corresponding partitions of A. Use notation (5.4). See also problem 5.1.

6.3 Determine whether the matrix below is the relation-matrix of an equivalence relation. Explain your answer.

	1	2	3	4	5	6	7	8
1	1					1		
2		1		1			1	
3			1					
4		1		1	1			1
5				1	1		1	
6	1					1		
7		1			1		1	
8				1				1

6.4 How many equivalence relations are there on a four-point set? See also problem 3.3.

6.5 Define the relation E on lists by the rule $a_1, a_2, \ldots E\ b_1, b_2, \ldots$ iff the sets are equal: $\{a_1, a_2, \ldots\} = \{b_1, b_2, \ldots\}$. Prove that E is an equivalence relation on any set of lists.

6.6 Suppose $C = \{1, 2, 3, 6\}$, and define R to be the following relation on $C \times C$: For all a, b, x, y in C, $(a, b)\ R\ (x, y)$ iff $ay = bx$. Prove that R is an equivalence relation. What is the partition associated with R?

6.7 Let R and R' be equivalence relations on a set A. Prove that $R \cup R'$ (consider R and R' as subsets of $A \times A$) is not necessarily an equivalence relation on A.

6.8 Give an example of two different equivalence relations on a set A, neither of which includes the other, such that their union is an equivalence relation on A.

***6.9** Let R and R' be two equivalence relations on the set A with associated partitions P and P'. Prove that $R \cup R'$ is an equivalence relation on A iff

$$\begin{gathered}\text{for all cells } C \text{ of } P \text{ and } C' \text{ of } P', \\ \text{if } C \cap C' \neq \varnothing, \\ \text{then either } C \subseteq C' \text{ or } C' \subseteq C.\end{gathered}$$

6.10 Let A be a set of five points, and let R be a relation on A consisting of 15 ordered pairs. Prove that R is not an equivalence relation. [Hint: Prove that the number of ordered pairs in any equivalence relation on A is 5, 7, 9, 11, 13, 17, or 25.]

Problems for Section 7

7.1 (*i*) Which of the following collections of sets is a partition of $\{1, 2, \ldots, 10\}$? Answer "yes" or "no" for each and explain.

$$P_1 = \{\{1,3,8\}, \{2,4,6\}, \{5,7,10\}, \{9\}\}.$$
$$P_2 = \{\{7,4,3,8\}, \{1,5,10,3\}, \{2,6\}\}.$$
$$P_3 = \{\{1,5,9\}, \{2,10,4,7\}, \{8,3,6\}\}.$$
$$P_4 = \{\{4,2\}, \{3,8\}, \{6\}, \{10,7\}, \{1\}, \{5\}, \{9\}\}.$$

(*ii*) Which of the partitions is a refinement of which other partitions (from the above list)? Explain.

7.2 List a partition P_1 of $\{1, \ldots, 8\}$ and a refinement P_2 of P_1.

7.3 The set $\{1, 2, 3, 4, 5, 6\}$ is partitioned as $\Pi = 1, 2, 3 \mid 4, 5 \mid 6$. How many refinements does Π have? You need not list them all if you can explain otherwise.

7.4 Let R be an equivalence relation on the set C. Let P be the partition of C associated with R. Let Q be a refinement of P, such that $Q \neq P$. Let S be the equivalence relation on C associated with Q. Prove one of the statements

$$R \subseteq S, \quad S \subseteq R$$

and disprove the other.

7.5[Ans] Let P be a refinement of the partition Q of the set A. Prove that for every cell C of Q there is a collection of cells of P that is a partition of C. Conversely, prove that if each cell of Q is partitioned, then the union of all those partitions is a refinement of Q.

7.6 Let $P(x)$ and $Q(x)$ be predicates with the same domain $D \neq \emptyset$. We can say that

$$\forall a, b \in D \quad \begin{array}{l} a\, R_P\, b \text{ iff } P(a) = P(b) \\ a\, R_Q\, b \text{ iff } Q(a) = Q(b), \end{array}$$

where $P(a)$ is true or false, and the same for $P(b)$, $Q(a)$, $Q(b)$.

(*i*) Prove that R_P and R_Q are equivalence relations on D.

*(*ii*) If the partition associated with R_Q is a refinement of that associated with R_P, what can you say about $P(x)$ and $Q(x)$? Explain.

7.7 [Recommended] Let A be any set and suppose that $P_1 = \{C_1, \ldots, C_m\}$ and $P_2 = \{D_1, \ldots, D_n\}$ are partitions of A. Prove that

$$Q := \{C_i \cap D_j;\ 1 \le i \le m, 1 \le j \le n\} - \{\emptyset\}$$

is a partition of A *and* that it is a refinement of both P_1 and P_2. [Careful! See Chapter 2, (19.10).]

7.8 Let P_1, P_2, and Q be as defined in the previous problem. Let P_3 be any partition of A that is a refinement of both P_1 and P_2. Prove that P_3 is a refinement of Q.

Test Questions

T1 Define $A := \{1,3,4,5,6,7,9,10\}$. Define the relation R on A by means of the predicate $P(a,b)$: a and b have a prime factor in common. $D_a := D_b := A$. List R as a set of ordered pairs. Then draw the graph of R. Then write the relation-matrix of R.

T2 Let A and B be sets. Let R be an equivalence relation on A. Let R' be an equivalence relation on B. Prove that $R \cap R'$ is an equivalence relation on $A \cap B$.

T3 Let R be a nontransitive relation on a set B. Define a relation R_1 on B as $R_1 := R \cup X$, where

$$X := \{(a,c);\ a, c \in B,\ \exists b \in B,$$
$$(a,b) \in R \text{ and } (b,c) \in R\}.$$

Give an example to show that R_1 is not necessarily transitive.

T4 For any relation R on a set A, define

$$R^c = \{(b,a);\ (a,b) \in R\}.$$

Prove that if R is transitive, then R^c is transitive.

T5 Define a relation S on the set $J := \{n;\ n \in \mathbf{N}, n \ge 2\}$ by the rule

$\forall m, n \in J$, mSn iff the set of all primes dividing m is the same as the set of all primes dividing n.

(*i*) Prove that S is an equivalence relation on J.

(*ii*) Find the equivalence class of the integer 12 and prove your answer.

T6 List without explanation all the refinements of the partition $P = \{\{1,2\}, \{3,4\}\}$. You may use the following shorthand: for example, represent the partition P as $1,2 \mid 3,4$.

T7 Write without explanation a partition P of the set $A = \{1,2,3,4,5\}$ having exactly two cells. Then write without explanation a refinement Q of P having exactly three cells.

T8 Let $E = \{1,2,3,4\}$. Write a partition P of E and a refinement Q of P such that $1 \mid 2 \mid 3 \mid 4 \neq Q \neq P \neq \{E\}$. No explanation is needed.

T9 Consider the relation R on the set G of integers $n \ge 2$ defined as follows:

$\forall a, b \in G$, $a\,R\,b$ iff the smallest prime dividing a equals the smallest prime dividing b.

Prove that R is an equivalence relation on G, and identify the equivalence class of 3.

T10 Consider the relation R on $\mathbf{Z}$ defined by the rule $\forall a, b \in \mathbf{Z}$, $a\,R\,b$ iff $a + b$ is a multiple of 3 (means $\exists k \in \mathbf{Z}$ such that $a + b = 3k$). Is R reflexive? Is R symmetric? Is R transitive? Explain.

T11 Let A be the set $\{1,2,\ldots,10\}$. Define partitions of A as:

$$P := 1,2,3 \mid 4,5 \mid 6,7,8,9 \mid 10$$
$$Q := 3,1,4 \mid 7,5 \mid 9,2,6 \mid 8,10$$

Using the same notation, write out the partition

$$\{C \cap D;\ C \in P,\ D \in Q\} - \{\emptyset\}.$$

T12 Consider the set C of all circular arrays of four 0's or 1's. [Example:

$$a = \begin{matrix} & 1 & \\ 0 & & 1 \\ & 0 & \end{matrix}$$

is one such.] Define a relation S on C via: $\forall a, b \in C$, aSb iff b becomes identical to a after b is rotated clockwise by zero, one, two, or three 90° rotations. Thus a above satisfies aSb for

$$b = \begin{matrix} & 0 & \\ 1 & & 0. \\ & 1 & \end{matrix}$$

Prove S is an equivalence relation. Find the equivalence classes. Explain.

T13 Consider a set of coin purses with coins in them. Define a relation R on this set as follows: For any two of the coin purses P and P', $P\ R\ P'$ if and only if every amount of money makable with the contents of P is makable with the contents of P'. Is R symmetric? Is R reflexive? Is R transitive? Explain.

T14 Four people are going to play a game while seated around a table. There are to be two teams of two partners each, and partners sit opposite each other. Consider the set S of all 24 possible seatings. Define a relation R on S as follows: for any two seatings b and b', we say bRb' iff the two teams of seating b are the same teams as those of seating b'. Example: The people are A, B, C, D. The places are fixed as 1, 2, 3, 4.

b_1: A at 1, B at 2, C at 3, D at 4. b_2: B at 1, A at 2, D at 3, C at 4.

$b_1\ R\ b_2$

(11.1F)

(*i*) Prove that R is an equivalence relation.
(*ii*) Determine the number of equivalence classes. Explain.

Answers to Practice Problems

(2.4P) We must prove that for all $x \in \mathbf{N}$, $x \leq x$. It is true by our definition of "less than or equal," $\leq$.

(3.2P) I'll do this for some of the possibilities in (2.10).

Example 1. $E(1) = \{1,3\}$, $E(2) = \{2\}$.

These are easy to read from the matrix.

Example 2. $E(1) = E(2) = \{1,2,5\}$.
$E(5) = \{1,2,3,5\}$.
Example 3. $E(1) = E(2) = E(3) = \{2,3\}$.
Example 4. $E(1) = \{1,2,3,4,5\}$
$= E(x)$ for all x in A.
Example 5. $E(1) = \{1\}$.
Example 6. $E(1) = \{1,2,3\}$, $E(4) = \{4,5\}$.

(3.5P) We assumed the antecedent of (3.4), that $E(x) \cap E(y) \neq \varnothing$. We concluded that $E(y) \subseteq E(x)$. If we interchange x and y in "$E(x) \cap E(y) \neq \varnothing$," we get the same statement. The hypothesis remains the same, so the conclusion must also hold after the interchange. Thus $E(x) \subseteq E(y)$.

(5.3P) The empty set has only the empty set of cells as a partition. A singleton set $\{a\}$ has only $\{\{a\}\}$ as a partition. A two-point set $\{a,b\}$ has two partitions. In the notation of (5.4), they are

$$a, b$$
$$a \mid b.$$

The set $\{a,b,c\}$ has five partitions

$$a, b, c$$
$$a \mid b, c \quad a, b \mid c \quad a, c \mid b$$
$$a \mid b \mid c\,.$$

By the main result of Section (6), these are the numbers of equivalence relations on the sets. For example, there are five equivalence relations possible on a set of three points.

(7.8P) (*i*) D is the cell with $a \longrightarrow b$ in it. The truth-table of a propositional formula determines what is in its cell. The truth-table of $a \longrightarrow b$ is

a	b	$a \longrightarrow b$
T	T	T
T	F	F
F	T	T
F	F	T

Every one-atom formula in a must be constant (in truth-value) on the first two lines of its table, since $a = T$ there. But $a \longrightarrow b$ is not constant on those lines. Similarly $a \longrightarrow b$ is not constant on lines 2 and 4, where $b = F$. Therefore there are no one-atom formulas in D.

(*ii*) The truth-table of a two-atom formula has four lines, with truth-values v_1, v_2, v_3, v_4. There are $2^4 = 16$ choices for these four truth-values (see Chapter 3, (11.5)). Thus there are 16 possible truth-tables, all of which are attainable via propositional formulas (by Chapter 2, (13.2)). Therefore there are 16 cells for $\Leftrightarrow$.

(*iii*) The one-atom formulas in a have $2^2 = 4$ possible truth-tables:

a	$f(a)$
T	v_1
F	v_2

with rows duplicated as b varies

a	b	$f(a)$
T	T	v_1
T	F	v_1
F	T	v_2
F	F	v_2

The same is true for one-atom formulas in b. Their tables are

a	b	$g(b)$
T	T	v_3
T	F	v_4
F	T	v_3
F	F	v_4

There is overlap only when $v_3 = v_4 = v_1 = v_2$, i.e., for tautologies or contradictions. Thus there are $4 + 4 - 2 = 6$ one-atom cells.

(*iv*) The answer is two—one for tautologies, one for contradictions.

(9.1P) Reflexivity is expressed as $\forall x \in A\ P(x,x)$; symmetry as $\forall x,y \in A,\ P(x,y) \longrightarrow P(y,x)$; transitivity as $\forall x,y,z \in A,\ P(x,y)$ and $P(y,z) \longrightarrow P(x,z)$.

CHAPTER 5

Functions

...[I]f...X depends on...x and on no other variable magnitude, it is customary to regard X as a function of x; and there is usually an implication that X is derived from x by some series of operations.

A. R. Forsyth, 1900

1. Functions

One of the most pervasive terms in mathematics is *function*. In computer science its meaning is close to that of *program* and *algorithm*, terms that you will agree are also widely used. You have already seen lots of functions, such as $x+2, x^2+1, (x-1)/(x+1)$, sines and cosines, and logarithms. Now look at a less familiar example. Consider the *length* function for strings over $\{0, 1\}$:

(1.1) If x is any finite string (Chapter 1, (19.9)) of 0's and 1's, $\ell(x) :=$ the total number of 0's and 1's used to form x. ℓ (empty string) $= 0$, $\ell(0) = \ell(1) = 1$, $\ell(00) = \ell(01) = \ell(10) = \ell(11) = 2$, and so on.

This ℓ seems to be a function, too. How can we define the concept to include the familiar functions of algebra and calculus and such functions as ℓ above? Mathematicians have answered that question as follows:

DEFINITION (1.2) A **function** is a rule f that, for every member x of a set X, selects exactly one element, denoted $f(x)$, of a set Y.

We represent the function and these two sets symbolically as

$$f : X \longrightarrow Y.$$

X is called the **domain**, and Y the **codomain**, of f.

Examples (1.3) Let $X = Y = \mathbf{Z}$, the set of all integers.

(*i*) f_1 is the rule that for each element x of X selects the element $x + 1$ of Y.

(*ii*) f_2 is the rule that for each element x of X selects the element x^2 of Y.

(*iii*) f_3 is the rule that for each element x of X selects the element 0 if x is even and 1 if x is odd.

We could define these functions instead by saying: for all x in X,

$$f_1(x) := x + 1$$
$$f_2(x) := x^2$$
$$f_3(x) := \begin{cases} 0 & \text{if } x \text{ is even} \\ 1 & \text{if } x \text{ is odd} \end{cases}$$

(*iv*) A different type of rule is one specified by a table. For example, we could define a function $f_4 : \{1,2,3,4\} \longrightarrow \mathbf{N}$ by the rule

x	1	2	3	4
$f_4(x)$	7	1	3	9

That is, $f_4(1) := 7, f_4(2) := 1$, and so on.

(*v*) Another example: A is a set, P is a partition of A, and $\pi : A \longrightarrow \mathcal{P}(A)$ is defined by the rule $\forall x \in A, \pi(x) :=$ the cell $C \in P$ such that $x \in C$.

Practice (1.4P) Give an example of the last example for $A = \{1,\ldots,5\}$.

A computer program that for each element of a set X of inputs produces an output is a representation of a function with domain X. The program is the rule; with it the computer transforms the input to the output. The codomain is for us to specify, as long as it has the set of outputs as a subset.

We often draw a Venn diagram of the domain and codomain with an arrow for the function.

(1.5F)

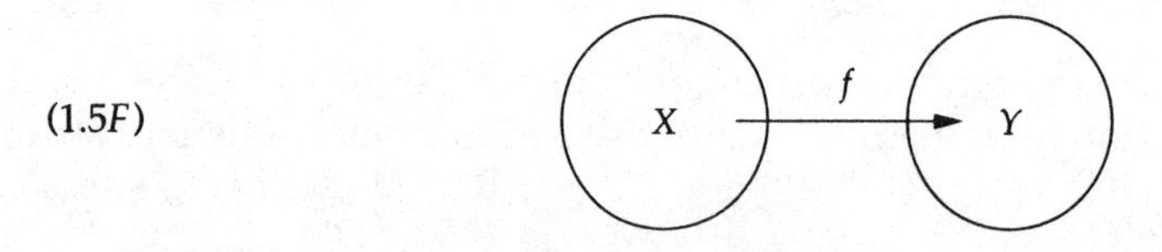

You can see that this definition of function includes many more types of functions than you studied in calculus.

2. Equality; Partial Functions

Two functions $f : X \longrightarrow Y$ and $g : U \longrightarrow V$ are **equal** (by definition) if and only if the domains are equal ($X = U$), the codomains are equal ($Y = V$), and

$$\forall x \in X, \qquad f(x) = g(x).$$

The last condition expresses the fact that the rules are the same.

If $f : X \longrightarrow Y$ is a function and W is a superset of X, we say that $f : W \longrightarrow Y$ is a **partial function**. That is, there may be points of W on which f is not defined. Division, for example, is a partial function from $\mathbf{R} \times \mathbf{R}$ to $\mathbf{R}$. For all $x,y \in \mathbf{R}$, we set $d(x,y) := x/y$ if $y \neq 0$. The function computed by a program is partial if for some inputs there is no output.

3. Some Terminology; $\widehat{f}$

We say: "f is a function **from** X **to** Y," "f **maps** X **into** Y," and "f **maps** b **to** $f(b)$." We sometimes call b the **argument** of the function. We may also say "f is (a function) **defined on** X, or simply "f is a function **on** X (to Y)."

A function is sometimes called instead a **mapping**, a **map**, a **transformation**, or a **correspondence**.

The point $f(x)$ of Y (for x in X) is called the **value of** f **at** x, or the **image** of x **under** f. Note that $f(x)$ is not the function f; $f(x)$ is the *value* of the function f at the point x.

Thus we made a syntax error when we spoke of the "function ... $x + 2$" in (1). Strictly, it is the function $f : \mathbf{R} \longrightarrow \mathbf{R}$ (say) defined by the rule $f(x) := x + 2$ for all $x \in \mathbf{R}$. This syntax error is common in calculus, however, where everyone knows that all domains and codomains are $\mathbf{R}$ or subsets of $\mathbf{R}$; it's not so serious in that case.

We also say "f **takes the value** $f(x)$ **at** x." The functions f_2 and f_3 in (1.3) show that not every point of the codomain need be a value of the function. That observation leads to the next definition.

DEFINITION (3.1) The **range** of the function $f : X \longrightarrow Y$ is the subset of all points of the codomain Y that are values of f.

That is, in symbols,

$$\text{range of } f := \{y;\ y \in Y, \exists x \in X, y = f(x)\}. \tag{3.2}$$

We usually compress this symbolism by writing

$$\text{range}\, f = \{f(x);\ x \in X\}. \tag{3.3}$$

Sometimes the range of f is called the **image** of f.

In the examples (1.3) above

$$\begin{aligned}
\text{range}\, f_1 &= \mathbf{Z} = Y \\
\text{range}\, f_2 &= \{0, 1, 4, 9, 16, 25, \ldots\} = \{n^2;\ n \in \mathbf{Z}\} \\
\text{range}\, f_3 &= \{0, 1\} \\
\text{range}\, f_4 &= \{1, 3, 7, 9\} \\
\text{range}\, \pi &= P.
\end{aligned}$$

We may diagram the range as follows:

(3.3F)

An extension of the idea of the range comes up all the time. It is the set of values of f on a subset of the domain of f. In symbols, we define for any subset A of X

(3.4) $$\widehat{f}(A) := \{ f(x);\ x \in A \}.$$

Since $\widehat{f}(A)$ is a subset of Y, (3.4) is the rule defining a function from $\mathcal{P}(X)$ to $\mathcal{P}(Y)$.

$$\widehat{f} : \mathcal{P}(X) \longrightarrow \mathcal{P}(Y);$$

after all, A can be any subset of X, so $A \in \mathcal{P}(X)$, the power set of X. We diagram $\widehat{f}(A)$ as follows:

(3.4F)

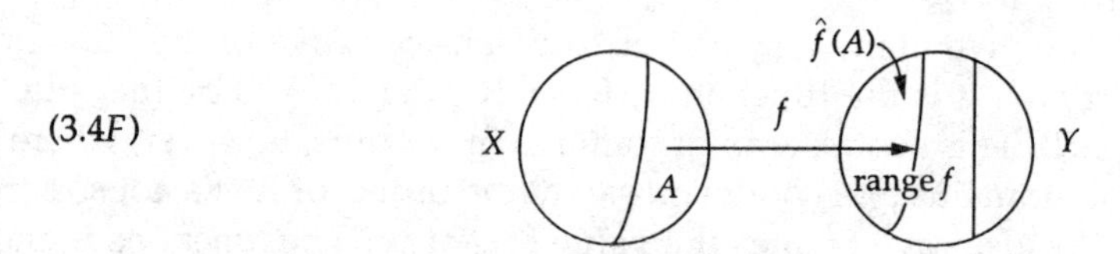

If A is a finite set $\{ a_1, \ldots, a_n \}$, we may list $\widehat{f}(A)$ as $\{ f(a_1), \ldots, f(a_n) \}$.

Example (3.5) For the function f_2 of (1.3) ($f_2(x) := x^2$ for all $x \in \mathbf{Z}$):

$$\widehat{f_2}(\{ -2, 1, 3 \}) := \{ f_2(-2), f_2(1), f_2(3) \} = \{ 1, 4, 9 \}$$
$$\widehat{f_2}(\{ -1, 0, 1 \}) = \{ 0, 1 \}$$
$$\widehat{f_2}(\{ 2 \}) = \{ 4 \}.$$

Warning: The standard notation for $\widehat{f}$ is f. People usually write $f(A)$ for what we've called $\widehat{f}(A)$. It is an abuse of notation because f maps X to Y; f is not defined on $\mathcal{P}(X)$. Nevertheless, when you see $f(k)$ for some k, it is almost always clear from the context whether k is in X or in $\mathcal{P}(X)$, so there is no ambiguity.

Trouble can arise when $X \cap \mathcal{P}(X) \neq \varnothing$. For example, take $X = \{ 1, \{ 1 \} \}$ and define

$$f(1) := 2, \qquad f(\{ 1 \}) := 3.$$

Here $\widehat{f}(\{ 1 \}) = \{ 2 \} \neq f(\{ 1 \})$.

The set $\widehat{f}(A)$ is called the **image** of A **under** f (no hat).

We'll use only the notation $\widehat{f}$ in the rest of this chapter. To acclimatize you to the prevailing practice, however, we'll gradually drop "$\widehat{f}(A)$" for "$f(A)$" in later chapters.

Notice, finally, how the definition of $\widehat{f}$ extends our set-builder notation of Chapter 2, (19.5). There we defined

$$T_Q := \{ x;\ Q(x) \},$$

where $Q(x)$ could be "$x \in D$ and $P(x)$." Now we allow ourselves to denote the set

(3.6) $$\{ f(x);\ Q(x) \},$$

where f is any function with domain including the truth-set T_Q of $Q(x)$. This set is the image of T_Q under f, namely, $\widehat{f}(T_Q)$. We could denote it with the original set-builder notation, as we did in (3.2) in defining the range of f, but that's cumbersome.

For the definition (3.4) we took $Q(x)$ to be "$x \in A$." But in general the set A of (3.4) might be specified as the truth-set of some predicate $Q(x)$, as in (3.6).

The range of f is $\widehat{f}(X)$.

4. What Is a "Rule"?

You may ask, what is the *rule* in the examples in (1.3) above? The rule f_1 is to add 1 to x to get $f_1(x)$; the rule f_2 is to square x to get $f_2(x)$; the rule f_3 is to see whether x is even and if so, assign 0; if not, assign 1. But what is a rule in general? It is, after all, a new term for us. The answer: *rule* is another undefined term. But you will soon see that we can explain it in terms of another of our undefined terms.

(4.1) We have seen in Example (1.3) that a rule can be specified by a general formula, as in f_1, f_2, and f_3. It could also be defined via a table for finite domains X, as in f_4 or f_5.

x	1	2	3	4	5
$f_5(x)$	-2	0	1	-7	0

Here, to complete the definition of the function f_5, we must specify some superset Y of the range of f_5 as the codomain. Let us define $Y :=$ range $f_5 = \{-2, 0, 1, -7\}$.

(4.2) Another way to define the rule, again when the domain is finite, is via a directed graph. For example let f_6 be defined by the arrows in this diagram:

(4.2F)

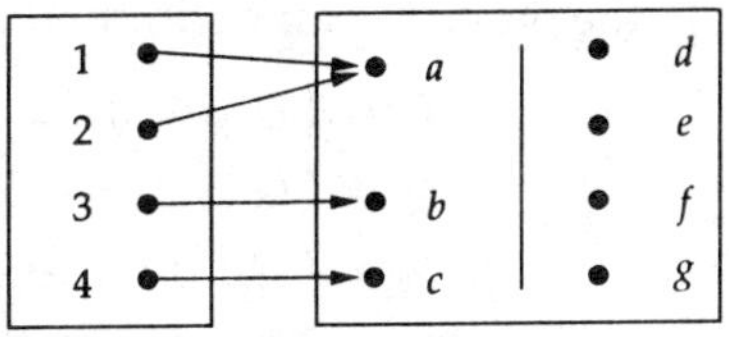

1 → a, 2 → a, 3 → b, 4 → c; d, e, f, g

This means that $f_6(1) = f_6(2) = a$, $f_6(3) = b$, $f_6(4) = c$. $X = \{1, 2, 3, 4\}$, and from the picture $Y = \{a, b, c, d, e, f, g\}$, and range $f_6 = \{a, b, c\}$.

In general a rule need not be predictable or "regular." It may even be unruly. Indeed if you put the output of a random-integer generator (used 100 times) in the table below (we have imagined the values 17, 330, ..., 47)

x	1	2	$\cdots$	100
$f(x)$	17	330	$\cdots$	47

the result would be a perfectly good rule defining a function f from $\{1, \ldots, 100\}$ to (say) $\mathbf{Z}$.

But to define *rule* properly we use the following older definition of function as a set of ordered pairs.

5. Alternate Definition of Function

(5.1) An older definition is this: A function f is a set of ordered pairs (a, b) with the property that no two different ordered pairs have the same first coordinate. In symbols

(5.2) $$\forall (b,c), (x,y) \in f \quad \text{if } b = x \quad \text{then} \quad c = y.$$

The rule here is simply that for every ordered pair $(b,c) \in f$, $f(b) = c$; the second coordinate of each ordered pair is the value of the function at the first coordinate. The restriction (5.2) accords with the idea of a rule; at each point of the domain a rule must select a value *unambiguously*. Condition (5.2) assures that for each first coordinate there is only one second coordinate. The familiar graph of a function in analytic geometry represents this definition of function as a set of points in the plane. The table in (4.1) used to define f_5 is a set of ordered pairs in an easily penetrated disguise.

There is no requirement that the second coordinates be unique. There may be any number of different ordered pairs with the same second coordinate. For example, f might be a constant function: $f_6 = \{(x,1);\ x \in \mathbf{R}\}$ is a perfectly good function. It is the rule that assigns to each real number the value 1; $f_6(x) = 1$ for every real x. In analytic geometry f_6 is graphed as the line $y = 1$.

(5.3F)

y

y=1

0

x

(5.4) In the ordered-pair definition, our examples (1.3) become

$$\begin{aligned} f_1 &= \{(x, x+1);\ x \in \mathbf{Z}\} \\ f_2 &= \{(x, x^2);\ x \in \mathbf{Z}\} \\ f_3 &= \{(2x, 0);\ x \in \mathbf{Z}\} \cup \{(2x+1, 1);\ x \in \mathbf{Z}\} \\ f_4 &= \{(1,7), (2,1), (3,3), (4,9)\}. \end{aligned}$$

Notice that defining a function as a set of ordered pairs does not specify any codomain. But the domain and range of the function can be recovered from the ordered-pair definition, as follows:

(5.5) $$\begin{aligned} \operatorname{domain} f &= \{b;\ \exists y\ (b,y) \in f\} \\ \operatorname{range} f &= \{c;\ \exists b\ (b,c) \in f\}. \end{aligned}$$

The domain is the set of first coordinates, and the range the set of second coordinates, of the set of ordered pairs defining f.

We usually use the first definition of function, but now you see that the term *rule* rests on the concept of set (of ordered pairs). If the set of ordered pairs is given as a list, then we have, in effect, defined our function by means of a table or directed graph, as in (4). If the ordered pairs are specified in the set-builder notation, then we may be able to define our function by a formula.

Example (5.6) In (5.4) we redefined the four functions f_1, f_2, f_3, f_4 as sets of ordered pairs in the set-builder notation. For f_1 and f_2 there are formulas, shown in (1.3), by which we could define the functions in the rule sense (1.2). For f_3 there's no nice formula, but there is a compact predicate allowing a definition of f_3 in the rule sense of (1.2). The table defining f_4 in (1.3) is functionally the same as a set of ordered pairs.

Example (5.7) We may construct another example using the telephone dial. Its grouping of letters with numbers (A B C with 2, D, E, F with 3, ..., W, X, Y with 9) suggests the function g:

(5.8) $$g = \{ (A,2),(B,2),(C,2),(D,3),\ldots,(W,9),(X,9),(Y,9) \}.$$

As we have defined g, it has domain $\mathscr{A} - \{ Q,Z \}$, where $\mathscr{A}$ is the set of all 26 letters of the Roman alphabet. We *could* define its domain to be $\mathscr{A}$, as

$$g : \mathscr{A} \longrightarrow \{ 0,1,\ldots,9 \},$$

using the same rule (5.8); then g would be a partial function. The range of g is $\{ 2,\ldots,9 \}$, and $\widehat{g}(\{ N,O,D,E \}) = \{ 3,6 \}$.

Practice (5.9P) Find $\widehat{g}(\{ V,A,G,U,E \})$. Find $\widehat{g}$ for some of your favorite words by first replacing the word by the set of letters used to spell it. Words with Q or Z are not allowed.

6. Functions as Relations

You might wonder how functions defined as sets of ordered pairs are related to relations. The answer is that a function is a special type of relation. We defined a relation on the set A in Chapter 4 as a subset of $A \times A$, but we could easily generalize that to a "relation from A to B": a subset of $A \times B$. Now that we have defined in (5) a function $f : X \longrightarrow Y$ as a subset of $X \times Y$, we can see that f is simply a relation from X to Y with the important restriction (5.2) that the first coordinates of the ordered pairs defining the function are unique. We also require that the domain *of the relation* be X. (The **domain** of a relation is the set of all first coordinates of the ordered pairs of the relation.) In terms of the directed graph the restriction (5.2) says that there is only one arrow going out from any point of X; in the relation matrix it says that every row has exactly one 1 in it. To force the domain to be X is to require that every point in the directed graph have an arrow going out from it (hence each point has exactly one outgoing arrow).

Examples (6.1) Relation $R_6 := \{ (1,a),(2,a),(3,b),(4,c) \}$. R_6 is the ordered-pair definition of the function f_6 defined in (4.2). Its relation matrix is

	a	b	c	d	e	f	g
1	1						
2	1						
3		1					
4			1				

It has a single 1 in each row, so it represents a function.

Consider relation $R_7 := \{(1,2),\ (1,3),\ (2,3)\}$. The graph and matrix of this relation are

(6.1*F*)

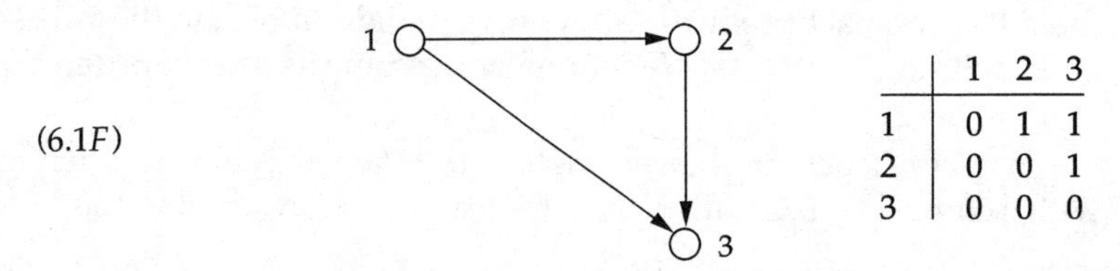

	1	2	3
1	0	1	1
2	0	0	1
3	0	0	0

Notice that the graph of R_7 has two arrows going out from a single point; the matrix has two 1's in the row of that point. R_7 is not a function, and we have seen why from three points of view. Neither is it a partial function.

7. Injectivity, Surjectivity, and Bijectivity

For the following two definitions, let f be a function with domain X and co-domain Y.

DEFINITION (7.1) We say that f is **injective** if and only if f takes a different value at every point of the domain X. In symbols

$$\forall a,b \in X, \quad f(a) = f(b) \text{ implies } a = b.$$

You might understand (7.1) better at first if you put it in (the equivalent) contrapositive form:

$$f \text{ is injective iff}$$

(7.2) $$\forall a,b \in X, \quad a \neq b \text{ implies } f(a) \neq f(b).$$

But (7.1) is easier to work with.

Examples (7.3) (*i*) Suppose we wanted to prove injectivity of the function $f : \mathbf{N} \longrightarrow \mathbf{N}$ defined by the rule $\forall n \in \mathbf{N}\ f(n) := n + 3$. How to proceed? Very simple. Just use (7.1) directly:

Let $a, b \in \mathbf{N}$ and assume $f(a) = f(b)$. Thus you are assuming $a + 3 = b + 3$. It follows immediately that $a = b$. Therefore f is injective.

(*ii*) Consider $g : \mathbf{N} \longrightarrow \mathbf{N}$ defined by the rule $g(x) := x^2 + x + 1$ for all $x \in \mathbf{N}$. We use (7.1) to prove that g is injective. The proof is a little more complicated than that for f.

Let $a, b \in \mathbf{N}$ and assume $g(a) = g(b)$. Thus

$$a^2 + a + 1 = b^2 + b + 1.$$

Therefore $a^2 - b^2 = b - a$. Factoring the left side we see that

(7.3.1) $$(a - b)(a + b) = -(a - b).$$

Now we use contradiction. We want to show that $a = b$, i.e., that $a - b = 0$. If that is not true, then we may cancel $a - b$ from both sides of (7.3.1), getting

$$a + b = -1,$$

which contradicts the fact that a and b are positive. Therefore $a = b$, and g is an injection.

Practice (7.4P) For the same g but with domain $\mathbf{Z}$, show that g is not injective. Notice that you must prove the negation of the quantified predicate in (7.1) or in (7.2):

$$\exists a, b \in \mathbf{Z} \qquad (a \neq b) \wedge (g(a) = g(b)).$$

See Chapter 2, (22) if you need to review that topic.

Thus an injective function (or **injection**) is as different from a constant function as possible. A constant function takes the same value at every point of X; an injective function does not do so at even two points of X.

An older term for *injective* is *one-to-one into*. That comes from the fact that

(7.5) An injective function is a one-to-one correspondence between its domain and range.

That is, if f is injective, then for every point y in the range of f, there is exactly one point b in the domain of f such that $f(b) = y$. Simply because f is a function, for every point p in the domain of f, there is exactly one point q in the range of f such that $f(p) = q$.

In a sketch of the function as a graph we can see this property clearly:

(7.5F)

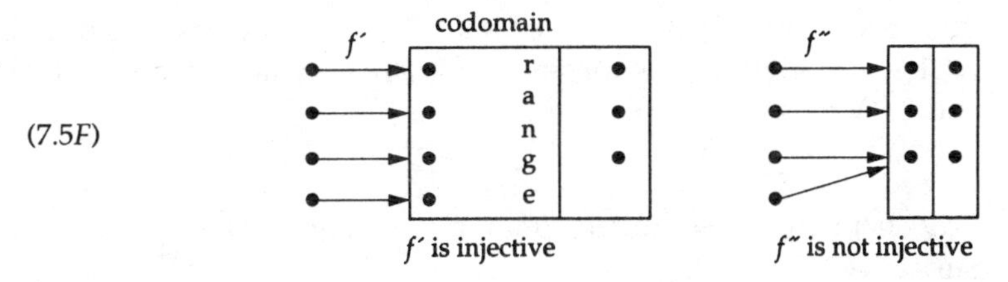

The obvious mapping from the set of people in the United States to the set of social security numbers is a partial function that is meant to be an injection.

(7.6) As a relation, a function is injective if and only if its relation-matrix has at most one 1 in each column.

Example The relation matrices M' for f' and M'' for f'' of (7.5F) are

(7.5F) $$M' = \left[\begin{array}{cccc|c} 1 & & & & \\ & 1 & & & \\ & & 1 & & \\ & & & 1 & \end{array}\right] \qquad M'' = \left[\begin{array}{ccc|c} 1 & & & \\ & 1 & & \\ & & 1 & \\ & & 1 & \end{array}\right]$$

(omitting the labeling of rows and columns). Since f' is injective we find no 1 or just one 1 in each column of M'. Since f'' is not injective it has a column with at least two 1's. In the matrices, blanks stand for 0's.

(7.7) A consequence of (7.5): If it is finite, the domain of an injective function consists of exactly the same number of points as its range.

DEFINITION (7.8) A function is called **surjective** (or a **surjection**) if and only if its range equals its codomain. In symbols, the function $f : X \longrightarrow Y$ is surjective iff

$$\forall y \in Y \quad \exists x \in X \quad f(x) = y.$$

The condition in symbols states that every point of the codomain Y is a point of the range $\{ f(b); b \in X \}$. That is, it states $Y \subseteq \text{range} f$. There is no need to state the reverse inclusion, because it holds by definition for all functions.

Injectivity is defined with reference only to the domain and range. But as you see, surjectivity does involve the codomain.

If f is any function, and if we redefine the codomain of f to be equal to the range of f, then f becomes surjective. Conversely, if f is surjective and we replace its codomain by a proper superset, then f is no longer surjective.

An older term for surjective is **onto.** Yes, *onto* is an adjective here: "f is an onto function." It also serves as a preposition: "f maps X onto Y" means $f : X \longrightarrow Y$ is surjective.

(7.9) Let us check the examples of (1.3) for these properties:

f_1 is both injective and surjective.
f_2 is neither injective nor surjective.
f_3 is neither injective nor surjective.

Since range f_2 does not contain 2, f_2 does not map **Z** onto **Z**; and range $f_3 = \{ 0, 1 \} \neq \mathbf{Z}$, so f_3 is not surjective. Why are they not injective? Because $f_2(-1) = f_2(1)$ and $f_3(1) = f_3(3)$.

You can check f_1.

Practice (7.10P) Is the "telephone" function $g : \mathcal{A} \longrightarrow \{ 0, \ldots, 9 \}$ of (5.7) injective? Surjective?

Examples (7.11) The sine function: Let f_7 be sin, that is, for all x in $X_7 = \{ r;\ r \in \mathbf{R}, 0 \leq r \leq 2\pi \}$, define $f_7(x) := \sin x$. Then f_7 is not injective because $f_7(0) = f_7(\pi) = 0$. The range of f_7 is the set $\{ r;\ r \in \mathbf{R}, -1 \leq r \leq 1 \} =: I$. We must define the codomain Y of f_7 to be a superset of I. If we say $Y := I$, then f_7 is surjective. If, for example, we say $Y := \mathbf{R}$ then f_7 is not surjective.

It will be helpful in discussing the sine function to have the following notation. For real a, b

(7.12) $$[a, b] := \{ t;\ t \in \mathbf{R},\ a \leq t \leq b \}.$$

Thus $X_7 = [0, 2\pi]$ and $I = [-1, 1]$. This notation is standard; you'll see more of it in books on analysis than here.

(7.13) Consider the sine function again, but now with domain $X_8 := [0, \pi/2]$. Also, $f_8(x) := \sin x$ for all x in X_8. This function is injective, as a look at its graph (in the sense of analytic geometry, not in our sense) shows. More discussion of this point is in (12.8).

(7.14F)

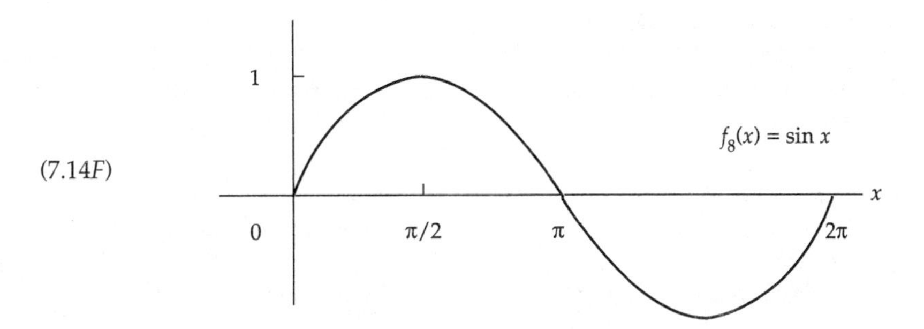

The range of f_8 is $[0,1] =: I'$. Again we must choose Y_8, the codomain of f_8, so that $I' \subseteq Y_8$. Then f_8 is surjective if and only if $Y_8 = I'$.

(7.15) As a relation, the function f is surjective if and only if its relation-matrix has at least one 1 in every column.

Example (7.16) Define $f_9(x) := 1 + x + x^2$ for $X := \{-1, 0, 1, 2\}$, and $Y := \{1, 3, 7\}$. Then the relation-matrix is

	1	3	7
−1	1	0	0
0	1	0	0
1	0	1	0
2	0	0	1

You see that each column has at least one 1. And of course, f_9 is surjective.

DEFINITION (7.17) A function $f : X \longrightarrow Y$ is called **bijective** if and only if f is both injective and surjective.

The older term for bijective is *one-to-one onto* or *1–1 onto*. Since an injective function is a 1–1 correspondence between domain and range, and since for a surjective function, the range is the codomain, the term *1–1 onto* is a good description of a bijective function.

Examples (7.18) Our first example, f_1, is a bijective function, as we remarked, in effect, in (7.9).

The sine function with domain $X = \{x;\ -\pi/2 \leq x \leq \pi/2\}$ and codomain $Y = \{y;\ -1 \leq y \leq 1\}$ is a bijection. See (7.14F).

To restate (7.5), any injective function is a bijection if its codomain is equal to its range. Thus any function f (on a finite domain) defined by a table is a bijection if all the entries on the $f(x)$ line of the table are different from each other and if those entries are all the points of the codomain. For example, the table

x	1	2	3	4	5
$f(x)$	12	13	14	15	11

gives the rule for an injective function f on the domain $\{1, \ldots, 5\}$; it is bijective if the codomain is $\{11, \ldots, 15\}$.

8. The Pigeonhole Principle

Suppose your school entertains prospective students by assigning each one a student host who is to show them the campus and answer questions about the intellectual and social life there. One day when there are ten hosts, eleven prospective students appear. Although your school likes to have a one-to-one relationship for this peer-tour, this time they can't. Two prospective students must be assigned to one host.

The foregoing is an instance of the **pigeonhole principle** at work.

(8.1) If you have distributed k pigeons to n pigeonholes and $k > n$, then some pigeonhole must hold more than one pigeon.

More prosaically:

(8.2) Let A and B be sets of exactly k and n elements, respectively. Let f be any function from A to B. If $k > n$, then there are at least two points $x, y \in A$ such that

$$x \neq y \quad \text{and} \quad f(x) = f(y).$$

In other words, there is no injection from A to B when $k > n$.

This perfectly obvious remark can be used to prove some unobvious results.

Practice (8.3P) Why is the last sentence of (8.2) perfectly obvious? Use something from (7) to explain it.

Examples (8.4)

(*i*) Suppose an eccentric person gave a large university a cash grant to be awarded in its entirety to one student. The donor requires that the winner be chosen by the following process: First, there is a random choice of a sequence of four integers a_1, a_2, a_3, a_4, such that $0 \leq a_i \leq 9$. The student with $a_1a_2a_3a_4$ as the last four digits of his or her social security number wins the entire grant. If there is no such student the process repeats.

What is wrong with this method? First, a large university has more than 10,000 students. Since there are only 10,000 possible four-digit numbers, there are at least two students with the same last four digits. If those were the digits chosen by the random process, it would not be possible to give the grant entirely to one student and be fair at the same time.

(*ii*) There are nine tennis teams of two players each for a round-robin doubles tournament. This means each team is to play each other team once. There is only one court available, and only five matches can be scheduled per day. Show that they are unable to complete the tournament in one week's time.

We first count the number of matches. If we represent the teams by the numbers $1, \ldots, 9$ we can represent the match between team i and team j by the ordered pair (i, j) where we take $i < j$. (That avoids double counting by ruling out (j, i), which represents the same match.) How many such pairs are there? They are the points above the diagonal $j = i$ in the 9×9 array

(8.5F)

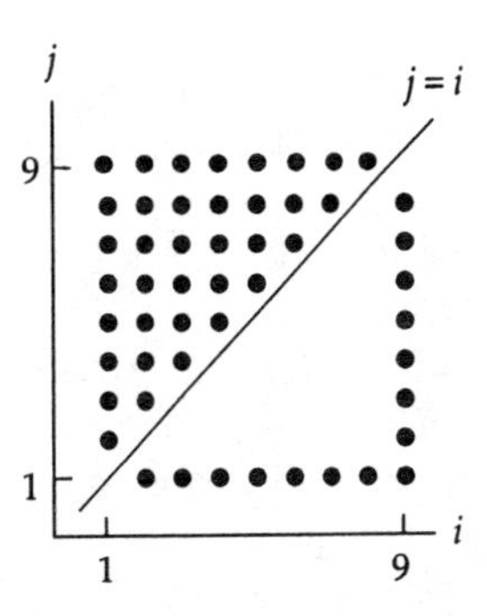

There are $9^2 = 81$ points in the whole array, and our subset is half of what's left after we take out the diagonal: $(9^2 - 9)/2 = 9 \cdot 8/2 = 36$.

Since there are seven days in a week there can be only 35 matches in a week.

Practice (8.6P) Count the number of matches differently by using a result from Chapter 3.

(8.7) Sometimes the pigeonhole principle can trip up its users. Consider this instance: " 'If there are more trees in the world than there are leaves on any one tree, then there must be at least two trees with the same number of leaves.'—this sentence records a thought."[†]

Practice (8.8P) What is wrong with the assertion in single quotes above?

Here is a generalization of the pigeonhole principle:

Proposition (8.9) Let $n \geq 1$. Let $x_1, \ldots, x_n$ be real numbers. Then

$$\max(x_1, \ldots, x_n) \geq \operatorname{avg}(x_1, \ldots, x_n) := \frac{x_1 + \cdots + x_n}{n}. \quad \blacksquare$$

By "$\max(x_1, \ldots, x_n)$" we mean a largest of the numbers $x_1, \ldots, x_n$.

Example

$$\max(3,1,4,1,5,9,2,6,5) = 9. \qquad \max(0,0,1,2,3,3) = 3.$$

Proof of Proposition (8.9) (8.10) It is intuitively obvious that the maximum of a sequence of numbers is at least as large as the average, but let's prove it anyway.

Let $m := \max(x_1, \ldots, x_n)$. Thus $m = x_k$ for some k, and $m \geq x_i$ for all $i = 1, \ldots, n$. Then

$$\sum_{1 \leq i \leq n} (m - x_i) \geq 0$$

[†]Jacques Barzun, *Teacher in America* (Boston: Little, Brown, 1945), page 59.

since $m - x_i \geq 0$ for all i. But

$$\sum_{1 \leq i \leq n} (m - x_i) = mn - \sum_{1 \leq i \leq n} x_i \geq 0.$$

Proposition (8.9) follows after we divide by n. QED

How does this result generalize the pigeonhole principle? We have k pigeons in n pigeonholes, and $k > n$.

Let x_i be the number of pigeons in the ith pigeonhole, for $i = 1, \ldots, n$. By (8.9) $\max(x_1, \ldots, x_n) \geq (1/n)(\sum x_i) = k/n > 1$. Since the maximum value of $x_1, \ldots, x_n$ is an integer and greater than 1, as we just saw, it must be at least 2.

Example (8.11) There is a famous old problem that (8.9) helps us solve. Suppose there are six points in space, no three in a line. Each two of the points are joined by a line-segment, of which there are necessarily $15 = (6^2 - 6)/2$. Now, each line segment is painted one of two colors, red or blue. Prove that no matter how the colors are chosen, there must be a one-color triangle (i.e., three points such that the line-segments joining them in pairs are all of the same color). You might want to try to solve this on your own before reading on.

The proof: Any point A is joined to five other points by five line-segments ("lines"). There are two colors available for these five lines, so there must be at least three lines of the same color. (This is a simple case of (8.9) with $n = 2$ and $x_1 + x_2 = 5$; $x_1[x_2]$ is the number of red [blue] lines from the point.)

So we have the picture

(8.12*F*)

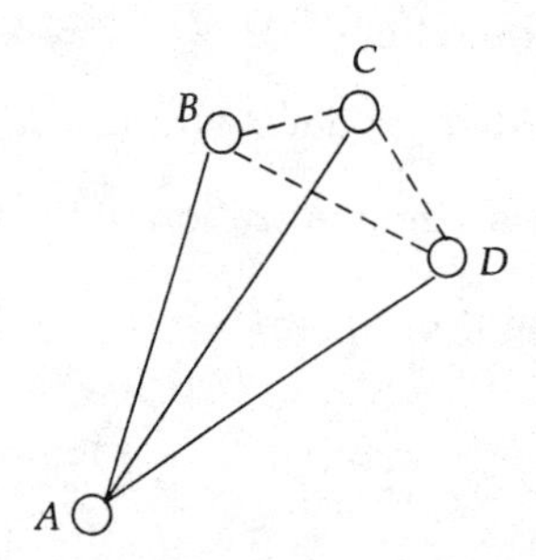

where the three lines from A of the same color, say red, go to B, C, D. Now look at the triangle BCD. If any one of its sides, say BC, is red, then there's a red triangle ABC. If none is red, then BCD is a blue triangle. QED

I admit that the reasoning here went beyond pigeons.

Example (8.13) For a final example we consider the problem of six points in space, 10 pairs of which are joined by distinct line segments ("lines"). We are to prove that, no matter how the lines are chosen, there is at least one triangle.

Let S be the set of six points and let L be the set of the 10 lines. Consider a set P of ordered pairs defined as follows:

$$P := \{(x, \ell);\ x \in S, \ell \in L, x \text{ is an endpoint of } \ell\}.$$

P has exactly 20 points, because there are 10 lines ℓ and each ℓ has two endpoints in S.

The average number of lines ending at a point of S is thus $20/6 = 3\ 1/3$. Therefore some point of S has at least four lines ending at it.

We just used (8.9) and the fact that the number of lines ending at a point of S is an integer. In more detail, if $x_1, \ldots, x_6$ are the points of S and if they have, respectively, $n_1, \ldots, n_6$ lines of L ending at them, then $n_1 + \cdots + n_6 = 20$. From (8.9) we see that $\max(n_1, \ldots, n_6) \geq 20/6$.

We now have two cases, that $\max(n_1, \ldots, n_6)$ is five or four. If it is five, then we have five lines as in the figure

(8.14*F*)

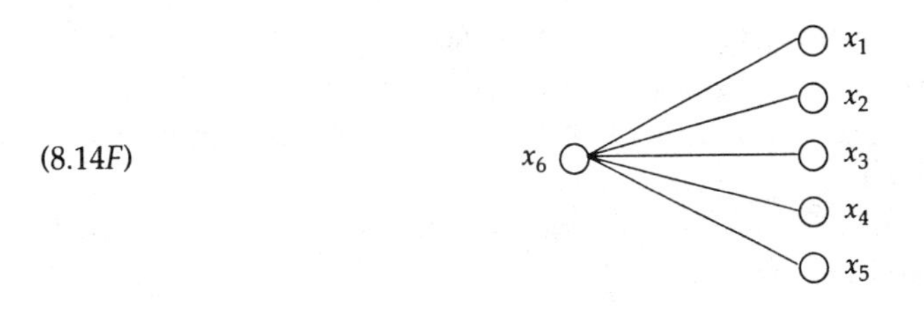

Any one of the five remaining lines forms a triangle since it must join two of $x_1, \ldots, x_5$.

Thus we now treat the other case, $\max(n_1, \ldots, n_6) = 4$. We have then

(8.15*F*)

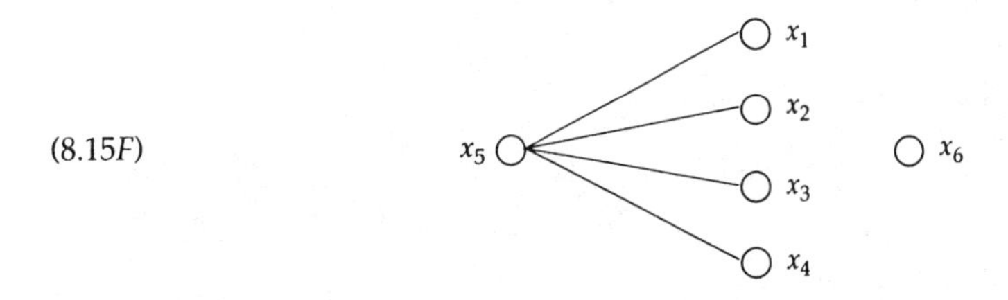

with no line between x_5 and x_6 but with six more lines joining some pairs. If there is a line between any two of $x_1, \ldots, x_4$ (say x_1, x_2), then there is a triangle x_5, x_1, x_2. If not, however, then all the remaining lines join x_6 and $x_1, \ldots, x_4$. But there can be only four such lines, contradicting the given condition that there are six more lines. Therefore there must be at least two lines with both end-points in $\{x_1, \ldots, x_4\}$. So there must be at least two triangles. (This problem is a special case of a "Putnam problem" from 1956. See A. M. Gleason, R. E. Greenwood, and L. M. Kelly, *The William Lowell Putnam Mathematical Competition—Problems and Solutions 1938–1964*, Mathematical Association of America, Washington, 1980. The general problem has $2n$ points (with $n > 1$) and $n^2 + 1$ lines and asks us to prove a triangle must exist.)

***Practice** (8.16P) Under the conditions of (8.13) prove that at least 3 triangles exist.

A more involved example of the pigeonhole principle appears in the *American Mathematical Monthly*, volume 89 (February, 1991), pages 160–162.

9. The Inverse "Function"

(9.1) Let $f : X \longrightarrow Y$ be a function. We define for any subset B of Y the following subset A of X : A is the set of all points a of X such that $f(a)$ is an element of B. Our notation for A, which depends on B, of course, is $\widehat{f}^{-1}(B)$. In symbols

$$A = \widehat{f}^{-1}(B) := \{ a;\ a \in X, f(a) \in B \}.$$

In words, A is the set of all points of the domain that map into B under f. The *all* here gets its usual heavy emphasis. See Chapter 2, (19.10).

We read $\widehat{f}^{-1}(B)$ as "f inverse of B." The set $\widehat{f}^{-1}(B)$ is called the **inverse image** of B, or the **counterimage** of B, whereas $\widehat{f}(A)$ is called the image of the set A, as we said in (3). This is a good place for another warning: that $\widehat{f}^{-1}$ is not a standard notation. Usually you will see simply f^{-1}. We'll use $\widehat{f}^{-1}$ consistently in this chapter and phase into f^{-1} later.

Our new $\widehat{f}^{-1}$ is not a function from Y to X, but it is a function from the power set $\mathcal{P}(Y)$ to the power set $\mathcal{P}(X)$.

$$\widehat{f}^{-1} : \mathcal{P}(Y) \longrightarrow \mathcal{P}(X)$$

When $B = \{ y \}$ is a singleton we customarily write $\widehat{f}^{-1}(y)$ instead of $\widehat{f}^{-1}(\{ y \})$, even though the latter is the correct form.

Example (9.2) Let f be defined as follows:

$$f(1) = f(2) = f(3) = a, \qquad f(4) = b,$$
$$f(5) = f(6) = c, \qquad f(7) = d,$$

where $X = \{ 1, \ldots, 7 \}$ and $Y = \{ a, b, c, d, e, g \}$. The function is graphed in (9.2F). The table shows B and $\widehat{f}^{-1}(B)$ for several subsets B of Y.

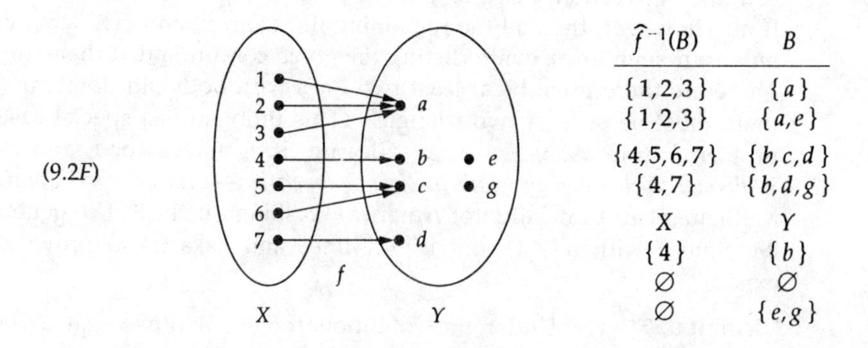

(9.2F)

$\widehat{f}^{-1}(B)$	B
$\{1,2,3\}$	$\{a\}$
$\{1,2,3\}$	$\{a,e\}$
$\{4,5,6,7\}$	$\{b,c,d\}$
$\{4,7\}$	$\{b,d,g\}$
X	Y
$\{4\}$	$\{b\}$
$\varnothing$	$\varnothing$
$\varnothing$	$\{e,g\}$

Example (9.3) Let's consider $\widehat{f}^{-1}$ for a familiar function from calculus, say $f(x) := \sin x$, with domain $X := [0, 2\pi]$ (the interval from 0 to 2π—see (7.12)) and codomain $Y := \mathbf{R}$.

(9.3F)

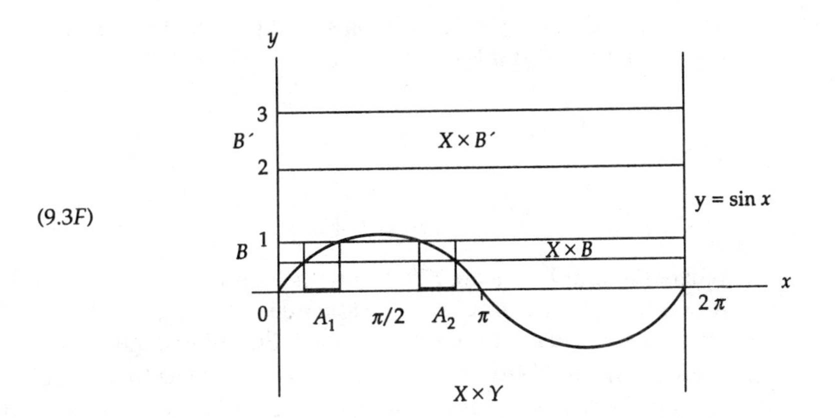

Consider an interval B, say the one in (9.3F). We find $\widehat{f}^{-1}(B)$ by looking at all points in $X \times Y$ with y-coordinate in B; that is the horizontal band $X \times B$ in (9.3F).

We find the intersection of the horizontal band with the graph of f.

We then project vertically from this intersection onto X, getting two intervals A_1 and A_2, the union of which is $\widehat{f}^{-1}(B)$:

$$\widehat{f}^{-1}(B) = A_1 \cup A_2.$$

If x is any point of A_1 or of A_2 then $f(x) \in B$; if $x \notin A_1 \cup A_2$, then $f(x) \notin B$.

For a second interval, take $B' := [2,3]$. Then $\widehat{f}^{-1}(B') = \varnothing$ because in (9.3F) the horizontal band $(X \times B')$ determined by B' is disjoint from the graph.

Finally, if $B'' := [1,2]$, then $\widehat{f}^{-1}(B'') = \{\pi/2\}$, since $x = \pi/2$ is the only argument for which $\sin x = 1$, and so the horizontal band $X \times B''$ (not shown) intersects the graph of f only at the point $(\pi/2, 1)$.

Notice that for all functions $f : X \longrightarrow Y$ and all $B \subseteq Y$,

(9.4)
$$\widehat{f}^{-1}(B) = \bigcup_{y \in B} \widehat{f}^{-1}(y).$$

Another, more interesting result is this:

(9.5) The set $\{\widehat{f}^{-1}(y);\ y \in \operatorname{range} f\}$ is a partition of X.

This fact is easy to prove and will be left as a problem. Actually this way of getting partitions is pervasive; we'll often refer to (9.5).

Examples (9.6)

(i) In example (9.2) above, the cells of the partition are

$$\{1,2,3\},\ \{4\},\ \{5,6\},\ \{7\}.$$

(ii) Consider logical equivalence on a set X of propositional formulas, all on some fixed finite set B of atoms. Define a function

$$t : X \longrightarrow Y$$

where Y is the set of all truth-tables on the set B of atoms. If $B = \{a, b\}$ then Y is the set of all truth-tables

a		b		
T		T		$\square$
T			F	$\square$
	F	T		$\square$
	F		F	$\square$

where the four boxes are filled in with T and F in all 16 possible ways. The rule for t is this: for any propositional formula $f \in X, t(f) :=$ the truth-table of f (with all atoms of B used to make the table). Since logical equivalence means having the same truth-table, if f in X has $t(f) = y$, we see that $\widehat{t}^{-1}(y)$ is the subset of X of all propositional formulas logically equivalent to f. Here is an example:

a		b		a		$(a \wedge b) \vee (a \wedge \sim b)$	
T		T		T		T	
T			F	T		T	
	F	T			F		F
	F		F		F		F

The formulas a and $(a \wedge b) \vee (a \wedge \sim b)$ have the same truth-table—call it y—so $a \in \widehat{t}^{-1}(y)$ and $(a \wedge b) \vee (a \wedge \sim b) \in \widehat{t}^{-1}(y)$.

Practice (9.7P) Why are there exactly 16 ways to fill the boxes with T and F? If B consists of n atoms (for $n \in \mathbf{N}$), how many truth-tables are there in Y?

(9.8) You have already seen the inverse function; the familiar arcsin or $\sin^{-1}$ is the same function that we have defined as the inverse of the sine. Normally the arcsin is evaluated at singletons, so you may have once been asked, for example, to find $\sin^{-1}(1)$, "the angle whose sine is 1." That is, $\alpha = \sin^{-1}(1)$ if and only if $\sin \alpha = 1$. That accords with our definition of the inverse function, except we would write $\{\alpha\} = \sin^{-1}(1)$. There are $0, 1, 2, \ldots,$ or infinitely many, values for α depending on our choice of domain X for the sine function. If we take $X = \mathbf{R}$, then

$$\sin^{-1}(1) = \{\pi/2 + 2\pi k;\ k \in \mathbf{Z}\},$$

an infinite subset of the domain. If we take $X := [0, \pi/4]$ then $\sin^{-1}(1) = \varnothing$. If we take $X = [-\pi/2, \pi/2]$ then for each y between -1 and 1 there is exactly one α in X such that $\{\alpha\} = \sin^{-1}(y)$. The latter domain X consists of the so-called principal values of the arcsine.

(9.9) The logarithm and exponential functions are inverses of each other. That is, if we define the function $\exp : \mathbf{R} \longrightarrow \mathbf{R}_{>0}$ by the rule

$$\forall a \in \mathbf{R} \qquad \exp(a) = 2^a,$$

then $\exp^{-1}(y) = \{\log_2(y)\}$ for each y in $\mathbf{R}_{>0}$ (which we define to be the set of positive real numbers), and $\log_2^{-1}(a) = \{\exp(a)\} = \{2^a\}$ for each $a \in \mathbf{R}$.

The notation $\widehat{f}^{-1}$ is imperfect, because it suggests more than it delivers. The function $\widehat{f}^{-1}$ is not always an inverse to the function $\widehat{f}$ (see (12.3) below for the definition of an inverse function). That is, there are functions $f : X \longrightarrow Y$ and subsets $A \subseteq X$ and $B \subseteq Y$ such that

(9.10) $$\widehat{f}(\widehat{f}^{-1}(B)) \neq B$$

and

(9.11) $$\widehat{f}^{-1}(\widehat{f}(A)) \neq A.$$

Sometimes, however, equality does hold in (9.10) or (9.11).

10. Composition of Functions

Suppose there are two functions $f : X \longrightarrow Y$ and $g : Y \longrightarrow Z$ in which the domain of g is the codomain of f. Then we denote the **composition** of these two functions as $g \circ f$, a function mapping X to Z defined as follows:

(10.1) $$\forall a \in X, (g \circ f)(a) := g(f(a)).$$

(10.1F)

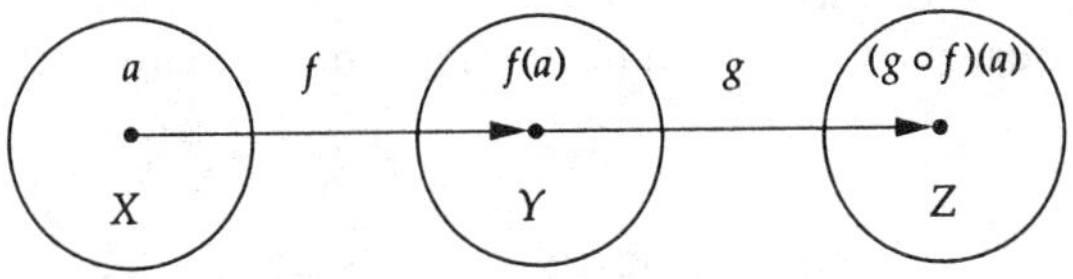

a f $f(a)$ g $(g \circ f)(a)$

X Y Z

(10.2) The notation $g \circ f$ may seem backward, but it is necessary because we put the argument of the function on the right-hand side. From the sketch (10.1F) you see that f is the function we must apply first. From a we go to $f(a)$. From $f(a)$ we go to $g(f(a))$. We call this function $g \circ f$ and write

$$g \circ f : X \longrightarrow Z.$$

(If, as some do, we wrote the arguments of our functions on the left, we would denote the same composition as $f \circ g$ and define it as $((a)f)g$. See (18).)

(10.3) You have seen the composition of functions before, as for example $\log(1+x)$, which is the composition of the logarithm function g with the function f defined by the rule $f(x) = 1 + x$ for all real $x > -1$. Here is another example: a program (f) that acts on the input (a) to produce a value ($f(a)$), which is then taken as the input to a subroutine (g).

The *chain rule* in calculus tells you how to get the derivative of a composition of functions $g \circ f$ in terms of the derivatives of the functions f and g. In our notation, if the prime $'$ denotes derivative, the chain rule is

$$(g \circ f)' = (g' \circ f) \cdot f'.$$

Thus to find $(\sin^2 x)'$ we have, for all $x \in \mathbf{R}$,

$$g(x) := x^2$$
$$f(x) := \sin x.$$

Then $(g \circ f)(x) := g(f(x)) = g(\sin x) = (\sin x)^2 =: \sin^2 x$. The derivatives: $g'(x) = 2x$ and $f'(x) = \cos x$. By the chain rule $(g' \circ f)(x) := g'(f(x)) = g'(\sin x) = 2\sin x$; multiply these by $f'(x) = \cos x$ to get the result $(\sin^2 x)' = 2\sin x \cos x$.

Proposition (10.4) The composition of injections [surjections] is injective [surjective].[†] ■

The proof, omitted here, is a nice exercise.

Corollary (10.5) The composition of bijections is a bijection.

Proposition (10.6) The inverse of a composition is the composition of the inverses in reverse order:

$$\widehat{(g \circ f)}^{-1} = \hat{f}^{-1} \circ \hat{g}^{-1}.$$

Proof Notice first that this claim makes sense, because $\widehat{(g \circ f)}^{-1}$, with the above choice of notation in (10.1F) for domains and codomains, is a function from $\mathcal{P}(Z)$ to $\mathcal{P}(X)$, and $\hat{g}^{-1}$ maps $\mathcal{P}(Z)$ to $\mathcal{P}(Y)$, and $\hat{f}^{-1}$ maps $\mathcal{P}(Y)$ to $\mathcal{P}(X)$.

Let us prove the claim (10.6) of equality between functions. We have just observed that both have the same domain and codomain. We now must show that

$$\forall C \subseteq Y, \widehat{(g \circ f)}^{-1}(C) = (\hat{f}^{-1} \circ \hat{g}^{-1})(C). \tag{10.7}$$

We can transform the left side of (10.7) into the right side more easily than you might think:

$$\begin{aligned}
\widehat{(g \circ f)}^{-1}(C) &:= \{a;\ a \in X, (g \circ f)(a) \in C\} \\
&:= \{a;\ a \in X, g(f(a)) \in C\} \\
&:= \{a;\ a \in X, f(a) \in \hat{g}^{-1}(C)\} \\
&:= \{a;\ a \in X, a \in \hat{f}^{-1}(\hat{g}^{-1}(C))\} \\
&:= (\hat{f}^{-1} \circ \hat{g}^{-1})(C).
\end{aligned}$$

QED

All we used were the definitions of composition and inverse of functions. The technique was to take apart the left side and reassemble it differently by replacing each key term by its definition. Thus, for example, $g(f(a)) \in C$ iff $f(a) \in \hat{g}^{-1}(C)$, by the definition of $\hat{g}^{-1}$. Here we used the definition of $\hat{g}^{-1}$ backward, so to speak. We also used Chapter 2, (19.10) throughout: $\{x; P(x)\}$ is the set of it all x for which $P(x)$ is true.

(10.8) The composition of functions is associative. That is, if domains and codomains are as indicated, namely,

$$f : U \longrightarrow V, \qquad g : V \longrightarrow W, \qquad h : W \longrightarrow X,$$

[†]This use of square brackets is a standard way in mathematics to compress two sentences into one. It means that on first reading you are to ignore what's inside the brackets. That yields, "The composition of injections is injective." It also means that you are then to reread the sentences and are to replace what's just before the brackets with the contents of the brackets. The result is, "The composition of surjections is surjective."

then

$$h \circ (g \circ f) = (h \circ g) \circ f.$$

This function maps U to X, as shown in the diagram (10.9F).

(10.9F)

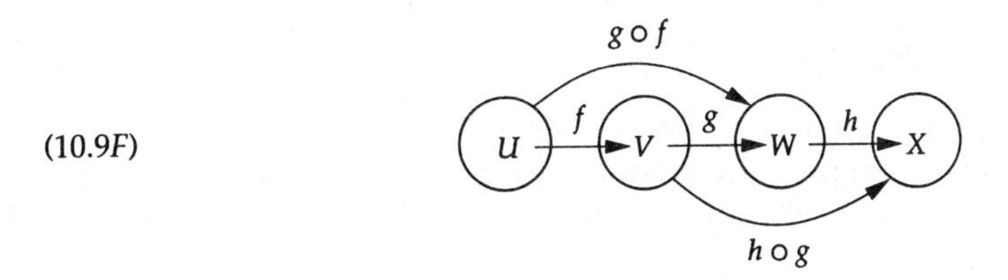

The result (10.8) says we end at the same place if we start with $g \circ f$ and then take h as if we start with f and then take $h \circ g$. We leave the proof to you in the problems.

11. Identity Functions

Let X be any set. We denote the **identity function** on X as

$$i_X : X \longrightarrow X; \tag{11.1}$$

we define it by the rule

$$\forall a \in X, \qquad i_X(a) := a.$$

As a relation, the identity function is nothing but the smallest reflexive relation. As a function it is both injective and surjective.

12. The Inverse of a Bijective Function

The graph of a bijective function is a helpful picture to have in mind.

(12.1F)

f: $1 \to a$, $2 \to b$, $3 \to c$, $4 \to d$, $5 \to e$

Notice that since every point of the codomain $Y := \{a, b, c, d, e\}$ is "hit" by exactly one arrow from the domain $X := \{1, 2, 3, 4, 5\}$, we could read these arrows backward to define a function g from Y to X. That is, $g(a) := 1$, $g(b) := 2$, and so on. Clearly, we could *go backward* this way for any bijective function f.

(12.2) *Comment:* If f were injective but not surjective, we could still define a function g by reading the arrows backward; but g would then be a partial function. For example,

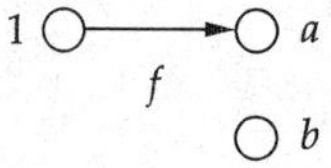

1 ○⟶○ a
f
○ b

Here $g(a) = 1$ but $g(b)$ is not defined.

If f were not injective then there would be ambiguity in trying to define g, so it would not be a function. For instance, in (12.2F), what is $g(a)$? Is it 1? Is it 2?

(12.2F)

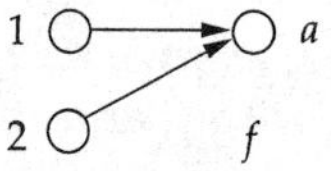

1 ○⟶○ a
2 ○↗ f

We now use the "one-to-one-ness" of bijectivity to define formally an *inverse function* for a bijective function.

DEFINITION (12.3) Let $f : X \longrightarrow Y$ be a function. An **inverse function** for f is any function g such that $g : Y \longrightarrow X$ and the compositions of g and f satisfy $g \circ f = i_X$ and $f \circ g = i_Y$, the identities on the domain and codomain, respectively.

The purpose of this section is to prove that every bijective function has a unique inverse function. Keep in mind the graph (12.1F); it makes our formal procedures transparent.

Theorem (12.4) Let $f : X \longrightarrow Y$ be bijective. Then there is a unique function $g : Y \longrightarrow X$ that is an inverse function for f. Furthermore, g is bijective.

Proof We are given a bijective f. Our first job is to define g. Then we must prove that g has the properties of (12.3).

To define $g : Y \longrightarrow X$ for a given bijective $f : X \longrightarrow Y$, we must set down a rule that assigns to each point of Y a point of X. But bijectivity makes it easy to do this: for each point y of Y, there is a *unique* point x of X such that $f(x) = y$. This x we take to be the value of g at y. Thus

(12.5) $$\forall x \in X, \quad \text{if } f(x) = y, \quad \text{then } g(y) := x.$$

(Here's where we go backward on the arrows.) To emphasize the bijectivity, we are taking x to be a point of $\widehat{f}^{-1}(y)$ and defining that x as $g(y)$. But by injectivity $\widehat{f}^{-1}(y)$ consists of only one point. So the rule is unambiguous. Furthermore the rule can be applied to every point of Y—that's where the surjectivity comes in.

Now we verify that $g \circ f = i_X$ and $f \circ g = i_Y$. Let x be any point of X. Then $(g \circ f)(x) = g(f(x)) = x = i_X(x)$, directly from our definitions of g and i_X. Therefore $g \circ f = i_X$.

For the other equation, let y be any point of Y. Then $(f \circ g)(y) = f(g(y)) = f(x)$, where x is, by definition of g, the unique element of X such that $f(x) = y$. That is, $g(y) = x$. But this equation tells us that $f(g(y)) = f(x) = y = i_Y(y)$. Therefore $f \circ g = i_Y$.

Now we prove that g is unique: Suppose that h is *any* function from Y to X satisfying

$$h \circ f = i_X.$$

(We don't need the other condition $f \circ h = i_Y$.) We'll prove $h = g$. Compose with g on the right to get

$$(h \circ f) \circ g = i_X \circ g = g.$$

But $(h \circ f) \circ g = h \circ (f \circ g)$, as you can easily verify; since $f \circ g = i_Y$, we find that $h = h \circ i_Y = g$.

Notice that when f is bijective, then its inverse g is also bijective, and $g^{-1} = f$.

QED

(12.6) NOTATION. An almost universal notation for this function g is f^{-1}.

Example (12.7) The function exp defined in (9.9) is a bijection from $\mathbf{R}$ to $\mathbf{R}_{>0}$. Its inverse function is $\log_2$. Another way to state Theorem (12.4) for exp is this: Every positive real number has a unique logarithm. For every positive y, $\exp(\log_2(y)) = y$ (that's $f \circ g = i_Y$). For every real x, $\log_2(\exp(x)) = x$ (that's $g \circ f = i_X$). In more familiar terms, if lg denotes $\log_2$, we have $2^{\lg(y)} = y$ and $\lg(2^x) = x$.
The arcsine is another familiar example of an inverse function. We usually choose $X := [-\pi/2, \pi/2]$ and $Y := [-1, 1]$ ($[a, b]$ is defined in (7.12)).

Example (12.8) The sine function $\sin : X \longrightarrow Y$ is a bijection. (How might you prove that claim? One good way is with calculus. The derivative is the cosine, which is positive on X except at the endpoints. But

$$\frac{d}{dx}\sin(x) > 0 \quad \text{for all } x \in X - \{\pm\pi/2\}$$

means $\sin x$ increases as x increases. That is, for all $x_1, x_2 \in X$

$$\text{if } x_1 < x_2 \quad \text{then } \sin(x_1) < \sin(x_2). \tag{12.9}$$

In particular, the sine is injective, for if $\sin(x_1) = \sin(x_2)$ and $x_1 \neq x_2$, then either $x_1 < x_2$ or $x_2 < x_1$. Either way (12.9) fails to hold, so x_1 must equal x_2.

To prove surjectivity we first note that $\sin(-\pi/2) = -1$ and $\sin(\pi/2) = 1$. Then by continuity the sine function takes every value between -1 and 1 as x runs over $X = [-\pi/2, \pi/2]$. Since $Y = [-1, 1]$, the function is surjective.)

There are two traditional notations for the inverse function here: arcsin and $\sin^{-1}$.

(12.10) Notice the relation between f^{-1} and $\widehat{f}^{-1}$ when f is bijective. It is this:

$$\forall y \in Y \qquad \{f^{-1}(y)\} = \widehat{f}^{-1}(\{y\}) =: \widehat{f}^{-1}(y). \tag{12.11}$$

In words, $f^{-1}(y)$ is a point of X; $\widehat{f}^{-1}(y)$ is the singleton set of that point.

13. The Restriction of a Function

Suppose $f : X \longrightarrow Y$ is a function and let $E \subseteq X$ be a subset of the domain of f. We may define a function $h : E \longrightarrow Y$ by the rule

$$\forall a \in E, \qquad h(a) := f(a). \tag{13.1}$$

The function h is called the **restriction** of f **to** E. One notation for h is $f|_E$. Thus

$$f|_E : E \longrightarrow Y$$

and for all $a \in E$,

$$f|_E(a) := f(a).$$

This idea arose because we sometimes want to focus on the action of a function on a subset of its domain. Notice that the range of $f|_E$ is $\widehat{f}(E)$.

(13.2) If the functions f and g are defined as sets of ordered pairs, then g is a restriction of f if and only if $g \subseteq f$.

Notice that $f|_X = f$.

Examples Let $f : \mathbf{R} \longrightarrow \mathbf{R}$ be defined by the rule

(*i*) $f(x) = x^2$. Then the restriction of f to $E = \{-1, 0, 1, 2\}$ is the function $g = f|_E$ defined by the table

x	-1	0	1	2
$g(x)$	1	0	1	4

(*ii*) $f(x) = \sin x$. Then the restriction of f to

$$E_1 = \{\pi n;\ n \in \mathbf{Z}\}$$

is a constant function;

$$f|_{E_1}(a) = 0 \quad \text{for all } a \in E_1.$$

We discussed in (12.7) the restriction of f to $E_2 = \{t;\ -\pi/2 \leq t \leq \pi/2\}$. Many of the familiar functions of calculus—polynomials, $\sin x$, e^x—are definable for complex numbers x. That is, letting $\mathbf{C}$ denote the set of all complex numbers, we may view them as functions $f : \mathbf{C} \longrightarrow \mathbf{C}$. In calculus we study them as real functions, i.e., we restrict them to $\mathbf{R}$ or to a subset of $\mathbf{R}$.

14. A Property of Injective Functions

We can characterize injective functions according to whether an inverse function exists "on one side." More precisely

Proposition Left (14.1) Let X be nonempty. The function $f : X \longrightarrow Y$ is injective if and only if there is a function $j : Y \longrightarrow X$ such that $j \circ f = i_X$.

Proof Let f be injective. Denote the range of f by V. Thus $V \subseteq Y$. Now $f : X \longrightarrow V$ is a bijection because f is injective. Let $j_1 : V \longrightarrow X$ be the inverse function for this bijection. Thus in particular $j_1 \circ f = i_X$. Now define a function j from Y to X by setting

(14.2) $$j(y) := \begin{cases} j_1(y) & \text{for all } y \in V \\ b & \text{for all } y \in Y - V \end{cases}$$

where b is any point of X. It is easy to check that $j \circ f = i_X$, because the arbitrarily chosen b is fenced out. A look at (14.2F) helps.

(14.2F)

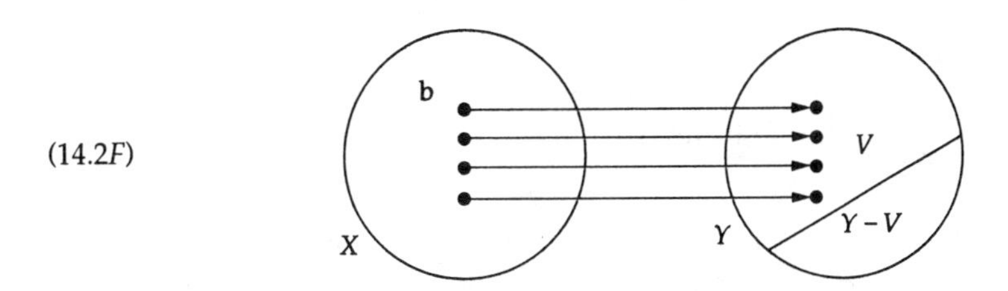

(14.3) Conversely, suppose there is a function j such that $j \circ f = i_X$. We prove that f is injective as follows: Let x and x' be any two points of X such that $f(x) = f(x')$. Apply j to that equation to get

$$j(f(x)) = j(f(x')).$$

But $j \circ f = i_X$. Therefore $x = x'$, and f is injective. QED

Practice (14.4P) Show that $j \circ f = i_X$ in (14.2).

15. A Property of Surjective Functions

There is an analogous result for surjective functions.

Proposition Right (15.1) The function $f : X \longrightarrow Y$ is surjective if and only if there exists a function $h : Y \longrightarrow X$ such that $f \circ h = i_Y$.

Proof Assume first that such a function h exists. We prove that f is surjective by picking any y in Y and finding an x in X such that $f(x) = y$. All we do is observe that

$$(f \circ h)(y) = f(h(y)) \\ = i_Y(y) = y.$$

Therefore $x = h(y)$ is a choice that works.

For the converse, assume f is surjective. That means in particular that for each y in Y, $\widehat{f}^{-1}(y)$ is a nonempty subset of X. For each y in Y, then, choose one point of $\widehat{f}^{-1}(y)$ and let that point be the value $h(y)$ of the function h that you are hereby defining at y.† It follows immediately from the definition of h that $f \circ h = i_Y$. QED

†See the caution under "The Axiom of Choice" in Section (23).

We have named these two propositions to accord with the existence of the left– and right–inverses that they describe. A left-inverse j of a function $f : X \longrightarrow Y$ is a function that satisfies $j \circ f = i_X$ (but not necessarily $f \circ j = i_Y$), and correspondingly for right-inverse. Here is a mnemonic device for remembering which is which: *Surjective* has an r in it, and a surjective function has a right inverse.

Notice also that both Propositions (14.1) and (15.1) are "if and only if" statements, i.e., $p \longleftrightarrow q$ for propositions p and q. Therefore each required two proofs, namely a proof of $p \longrightarrow q$ and a proof of the converse, $q \longrightarrow p$.

Practice (15.2P)

(i) Give an example of an injective function f and find its left-inverse j; do it so that j is not a right-inverse of f, and show that $f \circ j \neq i_Y$.

(ii) Do the analogue for a surjective function g.

(15.3) When f is bijective, then Proposition Left (14.1) gives us a left-inverse j for f, and Proposition Right (15.1) gives us a right-inverse h. But we earlier showed in (12.4) that f has a unique inverse function, something we defined there as a left- and right-inverse g (in our new terminology). In fact, when f is bijective it's easy to see that these three inverses are all equal:

$$f^{-1} = g = h = j.$$

We need prove only that $h = j$; since $g = f^{-1}$ is unique, and $h = j$ implies that h is a two-sided (ambidextrous) inverse, h will then have to equal g.

We now prove $h = j$. We know h and j satisfy $f \circ h = i_Y$ and $j \circ f = i_X$. Apply j on the left of the first of these to get

$$j \circ (f \circ h) = j \circ i_Y = j.$$

Now use associativity of composition of functions (10.8). The result is

$$(j \circ f) \circ h = j.$$

But $j \circ f = i_X$, so this is

$$i_X \circ h = h = j. \qquad \text{QED}$$

Practice (15.4P) Check that all the function-compositions here are *grammatical*, i.e., that in $\alpha \circ \beta$ the codomain of β is a subset of the domain of α. It helps to draw Venn diagrams of the sets involved with arrows for the functions.

(15.5) Propositions Left and Right, and (15.3), combine to make a converse of Theorem (12.4), that a bijective function has a unique inverse. That converse is

(15.6) If $f : X \longrightarrow Y$ has a right-inverse and a left-inverse, then f is bijective, the two inverses are equal, and the inverse is unique.

We may thus combine (12.4) and (15.6) into one "if and only if" result:

Theorem (15.7) The function $f : X \longrightarrow Y$ is bijective if and only if there is a function $g : Y \longrightarrow X$ such that

(15.8) $$f \circ g = i_y \quad \text{and} \quad g \circ f = i_X.$$

If g exists, it is unique. ∎

Example (15.9) Here is a simple example of the use of (15.7). Consider a set A and its power set $X := \mathcal{P}(A)$. Define the complementation function $c : X \longrightarrow X$ by the rule: $\forall B \subseteq A \quad c(B) := \overline{B} = A - B$. We know that $\overline{\overline{B}} = B$ for all B. Thus, for all $B \subseteq A$, $(c \circ c)(B) = B$, since $(c \circ c)B := c(c(B)) = c(\overline{B}) = \overline{\overline{B}} = B$. In other words, $c \circ c = i_X$. Since X is both domain and codomain for c we have satisfied both equations of (15.8). Therefore c is its own two-sided inverse and is a bijection from $\mathcal{P}(A)$ onto $\mathcal{P}(A)$, by (15.7).

(15.10) Now let's carry this kind of analysis further. Let A be a finite set of n elements and let k be an integer satisfying $0 \le k \le n$. Consider the subset S_k of $\mathcal{P}(A)$ consisting of all subsets of A having exactly k elements. Let S_{n-k} be the set of all subsets of A of size $n - k$. Now consider the function c restricted (see (13)) to S_k. Call it c_k.

$$c_k := c|_{S_k}.$$

Obviously c_k maps S_k into S_{n-k}; that is, $\widehat{c}_k(S_k) \subseteq S_{n-k}$. Why? Because for any set $B \in S_k, c_k(B) = \overline{B}$ has size $n - k$ since B has size k. Therefore

$$(\text{range of } c_k) \subseteq S_{n-k}.$$

We may therefore define a codomain for c_k as S_{n-k}. Thus we have

$$c_k : S_k \longrightarrow S_{n-k}.$$

The purpose of this section is to prove that c_k is a bijection.

Proposition (15.11) Complementation is a bijection between S_k and S_{n-k}.

Proof We do the obvious: define $c_{n-k} := c|_{S_{n-k}}$. Then by symmetry of argument with the preceding,

$$\widehat{c}_{n-k}(S_{n-k}) \subseteq S_k$$

since $n - (n - k) = k$. Therefore

$$c_{n-k} : S_{n-k} \longrightarrow S_k.$$

Now what? We simply prove that

$$c_k \circ c_{n-k} = i_{n-k} \qquad (:= \text{the identity on } S_{n-k})$$

and that

$$c_{n-k} \circ c_k = i_k \qquad (:= \text{the identity on } S_k).$$

But these claims are obvious because $\overline{\overline{B}} = B$ for any subset B of A. Remember, $c_k(B)$ is nothing but $\overline{B}$ (*if* $B \in S_k$), and $c_{n-k}(\overline{B})$ is $\overline{\overline{B}} = B$, and similarly on S_{n-k}. Therefore c_{n-k} is a two-sided inverse for c_k, so by (15.7) c_k is a bijection. QED

Maybe this proof belabors the obvious. At least, I think so; but students don't always leap to agree that Proposition (15.11) is obvious. In any case (15.9) and (15.10) are good practice in functions for you.

16. Permutations

(16.1) A bijection f a set X onto itself is called a **permutation** of X. It follows from (12.4) that f^{-1} is also a permutation of X. If g is another permutation of X, then $f \circ g$ is also a permutation of X, from (10.4). In particular, $f \circ f^{-1} = f^{-1} \circ f = i_X$, the identity function on X.

Examples

(*i*) Define $f_1 : \mathbf{Z} \longrightarrow \mathbf{Z}$ by the rule $\forall a \in \mathbf{Z} \quad f_1(a) := a + 1$. Then $f_1 \circ f_1$ maps a to $a + 2$, and f_1^{-1} maps a to $a - 1$.

(*ii*) Let $X_1 = \{1, 2, 3, 4, 5, 6\}$ and define f_2 by the table

x	1	2	3	4	5	6
$f_2(x)$	5	1	6	2	4	3

(*iii*) For $X_3 = \{t;\ t \in \mathbf{R}, 0 \le t \le 1\}$ define $f_3(t) := \sin(\pi t/2)$.

Practice (16.2P) Calculate $g = f_2 \circ f_2$ and then $g \circ g$.

If X is a finite set it is helpful to draw the graph of a permutation viewed as a relation. For Example (*ii*) above the graph is

(16.2F)

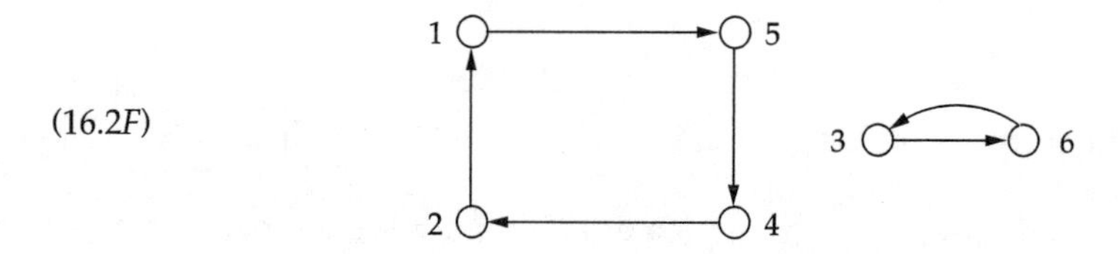

You see there are two *cycles*, one going from 1 to 5 to 4 to 2 and back to 1, the other from 3 to 6 and back to 3.

DEFINITION (16.3) A **cycle** on a finite set S is a permutation of S that, for some ordering $a_1, a_2, \ldots, a_n$ of the points of S, maps (for $1 \le i < n$) a_i to a_{i+1} and a_n to a_1.

In other words, the graph of a cycle looks like a circle:

(16.3F)

The situation of Figure (16.2F) holds generally.

Theorem (16.4) Let X be a finite set and f a permutation on X. Then there is a unique partition of X such that for each cell, the restriction of f to that cell is a cycle.

Proof We shall produce the cycles and the partition at the same time. We first define, recursively, permutations f^i that are powers of f:

$$f^0 := i_X$$
$$f^i := f \circ (f^{i-1}) \quad \text{for } i \in \mathbf{N}$$

In words, $f^0(a) := a$ for all $a \in X, f^1(a) := f(a), f^2(a) = f(f(a))$, and so on.

For each $a \in X$ define $S(a) := \{f^i(a); i = 0, 1, 2, \ldots\}$. Notice that

$$S(a) = \{a, f(a), f(f(a)), \ldots\}; \tag{16.5}$$

in particular $a \in S(a)$. Since $S(a) \subseteq X$, $S(a)$ is finite; therefore, there are exponents $0 \leq i < j$ such that $f^i(a) = f^j(a)$. Define k as the smallest such j. Thus $0 \leq i < k$, and $f^i(a) = f^k(a)$.

It follows that

$$f^0(a), \ldots, f^{k-1}(a)$$

are k different points of $S(a)$, and k is a positive integer. Now graph $S(a)$:

(16.6F)

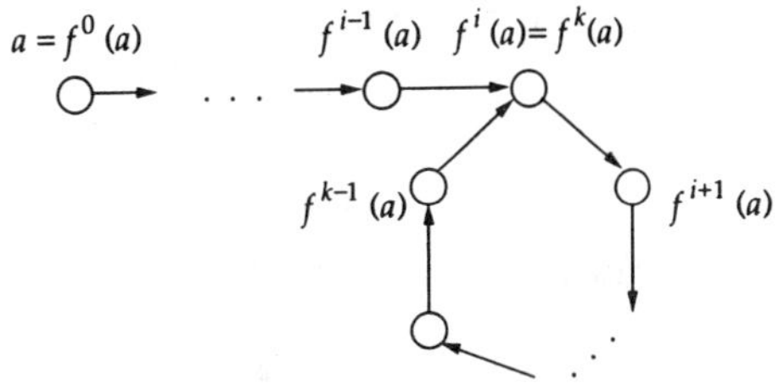

It is now obvious that $S(a) = \{f^0(a), \ldots, f^{k-1}(a)\}$. To show that f restricted to $S(a)$ is itself the cycle we seek, all we need do is show that the tail of (16.6F) does not exist; i.e., that $i = 0$. From (16.6F) you can see that if the tail existed, then i would be positive and two arrows would come into $f^i(a)$, one from $f^{k-1}(a)$ and one from $f^{i-1}(a)$. Since these two points are different, f would not be injective.

Therefore $i = 0$. Thus for every point a of X there is a cycle $S(a)$ of X containing a. Clearly, if b is any point of X, and $b \in S(a)$, then $a \in S(b)$; in fact, $S(b) = S(a)$. Thus

$$\forall a, b \in X \qquad S(a) = S(b) \quad \text{or} \quad S(a) \cap S(b) = \varnothing.$$

Therefore the set $\{ S(a); a \in X \}$ is a partition of X satisfying Theorem (16.4). It is unique because any cell of such a partition must include $S(a)$ for each a in the cell. QED

See problem 16.4 for an alternative to the last part of the foregoing proof.

You may have noticed that we used only the injectivity of f in the preceding proof, but note also that

(16.7) A function mapping a finite set to itself is injective if and only if it is surjective.

From the proof of Theorem (16.4) we see at once that when f is applied over and over again, as

$$a = f^0(a), f(a), f^2(a), \ldots,$$

we get a cycle as defined above, because we come back to the starting point a after a finite number of steps.

Example (16.8) (*i*) Consider $X = \{1, \ldots, 6\}$ and define g_1 by the table

x	1	2	3	4	5	6
$g_1(x)$	2	3	1	5	4	6

The graph of g_1 is

(16.8F)

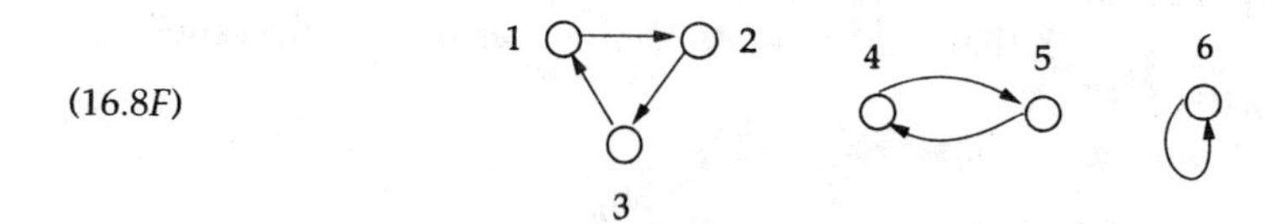

The cycle for $a = 1$ is $1, 2, 3, 1, 2, 3, 1, \ldots$; for $a = 2$ it is $2, 3, 1, 2, 3, 1, 2, \ldots$. For $a = 5$ it is $5, 4, 5, 4, 5, \ldots$ and for $a = 6$ it is $6, 6, 6, 6, \ldots$.

(*ii*) Define g_2 by the table

x	1	2	3	4	5	6
$g_2(x)$	2	3	4	1	6	5

The graph of g_2 is

(16.9F)

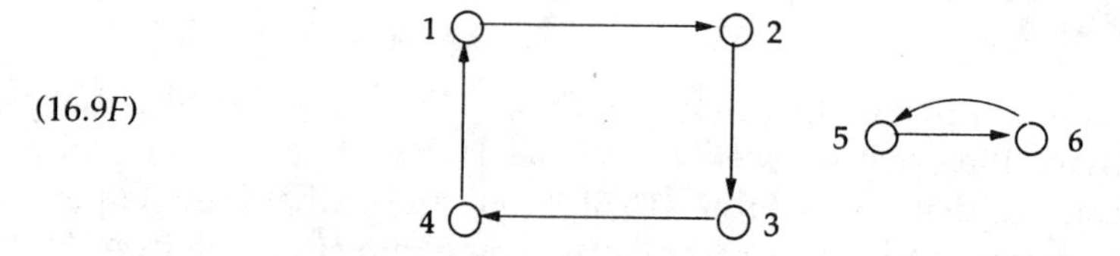

We apply this cycle construction to a point a and get the cycle with a in it. We then pick a point, if any, from X not in the cycle and repeat the process. In this way we exhaust X.

If b is not in the cycle of a, then nothing in the cycle of b can be in the cycle of a, because that would violate the bijectivity of f:

(16.10F)

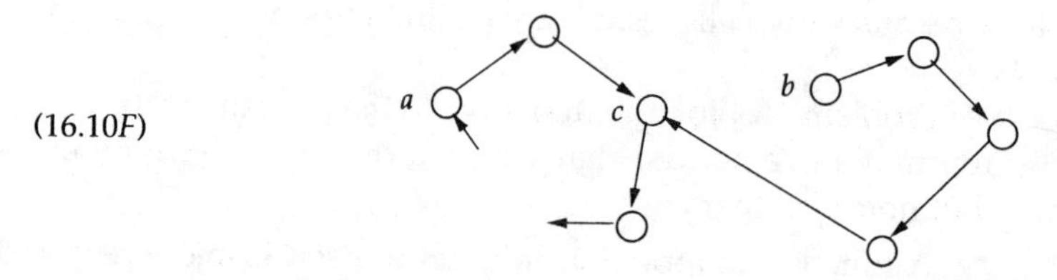

It would cause one point (c in (16.10F)) to have two incoming arrows.

17. The Disjoint-Cycle Notation for Permutations

Theorem (16.4) leads to a compact representation of permutations. Instead of defining the permutation f via a table, we use the cycle structure. The permutations g_1 and g_2 in (16.9) are thus represented as

$$g_1 = (1,2,3)(4,5)(6)$$
$$g_2 = (1,2,3,4)(5,6).$$

We agree that this notation means that g_1 sends 1 to 2, 2 to 3, and 3 to 1; it sends 4 to 5 and 5 to 4; it sends 6 to 6 (it **fixes** 6), and similarly for g_2.

(17.1) In general this **cycle representation** of a permutation f means that *adjacent* symbols $\ldots a, b \ldots$ within the same parentheses signify that $f(a) = b$; it also means that f maps the last symbol before a right parenthesis) to the first symbol after the preceding left parenthesis (.

Fixed points c may be written (c) or may be omitted.

Example (17.2)

(*i*) g_1 in the preceding paragraph could be written (1,2,3)(4,5). The fixed point, 6, is not written. And we re-emphasize: g_1 maps 3 to 1 and 5 to 4.

(*ii*) The table

x	1	2	3	4	5	6
$g_3(x)$	5	3	2	6	4	1

defines a permutation g_3 that can be denoted

$$g_3 = (1,5,4,6)(2,3).$$

(*iii*) Suppose f is the permutation of $A = \{1, \ldots, 20\}$ defined by the rule

$$\forall x \in A \qquad f(x) \in \{3x, 3x - 20, 3x - 40\} \cap A.$$

Then the cycle representation of f is (1,3,9,7)(2,6,18,14)(5,15)(8,4,12,16)(10)(11,13,19,17)(20).

18. Composition of Permutations

The disjoint-cycle representation allows rapid calculation of the composition $f \circ g$ of two permutations f and g on the same set. Before we explain how, it's necessary to point out that when we speak of the composition of *permutations* we think of the argument of the function as being on the left. That is, if f maps a to b we would write $(a)f = b$. (See (10.2).) Then in the composition $f \circ g$, f is applied first and g second:

(18.1F)

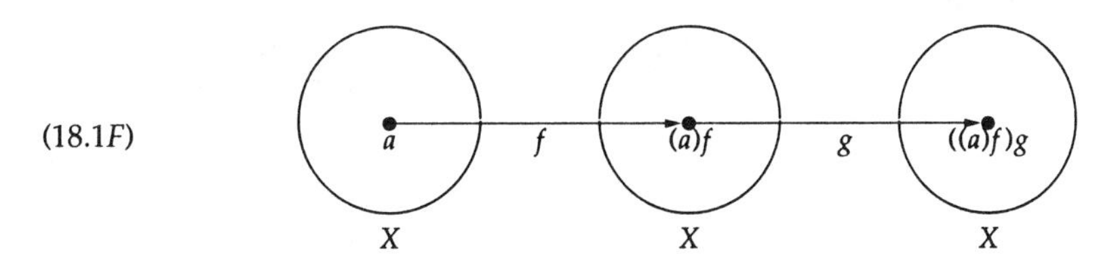

Consider $g_1 \circ g_2$, for example. We write their disjoint-cycle representations side by side, as

(18.2)
$$\begin{array}{cc} g_1 & g_2 \\ (1,2,3)(4,5) & (1,2,3,4)(5,6)\cdot \end{array}$$

We find the disjoint-cycle representation of $g_1 \circ g_2$ as follows:

(18.3) We write "(1", the first symbol of (18.2). Since 1 goes to 2 in the first cycle, we see where 2 goes in its *next* appearance. 2 goes to 3 in the third cycle. Since 3 does not appear *after* the third cycle, we write "(1, 3".

That is, g_1 sends 1 to 2 and g_2 sends 2 to 3, so $g_1 \circ g_2$ sends 1 to 3.

Next we see where $g_1 \circ g_2$ sends 3 by following the same procedure. Scanning from the left in (18.2) we see that 3 goes to 1 under g_1 and 1 goes to 2 under g_2, so 3 goes to 2 under $g_1 \circ g_2$. Thus we write

(18.4)
$$(1,3,2.$$

Continuing the process yields

$$g_1 \circ g_2 = (1,3,2,4,6,5).$$

The general procedure should be apparent from this example.

Practice (18.5P)
(*i*) Find (1,3,5)(1,2,3,4,5,6).

For (*ii*) and (*iii*), define f as (1,2,3,4,5,6) and g as (1,6,3,5,4,2). Then

(*ii*) Calculate $f \circ g$.
(*iii*) Calculate $g \circ f$.

Remarks We omit the commas in writing cycles when there is no ambiguity.

You can see why we must think of arguments of permutations as being on the left. If we had arguments on the right, then we'd have to calculate $f \circ g$ by writing the cycle representation of f to the right of that of g. That's too awkward, so we simply understand that for permutations $f \circ g$ means first do f, then do g. ■

19. Characteristic Functions

Let X be a set and A a subset of X. We define the **characteristic function** f_A **of** A as follows:

(19.1)
$$\forall x \in X, \quad f_A(x) := \begin{cases} 1 & \text{if } x \in A \\ 0 & \text{if } x \notin A. \end{cases}$$

Thus f_A has domain X and codomain $\{0,1\}$:

$$f_A : X \longrightarrow \{0,1\}.$$

Example (19.2) Let $X = \{1,2,3\}$ and consider all the characteristic functions defined on X. In tabular form they are represented as the eight rows of the table (the headers 1, 2, 3 go with each row.)

A	1	2	3
$\varnothing$	0	0	0
$\{3\}$	0	0	1
$\{2\}$	0	1	0
$\{2,3\}$	0	1	1
$\{1\}$	1	0	0
$\{1,3\}$	1	0	1
$\{1,2\}$	1	1	0
X	1	1	1

To any function g mapping X to $\{0,1\}$ there corresponds the subset B_g of X, where

$$B_g = \{x;\ x \in X, g(x) = 1\}.$$

B_g is simply the set of all points of X where g takes the value 1. Thus the set of *all* functions from X to $\{0,1\}$ corresponds naturally to the power set $\mathcal{P}(X)$, the set of all subsets of X. Thus $\mathcal{P}(X)$ is sometimes denoted 2^X, since the set of all functions from X to Y is denoted Y^X, and the "2" is a convenient symbolism for $\{0,1\}$.

Let us define "$\sim$" on $\{0,1\}$, and "$\vee$" and "$\wedge$" on $\{0,1\} \times \{0,1\}$, as follows:

$\sim$	0	1
	1	0

$\vee$	0	1
0	0	1
1	1	1

$\wedge$	0	1
0	0	0
1	0	1

This means, for example, that $0 \vee 1 = 1$ and $1 \wedge 0 = 0$. It is as if 0 were *false*, 1 were *true*, $\sim$ were *not*, $\vee$ were *or*, and $\wedge$ were *and*.

We now extend $\sim$, $\vee$, and $\wedge$ to characteristic functions:

$$(\sim f_A)(x) := \sim (f_A(x)).$$
$$(f_A \vee f_B)(x) := f_A(x) \vee f_B(x).$$
$$(f_A \wedge f_B)(x) := f_A(x) \wedge f_B(x).$$

Then for all subsets A, B of X

(19.3)
$$\begin{aligned} \sim f_A &= f_{\overline{A}} \\ f_A \vee f_B &= f_{A \cup B} \\ f_A \wedge f_B &= f_{A \cap B}, \end{aligned}$$

where $\overline{A} = X - A$ is the complement of A in X. (Compare Chapter 2, (29).)

Practice (19.4P) Prove (19.3).

(19.5) When X is finite, say $X = \{1, \ldots, n\}$, we may order the points of X as $1, 2, \ldots, n$ and then express the function f_A in tabular form as a vector or string in 0,1, as was done in Example (19.2).

If $n = 6$ the subset $\{2, 4, 6\}$ would have characteristic function 010101.

The set of all 2^n strings in 0,1 of length n then corresponds to the set $\mathcal{P}(X)$ of all 2^n subsets of X, an idea we raised in Chapter 3, (11.3). These strings are a means of representing subsets of X in a computer.

(19.6) A relation-matrix is the characteristic function (in tabular form) of its relation. Remember the relation on A is defined as a subset of $A \times A$, which we tabulate economically in square form (rather than linearly). See Chapter 4, where an entry is defined to be 1 in the relation-matrix if the corresponding ordered pair is in the relation. The entry is 0 if the pair is not in the relation.

20. Boolean Functions

Consider a circuit (in a computer, say) that has three *inputs* and one *output*. Each input and output can be either of 2 states; call the states 0 or 1. (They may be two different voltages, for example.) The circuit may be schematically represented as

(20.1*F*)

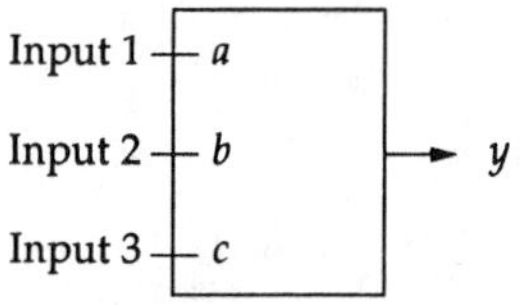

Input 1 — a
Input 2 — b → y
Input 3 — c

where a, b, c are the states of the three inputs, respectively, and y is the state of the output. What is a *circuit*? It is, in this example, a device that, for each of the $8 = 2^3$ possible inputs ($abc = 000, 001, 010, \ldots, 111$) produces an output $y = 0$ or 1. In other words, a circuit is a device that realizes a *function* from the set of eight input vectors or strings to the set $\{0, 1\}$.

Example For instance, $y = a \vee b \vee c$ defines such a function. It takes the value 0 at 000 and the value 1 at the seven other inputs. For another, $y' = a \wedge b$. And another: $y'' = 1$ iff a total of 2 or 3 of a, b, c are 1.

In general we define a **Boolean function of n variables** as a function from the set of all 2^n strings in 0,1 of length n to $\{0, 1\}$. Regarding the strings as vectors allows us to say that the domain of these functions is $\{0, 1\}^n$, by which we mean the Cartesian product of n copies of $\{0, 1\}$, defined in Chapter 1, (19.5). Then we can say

(20.2) A Boolean function of n variables is any function

$$f : \{0, 1\}^n \longrightarrow \{0, 1\}.$$

In particular, $\vee$ and $\wedge$ are Boolean functions of two variables. And $\sim$ is a Boolean function of one variable. A Boolean function is by definition a characteristic function. Since the domain has 2^n points, it has 2^{2^n} subsets (Chapter 3, (11.1)).

Therefore there is a total of 2^{2^n} Boolean functions of n variables. Please notice that 2^{2^n} is 2^e, where $e = 2^n$; it is not $(2^2)^n$, which is merely 2^{2n}.

It may at first seem surprising, but in fact it's easy to see that every Boolean function can be expressed in terms of the n input states and $\sim$, $\bigvee$, and $\bigwedge$.

For example, we can express the foregoing y'' as $(a \bigwedge b) \bigvee (a \bigwedge c) \bigvee (b \bigwedge c)$.

One simple way to express a Boolean function f is to use the *disjunctive normal form* (see Chapter 2, (13)).

First, notice that $a \bigwedge b \bigwedge c$ is 1 on only one string 111; similarly $\sim a \bigwedge b \bigwedge \sim c$ is 1 on only the string 010. In general let the n inputs be $x_1, \ldots, x_n$ and let y_i be one of x_i or $\sim x_i$ for each i. Then $y_1 \bigwedge \cdots \bigwedge y_n$ is 1 on only one input string.

Example If $y_2 = \sim x_2$ but $y_i = x_i$ for $i \neq 2$, then $y_1 \bigwedge \cdots \bigwedge y_n$ is 1 only at $101\ldots1$.

Second, we identify the subset A of all input strings on which the function f is 1. For each $t \in A$ we construct the term (function) $z_t = y_1 \bigwedge \cdots \bigwedge y_n$ that is 1 at t and 0 elsewhere.

Example If $t = 001101 \in A$ the corresponding term is

$$z_t = \sim x_1 \bigwedge \sim x_2 \bigwedge x_3 \bigwedge x_4 \bigwedge \sim x_5 \bigwedge x_6.$$

We then express f as

$$f = \bigvee_{t \in A} z_t, \qquad (20.3)$$

the disjunctive normal form (DNF) for f.

Practice (20.4P) Prove (20.3).

The DNF of y'' preceding is

$$(a \bigwedge b \bigwedge \sim c) \bigvee (a \bigwedge \sim b \bigwedge c) \bigvee (\sim a \bigwedge b \bigwedge c) \bigvee (a \bigwedge b \bigwedge c).$$

The DNF is not necessarily the simplest expression for a function, but it is simple in that it consists of the disjunction of terms of the same form. That fact allows the realization in hardware of the DNF of the Boolean function through a *programmable logic array*, studied in courses in digital logic design.

21. Propositions and Functions

Consider a propositional formula q in n atoms $a_1, \ldots, a_n$. For each assignment of truth-values to the atoms, q takes a truth-value. In other words, q has a truth-table of 2^n lines.

Example For instance, q could be $(a_1 \longrightarrow a_2) \bigvee a_3$ for $n = 3$. You may work out the truth-table, but note that q is false on only the line $a_1 = T, a_2 = F, a_3 = F$. The CNF of q is $\sim a_1 \bigvee a_2 \bigvee a_3$. The truth-table of q may be taken as the definition of a function B_q,

$$B_q : \{T, F\}^n \longrightarrow \{T, F\}. \qquad (21.1)$$

Looks familiar, doesn't it? We've spoken before of identifying 0 with F and 1 with T; if we do so here, we see that B_q is a Boolean function. The point is that "0" and "1" are just abstract symbols denoting two different things, and so are "F" and "T." What matters is that $\{0,1\}$ and $\{F,T\}$ are sets of two elements. So B_q is a Boolean function even if it is expressed in terms of F and T.

There are infinitely many propositional formulas in the n atoms $a_1,\ldots,a_n$ (for example, a_1, $a_1 \vee a_1$, $a_1 \vee a_1 \vee a_1,\ldots$), but each gives rise to one of the 2^{2^n} Boolean functions. If we recall that logical equivalence ($\Leftrightarrow$) of propositional formulas means that the formulas have the same truth-table, we see that $\Leftrightarrow$ is an equivalence relation (Chapter 4) on the set PF of all propositional formulas in $a_1,\ldots,a_n$. There are 2^{2^n} equivalence classes, one for each Boolean function.

22. The Cartesian Product Again

A more abstract definition of Cartesian product, and of *any* collection of sets, can be made with the concept of function. Suppose $\{A_i;\ i \in I\}$ is the collection of sets, with I as *index set*.

Examples Let us consider examples of index set. We might have

$$\{A_1, A_2\} = \{A_i;\ i \in \{1,2\}\},$$

or

$$\{A_1,\ldots,A_n\} = \{A_i;\ i \in \{1,\ldots,n\}\},$$

or

$$\{A_1, A_2,\ldots\} = \{A_i;\ i \in \mathbf{N}\},$$

or even

$$\{A_i;\ i \in \mathbf{R}\}.$$

In general I might be any set.

DEFINITION (22.1) The **Cartesian product** of $\{A_i;\ i \in I\}$ is the set of all functions $f: I \to \bigcup_{i\in I} A_i$ such that

$$\forall i \in I \qquad f(i) \in A_i.$$

NOTATION: $\prod_{i\in I} A_i$

It is necessary to *assume* that the Cartesian product is nonempty in general. See (23) for the reason.

This definition amounts to the same as that in Chapter 1 (for finite I), because, for example, if $I = \{1,2\}$, we can tabulate the functions as

1	2
$f(1)$	$f(2)$

and see immediately that there is a 1–1 correspondence between the set of all functions f in $\prod_{i\in\{1,2\}} A_i$ and the set of all ordered pairs

$$(a_1, a_2)$$

with $a_1 \in A_1$ and $a_2 \in A_2$. The correspondence is

f maps to the pair $(f(1), f(2))$;

The pair (a_1, a_2) maps to the function f defined by the rule $f(1) = a_1$, $f(2) = a_2$.

The new definition allows us to extend the definition to infinite index sets. It no longer requires that there be ordered pairs, triples, or whatever; the function concept finesses that need of the older definition.

23. The Axiom of Choice

It may surprise you to learn that the proof of (15.1), that every surjective function has a right-inverse, glossed over a fundamental difficulty. The trouble arises only when Y is infinite, so it is not so important for computer science. The problem is that it is impossible to prove that the procedure described in the proof—to choose a point from $\widehat{f}^{-1}(y)$ for each y in the *infinite* set Y—can actually be carried out! The word *impossible* here is used precisely: Kurt Gödel proved (in 1938) that if we *assume* that every surjective function has a right-inverse, then we do not introduce any contradictions into mathematics (supposing, that is, that mathematics is not contradictory already). Then Paul Cohen proved (in about 1962) that if we *assume* that some surjective functions do *not* have right-inverses, then again we introduce no contradictions into mathematics (supposing, that is, that mathematics is not contradictory already).

The statement involved here is called the *axiom of choice.*

(23.1) **The Axiom of Choice.** Every surjective function has a right-inverse.

The usual statement of the Axiom of Choice is this: the Cartesian product of nonempty sets is nonempty. Applied to our surjective function, this version says $\prod_{y\in Y} \widehat{f}^{-1}(y)$ is nonempty, which is to say that there is a function $h : Y \longrightarrow X$ such that $h(y) \in \widehat{f}^{-1}(y)$ for all $y \in Y$. That function h is a right-inverse for f.

As an exercise, verify that $f \circ h = i_Y$.

The results of Gödel and Cohen show that the Axiom of Choice is *independent* of the "standard" axioms of mathematics† (those that allow us to form sets, talk of membership, construct the integers, and so on). It has to be assumed as an axiom if one wants Proposition Right to be true for infinite codomains Y.

24. Factoring Functions

It is possible to express any function f as the composition of two functions, one injective (h) and one surjective (π). Since we can think of composition as being like multiplication, we say that we *factor* the function. The result: $f = h \circ \pi$.

†Our treatment in this text is so intuitive that we do not even speak of these axioms except in passing.

We shall now prove this claim. Let $f : X \longrightarrow Y$ be any function. We consider the partition of X into subsets on which f is constant:

(24.1) $$P_f = \{\widehat{f}^{-1}(y);\ y \in \text{range} f\}.$$

The cells of P_f are the subsets $\widehat{f}^{-1}(y)$ of X for all y in the range of f.

Example For the f of the table

x	1	2	3	4	5	6	7
$f(x)$	a	a	a	b	c	c	d

,

the cells are

$$\{1,2,3\} = \widehat{f}^{-1}(a) \qquad \{5,6\} = \widehat{f}^{-1}(c)$$
$$\{4\} = \widehat{f}^{-1}(b) \qquad \{7\} = \widehat{f}^{-1}(d).$$

The function f is not only constant on each of these subsets, but also if x and x' are in different cells of P_f, then $f(x) \neq f(x')$. That statement is true because $x \in \widehat{f}^{-1}(y)$ and $x' \in \widehat{f}^{-1}(y')$ imply $f(x) = y$ and $f(x') = y'$. If the cells are different, then $y \neq y'$.

Thus these cells are the *largest* subsets of X on which f is constant. *Largest* is used here in the sense of inclusion: there is no proper superset of $\widehat{f}^{-1}(y)$ on which f takes the constant value y—from the definition of $\widehat{f}^{-1}(y)$ as the set of *all* $x \in X$ for which $f(x) = y$.

Now define a mapping (function) $\pi : X \longrightarrow P_f$ by the rule

(24.2) $$\pi(b) := \text{ the unique cell of } P_f \text{ having } b \text{ as an element.}$$

This definition of π is unambiguous because P_f is a partition: each point b of X belongs to exactly one cell of P_f. Furthermore, π is surjective, because every cell is nonempty, and any point of a cell is mapped to that cell by π.

Now we define a mapping h from P_f to Y by the rule

(24.3) $$h(\widehat{f}^{-1}(y)) := y$$

for all y in the range of f.

Our definition of the cells of P_f allows only one value of f at all the points of a cell. That value is the value of h at that cell. Therefore h is defined unambiguously.

The function h is injective. We have shown this fact already in remarking that if x and x' are in different cells, then $f(x) \neq f(x')$. That is, h does not take the same value at two different points of its domain P_f. Alternatively, we can cast this proof into exactly the form of the definition of injectivity:

$$\text{If } h(\widehat{f}^{-1}(y)) = h(\widehat{f}^{-1}(y')), \quad \text{then } y = y',$$

since by the definition of h, the left-hand side is y and the other is y'. Therefore $\widehat{f}^{-1}(y) = \widehat{f}^{-1}(y')$, and h is injective.

Now it is a simple matter to see that $f = h \circ \pi$, because for any b in X, $\pi(b) = \widehat{f}^{-1}(y)$, where $y = f(b)$. And $h(\pi(b)) = h(\widehat{f}^{-1}(y)) = y = f(b)$. QED

The factorization we speak of here is simply

(24.4) $$f = h \circ \pi.$$

A picture makes these ideas clearer:

(24.4F)

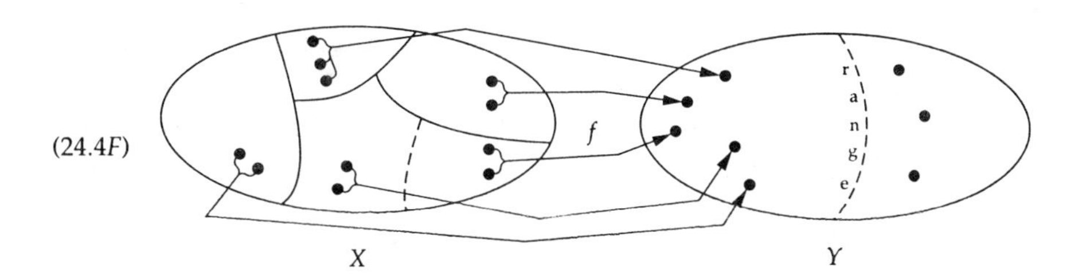

X is shown, partitioned into the cells of P_f. The mapping π maps X onto P_f by sending each point of X to the cell with that point in it. That cell is then mapped by h into the same point of Y that f sends its points to.

Another picture shows π and h:

(24.5F)

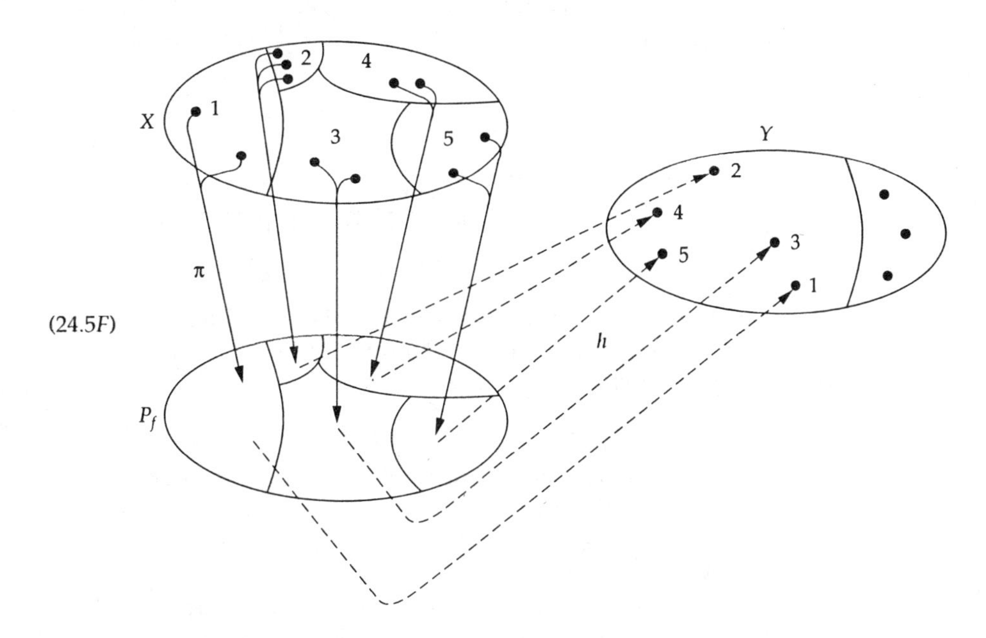

We omit the arrows for f to avoid clutter, but we have numbered the cells and their images correspondingly so that you may check that $h \circ \pi = f$ in the picture as well. That is, if x is any point of cell i, then $f(x) := i$.

Notice that in this pictorial example f is neither injective nor surjective, that we have expressed f as the composition of an injective function and a surjective function, and that, of course, such factorization of f does not make it either injective or surjective.

Example (24.6) An example of factoring functions arises in going from a list to the set it represents. Suppose we have a collection X of lists and a collection Y of sets, such that for each list in X, the set of the elements of that list is in Y. Define the function $f : X \to Y$ as follows:

$$\forall L \in X, \qquad f(L) = S,$$

where S is the set of the elements of the list. For example, if $L = 1,2,3,3,4$, then $f(L) = \{1,2,3,4\}$. The cell of all lists in X representing the same set S is $\widehat{f}^{-1}(S)$. The mapping π from X to the set P_f of cells $\widehat{f}^{-1}(S)$, for S in the range of f, is a surjection. The mapping h from cells to their sets is an injection. Suppose, to be more concrete, that $Y = \mathcal{P}(\{1,2,3,4\})$ and X is the following collection of lists:

$$\begin{aligned} &L_1 : 1,2,3,3,4 \\ &L_2 : 1,3,4,2 \\ &L_3 : 1,2,3,4 \\ &L_4 : 1,2,2,2 \\ &L_5 : 2,1,1 \end{aligned}$$

Then X is partitioned into just two cells, as $P_f = L_1, L_2, L_3 \mid L_4, L_5$. The range of f is $\{\{1,2,3,4\},\{1,2\}\}$. The surjection $\pi : X \to P_f$ is given by the rule

$$\begin{aligned} \pi(L_1) = \pi(L_2) = \pi(L_3) &= \{L_1, L_2, L_3\} \\ \pi(L_4) = \pi(L_5) &= \{L_4, L_5\}. \end{aligned}$$

The injection h is given by the rule

$$\begin{aligned} h(\{L_1, L_2, L_3\}) &= \{1,2,3,4\} \\ h(\{L_4, L_5\}) &= \{1,2\}. \end{aligned}$$

Example (24.7) Consider Boolean functions as in (21). We may define a mapping $\beta : PF \longrightarrow BF$, where BF is the set of all Boolean functions on n variables, and PF is defined as in (21), by the simple rule

$$\beta(q) := B_q, \tag{24.8}$$

where B_q is the Boolean function defined by the truth-table of q (for any $q \in PF$).

Applying the factorization (24.4) to $h = \beta$ would force us to discover logical equivalence between propositional formulas even if we'd never heard of it. Here's how.

The cells of the partition of the domain PF are, by (24.1), the $\beta^{-1}(B_q)$ for all $B_q \in BF$, since β is surjective. They are the largest subsets of PF on which β is constant. They form the partition P_β.

$$P_\beta := \{\beta^{-1}(f);\ f \in BF\}.$$

The equivalence relation E_{P_β} associated to the partition P_β (by the obvious process of Chapter 4, (6.2)) is precisely logical equivalence, $\Leftrightarrow$, as defined in Chapter 2, (7.9).

Practice (24.9P) Express the mappings π and h for this β.

Practice (24.10P) For $n = 2$ find four formulas in PF in different cells from each other and from those of $\sim a$, $a \vee b$, $a \wedge b$, and $a \rightarrow b$.

25. Numbers as Sets

Here is a brief indication, as promised at the end of Chapter 1, of how the positive integers may be defined in terms of sets.

DEFINITION (25.1) The integers 0, 1, 2, 3, ... are defined by

$$0 := \varnothing,$$
$$1 := \{0\} = \{\varnothing\},$$
$$2 := \{0,1\} = \{\varnothing,\{\varnothing\}\},$$
$$3 := \{0,1,2\} = \{\varnothing,\{\varnothing\},\{\varnothing,\{\varnothing\}\}\},$$

And so on.

This "and so on" is the biggest in the book. What happens in formal treatments is that an axiom for sets comes in here to *allow* the definition of an infinite set of "numbers" this way. Such an "and so on" always conceals an induction, we've learned; here it conceals the very axiom that makes induction possible.

Briefly, formal treatments define the *successor* of a set x to be the set

$$x^+ := x \cup \{x\}.$$

Then $0 = \varnothing$ by definition, and

$$0^+ := \varnothing \cup \{\varnothing\} = \{\varnothing\} = 1$$
$$1^+ := 1 \cup \{1\} = \{0,1\} = 2$$
$$2^+ := 2 \cup \{2\} = \{0,1,2\} = 3$$

and so on (again). The successor of a *number* is that number plus 1. The axiom is

> **The Axiom of Infinity**. There is a set containing 0 and containing the successor of each of its elements.

With that harmless assumption, one is formally off to the races.†

Practice (25.2P) Show that what we wrote in Chapter 1, (6.1) is true under definition (25.1), namely $1 \not\subseteq \{1,2\}$.

Practice (25.3P) Suppose you mistakenly used this definition of ordered pair:

$$(a,b) := \{a,\{a,b\}\}.$$

Compare Chapter 1, (17). What would $(0,0)$ turn out to be?

†See Paul R. Halmos (cited in Chapter 1, (20.2)) for more details.

26. A Confession

We sneaked functions into the text before Chapter 5. We had to; doing so was another inevitable start-up glitch. Union and intersection of sets are functions on $\mathcal{P}(V) \times \mathcal{P}(V)$ to $\mathcal{P}(V)$, where V is a universal set. Thus if $(A, B) \in \mathcal{P}(V) \times (\mathcal{P}(V)$, i.e., $A \subseteq V$ and $B \subseteq V$, then

$$\text{union } (A, B) := A \cup B \in \mathcal{P}(V)$$
$$\text{intersection } (A, B) := A \cap B \in \mathcal{P}(V).$$

Complementation c is a permutation of $\mathcal{P}(V)$, as we showed in (15.9):

$$c(A) := \overline{A} := V - A.$$

For any predicate $P(x)$ with domain X there is a function f from X to a set of propositions defined by the rule

$$\forall x \in X \qquad f(x) := P(x). \tag{26.1}$$

Conversely, for any such function f we may reverse the ":=" in (26.1) to define $P(x)$ as $f(x)$, thereby getting a predicate on X.

Quantifiers are functions from predicates to predicates (recall that propositions are predicates in zero variables).

Sequences of propositions are predicates with domain, say, **N**, hence are functions mapping **N** to sets of propositions.

27. Orders of Growth

For real-valued functions with domain **R** or **N** we sometimes want to know how fast the function grows as the argument gets larger and larger. Such questions come up in the analysis of algorithms, where we may find that this or that function measures the amount of time or space used in the execution of the algorithm. A principal benchmark of the performance of an algorithm is the time it requires as the input gets larger and larger. If $f(n)$ is the time it takes for inputs of size n, and $g(n)$ is that for another algorithm, the key comparison of the two algorithms is the ratio $f(n)/g(n)$ for large n. We'll give a brief introduction to the standard language for discussing these questions. We begin with a table of some values of common functions defined on **N**, in which *log* stands for $\log_2$.

(27.1F)

n	$\log(\log n)$	$\log n$	$n \log n$	n^2	n^3	2^n
10	1.73+	3.32+	33.22−	100	1000	1,024
20	2.11+	4.32+	86.44−	400	8000	1,048,576
30	2.29+	4.91−	147.2+	900	27000	1,073,741,824
40	2.41+	5.32+	212.9−	1600	64000	1,099,511,627,776
50	2.50−	5.64+	282.2−	2500	125000	1,125,899,906,842,624

This table gives you some idea of the different rates of growth possible.

Mathematicians and computer scientists have introduced several compact notations to compare the rates of growth of functions. For example, two functions growing at the same rate as each other are called **asymptotic**. The notation is $f \sim g$, or $f(n) \sim g(n)$. The precise definition is

$$f \sim g \quad \text{iff} \quad \frac{f(n)}{g(n)} \longrightarrow 1 \quad \text{as} \quad n \longrightarrow \infty. \tag{27.2}$$

Practice (27.3P) Prove that $\sim$ is an equivalence relation on the set of all functions $f : \mathbf{N} \longrightarrow \mathbf{R}$ such that for some k in $\mathbf{N}$ $f(n)$ is not zero for any $n > k$.

Examples

Define $f_1(n) := n^2 + n + 1$ and $g_1(n) := n^2$. Then $f_1 \sim g_1$, because

$$\frac{f_1(n)}{g_1(n)} = \frac{n^2 + n + 1}{n^2} = 1 + \frac{1}{n} + \frac{1}{n^2}, \tag{27.4}$$

which has a limit equal to 1 as n tends to infinity.

Let $f_2(n) := 2n + \log n$ and let $g_2(n) := 2n + 3$. Then

$$\frac{f_2(n)}{g_2(n)} = \frac{2n + \log n}{2n + 3} = \frac{2 + n^{-1}\log n}{2 + 3/n}. \tag{27.5}$$

Since both numerator and denominator have the same nonzero limit, namely 2, the quotient has limit $2/2 = 1$. Hence $f_2 \sim g_2$. (As you remember from calculus, $\log n/n$ tends to 0 as n tends to infinity—and the limit of a quotient is the quotient of the limits if both the latter exist and the denominator limit is nonzero.)

The factorial is our final example. Let $f_3(n) := n!$ (defined in Chapter 3, (13.1)). We know that $n!$ is the product of the n integers from 1 to n. Thus it grows even faster than 2^n, the product of n 2's, since $n!$ is the product of n integers, most of which are bigger than 2. A famous result called **Stirling's approximation** gives us the precise rate of growth of $n!$ in terms of simpler functions. It's not at all obvious, but

$$n! \sim \left(\frac{n}{e}\right)^n \sqrt{2\pi n}, \tag{27.6}$$

where $e = 2.71828+$ is the base of natural logarithms. I ask that you accept Stirling's approximation without proof. (Among many places where you may find a proof, I recommend William Feller, Vol. I [cited in Chapter 10, (17.1)].)

When you calculate the *ratio* of functions for large n you may substitute g for f if $f \sim g$. Thus to find a simpler asymptotic equivalent for $n^n/n!$, you could use (27.6) to get

$$\frac{n^n}{n!} \sim \frac{n^n}{(n/e)^n \sqrt{2\pi n}} = \frac{e^n}{\sqrt{2\pi n}}.$$

It could be wrong, however, to substitute an asymptotic equivalent in a sum.

Example

Take $h := n^2 + \log n$ and $f := n^2 - 2n + 3$. Then $h - f = 2n + \log n - 3 \sim 2n$. But if we noticed that $f \sim n^2$ and then substituted for f, we'd get, erroneously, $h - f \sim n^2 + \log n - n^2 = \log n$.

You may remember from analytic geometry that in plotting the graph of a rational function like, for example,

$$f(x) := \frac{x^3 + 3x^2 - 1}{2x^3 - x + 7} \tag{27.7}$$

you could determine the behavior of f at infinity by simply looking at the ratio of the terms of highest degree in numerator and denominator. In this case $f(x)$ "behaves like" $x^3/2x^3 = 1/2$ as $x \longrightarrow \infty$; i.e., $\lim_{x\to\infty} f(x) = 1/2$. The terms of degree lower than 3 are irrelevant for the behavior of f as $x \longrightarrow \infty$. Mathematicians have introduced a generalization of the idea that the behavior of a polynomial in x as $x \longrightarrow \infty$ is determined by its term of highest degree. They call it the *big Oh* notation. It's defined as follows: We say of two functions f and g mapping **N** to **R** that

$$\begin{gathered} f = O(g) \quad \text{iff} \quad \exists c > 0 \quad \text{and} \quad \exists k \in \mathbf{N} \\ \text{such that } n > k \text{ implies } |f(n)| \leq c|g(n)|. \end{gathered} \tag{27.8}$$

In a picture, (27.8) says that the graph of $f(n)/g(n)$ lies between the lines $y = c$ and $y = -c$ if $n > k$, as sketched in (27.8F).

(27.8F)

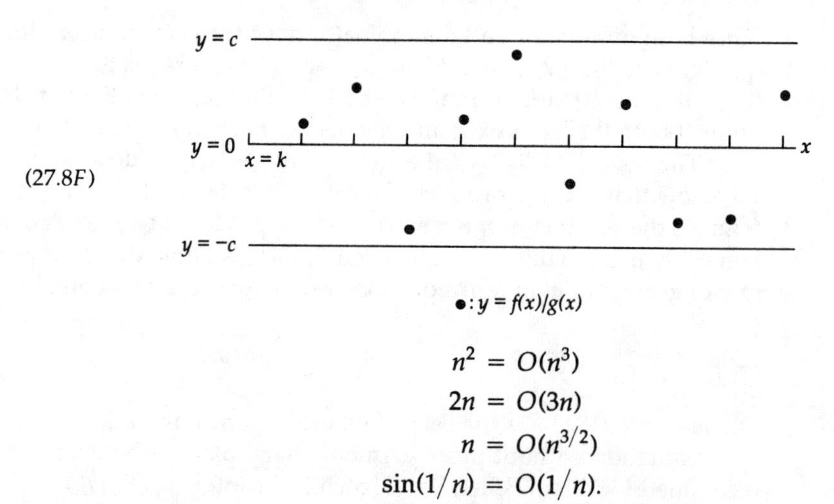

Examples (27.9)

$$\begin{aligned} n^2 &= O(n^3) \\ 2n &= O(3n) \\ n &= O(n^{3/2}) \\ \sin(1/n) &= O(1/n). \end{aligned}$$

The last of these holds because $(\sin x)/x \longrightarrow 1$ as $x \longrightarrow 0$; you know that from calculus. Thus $n \sin(1/n) \longrightarrow 1$ as $n \longrightarrow \infty$. Therefore, from the definition of limits (choose "ϵ" $= 1$), $|n \sin(1/n)| < 2$ for all sufficiently large n. Thus

$$|\sin(1/n)| < 2/n$$

for all large enough n.

In using the big Oh notation we usually don't care what the values of c and k are. It's enough to know they exist.

We also use the big Oh to simplify expressions by replacing something complicated by big Oh of something simple.

Example

In (27.7) the numerator $x^3 + 3x^2 - 1 = x^3 + O(x^2)$. (Big Oh is defined for functions mapping **R** to **R** in the same way as in (27.8).) This is mathematicians' way of saying $x^3 + 3x^2 - 1$ is x^3 plus a function $(3x^2 - 1)$, which "is $O(x^2)$." Since $3x^2 - 1$ is less than some constant c times x^2, and since we care only about very large x,

$$|x^3 + 3x^2 - 1| \leq |x^3| + |3x^2 - 1| \leq |x^3| + c|x^2| \tag{27.10}$$

for all large x. Thus we write, instead of all that in (27.10), $x^3 + 3x^2 - 1 = x^3 + O(x^2)$.

As the examples (27.9) show, the use of big Oh is not confined to polynomials but can be used for any functions where the relationship defined in (27.8) can be established.

Examples

Whenever f and g are asymptotic, we see that $f = O(g)$ and $g = O(f)$, for we may choose the constant c to be any number bigger than 1.

The sum of the first n integers, $f(n) := 1 + 2 + \cdots + n = O(n^2)$, since $f(n) = n(n+1)/2$, as we know from Chapter 3, (3.8). We may take c to be any constant bigger than $1/2$, and $k = 1$ (or any other integer).

The logarithm: $\log n = O(n)$, because $(\log n)/n$ tends to 0 as n tends to ∞. Thus $\log n < cn$ for any constant $c > 0$ for all values of n after a certain value (which depends on c).

If we say $f(n) = O(1)$, we mean that f is *bounded*; that is, for some constant $c > 0, | f(n) | < c$ for all n bigger than some fixed value. The logarithm, for all its slowness of growth, is not bounded; neither is the function d defined via the rule $d(n) :=$ the total number of distinct primes dividing n. But $(n + 1)/(n - 12)$ is bounded in our sense, which speaks only of the behavior of the function as n approaches infinity. So is $\sin n$.

Practice (27.11P) Prove that $d(n) = O(\log_2(n))$. This isn't too hard, but d appears to have a smaller order of growth, which may be tough to find.

To be more precise, we should say that $O(g)$ is really a *set*, namely, the set of all functions $f : \mathbf{N} \longrightarrow \mathbf{R}$ such that there are positive constants c and k for which $|f(n)| \leq c|g(n)|$ for all $n > k$. (The c and k depend on f.) Then instead of writing $f = O(g)$ we should write $f \in O(g)$. The latter is the precise meaning of the more casual and commonly found $f = O(g)$.

Example

We consider a polynomial $f(n) = n^3 + 2n^2 - 7n + 12$. All we care about is its rate of growth as n gets large, so we write

$$f(n) = n^3 + O(n^2).$$

This means that $f(n)$ equals n^3 plus some element of $O(n^2)$. Which element doesn't matter in this context, because

$$\text{for any } h \text{ in } O(n^2), \quad h(n)/n^3 \longrightarrow 0 \text{ as } n \longrightarrow \infty. \tag{27.12}$$

We write $f = n^3 + O(n^2)$ because it is briefer than $f = n^3 + h$, where $h \in O(n^2)$.

Practice (27.13P) Prove (27.12).

The equality sign used in $f = O(g)$, then, is not really equality but is instead a convenient shorthand for the membership relation $\in$. Thus we may not replace $O(g)$ in, say, $f = n^3 + O(g)$ with a function h satisfying $h = O(g)$ (unless it is the one function h from the infinite set $O(g)$ making $f = n^3 + h$ true).

Practice (27.14P) Prove that the relation S defined by $f \; S \; g$ iff $f = O(g)$ is not an equivalence relation.

We take up one more such idea, recently introduced by computer scientists, the *theta* notation. For functions $f, g : \mathbf{N} \longrightarrow \mathbf{R}$ we write

(27.15) $$f = \Theta(g) \quad \text{iff} \quad f = O(g) \quad \text{and} \quad g = O(f).$$

In other words, $f = \Theta(g)$ if and only if there are positive constants c, c', and k such that for all $n > k$

$$c|g(n)| \leq |f(n)| \leq c'|g(n)|.$$

Often we don't care about the value of the constants c and c'; we may wish to know only that the function is of order, say, n^2. Thus if the function were in $\Theta(n^2)$, we'd know that it grows as fast as cn^2 and no faster than $c'n^2$ for some constants c and c'. Again, if f and g are asymptotic, then certainly $f = \Theta(g)$, but they need not be asymptotic for the theta relationship to hold.

Practice (27.16P) Prove that the theta relationship defined by $f \; R \; g$ iff $f = \Theta(g)$ is an equivalence relation on the set of all functions mapping $\mathbf{N}$ to $\mathbf{R}$ that are never 0 after some point.

Example Let $f(n) := (-1)^n n$ and $g(n) := n$. Then $f(n)/g(n) = (-1)^n$ is alternately 1 and -1, so it has no limit. Therefore f is not asymptotic to g, but certainly $f = \Theta(g)$, since, for all $n > 0$, $|f(n)| = |g(n)|$.

28. Postscript

Here are a few comments and additions.

(28.1) *The definition of function:* The official definition combines both of ours. It says a function is a domain X, a codomain Y, and a **map**, where now a map is the set of ordered pairs defined in (5.1). Of course, the map should be a subset of $X \times Y$.

(28.2) *Binary operations:* These are functions. You've used them many times. A **binary operation on the set** A is a function from $A \times A$ to A.

Examples Addition is a binary operation on $\mathbf{N}$. So is multiplication. Subtraction is a binary operation on $\mathbf{Z}$.

If X is a set, union and intersection are binary operations on $\mathcal{P}(X)$.

(28.3) Notice that the implication in (7.1) in the definition of injectivity is the converse of that in (5.2) which assures the *single-valued* property of a function.

29. Further Reading

(29.1) Andrew M. Gleason, cited in Chapter 1, (20.1).

(29.2) Paul R. Halmos, cited in Chapter 1, (20.2).

(29.3) Israel Kleiner, Evolution of the function concept, *College Mathematics Journal*, Vol. 20 (1989), pages 282–300. (Our definitions appear on page 299.)

(29.4) Gregory H. Moore, *Zermelo's Axiom of Choice: Its Origins, Development, and Influence*, Vol. 8 in Studies in the History of Mathematics and Physical Sciences, Springer-Verlag, New York, 1982. Reviewed by Robert Bunn, *American Mathematical Monthly* Vol. 91, No. 10 (December, 1984), pages 654–662.

(29.5) L. E. Sigler, cited in Chapter 1, (20.3).

30. Problems for Chapter 5

Problems for Section 1

1.1 [Recommended] List all functions with domain $\{1,2\}$ and codomain $\{a,b\}$. Assume $a \neq b$.

1.2 List all functions with domain $\{a,b,c\}$ and codomain $\{1,2\}$.

1.3 Let X be nonempty. Find the set of (i) all functions $f : X \to \varnothing$. (ii) all functions $g : \varnothing \longrightarrow X$.

1.4 [Recommended] (Bunch) Define the "maximum" function $m : \mathbf{Z} \times \mathbf{Z} \longrightarrow \mathbf{Z}$ by the rule

$$(\star) \qquad \forall x,y \in \mathbf{Z} \qquad m(x,y) := \begin{cases} x & \text{if } x \geq y \\ y & \text{if } y > x \end{cases}$$

Here is an inductive "proof" that for all x,y in $\mathbf{N}, x = y = m(x,y)$. Find the error.

Define the predicate $P(n)$, with $D_n = \mathbf{N}$, as $P(n)$: $\forall x,y \in \mathbf{N},\ m(x,y) = n \longrightarrow x = y = n$. *Basis step.* $P(1)$ is true, because it says if the maximum of two *positive* integers is 1, then both those integers are 1. *Inductive step.* Let n be a fixed but arbitrary integer ≥ 1, assume $P(n)$, and try to prove $P(n+1)$,

$$(\star) \qquad P(n+1) : \forall x,y \in \mathbf{N} \qquad m(x,y) = n+1 \longrightarrow x = y = n+1.$$

We thus let x and y be any positive integers satisfying $m(x,y) = n+1$. That is, we set up the most general case in which the antecedent of the implication in $(\star)$ is true.

We subtract 1 from each argument in $(\star)$, getting $m(x-1, y-1) = n$. This is obviously correct. But now the inductive assumption tells us that $x - 1 = y - 1$. Thus $x = y$. This proves $\forall n \in \mathbf{N}, P(n) \longrightarrow P(n+1)$.

By the first theorem of induction (Chapter 3, (3.2)), $\forall n \in \mathbf{N}\ P(n)$.

Problems for Section 3

3.1 Find the range of each of the following functions. Each has domain $I := \{t;\ t \in \mathbf{R}, 0 \leq t \leq 1\}$, and codomain $\mathbf{R}$. The rules defining the functions are:

$$f(x) := 2x + 1;$$
$$g(x) := x^2 - x;$$
$$h(x) := 8(x-1)^3 + 6(x-1)^2 + 2.$$

For the following two problems, *prime* is defined in Chapter 3, (5.4). Note that 1 is not a prime.

3.2 Let $X := \{8, 9, 11, 12, 13, 15, 17, 21, 24, 30\}$. Define the function $f : X \longrightarrow \mathbf{N}$ by the rule

For all x in X $\quad f(x) :=$ the smallest prime dividing x.

Find the range of f. Explain.

3.3 For the same domain X as defined in the previous problem define the function $g : X \longrightarrow \mathcal{P}(\mathbf{N})$ by the rule

For all x in X $\quad g(x) := \{p;\ p \text{ prime}, p \text{ divides } x\}$.

Find the range of g. Explain.

3.4[Ans] Define Y as the set of all strings of positive length over $\{0,1\}$. Thus $Y = \{0, 1, 00, 01, 10, 11, 111, \ldots\}$.

Define $s : \mathbf{N} \longrightarrow Y$ by the rule $s(n) :=$ the base 2 representation of n (for all n in $\mathbf{N}$). For example, $s(2) = 10$ and $s(6) = 110$. List $\widehat{s}(\{7, 11, 15\})$.

3.5 Let f be a function $f : X \to Y$. Prove that for any subsets A, B of X

$$\widehat{f}(A \cap B) \subseteq \widehat{f}(A) \cap \widehat{f}(B).$$

Is the reverse inclusion true? If so, prove it; if not, give a counterexample.

3.6 Let A and B be subsets of X such that $A \subseteq B$. Prove that $\widehat{f}(A) \subseteq \widehat{f}(B)$.

Problems for Section 5

5.1 Determine whether the following relations are functions. Explain your answers. Also list the domain and range.

(*i*) $\{(0,1), (1,-3), (3,3), (4,-1), (6,-2), (7,1)\}$.
(*ii*) $\{(0,1), (1,4), (2,0), (3,3), (1,-3), (5,2)\}$.

5.2 Which of the following sets of ordered pairs are functions?

(*i*) $\{(a,b);\ a,b \in \mathbf{N}, a+b < 4\}$
(*ii*) $\{(a,b);\ a,b \in \mathbf{N}, a+b = 6\}$
(*iii*) $\{(a,b);\ a,b \in \mathbf{Z}, a = \pm b\}$
(*iv*) $\{(a,b);\ a,b \in \mathbf{Z}, b = a^2\}$
(*v*) $\{(a,b);\ a,b \in \mathbf{Z}, b = 3\}$.

Use the second definition of function (in Section (5), as a set of ordered pairs) and explain for each one why it is or is not a function.

5.3 [A test question] Express as sets of ordered pairs the functions f_1 and f_2 with domain $D := \{1,2,3,4\}$ and codomain $\mathbf{N}$:

(*i*) For all b in D, $f_1(b) := b^2$.

(*ii*) For all b in D, $f_2(b) :=$ the smallest prime that divides $3b+1$.

Problem for Section 6

6.1 Let A be any set. Consider relations on A and functions from A to A as sets of ordered pairs. Find all equivalence relations on A that are also functions from A to A. Explain.

Problems for Section 7

7.1 Which of these categories—injective; surjective; both; neither—fits the following functions? Explain.

(*i*) $g : \mathbf{R} \to \mathbf{R}$. The rule defining g is $\forall x \in \mathbf{R}$ $g(x) := 3x - 2$.
(*ii*) $h : \mathbf{N} \to \mathbf{N}$. Rule: $\forall y \in \mathbf{N}$ $h(y) := y^2$.
(*iii*) $j : \mathbf{N} \to \mathbf{N}$. Rule: $\forall z \in \mathbf{N}$ $j(z) := 3z$.
(*iv*) $k : \mathbf{Z} \to \mathbf{N}$. Rule: $k := \{(a, 1+a^2);\ a \in \mathbf{Z}\}$.
(*v*) $l : \mathbf{N} - \{1\} \to \mathbf{N}$. Rule: $\forall n \geq 2$ $l(n) :=$ the number of primes dividing n.

7.2 Which of the following functions are: injective, surjective, both, neither?

(*i*) $g : \mathbf{R} \to \mathbf{R}$. Rule is: $\forall x \in \mathbf{R}, g(x) := 2x + 1$.
(*ii*) $h : \mathbf{N} \to \mathbf{N}$. Rule is: $\forall x \in \mathbf{N}, h(x) := x^2 + 2$.

7.3 In problems 3.1, 3.2, 3.3 determine whether each of the functions is injective, surjective, both, or neither. Explain, except that for h in problem 3.1 you may give an informal discussion.

7.4 Prove that f is injective if and only if $\widehat{f}$ is injective; do the same for surjective.

7.5 (*i*) Find a surjection from $\mathbf{N}$ onto $\mathbf{N}$ that is not an injection. Prove your answer correct.
(*ii*) Same problem, but interchange *surjection* and *injection* (and for *onto* read *into*).

7.6 [A test question] Consider the function f defined by the table below. List $\widehat{f}(A)$ for $A := \{1,2,3,4\}$ and for $A := \{1,3,5,7\}$.

x	1	2	3	4	5	6	7	8
$f(x)$	b	a	d	a	e	c	e	c

7.7 Determine whether the function $f : \mathbf{N} \longrightarrow \mathbf{N}$ is surjective, where f is defined by the rule

$$f(x) := \begin{cases} x & \text{if } x \text{ is a multiple of 3} \\ x+1 & \text{if } x-1 \text{ is a multiple of 3} \\ x-1 & \text{if } x+1 \text{ is a multiple of 3} \end{cases}$$

Explain.

7.8 Let $f : X \to Y$ and assume that X and Y are finite sets. Specify the conditions injectivity, surjectivity, and bijectivity in terms of the relation-matrix of f.

Problems for Section 8

8.1 Prove that any subset of six elements of the set $\{1,2,\ldots,9\}$ must have at least two elements with sum equal 10.

8.2 Prove that at any party there must be two people who have shaken hands with the same number of others present.

8.3 Consider the problem of (8.11), the 6 points and 15 lines joining them in pairs. Prove that there must be two one-color triangles. (The color of one triangle need not be the same as that of the other.)

Problems for Section 9

9.1 [A test question] Each social security number (SSN) has nine digits as in $a_1a_2a_3 - a_4a_5 - a_6a_7a_8a_9$. Consider it as a function f,

$$f : \{1,\ldots,9\} \longrightarrow \{0,1,\ldots,9\},$$

defined by the rule

$$\forall i,\ 1 \le i \le 9, f(i) := a_i.$$

(*i*) Is anyone's SSN surjective?
(*ii*) List your own SSN, and calculate

$$\hat{f}^{-1}(\{1,3,5,7,9\})$$
$$\hat{f}^{-1}(\{0,2,4\})$$
$$\hat{f}^{-1}(\{1,3,5\}).$$

If you prefer to keep your SSN private, make up one at random.

9.2 Let $f : X \longrightarrow Y$ and let $x \in X$, $B \subseteq Y$. Prove that if $x \notin \hat{f}^{-1}(B)$, then $f(x) \notin B$.

9.3 Let $f : X \longrightarrow Y$ be any function. For any sets $B_1, B_2 \subseteq Y$ prove that

(*i*) $\hat{f}^{-1}(B_1 \cup B_2) = \hat{f}^{-1}(B_1) \cup \hat{f}^{-1}(B_2)$.
(*ii*) $\hat{f}^{-1}(B_1 \cap B_2) = \hat{f}^{-1}(B_1) \cap \hat{f}^{-1}(B_2)$.

9.4 [Recommended] Let $B \subseteq Y$ and $A := \hat{f}^{-1}(B)$. Prove that

$$\hat{f}(A) \subseteq B$$

and that in some cases equality does not hold.

9.5 Let B and B' be subsets of Y such that $B \subseteq B'$. Prove that $\hat{f}^{-1}(B) \subseteq \hat{f}^{-1}(B')$.

9.6 Let f be a function and let A be a subset of the domain of f. Assume that $\hat{f}^{-1}(\hat{f}(A)) \subseteq A$. Prove that $A = \hat{f}^{-1}(\hat{f}(A))$.

9.7 Why would it be wrong in general to say that for any function $f : X \longrightarrow Y$, the set $\{\hat{f}^{-1}(y);\ y \in Y\}$ is a partition of X?

9.8 Let g be a function with domain the six letters $\{a,\ldots,f\}$ and codomain $\{1,\ldots,6\}$. Suppose all you know about g is a few values of $\hat{g}^{-1}$, namely,

B	$\hat{g}^{-1}(B)$
$\{1,2,3\}$	$\{a,b,c,d\}$
$\{1,3\}$	$\{a,b\}$
$\{1,2,5\}$	$\{a,c,d,e,f\}$
$\{1,3,6\}$	$\{a,b\}$

Can you determine g? If so, explain how. If not, how would you prove that you couldn't?

9.9 Let A, X, Y be any sets such that $A \subseteq Y$, and let $f : X \longrightarrow Y$ be a function.

(*i*) Prove that $X = \hat{f}^{-1}(A) \cup \hat{f}^{-1}(Y - A)$.
(*ii*) Under what conditions is $\{\hat{f}^{-1}(A), \hat{f}^{-1}(Y - A)\}$ a partition of X?

9.10 Prove statement (9.5).

9.11 Let $f : X \longrightarrow Y$ be a function. Define a relation R on X as follows: for all $a, b \in X$

$$a\ R\ b \quad \text{iff} \quad f(a) = f(b).$$

In the notation of Chapter 4, (6), prove that P_R is the partition defined in (9.5). (Compare the previous problem.)

9.12 Let $f : X \longrightarrow Y$ and suppose $\{B_i;\ i \in I\}$ is a partition of the range of f.

(*i*) Prove that $\{\hat{f}^{-1}(B_i);\ i \in I\}$ is a partition of X.
(*ii*) If $\{A_j;\ j \in J\}$ is a partition of X, is $\{\hat{f}(A_j);\ j \in J\}$ a partition of $\hat{f}(X)$? Explain.

9.13 Let $f : X \longrightarrow Y$ be a function.

(*i*) Prove that if f is injective, and A is any subset of X, then $\hat{f}^{-1}(\hat{f}(A)) = A$. Explain where you use the injectivity.
(*ii*) Prove that if f is surjective and B is any subset of Y, then $\hat{f}(\hat{f}^{-1}(B)) = B$. Explain where you use the surjectivity.

9.14 Prove that f is injective if and only if for each y in the range of $f, \widehat{f}^{-1}(y)$ is a singleton.

9.15 Let $f : X \longrightarrow Y$ be the sine function, with $X = \{x;\ 0 \le x \le 2\pi, x \in \mathbf{R}\}$ and $Y = \mathbf{R}$, the set of all real numbers. For each of the following choices of subsets B_i of Y, find the set $\widehat{f}^{-1}(B_i)$. You may use a sketch to justify your answers.

$B_1 = \{1\}$
$B_2 =$ the set of all positive real numbers
$B_3 = \{x;\ x \in \mathbf{R},\ 0 \le x \le 1/2\}$
$B_4 = \{2\}$
$B_5 = \{0, \pi/2, \pi, 3\pi/2, 2\pi\}$.

9.16 Let $f : X \longrightarrow Y$ be any function. Let A be any subset of X, B any subset of Y. Prove or disprove: $\widehat{f}^{-1}(\widehat{f}(A)) = A$, and $\widehat{f}(\widehat{f}^{-1}(B)) = B$.

9.17 Let $A \subseteq X$. Prove or disprove:

$$(\widehat{f}^{-1} \circ \widehat{f})(A) = A$$
$$(\widehat{f} \circ (\widehat{f}^{-1} \circ \widehat{f}))(A) = \widehat{f}(A).$$

9.18[Ans] Let $B \subseteq Y$. Prove or disprove:

$$(\widehat{f} \circ \widehat{f}^{-1})(B) = B$$
$$(\widehat{f}^{-1} \circ (\widehat{f} \circ \widehat{f}^{-1}))(B) = \widehat{f}^{-1}(B).$$

***9.19** (*i*) Prove that f is injective if and only if $\widehat{f}^{-1}$ is surjective.
(*ii*) Prove that f is surjective if and only if $\widehat{f}^{-1}$ is injective.

9.20 Let A be a set and P a partition of A. Define a function $\pi : A \rightarrow \mathcal{P}(A)$ by the rule: for each $x \in A$, $\pi(x)$ is the cell of P containing x.

(*i*)[Ans] Prove that the range of π is P.
(*ii*) Under what conditions is π injective? surjective? bijective?
(*iii*) What is $\widehat{\pi}^{-1}(A)$? Let a be a point of A. What is $\widehat{\pi}^{-1}\{a\}$?

Problems for Section 10

10.1 Let the functions $f : X \longrightarrow X$ and $g : X \longrightarrow X$, where $X = \{1,2,3,4,5\}$, be defined by the table

x	1	2	3	4	5
$f(x)$	1	3	2	5	4
$g(x)$	5	1	3	1	5

Calculate $f \circ g$ and $g \circ f$.

10.2 Define the functions f and g by the ordered-pair method using tables, as follows:

x	1	2	3	4	5	6	7	8	9	10
$f(x)$	3	1	4	1	5	9	6	2	1	7
$g(x)$	7	8	9	10	11	12	11	10	9	8

For each of the four compositions $f \circ g, g \circ f, f \circ f, g \circ g$ give the table of values if the composition is defined. If it is not defined, explain why.

10.3 Let $f : X \longrightarrow X$, where $X = \{1,2,3,4\}$, be defined by the table

b	1	2	3	4
$f(b)$	2	3	3	1

What is the table for $f \circ f \circ f$?

10.4 Calculate $f \circ f$, where f is the function $g \circ h$, and g and h are defined by the tables

x	1	2	3	4	5	6
$g(x)$	4	2	3	6	5	1
$h(x)$	2	3	4	5	1	6

Enter $f \circ f$ as another row of the table.

10.5 Let f and g be injective functions, $f : X \longrightarrow Y$ and $g : W \longrightarrow X$, for sets W, X, Y. Prove that the composition $f \circ g$ is injective. (First verify that composing f and g makes sense.)

10.6 Let f and g be functions as in the previous problem. Prove two things:

(*i*) If $f \circ g$ is a surjection, then f is a surjection; and
(*ii*) If $f \circ g$ is an injection, then g is an injection.

10.7 Let f be the function on the set $\{1,2,3,4\}$ to itself defined by the table

x	1	2	3	4
$f(x)$	2	3	4	4

Let $g = f \circ f$. What is $\widehat{g}^{-1}(\{3,4\})$?

10.8 Prove Proposition (10.4) for surjectivity.

#10.9 Prove the associativity of function-composition: Assuming that domains and codomains are compatible, prove that

$$(f \circ g) \circ h = f \circ (g \circ h).$$

10.10 [A test question] Let D be the set $D = \{2,4,6,8,10,12,14\}$. Define $f : D \longrightarrow \mathcal{P}(\mathbf{N})$ by the rule

$$\forall n \in D, \quad f(n) = \{p;\ p \text{ prime}, p \mid n\}.$$

Define the partial function $g : \mathcal{P}(\mathbf{N}) \longrightarrow \mathbf{N} \cup \{0\}$ by the rule

if A is a finite subset of $\mathbf{N}$ then $g(A) := \sum_{x \in A} x$.

(*i*) Is f injective? Explain.
(*ii*) List the elements of the range of the composition $g \circ f$. Show your calculations.

10.11 [A test question] Give an example of two functions $f : D \longrightarrow Y$ and $g : Y \longrightarrow W$ such that D, Y, and W are finite sets and $g \circ f$ is bijective, but neither f nor g is bijective.

Problems for Section 16

16.1 If V is any set, prove that complementation is a permutation on $\mathcal{P}(V)$.

16.2 Let f be a permutation of a finite set X. For each $a \in X$ define $S(a) := \{f^i(a);\ i = 0,1,2,\dots\}$. Define the relation R on X by the rule: for all $a, b \in X$, aRb iff $b \in S(a)$. Prove that R is an equivalence relation on X.

[This problem allows an explanation of the apparent command "define," found here twice and in other settings from time to time. Some students think they are supposed to go somewhere and define something when they see this usage. Not so. *Define* is simply short for *Let us define*.]

16.3 [Recommended] Prove (16.7). (This problem also fits section (7).)

*16.4 See the proof of Theorem (16.4). After (16.5) i and k are defined as follows: k is the least positive integer such that for some i satisfying $0 \le i < k$, $f^i(a) = f^k(a)$. Shortcut the rest of the proof by using the inverse g of f guaranteed by (12.4).

Problems for Section 17

17.1 Define the permutations f and g by the tables here.

x	1	2	3	4	5	6	7	8	9	10
$(x)f$	7	2	4	1	9	5	3	10	6	8
$(x)g$	7	2	1	5	9	10	3	6	4	8

(*i*) Write f and g in disjoint-cycle form.
(*ii*) Since f and g are permutations of $\{1,\dots,10\}$ they have inverses f^{-1} and g^{-1}. Write disjoint-cycle versions f^{-1} and g^{-1}.

17.2 Suppose a permutation g is a cycle on its domain of n points. How many ways are there to write g in the disjoint-cycle notation of Section 17?

Problems for Section 18

18.1 Suppose f is the permutation (1,2,3,4)(5,6,7)(8,9) of the set $X = \{1,2,3,\dots,10\}$ and g is the permutation $g = (1,4)(2,7)(3,10)(5,9)(6,8)$ of X. Calculate $f \circ g$ and $g \circ f$. Also, what is $\widehat{f}^{-1}(\{1,4,9,10\})$?

*18.2[Ans] Let f and g be permutations on the same finite set X. Prove that whenever the disjoint-cycle representation of g has $\dots(\dots a, b, \dots)\dots$, then the disjoint-cycle representation of $f^{-1} \circ g \circ f$ has

$$\dots(\dots (a)f, (b)f, \dots)\dots.$$

We write the arguments of permutations on the left, as in Section 18. [In other words, to get the representation of $f^{-1} \circ g \circ f$, simply replace each point in the disjoint-cycle representation of g by its image under f. It follows that $f^{-1} \circ g \circ f$ has the same *shape* as g, in that for every cycle of g there corresponds a cycle of $f^{-1} \circ g \circ f$ of the same length.]

#18.3 A *transposition* is a permutation that interchanges two elements and fixes others.

(*i*) Calculate (12)(13)(14).
(*ii*) Prove that any permutation on a finite set is a product (i.e., composition) of transpositions.

#18.4 Suppose there are n people lined up in some order. Show that you can achieve any other ordering of them by successively having some two adjacent people trade places.

Problems for Section 20

20.1 How many Boolean functions of n variables take the value 1 at exactly an even number of points of the domain? Explain.

Problems for Section 24

24.1 Let $X = \{ n;\ n \in \mathbf{Z}, -5 \leq n \leq 10 \}$. Define a function $f : X \to \mathbf{Z}$ by the rule

$$\forall n \in X \qquad f(n) := n^2.$$

List the cells of the partition of X induced by f; tabulate the functions π and h such that $f = h \circ \pi$.

*24.2 [A test question] Define $X := \{1,2,3,\ldots,8\}$. Also define the partition P of X:

$$P := 1,2,3 \mid 4,5 \mid 6 \mid 7,8.$$

Define the function $h : X \longrightarrow P$ by the rule

$$h(x) := \text{the cell of } P \text{ with } x \text{ in that cell.}$$

(*i*) Prove that $\widehat{h}^{-1}(h(x)) = h(x)$ for all x in X.
(*ii*) State and prove a general result including that of (*i*) for any set X and any partition P of X.

Problems for Section 27

27.1 Consider the function $f_k : \mathbf{N} \longrightarrow \mathbf{R}$ defined as $f_k(n) := (n^2 + 3n + 1)/(2n^k - 3n + 7)$. Find a function $g_k(n) := (\text{constant}) \times n^a$ for some a in $\mathbf{Z}$ such that $f_k \sim g_k$ for $k = 1$. Then—solve the same problem for $k = 2$. Then—same for $k = 3$. Explain.

27.2 Find a simple asymptotic equivalent of $(2n)!/2^{2n}(n!)^2$. Explain.

27.3 Define $f(n) := n^3 + 2n^2 + 3n + 1$. True or false:

(*i*) $f(n) = n^3 + O(n^3)$?
(*ii*) $f(n) = n^3 + O(n^2)$?
(*iii*) $f(n) = n^2 + O(n^3)$?
(*iv*) $f(n) = n^2 + O(n^2)$?
(*v*) $f(n) = 1 + O(n)$?

Explain your answers to parts (*ii*) and (*iii*).

27.4 Define $g(n) := 3n^3 - 4n^2 + 1$. Are the following assertions true? False? Which? Explain each one individually.

(*i*) $g(n) \sim 9n^2$
(*ii*) $g(n) = 3n^3 + O(n^2)$
(*iii*) $g(n) = -4n^2 + O(1)$
(*iv*) $g(n) = O(n^4)$
(*v*) $g(n) = \Theta(n^3)$.

27.5 For each function f below, find the function $g(x) := x^a$ with the smallest *integral* a such that $f = O(g)$. [Note: *Integral* is the adjectival form of *integer*.]

$$f_1(x) := 7x^3 + 2x - 11$$
$$f_2(x) := -2x^4 \log x + x^3 \log x - 11$$
$$f_3(x) := 5x^{3/2} + 2x^{1/2} - x^{-1/2}$$

Justify your answer for f_2 or f_3.

27.6 For each function in the previous problem, find a simpler function h such that $f = \Theta(h)$. Justify your answer for f_3.

27.7 Prove that if $f = O(g)$ $[\Theta(g)]$, then for any real $c > 0$, $f = O(cg)$ $[\Theta(cg)]$.

27.8 Let A be the set of all functions $f : \mathbf{N} \longrightarrow \mathbf{R}$ such that f takes the value 0 only a finite number of times. Define a relation R on A as follows: $(f,g) \in R$ iff $f = O(g)$. Prove that R is reflexive and transitive but not symmetric.

31. Answers to Practice Problems

(1.4P) $A := \{1,2,3,4,5\}$ and $P := 1,2,3 \mid 4,5$. $C_1 := \{1,2,3\}$ and $C_2 := \{4,5\}$. Then the table for π is

x	1	2	3	4	5
π (x)	C_1	C_1	C_1	C_2	C_2

(5.9P) $\widehat{g}(\{V,A,G,U,E\}) = \{8,2,3,4\}$, since $g(V) = 8 = g(U), g(A) = 2$, and so on. My favorite word is *dinner*. $\widehat{g}(\{D,I,N,E,R\}) = \{3,4,6,7\}$.

(7.4P) $g(x) := x^2 + x + 1$ and $g : \mathbf{Z} \longrightarrow \mathbf{Z}$. We seek a and b in $\mathbf{Z}$ such that $a \neq b$ but $g(a) = g(b)$. We see from (7.3.1) that $a \neq b$ and $g(a) = g(b)$ imply $a + b = -1$. So try that. Take $a = 0$ and $b = -1$. We get $g(0) = 1$ and $g(-1) = (-1)^2 - 1 + 1 = 1$. One counterexample is all we need.

(7.10P) We did not specify the codomain for g in (5.7). We don't need a codomain to tell whether functions are injective. This one is not injective, because $g(A) = g(B)$. As we said, range $g = \{2,\ldots,9\}$. g is surjective iff we define the codomain to be $\{2,\ldots,9\}$.

(8.3P) A has exactly k points, and B exactly n. We are to explain why it is perfectly obvious that there is no injection from A to B if $k > n$. The answer is in (7.7): The domain of an injective function has the same number of points as the range. Here the domain has k points; but the range has at most n points, since it's a subset of B. So the range has fewer points than the domain. Therefore the function is not injective.

This argument is an example of the contrapositive at work. In sketch form, (7.7) says,

If injective, then size of domain equals size of range.

(8.6P) Problem 37 of Chapter 3 says that the total number of subsets of exactly two elements from a set of n elements is $n(n-1)/2$. There is a 1–1 correspondence between the set of such subsets of $X := \{1,\ldots,n\}$ and the set of ordered pairs (i,j) from X with $i < j$. (Each subset consists of two unequal points; we put the smaller as the first coordinate.)

(8.8P) The statement is wrong. A world with just one tree having no leaves is a counterexample. Maybe the originator of the statement forgot that zero is a counting number.

(8.16P) We use the results in the text. In the first case, when $\max(n_1,\ldots,n_6) = 5$, we have five triangles, as mentioned after (8.14F). In the second case $\max(n_1,\ldots,n_6) = 4$. We saw from (8.15F) that there must be at least two triangles by showing there are at least two lines (say L_1, L_2) with both endpoints in $\{x_1,\ldots,x_4\}$. Now we go further by dividing this case into subcases. There are four lines to place somehow in (8.15F)—call them $M_1,\ldots,M_4$—aside from L_1, L_2, and the four shown there.

Subcase 1. L_1 and L_2 are the only lines with both endpoints in $\{x_1,\ldots,x_4\}$: Then each of $M_1,\ldots,M_4$ has x_6 as one endpoint, and $M_1,\ldots,M_4$ must have $x_1,\ldots,x_4$ as their other endpoints. In this subcase there are four triangles (one example is in (31.1F)),

(31.1F)
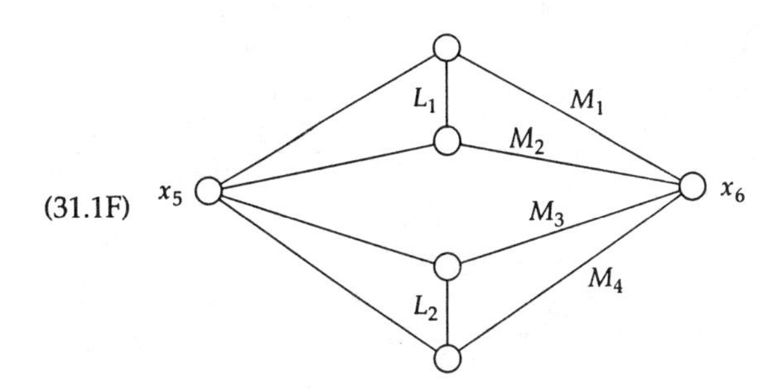

which we may specify as (L_1, x_5), (L_1, x_6), (L_2, x_5), and (L_2, x_6), writing the base and the third vertex.

Subcase 2. L_1 and L_2 are not the only lines with both endpoints in $\{x_1,\ldots,x_4\}$: Say that M_1 is another such line. Then there are the three triangles (L_1, x_5), (L_2, x_5), and (M_1, x_5). QED

For sport, prove that there need not be four triangles.

(9.7P) We have truth-tables of four lines. There are four boxes to be filled. The filling corresponds to a string of length four over $\{T, F\}$. From Chapter 3, (11.3ff), we know that the total number of such strings is $2^4 = 16$.

In the general case there are n atoms. Hence there are 2^n lines in the truth-table, because each line of the table corresponds to a string of length n over $\{T, F\}$, indicating the truth-values of the n atoms. Now there are 2^n boxes to be filled with T or F, one box for each line of the table. By the same reasoning there are 2^{2^n} strings over $\{T, F\}$ of length 2^n.

$$\text{For } n = 5, \qquad 2^{2^n} = 2^{32} > 4{,}000{,}000{,}000.$$

(14.4P) Let x be any point of X. We are to show that $(j \circ f)(x) = i_X(x) = x$.

$(j \circ f)(x) = j(f(x))$	definition of composition
$f(x) \in V$	definition of V as range f
$j(f(x)) := j_1(f(x))$	from (14.2)
$j_1(f(x)) = (j_1 \circ f)(x)$	definition of composition
$= i_X(x)$	definition of j_1
$= x$	definition of i_X

(15.2P) All we need is to make sure our examples are not bijective.

(i) Let us consider the injective function
$$f_1 : \{1\} \longrightarrow \{a, b\}$$
defined by the rule $f(1) := a$. The left-inverse has to be
$$j : \{a, b\} \longrightarrow \{1\}$$
with $j(a) := j(b) := 1$; that is the only function from $\{a, b\}$ to $\{1\}$. Indeed it satisfies
$$j \circ f_1 = i_{\{1\}},$$
because $j \circ f_1(1) = j(f_1(1)) = j(a) = 1$. But $(f_1 \circ j)(b) = f_1(j(b)) = f_1(1) = a$, so j is not a right-inverse of f_1.

(ii) For surjectivity we may read part (i) backward, taking j as our surjective function and f_1—as we just showed—as its right-inverse.

(15.4P) We have $f : X \longrightarrow Y$, and the left-inverse $j : Y \longrightarrow X$, and the right-inverse $h : Y \longrightarrow X$.

(31.2F)

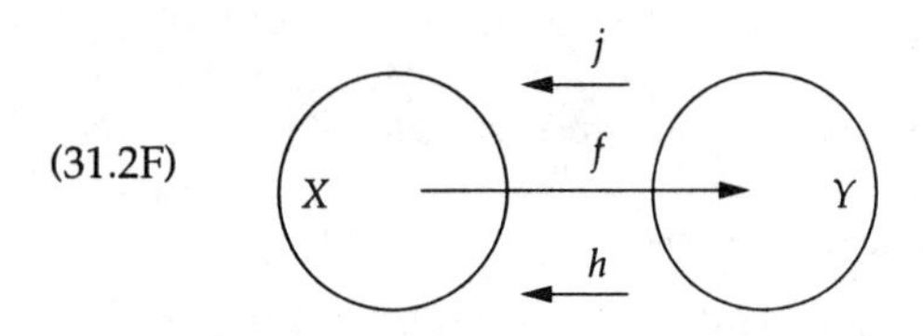

What compositions did we write? The first was $j \circ (f \circ h)$. The composition $f \circ h$ maps Y to X via h and then X to Y via f. So $f \circ h : Y \longrightarrow Y$.

Since j maps Y to X, $j \circ (f \circ h)$ makes sense, because the codomain of $f \circ h = Y =$ domain of j. Next we applied associativity to $j \circ (f \circ h)$. We checked in (10.8) that that's OK.

Finally we wrote $i_X \circ h$. Since $h : Y \longrightarrow X$, codomain $h =$ domain i_X; again the composition makes sense.

(16.2P) To calculate $g := f_2 \circ f_2$ we may add a new line to the table defining f_2. We find $g(1) = 4$, for example, since $g(1) = f_2(f_2(1)) = f_2(5)$; we look up $f_2(5) = 4$ in the table and write 4 under the 5 in line 2.

x	1	2	3	4	5	6
$f_2(x)$	5	1	6	2	4	3
$f_2 \circ f_2(x)$	4	5	3	1	2	6

To find $g \circ g$ we ignore line 2 of the table and repeat the above process for lines 1 and 3. The entries in the $g \circ g$ line are then $g(g(1)) = g(4) = 1$; $g(g(2)) = g(5) = 2$; and so on. $g \circ g$ is the identity. (That means g is its own inverse.)

(18.5P) (*i*) $(1,3,5)\,(1,2,3,4,5,6) = (1,4,5,2,3,6)$.

(*ii*) $f \circ g = (1)\,(2,5,3)\,(4)\,(6)$. You may also write it as $(2,5,3)$.

(*iii*) $g \circ f = (1)\,(2)\,(3,6,4)\,(5) = (3,6,4)$.

(19.4P) $\sim f_A$ is the function mapping X to $\{0,1\}$ defined by the rule $\forall x \in X$,

$$(\sim f_A)(x) := \sim (f_A(x)) := \begin{cases} 0 & \text{if } f_A(x) = 1 \\ 1 & \text{if } f_A(x) = 0. \end{cases}$$

Therefore $\sim f_A$ takes the value 1 on precisely the points of X outside of A, i.e., on $\overline{A}$. It is, therefore, the characteristic function of $\overline{A}$, $f_{\overline{A}}$.

The rule $\forall x \in X$ defines the function $f_A \bigvee f_B$.

$$(f_A \bigvee f_B)(x) := f_A(x) \bigvee f_B(x) :=$$

$$\begin{cases} 1 & \text{if } f_A(x) = 1 \text{ or } f_B(x) = 1 \\ 0 & \text{if } f_A(x) = f_B(x) = 0. \end{cases}$$

Thus $f_A \bigvee f_B$ takes the value 1 precisely on the points of $A \cup B$. And so forth.

(20.4P) We've just about proved this statement in the text. The function f takes the value 1 at exactly the strings in A. For each such string $t = t_1 \ldots t_n \in A$ we construct the Boolean function

$$z_t := {}^{t_1}x_1 \bigwedge {}^{t_2}x_2 \bigwedge \cdots \bigwedge {}^{t_n}x_n,$$

where we define ${}^0x := \sim x$, and ${}^1x := x$ for any variable x. It's clear now that z_t takes the value 1 on just one string, namely, $t = t_1 \ldots t_n$. Why? Because

$${}^0x := \sim x = \begin{cases} 1 & \text{if } x = 0 \\ 0 & \text{if } x = 1 \end{cases}$$

and similarly ${}^1x = 1$ iff $x = 1$. So even a single variable $x_i \neq t_i$ makes z_t take the value 0.

Therefore the *or* of all the z_t's over all t in A takes the value 1 at each point of A. If $s = s_1 \ldots s_n \in \{0,1\}^n$ is not in A, then for all $t \in A$, $s \neq t$; thus $z_t(s) = 0$, and so

$$\bigvee_{t \in A} z_t(s) = \bigvee_{t \in A} 0 = 0.$$

So this function equals f.

(24.9P) Our domain is PF, the set of all propositional formulas in n atoms $a_1, \ldots, a_n$. The mapping $\pi : PF \longrightarrow P_\beta$ sends each Boolean function f to the cell of the partition of which f is an element.

$\pi(f)$ = the set of all propositional formulas in $a_1, \ldots, a_n$ which are logically equivalent to f.

The mapping $h : P_\beta \longrightarrow BF$ is defined as follows. An element of P_β is a subset C of PF consisting of propositional formulas, all logically equivalent to each other. Since they have the same truth-table, we define $h(C)$ to be the unique Boolean function in BF with that truth-table. Indeed $\beta = h \circ \pi$.

(24.10P) We can easily solve this problem if we write four different truth-tables also different from those of the given formulas. We can even take a bit of a shortcut if we notice that except for $\sim a$ the three givens have a 3–1 distribution of truth-values. We can find four 2–2 distributions different from $\sim a$, such as

a	b	$\sim a$	f_1	f_2	f_3	f_4
T	T	F	T	T	T	F
T	F	F	F	T	F	T
F	T	T	T	F	F	T
F	F	T	F	F	T	F

Now we might use DNF to make them into formulas, but we can say right off that $f_1 = b, f_2 = a, f_3 = (a \wedge b) \vee (\sim a \wedge \sim b)$, and $f_4 = \sim f_3 = \sim (a \wedge b) \wedge \sim (\sim a \wedge \sim b) = (\sim a \vee \sim b) \wedge (a \vee b)$.

(25.2P) Since $1 := \{0\}$ and $2 := \{0, 1\}$, we see that $\{1, 2\} = \{\{0\}, \{0, 1\}\}$. Now

$$1 = \{0\} \not\subseteq \{\{0\}, \{0, 1\}\} = 2,$$

because the only element of 1, namely, 0, is not an element of 2.

(25.3P) Since

$$\begin{aligned} 0 &:= \varnothing, \\ 1 &:= \{\varnothing\}, \\ 2 &:= \{\varnothing, \{\varnothing\}\}, \end{aligned}$$

we'd have

$$\begin{aligned} (0, 0) &:= \{0, \{0, 0\}\} = \{0, \{0\}\} \\ &= \{\varnothing, \{\varnothing\}\} \\ &= 2. \end{aligned}$$

(27.3P) The condition ruling out the value 0 for all large enough n allows us to divide freely. So any function f is asymptotic to itself because $f(n)/f(n) = 1$ for all large enough n. If for f and g

$$\lim_{n \to \infty} \frac{f(n)}{g(n)} = 1,$$

then

$$\lim_{n \to \infty} \frac{g(n)}{f(n)} = 1$$

also, as you know from calculus. Therefore $\sim$ is symmetric.

For transitivity we assume both ratios $\frac{f(n)}{g(n)}$ and $\frac{g(n)}{h(n)}$ have limit 1. It follows that

$$\frac{f(n)}{g(n)} \cdot \frac{g(n)}{h(n)} = \frac{f(n)}{h(n)}$$

also has limit 1, so $\sim$ is transitive.

(27.11P) Let $n = p_1^{e_1} \dots p_r^{e_r}$ with $r \geq 1$, all the p_i's distinct, and $e_i \geq 1$ for $i = 1, \dots, r$. Then $d(n) := r$, and since $2 \leq p_i$,

$$2^r \leq P_1 \dots p_r \leq n.$$

So $r \leq \log_2 n$. Therefore, $d(n) = O(\log_2 n)$.

(27.13P) We know that for some constant c and all large n,

$$|h(n)| \leq cn^2.$$

Divide by n^3 to find for all large n that

$$\left| \frac{h(n)}{n^3} \right| \leq \frac{c}{n}.$$

The result follows, since c/n tends to 0.

(27.14P) The relation S is not symmetric. For example, $1 = O(n)$ but $n \neq O(1)$. (You may check that claim.)

(27.16P) Every function f in the given set satisfies $f = O(f)$, so R is reflexive. Since Definition (27.15) is symmetric in f and g, R is symmetric. To prove transitivity, let $f = \Theta(g)$ and $g = \Theta(h)$. Then since $f = O(g)$ and $g = O(h)$, there exist constants c, c' such that for all large n, $|f(n)/g(n)| < c$ and $|g(n)/h(n)| < c'$. We multiply these two inequalities to see that $|f(n)/h(n)| < cc'$. Thus $f = O(h)$.

Starting from $h = O(g)$ and $g = O(f)$, the same argument shows that $h = O(f)$. Thus R is transitive.

CHAPTER

6

Divisibility in the Integers

We now look at some basic properties of the integers. Divisibility is the foundation for Chapter 7, on the integers modulo m. Chapters 6 and 7 constitute an introduction to number theory, a subject once thought to be nothing but art for art's sake. It now turns out to have several important applications. (We present one at the end of Chapter 7.) Chapters 6 and 7 also are basic for the theory of groups and of error-correcting codes, two topics we recommend to you for their beauty and interest, although we don't include them in this book.

1. Basic Ideas

When discussing integers, everyone knows that "a is a multiple of b" means that $a = bc$ for some integer c. Thus 12 is a multiple of 3, 14 is a multiple of 2 (and of 7), -24 is a multiple of 8.

The converse relationship is that of divisibility.[†] To say "b divides a" is to say that a is a multiple of b; i.e.,

DEFINITION (1.1) For all $a, b \in \mathbf{Z}$ we say b **divides** a iff $\exists c \in \mathbf{Z}$ such that $a = bc$.

NOTATION: $b \mid a$.

Thus 3 divides 12, 2 divides 14, 8 divides -24, and 23 divides $2^{11} - 1$. When b divides a we say that b is a **divisor** of a. Divisors always come in pairs; that is, if b divides a,

[†]Note that *divisibility* is defined entirely in terms of multiplication.

then so does c, where $a = bc$. But sometimes $b = c$: $4 = 2 \times 2$. The positive divisors of 4 are 1, 2, and 4; 1 and 4 are paired, and so are 2 and 2. The positive divisors of 6 are 1, 2, 3, and 6; they are paired as 1,6 and 2,3.

Divisibility offers a strategy for proving integers equal.

Lemma (1.2) Let a and b be *positive* integers. If a divides b and b divides a, then $a = b$. ■

The proof is left as an exercise.

The **absolute value** of any real number t is defined to be

(1.3)
$$|t| := \begin{cases} t & \text{if } t \geq 0 \\ -t & \text{if } t < 0. \end{cases}$$

Thus $|t| = |-t| \geq 0$ for all real t.

Example $|-3| = 3 = |3|$, $|1-2+3| = |-1+2-3|$, and $|7-11| = |11-7|$.

Notice that for all real numbers t, t'

(1.4)
$$|tt'| = |t||t'|,$$

and

(1.5)
$$||t| - |t'|| \leq |t + t'| \leq |t| + |t'|.$$

We shall also use the *floor* function and the *ceiling* function. If t is any real number, the value $\lfloor t \rfloor$ of the floor function at t is defined by the rule

(1.6) **floor**: $\lfloor t \rfloor$ is the *largest* integer a such that $a \leq t$.

The value $\lceil t \rceil$ of the ceiling function at t is defined by the rule

(1.7) **ceiling**: $\lceil t \rceil$ is the *smallest* integer b such that $t \leq b$.

Here *largest* and *smallest* are in the algebraic sense of being rightmost or leftmost, respectively, on the number line. Thus if $t \in \mathbf{Z}$ then $a = \lfloor t \rfloor = \lceil t \rceil = b$. But if t is not an integer, then $\lfloor t \rfloor < t < \lceil t \rceil$, and $\lceil t \rceil - \lfloor t \rfloor = 1$.

(1.8*F*)

$$a = \lfloor t \rfloor < t < \lceil t \rceil = b = 1 + a$$

That is, if $t \notin \mathbf{Z}$, then t lies between two consecutive integers; the smaller is $\lfloor t \rfloor$ and the larger $\lceil t \rceil$. Thus

(1.9)
$$\lceil t \rceil - \lfloor t \rfloor = \begin{cases} 0 & \text{if } t \in \mathbf{Z} \\ 1 & \text{if } t \notin \mathbf{Z}. \end{cases}$$

(1.10) Another way to put it is this. Suppose x is any real number; then $x = m + s$ for some $m \in \mathbf{Z}$ and $s \in \mathbf{R}$ with $0 \leq s < 1$. Then

$$\lfloor m + s \rfloor = \lfloor x \rfloor = m,$$

$$\lceil m + s \rceil = \lceil x \rceil = \begin{cases} m & \text{if } s = 0 \\ m + 1 & \text{if } s \neq 0. \end{cases}$$

Similarly, if $y = m - s$, with the same conditions on m and s, then

$$\lfloor m - s \rfloor = \lfloor y \rfloor = \begin{cases} m & \text{if } s = 0 \\ m - 1 & \text{if } s \neq 0, \end{cases}$$

$$\lceil m - s \rceil = \lceil y \rceil = m.$$

Here is a graph of the absolute value and the floor and ceiling functions:

(1.10F)

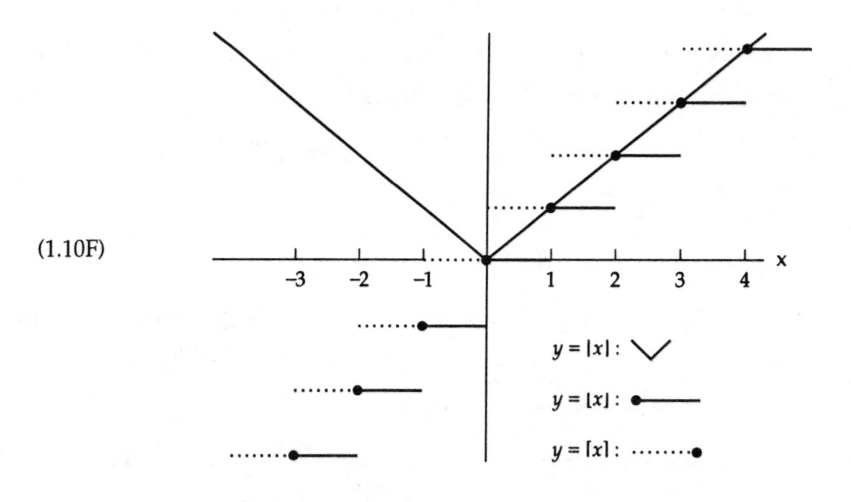

Examples (1.11)

$$\lfloor 3 \rfloor = \lceil 3 \rceil = 3$$

$$3 = \lfloor 3.1 \rfloor, \qquad \lceil 3.1 \rceil = 4$$

$$3 = \lfloor \pi \rfloor, \qquad \lceil \pi \rceil = 4$$

$$-2 = \lfloor -1.5 \rfloor, \qquad \lceil -1.5 \rceil = -1$$

$$\forall x \in \mathbf{R}, \lfloor x \rfloor = \lfloor x^2 \rfloor \quad \text{iff } 0 \leq x < \sqrt{2}.$$

(1.12) *Postage rates.* In 1987 for first-class mail of weight $w \leq 12$ ounces from the United States to Canada, Mexico, or the United States, the postage (in U.S. cents) was

$$17\lceil w \rceil + 5 + e(\lceil w \rceil - 1),$$

where $e := 0$ for U.S. addresses and $e := 1$ for Canadian or Mexican addresses.

Practice (1.13P) At these rates what was the first-class postage for a 1.5-ounce letter from the United States to Mexico? To the United States?

(1.14) Most people state their age as that at their last birthday. That is, if $x \in \mathbf{R}$ is their true age they say they are $\lfloor x \rfloor$ years old. Some insurance companies base their rates on the insured's age at the *nearest* birthday, namely,

$$\lfloor x \rfloor \quad \text{if } x - \lfloor x \rfloor < 1/2, \qquad \lceil x \rceil \quad \text{if } x - \lfloor x \rfloor \geq 1/2.$$

There is an interesting relationship between the floor and ceiling functions, namely,

$$\forall t \in \mathbf{R}, \qquad \lfloor t \rfloor = -\lceil -t \rceil. \tag{1.15}$$

We leave the proof to the problems.

2. Division

Division of any integer a by any nonzero integer b is a familiar process; it leads out of the integers $\mathbf{Z}$ in general, yielding a rational number. There is a way to express this process without speaking of rational numbers, however: the familiar partial quotient and remainder on division of a by b are the terms employed. It is known as the division algorithm, although it isn't an algorithm—it's a theorem.

Theorem (2.1) *The division algorithm.* Let $a, b \in \mathbf{Z}$ with $b \neq 0$. Then there exist unique integers q, r such that

$$a = bq + r, \quad \text{and}$$
$$0 \leq r < |b|.$$

The integer q is called the **partial quotient** of a by b, and r the **remainder**. (The **quotient** is a/b.)

Example (2.2)

(*i*) For $a = 29$ and $b = 8$ we have $29 = 8 \cdot 3 + 5$; q is 3 and r is 5.

(*ii*) If we change b to -8 we have almost the same equation: $29 = (-8)(-3) + 5$; q becomes -3.

Proof For clarity we assume $b > 0$. We first prove that if q and r exist, then they are unique. So we know that

$$a = bq + r \quad \text{and} \quad 0 \leq r < b \tag{2.3}$$

for some q and r in $\mathbf{Z}$. Then

$$\frac{a}{b} = q + \frac{r}{b}.$$

We divide the inequality $0 \leq r < b$ by b to see that

$$0 \leq \frac{r}{b} < 1.$$

Therefore $q = \lfloor a/b \rfloor$, by (1.10). Thus if (2.3) holds, then q is uniquely determined as the floor of a/b. And from (2.3) it follows that

$$r = a - bq$$

is also unique.

Now we prove the existence of q and r satisfying (2.3). (We've almost done it already, because we just showed what they must be if they do exist.) We set

$$q := \left\lfloor \frac{a}{b} \right\rfloor, \tag{2.4}$$

and

$$r := a - bq := a - b\left\lfloor \frac{a}{b} \right\rfloor. \tag{2.5}$$

Obviously q and r, so defined, are integers. Because of (2.5) they satisfy the equation of (2.3). It only remains to show that r satisfies the inequalities of (2.3).

To prove $r \geq 0$ we multiply

$$\left\lfloor \frac{a}{b} \right\rfloor \leq \frac{a}{b}$$

by b to get

$$b\left\lfloor \frac{a}{b} \right\rfloor \leq a.$$

We multiply by -1, switching the inequality to $\geq$, and add a :

$$r := a - b\left\lfloor \frac{a}{b} \right\rfloor \geq a - a = 0.$$

To prove $r < b$ we use

$$\frac{a}{b} < 1 + \left\lfloor \frac{a}{b} \right\rfloor.$$

Again we multiply by $-b$; we get

$$-a > -b - b\left\lfloor \frac{a}{b} \right\rfloor.$$

It follows immediately that

$$r := a - b\left\lfloor \frac{a}{b} \right\rfloor < b.$$

This completes the proof of Theorem (2.1) for the case when $b > 0$.

Notice that we could combine these two proofs by starting with

$$\left\lfloor \frac{a}{b} \right\rfloor \leq \frac{a}{b} < 1 + \left\lfloor \frac{a}{b} \right\rfloor.$$

We must mop up: If $b < 0$ we could repeat the previous proof, making the necessary changes; but it's slicker to piggyback on it, as follows. Suppose $b < 0$; then $-b > 0$ and we've just solved the problem for a and $-b$. Thus there are unique $q, r \in \mathbf{Z}$ such that

$$a = (-b)q + r, \qquad 0 \le r < |b|.$$

But this means that

$$a = b(-q) + r \quad \text{and} \quad 0 \le r < |b|.$$

Therefore $-q$ and r satisfy (2.3) for a, b if q and r satisfy (2.3) for a and $-b$. We determine $-q$ and r now by formulas like (2.4) and (2.5) that will show they are unique.

From (2.4) we see that

$$q = \left\lfloor \frac{a}{-b} \right\rfloor.$$

Therefore,

$$-q = -\left\lfloor -\frac{a}{b} \right\rfloor = \left\lceil \frac{a}{b} \right\rceil \tag{2.6}$$

by (1.15). Now

$$r = a - b(-q) = a - b\left\lceil \frac{a}{b} \right\rceil. \tag{2.7}$$

QED

Comment: Purists prefer proofs that stay within **Z**. See, for example, (16.3, page 30) and (16.4, page 5). We went outside **Z** by considering a/b. There is, however, a simple proof by induction that stays in **Z**. See section (17), problem 2.3.

We may summarize the findings of our proof of Theorem (2.2) as follows: Equations (2.4), (2.5), (2.6), and (2.7) tell us that

For $a, b \in \mathbf{Z}$ with $b \neq 0$, the unique $q, r \in \mathbf{Z}$ satisfying $a = bq + r$, $0 \le r < |b|$, are

$$q = \begin{cases} \lfloor \frac{a}{b} \rfloor & \text{if } b > 0 \\ \lceil \frac{a}{b} \rceil & \text{if } b < 0 \end{cases} \tag{2.8}$$

and $r = a - bq$.

Examples (2.9) Let's use $a = \pm 29$ and $b = \pm 8$ to follow up (2.2). Since $3 < 29/8 < 4$, we have for the four cases

$$\begin{array}{ll} 29 = 8 \cdot 3 + 5 & \lfloor \frac{29}{8} \rfloor = 3 \\ 29 = (-8)(-3) + 5 & \lceil \frac{29}{-8} \rceil = -3 \\ -29 = 8(-4) + 3 & \lfloor \frac{-29}{8} \rfloor = -4 \\ -29 = (-8)4 + 3 & \lceil \frac{-29}{-8} \rceil = 4. \end{array}$$

Practice (2.10P) Work out the analogous four cases for $a = \pm 17$ and $b = \pm 6$.

The division algorithm is really simple, but you will be surprised to see how useful it can be. For one thing it tells us that, in the notation of (2.1), b divides a if and only if $r = 0$.

Example (2.11)

	a	b	q	r
$18 = 5 \cdot 3 + 3$	18	5	3	3
$18 = 3 \cdot 6$	18	3	6	0
$27 = 6 \cdot 4 + 3$	27	6	4	3
$11 = 2 \cdot 5 + 1$	11	2	5	1

3. The Division Algorithm on a Computer

How do we get the partial quotient and remainder using division in a computer or hand calculator? The answer leaps at us from (2.8). The floor function is available in most programming languages, for example, indirectly via integer arithmetic in Fortran, or as a separate function in APL. The ceiling function is determined by the floor function via (1.15).

Thus we can see that the following algorithm produces the partial quotient and remainder.

(3.1) **The division algorithm (div alg)**

Input: $n, b \in \mathbf{Z}$, $b \neq 0$

Output: (q, r) such that $q, r \in \mathbf{Z}$, $n = bq + r$, and $0 \leq r < |b|$.

$s := n/b$

If $b > 0$ $\quad q := \lfloor s \rfloor$

If $b < 0$ $\quad q := \lceil s \rceil$

$r := n - bq$

End.

Suppose we want only r and not q. First of all, r is the same for n and b as for n and $-b$ (why?). Therefore we assume b positive.

Using the notation just preceding, we know that the fractional part of the quotient n/b is

$$fp(n/b) := n/b - \lfloor n/b \rfloor = r/b.$$

If we multiply by b we get r—almost! There may be a truncation error. To get rid of it, we add $\frac{1}{2}$ and take the floor function.

Practice (3.2P) Prove that for any real x, $\lfloor x + \frac{1}{2} \rfloor$ is the nearest integer to x (unless x is half an odd integer, in which case $\lfloor x + \frac{1}{2} \rfloor$ is one of two integers nearest to x).

Here is the algorithm.†

(3.3) **rem** n **by** b

Input: $n, b \in \mathbf{Z},\ b \neq 0$

Output: r such that $\exists q \in \mathbf{Z}, n = bq + r, 0 \leq r < |b|$.

$b := |b|$

$s := n/b$

$t := fp(s) := s - \lfloor s \rfloor$

$u := bt$

$r := \lfloor u + 1/2 \rfloor$

End.

This algorithm *rem n by b* is well suited to hand calculators; we round by eye.

Example (3.4) We reproduce the hand-calculation of rem 129 by 17:

$$\begin{aligned} s &= 7.5882352 \\ t &= s - 7 \\ bt &= 17(.5882352) = 9.9999984 \\ \text{rem } 129 \text{ by } 17 = r &= 10 \end{aligned}$$

(by eye, instead of adding $\frac{1}{2}$ to bt and then applying $\lfloor\ \rfloor$).

Practice (3.5P) With most hand calculators we will always round upward. Why?

4. Integers in Base *b*

One application of (2.1) is in representing integers in the base b, for any integer $b \geq 2$. If $n > 0$ is an integer, and if there are integers $n_0, n_1, \ldots, n_k$ all in $\{0, 1, \ldots, b-1\}$ such that $n_k \neq 0$ and

(4.1) $$n = n_k b^k + n_{k-1} b^{k-1} + \cdots + n_1 b + n_0$$

then the base b representation of n is the symbol formed by juxtaposing the numbers $n_k, n_{k-1}, \ldots, n_0$. The subscript b and parentheses may be present.

$$n = (n_k n_{k-1} \ldots n_1 n_0)_b.$$

The base b representation of 0 is usually taken to be just 0.

Examples (4.2) 341 is the base 10 representation of 341, as we usually call it, since

$$341 = 3 \cdot 10^2 + 4 \cdot 10^1 + 1 \cdot 10^0.$$

We might write it as $(341)_{10}$.

†In APL rem n by b is just $(|b|)|n$.

The base 2 representation of $(13)_{10}$ is 1101, since $(1101)_2$ stands for, by (4.1),

$$1 \cdot 2^3 + 1 \cdot 2^2 + 0 \cdot 2^1 + 1 \cdot 2^0 = 8 + 4 + 1 = 13.$$

We now prove such a representation exists for every positive integer and every base b. Look first at (4.1) again. It says

$$n = bq + n_0, \qquad \text{and} \quad 0 \leq n_0 < b$$

(where $q = n_k b^{k-1} + n_{k-1} b^{k-2} + \cdots + n_1$). By the uniqueness part of the division algorithm, n_0, the least significant term in "n base b," is the remainder of n on division by b. That observation tells us how to find all the coefficients in the base b representation, by induction.

Theorem (4.3) Let b be an integer ≥ 2. Then every positive integer n has a unique base b representation (4.1).

Proof We use the second form of induction, giving a streamlined proof. Our predicate $S(n)$ is (4.1) with the conditions $0 \leq n_i < b$ and $n_k \neq 0$.

First, if $n < b$ then $n_0 := n$ is unique and n is its own base-b representation. (Here $k = 0$, and $n_k = n_0 \neq 0$.)

(4.3.1) If $n \geq b$ use the division algorithm (2.1) to find q and r such that

$$n = bq + r, \qquad 0 \leq r < b,$$

where q and r are unique by (2.1). Set $n_0 := r$.

(4.3.2) Since $1 \leq q < n$, q has a unique base b expression, by induction. Let us say it is, in the form (4.1),

$$q = n_k b^{k-1} + \cdots + n_1. \tag{4.3.3}$$

Note that, also by induction, $n_k \neq 0$. Then bq has a base b expression, namely,

$$bq = n_k b^k + \cdots + n_1 b.$$

This base b expression for bq is also unique, because any such expression for bq yields on division by b the unique such expression for q.

Therefore the expression (4.1) for n is

$$n = bq + r = n_k b^k + \cdots + n_1 b + n_0,$$

in which the n_i's are unique since n_0 is unique by (4.3.1) and the rest are unique by (4.3.2). QED

The proof just concluded suggests a procedure for finding the base b expression of an integer by repeated use of the division algorithm. That is, $n_0 =$ **rem** n **by** b, and (see (4.3.3)) $n_1 =$ **rem** q **by** b, and so on.

Example (4.4) Let $b = 2$. Then for $n = 23$ we have

$$\begin{aligned} q_{-1} &= 23 = 2 \cdot 11 + 1 & i &= -1 \\ q_0 &= 11 = 2 \cdot 5 + 1 & i &= 0 \\ q_1 &= 5 = 2 \cdot 2 + 1 & i &= 1 \\ q_2 &= 2 = 2 \cdot 1 + 0 & i &= 2 \\ q_3 &= 1 = 2 \cdot 0 + 1 & i &= 3 \end{aligned}$$

Since $q_4 = 0$ we stop. Thus $(10111)_2 = (23)_{10}$. We found the partial quotient, divided it by 2, and repeated the process.

Thus we are led to this simple algorithm or program for computing the base b representation of n:

(4.5) n **base** b

Input: integers $n \geq 1$, $b \geq 2$.
Output: the base b representation of n, $(n_i n_{i-1} \ldots n_0)_b$.

Initial step. Set $q_{-1} := n$; $i := -1$.
While $q_i \neq 0$ **do**

$(q_{i+1}, n_{i+1}) :=$ div alg (q_i, b)
$i := i + 1$

End.

The understanding here is, of course, that when $q_i = 0$ you go to the next step, the *End* in this case. (Here the juxtaposition of numerals does not mean that they are to be multiplied together.) Notice that the algorithm claims that k is the value of i when the algorithm stops.

Example (4.4) shows this procedure. It is marked to show the values of i and q_i. Actually $i = 4$ when the algorithm stops, because q_3 being 1 causes the **while** loop to be done: q_4 is calculated (as 0) and i is increased from 3 to 4. In other words, the algorithm is just a formal statement of the procedure that anyone would find after a few minutes' search.

Practice (4.6P) Find the base 2 representation of $(2345)_8$.

5. Proof of Correctness

You probably accept algorithm (4.5) as correct, but here is a formal proof in the spirit of current program-proving. The idea is to introduce a *comment*, called an **invariant**, at the while loop and prove it is correct just before each execution of the test "$q_i \neq 0$" in the while statement. At the end of the algorithm the comment should lead to a proof of correctness. In general, picking the right invariant can be tricky; here, however, it is transparently simple. We repeat the algorithm and insert the comment:

(5.1) n **base** b

$q_{-1} := n$; $i := -1$
[*Comment*: $n = q_i b^{i+1} + \Sigma_{0 \leq j \leq i} n_j b^j$]

while $q_i \neq 0$ **do** $\begin{cases} (q_{i+1}, n_{i+1}) := \textbf{div alg } (q_i, b) \\ i := i + 1 \end{cases}$

End.

The comment is true just before the first execution of the test, because q_{-1} is defined to be n, and i is -1. The sum over the empty set $\{j;\ 0 \le j \le -1\}$ is 0; thus

$$n = q_{-1}b^{-1+1} + 0.$$

Now we apply mathematical induction to prove that the comment, which we take to be a predicate $S(i)$ of the one variable i, is true for all i such that $q_i \neq 0$.

Let i be any integer such that $q_i \neq 0$. Assume that $S(i)$ is true. We try to prove that $S(i+1)$ is true. Our assumption $S(i)$ reads:

$$S(i):\ n = q_i b^{i+1} + n_i b^i + \cdots + n_0 b^0. \tag{5.2}$$

Since $q_i \neq 0$, the test is satisfied and we do the division algorithm, getting

$$q_i = q_{i+1}b + n_{i+1}. \tag{5.3}$$

We substitute q_i from (5.3) into (5.2) to transform (5.2) into precisely $S(i+1)$.

This completes the inductive proof that the comment is true just before each test whether q_i is 0. In particular it is true just before the last test, when $q_i = 0$, because q_{i-1} is not 0. That is, for this final value of i

$$S(i):\ n = 0 \cdot b^{i+1} + n_i b^i + \cdots + n_0 b^0 \tag{5.4}$$

is true. And we note that $n_i \neq 0$ because, by (5.3),

$$0 \neq q_{i-1} = q_i b + n_i = 0 + n_i = n_i.$$

Now we see that (5.4) yields our base b expression for n. QED

Remarks (5.5) (*i*) Notice that in doing our inductive proof we did not prove $q_i \neq 0 \longrightarrow q_{i+1} \neq 0$. Instead we proved $q_i \neq 0 \longrightarrow S(i+1)$.

(*ii*) If $n = 0$, the algorithm n **base** b as it stands produces the empty sequence. Since we customarily represent zero as 0 in any base, perhaps a new first line of the algorithm, reading

if n = 0, output 0

would be appropriate (with the allowance of $n = 0$ in the input).

(*iii*) A comment on the induction: This proof is the first one we've seen that does induction over a finite set. Clearly that restriction is no restriction, for we may either re-do the proof of induction for finite segments of the integers, or notice that to prove the predicate $P(m)$ only for $m = 1, \ldots, k$, we may simply *define* $P(k+1), P(k+2), \ldots$ to be true and then apply either theorem on induction from Chapter 3. ■

Example (5.6) Take $b = 6$ and $n = (248)_{10}$.

$$\begin{aligned} 248 &= 6 \cdot 41 + 2 \\ 41 &= 6 \cdot 6 + 5 \\ 6 &= 6 \cdot 1 + 0 \\ 1 &= 6 \cdot 0 + 1 \end{aligned}$$

Thus $(248)_{10} = (1052)_6$. We do not stop when a remainder is 0, but only when the partial quotient becomes 0.

6. Analysis of the Algorithm *n* base *b*

The main question to ask in analyzing the algorithm just described is this: how many calls of the subroutine **div alg** does it make? The answer is $k+1$ for k as defined in (4.1); but how is k related to the input n? Since

$$b^k \leq n < b^{k+1},$$

we see that

$$k \leq \log_b n < k+1.$$

Therefore $k+1 = \lceil \log_b n \rceil$ unless $n = b^k$.

We usually speak of this result by saying that the complexity of this algorithm to express n in the base b is approximately $\log n$.[†] This manner of speaking deliberately ignores the base of the logarithm, because

$$\log_b n = (\log_b c) \log_c n$$

for any two bases b and c. Thus the logarithm of n to one base is a constant times its logarithm to another base. In other words, the complexity here is $O(\log n)$, in the terms of Chapter 5, (27).

If we wanted to analyze n **base** b further, we could time the division and other operations in **div alg**. But those are machine-dependent and tedious to do.

Now you've seen a simple example of the analysis of algorithms. Later ones will be more demanding.

7. Greatest Common Divisor

Suppose a and b are integers, not both 0. Consider the set D of integers that divide *both a and b*. If we want to be detailed, we may say that $D = D(a) \cap D(b)$, where we define $D(x)$ for each integer x to be the set of all divisors of x. If $x \neq 0$, then $D(x)$ is finite; therefore D is finite.

D is called the set of **common divisors** of a and b. By our definition both negative and positive divisors are in D; our focus, however, is on the positive divisors.

DEFINITION (7.1) Let $a, b \in \mathrm{Z}$. Let D be the set of common divisors of a and b. Then we define the **greatest common divisor** d of a and b as

$$d := \begin{cases} \text{the largest element of } D & \text{if } (a,b) \neq (0,0) \\ 0^{\ddagger} & \text{if } (a,b) = (0,0) \end{cases}$$

[†] Since the *representation* of n in the "input" base c takes about $\lceil \log_c n \rceil$ bits, we might instead say that the complexity of this algorithm to express n in the base b is proportional to the input.

[‡] This seemingly arbitrary definition will be tied to the non "zero, zero" case in (7.13).

$$\text{NOTATION:} \qquad d = \gcd(a,b).$$

(In some books you see $\gcd(a,b)$ denoted simply as (a,b), but not here.)

Since 1 divides every integer, $1 \in D$; therefore,

$$1 \le \gcd(a,b) \quad \text{if } (a,b) \neq (0,0).$$

Notice that gcd is a function from $\mathbf{Z} \times \mathbf{Z}$ to $\mathbf{Z}$ with range $\mathbf{N} \cup \{0\}$.

Examples (7.2)

$$\begin{aligned} \gcd(1,n) &= 1 \qquad \text{for all } n \in \mathbf{Z} \\ \gcd(a,0) &= |a| \qquad \text{for all } a \text{ in } \mathbf{Z} \\ \gcd(2,6) &= 2 \\ \gcd(-2,\pm 6) &= 2 \\ \gcd(30,42) &= 6 \\ \gcd(3,5) &= 1 \\ \gcd(a,b) &= \gcd(b,a) \end{aligned}$$

DEFINITION (7.3) The integers a and b are called **relatively prime** if and only if

$$\gcd(a,b) = 1.$$

(7.4) *Terminology*. Sometimes we say a **is prime to** b instead of *a and b are relatively prime*. In the latter form we *should* say *mutually relatively prime or relatively prime to each other*, but we rarely do so. You may even see "a is **coprime** to b" for this idea.

We now prove that the gcd, as we often call it, can be expressed in a way that we'll use over and over to study this function.

Theorem (7.5) Let $a, b \in \mathbf{Z}$. Then there are integers x, y such that

$$d = \gcd(a,b) = ax + by.$$

Proof If $(a,b) = (0,0)$, then (7.5) is obvious. So assume not both a, b are 0. We prove this result by investigating the set S of all such linear combinations of a and b with integral coefficients. Define

(7.6) $$S := \{\, au + bv;\ u,v \in \mathbf{Z}\,\}.$$

By definition, since d divides both a and b, there are integers a' and b' such that $a = da'$ and $b = db'$. It follows that every element of S is a multiple of d, since

$$au + bv = d(a'u + b'v).$$

All we have to do is show that d itself is in S, since $d \in S$ is what (7.5) says.

To show $d \in S$, let s be the smallest[†] positive element in S. (Since $a^2 + b^2 \in S$ (why?), S does have positive elements.) We first show that s divides *every* element

[†]Here we tacitly use well ordering on the subset S^+ of all positive integers in S.

$c = au + bv$ of S: Using the division algorithm (2.1) we find integers q, r such that

$$c = qs + r, \quad \text{and} \quad 0 \leq r < s.$$

To show $s \mid c$ we'll show $r = 0$. Now s has an expression

$$s = au_0 + bv_0$$

for certain integers u_0, v_0. Therefore

$$\begin{aligned} r = c - qs &= au + bv - q(au_0 + bv_0) \\ &= a(u - qu_0) + b(v - qv_0); \end{aligned}$$

and so r is also in S. Thus we have shown that

(*i*) $r \in S$

(*ii*) $0 \leq r < s$

(*iii*) s is the least positive element in S.

It follows that $r = 0$. Thus s divides c, and in particular both a and b. (Why?)

We will now prove that in fact $s = d$, the greatest common divisor of a and b. We have shown that s is a common divisor of a and b. Hence $s \leq d$.

From

$$s = au_0 + bv_0 = d(a'u_0 + b'v_0)$$

we see that d divides s, forcing $d \leq s$ (since both are positive). Therefore $s = d$. QED

There are a number of remarks to make about this result and the proof just given.

First, the proof is *nonconstructive*. It proves the *existence* of x and y ($x = u_0$ and $y = v_0$, incidentally) without giving you a clue as to how to find such x and y. An *algorithm* producing x and y satisfying $\gcd(a, b) = ax + by$ would be a *constructive* proof of Theorem (7.5). (Hang on; two algorithms are coming.)

Second, the proof illustrates a good way to prove that one integer, that is, s divides another, c: Apply the division algorithm (2.1) to get $c = qs + r$ and then prove $r = 0$. (Of course, this technique may fail, especially if s does not divide c.)

Third, this theorem is our first nonobvious result. To prove things about the gcd, we will use it frequently—instead of going back to the definition of gcd. We state more remarks as corollaries.

Corollary (7.7) Define $S := \{ au + bv;\ u, v \in \mathbf{Z} \}$, as in (7.6). Then $S = \{ kd;\ k \in \mathbf{Z} \}$.■

This is a corollary of the proof of Theorem (7.5), where we showed $S \subseteq \{ kd;\ k \in \mathbf{Z} \}$. We leave the reverse inclusion as a problem. Corollary (7.7) says that the set of all integral linear combinations of a and b equals the set of all integral multiples of $\gcd(a, b)$.

Corollary (7.8) Every common divisor of a and b divides $d = \gcd(a, b)$.

Proof We know that for some $x, y \in \mathbf{Z}$, $d = ax + by$. Therefore, if d_1 divides both a and b, say $a = d_1a_1$ and $b = d_1b_1$, then

$$d = d_1(a_1x + b_1y).$$

So d_1 divides d.

Corollary (7.9) If $\exists x, y : ax + by = 1$, then $\gcd(a, b) = 1$. ■

This corollary is obvious even without the Theorem or its proof, because if $d = \gcd(a, b)$, then d divides every linear combination of a and b, including $ax + by = 1$. Thus $1 \le d$ and $d \mid 1$. Therefore $d = 1$.

Corollary (7.10) Let a and b be relatively prime. For all c, if b divides ac, then b divides c. In symbols: $\forall a, b, c \in \mathbf{Z}, [\gcd(a, b) = 1 \wedge b \mid ac] \longrightarrow b \mid c$.

Proof From (7.5) we know there are x, y such that $ax + by = 1$. Then $acx + bcy = c$. Now $b \mid (ac)$ by hypothesis. Therefore $b \mid (acx)$, and of course $b \mid (bcy)$. Therefore b divides the sum of these two, which is c. QED

Notice that we began the proof by invoking (7.5); that's usually helpful in approaching problems involving gcd's.

Examples (7.11)

$$\begin{aligned} \gcd(10, 15) &= 5 = 10(-1) + 15(1) \\ \gcd(30, 77) &= 1 = 30(18) + 77(-7) \\ \gcd(15, 36) &= 3 = 15(-7) + 36(3) \\ \gcd(5, 3) &= 1 = 5(2) + 3(-3) \\ \gcd(5313, 2047) &= 23 = 5313(42) + 2047(-109) \end{aligned}$$

NOTE: $5313 = 23 \cdot 231,\ 2047 = 23 \cdot 89$.

Practice (7.12P) Notice that we cannot push Corollary (7.9) much further. That is, if for some x, y, $ax + by = c$, it is not necessarily true that $\gcd(a, b) = c$. But this equation says something about $\gcd(a, b)$—what?

(7.13) We now connect the definition "$\gcd(0, 0) = 0$" to the definition of $\gcd(a, b)$ where not both a and b are 0. There are two ways to do it, with (7.7) or (7.8).

Let a, b be any two integers. Define $S = \{au + bv;\ u, v \in \mathbf{Z}\}$. We may immediately interpret Corollary (7.7) as follows:

(7.14) $d = \gcd(a, b)$ is the unique integer ≥ 0 such that $S = \{kd;\ k \in \mathbf{Z}\}$.

Why so? The claim (7.14) is immediate if $a = b = 0$, for then $S = \{0\}$ and d is 0 by definition. In case a or b is not 0, it's just a restatement of (7.7). So this is a *characterization*† of gcd that ties the two cases together.

We can use (7.8) to connect the two cases as follows: Let us consider $(a, b) \ne (0, 0)$. By (7.8) we see that

(7.15) $d = \gcd(a, b)$ is the unique common divisor ≥ 0 of a and b such that every common divisor of a and b divides d.

†A *characterization* of a mathematical object, say an integer x, is a theorem that says, for example, "The integer y equals x *if and only if* y has properties p and q." In other words, x is defined one way, say "x is the gcd (or whatever) if and only if it has properties r and s"; and then you prove that you could have defined it with properties p and q instead. A characterization is a theorem allowing a different but equivalent definition of the thing characterized. You could define a set as the truth-set of a predicate $P(x)$ and then *characterize* it by proving it is the truth-set of another predicate $Q(x)$.

Indeed (7.8) states all of this claim but the uniqueness, which we leave for you in problem 7.9. But now we notice that (7.15) is still true if $a = b = 0$: The set of common divisors of 0 and 0 is **Z**, so there is no largest. But since 0 divides 0, 0 is a common divisor of 0 and 0; and it is the *only* common divisor divisible by all common divisors. So (7.15) gives another characterization of the gcd.

Practice (7.16P) Let $a, b, c \in \mathbf{Z}$. If a is prime to b and b is prime to c, is a necessarily prime to c?

8. The Euclidean Algorithm

We now present two algorithms, one to find $d = \gcd(a, b)$; from the first we get the second, which finds x, y in **Z** such that $d = ax + by$.

The Euclidean algorithm is a recipe for finding $d = \gcd(a, b)$, where a and b are any two integers not both 0. It consists of repeated use of the division algorithm. For definiteness we assume $b \neq 0$. Here is the procedure:

(8.1)

$$\begin{array}{lrl}
\text{Step 1.} & a = |b|q_1 + r_1 & 0 \le r_1 < |b| \\
& \swarrow \quad & \\
\text{Step 2.} & |b| = r_1q_2 + r_2 & 0 \le r_2 < r_1 \\
& \swarrow \quad \swarrow & \\
\text{Step 3.} & r_1 = r_2q_3 + r_3 & 0 \le r_3 < r_2 \\
& \swarrow \quad & \\
& \vdots & \\
& \swarrow \quad \swarrow & \\
\text{Step } n. & r_{n-2} = r_{n-1}q_n + r_n & 0 \le r_n < r_{n-1} \\
& \swarrow \quad \swarrow & \\
\text{Step } n+1. & r_{n-1} = r_nq_{n+1} & (\text{i.e., } r_{n+1} = 0).
\end{array}$$

Conclusion: $r_n = \gcd(a, b)$.

Concentrate for now on the idea of the algorithm—to divide the divisor by the remainder, over and over again, as long as the remainder is not 0. The arrows help to understand the procedure: the divisor at one step is divided at the next step by the remainder. Notice that if $r_1r_2 \neq 0$ then r_1 appears three times: first as a remainder, second as a divisor, and last as something divided by r_2. The same is true for $r_2, r_3, \ldots, r_{n-1}$. The gcd r_n appears only twice.

(8.2) The procedure must end as advertised, because the remainders make a strictly decreasing sequence of nonnegative integers:

$$|b| > r_1 > r_2 > \cdots > r_n > r_{n+1} = 0.$$

Therefore for some n, r_{n+1} must be 0 (certainly $n + 1 \le |b|$); there the algorithm stops.

(8.3) Why is $r_n = d = \gcd(a, b)$? We prove that $r_n = d$ by showing that r_n is a common divisor of a and b, and then that r_n is a multiple of d. Since $r_n > 0$, r_n must then be d. (That is, we'll use Corollary (7.8) and then Lemma (1.2).)

Thus r_n divides a and b: from Step $n+1$, $r_n \mid r_{n-1}$. From Step n, $r_n \mid r_{n-2}$. and so on (a disguised induction). We find that $r_n \mid r_1$, and from Step 2, $r_n \mid b$. Finally, Step 1 tells us that $r_n \mid a$. Thus r_n is a common divisor of a and b. By (7.8), $r_n \mid d$.

Any divisor d' of a and b divides r_n: $d' \mid r_1$, from Step 1; $d' \mid r_2$, from Step 2,..., $d' \mid r_n$, from Step n. Therefore, in particular, $d \mid r_n$. By Lemma (1.2) $r_n = d$, and the Euclidean algorithm (8.1) is indeed correct.

Examples (8.4) (*i*) gcd(33, 93) = 3, as we can tell by inspection. The Euclidean algorithm produces 3 in the following steps:

Step 1. $93 = 33 \cdot 2 + 27$

Step 2. $33 = 27 \cdot 1 + 6$

Step 3. $27 = 6 \cdot 4 + 3$

Step 4. $6 = 3 \cdot 2 \quad \therefore 3 = \gcd(33, 93)$.

(*ii*) gcd(5313, 2047) = ?

Step 1. $5313 = 2047 \cdot 2 + 1219$

Step 2. $2047 = 1219 + 828$

Step 3. $1219 = 828 + 391$

Step 4. $828 = 391 \cdot 2 + 46$

Step 5. $391 = 46 \cdot 8 + 23$

Step 6. $46 = 23 \cdot 2$

Thus 23 = gcd (5313, 2047).

In these two examples, at least, the numbers of steps required was far fewer than the smaller of $|a|$ and $|b|$. Incidentally, the algorithm "doesn't care" which one of $|a|$ and $|b|$ you divide by at Step 1 (if both are nonzero), for if you divide by the larger one at Step 1, then Step 2 is the division of the larger by the smaller.

Example Let us suppose we started "wrong" with the last example. We'd have

Step 1. $2047 = 5313 \cdot 0 + 2047$

Step 2. $5313 = 2047 \cdot 2 + 1219$

$\vdots$

which would be just one more step.

Contrast this procedure (8.1) with that for n **base** b in (5.1). There we always divide by b, and it's the quotient that gets divided. Here, after Step 1 it's the remainder that we divide by; after Step 2 it's the previous remainder that gets divided.

Practice (8.5P) Find gcd(14, 36) by the Euclidean algorithm.

(8.6) *The formal version of the Euclidean algorithm*

Euclid (a, b)

Input: any two integers a, b

Output: d = **Euclid** $(a, b) = \gcd(a, b)$.

If $ab = 0$, then **Euclid** $(a, b) := |a + b|$. End.

Otherwise,

do initial step: $r_{-1} := a$

$r_0 := |b|$

$n := 0$

while $r_n \neq 0$ **do** $r_{n+1} :=$ **rem** r_{n-1} **by** r_n

$n := n + 1$

$d := r_{n-1}$

End.

You can supply a proof of correctness in the style of that given for **div alg**. Which of (8.1) and (8.6) do you find easier to understand?

9. Finding the Coefficients

Now we take up the question of how to find integers x and y such that

$$d = \gcd(a, b) = ax + by.$$

It is easy enough for small a, b to use inspection. Thus

$$1 = \gcd(3, 5) = 3(2) + 5(-1) = 3(-3) + 5(2).$$

But for larger a, b it would be tedious to try value after value of x and y until we hit on a correct solution. We know it exists—how do we find it efficiently?

(9.1) One answer is to drive the Euclidean algorithm (8.1) backward, solving for d in terms of a and b. Consider the example:

$$2 = \gcd(38, 16)$$

Step 1.	$38 = 16 \cdot 2 + 6$
2.	$16 = 6 \cdot 2 + 4$
3.	$6 = 4 \cdot 1 + 2$
4.	$4 = 2 \cdot 2$

We find coefficients x, y so that $2 = 38x + 16y$ as follows:

Using Step	*we get*
#3	$2 = 6 - 4$
#2	$2 = 6 - (16 - 6 \cdot 2)$
	$= 6 \cdot 3 - 16$
#1	$2 = (38 - 16 \cdot 2)3 - 16$
	$= 38(3) + 16(-7)$

Thus $x = 3, y = -7$ is a solution. The idea of this procedure is to substitute the value of the remainder from a given step into the formula for d obtained from the steps below: We solved for the remainder in #3; we then substituted for the value of the remainder 4 obtained from #2; after collecting terms we substituted for the value of the remainder 6 from #1.

It is easier to understand in the general case. Suppose we assume a problem for which $n = 3$.

(9.2)
$$\begin{array}{llll} \#1. & a & = & |b|q_1 + r_1 \\ \#2. & |b| & = & r_1q_2 + r_2 \\ \#3. & r_1 & = & r_2q_3 + r_3 \\ \#4. & r_2 & = & r_3q_4. \end{array}$$

Then $d = r_3$, and

$$\begin{array}{rcll} r_3 & = & r_1 - r_2q_3 & \text{(use \#3)} \\ & = & r_1 - (|b| - r_1q_2)q_3 & \text{(use \#2)} \\ & = & r_1(1 + q_2q_3) - |b|q_3 & \text{(collect terms)} \\ & = & (a - |b|q_1)(1 + q_2q_3) - |b|q_3 & \text{(use \#1)} \\ & = & a(1 + q_2q_3) - |b|(q_1(1 + q_2q_3) + q_3) & \text{(collect terms)} \end{array}$$

Thus $x = 1 + q_2q_3$ and $y = \pm(q_1 + q_3 + q_1q_2q_3) = \pm(q_3 + q_1x)$ (use + if $b < 0$, − if $b > 0$).

The same procedure works in all cases. Not surprisingly, I used it to find

(9.2.1) $$23 = \gcd(5313, 2047) = 5313(42) + 2047(-109)$$

in (8.4). Here $n = 5$, so the playback (9.2) took 9 lines (and several tries to get it right). An algorithm to find the coefficients is presented in Reference (16.3, pages 145–147), but we shall give a different one.

You may have noticed that the formal algorithm (8.6) for the gcd did not preserve the partial quotients and remainders, so we'd have to modify it if we wanted to find the coefficients. In fact, there is an algorithm that finds the coefficients (*extends* the algorithm) as it is finding the gcd. We present it now, calling it **EE** for *extended Euclid*. Notice it is defined for positive inputs.

(9.3) **EE** (a, b)†

Input: $a, b \in \mathbf{N}$

Output: **EE**$(a, b) := (d, x, y)$, where $d = \gcd(a, b)$ and $x, y \in \mathbf{Z}$ satisfy $ax + by = d$.

Initial Step: $x' := y := 1,\ y' := x := 0$

$c := a,\ d := b$

[*Comment*: $ax' + by' = c$, and $ax + by = d$],

$(q, r) :=$ **div alg** (c, d)

†Adapted from D. E. Knuth, page 14, cited in Chapter 3, (15.2).

```
while r ≠ 0 do
    c := d
    d := r
    t := x'
    x' := x
    x := t − qx
    t := y'
    y' := y
    y := t − qy
  [Same comment.]
  (q,r) := div alg (c,d)
End.
```

We have introduced the comment

$$ax' + by' = c \text{ and } ax + by = d \tag{9.4}$$

before each call to **div alg** as an invariant to help with the following proof.

Proof of correctness of **EE** How does **EE** work? It uses variables c and d, initially set to a and b, respectively, and follows the Euclidean algorithm (8.1) in resetting c and d until the end. This resetting occurs in the **div alg** steps and in the instructions $c := d$ and $d := r$. It stops when it finds $r = 0$, so it doesn't set d to 0; d is the last nonzero remainder, so we know from (8.3) that it's $\gcd(a, b)$.

The coefficients x and y initially make $ax + by = d$ (trivially). Our job is to show that as d changes, x and y change so as to preserve this equation.

We do it by using the x' and y' that someone cleverly introduced into the algorithm to make, initially,

$$ax' + by' = c.$$

Here's how we proceed by induction on the number n of times we perform the **div alg** step:

Basis: For $n = 0$ we have just done the initial step, and the comment is true.

Inductive step: Let $n \geq 0$ be a fixed but arbitrary integer. Assume the comment is true just before the $(n + 1)^{\text{st}}$ execution of **div alg**. Try to prove it is true just after the next round of resettings (if r remains nonzero).

So—we are going through the algorithm, and the comment is true. We then do

$$(q, r) := \textbf{div alg}\,(c, d) \tag{9.5}$$

and reset variables if $r \neq 0$. (Notice if $r = 0$ we do not reset anything in (9.4), so it remains true.) So assume $r \neq 0$.

Notice that

$$c = qd + r \tag{9.6}$$

is the equation underlying (9.5). Equation (9.6) gives us the new values of q and r, in terms of the old values of c and d. In terms of the old values of x, x', y, and y' the comment (9.4) is true:

$$\begin{aligned} ax' + by' &= c \\ ax + by &= d. \end{aligned} \tag{9.7}$$

Let's use capital letters for the new values. We have, from the reset instructions in (9.3),

$$\begin{aligned} C &= d \\ D &= r \\ X &= x' - qx \\ Y &= y' - qy \\ X' &= x \\ Y' &= y. \end{aligned}$$

Then $aX' + bY' = ax + by = d = C$, and

$$aX + bY = a(x' - qx) + b(y' - qy) = c - qd,$$

as we see if we multiply the second equation of (9.7) by q and subtract it from the first. But, from (9.6), $c - qd = r = D$. This completes our inductive step and hence our proof of correctness.
QED

Example (9.8) Let $a = 128{,}520$ and $b = 493$. We reproduce the key values at each step of **EE** in the following table:

x'	x	y'	y	c	d	q	r
1	0	0	1	128,520	493	260	340
0	1	1	−260	493	340	1	153
1	−1	−260	261	340	153	2	34
−1	3	261	−782	153	34	4	17
3	−13	−782	3389	34	17	2	0

Accordingly, we say that $17 = \gcd(128520, 493) = d$ and that

$$(128{,}520)(-13) + (493)(3389) = 17.$$

In case you doubt that $d = 17$, you can verify the latter equation directly and then check that 17 divides both 128,520 and 493. It follows that 17 is both a multiple and a divisor of d, hence $17 = d$.

Practice (9.9P) Drive the Euclidean algorithm backward to find x, y such that

$$2 = 36x + 14y.$$

10. Application: Pouring Problems

Suppose at the end of a tough day of backpacking, needing exactly one cup of water to make crêpes suzettes, you have only a 3-cup and a 5-cup measure. The nearby stream has plenty of water, but how can you measure just one cup? Answer: fill the 3-cup measure, empty it into the other, fill the 3-cup again, and then fill the other from it. There remains just one cup in the 3-cup measure.

That is a simple example of a pouring problem. More generally you might have measures A of a units and B of b units, with the problem being to measure e units into some container C. C could be A, if $e \leq a$, or B if $e \leq b$. But if not, then C is some container of unknown volume greater than e, a, and b. *The only moves allowed are to fill or empty A or B.* A reservoir of plenty of water is assumed.

After poring over many such problems I found a systematic approach using the division algorithm. The rest of this section presents that approach.

Proposition (10.1) The pouring problem just stated is soluble if and only if e is a multiple of gcd (a, b). ■

We prove first a lemma:

Lemma (10.2) After each move, the contents of each of A, B, and C is a multiple of $d = \gcd(a, b)$.

Proof Let $v(X)$ stand for the amount of water in X, for $X = A, B$, or C. At the start all are empty. We assume that after n moves the lemma is true (for $n \geq 0$). If our next move is to empty A (into B or C), then we've added $v(A)$, a multiple of d, to $v(B)$ or $v(C)$ and $v(A)$ is now 0. The conclusion of the lemma still is true. If our next move is to fill B from A, then $v(B)$ is now b and $v(A)$ is now (old $v(A)$) $-$ $(b - v(B))$. This argument proves the lemma, by induction.

Proof of Proposition (10.1) If the pouring problem is soluble, then by Lemma (10.2), e is a multiple of $d = \gcd(a, b)$. If, conversely, $e = kd$ for some k, we first divide e by a, getting $e = aq + r$ with $0 \leq r < a$. We fill A q times, emptying it each time into C; the state of things is now

X	A	B	C
$v(X)$	0	0	aq

All we need do now is measure r units, and we will show that we can do it *without using* C. Of course, if $r = 0$, we're done. So assume now $r \geq 1$.

Notice that if a or b equals d, then it is trivial to achieve $kd = e$, so we may also assume both a and b are greater than d. Since r is a multiple of d, there are *positive* integers x, y such that

$$ax - by = r.$$

Since $r < a$ the coefficient of b must be negative. If we were using C, the proof now would be trivial: we would fill A (from the reservoir) and empty it into C x times; then we would fill B from C and empty B onto the ground y times. The net increase in C would be r, and we'd be done. But many pouring problems have only A and B with $r < a$, so we do it that way. The idea is still the same—fill A x times and empty it into B y times.

(10.3) We have $ax > by$. Thus in filling A x times we are able to pour from A into B enough to fill B exactly y times. (Each time we fill B we discard the contents.) When we finish, there remain r units in A. QED

Sometimes the problem is posed with a small reservoir R of $a + b$ units and target of $r < \max\{a, b\}$ units. To solve it, we follow the same method except that we pour from B back into R.

This approach helps to solve some of the more complicated pouring problems, but now we have to think first

Example (from Sam Loyd (16.1)) Consider two full reservoirs R and S of 10 units each, and containers A of 5 units and B of 4 units. Problem: Measure exactly 2 units into each of A and B.

Remember that the only legal move is to fill or empty a container.

(10.4) Therefore any reachable state of this system must have at least one container full or empty.

Therefore the final state must be

R	S	A	B
10	6	2	2

since taking 4 units from the 20 in R and S cannot empty either one. The next thought required is to list the possible penultimate states, those just before the final state. Since in the final state nothing is empty and only R is full, the last move must have been to fill R. If no move was wasted, R must have been filled from A or B. Thus there are only two possible penultimate states, by (10.4), namely,

(10.5)

	R	S	A	B
(*i*)	8	6	2	4
		or		
(*ii*)	7	6	5	2

By the previous method (10.3) we can easily get 2 units in A or B using just R, A, and B. If we thus put 2 units in A we have the state

R	S	A	B
8	10	2	0

We fill B from S to achieve state (10.5*i*). Now we solve the problem by filling R from B. (It is not clear whether the other penultimate state (10.5*ii*) is reachable.) Here is the full sequence of steps, in which a blank indicates no change:

	R	S	A	B	
Capacity	10	10	5	4	
Target	10	6	2	2	
Start	10	10	0	0	
Intermediate target	8	10	2	0	Use $2 = 5 \cdot 2 - 4 \cdot 2$
Start	10	10	0	0	
	5		5		
			1	4	
	9			0	
			0	1	
	4		5		
			2	4	
	8	10	2	0	
		6		4	
	10	6	2	2	

For another approach to pouring problems, see (16.2).

There are "pouring" algorithms for finding the gcd. That is, they use no division, only subtraction. Some go back to the time of Euclid; more modern ones with improvements are faster than the Euclidean algorithm.†

Practice (10.6P) Suppose you have containers A and B holding 7 pints and 3 pints, respectively. You have a large supply of water. How can you measure exactly 5 pints?

11. Primes

Now we discuss prime numbers (*primes*) and their most elementary properties. We first recall the useful result on divisibility, Corollary (7.10).

Another result we'll need is

Lemma (11.1) For all $n \geq 1$, if $b_1, \ldots, b_n$ are integers to each of which the integer a is relatively prime, then a is relatively prime to $b_1 \cdots b_n$.

Proof For $i = 1, \ldots, n$ there are integers x_i, y_i such that

(11.2) $$1 = ax_i + b_iy_i,$$

†Graham H. Norton, Extending the Binary GCD Algorithm, in Jacques Calmet (ed.), *Algebraic Algorithms and Error-Correcting Codes*, Lecture Notes in Computer Science 229, (Berlin: Springer-Verlag, 1986), pages 363–372.

Donald E. Knuth, *The Art of Computer Programming*, Vol. 2, Seminumerical Algorithms (2d ed:), (Reading, Addison-Wesley, 1981), pages 319–324.

as we see from the hypothesis and Theorem (7.5). Multiply these all together to get

$$(11.3) \qquad \begin{aligned} 1 &= \prod_{1 \le i \le n} (ax_i + b_i y_i) \\ &= (b_1 b_2 \cdots b_n)(y_1 y_2 \cdots y_n) + aB, \end{aligned}$$

where B is some integer (a polynomial in a and in the x_i's, b_i's, and y_i's; the exact formula for B is irrelevant). This expression is true because when the product in (11.3) is expanded, every term has a factor of a in it except the one obtained from the choice of $b_i y_i$ in every $(ax_i + b_i y_i)$. Thus we have a linear combination of a and $b_1 \cdots b_n$ equal to 1. By (7.9) a and $b_1 \cdots b_n$ are relatively prime. QED

DEFINITION (11.4) A **prime** is an integer greater than 1 that has no positive divisors except itself and 1.

Integers greater than 1 that are not prime are called **composite.**

Examples The smallest few primes are 2, 3, 5, 7, 11, 13, 17, 19, 23, 29, 31, 37, 41, 43, 47. Notice that 1 is *not* a prime, that 2 *is* prime; and that 4, 6, 8, and 9 are composite.

We use the word *prime* as both a noun and an adjective. For example, in 1964, at least one postage meter at the University of Illinois printed on each envelope next to the postage "$2^{11,213} - 1$ is prime." At the time it was the largest known prime, a fact proved at Illinois. (See Section (13)).

We have already introduced another use of the term: a is prime to b means $\gcd(a, b) = 1$, a relationship between a and b. Prime-ness, on the other hand, is a property of (some) single integers.

Primes have fascinated people interested in mathematics since the time of Euclid (if not before). They have an irregular distribution among the integers. One famous problem, posed in antiquity but still unsolved, is to prove or disprove that there are infinitely many *twin primes* (two primes differing by 2, such as 3 and 5, 5 and 7, 11 and 13, 41 and 43). We confine ourselves to the simplest of results on primes, however.

Lemma (11.5) Let p be a prime. For all integers n, p is prime to n or p divides n.

Proof $\qquad \gcd(n, p) = 1 \quad \text{or} \quad p.$

Lemma (11.6) If a and b are any integers, and if the prime p divides ab, then p divides a or p divides b:

$$p \mid ab \quad \longrightarrow \quad p \mid a \ \ or \ \ p \mid b.$$

Proof By the previous lemma, p divides a or p is relatively prime to a. If the former, we are done. If the latter, then we use (7.10) to conclude that $p \mid b$. QED

Corollary If p divides $b_1 \cdots b_n$ then p divides b_i (for some $i = 1, \ldots, n$).

Lemma (11.7) If $q_1, \ldots, q_r$ are $r \geq 1$ primes, and if p is prime, then

$$p \mid (q_1 \cdots q_r) \longleftrightarrow \exists i : p = q_i.$$

Proof In the direction "$\longleftarrow$" the result is obvious: If $p \in \{q_1, \ldots, q_r\}$, then p divides $q_1 \cdots q_r$.

For the converse, suppose $p \mid (q_1 \cdots q_r)$. Then by the previous result, p divides q_i for some i. But since p and q_i are prime, they are equal. (Why?)

Theorem (11.8) *(Unique Factorization).* Every integer $n \geq 2$ is a product of primes:

$$\forall n \geq 2 \; \exists \text{ primes } p_1, \ldots, p_r \quad \text{such that } n = p_1 \cdots p_r.$$

Apart from their order in the list $p_1, \ldots, p_r$, the primes are unique. ■

Examples $30 = 2 \cdot 3 \cdot 5, \quad 24 = 2 \cdot 2 \cdot 2 \cdot 3, \quad 31 = 31, \quad 36 = 2 \cdot 2 \cdot 3 \cdot 3, \quad 2^{11} - 1 = 23 \cdot 89.$

Proof of (11.8) We proved the existence of the primes $p_1, \ldots, p_r$ by induction on n in Chapter 3, (5.4). For the uniqueness, suppose for some n that there are two factorizations

$$p_1 \cdots p_r = n = q_1 \cdots q_s.$$

Since p_1 is a prime dividing the left side, it divides the right side; and Lemma (11.7) applies to tell us that $p_1 = q_i$ for some i. Let us choose the notation so that $p_1 = q_1$. Cancel p_1 and we have

$$p_2 \cdots p_r = q_2 \cdots q_s.$$

Repeat this argument until all the primes are canceled from one side or the other. This would leave either $1 = 1$, in which case we would have proved the desired uniqueness; or we would have $1 = \cdots q_s$. The last is impossible because q_s divides the right-hand side but not the left-hand side. (Or it could be $\cdots p_r = 1$: same consideration applies.) QED

This theorem allows us to write any integer $n \geq 2$ uniquely as

$$n = p_1^{e_1} p_2^{e_2} \cdots p_r^{e_r}$$

where $r \geq 1$, the p_i's are mutually distinct, and $e_i \geq 1$. Thus

$$24 = 2^3 \cdot 3, \qquad 30 = 2^1 \cdot 3^1 \cdot 5^1, \qquad 108 = 2^2 \cdot 3^3.$$

We can even push this idea a little further and say that for all $n \geq 1$ we may express n uniquely as

$$n = \prod_{p \text{ prime}} p^{e_p},$$

where the exponent e_p is now allowed to be 0 and the product is over *all* primes p. Of course p^0 is 1, so if $p \nmid n$ we set $e_p = 0$. (The symbol $\nmid$ is the negation of $\mid$: $p \nmid n$

means "p does not divide n.") The product is then a finite product of integers greater than 1 times an infinite product of 1's (which is 1). Using this notation we state

Proposition (11.9) For all $a, b \in \mathbf{N}$

$$\text{if } a = \prod p^{e_p} \text{ and } b = \prod p^{f_p},$$
$$\text{then } \gcd(a,b) = s = \prod p^{\min\{e_p, f_p\}}.$$

The products are over all primes p.

The exponent $\min\{e_p, f_p\}$ is the smaller of the two exponents. Thus $\gcd(2^2 \cdot 3^1 \cdot 5^4 \cdot 7^3, 2^4 \cdot 5^1 \cdot 7^2) = 2^2 \cdot 5^1 \cdot 7^2$.

Proof of (11.9) With s defined as the product above, clearly s divides each of a and b. Conversely, if $t \geq 1$ is a common divisor of a and b, then $t|a$ implies $g_p \leq e_p$ for all p, where

$$t = \prod_{p \text{ prime}} p^{g_p},$$

and similarly for b. Thus $g_p \leq e_p$ and $g_p \leq f_p$, so $g_p \leq \min\{e_p, f_p\}$. Therefore $t \mid s$. By (7.15), $s = \gcd(a,b)$. QED

Just as *being a multiple of* is the converse relation to *being a divisor of*, there is a converse concept to that of greatest common divisor. It is called *least common multiple*.

DEFINITION (11.10) The **least common multiple** of nonzero integers a, b is the smallest positive integer that is a multiple of a and of b.

NOTATION: $\operatorname{lcm}(a,b)$.

Proposition (11.11) With notation as in (11.9),

$$\operatorname{lcm}(a,b) = \prod_{p \text{ prime}} p^{\max\{e_p, f_p\}}. \blacksquare$$

Corollary (11.12) For all $a, b \in \mathbf{N}$

$$\gcd(a,b)\operatorname{lcm}(a,b) = ab. \blacksquare$$

Corollary (11.12) ties greatest common divisor and least common multiple together in a simple way. They are *dual* to each other. Here's an example of what *dual* means: From (11.9) we can see—what we already knew from (7.8)—that every common divisor of a and b divides $\gcd(a,b)$. From (11.11) we can see the dual property of lcm, that every common multiple of a and b is a multiple of $\operatorname{lcm}(a,b)$.

Example In adding two fractions we often find the lcm of the denominators. For example,

$$\frac{11}{30} + \frac{1}{105} = \frac{77}{30 \cdot 7} + \frac{2}{2 \cdot 105} = \frac{79}{210}.$$

In this process the lcm often gets some wrong name, like *greatest common denominator*.

Proposition (11.13) (The square-root bound) For all $n \geq 4$, if n is composite, then there is a prime $p \leq \sqrt{n}$ that divides n.

Proof By hypothesis, $n = ab$ for integers a, b both greater than 1. If $a > \sqrt{n}$ and $b > \sqrt{n}$ then $ab > \sqrt{n} \cdot \sqrt{n} = n$, a contradiction. So one of them, say a, is at most $\sqrt{n}$. Since a is divisible by some prime p, that prime is also at most $\sqrt{n}$. QED

Theorem (11.14) There are infinitely many primes.

Proof Suppose not. Then for some integer n there are only n primes. Say they are $p_1, p_2, \ldots, p_n$. Then $p_1 p_2 \cdots p_n + 1$ is not divisible by any of the primes $p_1, \ldots, p_n$, so it must be a prime, contradicting our supposition that $p_1, \ldots, p_n$ are all the primes. QED

In fact, if we multiply the first n primes together, and add 1, as $p_1 \cdots p_n + 1 =: q_n$, we do get primes for $n = 1, \ldots, 5$ but not for $n = 6$.

$$\begin{aligned}
2 + 1 &= 3 \\
2 \cdot 3 + 1 &= 7 \\
2 \cdot 3 \cdot 5 + 1 &= 31 \\
2 \cdot 3 \cdot 5 \cdot 7 + 1 &= 211 \text{ (prime)} \\
2 \cdot 3 \cdot 5 \cdot 7 \cdot 11 + 1 &= 2311 \text{ (prime)} \\
2 \cdot 3 \cdot 5 \cdot 7 \cdot 11 \cdot 13 + 1 &= 30{,}031 = 59 \cdot 509
\end{aligned}$$

That 3, 7, and 31 are primes can be "explained" by the square-root bound (11.13), since $\lfloor \sqrt{q_n} \rfloor \leq p_n$ for $n = 1, 2, 3$. For $n \geq 3$ there are more and more primes between p_n and $\sqrt{q_n}$, and it is hard to predict whether they divide q_n.

Practice (11.15P) Prove that for any $n \in \mathbf{N}$, if no prime $p \leq n^{1/3}$ divides n, then n is prime or n is the product of two primes.

12. The Sieve of Eratosthenes

Suppose we write out the first n^2 integers in their natural order, delete 1, and then, counting from the first number, 2, in the list, check off every second number (indicated here by putting them in italics)

(12.1)

$$\mathbf{2}\ \mathbf{3}\ \mathit{4}\ \mathbf{5}\ \underline{\mathit{6}}\ 7\ \mathit{8}\ \underline{9}\ \mathit{10}'\ 11\ \underline{\mathit{12}}\ 13\ \mathit{14}\ \underline{15'}\ \mathit{16}\ 17\ \underline{\mathit{18}}\ 19\ \mathit{20}'$$

$$\underline{21}\ \mathit{22}\ 23\ \underline{\mathit{24}}\ 25'\ \mathit{26}\ \underline{27}\ \mathit{28}\ 29\ \underline{\mathit{30}'}\ 31\ \mathit{32}\ \underline{33}\ \mathit{34}\ 35'\ \underline{\mathit{36}}.$$

We show the process in (12.1) for $n = 6$. Make the 2 boldface. Now go to the first unmarked number, 3, make it boldface, and check off every third number after it. (Done with underlines above.) Now the first untouched number in the list is 5. Make it boldface and check off (here with primes ′) every fifth number after it. Once every $i \leq n$ is either checked or boldfaced, stop. Now the boldfaced, and also, by (11.13), the unchecked numbers are all the primes up to n^2. So we find that not only are 2, 3, and 5 prime, but also 7, 11, 13, 17, 19, 23, 29, and 31.

13. Mersenne and Fermat Primes

Special types of primes have been heavily studied. Two such are Mersenne and Fermat primes.

Mersenne primes. A Mersenne prime is by definition any prime p of the form $2^n - 1$. The first few Mersenne primes follow:

$$\begin{aligned} 3 &= 2^2 - 1 \\ 7 &= 2^3 - 1 \\ 31 &= 2^5 - 1 \\ 127 &= 2^7 - 1. \end{aligned}$$

It is easy to see that if $2^n - 1$ is prime, then n is prime. The converse is false, however, because

$$2^{11} - 1 = 23 \cdot 89.$$

The largest prime known at any given time has usually been a Mersenne prime. Let's say that m denotes the largest exponent for which $2^m - 1$ is known to be prime. Of course, m varies with time. Here is a selection of values of m, the earliest from 1750, when Euler showed that $m = 31$. In 1876, m was 127. It remained 127 until 1952, when Robinson found five new values, the largest being $m = 2,281$. In 1963, m was 11,213, as mentioned near the beginning of (11). In 1971, Tuckerman found $m = 19,937$. By 1983, Slowinski found $m = 132,049$, and in 1985 he showed $m = 216,091$. At this point a nonMersenne prime intervenes: In 1989, $p := 391,581 \times 2^{216,193} - 1$ was proved to be prime. But in 1992, Slowinski riposted with $m = 756,839$. The Mersenne prime for that m is much larger than p.

There is a fascinating account of primes, including the pre-1990 records above, in (16.4). That book is advanced, however, for readers of this book.

Fermat primes. A Fermat prime is by definition a prime of the form $2^{2^m} + 1$. The only known Fermat primes are those for $m = 0, 1, \ldots, 4$:

$$\begin{aligned} 2^1 + 1 &= 3 \\ 2^2 + 1 &= 5 \\ 2^4 + 1 &= 17 \\ 2^8 + 1 &= 257 \\ 2^{16} + 1 &= 65,537. \end{aligned}$$

The Fermat number $2^{32} + 1$ is, however, not prime, as Euler showed in 1739 (see 16.3, page 74); he proved 641 divides it.

Still, the Fermat primes have been proposed for applications in signal processing using a fast Fourier transform.

It is not hard to show that if $2^h + 1$ is an odd prime, then h is a power of 2. As we just remarked, Euler showed the converse false.

14. Fermat's Method of Descent: An Example

We close with the example promised in Chapter 3, Section 15. Consider the claim

(14.1) There are no odd positive integers x, y such that $3x^2 + y^2 = 2^{2k}$ for any integer k, except $x = y = k = 1$.

We now give a proof (due to C. L. Chen) of (14.1) by Fermat's method of descent. We do induction on k. For $k = 2$,

$$3x^2 + y^2 = 16$$

has no such solution because x has to be 1, forcing $y^2 = 13$. For the inductive step we prove instead of $(\bigwedge_{2 \le j < k} S(j)) \longrightarrow S(k)$ the contrapositive

$$\sim S(k) \longrightarrow \sim S(k-1) \vee \cdots \vee \sim S(2).$$

That is, we prove:

(14.2) If there is a solution to (14.1) for $k > 2$ then there is one for a smaller value of k.

So suppose that $k > 2$ and that $\exists$ odd $x, y \ge 1$ such that

$$3x^2 + y^2 = 2^{2k}.$$

Then x and y are relatively prime because $\gcd(x, y)$ divides 2^{2k} (why?) and is thus a power of 2. But x and y are odd, so their gcd is 2^0. Then

$$3x^2 = (2^k - y)(2^k + y).$$

Now $d := \gcd(2^k - y, 2^k + y) = 1$, because the sum of these numbers is 2^{k+1}—so d is a power of 2, by (7.7). And 2 does divide 2^k, since $k > 0$; but y is odd, so $2 \nmid (2^k - y)$.

Therefore, by prime factorization, there are integers r, s such that

$$3x^2 = 3r^2 \cdot s^2 \quad \text{and} \quad \{3r^2, s^2\} = \{2^k - y, 2^k + y\}.$$

It follows that

$$3r^2 + s^2 = 2^{k+1}. \tag{14.3}$$

Both r and s are odd, so if we can show that $k+1$ is even $(= 2t)$ we shall have finished the inductive step, because $k > 2$ implies $1 < (k+1)/2 = t < k$. But if k is even, then it is easy to show that

(14.4) 2^j has remainder 2 on division by 3 if $j \ge 1$ is odd.

But

(14.5) $3r^2 + s^2$ has remainder 1 on division by 3 if s is not divisible by 3.

Clearly 3 does not divide s because $3 \mid s \longrightarrow 3 \mid 2^{k+1}$, from (14.3), a contradiction. Therefore $k+1$ cannot be odd, so (14.3) is the *smaller* solution claimed in (14.2). Thus (14.1) is proved. QED

As we said in Chapter 3, we shortcut this induction proof in Chapter 7.

15. Summary

The division algorithm (2.1) is simple, yet it is the foundation for all of this chapter. We use it to prove (7.5), that the greatest common divisor d of a and b can be represented as a linear combination $d = ax + by$. Theorem (7.5) is only our second nonobvious basic result (the first being (11.1) of Chapter 3). It is as much a lemma as a theorem, since we use it to prove more interesting results.

We use (7.5) to prove Corollary (7.10), that if $\gcd(a, b) = 1$ and b divides ac, then b divides c. As we proved easily by induction in Chapter 3 (5.4), every integer $n > 1$ is prime or is the product of primes. Now with (7.10) we prove that n determines these primes uniquely. Unique factorization of integers into primes seems intuitively obvious—until we try to prove it and see that to do so we need the division algorithm and its consequences outlined above.

Here is an example of a system in which unique factorization does not hold. Consider the subset I of all positive integers having remainder 1 on division by 5:

$$\begin{aligned} I &:= \{1, 6, 11, 16, 21, 26, 31, 36, \dots\} \\ &= \{5k + 1;\ k \in \mathbf{Z}, k \geq 0\} \end{aligned}$$

The product of any two integers in I is in I. We may define a "prime" in I as any element of I which is not the product of any two elements of I except itself and 1 (and is not 1). Then 6, 11, 16, and 21 are "prime," for example. In fact, all elements of I less than 100 except 36, 66, and 96 are "prime" in I. These three do have unique factorizations: $36 = 6 \cdot 6, 66 = 6 \cdot 11, 96 = 6 \cdot 16$. But $6 \cdot 56 = 16 \cdot 21$, so some elements in I are products of "primes" in more than one way.

We also studied the Euclidean algorithm, used to compute the gcd, and we showed how to extend it to compute the x and y of (7.5).

For fun we looked at pouring problems as an application of divisibility. For your cultural enrichment there is also the square-root bound (11.13), the sieve of Eratothsenes, Mersenne and Fermat primes, and an example of Fermat's method of descent.

16. Further Reading

(16.1) Martin Gardner, ed. *Mathematical Puzzles of Sam Loyd*, Dover, New York, 1961.

(16.2) T. H. O'Beirne, *Puzzles and Paradoxes*, Dover, New York, 1965.

(16.3) Oystein Ore, *Number Theory and Its History*, McGraw-Hill, New York, 1948. (Reprinted by Dover, New York, 1989.)

(16.4) Paulo Ribenboim, *The Little Book of Big Primes*, Springer-Verlag, New York, 1991.
(16.5) André Weil, *Number Theory for Beginners*, Springer-Verlag, New York, 1979.

17. Problems for Chapter 6

Problems for Section 1

1.1 Consider your social security number. Represent it as $a_1a_2\dots a_9$. Calculate the floor function of the four ratios a_1/a_2, a_2/a_3, a_3/a_4, and a_4/a_5. Also find the ceiling function of the next four ratios. If you have zero as a digit, delete any ratio calling for division by 0.

1.2 For what integers a is $|2^a| = |2^{-a}|$?

1.3[Ans] Prove that for all $a \in \mathbf{Z}$, a divides 0.

1.4 Prove that for all $x \in \mathbf{R}$, $\lfloor x + \frac{1}{2} \rfloor$ is the unique nearest integer to x unless $x + \frac{1}{2} \in \mathbf{Z}$.

1.5 For all real t prove that $\lceil t \rceil = -\lfloor -t \rfloor$, and that $\lfloor t \rfloor = -\lceil -t \rceil$.

1.6 Consider the inequality (1.5). Prove: if $tt' \le 0$, then the left-hand inequality may be replaced by equality. If $tt' \ge 0$, the same may be done for the right-hand inequality.

1.7 (*i*) Find a neater formula than the one in (1.14) for the insurance age.
(*ii*) Solve the same problem, except if $x - \lfloor x \rfloor = \frac{1}{2}$, make the insurance age $\lfloor x \rfloor$.

1.8 Let x and y be any real numbers such that $x + y = n \in \mathbf{Z}$. Prove that

$$\lfloor x \rfloor + \lceil y \rceil = n.$$

1.9[Ans] Prove that if n is an integer for which Fermat's equation $x^n + y^n = z^n$ (Chapter 1, (7)) has an integral solution, then it also has one for all positive divisors of n. Hence it would be enough to prove Fermat's theorem only for $n = 4$ and for odd primes n—explain why.

1.10 Consider the functions defined by the rules $g(x) := x - \lfloor x \rfloor$ and $h(x) := x - \lceil x \rceil$. For which $x \in \mathbf{R}$ is $|g(x)| = |h(x)|$? Prove your answer.

1.11 Prove that for all odd integers n, 8 divides $n^2 - 1$.

Problems for Section 2

2.1 Find integers q and r satisfying (2.1) for
(*i*) $a = 297$ and $b = 18$; (*ii*) $a = 171$ and $b = -12$.

2.2 Use induction on $j \ge 1$ to prove that on division by 3, 2^j has remainder 1 if j is even, remainder 2 if j is odd.

2.3 Let $n \ge 0$ be an integer. Prove by induction or otherwise that 5 divides $n(n^4 - 1)$.

2.4 Let $b \ne 0$ be an integer. Prove by induction on a that for all $a \in \mathbf{Z}$ there are unique $q, r \in \mathbf{Z}$ such that $a = bq + r$ and $0 \le r < |b|$.

2.5 Let x be any real number.

(*i*) Prove that $\lfloor \frac{1}{2}\lfloor x \rfloor \rfloor = \lfloor \frac{x}{2} \rfloor$.
(*ii*) For any $m \in \mathbf{N}$, prove that

$$\left\lfloor \frac{1}{m}\lfloor x \rfloor \right\rfloor = \left\lfloor \frac{x}{m} \right\rfloor.$$

(*iii*) Prove the analogues of (*i*) and (*ii*) for the ceiling function.

2.6 Let $m \in \mathbf{N}, a \in \mathbf{Z}$. Prove that

$$\left\lceil \frac{a}{m} \right\rceil = \left\lfloor \frac{a + m - 1}{m} \right\rfloor.$$

2.7 Let $t \in \mathbf{Z}$.

(*i*) Prove

$$\left\lceil \frac{1}{2}\left\lfloor \frac{t}{2} \right\rfloor \right\rceil = \left\lfloor \frac{t+2}{4} \right\rfloor.$$

(*ii*) Find and prove an analogous expression for $\lfloor \frac{1}{2} \lceil t/2 \rceil \rfloor$.
(*iii*) Prove that the results of (*i*) and (*ii*) hold for any $t \in \mathbf{R}$.

***2.8**[Ans] Generalize the preceding problem as follows. Let $x \in \mathbf{R}$ and let $m, n \in \mathbf{N}$. Then find and prove an expression like that in Problem 2.7 (*i*) for

$$\left\lceil \frac{1}{m}\left\lfloor \frac{x}{n} \right\rfloor \right\rceil.$$

***2.9**[Ans] Egyptian mathematicians in 1800 B.C. represented rational numbers between 0 and 1 as sums of unit fractions $1/a + 1/b + \cdots + 1/k$ where $a, b, \dots, k$ were distinct positive integers. For example, they wrote $\frac{2}{5}$ as $\frac{1}{3} + \frac{1}{15}$. Prove that it is always possible to do this in a systematic way: If $0 < m/n < 1$, then

$$\frac{m}{n} = \frac{1}{q} + \left\{ \text{representation of } \frac{m}{n} - \frac{1}{q} \right\}, \text{ where } q := \left\lceil \frac{n}{m} \right\rceil.$$

This is Fibonacci's algorithm, dating from 1202 A.D. (Graham, Knuth, and Patashnik). [What you must do is show that the procedure terminates after a finite number of steps.]

***2.10** Ans Prove that for all $n \in \mathbf{N}$

$$n + \left\lfloor \frac{n + \lfloor \frac{n}{5} \rfloor}{2} \right\rfloor = \left\lfloor \frac{8n}{5} \right\rfloor.$$

2.11 We carry problem 2.5 further.

(i) Disprove: $\forall r,\ x \in \mathbf{R}$, if $0 \le r < 1$, then $\lfloor r \lfloor x \rfloor \rfloor = \lfloor rx \rfloor$.

(ii) Is part *(ii)* of problem 2.5 true if m is a negative integer? Explain.

**(iii)*Ans A converse of part *(ii)* of problem 2.5: Let $r \in \mathbf{R}$. Prove that if

$$\forall x \in \mathbf{R}\ \lfloor r \lfloor x \rfloor \rfloor = \lfloor rx \rfloor,$$

then $r = 0$ or $r = 1/m$ for some $m \in \mathbf{N}$.

Problem for Section 3

3.1 Use **rem** n **by** b to get r for $n = -11, b = 4$.

Problems for Section 4

4.1 Find the base 8 representation of 100 (base 10) and of 1000 (base 9). Show how you solve these problems.

4.2 Express 197 (base 10) in the base 7. Show your steps.

4.3 Prove the uniqueness of the base b representation (4.1) of an integer.

4.4 Ans Let $b \ge 2$ be an integer. Write a recursive algorithm b-**ary** n to find, for any $n \in \mathbf{N}$, the base b expression of n. [The latter is often called the b-ary expression of n. ("Ary" rhymes with "Larry.")]

Problems for Section 7

7.1 Prove that as a function (see (7.1)) gcd has range $\mathbf{N} \cup \{0\}$.

7.2 Ans Let $a, b, m \in \mathbf{Z}$. Prove that

$$\gcd(ma, mb) = |m| \gcd(a, b).$$

7.3 Define S as in (7.6). Prove that $a, b \in S$.

7.4 For the same S, prove that $\{ kd;\ k \in \mathbf{Z} \} \subseteq S$.

7.5 Define S as in (7.6). Prove $a^2 + b^2 \in S$.

7.6 If a and b are integers satisfying $66a - 51b = 21$, what can you say about the value of $\gcd(a, b)$?

7.7 Give an example of integers a, b such that there are integers x, y making $ax + by = 2$, but $\gcd(a, b) \ne 2$. Explain.

7.8 Let $d \ge 0$. Define

$$(\star) \qquad S = \{ kd;\ k \in \mathbf{Z} \}.$$

Prove that d is the only nonnegative element of S for which $(\star)$ is true. That is, if $e \ge 0$ and $S = \{ ej;\ j \in \mathbf{Z} \}$, prove $e = d$.

7.9 Prove the uniqueness claimed in (7.15).

7.10 [Recommended] With $d = \gcd(a, b) \ne 0$, prove that a/d and b/d are relatively prime.

7.11 [Recommended] Let a, b, x, and y be integers such that $\gcd(a, b) = ax + by$. Prove that x and y are relatively prime.

7.12 Let $a, b, d \in \mathbf{Z}$. Suppose that d is a common divisor of a and b, but that $d \ne \pm \gcd(a, b)$. Prove that there are no integers x, y such that $d = ax + by$.

7.13 Ans Let a, b be relatively prime integers. Suppose m is any integer such that $a \mid m$ and $b \mid m$. Prove that $ab \mid m$.

***7.14** Write a recursive algorithm to calculate $\gcd(a, b)$.

7.15 Ans (Weil) Prove that in the Fibonacci sequence $1, 2, 3, 5, 8, 13, \ldots$, in which each term after the second is the sum of the preceding two terms, every two consecutive terms have gcd 1.

***7.16** (Weil) If p, q, r, s are integers such that $ps - qr = \pm 1$, and a, b, a', b' are integers such that

$$(\star) \qquad \begin{aligned} a' &= pa + qb \\ b' &= ra + sb, \end{aligned}$$

prove that $\gcd(a, b) = \gcd(a', b')$. [Hint: Solve $(\star)$ for a, b.]

Problems for Section 8

8.1 Find by Euclid's algorithm $\gcd(242, 165)$ and $\gcd(17296, 18416)$.

8.2 Use the Euclidean algorithm to find the gcd of *(i)* 726 and 946; *(ii)* 1247 and 98.

8.3 [Ans] Prove that the division algorithm (2.1) and the Euclidean algorithm (8.1) remain true if we use the *least absolute* remainder instead of the one in (2.1). That is, we divide a by b as $a = bq' + l$, where $-b/2 < l \leq b/2$.

Problems for Section 9

9.1 Run the Euclidean algorithm backward to find $x, y \in \mathbf{Z}$ such that

$$\gcd(242, 165) = 242x + 165y.$$

9.2 Use the algorithm **EE** to find the coefficients x, y in $\mathbf{Z}$ such that $243x + 712y = 1$. Show your work in a table like that in (9.8).

9.3 Find integers x and y such that $\gcd(132, 210) = 132x + 210y$.

9.4 Write an algorithm that calls **EE** and produces $\gcd(a, b)$ and the extension for all a, b in $\mathbf{Z}$.

9.5 Why is t needed in **EE**? Why not simply set $x := x' - qx$ and $y := y' - qy$?

9.6 In our main theorem (7.5) on the gcd we proved that for all $a, b \in \mathbf{Z}$ not both 0, $\exists x, y \in \mathbf{Z}$ such that

$$\gcd(a, b) = ax + by.$$

How many solutions (x, y) are there to this equation?

*9.7 As a project, read Daniel A. Marcus, An alternative to Euclid's algorithm, *American Mathematical Monthly*, Vol. 88 (April 1981), pages 280–283, and present this extended algorithm to your class.

*9.8 Suppose $a, b \in \mathbf{N}$ and $\gcd(a, b) = d$. Suppose not both a and b equal d. Assuming Theorem (7.5), use a simple algebraic argument to prove that there are coefficients $x, y \in \mathbf{Z}$ such that

$$ax + by = d \quad \text{and} \quad |x| < b/d, \ |y| < a/d.$$

Problem for Section 10

10.1 The following problems have been posed since the sixteenth century A.D. (15.2, page 49).

(i) With containers A, B, C of capacity 8, 5, and 3 units, respectively, divide 8 units of water into two equal parts.

(ii) With containers W, X, Y, Z with respective capacity 5, 11, 13, and 24 units, divide 24 units of water into three equal parts.

Instead of just writing down your series of steps, explain your approach.

Problems for Section 11

11.1 Prove or disprove: for all integers $n \geq 0, n^2 - n + 41$ is prime.

11.2 Prove Lemma (11.1) by induction on n.

11.3 For all $n \in \mathbf{N}$, n is the square of an integer if and only if the total number of positive divisors of n is odd.

11.4 Prove that if $2^n - 1$ is prime, then n is a prime. Is the converse true?

11.5 Let a and b be integers different from 0. Define sets A, B, α, β as follows:

$$A = \{x;\ x \mid a,\ x \geq 1,\ x \in \mathbf{Z}\}$$
$$B = \{y;\ y \mid b,\ y \in \mathbf{Z},\ y \geq 1\}$$
$$\alpha = \{u;\ u \in \mathbf{Z},\ u \geq 1,\ a \mid u\}$$
$$\beta = \{v;\ v \in \mathbf{Z},\ v \geq 1,\ b \mid v\}$$

Prove: $\gcd(a, b) = \max A \cap B$, and $\operatorname{lcm}(a, b) = \min \alpha \cap \beta$, where $\max X$ is the largest integer in the set X, if there is one; and $\min Y$ is the smallest integer in the set Y, if there is one.

11.6 Find the prime factorization of

(i) 33,815,202.
(ii) 35,225,372.
(iii) 80,353,292.

[Hint: One of these is straightforward. The other two are nasty, but each in the same way as the other. Keep (11.13) in mind.]

11.7 Prove Proposition (11.11) and Corollary (11.12).

11.8 Let $n > 1$ be an integer that is not an integral power of 2. What can you say about the prime factorization of n?

Problem for Section 13

13.1 Prove that if $2^h + 1$ is an odd prime, then h is an integral power of 2.

Problem for Section 14

14.1 Prove the assertion (14.5).

Some Test Questions

T1. Consider the relation R on the set G of integers $n \geq 2$ defined as follows:

$\forall a, b \in G$, $a\,R\,b$ iff the smallest prime dividing a equals the smallest prime dividing b

Prove that R is an equivalence relation on G, and *identify* the equivalence class of 3.

T2. What is the range of the function $g : \mathbf{Z} \longrightarrow \mathbf{N}$ defined by the rule

$$n \in \mathbf{Z},\ g(n) = \gcd(n-1, n+1)?$$

Prove your answer. Also, for each point $y \in$ range g, identify the set $\overleftarrow{g}^{-1}(y)$.

T3. Use the Euclidean algorithm (the hand computation version) to find the gcd of 1986 and 1878. Check that your answer is a common divisor.

T4. What is the least common multiple m of 30 and 42? Explain your answer in terms of prime factorization.

T5. What is the set of primes that divide 120? In the expression $120 = p^{e_1} \ldots p_r^{e_r}$, where the p_i are distinct primes and the e_i are positive integers, identify r, and for all $i = 1, \ldots, r$, identify p_i and e_i.

T6. Find the greatest common divisor of 1988 and 1778 by using the Euclidean algorithm.

T7. Represent 297 (base 10) in the base 3. Show your procedure step by step.

T8. Suppose you are told that four integers a, b, x, and y satisfy the equation $ax + by = 6$. What can you say about the value of $\gcd(a, b)$? Explain.

T9. Use induction on $j \geq 1$ to prove that 4^j has remainder 1 on division by 5 if j is even and remainder 4 if j is odd.

T10. Prove or disprove:

$$\forall x, y \in \mathbf{R} \qquad \lceil xy \rceil = \lceil x \rceil \lceil y \rceil.$$

T11. Assume $x - \frac{1}{2}$ is not an integer. Prove that $\lceil x - \frac{1}{2} \rceil$ is the nearest integer to x.

T12. Let n be any integer. Prove that each of the integers $n, n+1$, and $2n+1$ is relatively prime to the other two.

T13. Prove or disprove that for all $a, b, m \in \mathbf{N}$, if $m \mid ab$, then $m \mid a$ or $m \mid b$.

General Problems

*G1. Prove that gcd is an associative binary operation on $\mathbf{Z}$. (See Chapter 5, (28.2) for the definition of *binary operation*.)

*G2.[Ans] (Weil) Prove that any positive integer n that is not an integral power of 2 is the sum of (two or more) positive consecutive integers.

*G3. For all $n \in \mathbf{N}$ define

$$a_n := \sum_{1 \leq i} \left\lfloor \frac{n}{2^i} \right\rfloor.$$

For all $n \in \mathbf{N}$ prove

(*i*) $a_{2n+1} = a_{2n}$;

(*ii*) $a_{2n} = n + a_n$;

(*iii*) $a_n = n - wt_2(n)$, where $wt_2(n)$, the *binary weight* of n, is defined to be the total number of 1's in the base 2 representation of n;

(*iv*) 2^{a_n} divides $n!$ exactly. This means that $n! = 2^{a_n} \times$ (odd integer).

*G4. (Moen) Define the function $s : \mathbf{N} \longrightarrow \mathbf{N}$ by the rule

$$\forall n \in \mathbf{N} \qquad s(n) := \left\lfloor (n + \sqrt{n})^{1/2} \right\rfloor + n.$$

Find the range of s.

17. Answers to Practice Problems

(1.13P) Since $w = 1.5$, $\lceil w \rceil = 2$. The formula yields (in cents)

$$17 \cdot 2 + 5 + \begin{cases} 1 \cdot (2-1) = 40 \text{ (Mexico)} \\ 0 \cdot = 39 \text{ (United States)} \end{cases}$$

(2.10P)

$$\begin{aligned} 17 &= 6 \cdot 2 + 5; & \left\lfloor \tfrac{17}{6} \right\rfloor &= \lfloor 2+ \rfloor &&= 2 \\ 17 &= (-6)(-2) + 5; & \left\lceil \tfrac{17}{-6} \right\rceil &= \lceil -(2+) \rceil &&= -2 \\ -17 &= 6(-3) + 1; & \left\lfloor \tfrac{-17}{6} \right\rfloor &= \lfloor -(2+) \rfloor &&= -3 \\ -17 &= (-6)3 + 1; & \left\lceil \tfrac{-17}{-6} \right\rceil &= \lceil 2+ \rceil &&= 3 \end{aligned}$$

By "2+" I mean a number x such that $2 < x < 3$.

(3.2P) Suppose $x = \lfloor x \rfloor + s$, where $0 \le s < 1$. Then the nearest integer to x is either $\lfloor x \rfloor$ or $\lceil x \rceil$, depending on whether $s < \frac{1}{2}$ or $s > \frac{1}{2}$. (If $s = \frac{1}{2}$, then x is the distance from each one, and $\lfloor x + \frac{1}{2} \rfloor = \lceil x \rceil$. If $s = \frac{1}{2}$, then x is half an odd integer, and conversely.) If $s < \frac{1}{2}$, then $x + \frac{1}{2} < \lceil x \rceil$, so $\lfloor x + \frac{1}{2} \rfloor = \lfloor x \rfloor$, the nearest integer to x. If $s > \frac{1}{2}$, then $\lceil x \rceil < x + \frac{1}{2} < 1 + \lceil x \rceil$, so $\lfloor x + \frac{1}{2} \rfloor = \lceil x \rceil$.

(3.5P) Most hand calculators truncate rather than round. Thus the s they give you is less than the exact value. When you subtract $\lfloor s \rfloor$ and then multiply by b, the result will be less than the exact value. Since $129 \div 17 = 7.588235294+$, our calculator truncated the 294 to 2 instead of rounding to 3. That's an error of about 9.4×10^{-8}. When we multiply by 17 in the next step we should find an error of 159.8×10^{-8}, or about 16×10^{-7}. That's exactly the amount by which our calculator is too low.

(4.6P) This one is cute, because 8 is a power of 2. Thus

$$(2345)_8 = 2 \cdot 8^3 + 3 \cdot 8^2 + 4 \cdot 8^1 + 5 \cdot 8^0.$$

Since $8 = (1000)_2$, and

$$2 = (010)_2,\ 3 = (011)_2,\ 4 = (100)_2,\ 5 = (101)_2,$$

we may string the coefficients 2, 3, 4, and 5 in base 2 together:

$$(2345)_8 = 10\ 011\ 100\ 101$$

The "100" for 4, for example, is $100 \times 1000 = 100000$ (i.e., 4×8) in the binary representation.

(7.12P) I leave this one to you.

(7.16P) Not necessarily. Consider $a, b, c = 2, 3, 4$.

(8.5P)

$$\begin{aligned} 36 &= 14 \cdot 2 + 8 \\ 14 &= 8 \cdot 1 + 6 \\ 8 &= 6 \cdot 1 + 2 \\ 8 &= 2 \cdot 3 \end{aligned}$$

Therefore gcd $(14, 36) = 2$.

(9.9P) Using the answer to (8.5P), we get

$$\begin{aligned} 2 &= 8 - 6 \\ &= 8 - (14 - 8) = 8 \cdot 2 - 14 \\ &= (36 - 14 \cdot 2)2 - 14 \\ &= 36 \cdot 2 - 14 \cdot 5. \end{aligned}$$

(10.6P) We first check that 5 is a multiple of gcd $(7, 3)$. Now we must represent 5 as $7x + 3y$. We find a way:

$$5 = 7 \cdot 2 + 3(-3).$$

This expression tells us to fill A twice and to fill B from A three times. Here are the steps:

Amount in A:	7	4	4	1	1	0	7	5
Amount in B:	0	3	0†	3	0†	1	1	3

†We have emptied B onto the petunias.

(11.15P) If the conclusion is false, then n is the product of three integers (not necessarily prime) all bigger than 1. Say they are a, b, c. Thus $n = abc$. Suppose $a \le b \le c$. Then $a^3 \le n$, so $a \le n^{1/3}$. Some prime divides a, however, contradicting the hypothesis.

CHAPTER 7

Congruences

1. Introduction

We now take up the idea of *congruence modulo m*, a far-reaching technique for studying the integers. First published by Gauss in 1801, this approach is still basic today. It has many applications, for example, in computers, cryptography, error-correcting codes, and signal processing. It is the basis of group theory and number theory.

DEFINITION (1.1) Let $a, b, m \in \mathbf{Z}$. We say a **is congruent to** b **modulo** m, in symbols,

$$a \equiv b \pmod{m},$$

if and only if $a - b$ is a multiple of m.

(1.2) If we use the definition of *multiple* we get

$$a \equiv b \pmod{m} \quad \text{iff} \quad \exists k \in \mathbf{Z} \quad \text{such that} \quad a - b = km.$$

Examples

$17 \equiv 7 \pmod{2}$ because $17 - 7 = 5 \cdot 2$

$23 \equiv 20 \pmod{3}$ because $23 - 20 = 3$

$23 \equiv 17 \pmod{3}$ because $23 - 17 = 2 \cdot 3$

$25 \equiv 3 \pmod{11}$ because $25 - 3 = 2 \cdot 11$

$$12 \equiv 0 \pmod 4 \text{ because } 4 \mid 12.$$
$$5280 \equiv 168 \pmod{36} \text{ because } 5280 - 168 = 142 \cdot 36$$
$$-12 \equiv 14 \pmod{13} \text{ because } -12 - 14 = (-2) \cdot 13$$
$$-57 \equiv -29 \pmod 7 \text{ because } -57 + 29 = (-4)7$$
$$125 \equiv -3 \pmod{16} \text{ because } 125 + 3 = 8 \cdot 16$$

Practice (1.3P) Prove that for any $n, k \in \mathbf{Z}$, $-(n-k) \equiv n + k \pmod n$.

We emphasize an important special case of Definition (1.1):

(1.4) $$\forall a, m \in \mathbf{Z},\ a \equiv 0 \pmod m \longleftrightarrow m \mid a.$$

Because for all $a, b, m \in \mathbf{Z}$, $a \equiv b \pmod m$ iff $a \equiv b \pmod{-m}$, we usually take $m \geq 0$.

(1.5) Congruence mod m, as we often speak of it, is a binary relation on $\mathbf{Z}$, and $\equiv \pmod m$ is its infix notation. In fact, it is an equivalence relation, as we now show.

Reflexivity: $\forall a \in \mathbf{Z}, a \equiv a \pmod m$ because $a - a = 0 \cdot m$.

Symmetry: $\forall a, b \in \mathbf{Z},\ a \equiv b \pmod m \longrightarrow b \equiv a \pmod m$, because $(\exists k \in \mathbf{Z},\ a - b = k \cdot m) \longrightarrow b - a = (-k)m$.

Transitivity: $\forall a, b, c \in \mathbf{Z},\ [a \equiv b \pmod m$ and $b \equiv c \pmod m]$ implies $a \equiv c \pmod m$, because if $\exists k, k' \in \mathbf{Z}$ such that $a - b = km$ and $b - c = k'm$, then $a - c = (k + k')m$.

Therefore, *congruence mod m* is an *equivalence relation* on $\mathbf{Z}$ (different for different m, of course).

It is easy to find the cells of the associated partition. We denote the equivalence class containing a by $\lfloor a \rfloor_m$, or $\lfloor a \rfloor$ if the m is understood from the context. Recall from Chapter 4, (3.1) that the equivalence class of a is

(1.6) $$\lfloor a \rfloor_m := \{x;\ x \in \mathbf{Z},\ x \equiv a \pmod m\}.$$

The equivalence class $\lfloor a \rfloor_m$ of a is the set of all integers x congruent to a mod m. The notation $\lfloor a \rfloor_m$ is not standard. In fact, there is no standard notation, though $\bar{a}$ is sometimes used for $\lfloor a \rfloor_m$. Because it omits m, the notation $\bar{a}$ can be unsuitable.

DEFINITION (1.7) $\lfloor a \rfloor_m$ is called the **congruence class of** a **mod** m, or the **residue class of** a **mod** m. The set of all congruence classes mod m is denoted $\mathbf{Z}_m$.

(1.8) What are the congruence classes mod 2? The cell $\lfloor 0 \rfloor_2$ is the set of all integers congruent to 0 mod 2, that is, divisible by 2: $x \equiv 0 \pmod 2$ if and only if x is even. This is a special case of (1.4), but let's work it out from the definition (1.2): $x \equiv 0 \pmod 2$ iff $\exists k \in \mathbf{Z}$ such that $x - 0 = 2k$ iff x is even.

$$\lfloor 0 \rfloor_2 = \text{the set of all even integers.}$$
$$\lfloor 1 \rfloor_2 = \text{the set of all odd integers.}$$

Why the latter? Because

$$\lfloor 1 \rfloor_2 := \{x;\ x \in \mathbf{Z},\ x - 1 \text{ is even}\}.$$

Thus there are only the two cells mod 2, $\lfloor 0 \rfloor_2$ and $\lfloor 1 \rfloor_2$ mod 2. And $\mathbf{Z} = \lfloor 0 \rfloor_2 \cup \lfloor 1 \rfloor_2$, and $\lfloor 0 \rfloor_2 \cap \lfloor 1 \rfloor_2 = \varnothing$, in agreement with Theorem (3.3) of Chapter 4.

Notice that $\lfloor 0 \rfloor_2 = \lfloor 2 \rfloor_2 = \lfloor 4 \rfloor_2 = \lfloor -6 \rfloor_2$, that if $b \in \lfloor a \rfloor_m$ then $a \in \lfloor b \rfloor_m$; and the latter is true for the equivalence classes of any equivalence relation.

We describe this equivalence relation by listing its cells in the following result.

Theorem (1.9) Let $m \geq 1$. There are exactly m distinct congruence classes mod m. They may be denoted

$$\lfloor 0 \rfloor_m, \lfloor 1 \rfloor_m, \ldots, \lfloor m-1 \rfloor_m.$$

They are the cells of a partition of **Z**; that is,

(1.9.1) $$\forall i \in \mathbf{Z},\ \lfloor i \rfloor_m \neq \varnothing$$

and

(1.9.2) $$\lfloor 0 \rfloor_m \cup \lfloor 1 \rfloor_m \cup \cdots \cup \lfloor m-1 \rfloor_m = \mathbf{Z}.$$

Also

(1.9.3) $$0 \leq i < j < m \longrightarrow \lfloor i \rfloor_m \cap \lfloor j \rfloor_m = \varnothing.$$

Thus

(1.9.4) $$\mathbf{Z}_m = \{\lfloor 0 \rfloor_m, \ldots, \lfloor m-1 \rfloor_m\}. \blacksquare$$

Examples (1.10) We have just verified this claim for $m = 2$. Let's do it for $m = 3$ before giving the general proof.

Certainly $\lfloor 0 \rfloor_3$, $\lfloor 1 \rfloor_3$, and $\lfloor 2 \rfloor_3$ are three different cells, because no two of 0, 1, 2 are congruent mod 3. But

(1.10.1) for any integer a, one of $a - 1$, a, $a + 1$ is a multiple of 3. If it is $a - 1$, then $a \equiv 1 \pmod 3$ and so $a \in \lfloor 1 \rfloor_3$. If it is a, then $a \equiv 0 \pmod 3$ and $a \in \lfloor 0 \rfloor_3$. If it is $a + 1$, then $a \equiv -1 \pmod 3$, so $a \in \lfloor -1 \rfloor_3$. But $\lfloor -1 \rfloor_3 = \lfloor 2 \rfloor_3$, because $-1 \equiv 2 \pmod 3$. Thus we've proved

$$\mathbf{Z} = \lfloor 0 \rfloor_3 \cup \lfloor 1 \rfloor_3 \cup \lfloor 2 \rfloor_3$$

and that these three are all the cells mod 3.

Practice (1.11P) Prove (1.10.1).

Proof of (1.9) For each i in **Z**, $\lfloor i \rfloor_m \neq \varnothing$, because $i \in \lfloor i \rfloor_m$.

We use the division algorithm, Chapter 6, (2.1), to prove (1.9.2). Let $a \in \mathbf{Z}$. Then there are unique integers q, r such that

$$a = mq + r \qquad \text{with } 0 \leq r < m.$$

Thus $a \equiv r \pmod{m}$, so $a \in \lfloor r \rfloor_m$. Since r is one of $0, \ldots, m-1$, we have proved that

$$\mathbf{Z} \subseteq \lfloor 0 \rfloor_m \cup \cdots \cup \lfloor m-1 \rfloor_m. \tag{1.12}$$

The reverse inclusion is automatic, because each residue-class is a subset of $\mathbf{Z}$ by definition. Therefore (1.9.2) is true.

The uniqueness of r proves (1.9.3). Every integer is in the residue-class of its remainder r, and r is only one element of $\{0, \ldots, m-1\}$; thus no integer can be in two of these residue-classes. QED

The use of the remainder in the proof just ended is important enough to be recorded separately. Let's state the idea this way:

Lemma (1.13) For $m \geq 1$ define $M := \{0, 1, \ldots, m-1\}$. Then, for all $a \in \mathbf{Z}$, a is congruent mod m to exactly one element of M, namely, to the remainder of a on division by m.

Proof Let $a \in \mathbf{Z}$. Divide a by m, using the division algorithm. Thus there are unique integers q and r such that

$$a = qm + r, \quad \text{and} \quad 0 \leq r < m.$$

It follows, of course, that $a \equiv r \pmod{m}$. To prove the uniqueness: If $a \equiv s \pmod{m}$ and $s \in M$, then for some $k \in \mathbf{Z}, a = km + s$, and $0 \leq s < m$. By the uniqueness of q and r, we see that $s = r$ (and $k = q$). QED

Comment: We already proved in Chapter 4 that the equivalence classes of an equivalence relation on A partition the set A. The whole burden of the present proof is to show that the number of equivalence classes is exactly m, and to identify them explicitly. In other words, since we knew that $\lfloor 0 \rfloor_m, \ldots, \lfloor m-1 \rfloor_m$ were equivalence classes, we merely had to show that they were all of them (that was (1.12)) and that there were no duplications (that was (1.9.3)). To prove (1.9.1) and (1.9.2) took little extra effort, so we also gave a proof from scratch of Theorem (3.3) of Chapter 4 in this case.

DEFINITION (1.14) The set $\{0, 1, \ldots, m-1\}$, or any set of m integers $a_0, \ldots, a_{m-1}$ such that

$$\mathbf{Z} = \lfloor a_0 \rfloor_m \cup \cdots \cup \lfloor a_{m-1} \rfloor_m,$$

CAUTION

is called a **complete system of residues mod** m.

In this situation we may choose the notation so that for all i, $\lfloor a_i \rfloor_m = \lfloor i \rfloor_m$, that is, $a_i \equiv i \pmod{m}$. For example, $\{10, 1, -3, 28, -11\}$ is a complete system of residues mod 5. Thus any integer is congruent mod 5 to exactly one of 10, 1, -3, 28, or -11.

Example (1.15) This idea is easier than it may appear to be at first. For $m = 5$, imagine the five equivalence classes $\lfloor 0 \rfloor_5, \ldots, \lfloor 4 \rfloor_5$ listed as five infinite columns:

⋮	⋮	⋮	⋮	⋮
−15	−14	−13	−12	−11
−10	−9	−8	−7	−6
−5	−4	−3	−2	−1
0	1	2	3	4
5	6	7	8	9
10	11	12	13	14
15	16	17	18	19
⋮	⋮	⋮	⋮	⋮

A complete system of residues mod 5 is nothing more or less than a set of five integers chosen one from each column.

In particular,

(1.16) Any set of m consecutive integers is a complete system of residues mod m.

(1.17) If $\{a_0, \ldots, a_{m-1}\}$ is a complete system of residues mod m and s is any integer, then

$$\{a_0 + s, \ldots, a_{m-1} + s\}$$

is also a complete system of residues mod m.

(1.18) The set of exactly m integers $c_1, \ldots, c_m$ is a complete system of residues mod m if and only if for each $i \in \{0, 1, \ldots, m-1\}$ there is a $j \in \{1, \ldots, m\}$ such that $c_j \equiv i \pmod{m}$.

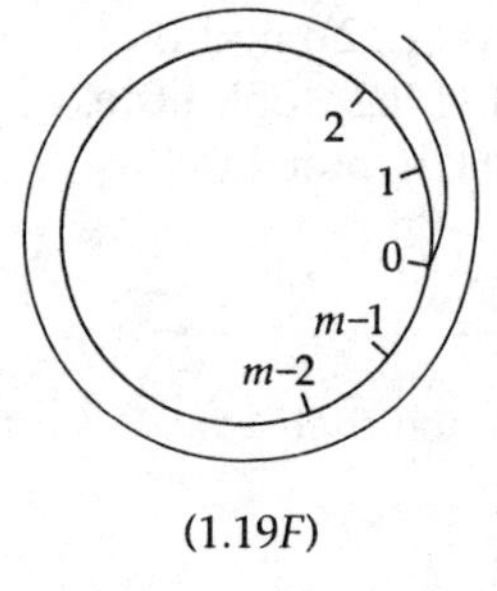

(1.19F)

Lemma (1.13) allows an analogy with the trigonometric functions (sometimes called the circular functions): We imagine a circle of circumference m, with distances $0, 1, \ldots, m-1$ marked on it from one fixed point. To find the remainder r on division of an integer a by m imagine a string of length $|a|$. Starting at 0, wrap it snugly around the circle in the positive [negative] direction if $a \geq 0$ [< 0]. Where it ends is r. (The number of full wraps is $|q|$.) An example is drawn in (1.19F). A clock keeps time modulo 12 hours (24 hours for some, one full year for some).

2. An Application to Computers

The adder in a computer operates modulo m. We first consider the most common choice for m, namely $m = 2^n$. The adder works as follows. Its input is two strings of 0's and 1's

$$a_1 a_2 \ldots a_n, \qquad b_1 b_2 \ldots b_n.$$

We let the strings stand for the base 2 representations of integers:

$$a_1a_2\ldots a_n \quad \text{stands for} \quad a = a_1 2^{n-1} + a_2 2^{n-2} + \cdots + a_n.$$

Thus the adder deals with strings standing for the 2^n integers

(2.1) $$0, 1, 2, \ldots, 2^n - 1.$$

It adds two input strings $a_1 \ldots a_n$ and $b_1 \ldots b_n$ of 0's and 1's by adding and carrying, just as we would add $a_1 2^{n-1} + \cdots + a_n$ to $b_1 2^{n-1} + \cdots + b_n$. There is one exception: when a 1 would be carried to the 2^n-place the adder ignores it. There *is* no 2^n-place, after all. For example, consider the sum $a + b$, where

	2^n	2^{n-1}	2^{n-2}	$\ldots$		2^0
$a =$		1	0	$\ldots$	0	1
$b =$		1	0	$\ldots$	0	0
$a + b =$	1	0	0	$\ldots$	0	1

As shown, there is a carry of 1 to the 2^n-place. *The adder throws that 1 away* and produces as the sum $a \dot{+} b$ only what is in its n places (here $00\ldots01$). That rule holds in general: any carry to the 2^n-place is ignored;† the least n bits of the ordinary sum $a + b$ are all that remain.

(2.2) In other words, the adder treats 2^n as if it were 0! The computer sum $a \dot{+} b$ of two integers is not necessarily $a + b$, but it is *congruent* to $a + b$ mod 2^n. If $a + b \geq 2^n$, then $a \dot{+} b = a + b - 2^n$. Since $2^n \equiv 0 \pmod{2^n}$

$$a + b \equiv a \dot{+} b \pmod{2^n}.$$

Here are the strings that the computer works with:

(2.3)

n columns				
0	0	$\cdots$	0	
0	0	$\cdots$	1	
		$\vdots$		2^n rows
1	1	$\cdots$	1	

Now—what numbers does the computer represent with these strings of n 0's or 1's in (2.3)? Certainly not the integers $0, 1, 2, \ldots, 2^n - 1$; that representation would not allow any negative numbers. It represents

(2.4) $$-2^{n-1}, -2^{n-1} + 1, \ldots, -1, 0, 1, \ldots, 2^{n-1} - 1,$$

each integer from -2^{n-1} to $2^{n-1} - 1$. But how? The only way consistent with the adder is to use congruence mod 2^n.

†In practice such carry is recorded in a system apart from the adder for use in, say, multiple precision.

(2.5) Thus each negative number in (2.4) is represented by the number congruent to it mod 2^n among $0, 1, \ldots, 2^n - 1$.

The following circular diagram (2.5F) shows this representation for $m = 16$ $(n = 4)$.

(2.5*F*)

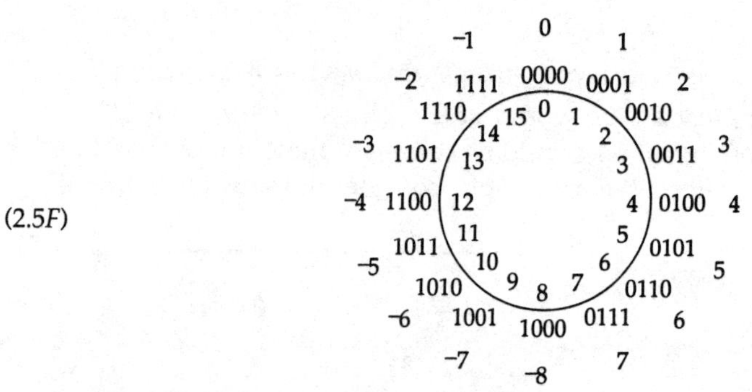

Two's Complement

The numbers in the inner ring are the base 10 values of the base 2 numbers in the middle ring. Each number in the outer ring is congruent mod 16 to its nearest neighbors in the other two rings. The numbers in the outer ring, (2.4) for $n = 4$, are what the programmer "sees," or works with. Those in the middle ring, (2.3) for $n = 4$, are what the computer "sees." The inner ring is (2.1) for $n = 4$. The numbers in each ring form a complete system of residues mod 16.

(2.6) *The sign bit*. You can see that the negative numbers are represented in the middle ring with a 1 in the 2^3-place. The positives have a 0 in that place. The value of the bit in the 2^{n-1}-place determines the sign of the number from (2.3) that it stands for. For that reason it is called the *sign bit*.

(2.7) *Overflow*. If we add 7 and 3 we get 10. In our mod 16 adder, 10 stands for -6.

$$7 \leftrightarrow 0111$$

$$3 \leftrightarrow 0011$$

$$7 + 3 \leftrightarrow 1010$$

Notice that the sign bits of the two numbers 7 and 3 were equal to each other but that their sum has a different sign bit. (Both were 0, but in the sum the bit is 1.) Any time this happens we say that overflow has occurred, because we are trying to represent a number too big (in absolute value) for our system. We've gone past the 8 (in general, the 2^{n-1}) at the bottom of the circle.

If we add two numbers of different signs no overflow can occur, because their sum is closer to 0 than the farthest of them is. For example, $5 = 7 \dot{+} (-2)$ is 2 units *toward 0* from 7. It's impossible to pass the bottom of the circle.

(2.8) **Two's complement arithmetic**. This term is the name of the setup we've been discussing. Here now is the algorithm used to find $-a$. First notice that for any number a (in any representation)

$$-a = (-1 - a) + 1.$$

Let a be any n-bit string from (2.3).

> We find $-1 - a$ by complementing the string a (result: $\bar{a}$), because -1 is represented by the string $11\ldots1$ of n 1's. We then add $0\ldots01$ to the string for $-1 - a$.

Example (2.9)

$$a = 0110$$
$$\bar{a} = 1001$$
$$\bar{a} + 0001 = 1010 = -a.$$

Here $a = 6$, and $(1010)_2 = (10)_{10} \equiv -6 \pmod{16}$, so 1010 is indeed -6 according to (2.5).

Practice (2.10P) Define a function f on the set of strings over 0,1 of length n by the rule

$$f(a_1 \cdots a_n) = -a_1 2^{n-1} + a_2 2^{n-2} + \cdots + a_n.$$

Prove that the range of f is the set of numbers (2.4) and that the image of each string under f is the number represented in the computer by that string.

(2.11) In **1's complement arithmetic** the adder operates modulo $2^n - 1$. Since the same n bits are used to represent integers, zero is represented in two ways, as $00\ldots0$ and as $11\ldots1 \leftrightarrow 2^n - 1$. Overflow is handled as in 2's complement arithmetic. The slight change in the adder forces the following changes in the system. Here is a sketch for $n = 4$:

(2.11*F*)

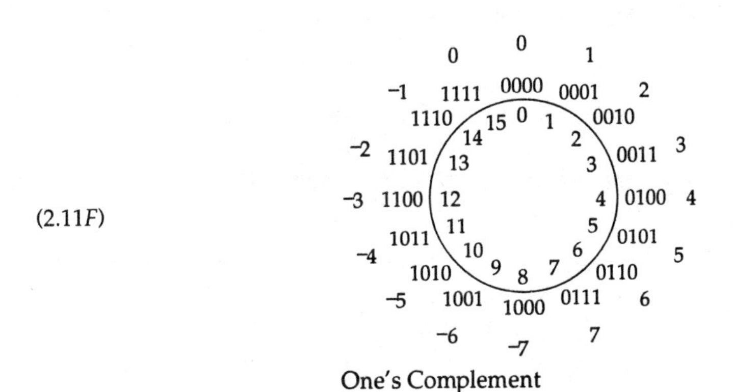

One's Complement

The inner numbers are the base 10 values of the 4-tuples considered as base 2 numbers. The 4-tuples are what the computer "sees." The outer numbers are what the programmer "sees"; each is congruent mod 15 to the corresponding 4-tuple.

Addition is done as before, with the usual carry on the n-tuples. But a carry to the 2^n-position, if any, is not thrown away; it is *added* to the 1's place. Why? It *has* to be so, simply because $2^n \equiv 1 \pmod{2^n - 1}$.

Example (2.12) To compute $(-1) + (-2)$:

	Computer		Programmer
	1 1 1 0	↔	−1
	1 1 0 1	↔	−2
The carry 1:	1 0 1 1		−3
is added to	0 0 0 1		
the sum	1 0 1 1		
Result:	1 1 0 0	↔	−3

Because the highest carry is added to the 1's place, we may think of the n bit positions as arranged in a circle, with each one sending its carry to be added to the one at its right:

(2.12*F*)

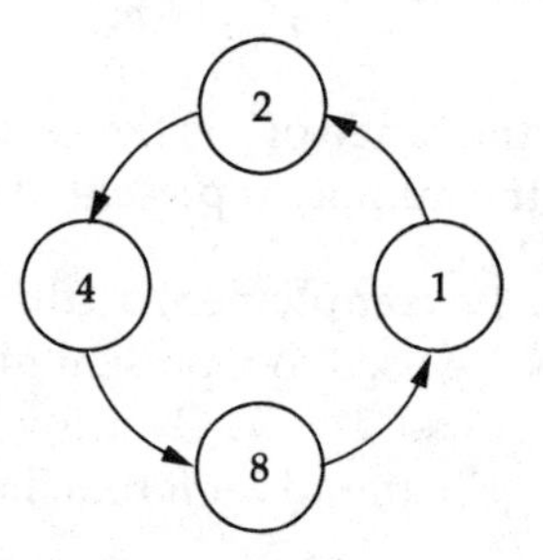

The system represents the numbers

(2.13) $$-(2^{n-1} - 1), \ldots, 0, \ldots, 2^{n-1} - 1.$$

This is the programmer's complete system of residues modulo $2^n - 1$. The computer represents them in base 2 via congruence mod $2^n - 1$. The extra representation $11\ldots1$ of 0 is called the *negative*, or *dirty, zero*. The name *1's complement* comes from the fact that

(2.14) the negative of any integer in the system is the bit-by-bit complement of that integer.

Example

$$6 \leftrightarrow 0110$$
$$-6 \leftrightarrow 1001.$$

Practice (2.15P) Check that addition of $11\ldots1$ does not change any integer in the system, that is, the dirty zero is mathematically clean.

Practice (2.16P) Prove (2.14).

(2.17) You can see that the sets S of integers to be represented—(2.4) for a mod-2^n adder, and (2.13) for a mod-$(2^n - 1)$ adder—are forced on us by the adder and by our natural wish to satisfy the requirements that:

- The integers should be consecutive. That is, if $x, y \in S$, then all integers between x and y should be in S.
- The negative of each element of S should be in S (as far as possible).

You have already seen how the adder forces us to handle the 1 in the 2^n-place in both systems, and the all-1 string in 1's complement.

3. Basic Properties of Congruence mod *m*

We have already established some basic properties of congruence mod m, namely reflexivity, symmetry, and transitivity. They imply in particular that integers congruent to the same integer are congruent to each other, all mod m.

And we have seen the important fact

$$\lfloor a \rfloor_m = \lfloor b \rfloor_m \quad \text{iff} \quad a \equiv b \pmod{m}. \tag{3.1}$$

This expression is just a restatement of the definition of congruence class. Now we get down to brass tacks.

Proposition (3.2) Let $m \geq 1$. Let a, b, x, y be any integers such that

$$a \equiv b \pmod{m} \quad \text{and} \quad x \equiv y \pmod{m}.$$

Then

$$a \pm x \equiv b \pm y \pmod{m} \tag{3.3}$$

$$ax \equiv by \pmod{m} \tag{3.4}$$

$$\forall\, d \mid m,\ a \equiv b \pmod{d} \tag{3.5}$$

(3.6) If e is a common divisor of a, b, and m, then

$$\frac{a}{e} \equiv \frac{b}{e} \left(\operatorname{mod} \frac{m}{e}\right).$$

Proof The first two seem obvious because they are also properties of equality. And they are easy to prove: for (3.3)

$$a \pm x - (b \pm y) = a - b \pm (x - y).$$

Since both $a - b$ and $x - y$ are multiples of m by hypothesis, so is their sum or difference.

For (3.4)

$$\begin{aligned} ax - by &= ax - bx + bx - by \\ &= (a - b)x + b(x - y); \end{aligned}$$

now proceed as in (3.3).

Although (3.5) does not have an analogue with equality, it is still easy to prove: Since $a - b = jm$ for some j, and $m = dm'$ for some m', $a - b = jm'd$ is a multiple of d.

The last one, (3.6), is also straightforward. We write $a = ea'$, $b = eb'$, and $m = em'$. Then $\exists j: a - b = mj \longrightarrow e(a' - b') = em'j$. Use the cancellation law in **Z** to conclude $a' - b' = m'j$; that is, $a' \equiv b' \pmod{m'}$. In other words, cancellation of a common factor on both sides of a congruence is all right if that factor divides m and is also canceled from m. (We present a stronger cancellation law in (3.12).) QED

Example of (3.3) $29 \equiv -1 \pmod 6$ and $13 \equiv 1 \pmod 6$. Therefore, $29 + 13 \equiv -1 + 1 = 0 \pmod 6$, and $29 - 13 \equiv -1 - 1 = -2 \equiv 4 \pmod 6$.

Example of (3.4) $25 \equiv 14 \pmod{11}$ and $79 \equiv 2 \pmod{11}$. Therefore $25 \cdot 79 = 1975 \equiv 14 \cdot 2 = 28 \pmod{11}$. Of course $28 \equiv 6 \pmod{11}$, since, for example, $14 \equiv 3 \pmod{11}$.

Example of (3.5) Since $83 \equiv 35 \pmod 6$ and $3 \mid 6$, $83 \equiv 35 \pmod 3$.

Example of (3.6) Since $66 \equiv 30 \pmod{18}$ we may cancel the common divisor 6 from all three numbers to get

$$11 \equiv 5 \pmod 3.$$

Corollary (3.7)

(3.7.1) $a \equiv b \pmod m \longrightarrow \forall n \geq 0, a^n \equiv b^n \pmod m$

(3.7.2) $a \equiv b \pmod m \longrightarrow \forall k \in \mathbf{Z}, ka \equiv kb \pmod m$

(3.7.3) For any polynomial $f(x) = a_0 + a_1x + \cdots + a_nx^n$ with coefficients a_i in **Z**,

$$\forall s \in \mathbf{Z},\ f(s) \equiv b_0 + b_1t + \cdots + b_nt^n \pmod m$$
$$\text{if} \quad s \equiv t \pmod m \text{ and } \forall i\ a_i \equiv b_i \pmod m.$$

Proof (3.7.1) follows from (3.4) by a simple induction on n. For $n \geq 1$ we set $x = a^{n-1}$ and $y = b^{m-1}$. The induction assumption is $x \equiv y \pmod m$.

The rule (3.7.2) is the special case of (3.4) with $x = y = k$.

Before we do (3.7.3), here is another bit of terminology: If we start with an integer a and find an integer b such that $a \equiv b \pmod m$, we may say that we have **reduced** a mod m (even if $|a| < |b|$). We need a verb for this process, and *reduce* has become it.

The claim (3.7.3) on the polynomial uses (3.3), (3.4) and (3.7.1). It says we can reduce a polynomial mod m by reducing each term and each factor in each term mod m. QED

Example of (3.7.1) (3.8) Find the remainder of $(795)^{25}$ on division by 11. $795 \equiv 3 \pmod{11}$. Therefore $(795)^{25} \equiv 3^{25} \pmod{11}$. We reduce 3^{25} mod 11 with a bit more effort. We use the same rules, and the law of exponents. $3^2 = 9 \equiv -2 \pmod{11}$. Therefore, by (3.3),

$$3^4 = (3^2)^2 \equiv (-2)^2 = 4 \pmod{11}$$
$$3^8 = (3^4)^2 \equiv 4^2 = 16 \equiv 5 \pmod{11}$$
$$\therefore 3^{16} = (3^8)^2 \equiv 5^2 = 25 \equiv 3 \pmod{11}.$$

Since $25 = 16 + 8 + 1$, the law of exponents tells us that $3^{25} = 3^{16} \cdot 3^8 \cdot 3^1$. Applying (3.4), we see that $3^{25} \equiv 3 \cdot 5 \cdot 3 = 15 \cdot 3 \equiv 4 \cdot 3 \equiv 1 \pmod{11}$—all that work just to get a number between 0 and 10. Still, it's far better than computing every power $2, 3, 4, \ldots, 24, 25$ of 795, even reduced mod 11 each time. Later we shall see more efficient ways to make computations like this one.

This example does illustrate a useful tactic, however. To compute a^n, represent n in base 2; calculate a^{2^i} for $i = 1, \ldots, \lfloor \log_2 n \rfloor$ by repeated squaring; then multiply together a^{2^i} for appropriate i to get a^n:

$$\begin{array}{llll} & \text{Find } n_i: & n = \sum_{0 \le i \le k} n_i 2^i & n_i = 0 \text{ or } 1; \\ (3.9) & \text{Do:} & a^2 = a \cdot a,\ a^4 = a^2 \cdot a^2, \ldots,\ a^{2^k} = a^{2^{k-1}} \cdot a^{2^{k-1}}; & \\ & \text{Then:} & a^n = \prod_{n_i = 1} a^{2^i}.^{\dagger} & \end{array}$$

This tactic is useful in **Z** or **R**; it works mod m too, where sometimes there are even quicker methods.

Examples (3.10)

(*i*) (3.7.2): $29 \equiv 17 \pmod 6 \longrightarrow 58 \equiv 34 \pmod 6$.

(*ii*) (3.7.3): Let $f(x) = 2x^3 + 11x^2 - 4x - 13$. We reduce $f(5)$ mod 12 as follows:

$$\begin{aligned} 5^2 &= 25 \equiv 1 \pmod{12} \\ 5^3 &\equiv 5 \pmod{12} \\ f(5) &\equiv 2 \cdot 5 - 1 \cdot 1 - 4 \cdot 5 - 1 \\ &\equiv 9 - 21 \equiv -12 \equiv 0 \pmod{12}, \text{ using (3.7.2) and } 5^2 \equiv 1. \end{aligned}$$

We don't know the value of $f(5)$ from this, but we know that 12 divides $f(5)$.

(*iii*) Since $9 \equiv -4 \pmod{13}$, we see from (3.7.1) that $9^2 \equiv (-4)^2 = 16 \equiv 3 \pmod{13}$. From $9^2 \equiv 3 \pmod{13}$ we get, using (3.7.2), $9^3 \equiv 3 \cdot 9 = 27 \equiv 1 \pmod{13}$.

(Why the last congruence? $27 = 2 \cdot 13 + 1$, and $13 \equiv 0 \pmod{13}$, so we throw away any multiple of 13. Or, simply, 1 is the remainder of 27 on division by 13—see Lemma (1.13).)

(*iv*) When doing these calculations by hand it can be helpful to list several multiples of m. They are all $\equiv 0 \pmod m$. For example, for $m = 18$ you could list $-72, -54, -36, -18, 0, 18, 36, 54, 72$. Then if you came upon 57, say, you could quickly find $57 \equiv 3 \pmod{18}$ by adding -54; you used (3.3).

Practice (3.11P) What is the quickest way to show in (3.10*ii*) that, although $f(5) \equiv 0 \pmod{12}$, $f(5) \neq 0$?

[†] Notice carefully the symbolism. The large Π stands for product (multiplication in **Z** or **R**). The subscript $n_i = 1$ tells us to choose each i in the truth-set of the predicate $n_i = 1$. *For those* i we are to evaluate the terms a^{2^i} and multiply them together. (A more complete notation uses $i : n_i = 1$ for the subscript, but usually we understand the variable of the predicate from the context.)

Proposition (3.12) The Cancellation Law for Congruences. Let $m \in \mathbf{N}$. Let $a, b, k \in \mathbf{Z}$ and set $d := \gcd(k, m)$. Suppose that

$$ak \equiv bk \pmod{m}.$$

Then

$$a \equiv b \left(\text{mod } \frac{m}{d}\right).$$

Proof This result reminds us of (3.6), which would yield

$$ak' \equiv bk' \pmod{m'} \tag{3.13}$$

after we canceled the common factor d, setting $k = dk'$, $m = dm'$. The present Cancellation Law (3.12) claims that we can go further, by canceling k' from both sides of (3.13). Why can we? The answer is in the *gcd* part of the hypothesis, which we haven't fully used. What we've done so far would be true for *any* common divisor d; but only for the greatest common divisor d (or $-d$) are there integers x, y such that

$$d = kx + my. \tag{3.14}$$

That result is Theorem (7.5) of Chapter 6. We divide (3.14) by d to get

$$1 = k'x + m'y,$$

showing that $\gcd(k', m') = 1$ by Chapter 6, (7.9). The definition of congruence applied to (3.13) tells us that

$$m' \mid k'(a - b).$$

Since m' is prime to k', the key result of Chapter 6, (7.10) implies m' divides $a - b$. Thus $a \equiv b \pmod{m'}$. QED

Corollary (3.15) Let $\gcd(k, m) = 1$. If $ak \equiv bk \pmod{m}$, then

$$a \equiv b \pmod{m}. \blacksquare$$

Examples of (3.12) The Cancellation Law (3.12) is especially useful. If we know that for some integer x,

$$3x \equiv 15 \pmod{23}, \tag{3.16}$$

then we can immediately conclude

$$x \equiv 5 \pmod{23},$$

because we may cancel the common factor 3 from both sides of the congruence (3.16). Since $\gcd(3, 23) = 1$, we may leave the modulus 23 alone.

If we had

$$3x \equiv 15 \pmod{24} \tag{3.17}$$

then we'd have to change the modulus: (3.17) implies

$$x \equiv 5 \pmod{8},$$

because $\gcd(3,24) = 3$. We used (3.6). Canceling the 3 without dividing the modulus by 3 in (3.17) could lead to an error. For example, $x = 13$ satisfies (3.17), but it does not satisfy $x \equiv 5 \pmod{24}$.

From the congruence $145 \equiv 80 \pmod{13}$ we may cancel 5 because $\gcd(5,13) = 1$. We get $29 \equiv 16 \pmod{13}$.

From the congruence

$$222 \equiv 96 \pmod{14}$$

we may cancel the common factor 6 as follows:

$$222 = 6 \cdot 37, \quad 96 = 6 \cdot 16$$
$$\gcd(6,14) = 2$$
$$\therefore 37 \equiv 16 \pmod{7}.$$

It may seem pointless to go through an example like the one just concluded, since after all it's perfectly obvious that $37 \equiv 16 \pmod{7}$. The real use of (3.12) is in solving equations mod m, as we showed in the first examples.

Example of (3.15) We used (3.15) in some of the examples of (3.12).

Proposition (3.18) Let $m_1, m_2 \in \mathbf{N}$. If $a \equiv b \pmod{m_1}$ and $a \equiv b \pmod{m_2}$, then

$$a \equiv b \pmod{\operatorname{lcm}(m_1, m_2)}.$$

Proof This proof is almost self-evident. By definition $a - b$ is a common multiple of m_1 and m_2. Every common multiple is a multiple of the least common multiple. (This follows immediately from Chapter 6, (11.11), as we remarked there.) QED

Corollary (3.19) Let $m_1, \ldots, m_r \in \mathbf{N}$ and assume $\forall i \neq j \; \gcd(m_i, m_j) = 1$. Then

$$[\forall i \; a \equiv b \pmod{m_i}] \longrightarrow a \equiv b \pmod{m_1 \cdots m_r}. \blacksquare$$

Example of (3.18)

$$751 \equiv 223 \pmod{11}$$
$$751 \equiv 223 \pmod{3}$$
$$\therefore 751 \equiv 223 \pmod{33}.$$

In fact, since $751 - 223 = 528 = 2^4 \cdot 3 \cdot 11$ we may say that in addition

$$\forall n = 1,2,3,4,^{\dagger} \qquad 751 \equiv 223 \pmod{2^n}.$$

We then conclude that for the same n

$$751 \equiv 223 \pmod{(2^n \cdot 33)},$$

an example of (3.19) with $r = 3$.

Examples (3.20) Let's do more examples of reducing a^n mod m.

† This is a common way to say $\forall n \in \{1,2,3,4\}$.

(*i*) Consider $7^{29} \pmod{19}$. We use the earlier tactic (3.9):

$$\begin{aligned}
29 &= 16 + 8 + 4 + 1 \\
7^2 &= 49 \equiv 11 \pmod{19} \\
(7^2)^2 &= 7^4 \equiv (11)^2 = 121 \equiv 7 \pmod{19} \\
(7^4)^2 &= 7^8 \equiv (7)^2 \equiv 11 \pmod{19} \\
(7^8)^2 &= 7^{16} \equiv (11)^2 \equiv 7 \pmod{19}. \\
7^{29} &= 7^{16} \cdot 7^8 \cdot 7^4 \cdot 7^1 \equiv 7 \cdot 11 \cdot 7 \cdot 7 \\
&\equiv 7 \cdot 11 \cdot 7^2 \equiv 7 \cdot 11 \cdot 11 \\
&\equiv 7 \cdot 11^2 \equiv 7 \cdot 7 \equiv 11 \pmod{19}.
\end{aligned}$$

So we may say that $7^{29} \equiv 11 \pmod{19}$.

(*ii*) Let us reduce $2^{1,000,000}$ mod 17. Since $2^{19} = 524,288$ we would have to do 19 successive squarings and then several multiplications if we followed (3.9), the preceding approach. But there's a shortcut: Calculate

$$\begin{aligned}
&2^2 = 4, \therefore\ 2^4 = 16 \equiv -1 \pmod{17} \\
&2^8 \equiv (-1)^2 = 1 \pmod{17}.
\end{aligned}$$

—*to be continued.*

(3.21) We wish to reduce a^n mod m. In general, as soon as we find an exponent e such that $a^e \equiv 1 \pmod{m}$, we may cut our work drastically, as follows. We use the division algorithm, Chapter 6, (2.1), to write

$$n = qe + r \qquad 0 \le r < e.$$

Then $a^n = a^{qe+r} = (a^e)^q \cdot a^r$ by the law of exponents. But since $a^e \equiv 1 \pmod{m}$ by assumption, $(a^e)^q \equiv 1^q \equiv 1 \pmod{m}$. Therefore,

(3.22) $$a^n \equiv a^r \pmod{m}.$$

Notice that all we want is the value of the remainder r. That is, by (1.13), we want to *reduce the exponent* mod e.

Example Returning to Example (3.20*ii*), we have $2^{10^6} \pmod{17}$, and we know that $2^8 \equiv 1 \pmod{17}$. So $e = 8$. Since $10 \equiv 2 \pmod 8$, $10^6 \equiv 2^6 \pmod 8$ by (3.7.1), so $r = 0$, and $2^{10^6} \equiv 1 \pmod{17}$. (Why is $r = 0$?)

Rules (3.2) to (3.7), (3.9), (3.12), (3.15), (3.18), and (3.21) are all the properties of congruence mod m you will need to do a great many problems.

Practice (3.23P) Identify the rules of this chapter that justify the preceding steps $(a^e)^q \equiv 1^q \equiv 1 \pmod{m}$ and the conclusion (3.22). Reduce 7^{29} mod 19 using (3.22).

4. Some Applications

Consider the fact that $10 \equiv 1 \pmod 9$. A simple result, but it is the basis of the old standby called *casting out nines.*

Proposition (4.1) Let $n = n_0 + n_1 10 + \cdots + n_k 10^k$ where $0 \leq n_i < 10$ for all i. Then

$$n \equiv n_0 + n_1 + \cdots + n_k \pmod 9.$$

Proof Use (3.7.3). QED

Corollary (4.2) $n \equiv \sum n_i \pmod 3$. ■

The Corollary follows from (3.5). (Unless stated otherwise, a corollary has the same hypotheses as the result of which it is a corollary.)

Examples To illustrate the use of (4.1) in checking hand calculations, consider these addition and multiplication problems:

(*i*)

$$\begin{array}{rrll} & 128 & \equiv 1+2+8 \equiv & 2 \\ + & 193 & \equiv 1+9+3 \equiv & 4 \pmod 9 \\ + & \underline{204} & \equiv 2+0+4 \equiv & \underline{6} \\ & 525 & & 12 \equiv 1+2 \equiv 3 \pmod 9 \end{array}$$

$$525 \equiv 5+2+5 = 12 \equiv 3 \pmod 9.$$

So 525 *may* be right. This check alone doesn't prove it's right, but it will detect some errors. For example, if we had 515 instead of 525, we would get 2 as its value mod 9, so we'd know there was an error somewhere.

(*ii*) 536×142:

$$\begin{array}{rl} 536 & \equiv 5+3+6 \equiv 5 \\ \underline{\times 142} & \equiv 1+4+2 \equiv 7 \equiv -2; \quad 5(-2) \equiv -1 \pmod 9 \\ 1172 & \\ 2144 & \\ \underline{536} & \\ 76{,}212 & \equiv 7+6+2+1+2 \equiv 0 \pmod 9 \end{array}$$

Since $-1 \not\equiv 0 \pmod 9$, there is an error somewhere.

This procedure is called *casting out nines* because one easily evaluates $\sum n_i$ mod 9 by disregarding 9's and numbers that sum to 9, as in 128; $1+8 = 9$ so $128 \equiv 2 \pmod 9$.

One type of error that it can't detect is transposition of digits. $1234 \equiv 3124 \equiv 1324 \pmod 9$. The following check will flag as an error the interchange of two adjacent digits.

Proposition (4.3) If $n = \sum n_i 10^i$ with $0 \leq n_i < 10$, then

$$n \equiv \sum_i (-1)^i n_i \pmod{11}.$$

Proof Since $10 \equiv -1 \pmod{11}$, (4.3) follows from (3.7.3).

This check could be used, as casting out nines is used, to check additions and multiplications. It also detects the interchange of only two adjacent digits, because the signs are altered. If we interchange n_0 and n_1, say, we get

$$n \equiv n_1 - n_0 + n_2 - n_3 + - \cdots \pmod{11}$$

instead of the correct

$$n \equiv n_0 - n_1 + n_2 - n_3 + - \cdots \pmod{11}.$$

These can be the same only if

$$n_1 - n_0 \equiv n_0 - n_1 \pmod{11}$$

This result implies

$$2n_0 \equiv 2n_1 \pmod{11}$$

and now the Cancellation Law (3.15) implies $n_0 \equiv n_1 \pmod{11}$. But $0 \leq n_i < 10$, so $n_0 = n_1$; thus no error occurred!

But if we interchange two adjacent pairs we might fool ourselves. For example,

$$136709 \equiv 316790 \equiv 1 \pmod{11}.$$

(4.4) If both checks (4.1) and (4.3) work out, then you know that your computation is correct mod 99. (Why?) And if it's also correct mod 10, then it's correct mod 990.

We could call (4.3) casting out elevens. At this point we can generalize these ideas to

Proposition (4.5) If $n = \sum n_i b^i$, then

$$n \equiv \sum n_i \pmod{b-1}$$

and

$$n \equiv \sum (-1)^i n_i \pmod{b+1}. \blacksquare$$

The proof is omitted.

(4.6) *Applications.* These ideas are in current use to detect errors in certain commercial systems. For example, the U.S. Postal Service uses casting out nines. They identify money orders with a 10-digit number and append a check-digit consisting of the sum of the ten digits mod 9. This scheme detects most, but not all, single errors (those affecting only one digit). It fails to detect most transposition errors (defined as interchanges of two digits, not necessarily adjacent).

Practice (4.7P) Which single errors does the Postal Service system fail to detect? Which transposition errors does it detect?

For its traveler's checks the VISA company appends the negative (mod 9) of the sum of the digits. This system is worse than the prior scheme because it can't detect *any* transpositions.

United Parcel Service uses a nine-digit number, say n, with one check-digit appended. It is n reduced mod 7. The scheme detects 94% of transposition errors.

Practice (4.8P) Which single errors does the UPS scheme fail to detect?

(4.9) The International Standard Book Number (ISBN) on books is another example of an error-detection code based on modular arithmetic. It works mod 11. There are nine digits $x_1, \ldots, x_9$ and one check-digit, x_{10}. The check digit is chosen to satisfy the congruence

$$\sum_{1 \le i \le 10} i\, x_i \equiv 0 \pmod{11}.$$

Since $10 \equiv -1 \pmod{11}$, we may say that

$$x_{10} \equiv \sum_{1 \le i \le 9} i\, x_i \pmod{11}.$$

When x_{10} turns out to be 10, they print X instead. This scheme detects each single error and each transposition error. (See Joseph A. Gallian and Steven Winters, Modular arithmetic in the marketplace, *Amer. Math. Monthly*, Vol. 95 (1988), pages 548–551.)

5. Solving Linear Congruences

Suppose we consider, for $m \ge 1$ and some integers a, b, the congruence

(5.1) $$ax \equiv b \pmod{m}.$$

Does it have a solution, and if so, how do we find it, or them?

Examples (5.2) (*i*) Let us consider the congruence

$$2x \equiv 3 \pmod{12}.$$

It has no solution x, because for all integers x, $2x - 3$ is odd and thus not divisible by 12.

(*ii*) Now look at

(5.2.1) $$3x \equiv 23 \pmod{16}.$$

This congruence has the solution $x = -3$, since $3 \cdot (-3) - 23 = -32 \equiv 0 \pmod{16}$. And if x_0 is any solution to (5.2.1), then $x_0 \pm 16$ is obviously also a solution. Thus by a trivial induction we see that the set of all solutions includes $\lfloor -3 \rfloor_{16}$. (Or we can simply observe that $x_0 + 16j$ is a solution to (5.2.1) for all $j \in \mathbf{Z}$.)

Where did the -3 come from? Answer: inspection. But we discuss systematic methods below.

The entire set of solutions to (5.2.1) is $\lfloor -3 \rfloor_{16}$, as explained in (5.12) below.

(*iii*) Finally, what about

(5.2.2) $$21x \equiv 18 \pmod{12}?$$

It does have a solution x, namely $x = 2$, since $21 \equiv 9 \pmod{12}$. Thus by (3.7.2) an *equivalent* congruence is $9x \equiv 18 \pmod{12}$, which certainly has $x = 2$ as one solution. Another solution is $x = -2$, since $-18 \equiv 18 \pmod{12}$. And any integer congruent mod 12 to a solution is also a solution: If x_0 is an integer satisfying $21x_0 \equiv 18 \pmod{12}$, and if $x_1 \equiv x_0 \pmod{12}$, then $21x_1 \equiv 18 \pmod{12}$ also, because by (3.4), $21x_1 \equiv 21x_0 \pmod{12}$. Thus 10 and 14 are also solutions, for example.

Practice (5.3P) Explain in detail why (5.2.2) is equivalent to $9x \equiv 18 \pmod{12}$. First explain what *equivalent* means here.

Practice (5.4P) Before reading on, try to find all x such that $17x \equiv 45 \pmod{11}$. Try to use properties of congruence developed so far and not just trial and error.

The general situation is described in the following result.

Theorem (5.5) Let $m \geq 1$, and $a, b \in \mathbf{Z}$. The congruence

(5.6) $$ax \equiv b \pmod{m}$$

has a solution x if and only if $d := \gcd(a, m)$ divides b.

If d does divide b, define

$$a =: da', \qquad b =: db', \qquad m =: dm'.$$

Then the modified congruence

(5.7) $$a'x \equiv b' \pmod{m'}$$

is equivalent to (5.6). That is, both (5.6) and (5.7) have the same set of solutions. If x_0 is any solution (to either congruence), then the set of all solutions to either congruence is

(5.8) $$\lfloor x_0 \rfloor_{m'}. \blacksquare$$

NOTE The congruence class (5.8) is mod m', not mod m (unless $d = 1$, of course).

Corollary (5.9) We may partition the solution-set (5.8) to express it as a union of d congruence classes mod m. We get

(5.10) $$\lfloor x_0 \rfloor_{m'} = \lfloor x_0 \rfloor_m \cup \lfloor x_0 + m' \rfloor_m \cup \cdots \cup \lfloor x_0 + (d-1)m' \rfloor_m.$$

In words, any solution x to (5.6) is congruent mod m to just one of the d solutions

(5.11) $$x_0,\ x_0 + m',\ x_0 + 2m',\ \ldots,\ x_0 + (d-1)m'. \blacksquare$$

We leave the proof of the corollary to the problems. It is neater to express the solution-set as just one congruence class. We now give examples of the theorem before proving it.

Examples (5.12) To begin, look at the first congruence of Example (5.2), $2x \equiv 3 \pmod{12}$. We already know it has no solutions; the theorem draws that conclusion, too, because $d := \gcd(2, 12) = 2$ does not divide $3 =: b$.

For the second congruence of (5.2), $3x \equiv 23 \pmod{16}$, $d := \gcd(3,16) = 1$ does divide $23 =: b$; so, according to the theorem, the set of all solutions is the congruence class mod 16 of any one solution. We found $x_0 = -3$ as one solution in (5.2*ii*), so $\lfloor -3 \rfloor_{16}$ is the entire solution-set.

For (5.2*iii*) the gcd of a and m does divide b; $d = \gcd(21,12) = 3$. The modified congruence is $7x \equiv 6 \pmod 4$. This can be rewritten as $-x \equiv 2 \pmod 4$, i.e., $x \equiv 2 \pmod 4$. It has solved itself! Indeed the solution is unique mod 4. Now the Theorem claims that the solution-set is $\lfloor 2 \rfloor_4$. Note that $m' = 4$ and that the corollary claims that

(5.13) Any solution to $21x \equiv 18 \pmod{12}$ is congruent to one of 2, $2 + 4 = 6$, or $2 + 2 \cdot 4 = 10$ mod 12.

We found solutions 2 and -2 in (5.2*iii*). The latter "is" the 10 we get from the theorem, since $-2 \equiv 10 \pmod{12}$; but we did not observe in (5.2) that 6 is a solution. We can check it, however:

$$21 \cdot 6 \equiv (-3) \cdot 6 = -18 \equiv 18 \pmod{12}.$$

That (5.13) is true follows from (3.6): We divide by 3, arguing that any x satisfying (5.2.2) must satisfy

$$7x \equiv 6 \pmod 4.$$

As before, we see that this condition holds iff $x \equiv 2 \pmod 4$. The only residue-classes $\lfloor x \rfloor_{12}$ mod 12 that satisfy $x \equiv 2 \pmod 4$ are $\lfloor 2 \rfloor_{12}$, $\lfloor 2+4 \rfloor_{12}$, and $\lfloor 2+8 \rfloor_{12}$. (As you will see, we have almost imitated the general proof of (5.5).)

Proof of (5.5) With the machinery we've developed so far, we can make the proof briefer than the statement. Let the colon (:) here stand for *such that*.

$$\begin{aligned} &\exists x \in \mathbf{Z} : ax \equiv b \pmod m \\ &\longleftrightarrow \exists x, j \in \mathbf{Z} : ax - b = mj \\ &\longleftrightarrow \exists x, j \in \mathbf{Z} : ax - mj = b \\ &\longleftrightarrow d \mid b, \text{ by Corollary (7.7) of Chapter 6} \end{aligned}$$

This proves the existence of at least one solution if and only if d divides b.

Assume now $d \mid b$, so (5.6) does have solutions. We first show that any two of these solutions, x and y, are congruent to each other mod m'. Thus

$$da'x \equiv db' \pmod{dm'}$$

and

$$da'y \equiv db' \pmod{dm'}.$$

Canceling d from these, by (3.6), we get

$$a'x \equiv b' \equiv a'y \pmod{m'}. \tag{5.14}$$

Now we use the stronger Cancellation Law (3.12) to cancel a' from (5.14). We may do so because $d := \gcd(a,m)$, and, therefore, $a' = a/d$ and $m' = m/d$ are relatively prime. It follows that

$$x \equiv y \pmod{m'}.$$

Therefore, the set of all solutions to (5.6) is a single congruence class mod m'.

We can easily see that (5.7) is equivalent to (5.6) as follows: We've just shown that every solution of (5.6) is a solution of (5.7). If (5.7) is true for x, then $da'x \equiv db'$ (mod dm') is true, by an obvious converse to (3.6) that we have not previously mentioned. But the latter congruence is (5.6).

To summarize, we showed (5.6) has a solution iff d divides b, and that in that case the solution-set is the same as that for (5.7), namely, one congruence class mod m'.

QED

We'll discuss how to find the solutions shortly.

Corollary (5.15) If $\gcd(a, m) = 1$, then for all b in $\mathbf{Z}$ there is an integer x satisfying the congruence

$$ax \equiv b \pmod{m},$$

and any two such x are congruent mod m. [More briefly: The congruence has a solution x that is unique mod m.] ∎

Examples To solve

(5.16)
$$6x \equiv 22 \pmod{40}.$$

we first divide by $2 = \gcd(6, 40)$.

$$3x \equiv 11 \pmod{20}.$$

With small numbers like 3 for a' we may try the following quick approach. Replace $b' = 11$ by $b' + km' = 11 + 20k$ for various k (use (1.5), transitivity) until we get a number divisible by 3. Since $k = -1$ works, an equivalent problem is

$$3x \equiv -9 \pmod{20}.$$

Now use cancellation. Thus $x \equiv -3 \pmod{20}$ is the unique solution mod 20, by (3.15) or (5.5). In terms of Corollary (5.9), any solution x to the original problem satisfies

$$x \equiv -3 \text{ or } x \equiv 17 \pmod{40}.$$

Consider the problem

(5.17)
$$21x \equiv 87 \pmod{93}.$$

Here $3 = \gcd(21, 93)$ and $3 \mid 87$, so there is a solution. We divide by d. Our problem now is to solve $7x \equiv 29 \pmod{31}$, which we rewrite as

$$7x \equiv -2 \pmod{31}.$$

The coefficient $a' = 7$ is still fairly small, so we may do trial and error as before. We look for k such that 7 divides $-2 + 31k$. (That is, we solve the congruence $31k \equiv 2$ (mod 7).) To do this we note $31 \equiv 3 \pmod 7$, so we solve

$$3k \equiv 2 \pmod{7}.$$

We see by inspection that $3 = k$ is a solution. Now we put $k = 3$ in $-2 + 31k$ to get

$$7x \equiv 91 \pmod{31},$$
$$x \equiv 13 \pmod{31}.$$

The set of all solutions to (5.17) is $\lfloor 13 \rfloor_{31}$. Thus there are $3 = d$ solutions mod 93 to the original congruence:

$$x \equiv 13, \qquad x \equiv 44, \qquad x \equiv 75 \pmod{93}.$$

(5.18) The congruence

$$30x \equiv 77 \pmod{265}$$

has no solution, because $\gcd(30, 265) = 5$ does not divide $b = 77$.

(5.19) Recall example (8.4*ii*) of Chapter 6: $\gcd(2047, 5313) = 23$. If we tried to solve

$$2047x \equiv 46 \pmod{5313}$$

by the preceding methods we'd have a tedious time of it. They should work, but the repeated change of the congruence to be solved as we reduce a' until we can solve the problem by inspection would lead to so much bookkeeping that we would likely make a mistake. Instead we could use the extended Euclidean algorithm, as shown in the next section.

6. An Algorithm to Solve $ax \equiv b \pmod{m}$

(6.1) **Solve (a, b, m).**

Input: $a, b \in \mathbf{Z}, m \in \mathbf{N}$.

Output: integer x such that $ax \equiv b \pmod{m}$ if such x exists.

Step 1. $(d, u, v) := \mathbf{EE}(|a|, m)$

Step 2. Check whether $d \mid b$. If not, stop. If so, do

Step 3. $x := \pm ub/d$ (+ if $a \geq 0$; − if $a < 0$).

End.

(6.2) Why does this algorithm (6.1) work? When $a > 0$, algorithm **EE** (Chapter 6, (9.3)) produces $d := \gcd(a, m)$ and integers u, v satisfying

$$au + mv = d. \tag{6.3}$$

If we multiply (6.3) by $b/d =: b'$ (when b' is an integer) we get

$$aub' + mvb' = db' = b;$$

and if we reduce this mod m it becomes

$$a(ub') \equiv b \pmod{m},$$

showing that $x = ub'$ is a solution to the congruence $ax \equiv b \pmod{m}$.

REMARK (6.4) The foregoing proof that Algorithm (6.1) is correct is an alternate, constructive proof of the existence part of Theorem (5.5) when a is positive. ■

Example (6.5) The values of d $(= 23)$ and u $(= -109)$ for Example (5.19) appear already in Chapter 6, (9.2.1). Thus $x = -218$ is a solution of (5.19). Since $m' := m/d = 5313/23 = 231$, we know from Theorem (5.5) that the set of all solutions is $\boxed{-218}_{231}$. Since $-218 + 231 = 13$, we see that 13 is a solution, and the solution-set is $\boxed{13}_{231}$.

(6.6) You could add another step to find all solutions mod m.

Step 4. $i := 1, x_0 := x$

while $i < d$ **do**

$$x_i = x_{i-1} + \frac{m}{d}; i := i + 1$$

End.

7. The Chinese Remainder Theorem

In an old parlor game, A calls for B to think of a number n between 1 and 60. A then asks B to tell the remainder of n on division by 3, 4, and 5 respectively. For example, these remainders might be 1, 2, and 4. Then A tells B what the number is. In this example it is 34. How does A know? Simple—A uses the Chinese remainder theorem. We solve the example and then do the general case.

(7.1) The "secret" number n is, by (1.17), a solution to the *simultaneous* congruences

$$\begin{aligned} n &\equiv 1 \pmod 3 \\ n &\equiv 2 \pmod 4 \\ n &\equiv 4 \pmod 5. \end{aligned}$$

Each of these individual congruences is already solved, but how do we (indeed, can we) solve all three at once? Let us take a laborious approach. The most general solution to the first congruence is $n = 1 + 3u$, for any $u \in \mathbf{Z}$. We substitute this n into the second congruence and solve for u:

$$\begin{aligned} 1 + 3u &\equiv 2 \pmod 4 \\ -u &\equiv 1 \pmod 4 \\ u &\equiv -1 \pmod 4. \end{aligned}$$

Thus $u = -1 + 4v$ for any $v \in \mathbf{Z}$, making

$$\begin{aligned} n = 1 + 3u &= 1 + 3(-1 + 4v) \\ &= -2 + 12v. \end{aligned}$$

Now we put this n into the last congruence and solve for v. We have

$$\begin{aligned} n = -2 + 12v &\equiv 4 \pmod 5 \\ 12v &\equiv 6 \pmod 5 \\ 2v &\equiv 1 \pmod 5). \end{aligned}$$

The last cancellation of 6 is allowed because $\gcd(6,5) = 1$. By inspection we may take $v = 3 + 5w$ for any $w \in \mathbf{Z}$. Thus $n = -2 + 12v = -2 + 12(3 + 5w) = 34 + 60w$, for any $w \in \mathbf{Z}$, is a solution to the congruences. The only one of these solutions between 1 and 60 is 34. Notice that any two solutions are congruent mod 60 and that $60 = 3 \cdot 4 \cdot 5$, the product of the moduli.

(7.2) Another way to say what A did is this: The last congruence $n \equiv 4 \pmod 5$ says that n is one of the 11 numbers

$$4,\ 9,\ \underline{14},\ 19,\ 24,\ \underline{34},\ 39,\ 44,\ 49,\ \underline{54},\ 59. \tag{7.3}$$

The next congruence $n \equiv 2 \pmod 4$ says that n is among every fourth one of the list (7.3), starting at 14. We underlined them. Those are the only ones congruent to 2 mod 4. Finally, $n \equiv 1 \pmod 3$ picks out 34, just one of the three marked. (Compare the sieve of Eratosthenes in Chapter 6, Section 12.)

The general question here is that of simultaneous congruences

$$\begin{aligned} x &\equiv b_1 \pmod{m_1} \\ x &\equiv b_2 \pmod{m_2} \\ &\ \ \vdots \\ x &\equiv b_n \pmod{m_n}. \end{aligned} \tag{7.4}$$

We assume that the moduli are relatively prime in pairs, i.e.,

$$\forall i, j \qquad i \neq j \longrightarrow \gcd(m_i, m_j) = 1.$$

The b_i's may be any integers.

Theorem (7.5) *The Chinese Remainder Theorem* (CRT). The simultaneous congruences (7.4) with moduli relatively prime in pairs have a solution, and it is unique mod $M := m_1 m_2 \cdots m_n$. That is, the set of all solutions is a congruence class mod M.

Proof We use induction on n. If $n = 1$ the result is self-evident, so assume it for fixed but arbitrary $n \geq 1$. Let x_0 be a solution to the first n congruences. Then the uniqueness part of the induction assumption implies that x satisfies the congruence

$$x \equiv x_0 \pmod{m_1 \cdots m_n} \tag{7.6}$$

if and only if the same x satisfies the n simultaneous congruences $x \equiv b_1 \pmod{m_1}, \ldots,$ $x \equiv b_n \pmod{m_n}$. We pair the new congruence (7.6) with $x \equiv b_{n+1} \pmod{m_{n+1}}$. Using the same idea as in (7.1) we say $x = b_{n+1} + j \cdot m_{n+1}$. Put this in (7.6) and solve for j:

$$j m_{n+1} \equiv x_0 - b_{n+1} \pmod{m_1 \cdots m_n}. \tag{7.7}$$

A value of j satisfying (7.7) exists because $\gcd(m_{n+1}, m_1 \cdots m_n) = 1$ (Corollary (5.15)). And j is unique mod $m_1 \cdots m_n$. Therefore, $x = b_{n+1} + j m_{n+1}$ is unique mod $m_1 \cdots m_n m_{n+1}$. This completes the induction step and proves the result. QED

Practice (7.8P) Explain why j is unique mod $m_1 \cdots m_n$ and why x is unique mod $m_1 \cdots m_n m_{n+1}$.

8. An Algorithm for the CRT

Here is an algorithm to solve (7.4).

(8.1) Algorithm CRT

Input: $n \in \mathbf{N}$, integers $b_1, \ldots, b_n$, and positive integers $m_1, \ldots, m_n$ relatively prime in pairs.

Output: an integer x satisfying the simultaneous congruences (7.4).

Step 0. Set $M. = m_1 \cdots m_n$.

Step 1. $\forall i = 1, \ldots, n$ solve for c_i in

$$c_i \frac{M}{m_i} \equiv 1 \pmod{m_i}. \tag{8.2}$$

Step 2. $x := \sum_i b_i c_i M/m_i$ is a solution to (7.4).

End.

Proof Consider any (fixed) i. Then for all $j \neq i$, $M/m_i \equiv 0 \pmod{m_j}$. Therefore,

$$\sum_i b_i c_i \frac{M}{m_i} \equiv b_j \pmod{m_j}$$

for all $j = 1, \ldots, n$. Why? Because the only nonzero term mod m_j is $b_j c_j M/m_j$, which is $\equiv b_j \pmod{m_j}$ by the choice of c_j. QED

Practice (8.3P) Why is it possible to solve for c_i in (8.2)?

One advantage of (8.1) over the algorithm suggested in (7.1) is that (8.1) saves labor if more one set of b_i's is involved for the same m_i's. Even for hand computations (8.1) involves less bookkeeping and is less error-prone than the other.

Example (8.4) Let's use (8.1) to redo (7.1). $M = 60$, $m_1 = 3$, $m_2 = 4$, $m_3 = 5$.

$$\begin{aligned} c_1 \frac{60}{3} &= 20c_1 \equiv 1 \pmod 3 \\ -c_1 &\equiv 1 \pmod 3 \\ c_1 &= -1. \end{aligned}$$

$$\begin{aligned} c_2 \frac{60}{4} &= 15c_2 \equiv 1 \pmod 4 \\ -c_2 &\equiv 1 \pmod 4 \\ c_2 &= -1. \end{aligned}$$

$$\begin{aligned} c_3 \frac{60}{5} &= 12c_3 \equiv 1 \pmod 5 \\ 2c_3 &\equiv 1 \pmod 5 \\ c_3 &= 3. \end{aligned}$$

Any solution for the c_i's will do, although it's often helpful to have them not all of the same sign. Now (7.5) tells us that

$$\begin{aligned} n &\equiv 1 \cdot (-1) \cdot 20 + 2 \cdot (-1)15 + 4 \cdot 3 \cdot 12 \pmod{60} \\ &\equiv 94 \equiv 34 \pmod{60}. \end{aligned}$$

9. Applications of the CRT

(9.1) *Multiple precision.* Suppose an integer t is to be calculated to a precision that is difficult to achieve on the computer available. Suppose you know $1 \leq t < M$. One tactic is to use the CRT by choosing n primes $p_1, \ldots, p_n$ with product M (or more). Calculate the remainder r_i of $t \pmod{p_i}$ for each $i = 1, \ldots, n$. This step can be easy and error-free. Now you know

$$\forall i = 1, \ldots, n \quad t \equiv r_i \pmod{p_i}.$$

You carry out algorithm (8.1), which requires as multiple-precision steps only the multiplication of b_i by $c_i M / m_i$ and the addition of those quantities.

Example (9.2) Consider the problem of calculating 2^{25} exactly (in base 10, that is; you know it exactly in base 2). Suppose your calculator can handle no bigger than five-digit numbers. We know that 2^{25} is about 32,000,000, so we choose the mutually relatively prime moduli (they don't *have* to be primes)

$$\begin{aligned} m_1 &= 400 \\ m_2 &= 401 \\ m_3 &= 403, \end{aligned}$$

since $M = m_1 m_2 m_3 > 4^3 \cdot 10^6 = 64{,}000{,}000$. We use the CRT to determine $x = 2^{25}$ uniquely mod M. Since $0 < x < M$ we will have x exactly. Since $2^8 = 256$ and $2^{16} = (256)^2 = 65{,}536$ we can get these exactly on our calculator. We reduce mod m_i:

$$\begin{aligned} 2^{16} &\equiv 336 \pmod{400} \\ &\equiv 173 \pmod{401} \\ &\equiv 250 \pmod{403}. \end{aligned}$$

Now $2^{24} = 2^8 \cdot 2^{16}$, so

$$\begin{aligned} 2^{24} &\equiv 256 \cdot 336 = 86{,}016 \equiv 16 \pmod{400} \\ &\equiv 256 \cdot 173 = 44{,}288 \equiv 178 \pmod{401} \\ &\equiv 256 \cdot 250 = 64{,}000 \equiv 326 \pmod{403}. \end{aligned}$$

Finally, $x = 2^{25}$ is congruent to the doubles of these residues, so we solve the CRT problem

$$\begin{aligned} x &\equiv 32 \pmod{400} \\ x &\equiv 356 \equiv -45 \pmod{401} \\ x &\equiv 249 \equiv -154 \pmod{403}. \end{aligned}$$

To find c_1, c_2, c_3:

$$401 \cdot 403 \cdot c_1 \equiv 3c_1 \equiv 1 \pmod{400}.$$

We do this "by eye:" $3 \mid 399$ and $399 \equiv -1 \pmod{400}$, so $c_1 \equiv -133 \pmod{400}$.

$$\begin{aligned} 400 \cdot 403c_2 &\equiv -1 \cdot 2c_2 \equiv 1 \pmod{401} \\ \text{that is,}\quad 2c_2 &\equiv -1 \pmod{401} \\ &\equiv 400 \pmod{401} \\ \therefore c_2 &\equiv 200 \pmod{401}. \\ 400 \cdot 401c_3 &\equiv (-3)(-2)c_3 \equiv 1 \pmod{403} \\ \text{that is,}\quad 6c_3 &\equiv 1 \pmod{403} \\ &\equiv -402 \pmod{403} \\ c_3 &\equiv -67 \pmod{403}. \end{aligned}$$

Now we know that

(9.3) $x \equiv 32(-133 \cdot 401 \cdot 403) - 45 \cdot (200 \cdot 400 \cdot 403) - 154(-67 \cdot 400 \cdot 401) \pmod{M}$.

So far we have not had to exceed five digits in any calculations; we have not even needed to find M. At this point, however, we have to do *multiple-precision* work, in our sense. If we work out the terms in (9.3) as they stand, we will need nine digits: the first term is $-687{,}782{,}368$, for example. M is eight digits: $M = 67{,}641{,}200$. We can restrict our calculations to eight digits if we reduce the individual terms of (9.3) mod M. (That's OK because our answer is correct only mod M anyway.) This step is easy:

$$32 \cdot (-133) \equiv 144 \pmod{400}$$

Therefore,

$$32 \cdot (-133) \cdot 401 \cdot 403 \equiv 144 \cdot 401 \cdot 403 \pmod{M}.$$

(Why?) Doing the analogous reduction on the other two terms of (9.3) and then combining the first two terms yields

$$-5{,}422{,}768.$$

The third term becomes

$$38{,}977{,}200$$

and the sum is 33,554,432. It is 2^{25} because it is between 0 and M.

If we left the terms in (9.3) alone and combined them we'd get

$$x \equiv -483{,}575{,}168$$

which is indeed $\equiv 33{,}554{,}432 \pmod{M}$, but it takes more work to find that out. The point of this example is to illustrate the process, not to calculate 2^{25} per se.

(9.4) Here is a modification: suppose t is the difference of two large integers; $t = a - b$. If you try to find t directly you can lose t when the most significant figures of a and b are the same. For example, $1234567 - 1234456 = (0000)111$. If the calculation of the less significant figures in a and b is not reliable you can get an error when you try to find t. But if you can reliably calculate the residues of a and b mod m

for some $m > t$ then you can get t exactly, if t is positive. In general you need $m > 2|t|$, as we explain in example (9.5).

This modification uses merely congruence mod m. If a and b are so huge that even t is beyond the precision of the machine at hand, then you can bring in the CRT approach of the previous application.

Example (9.5) Here we use congruences to calculate the difference of two large numbers. Problem: Find $a - b$, where

$$a := 2^{33} + 3^{17},$$
$$b := 2^{29} + 2 \cdot 3^{20} + 3^{19} + 3^{16} + 2 \cdot 3^{13} + 2 \cdot 3^{10} + 3^{9}.$$

Using logarithms or crude estimates we can see that both a and b are approximately 8×10^9. In particular, they are 10-digit numbers. All that tells us about $a - b$ is that $|a - b|$ is less than 10^9. (Why?)

We can calculate $a - b$ mod m for $m = 1000$, 1001, and 1003, three moduli relatively prime in pairs with product greater than 10^9. We then apply the Chinese Remainder Theorem (7.5) to find the value of $a - b$ correct modulo $M = m_1 m_2 m_3 = 1000 \cdot 1001 \cdot 1003 = 1{,}004{,}003{,}000 > 10^9$.

Using the techniques of Section (3) we find the residues mod m_1, m_2, and m_3 of the various powers of 2 and 3. For example,

$$\begin{aligned} 2^{10} &= 1{,}024 \equiv 24 \pmod{1000} \\ \therefore 2^{20} &\equiv (24)^2 = 576 \pmod{1000} \\ \therefore 2^{30} &\equiv (24) \times 576 \equiv 824 \pmod{1000} \\ \therefore 2^{33} &= 8 \cdot 2^{30} \equiv 8 \cdot 824 \equiv 592 \pmod{1000}. \end{aligned}$$

Calculating the same way mod 1001 and 1003, we begin our table of intermediate results as

(9.6)

		mod 1000	mod 1001	mod 1003
2^{33}	$\equiv$	592	239	869
3^{17}	$\equiv$	163	152	904
$a = 2^{33} + 3^{17}$	$\equiv$	755	391	770

		mod 1000	mod 1001	mod 1003
2^{29}	$\equiv$	912	578	117
3^{9}	$\equiv$	683	664	626
$2 \cdot 3^{10}$	$\equiv$	98	981	747
$2 \cdot 3^{13}$	$\equiv$	646	461	109
3^{16}	$\equiv$	721	718	970
3^{19}	$\equiv$	467	367	112
$2 \cdot 3^{20}$	$\equiv$	802	200	672
b	$\equiv$	4329	3969	3353
b	$\equiv$	329	966	344
a	$\equiv$	755	391	770
$-b$	$\equiv$	-329	-966	-334
$a - b$	$\equiv$	426	426	426

The fact that $a - b$ is congruent to the same number (426) modulo each of the three moduli means that we need not go through the rest of the CRT procedure;

(9.7) $$a - b \equiv 426 \pmod{M = 1{,}004{,}003{,}000}.$$

Congruence (9.7) follows from (3.19).

We are now almost ready to say that $a - b = 426$. We know that $|a - b| < M$, but $a - b = 426 - M$ is also a solution to the congruences, (9.6), and $|426 - M| < M$. So—is $a - b$ positive, or is it negative?

If we had total confidence in our estimate $|a - b| < 10^9$ we could conclude from (9.5) that $a - b$ is positive, because $|426 - M| > 10^9$. But if we had found $a - b \equiv 4{,}100{,}000 \pmod{M}$, for example, we'd still be in the soup.

Actually, "Is $a - b$ positive?" is the wrong question, because if it is, to find that out we'd have to determine not only the leading digit, 8, but also perhaps the next seven digits.

What we do instead is introduce a fourth modulus to increase the range, or value of M, so as to remove the ambiguity. Here there's an ideal new modulus, namely, 3. It is relatively prime to M, and it's trivial to reduce a and b mod 3:

$$a = 2^{33} + 3^{17} \equiv 2^{33} \equiv 2 \pmod{3}$$
$$b \equiv 2^{29} \equiv 2 \pmod{3}$$
$$\therefore a - b \equiv 0 \pmod{3}.$$

Since $426 \equiv 0 \pmod{3}$,

(9.8) $$a - b \equiv 426 \pmod{3M}.$$

Since $M \not\equiv 0 \pmod{3}$, $426 - M \not\equiv 0 \pmod{3}$. Therefore, $a - b = 426$.

Practice (9.9P) Justify (9.8) by treating it as a CRT problem with two congruences, one mod 3, one mod M.

The choice of the three moduli 1000, 1001, and 1003 was, therefore, not good enough, even though their product exceeds the absolute value of $a - b$. You should

(9.10) choose moduli so that their product M is at least *twice* the upper bound B on $|a - b|$.

Why will this work? Because if $M > 2B$ and $a - b \equiv x \pmod{M}$ and $0 < x < B$, then

$$|x - M| > B,$$

because

$$|x - M| = M - x > 2B - x > B,$$

since $x < B$. (Similarly, if $|x - M| < B$, then $x > B$.) So only one of the two solutions to the congruence satisfies the bound on the absolute value.

We eliminated $426 - M$ as a possible value of $a - b$ in our example by use of congruence mod 3 instead of by use of (9.10). If we wanted to use the argument of (9.10), we could say: (9.8) implies either $a - b = 426$, or $a - b = 426 - 3 \cdot (1{,}004{,}003{,}000)$. But the latter is impossible because its absolute value is greater than 3×10^9.

10. A Generalization of the CRT

Suppose we had simultaneous linear congruences in which the coefficients of the unknown x were not all 1.

(10.1)
$$\begin{aligned} a_1x &\equiv b_1 \pmod{m_1} \\ a_2x &\equiv b_2 \pmod{m_2} \\ &\vdots \\ a_nx &\equiv b_n \pmod{m_n}. \end{aligned}$$

Is there a solution x, and if so, how could we find it?

Certainly if there is a solution there must be one for each individual congruence $a_ix \equiv b_i \pmod{m_i}$. And this fact suggests an approach to problem (10.1). Namely,

(10.2) Solve (if possible) each congruence

$$a_ix \equiv b_i \pmod{m_i}$$

as

$$x \equiv t_i \pmod{m_i'},$$

where $m_i' = m_i/\gcd(a_i, m_i)$ and t_i is the solution given by (5.5). If for some i, $\gcd(a_i, m_i) \nmid b_i$, then (10.1) has no solution.

If the m_i' are relatively prime in pairs we may apply (7.5) to find a solution x_0, unique mod $M' = m_i' \cdots m_n'$, to the reduced simultaneous congruences

(10.3)
$$\begin{aligned} x &\equiv t_1 \pmod{m_1'} \\ &\vdots \\ x &\equiv t_n \pmod{m_n'}. \end{aligned}$$

The congruences (10.3) are equivalent to (10.1). If we define

$$M' := m_1' \cdots m_n'$$

and let x_0 be any solution to (10.3), then the solution-sets of (10.1) and (10.3) are

$$\lfloor x_0 \rfloor_{M'}.$$

Example (10.4) Consider the problem

(10.5)
$$\begin{aligned} 2x &\equiv 2 \pmod{6} \\ 2x &\equiv 4 \pmod{8} \\ 7x &\equiv 21 \pmod{35} \end{aligned}$$

The gcd's are 2, 2, and 7, respectively. The reduced simultaneous congruences are

$$\begin{aligned} x &\equiv 1 \pmod{3} \\ x &\equiv 2 \pmod{4} \\ x &\equiv 3 \pmod{5}. \end{aligned}$$

We solve the latter as before, finding

$$x \equiv 58 \equiv -2 \pmod{60}.$$

Thus $x_0 = -2$ and $M' = 60$. The solution-set to (10.5) is $\lfloor -2 \rfloor_{60}$.

(10.6) *A different generalization.* It is even possible to relax the requirement that the moduli m_i be relatively prime in pairs. The result is that (7.4) has a solution if and only if for all i, j, $b_i \equiv b_j \pmod{\gcd(m_i, m_j)}$. The solution is unique modulo the lcm of $m_1, \ldots, m_n$. We shall, however, content ourselves with the special cases presented previously. For an account of this generalization see (16.1, pages 244 ff.).

11. Fermat's Theorem

We now present Euler's proof of a famous result discovered by Fermat in 1640. You are asked to find another proof in problem G10 of Chapter 8.

Fermat's Theorem (11.1) Let p be a prime and a any integer not divisible by p. Then

$$a^{p-1} \equiv 1 \pmod{p}.$$

Proof Consider the two products

$$1 \cdot 2 \cdot 3 \cdots (p-1) = u$$

and

$$(1 \cdot a)(2 \cdot a)(3 \cdot a) \cdots ((p-1)a) = v.$$

Notice that no two factors in v are the same mod p; that is,

$$ia \not\equiv ja \pmod{p}$$

if $i \not\equiv j$ mod p, because

$$ia - ja = (i-j)a$$

is divisible by the *prime* p if and only if p divides one of the factors $i - j$ or a. That result is derived from Lemma (11.6), Chapter 6. (We may instead use the Cancellation Law (3.12).) Therefore, since none of the factors in v is 0 mod p either, those $p - 1$ factors must be congruent mod p to $1, 2, \ldots, p-1$ in some order. Therefore,

$$u \equiv v \pmod{p}.$$

But $v = a^{p-1}u$ by commutativity of multiplication. Thus

$$a^{p-1}u \equiv u \pmod{p};$$

and since u is relatively prime to p we may cancel u, by (3.15). QED

Examples

$$2^{10} \equiv 1 \pmod{11}$$
$$7^{12} \equiv 1 \pmod{13}$$
$$5^{100} \equiv 1 \pmod{101}$$
$$48^{72} \equiv 1 \pmod{73}$$

(11.2) We can also use Fermat's theorem to shorten such calculations as those in (3.8), (3.9), and (3.20): To find the least residue of $a^n \pmod{p}$ for the prime p we may divide n by $p-1$; thus

$$n = q(p-1) + r, \qquad 0 \leq r < p-1.$$

Then $a^n = (a^{p-1})^q a^r \equiv a^r \pmod{p}$ by Fermat's theorem. Since r is less than p, this approach can save a lot of effort.

Example To reduce $29^{1,000} \pmod{37}$ to its least positive residue, we find, using (11.2), that

$$1000 \equiv 28 \pmod{36};$$
$$\therefore 29^{1000} \equiv 29^{28} \pmod{37}.$$

Since $29 \equiv -8 \pmod{37}$ we can simplify further.

$$29^{28} \equiv (-8)^{28} = (-2^3)^{28}$$
$$\equiv 2^{84} \pmod{37}$$

We use (11.2) again:

$$84 \equiv 12 \pmod{36}$$
$$\therefore 2^{84} \equiv 2^{12} \pmod{37}$$

The quantity 2^{12} is a little easier to calculate than 29^{28} mod 37.

Since $12 = 8 + 4$, and $2^4 = 16$, $2^8 = 256 \equiv 34 \equiv -3 \pmod{37}$. Thus $2^{12} = 2^8 \cdot 2^4 \equiv (-3)(16) = -48 \equiv -11 \equiv 26 \pmod{37}$.

Notice that to find r we reduce n mod $p-1$.

Fermat's theorem is the basis of probabilistic methods for testing whether an integer is prime. See, for example, reference (16.4) of Chapter 6, page 96.

12. The Euler Phi Function

The Euler φ function maps $\mathbf{N}$ to $\mathbf{N}$. It is defined by the rule

(12.1) $\forall m \in \mathbf{N}$, $\varphi(m)$ is the total number of residue-classes in $\mathbf{Z}_m$ that are relatively prime to m.

For this definition to make sense we need to define gcd $(m, \lfloor a \rfloor_m) := \gcd(m, a)$, as you might expect. And the latter definition makes sense because for all $x \in \lfloor a \rfloor_m$, $\gcd(x, m) = \gcd(a, m)$; therefore, $\gcd(m, \lfloor a \rfloor_m)$ does not vary with the choice of elements from $\lfloor a \rfloor_m$.

Example The statement $\varphi(3) = 2$ is true because among the elements of $\mathbf{Z}_3$, $\lfloor 0 \rfloor_3$ is not relatively prime to 3 but $\lfloor 1 \rfloor_3$ and $\lfloor 2 \rfloor_3$ are. Also, $\varphi(1) = 1 = \varphi(2)$.

Practice (12.2P) Work out $\varphi(4)$ and $\varphi(5)$.

Another name for the phi function is the Euler *totient*.

Now consider the following statement: For a set S,

(12.3) $\varphi(m)$ is the total number of integers in S that are relatively prime to m.

Then (12.1) is equivalent to (12.3) for each of the following definitions of S:
(*i*) $S = \{0,1,\ldots,m-1\}$,
(*ii*) $S = \{1,2,\ldots,m\}$,
(*iii*) S is any complete system of residues mod m.

From (12.3 *i* or *ii*) we see at once that

(12.4) $$\varphi(p) = p - 1 \quad \text{if } p \text{ is prime,}$$

since any integer is relatively prime to p iff it is not divisible by p.

It is easy enough to see that

(12.5) $$\varphi(p^2) = p^2 - p \quad \text{if } p \text{ is prime;}$$

we use (12.3*ii*) and identify all the integers in $\{1,\ldots,p^2\}$ that *are* divisible by p. They are just the set of p elements

$$\{1\cdot p, 2p, 3p, \ldots, p\cdot p\}.$$

Thus $\varphi(9) = \varphi(3^2) = 9 - 3 = 6$; the 3 "undesirables" are 3, 6, 9.

The same technique shows that

(12.6) If p is prime and $e \geq 1$, then $\varphi(p^e) = p^e - p^{e-1}$.

Thus, for example, $\varphi(27) = 27 - 9 = 18$, and $\varphi(2^n) = 2^{n-1}$ for all $n \geq 1$.

The following result is one of the key properties of the Euler phi function. With it and (12.6) and unique factorization (Chapter 6, (11.8)), we can in principle calculate $\varphi(n)$ for any integer $n \geq 1$.

Theorem (12.7) For all positive integers a, b, if gcd $(a,b) = 1$ then $\varphi(ab) = \varphi(a)\varphi(b)$.

Proof We first define for any $m \geq 1$ the set Φ_m of all residue-classes in $\mathbf{Z}_m$ that are prime to m. We next define a mapping from Φ_{ab} to the Cartesian product of Φ_a and Φ_b,

$$f : \Phi_{ab} \longrightarrow \Phi_a \times \Phi_b;$$

and then we'll finish by showing that f is a bijection.

(12.8) For all $\lfloor x \rfloor_{ab} \in \Phi_{ab}$, $f(\lfloor x \rfloor_{ab}) := (\lfloor x \rfloor_a, \lfloor x \rfloor_b) \in \Phi_a \times \Phi_b$.

We have to make our usual check that the definition makes sense. That is, suppose $x' \in \lfloor x \rfloor_{ab}$; show that $\lfloor x' \rfloor_a = \lfloor x \rfloor_a$ and $\lfloor x' \rfloor_b = \lfloor x \rfloor_b$. But since $x \equiv x'$ (mod ab), it follows immediately from (3.5) that $x \equiv x'$ (mod a) and (mod b). Therefore, $\lfloor x' \rfloor_a = \lfloor x \rfloor_a$ and $\lfloor x' \rfloor_b = \lfloor x \rfloor_b$. Another point we must check is that the range of f is really a subset of the codomain. But since x is prime to ab, x is prime to each of a and b; thus $\lfloor x \rfloor_a$ is in Φ_a and $\lfloor x \rfloor_b$ in Φ_b.

Now we prove that f is a bijection. An element $(\lfloor y \rfloor_a, \lfloor z \rfloor_b)$ in the codomain is the value of f at some point $\lfloor x \rfloor_{ab}$ of the domain if and only if

$$f(\lfloor x \rfloor_{ab}) := (\lfloor x \rfloor_a, \lfloor x \rfloor_b) = (\lfloor y \rfloor_a, \lfloor z \rfloor_b).$$

But that holds iff

$$x \equiv y \pmod{a} \quad \text{and} \quad x \equiv z \pmod{b}.$$

By the CRT (7.5) there is always such an x ($\therefore f$ is surjective) since a and b are relatively prime. The CRT also tells us that such solutions x are unique mod ab ($\therefore f$ is injective). QED

Example (12.9) We see that $\varphi(2^3 \cdot 3 \cdot 5^2 \cdot 7) = \varphi(2^3)\varphi(3)\varphi(5^2)\varphi(7) = 4 \cdot 2 \cdot 20 \cdot 6 = 960$. It is wrong to say that $\varphi(25) = \varphi(5 \cdot 5) = 16$, because 5 is not prime to itself. Similarly it is wrong to say $\varphi(18) = \varphi(3 \cdot 6) = \varphi(3)\varphi(6) = 2 \cdot 2 = 4$, because 3 is not prime to 6.

Practice (12.10P) What is $\varphi(25)$? $\varphi(18)$?

(12.11) As mentioned in example (12.9), you need the hypothesis $\gcd(a, b) = 1$ before you can use the conclusion $\varphi(ab) = \varphi(a)\varphi(b)$ of Theorem (12.7). It may even be an "if and only if" result. That is, perhaps the converse is true: if $\varphi(ab) = \varphi(a)\varphi(b)$, then $\gcd(a, b) = 1$. See problem G10.

Now comes a generalization of Fermat's theorem, found by Euler ca. 1750.

Euler's Theorem (12.12) Let m be a positive integer and let $a \in \mathbf{Z}$ be prime to m. Then

$$a^{\varphi(m)} \equiv 1 \pmod{m}.$$

Proof The proof, just like that of Fermat's theorem (11.1), is due to Euler. Let $n = \varphi(m)$ and let $b_1, \ldots, b_n$ be all the integers prime to m among $1, \ldots, m$. Then

$$(ab_1)(ab_2)\cdots(ab_n) = a^n b_1 b_2 \cdots b_n$$

is also congruent mod m to $b_1 b_2 \cdots b_n$ because $ab_1, ab_2, \ldots, ab_n$ are all different mod m (by the Cancellation Law (3.12)) and so are congruent to $b_1, b_2, \ldots, b_n$ in some order, since a is prime to m by hypothesis. QED

Example The quantity $2^{90} \equiv 2^{10} \pmod{75}$ because $\varphi(75) = \varphi(3 \cdot 5^2) = 2 \cdot 20 = 40$. Thus $2^{40} \equiv 1 \pmod{75}$, so $2^{80} \equiv 1 \pmod{75}$.

(12.13) Euler's theorem offers another way to solve the linear congruences $ax \equiv b \pmod{m}$ of (5). First dividing a, m, and b by $\gcd(a, m)$, we may assume $\gcd(a, m) = 1$. Then we multiply both sides by $a^{\varphi(m)-1}$. That process solves the congruence; $x \equiv ba^{\varphi(m)-1} \pmod{m}$.

Examples (12.14) (*i*) $12x \equiv 10 \pmod{26}$. Divide by $2 = \gcd(12, 26)$ to get the equivalent problem $6x \equiv 5 \pmod{13}$. Since $\varphi(13) = 12$, the quantity $6^{12} \equiv 1 \pmod{13}$, so $6^{11} \cdot 6x \equiv 6^{11} \cdot 5 \pmod{13}$ or $x \equiv 6^{11} \cdot 5 \equiv 11 \cdot 5 \equiv 55 \equiv 3 \pmod{13}$. [Check that $6^{11} \equiv 11 \pmod{13}$.]

(*ii*) Consider also $12x \equiv 8 \pmod{70}$. Again $\gcd(12, 70) = 2$, so we solve the equivalent problem $6x \equiv 4 \pmod{35}$. Since $\varphi(35) = \varphi(5)\varphi(7) = 4 \cdot 6 = 24$, we can multiply both sides by 6^{23} to find x. We get

$$x \equiv 6^{23} \cdot 4 \equiv 6 \cdot 4 = 24 \pmod{35}.$$

Practice (12.15P) Verify the easy way that $6^{23} \equiv 6 \pmod{35}$.

Sometimes an exponent smaller than $\varphi(m)$ "turns a into 1," as in the last example, where $6^2 \equiv 1 \pmod{35}$. All we needed to do was multiply by 6. So it may pay to calculate (mod m) successively $a^2, a^3, a^4, \ldots$ instead of going straight to $a^{\varphi(m)-1}$. Here is an observation that shortens this approach by telling us to look only at n dividing $\varphi(m)$:

(12.16) Let $a \in \mathbf{Z}$ and $m \in \mathbf{N}$. If n is the smallest positive integer such that $a^n \equiv 1 \pmod{m}$, then n divides $\varphi(m)$.

The proof is left to the problems. In working example (12.14*i*) this way, we'd check 6^n for $n = 2, 3, 4$, and 6, finding $6^6 \equiv -1 \pmod{13}$. So we could stop here and multiply by 6^5 on both sides:

$$-x \equiv 6^5 \cdot 5 \equiv 2 \cdot 5 \equiv 10 \equiv -3 \pmod{13}, \quad \text{or} \quad x \equiv 3 \pmod{13}.$$

13. Application: A Cryptographic Scheme

We now discuss briefly the RSA algorithm for sending and deciphering secret messages. The name is the initials of the codiscoverers Rivest, Shamir, and Adelman. The receiver constructs his or her own system, publishing part of the key and keeping the rest secret. The key consists of the following elements

Public key	*Secret key*
pq	prime p
e	prime $q \neq p$
	f such that $ef \equiv 1 \pmod{\varphi(pq)}$.

To send a *message*, i.e., a number a prime to pq, the sender reduces a^e mod pq, and sends the result r. The receiver decodes r by calculating $r^f \pmod{pq}$. The result is a, because

$$r^f \equiv a^{ef} \equiv a \pmod{pq}$$

by Euler's theorem, (12.12). (Remember, $ef = 1 + k\varphi(pq)$, for some integer k, so $a^{ef} = a \cdot (a^{\varphi(pq)})^k \equiv a \cdot 1^k \equiv a \pmod{pq}$.)

From (12.7) we know that $\varphi(pq) = (p-1)(q-1)$.

Typically, p and q each have 50 to 100 digits. The belief that the RSA scheme is cryptographically strong (invulnerable to an adversary's attempt to discover f) is based on the presumed difficulty of factoring large numbers. Mathematicians have tried unsuccessfully for centuries to find efficient algorithms to factor integers. In 1988, however, a group of about 10 mathematicians using several hundred computers succeeded in factoring the 100-digit number $(11^{104} + 1)/(11^8 + 1)$. They found that it is the product of two primes, one of 41 digits, one of 60 digits. They were clever in their use of the computers, as they had to be, because brute-force methods would take an incredibly long time. Specifically, suppose they had known that one factor was a 41-digit prime. There are about about 10^{39} such primes, by the *prime number theorem*, the proof of which is way beyond us. Suppose they tried these primes, one after the other, to see if one divided the number. Say they could test a million primes each second. How long would it take? Answer: If the universe is 4×10^9 years old, then it is less than 10^{24} microseconds old. So, in a trillion lifetimes of the universe they could check about 0.1% of these primes.

In 1990 several hundred mathematicians, using about 1000 computers, were able to factor a certain number of 155 digits. It was the largest on a list called The 10 Most Wanted Numbers, set up to challenge factoring experts. Their feat will tend to make RSA users choose larger primes.

Practice (13.2P) Use congruences or algebra, not brute force, to show that $11^8 + 1$ divides $11^{104} + 1$.

Practice (13.3P) The prime number theorem says that the total number of primes less than x is asymptotic (as $x \longrightarrow \infty$) to $x/\ln x$, where $\ln x$ is the natural logarithm of x.

(*i*) Use that result to estimate that the total number of primes having 41 digits (in base 10) is about 9.5×10^{38}.

(*ii*) About how many primes of 100 digits are there?

The new feature of this cryptosystem (and of some others) is the public key. It's like a telephone listing in that anyone can "call" the receiver, but with a message enciphered so that only the receiver can decipher it.

Example (13.4) Using small primes, let us take $p = 47$ and $q = 53$. Then $pq = 47 \cdot 53 = 2491$, and $\varphi(pq) = 46 \cdot 52$. We choose $e = 45$ (e must be relatively prime to $\varphi(pq)$). We find f by solving

$$45f \equiv 1 \pmod{46 \cdot 52 = 8 \cdot 13 \cdot 23}. \tag{13.5}$$

Although we could use algorithm EE, Chapter 6, (9.3), an easy way by hand is to use the algorithm CRT (8.1). Thus we want to solve

$$\begin{aligned} 45f &\equiv -3f \equiv 1 \pmod 8 \\ 45f &\equiv \quad 6f \equiv 1 \pmod{13} \\ 45f &\equiv \quad -f \equiv 1 \pmod{23}. \end{aligned}$$

We put these in the standard form (7.4) by solving the individual congruences for f:

$$\begin{aligned} -3f \equiv 1 \pmod{8} &\quad \text{iff} \quad f \equiv -3 \equiv 5 \pmod 8 \\ 6f \equiv 1 \pmod{13} &\quad \text{iff} \quad f \equiv -2 \pmod{13} \\ -f \equiv 1 \pmod{23} &\quad \text{iff} \quad f \equiv -1 \pmod{23}. \end{aligned}$$

Now we solve the auxiliary congruences (8.2): We have $c_1 \cdot 13 \cdot 23 \equiv 1 \pmod 8$, or $3c_1 \equiv 1 \pmod 8$. Thus $c_1 = 3$. Next $c_2 \cdot 8 \cdot 23 \equiv 1 \pmod{13}$, or $-5(-3)c_2 = 15c_2 \equiv 2c_2 \equiv 1 \pmod{13}$. Thus $c_2 = 7$. Finally $c_3 \cdot 8 \cdot 13 \equiv 1 \pmod{23}$, or $-80c_3 \equiv 12c_3 \equiv 1 \pmod{23}$; $c_3 = 2$. Therefore, we may take $f \equiv 5 \cdot 3 \cdot 13 \cdot 23 - 2 \cdot 7 \cdot 8 \cdot 23 - 1 \cdot 2 \cdot 8 \cdot 13 \pmod{8 \cdot 13 \cdot 23}$. This operation makes $f = 1701$.

We publish 2491 and 45 and keep p, q, and f secret.

Someone wanting to send us the message 1059, for example, would encipher it by reducing $(1059)^{45}$ mod 2491, using in real life a computer. The result:

$$(1059)^{45} \equiv 1536 \pmod{2491}.$$

Thus we receive the transmission "1536." We decipher it by reducing 1536^{1701} mod 2491. Indeed the result is 1059.

14. The Euler Triangle

Now we present a fascinating graphic of the Euler phi function due to William J. Jones. It leads to questions and guesses about this function.

(14.1F)

The small circles stand for integers relatively prime to the row index n, which starts at $n = 2$. Thus the circle at the apex stands for 1, prime to 2. On the next line are the circles for 1 and 2 prime to $n = 3$. On the next line are circles for 1 and 3, prime to $n = 4$. There is a blank for 2 because 2 is not relatively prime to 4. The row with index n has space for the $n-1$ integers $1, \ldots, n-1$. The bottom row is for $n = 59$.

15. A Short Descent

In this section we give the shortcut bypassing Fermat's method of descent for the Diophantine problem in Chapter 6, Section 14. The result to be re-proved is this:

For integral $k > 1$, there are no odd integers x, y such that $3x^2 + y^2 = 2^{2k}$.

If $k \geq 2$ then $2^{2k} \equiv 0 \pmod 8$. For any odd integer z, $z^2 \equiv 1 \pmod 8$ (see Problem 3.13). Therefore, if there is a solution, then

$$3x^2 + y^2 \equiv 3 + 1 = 4 \pmod 8.$$

But $4 \not\equiv 0 \pmod 8$, so we have a contradiction.

16. Further Reading

(16.1) Øystein Ore [cited in Chapter 6, (16.3)].

(16.2) Harold M. Stark, *An Introduction to Number Theory*, Markham, Chicago, 1970, and The MIT Press, Cambridge, 1978.

(16.3) André Weil [cited in Chapter 6, (16.5)].

17. Problems for Chapter 7

Problems for Section 1

1.1 Find all integers congruent to 6 modulo 1.

1.2 Find all $m \geq 1$ such that $27 \equiv 9 \pmod m$.

1.3 Set $A := \{687, 589, 931, 847, 527\}$. Which of the elements of A are congruent to which other elements mod 3? mod 7? Explain.

1.4 For each subset $\{x, y\}$ of two elements from A in the preceding problem, find an integer $m > 2$ such that $x \equiv y \pmod m$. Explain.

1.5 What does it mean to say of integers a, b that $a \equiv b \pmod 0$? Explain.

1.6 For each pair (x, m) in B find the least positive integer r such that $x \equiv r \pmod m$. Explain.

$$B := \{(19,2), (131,5), (84,14), (141,17)\}.$$

1.7 (*i*) Prove or disprove that

$$37 \equiv 16 \pmod 7,$$
$$16 \equiv -11 \pmod{18},$$
$$101 \equiv 1987 \pmod{41}.$$

(*ii*) Find all $m \geq 0$ such that $23 \equiv 17 \pmod m$.

1.8 (*i*) Find by trial and error two different integers x such that $3x \equiv 7 \pmod{11}$.

(*ii*) Show that these two solutions are congruent to each other modulo 11.

1.9 Solve for $x : 5x \equiv 9 \pmod{17}$. That is, find all x satisfying the congruence, and also explain why your answer is correct.

1.10 Do the same for the congruence $12x \equiv 20 \pmod{52}$.

1.11[Ans] Prove that any integer z satisfies at least one of the following congruences:

$$z \equiv 0 \pmod 2 \qquad z \equiv 0 \pmod 3 \qquad z \equiv 1 \pmod 4$$
$$z \equiv 1 \pmod 6 \qquad z \equiv 3 \pmod 8 \qquad z \equiv 11 \pmod{12}$$

1.12 Define three subsets of the integers as follows:

$$A := \{\, x;\ 0 \le x \le 10,\ x \equiv 1 \pmod 2 \text{ and } x^2 \equiv 1 \pmod 5 \,\}$$
$$B := \{\, y;\ 0 \le y \le 10,\ y \equiv 1 \pmod 2 \text{ and } y^2 \equiv -1 \pmod 5 \,\}$$
$$C := \{\, z;\ 0 \le z \le 10,\ z \equiv 0 \pmod 6 \text{ or } z \equiv 2 \pmod 6 \text{ or } z \equiv -2 \pmod 6 \,\}.$$

Prove or disprove: $\{\, A, B, C \,\}$ is a partition of the set $\{\, 0, 1, 2, \dots, 10 \,\}$.

1.13 Do the numbers 19, 8, −3, −5, 10, 5 form a complete residue system modulo 6? Explain.

1.14 Is $\{\, 49, 17, 75, 92, 46, 33, 70, 58, 86, 54 \,\}$ a complete system of residues mod 9? mod 10? mod 11? Explain.

1.15 (*i*) Determine whether the set $\{\, 0, 1, 2, 3, 4, 5, 6, 7 \,\}$ is a complete system of residues mod 7. Explain your answer.

(*ii*) Same question for the set $\{\, 17, -13, 25, -35, -9, 27, 9 \,\}$.

(*iii*) Ditto for $\{\, -8, 105, 1111, 117, 32, -3, 21, -1 \,\}$ but now mod 9.

(*iv*) Ditto for $\{\, 1013, 2570, 29, -1023, 7238, 1111, -138, 5656, 1775, 1984 \,\}$ but now mod 10.

1.16 Find which congruence class among $\lfloor i \rfloor_6$, for $i = 0, 1, \dots, 5$, each of the following integers belongs to: 18, 71, 125, −10, −5, 1.

1.17 What relations of inclusion or equality are there between the following congruence classes? Explain your answer.

$$\lfloor 4 \rfloor_2,\ \lfloor 4 \rfloor_7,\ \lfloor 0 \rfloor_{10},\ \lfloor 15 \rfloor_5,\ \lfloor 5 \rfloor_6,\ \lfloor 2 \rfloor_3,\ \lfloor -2 \rfloor_2,\ \lfloor 3 \rfloor_{21}$$

Ignore inclusion if equality holds, and ignore the fact that $\lfloor a \rfloor_m = \lfloor a \rfloor_m$.

1.18 Which, if any, $m \in \mathbf{N}$ makes the set T a complete system of residues mod m? Explain.

(*i*) $T = \{\, x;\ x \in \mathbf{Z}\ -\lfloor \frac{m}{2} \rfloor \le x \le \lfloor \frac{m}{2} \rfloor \,\}$

(*ii*) $T = \{\, x;\ x \in \mathbf{Z}\ -\lfloor \frac{m}{2} \rfloor < x < \lfloor \frac{m}{2} \rfloor \,\}$

(*iii*) $T = \{\, x;\ x \in \mathbf{Z}\ -\lfloor \frac{m}{2} \rfloor < x \le \lfloor \frac{m}{2} \rfloor \,\}$

(*iv*) $T = \{\, x;\ x \in \mathbf{Z}\ -\lceil \frac{m}{2} \rceil \le x \le \lfloor \frac{m}{2} \rfloor \,\}$

(*v*) $T = \{\, x;\ x \in \mathbf{Z}\ -\lceil \frac{m}{3} \rceil \le x < \lceil \frac{2m}{3} \rceil \,\}$.

1.19 Prove that if $S = \{\, b_0, b_1, \dots, b_{m-1} \,\}$ is a set of integers such that $\mathbf{Z} = \lfloor b_0 \rfloor_m \cup \lfloor b_1 \rfloor_m \cup \dots \cup \lfloor b_{m-1} \rfloor_m$, then S is a complete system of residues mod m, and **conversely**. Use this definition: The set $\{\, a_0, a_1, \dots, a_{m-1} \,\}$ is a complete system of residues mod m iff it can be ordered so that for all i in $\{\, 0, 1, \dots, m-1 \,\}$ $a_i \equiv i \pmod m$.

1.20 Let $m \in \mathbf{N}$ and $s \in \mathbf{Z}$. Let C be any complete system of residues mod m. Prove that

$$C + s := \{\, c + s;\ c \in C \,\}$$

is also a complete system of residues mod m. [This problem asks you to prove (1.17). It can be difficult if assigned early in this chapter.]

1.21[Ans] (*i*) Prove (1.16).

(*ii*) Prove (1.18).

1.22 Suppose U is a set of integers such that no two elements of U are congruent to each other mod m. Under what further condition is U a complete system of residues mod m?

1.23 Let $m \in \mathbf{N}$. Suppose R is a set of exactly m integers such that

$$\forall a \in \mathbf{Z}\ \exists r \in R\ \ a \equiv r \pmod m.$$

Prove that R is a complete system of residues mod m.

1.24 Let $x \bmod m :=$ the least nonnegative integer u such that $x \equiv u \pmod m$. Thus $x \bmod m = u =$ rem x by m. Let C be a subset of exactly m elements of $\mathbf{Z}$.

(*i*) Prove that C is a complete system of residues (CSR) mod m iff the function $f : C \longrightarrow \mathbf{Z}$ defined by the rule $f(x) := x \bmod m$ (for all x in C) is an injection.

(*ii*) Define $r : \mathbf{Z} \longrightarrow \mathbf{Z}$ by the rule $r(x) := x \bmod m$ for all $x \in \mathbf{Z}$. Prove C is a CSR iff $r|_C$ is an injection.

1.25 Let $m, n \ge 1$ and $a \in \mathbf{Z}$. Prove that if $a^n \equiv 1 \pmod m$, then $\gcd(a, m) = 1$.

***1.26** Suppose a is an integer such that $a^{10} \equiv 10 \pmod{26}$. Find $\gcd(a, 26)$. Prove your answer. (It is not necessary to find a.)

Problems for Section 3

3.1 Cancel correctly the common factor on both sides of the congruences: $264 \equiv 1224 \pmod{48}$; $45 \equiv 150 \pmod 7$; $168 \equiv -48 \pmod{72}$.

3.2 Cancel the biggest possible common factor correctly in the following congruences: $280 \equiv 632 \pmod{88}$, and $1075 \equiv 175 \pmod{30}$.

3.3[Ans] Prove that the square of any integer is congruent to one of -2, 0, 1, 4 mod 9.

3.4 (*i*) Find three distinct integers x such that $x \equiv 3 \pmod 8$.
(*ii*) Find all integers y such that $2y \equiv 6 \pmod 8$. Explain.
(*iii*) Find all integers z such that $2z \equiv 12 \pmod{23}$. Explain.

3.5 Find all integers x such that $6x \equiv 3 \pmod{27}$. Explain carefully your procedures, which may be either of your own invention or from somewhere in the text.

3.6 Prove (3.9).

Show your work and explain your steps in the next four problems.

3.7 Find the least positive residue of $(15)^{35}$ mod 19.

3.8 Find the least positive residue of 17^{158} mod 23.

3.9 Calculate the least positive residue of (*i*) $(331)^{51}$ mod 49 and also (*ii*) $(1025)^{500}$ mod 83.

3.10 Find the least positive residue of $(29)^{36}$ mod 17.

3.11 Verify Euler's observation that

$$2^{2^5} + 1 \equiv 0 \pmod{641}.$$

You may use a calculator; but if so, record your steps.

3.12 As x runs through $\mathbf{Z}$ what is x^2 congruent to mod 3? mod 4? mod 5?

3.13 Prove that the square of any integer is congruent to 0, 1, or 4 mod 8. What is your conclusion if the integer is odd?

3.14 Let t be any integer relatively prime to m, and let $\{a_0, \ldots, a_{m-1}\}$ be any complete system of residues mod m. Prove that $\{ta_0, \ldots, ta_{m-1}\}$ is also a complete system of residues mod m.

3.15[Ans] Show without appealing to calendars that in any nonleap year January and October have the same calendar. Show that January and July have the same calendars in a leap year.

3.16 Without referring to calendars show that the calendar for December 1976 is the same as that for July 1987.

3.17 Let m and n be positive integers. Prove that the partition of $\mathbf{Z}$ given by congruence mod m is a refinement of that for congruence mod n iff m is a multiple of n.

***3.18** Define a sequence $a_0, a_1, a_2, \ldots$ of integers by the recursion $a_{n+2} = a_{n+1} + a_n$ (for all $n \geq 0$) and the initial conditions $a_0 = 0$ and $a_1 = 1$. Prove (by induction?) that $a_{5k} \equiv 0 \pmod 5$ for all $k \geq 0$. [Hint: Generalize the problem by setting $a_1 := a :=$ any integer.]

3.19 Let $f(x) = 13x^3 - 5x^2 + 14x - 10$. Compute the least positive residue of $f(12)$ mod 7.

3.20 Let $m \geq 1$. Prove that for all a, $x \in \mathbf{Z}$, if $x \in \lfloor a \rfloor_m$, then $\gcd(x, m) = \gcd(a, m)$.

3.21 (Weil) Prove that if a, b, c are integers such that $a^2 + b^2 = c^2$, then $abc \equiv 0 \pmod{60}$. [Hint: Do problems 3.12 and 3.13 first.]

Problems for Section 4

#4.1 Prove that any integer ≥ 0 is congruent mod 3 to the sum of its base 10 digits.

4.2 Find a formula for reducing a nine-digit number (base 10) mod 7.

4.3 Prove that the ISBN coding scheme of (4.9) detects all single errors and all transpositions. (A transposition is an interchange of two adjacent digits. You assume that a single error occurs, or a transposition—not both—and no other error.)

Problems for Section 5

5.1 Solve for x: $18x \equiv 8 \pmod{15}$.

5.2 Find all solutions mod m for the congruence $ax \equiv b \pmod m$ or prove that there are no solutions.

	a	b	m
(*i*)	15	9	36
(*ii*)	1	2	3
(*iii*)	6	210	20
(*iv*)	492	120	615
(*v*)	495	120	615

5.3 Prove Corollary (5.9).

Problems for Section 6

6.1 Prove that algorithm (6.1) is correct (*i*) when $a = 0$. If solutions exist, find the set of all solutions. (*ii*) when a is negative.

In the next two problems find *all* integers x, y, or z satisfying the given congruences. Express the set of solutions as a single congruence class for a divisor of the original modulus *and* as a union of congruence classes for the original modulus. Prove your answers.

6.2 $10x \equiv 25 \pmod{35}$.

6.3 (*i*) $15y \equiv 21 \pmod{48}$,
(*ii*) $166z \equiv 46 \pmod{22}$.

Problems for Section 7

7.1 [Recommended for its recall of Chapter 2.] Suppose a and b are integers such that

$$\forall n \geq 1 \; \exists \text{ prime } p > n : a \equiv b \pmod{p}.$$

What can you conclude about a and b?

7.2 Use the CRT to prove that there are five consecutive integers each divisible by the square of a prime. Generalize.

7.3 (*i*) Define addition $\oplus$ in $\mathbf{Z}_m$ by the rule $\lfloor x \rfloor_m \oplus \lfloor y \rfloor_m := \lfloor x+y \rfloor_m$. Prove that this definition is independent of the choice of x in $\lfloor x \rfloor_m$ and y in $\lfloor y \rfloor_m$.
(*ii*) Consider $\mathbf{Z}_{60}$. Find a mapping $g : \mathbf{Z}_{60} \to \mathbf{Z}_3 \times \mathbf{Z}_4 \times \mathbf{Z}_5$ and prove it is a bijection.
(*iii*) For $i = 1, 2$ let (a_i, b_i, c_i) be in $\mathbf{Z}_3 \times \mathbf{Z}_4 \times \mathbf{Z}_5$. Define addition here as addition on the coordinates; thus

$$(a_1, b_1, c_1) \oplus (a_2, b_2, c_2) := (a_1 \oplus a_2, b_1 \oplus b_2, c_1 \oplus c_2).$$

Inside the parentheses, the $\oplus$ means addition mod 3 for the a's, mod 4 for the b's, and mod 5 for the c's. Prove that for all X, Y in $\mathbf{Z}_{60}$, $g(X \oplus Y) = g(X) \oplus g(Y)$.
(*iv*) Do the analogues of (*i*) and (*iii*) for multiplication.
(*v*) Generalize (*ii*), (*iii*), and (*iv*) to $\mathbf{Z}_m$, for $m \geq 2$, in place of $\mathbf{Z}_{60}$.

Problems for Section 8

8.1 Solve the *simultaneous* congruences (systematically, not by trial and error) $x \equiv 2 \pmod 5$, $x \equiv 7 \pmod{11}$, $x \equiv 11 \pmod{13}$.

8.2 Solve the simultaneous congruences

$$x \equiv 3 \pmod 5$$
$$x \equiv 2 \pmod 6$$
$$x \equiv 3 \pmod 7.$$

Find the smallest positive solution, check your answer, and prove it is the smallest.

8.3 Find all integers y such that $y \equiv 5 \pmod{16}$, $y \equiv 1 \pmod 9$, $y \equiv 3 \pmod{25}$.

8.4 Solve the simultaneous congruences

$$x \equiv 2 \pmod 5$$
$$x \equiv 7 \pmod{13}$$
$$x \equiv 11 \pmod 8$$

8.5 Solve the simultaneous congruences

$$y \equiv 7 \pmod{16}$$
$$y \equiv 1 \pmod 9$$
$$y \equiv 2 \pmod{25}$$

8.6 Prove $\sum_{1 \leq i \leq n} c_i M/m_i \equiv 1 \pmod M$ in the notation of the CRT in (8.1).

8.7 A professor has a pot of at most 100 semicolons to be distributed to students. If the semicolons are divided into seven equal piles, there are two left over. If they are divided into four equal piles, there is one left over. If they are divided into five equal piles, there are three left over. How many semicolons are in the pot?

8.8 A computer has $N < 100{,}000$ memory bytes. When it runs several jobs at once it divides the memory equally among the jobs without using the leftovers. It runs 43 jobs and has 4 bytes left over; it runs 47 jobs and has 4 bytes left over; it runs 51 jobs and has 44 bytes left over. Find N.

Problems for Section 10

10.1 Find all integers x such that both congruences hold simultaneously:

$$6x \equiv 10 \pmod{20}$$
$$5x \equiv 2 \pmod{11}.$$

10.2 Find the set of all x satisfying simultaneously the congruences $2x \equiv 12 \pmod{14}$, $6x \equiv 22 \pmod{51}$, and $x \equiv 4 \pmod{11}$. Explain your answer.

10.3 Same question for $2y \equiv 5 \pmod 9$, $3y \equiv 1 \pmod{13}$, and $y \equiv 2 \pmod 8$.

Problems for Section 11

11.1 Calculate the least positive residue of (*i*) $2^{14} \pmod{17}$, (*ii*) $3^{100} \pmod 5$, (*iii*) $7^{35} \pmod{11}$.

11.2 Prove that for all integers $m \geq 0$, $(2^m - 1)(2^m - 2)(2^m - 4) \equiv 0 \pmod 7$.

11.3[Ans] Find all $n \geq 0$ for which

$$3^n + 4^n \equiv 0 \pmod 7.$$

11.4[Ans] Let n be composite (i.e., $n = st$, where s and t are integers greater than 1). Prove that $\exists a \in \mathbf{Z} : n \nmid a$ and $a^{n-1} \not\equiv 1 \pmod n$. What is the logical relationship between this result and Fermat's Theorem (11.1)?

11.5 (*i*) (Ore, ref. 16.1) If p and p_1, p_2, p_3 are primes such that $p = p_1^2 + p_2^2 + p_3^2$, prove that one of p_1, p_2, or p_3 is equal to 3.

*(*ii*) Generalize.

Problems for Section 12

12.1 Find φ (210), φ (18), φ (40), and φ (72).

12.2 Prove that $2^{20} \equiv 1 \pmod{75}$ by using (3.19) and Euler's theorem (12.12).

12.3 Prove that if $m \geq 3$ then $\varphi(m)$ is even. [Hint: Use the complete system of residues $\{x; -\lfloor \frac{m}{2} \rfloor \leq x < \lceil \frac{m}{2} \rceil\}$.]

12.4 Prove that for any $m \geq 1$,

$$\varphi(m) = m \prod_{p|m} (1 - \frac{1}{p}).$$

Here $p \mid m$ means that p is a *prime* dividing m. The product is taken over all such primes.

12.5[Ans] (*i*) Prove that if n is odd, then $\varphi(2n) = \varphi(n)$.

(*ii*) Is there an $n \geq 1$ for which $\varphi(3n) = \varphi(n)$?

(*iii*) Prove that if n is even, then $\varphi(2n) = 2\varphi(n)$.

12.6 Prove that for all $n, k \in \mathbf{N}$, $\varphi(n^k) = n^{k-1}\varphi(n)$.

12.7 Find all n such that $\phi(n) = 8$. Prove your answer.

12.8[Ans] (Ore, ref. 16.1) Find all integers $n \geq 1$ such that $\varphi(n)$ divides n.

*12.9 Prove (12.16). Hint: Let t be the smallest positive integer such that $a^t \equiv 1 \pmod m$, and apply the division algorithm (Chapter 6, Section 2).

*12.10 Prove that for all $n \in \mathbf{N}$,

$$\sum_{d|n} \varphi(d) = n.$$

The sum is taken over all the positive divisors d of n. Hint: Consider the equivalence relation R on $\{1, \ldots, n\}$ defined by xRy iff $\gcd(x, n) = \gcd(y, n)$.

Problem for Section 13

13.1 Find a "secret" decoding value for f for the cryptographic scheme of (13) for the primes $p = 29$ and $q = 47$ with $e := 33$. Explain.

Problems for Section 14

14.1 What accounts for the blank-free horizontal lines of circles occurring in Figure (14.1*F*)?

14.2 Describe precisely and explain the recurring pattern of "circles of six" down the center line of Figure (14.1*F*).

*14.3 Can you find, describe, and explain other recurring patterns in (14.1*F*)?

General Problems

G1. In the base 10, a four-digit number divisible by 9 has for three of its digits 3, 4, and 5, in some order. What is the other digit? Explain.

G2. For integral m, suppose $2^m - 1$ is prime. Prove that m is prime. Give an example to show that the converse is false.

G3.[Ans] Prove that any positive integer n divisible neither by 2 nor by 3 satisfies $n^2 \equiv 1 \pmod{24}$.

G4. " Sometimes a very simple question in elementary arithmetic will cause a good deal of perplexity. For example, I want to divide the four numbers 701, 1,059, 1,417, and 2,312 by the largest number possible that will leave the same remainder in every case. How am I to set to work? Of course, by a laborious system of trial one can in time discover the answer, but there is quite a simple method of doing it if you can only find it." (H. E. Dudeney)

G5. Let $a, b, \ldots, f$ be integers such that (α) $ad - bc = 1$. Consider the simultaneous equations

$$(\beta) \qquad \begin{aligned} ax + by &= e \\ cx + dy &= f \end{aligned}$$

The condition (α) means that there is a unique solution (x,y) to (β) and that both x and y are in $\mathbf{Z}$.

Let p be a prime, and let (u,v) be the unique solution mod p to the simultaneous congruences

$$(\gamma) \qquad \begin{aligned} au + bv &\equiv e \pmod p \\ cu + dv &\equiv f \pmod p. \end{aligned}$$

(*i*) Assuming the uniqueness stated for (β) and (γ), prove that

$$(\delta) \quad x \equiv u \pmod p \quad \text{and} \quad y \equiv v \pmod p.$$

[The Chinese Remainder Theorem could be applied to this situation: You would solve (γ) for various primes p and then get x and y by use of (8.1) on the two sets of simultaneous congruences (δ). Thus you can solve simultaneous equations with solutions that are too large for your computer with a minimum of multiple precision steps, just as in (9).

The restriction $ad - bc = 1$ is not essential. You need $ad - bc \neq 0$. Also, the idea carries over to n equations in n unknowns, for $n \in \mathbf{N}$. We omit the generalization because it is somewhat technical.]

(*ii*) An algebraic problem: prove the uniqueness claimed about (β) and (γ).

G6. Let h be any integer such that $2^h + 1$ is an odd prime. Prove that h is a power of 2.

G7. Find examples of (12.16) by calculating mod 13 $a, a^2, a^3, \ldots$ until you first reach 1, for $a = 1, 2, 3, \ldots, 12$.

G8. (Ore, ref. 16.1) Reduce $(n-1)!$ mod n for $n = 2, \ldots, 11$ and try to guess a general rule.

***G9.** Let p be prime. Prove that $1 \cdot 2 \cdots (p-1) \equiv -1 \pmod p$. (This Eighteenth-century result is known as Wilson's theorem.)

***G10.** (*i*) Investigate the converse to Theorem (12.7). Is it true?

(*ii*) Go farther; find a formula for $\varphi(ab)$ in terms of $\varphi(a)$ and $\varphi(b)$ that is good for any $a, b \in \mathbf{N}$. [Hint: To answer (*i*) you might have to solve (*ii*).]

G11. (Ireland and Rosen) Prove that there are no integers x, y such that

(*i*) $3x^2 + 2 = y^2$;

(*ii*) $7x^3 + 2 = y^3$.

G12. Let p be prime.

(*i*) Consider the set S_1 of all nonzero squares mod p. Thus

$$S_1 := \{\lfloor x^2 \rfloor_p;\ 0 < x < p,\ x \in \mathbf{Z}\}.$$

Prove that S_1 has exactly $(p-1)/2$ elements.

*(*ii*) Let S_2 be the set of all squares mod p^2. Thus

$$S_2 := \{\lfloor y^2 \rfloor_{p^2};\ 0 < y < p^2,\ y \in \mathbf{Z}\}.$$

Prove that S_2 has exactly $(p^2 - p + 2)/2$ elements.

*(*iii*) Prove that if r is a square mod p, then it is a square mod p^2. This means:

$$\begin{aligned} &\text{If} && \exists x \in \mathbf{Z} \ : \ r \equiv x^2 \pmod p, \\ &\text{then} && \exists y \in \mathbf{Z} \ : \ r \equiv y^2 \pmod{p^2}. \end{aligned}$$

Some Test Questions

T1. Let $x, y \in \{49, 35, 23, 16\}$. For which x, y with $x > y$ is it true that $x \equiv y \pmod 7$? Explain.

T2. Show without waving the calendar that in any year March and November start on the same day of the week. Explain.

T3. Suppose $\{a_1, a_2, \ldots, a_5\}$ is a complete system of residues mod 5. Prove that $\{7a_1, 7a_2, \ldots, 7a_5\}$ is also a complete system of residues mod 5.

T4. Consider the congruences $15x \equiv 9 \pmod{42}$ and $15x \equiv 10 \pmod{42}$. For *each* of the two congruences answer these questions:

(*i*) Is there a solution? Explain.

(*ii*) If there is a solution, what is the set of all solutions? Explain your answer by finding that set. Show your procedure.

T5. Find all integers x such that $3x \equiv 21 \pmod{11}$. Explain.

T6. From among the following integers choose a complete system of residues mod 5: 93, −37, 12, −71, 55, −23, 50, −97, 67, −59. Give a reason for choosing each of your numbers.

T7. Partition the congruence class $\lfloor 1 \rfloor_4$ into cells consisting of congruence classes mod 12. Explain.

T8. Let m be a positive integer. Prove that any m consecutive integers form a complete system of residues mod m.

T9. Find all integers x satisfying the congruence

$$5x \equiv 3 \pmod{31}$$

and explain your procedure.

T10. Solve the following simultaneous congruences:

$$x \equiv b \pmod 5$$
$$x \equiv b+1 \pmod 7$$
$$x \equiv b+2 \pmod p,$$

where b is the last (i.e., rightmost) digit in your social security number (SSN), and p is the smallest prime satisfying

$$p > 7 \quad \text{and} \quad p \geq a+b,$$

where a is the next-to-last digit in your SSN.

T11. Find the least positive residue of $(5)^{297}$ mod 23.

T12. Find $\varphi(7)$, $\varphi(8)$, $\varphi(9)$, and $\varphi(99)$. You should use theorems proved in the text and *quote* them.

T13. Use the division algorithm to show that if p is prime, and if a is any integer relatively prime to p, and if t is the smallest positive integer such that $a^t \equiv 1 \pmod p$, then t divides $p-1$.

T14. For any positive integer m, prove that $\varphi(m^2) \equiv 0 \pmod m$.

T15. Define $f(n) := n^2 + n$. Prove that f is symmetric mod 16 about $7\frac{1}{2}$. That is, prove

$$\forall x \in \mathbf{Z}, \quad f(8+x) \equiv f(7-x) \pmod{16}.$$

18. Answers to Practice Problems

(1.3P) From the definition (1.1),

$$(\star) \qquad -(n-k) \equiv n+k \pmod n$$

iff $-(n-k)-(n+k)$ is a multiple of n. Since it equals $-2n$, the congruence $(\star)$ holds.

(1.11P) We are asked to prove that one of any three consecutive integers is a multiple of 3. We use the division algorithm. If the first integer (here $a-1$) is not divisible by 3, then it has remainder 1 or 2 on division by 3.

Case 1. $a-1 = 3q+1$. Then $a+1 = 3q+3$ is a multiple of 3.

Case 2. $a-1 = 3q+2$. Then $a = 3q+3$ is a multiple of 3.

(2.10P) (*i*) If $a_1 = 0$, then $f(a_1 \ldots a_n)$ is the number in **N** with binary representation equal to $a_1 \ldots a_n$.

(*ii*) If $a_1 = 1$, then $f(a_1 \ldots a_n)$ is that number less 2^n (since $2^{n-1} - 2^n = -2^{n-1}$).

Thus in each case $f(a_1 \ldots a_n)$ is congruent (mod 2^n) to $(a_1 \ldots a_n)_2$. From (2.5) we see that $f(a_1 \ldots a_n)$ is the number represented in the computer by $a_1 \ldots a_n$.

The positive members of the range are $1, \ldots, (01\ldots1)_2 = 2^{n-1} - 1$. The range has 0; and its negative elements, from (*ii*), are $0, 1, \ldots, 2^{n-1} - 1$, all less 2^{n-1}, namely, $-2^{n-1}, -2^{n-1}+1, \ldots, -1$.

(2.15P) It's easiest to say

- $(11\ldots1)_2 = 2^n - 1 \equiv 0 \pmod m$.
- Addition is in **Z** followed by reduction mod $2^n - 1$ (or: Addition is mod $2^n - 1$).
- Since $2^n - 1 \equiv 0 \pmod m$, adding it makes no change.
- The fact that the numbers are in base 2 does not affect their sum. The sum of integers is well defined and has nothing to do with representations of them in base 2 or any other base.

(2.16P) Since 0 can be represented as the all-1 string, and $-a = 0 - a$ for all a, we may subtract a from the integer $(11\ldots1)_2$. The result is expressed in (2.14). It's better expressed in terms of congruences, since the adder works mod m:

$$-a \equiv 0 - a \equiv 11\ldots1 - a \pmod m.$$

(3.11P) If an integer is 0, then it is congruent to 0 mod m for all m. But $f(5) \equiv -13 \equiv 2 \pmod 5$, by (3.7.3). We chose a convenient modulus.

(3.23P) We're given $a^e \equiv 1 \pmod m$. Then $(a^e)^q \equiv 1^q \pmod m$ for all $q \geq 0$ by (3.7.1). But $1^q = 1$. Since $n = eq + r$, $a^n = a^{eq+r} = (a^e)^q a^r$ (law of exponents). $\therefore a^n \equiv a^r \pmod m$.

For 7^{29} mod 19, we first try to find e so that $7^e \equiv 1 \pmod{19}$. We compute

$$7^2 = 49 \equiv 11 \pmod{19}$$
$$7^3 = 77 \equiv 1 \pmod{19}.$$

Thus $e = 3$. Now $29 = 3 \cdot 9 + 2$, so $r = 2$. $\therefore 7^{29} \equiv 7^2 \equiv 11 \pmod{19}$.

(4.7P) With respect to single errors, the U.S. Postal Service system fails to detect any zero in place of nine, or vice versa. (This error could not happen at the check-digit, because nine is never used there.) Any other single error would change the value of the sum mod 9 (or would make the check-digit wrong).

It obviously fails to detect any transposition on the first 10 digits, by (4.1). But any interchange of the check-digit with one of the first 10 will be detected (unless they are equal), because of the following argument:

Say the ith digit n_i is interchanged with the check-digit c. Let s be the sum of the nine digits other than n_i. Then by definition

$$s + n_i \equiv c \pmod 9.$$

We see, however, the number with c and n_i interchanged. We check it by asking

$$\text{Is } s + c \equiv n_i \pmod 9?$$

The congruence cannot hold unless $c = n_i$, because if it were so, then we could subtract one congruence from the other, by (3.3), getting

$$n_i - c \equiv c - n_i \pmod 9.$$

This congruence implies $2(n_i - c) \equiv 0 \pmod 9$, and by the cancellation law (3.15), since $\gcd(2,9) = 1$, we see that

$$n_i - c \equiv 0 \pmod 9,$$

that is,

$$n_i \equiv c \pmod 9.$$

But n_i and c lie between 0 and 8, so they are equal.

(4.8P) Because of (3.7.3), if any one digit is changed to something congruent to it mod 7, the check-digit would be the same, so the error would be undetected. So any change of 0 to 7, 1 to 8, or 2 to 9, or vice versa, would be an undetected error.

(5.3P) For all $x \in \mathbf{Z}$, $21x \equiv 9x \pmod{12}$. That result derives from (3.4). Then

$$\forall x \in \mathbf{Z} \quad 21x \equiv 18 \pmod{12}$$
$$\text{iff} \quad 9x \equiv 18 \pmod{12},$$

because of transitivity of congruence mod m. ($18 \equiv 21x$ and $21x \equiv 9x$; $\therefore\ 18 \equiv 9x$, and conversely.)

Equivalent means they have the same solution-set, as just explained.

(5.4P) As given, $17x \equiv 45 \pmod{11}$. First, reduce the coefficients: An equivalent congruence is

$$6x \equiv 1 \pmod{11}.$$

Now by inspection we see that $x = 2$ is a solution. Are there more solutions? Yes, all integers in $\lfloor 2 \rfloor_{11}$ are solutions. Are there still more? No, because if $y \in \mathbf{Z}$ satisfies $6y \equiv 1 \pmod{11}$, then y satisfies $6y \equiv 12 \pmod{11}$, since $1 \equiv 12 \pmod{11}$. Now by the strong cancellation law (3.12), we may cancel the common factor 6, since $\gcd(6,11) = 1$. We find that $y \equiv 2 \pmod{11}$.

(7.8P) For j : j is unique mod $m_1 \cdots m_n$ by the inductive hypothesis.

For x : x is defined as $b_{n+1} + jm_{n+1}$, in which b_{n+1} and m_{n+1} are unique because they are given. The only variability is in j, which runs over some one congruence class mod $m_1 \cdots m_n$. Thus $j \in \{j' + km_1 \cdots m_n;\ k \in \mathbf{Z}\}$ for some fixed j'. Then, in full generality,

$$x \in \{b_{n+1} + j'm_{n+1} + km_1 \cdots m_n \cdot m_{n+1};\ k \in \mathbf{Z}\}.$$

Since j' is unique this set is a congruence class mod $m_1 \cdots m_{n+1}$.

(8.3P) Since m_i is relatively prime to m_j if $j \neq i$, m_i (the modulus) is relatively prime to $m_1 \cdots m_{i-1} m_{i+1} \cdots m_n = M/m_i$, the coefficient of the unknown c_i.

(9.9P) We know $a - b \equiv 426 \pmod M$ and $a - b \equiv 0 \pmod 3$. So we have the CRT problem

$$X \equiv 426 \pmod M$$
$$X \equiv 0 \pmod 3.$$

Since $426 \equiv 4 + 2 + 6 \equiv 0 \pmod 3$ (see problem 4.1), it is obvious that 426 is one solution to these simultaneous congruences. (We need not work through the algorithm (8.1).) Since $\gcd(3, M) = 1$, we know that the set of all solutions is exactly a congruence class mod $3M$; that conclusion comes from (7.5), the CRT. Therefore, the solution-set S is $\lfloor 426 \rfloor_{3M}$, since 426 is in S.

(12.2P) $\mathbf{Z}_4 = \{\lfloor 0 \rfloor_4, \lfloor 1 \rfloor_4, \lfloor 2 \rfloor_4, \lfloor 3 \rfloor_4\}$. Here 1 and 3 are prime to 4 (and 0 and 2 are not), so $\varphi(4) = 2$.

For $\varphi(5)$ we just observe that because 5 is prime, only 0 is not prime to 5 among $0, 1, \ldots, 4$. Therefore, $\varphi(5) = 4$.

(12.10P) By (12.5), $\varphi(25) = \varphi(5^2) = 5^2 - 5 = 20$. It is *not* $\varphi(5)\varphi(5) = 16$. To find $\varphi(18)$ we use (12.7), and (12.5):

$$18 = 2 \cdot 9 \quad \text{and} \quad \gcd(2, 9) = 1;$$
$$\therefore \varphi(18) = \varphi(2)\varphi(9)$$
$$= 1 \cdot (9 - 3) = 6.$$

(12.15P) Since $6^2 = 36 \equiv 1 \pmod{35}$, $6^{23} = 6^{22} \cdot 6 = (6^2)^{11} \cdot 6 \equiv 1^{11} \cdot 6 = 6 \pmod{35}$.

(13.2P) We use algebra. Since $104 \equiv 0 \pmod 8$, that is, $104 = 8 \cdot 13$, we use the identity, holding for odd n,

$$x^n + 1 = (x+1)(x^{n-1} - x^{n-2} + x^{n-3} - + \cdots + 1),$$

which you can prove by induction. Now set $n = 13$ and $x = 11^8$.

(13.3P) (*i*) The 41-digit numbers are those between 10^{40} and 10^{41}. Let's say

$\pi(x) :=$ the total number of primes less than x.

We wish to approximate $\pi(10^{41}) - \pi(10^{40})$.

Since $\ln 10 = 2.3+$ we find $\pi(10^{41}) \doteq 10^{41}/(41 \ln 10) = 10^{41}/41(2.3) =: A$ and $\pi(10^{40}) \doteq 10^{40}/40(2.3) =: B$. The denominators are 94.3 and 92;

$$A = 10.6 \times 10^{38}$$
$$B = 1.1 \times 10^{38}.$$

(*ii*) Similarly we find $\pi(10^{100} - \pi(10^{99}) \doteq (0.039+) \times 10^{99}$, so in round numbers there are 4×10^{97} primes of 100 digits.

CHAPTER

8

The Binomial Theorem

I never met an exponent I didn't like.
Isaac Newton

Introduction

In computer science the main application of the topic of this chapter is to the analysis of algorithms, of which we shall give two examples at the end. Along the way, however, there will be plenty of chances for you to experience new ideas, new forms of definition, applications of mathematical induction, and some challenging problems. About the binomial theorem, discovered by Isaac Newton in the seventeenth century, the mathematical literature is by now teeming with a lot o' news; but we shall confine ourselves to a barebones introduction to it, leaving you (if interested) to pursue it further on your own.† The central fact for us will be the counting property of the binomial coefficients.

Example In baseball the World Series is a best-of-seven-games match. Call the teams A and N. Record the outcome of each game as W if A wins, L if A loses (ties are impossible). Juxtapose these symbols for the Series to form a string of length four to seven over $\{W, L\}$.

For example, if A wins the series in four games, the string is $WWWW$. If N wins the Series by losing only the first and third games, the string is $WLWLLL$.

How many such strings can there be? You will find this an easy question before you finish this chapter.

1. Polynomials

You have seen many polynomials, that is, expressions like $1 + x$, $2 - x + xy$, $11 + 3x - 5x^2$, and so on. For the most part, our polynomials have only one variable x, like the first and third of the examples just given. Consider the term **variable**: A variable is a

†You could start with (10.2) and (10.4).

placeholder. As we did when we discussed the predicate calculus, we may attach to the variable x a domain D, the set of all numbers that we allow ourselves to put in place of x—if we wish to make a substitution. Often we don't mention the domain because it is known from the context. Let us agree now that in this chapter our domain D will always be **Z**, or sometimes **R**, the sets of all integers and real numbers, respectively.

Before defining *polynomial*, we alert you to two important points. One, mentioned before, is that numbers can be substituted for the variable. The other is that interesting facts can arise when we do not substitute anything for the variable. More on these points later. And now to the definition.

DEFINITION (1.1) A **polynomial** is a special sum, namely, a power series

(1.2) $$f(x) = \sum_{0 \le i} a_i x^i,$$

where for only a finite number of i is the real number a_i different from 0. Another way to express this condition is "For some $N \in \mathbf{N}$ and for all $i > N, a_i = 0$." In symbols

(1.3) $$\exists N \in \mathbf{N}\ \forall i > N \qquad a_i = 0.$$

Still another way: A polynomial is a power series with a zero tail.

Example (1.4) For the polynomial $1 + 2x$ represented in the form (1.2), $a_0 = 1, a_1 = 2$, and $a_i = 0$ for $i \ge 2$. For $x^2 - 1$ we have $a_0 = -1, a_1 = 0, a_2 = 1$, and $a_i = 0$ for all $i > 2$.

DEFINITION (1.5) We define **equality** between polynomials in this way: if $g(x) = \sum b_i x^i$ is another polynomial in the form (1.2), then $f(x) = g(x)$ if and only if for all $i, a_i = b_i$.

That is, two polynomials are equal if and only if coefficients of like powers of x are equal.

(1.6) Another way to express the polynomial (1.2) is this: It is an infinite vector in which all but a finite number of coordinates are zero. For example,

(1.7) $$(a_0, a_1, \ldots, a_i, \ldots, a_n, 0, 0, 0, 0, \ldots).$$

(If you don't like *infinite vector*, think of (1.7) as a function,

$$a : \mathbf{N}_0 \longrightarrow \mathbf{R},$$

that takes the value zero at all but a finite number of points of its domain $\mathbf{N}_0$, the set of all integers ≥ 0.) The representation (1.7) uses explicit coordinate places instead of the placeholders x^0, x^1, and so on, of (1.2). For example, see (1.10) below.

DEFINITION (1.8) The **degree** of the polynomial (1.2) is defined as follows: The degree of $f(x)$ is -1 if $f(x)$ is the zero polynomial, that is, if for all i, $a_i = 0$; otherwise the degree of $f(x)$ is n, where n is the largest value of i such that a_i is not 0.

If $a_n \neq 0$, then the degree of the polynomial in (1.7) is n.

Example (1.9) The degree of $1 + x + x^3$ is 3, of $100 - x^2 + x$ is 2, and of $1 + x + x^4 - x^n$ is $\max\{4, n\}$ unless $n = 4$, in which case the degree is 1. In terms of (1.7) the degree is determined by the position of the rightmost nonzero entry which in this expression,

$$(1,0,3,1,-4,0,0,1,2,0,0,\ldots), \tag{1.10}$$

is the 2. We count from the left to find that $2 = a_8$, so the polynomial represented by (1.10) has degree 8. Our customary representation of (1.10) is $1 + 3x^2 + x^3 - 4x^4 + x^7 + 2x^8$.

Practice (1.11P) Suppose N satisfies (1.3) for a given polynomial $f(x)$. Prove that (degree of $f(x)$) $\leq N$ and show that equality need not hold.

Although a polynomial is, strictly speaking, not a function, for every polynomial $f(x)$ we may *define* a function F from $\mathbf{R}$ to $\mathbf{R}$. We use substitution. If $f(x)$ is given by (1.2), we define F by the rule

$$\forall r \in \mathbf{R}, \quad F(r) := f(r) := \sum_{0 \leq i} a_i r^i. \tag{1.12}$$

There is no problem of convergence, because the sum is finite: a polynomial has only a finite number of nonzero terms. The polynomial *form* (1.2) expresses the rule defining the function F. So $f(x)$ is close to, but—in part because x is a variable and not a real number—not the same as, F. It would be much more convenient to denote F by f, but doing so at the outset would risk confusing $f(x)$ with F.

Example (1.13) The polynomial $f(x) = 1 + x^2$ determines by (1.12) the function F mapping r to $1 + r^2$ for each $r \in \mathbf{R}$. The graph of $y = F(r)$ in the (y, r) plane is a familiar parabola.

Here is an obvious but important property of polynomials:

(1.14) If polynomials $f(x)$ and $g(x)$ are equal, then

$$\forall r \in \mathbf{R}, \quad f(r) = g(r). \tag{1.15}$$

(1.16) The converse of (1.14) is also true: If polynomials $f(x)$ and $g(x)$ satisfy (1.15), then $f(x) = g(x)$. We leave the proof to the problems.

There is a stronger version of (1.16):

(1.17) If $f(x)$ and $g(x)$ agree at N points in $\mathbf{R}$, where N is bigger than the degree of either polynomial, then $f(x) = g(x)$.

Although we prove (1.17) only in part—since a decent proof requires some algebra—we propose it as a problem in Chapter 11. Here is the partial proof.

Proof We prove (1.17) in the special case that $f(x)$ and $g(x)$ have degrees at most 2. Then for some unknown coefficients $f(x) = a_0 + a_1x + a_2x^2$ and $g(x) = b_0 + b_1x + b_2x^2$. We allow the possibility that a_2 or b_2 is zero. Say these polynomials are equal at the $N = 3$ points $0, 1, 2$ (for definiteness—any three distinct real numbers would do). We want to prove that $a_0 = b_0$, $a_1 = b_1$, and $a_2 = b_2$. So we'll prove that

$$h(x) := f(x) - g(x) = (a_0 - b_0) + (a_1 - b_1)x + (a_2 - b_2)x^2$$

is the zero polynomial.

Since $f(0) = g(0), f(1) = g(1)$, and $f(2) = g(2)$, we see that $h(0) = h(1) = h(2) = 0$. Set $c_i = a_i - b_i$ for ease of writing. Then $h(x) = c_0 + c_1x + c_2x^2$, and

$$\begin{aligned} h(0) &= c_0 + c_1 \cdot 0 + c_2 \cdot 0^2 = c_0 = 0 \\ h(1) &= c_0 + c_1 \cdot 1 + c_2 \cdot 1^2 = c_0 + c_1 + c_2 = 0 \\ h(2) &= c_0 + c_1 \cdot 2 + c_2 \cdot 2^2 = c_0 + 2c_1 + 4c_2 = 0. \end{aligned} \tag{1.18}$$

You can easily see that the only solution to the linear equations (1.18) is $c_0 = c_1 = c_2 = 0$, which is what we wanted to show. QED

Another way to state (1.17) is this: Any polynomial of degree d is determined by its values at any $d + 1$ points.

You will see how x and its powers serve as placeholders as we define addition and multiplication of polynomials.

DEFINITION (1.19) With $f(x)$ and $g(x)$ defined as in (1.2) and (1.5), the **sum** of $f(x)$ and $g(x)$ is defined to be

$$f(x) + g(x) := \sum_{0 \le i} c_i x^i, \tag{1.20}$$

where for each $i \ge 0$, c_i is defined to be $a_i + b_i$.

The latter addition is defined because we know how to add the two real numbers a_i and b_i. Our definition of addition of polynomials, then, comes down to just this: add the corresponding coefficients. Here you see that a polynomial is nothing but the sequence (1.7) of its coefficients: $1 + 2x + x^3$ is the sequence (1,2,0,1). We suppressed the infinite tail of 0's in the sequence. When we add two polynomials we just add the corresponding terms in their sequences. When we use the notation (1.2), the powers of x keep the coefficients in their proper places.

Example (1.21) $(1 + 2x + x^3) + (-2 + x + (-3)x^2 + x^3 + (-7)x^4) = -1 + 3x - 3x^2 + 2x^3 - 7x^4$

Notice that $(-a)x^i = -ax^i$, and that $x^0 = 1$ and $x^1 = x$.

(1.22) We also operate on polynomials by **scalar multiplication**, in which we multiply a polynomial by a real number c:

$$cf(x) := c\sum a_i x^i := \sum ca_i x^i. \tag{1.23}$$

Here we understand that ca_i is the product in the real numbers of c and a_i. Having this, we can of course subtract one polynomial from another by first multiplying it by -1 and then adding the two by our rule above.

Multiplication of polynomials is a more interesting operation. We define it as follows:

DEFINITION (1.24) Let $f(x)$ and $g(x)$ be polynomials defined as in (1.2) and (1.5). The **product** $f(x)g(x)$ is the polynomial in which for each nonnegative integer i, the coefficient of x^i is

$$a_i b_0 + a_{i-1} b_1 + \cdots + a_1 b_{i-1} + a_0 b_i = \sum_{j+k=i} a_j b_k. \tag{1.25}$$

The sum in (1.25) is understood to be over all pairs (j, k) satisfying $j + k = i$.

Another way to write the sum (1.25) is this:

$$\sum_{0 \le j \le i} a_j b_{i-j}.$$

Notice that this expression is a finite sum, having at most $i + 1$ nonzero terms. By working it out you see that you obtain the product of two polynomials by applying the distributive law and the law of exponents blindly to the formal product of them. In fact, these laws for polynomials follow logically from the definition (1.24). For example, you can prove immediately from the new standpoint, using (1.25), that $(x^i)(x^j) = x^{i+j}$ for all nonnegative integers i and j.

If the polynomial is in the form (1.7), multiplying it by x shifts all the coefficients one place to the right and puts a zero in place of a_0:

Example (1.26) The best way to multiply the polynomial (1.2) by $x + 2x^3$ by hand might be to set up the array

$$\begin{array}{rcccccccc} xf(x) & : & 0 & a_0 & a_1 & a_2 & a_3 & \ldots \\ 2x^3 f(x) & : & 0 & 0 & 0 & 2a_0 & 2a_1 & \ldots \end{array}$$

and then add the columns to get the product in the form (1.7).

Example (1.27) Another way is to write one of the sequences (polynomials) backward, as $(\ldots, 0, 0, 2, 0, 1, 0)$ for $x + 2x^3$. Then slide it along under the first sequence, stopping first where the constant terms are lined up vertically, thus:

$$\begin{array}{cccccccc} \ldots & & a_{-1} & a_0 & a_1 & a_2 & a_3 & \ldots \\ 2 & 0 & 1 & 0 & & & & \end{array}$$

Then multiply term by term and add the resulting products. The constant term is $0 \cdot 2 + 0 \cdot 1 + a_0 \cdot 0 = 0$, since a_{-1}, and so on, are zero for all polynomials. We slide once more to get the coefficient of x:

$$\begin{array}{cccccc} \ldots & a_{-1} & a_0 & a_1 & a_2 \ldots \\ & 0 & 1 & 0 & \end{array}$$

The sum of products is $a_{-1} \cdot 0 + a_0 \cdot 1 + a_1 \cdot 0 = a_0$. And so on. You can see easily why this works: (1.25) shows that as one sequence of coefficient indices increases, the other decreases.

If we slide over two spaces to get the coefficient of x^3 we have

$$\begin{array}{cccccc} & & a_0 & a_1 & a_2 & a_3 & \ldots \\ \ldots & 0 & 2 & 0 & 1 & 0 & \end{array}$$

and the sum of products is $2a_0 + a_2$. These and the previous results are, of course, the same.

A more compact way to organize polynomial multiplication by hand is to use a rectangular array. For example, to multiply $(2 - 2x + x^2)(2 + 2x + x^2)$ you'd set up this array,

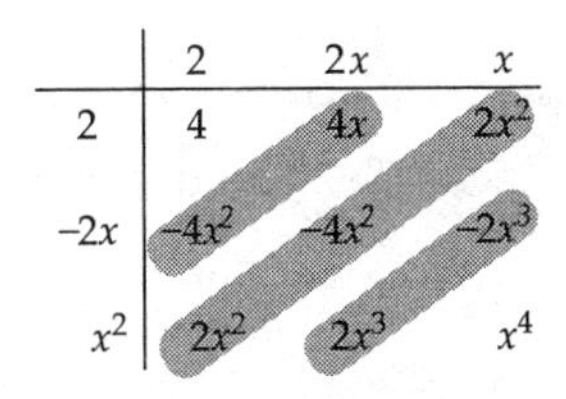

where you fill in the spaces with the product of the row and column entries. Then you add all the entries. As you see, like powers of x occur on diagonals as indicated, so if you're careful you can leave out the x's. This product is $4 + x^4$, which should have been obvious, since it is $[(2 + x^2) - 2x]\,[(2 + x^2) + 2x]$.

Nowadays computer algebra packages like Maple and Macsyma will do these and lots of other manipulations for you.

Since it is not our purpose to review high school algebra, we allow you to work out further examples of these operations. Let us say for now that the sum in (1.25), the coefficient of x^i in the product of the two polynomials with coefficient sequences $(a_0, a_1, \ldots)$ and $(b_0, b_1, \ldots)$, respectively, is sometimes called the **convolution with shift** i of these two sequences. It is heavily studied because of its importance in applications such as signal processing.

Notice that for all *nonzero* polynomials $f(x)$ and $g(x)$,

$$\text{(1.28)} \qquad \text{degree of } (f(x)g(x)) = (\text{degree of } f(x)) + (\text{degree of } g(x)).$$

To make (1.28) hold even when $f(x) = 0$ or $g(x) = 0$, some authors define the degree of the zero polynomial to be $-\infty$. Still other authors don't assign any degree to the zero polynomial.

Practice (1.29P) Prove (1.28).

Notice that we may regard any real number r as a polynomial, called a **constant polynomial**. In the form (1.7) it is $(r, 0, 0, 0, \ldots)$. The rules (1.20) and (1.25) for addition and multiplication of constant polynomials are

$$(r, 0, 0, \ldots) + (s, 0, 0, \ldots) = (r + s, 0, 0, \ldots)$$
$$(r, 0, 0, \ldots)(s, 0, 0, \ldots) = (rs, 0, 0, \ldots),$$

showing that the constant polynomials mimic the behavior of the real numbers.

Practice (1.30P)

(*i*) What is the degree of a constant polynomial?

(*ii*) Find the product of $1 - x$ and $1 + x + x^2 + \cdots + x^n$ by the two methods (1.26) and (1.27). Let $n \geq 0$.

(*iii*) Find the product $(1 - x + x^2)(1 + x + x^2)$. Then multiply it by $(1 - x)(1 + x)$.

2. Binomial Coefficients

We now use these ideas on polynomials to define the focus of this chapter.

DEFINITION (2.1) For all integers $n \geq 0$ and for all k in **Z**, we define the **binomial coefficient** $\binom{n}{k}$ (called "n choose k") as the coefficient of x^k in the expression of $(1 + x)^n$ as a polynomial in x.

To explain: $(1 + x)^n$ is the product of n polynomials $1 + x$; we know how to multiply these together to find the coefficient of each power of x in the polynomial that is the product; each such coefficient we call a binomial coefficient, give it this strange notation, and, for reasons to be explained, read it "n choose k." The first few of these polynomials are

$$(1 + x)^0 = 1$$
$$(1 + x)^1 = 1 + x$$
$$(1 + x)^2 = 1 + 2x + x^2$$
$$(1 + x)^3 = 1 + 3x + 3x^2 + x^3.$$

These four polynomials furnish in effect a table of values of $\binom{n}{k}$ for $n = 0, 1, 2$, and 3 and for all k. The first one tells us that $\binom{0}{0} = 1$ and that $\binom{0}{k} = 0$ for all $k \neq 0$, because the coefficient of x^0 is 1, and for each $k \neq 0$, the coefficient of x^k is 0 in $(1 + x)^0 = 1$. Similarly we see that

$$\binom{1}{0} = \binom{1}{1} = 1 \quad \text{and for all } k > 1, \quad \binom{1}{k} = 0.$$

Continuing, we get

$$\binom{2}{0} = 1, \quad \binom{2}{1} = 2, \quad \binom{2}{2} = 1$$

$$\binom{3}{0} = 1, \quad \binom{3}{1} = 3, \quad \binom{3}{2} = 3, \quad \binom{3}{3} = 1.$$

$$\text{NOTATION: } (1+x)^n = \binom{n}{0} + \binom{n}{1}x + \binom{n}{2}x^2 + \cdots + \binom{n}{k}x^k + \cdots + \binom{n}{n}x^n$$

$$= \sum_{0 \le k \le n} \binom{n}{k} x^k, \tag{2.2}$$

The sum could equally well be taken over all $k \geq 0$, because $\binom{n}{k} = 0$ whenever $k > n$. (See (2.5).)

For emphasis, we repeat: what we have introduced here is the definition

$$\binom{n}{k} \text{ is the coefficient of } x^k \text{ in } (1+x)^n. \tag{2.3}$$

Our procedure to define the binomial coefficients is the reverse of the usual. Normally we think of the coefficients as defining the polynomial, but here we let the polynomial define its coefficients. That is, our definition (1.25) for multiplication of polynomials allows us to *start* from $(1+x)^n$ as something well defined.

Practice (2.4P) Find $\binom{4}{k}$ for all $k \geq 0$ by multiplying $1 + 3x + 3x^2 + x^3$ by $1 + x$.

Notice how useful it is to think of a polynomial as an infinite sum in which all but a finite number of terms are zero. For example, this concept leads immediately to our first general fact about binomial coefficients: When n is a nonnegative integer, for all integers k

$$\text{If } k > n \quad \text{or} \quad \text{if } k < 0, \quad \text{then } \binom{n}{k} = 0. \tag{2.5}$$

This statement is true because the coefficient of x^k in $(1+x)^n$ is 0 if $k > n$ (Problem 1.2); ditto if $k < 0$, simply because there are no negative powers of x in a polynomial in x.

Now let us find some more values for $\binom{n}{k}$ before deriving the basic recursion satisfied by the binomial coefficients. For example,

$$\binom{n}{0} = \binom{n}{n} = 1 \quad \text{for all integers } n \geq 0. \tag{2.6}$$

That $\binom{n}{0} = 1$ follows immediately from setting $x = 0$ in the defining relation (2.2): $(1+x)^n = \sum \binom{n}{k} x^k$. That $\binom{n}{n} = 1$ can be proved by a simple induction (Problem 2.2), or by an intuitive argument that there is only one choice of term in each of the n factors $(1 + x)$ that will produce x^n in the polynomial. We shall later see other proofs that $\binom{n}{n} = 1$.

Notice that we substituted a value for the variable in order to find out something about the coefficients of the polynomial.

3. The Basic Recursion

Proposition (3.1) For all integers $n \geq 0$ and all integers k, the binomial coefficients satisfy the following relationship:

$$\binom{n}{k} + \binom{n}{k+1} = \binom{n+1}{k+1}. \tag{3.2}$$

Proof We use the definition of $\binom{n}{k}$. We multiply the defining relation (2.2) for the binomial coefficients by $1 + x$ on both sides:

$$(1+x)^{n+1} = (1+x)(1+x)^n = \sum_{0 \leq k} \binom{n}{k} x^k + x \sum_{0 \leq k} \binom{n}{k} x^k. \tag{3.3}$$

We multiply the "outside" x by each term of the last sum: The right-hand side of (3.3) is then

$$\sum_{0 \leq k} \binom{n}{k} x^k + \sum_{0 \leq k} \binom{n}{k} x^{k+1}. \tag{3.4}$$

We now collect terms, that is, we add together all the coefficients of like powers of x. For each k, the coefficient of x^{k+1} in the first sum of (3.4) is $\binom{n}{k+1}$; in the second sum it is $\binom{n}{k}$. Therefore, the coefficient of x^{k+1} in (3.4) is

$$\binom{n}{k+1} + \binom{n}{k}.$$

But this polynomial in (3.4) is equal to that in (3.3), namely, $(1 + x)^{n+1}$, which by definition (2.1) is equal to

$$\sum_{0 \leq j} \binom{n+1}{j} x^j.$$

Setting $j = k + 1$ yields the result of the Proposition, since equal polynomials have by definition equal coefficients. QED

Notice how often we used the definitions of polynomial sum and product and of equality of polynomials in the proof just concluded. Notice also that the basic recursion gives you a concrete link between the binomial coefficients for n and those for $n + 1$. It should thus allow you to go from n to $n + 1$ in proofs by induction.

4. Pascal's Triangle

The recursion (3.1) and the boundary conditions (2.6) determine the famous triangle of Pascal. It is a triangular array of the nonzero binomial coefficients $\binom{n}{k}$ for all integral $n \geq 0$. The horizontal rows are numbered with the value of n, starting with zero. The diagonal "files" starting from the left-hand edge are numbered with the value of k. The entry in row n and file k is the value of $\binom{n}{k}$. The first few rows of Pascal's triangle are

$$\begin{array}{ll} n=0 & 1 \\ n=1 & 1\quad 1 \\ n=2 & 1\quad 2\quad 1 \\ n=3 & 1\quad 3\quad 3\quad 1 \\ n=4 & 1\quad 4\quad 6\quad 4\quad 1 \end{array} \tag{4.1}$$

Thus we see that $\binom{4}{2} = 6$, and using (3.1) we can see immediately that $\binom{5}{2} = 4+6 = 10$. Going further, we notice that the result of Proposition (3.1) appears in the triangle as the property that

(4.2) Every entry is the sum of the two entries above it and on either side.

This property holds even at the boundaries, because the other binomial coefficients are zero, i.e., $\binom{n}{k} = 0$ if k is negative or if $k > n$. Using this property makes it an easy matter to produce row $n+1$ from row n (if n is not too large). For example, we could write down row 5 at top speed by merely adding pairs of entries from row 4, getting 1 5 10 10 5 1.

Practice (4.3P) Write out the Pascal triangle down to the row for $n = 10$. (A poster-sized version of the Pascal triangle should be a hands-down winner in any dormitory's room-decorating contest.)

The Pascal triangle may be calculated from the boundary conditions

$$\binom{n}{0} = \binom{n}{n} = 1 \quad \text{for all } n = 0, 1, 2, \ldots$$

and the basic recursion (3.2). In principle, therefore, you should be able to derive any fact about binomial coefficients by starting from the basic recursion. Sometimes you'll be able to bypass it, but if you don't know how to solve a problem in binomial coefficients a good strategy is to try the basic recursion.

Practice (4.4P) What is wrong with the statement "The Pascal triangle has all the binomial coefficients in it"?

5. Symmetry

You have probably noticed that each row of the Pascal triangle is a palindrome, that is, it reads the same backward as forward. In other words, (so far) the Pascal triangle is symmetric about the vertical line through the center. In fact this symmetry exists in every row:

Lemma (5.1) For all nonnegative integers n and all integers k

$$\binom{n}{k} = \binom{n}{n-k}. \blacksquare$$

In row n of the Pascal triangle, the index k counts from the left, and $n-k$ counts from the right.

Another way to express this result is to say that for all integers a and b such that $a+b \geq 0$, the two binomial coefficients $\binom{a+b}{a}$ and $\binom{a+b}{b}$ are equal.

Before giving the proof we show by example the key procedure used in it.

Example (5.2) Consider the polynomial $p(x) = 3x^2+x-4$ and the polynomial $q(x) = -4x^2+x+3$ obtained by "reversing" the sequence of (nonzero) coefficients of $p(x)$. There is a close algebraic relation between $p(x)$ and $q(x)$, namely,

$$q(x) = x^2 p\left(\frac{1}{x}\right). \tag{5.3}$$

Why? Because

$$p\left(\frac{1}{x}\right) = \frac{3}{x^2} + \frac{1}{x} - 4$$

and now (5.3) is obvious.

Proof of (5.1) We start directly from the definition. In the equation $(1+x)^n = \sum \binom{n}{k} x^k$ we substitute $1/x$ for x, and then multiply by x^n, getting

$$x^n(1+1/x)^n = \sum_{0\leq k} \binom{n}{k} x^{n-k}, \tag{5.4}$$

since we used the fact that $1/x = x^{-1}$. Here we have made a brief excursion out of the realm of polynomials. This operation—replacing x by $1/x$ and then multiplying by x^n—applied to any polynomial of degree n produces the polynomial with its sequence of coefficients the reverse of the original. The new constant term is the old coefficient of the term of degree n (Problem 1.3). We can certainly see that such an outcome has occurred with our polynomial, for on the right we now have $\binom{n}{k}$ as the coefficient of x^{n-k}. But now we multiply on the left to see that we are right back where we started: $x^n(1+1/x)^n = (x+1)^n$, which equals $(1+x)^n$ by commutativity. But from our definition, (2.2), the coefficient of x^{n-k} in $(1+x)^n$ is $\binom{n}{n-k}$. Since the two polynomials in (5.4) are equal, their coefficients are equal, so the symmetry lemma must be true. QED

Notice that this proof went directly from our definition (2.1) of binomial coefficients to the conclusion. Several other proofs of this result are possible; see Problems 5.1 and 6.6.

6. The Counting Property

We shall now develop the property of binomial coefficients that is most useful in computer science. A bit of terminology will be helpful: if m is a nonnegative integer,

(6.1) the statement "X is an m-**set**" means that X is a set having exactly m elements.

NOTATION: $|X| = m$.

Thus $\{1,2,3\}$ is a 3-set, $\{x\}$ is a 1-set, $\varnothing$ is a 0-set, and $\{1,3\}$ is a 2-subset of $\{1,2,3,4\}$. We may write $|\{1,2,3\}| = 3$, $|\varnothing| = 0$, and so on.

We may regard $|\ \ |$ as a function from the set of all finite sets to $\mathbf{N} \cup \{0\}$.

We shall now prove

Theorem (6.2) For all nonnegative integers n and all integers k, $\binom{n}{k}$ is the total number of k-subsets of an n-set. ■

Before giving the proof we do an example in the case $n = 4$.

Example (6.3) Consider the 4-set $\{1,2,3,4\}$. We list its 16 subsets by size:

0-subsets	1-subsets	2-subsets	3-subsets	4-subsets
$\varnothing$	$\{1\}$	$\{1,2\}$	$\{1,2,3\}$	$\{1,2,3,4\}$
	$\{2\}$	$\{1,3\}$	$\{1,2,4\}$	
	$\{3\}$	$\{1,4\}$	$\{1,3,4\}$	
	$\{4\}$	$\{2,3\}$	$\{2,3,4\}$	
		$\{2,4\}$		
		$\{3,4\}$		

We see that there is exactly one 0-subset, and $\binom{4}{0} = 1$. There are exactly four 1-subsets, and $\binom{4}{1} = 4$. There are six 2-subsets, and $\binom{4}{2} = 6$, and so forth. Notice also that there are zero 5-subsets, and $\binom{4}{5} = 0$.

We should also remark that each two n-sets have the same number of k-subsets as each other. If this claim is not obvious, imagine the two sets to be $\{1,2,\ldots,n\}$ and $\{1',2',\ldots,n'\}$.

***Proof** of the Theorem* We introduce n new variables $x_1,\ldots,x_n$ and consider the product

$$(1 + x_1)(1 + x_2)\ldots(1 + x_n) = \prod_{1\leq i\leq n}(1 + x_i) =: B. \tag{6.4}$$

We now work out in detail what this polynomial B in the several variables is. Let T stand for the set $\{1,2,\ldots,n\}$ of subscripts. T is an n-set. From the definition of multiplication of polynomials, B consists of the sum of all monomial terms of the following form:

$$x_{i_1}x_{i_2}\cdots x_{i_r}$$

where $\{i_1,i_2,\ldots,i_r\}$ is a subset of T. In other words, in forming a term of B, we make a choice in each of the terms $1+x_i$ of either the 1 or the x_i, and then multiply together the n choices. The result is the product of the x_i's chosen, and we identify the choice simply by specifying the set of subscripts on the chosen variables. Of course we must run through all possible choices to form B correctly. A compact way to state all this is the equation

$$B = \prod_{i\in T}(1+x_i) = \sum_{S:S\subseteq T}\left(\prod_{h:h\in S} x_h\right). \tag{6.5}$$

Think it over. The symbols $S : S \subseteq T$ under the summation sign mean that for each object S satisfying $S \subseteq T$—thus S must be a subset of T—the quantity at the right of the summation sign is to be calculated. The summation sign of course also means that we must then sum all these quantities. What are they?

The symbols $h : h \in S$ under $\prod x_h$ mean that we are to take the product of the elements of the set $\{x_h;\ h \in S\}$. For example, if $S = \{1,3,5\}$, then $\{x_h;\ h \in S\} = \{x_1,x_3,x_5\}$ and the product quantity in question is $x_1x_3x_5$.

Notice that if S is the empty subset of T, then

$$\prod_{h\in\varnothing} x_h = 1, \tag{6.6}$$

for the empty subset arises exactly when your n choices are all 1's, i.e., when you choose no x_i's at all. (Compare Chapter 1, (10.14).)

We now set for each $i = 1,\ldots,n$, $x_i := x$. Then the term $x_{i_1}\cdots x_{i_r}$ becomes simply x^r. Thus for each subset S of T

$$\prod_{h:h\in S} x_h = x^{|S|}.$$

Our equation (6.5) now becomes

$$(1+x)^n = \sum_{S:S\subseteq T} x^{|S|},$$

in which the left-hand side is the now-familiar $\sum\binom{n}{k}x^k$. But look at the right-hand side: the coefficient of x^k is the total number of k-subsets of T, since there are just that many terms x^k on the right, one for each S such that $|S| = k$. Since T is an n-set, $\binom{n}{k}$ is the total number of k-subsets of an n-set.

QED

Comments (6.7)

(*i*) Perhaps the main virtue of this proof is that is has enlarged your repertoire of uses of the summation and product signs, especially in that we took these operations over all elements of sets (like S and $\mathcal{P}(T)$) other than the standard $\{0,1,\ldots,m\}$. After all, it is easy enough to convince yourself of the truth of the Theorem by simply starting with $(1+x)^n$ and then waving your hands a bit.

(*ii*) Now it becomes clear why we read $\binom{n}{k}$ as "n choose k."

(*iii*) This proof too went directly from the definition (2.1) to the conclusion. Other proofs are possible, say via the basic recursion (3.2). In the other direction, you can use the counting theorem to prove symmetry and the basic recursion.

To nail down your understanding you might wish to work through the preceding proof for the case $n = 3$ or $n = 4$.

Example (6.8) According to (6.2) there are exactly $\binom{5}{2}$ 2-subsets of a 5-set and $\binom{5}{1}$ 1-subsets. Let's list these subsets of $\{1,2,3,4,5\}$ and verify (6.2) for this example.
The 1-subsets are $\{1\}, \{2\}, \{3\}, \{4\}$, and $\{5\}$; of course there are exactly five of them, and $5 = \binom{5}{1}$.

Omitting braces, the 2-subsets are as follows:

$$\begin{array}{llll} 1,2 & 2,3 & 3,4 & 4,5 \\ 1,3 & 2,4 & 3,5 & \\ 1,4 & 2,5 & & \\ 1,5 & & & \end{array}$$

and there are exactly $10 = \binom{5}{2}$, as we know from the Pascal triangle. In a moment we'll see that $\binom{5}{2}$ can be calculated by the formula $\binom{5}{2} = 5 \cdot 4/2 \cdot 1 = 10$.

Practice (6.9P) (*i*) In a class of 40 students two are to be chosen sergeants-at-arms. How many such choices are possible? (*ii*) A record store has a sale. Purchase of any three records (no two the same) from the current top ten hits costs \$15. Suppose n different customers each buy three records in this sale but no two customers choose the same set of three records. What can you say about n?

7. The Binomial Theorem

We've proved several properties of binomial coefficients just using the definition (2.1). The basic recursion (3.1) is even an algorithm for calculating the value of $\binom{n}{k}$. The recursion by itself, however, could be cumbersome in some theoretical settings, not giving us any insight into the size of $\binom{n}{k}$, for example.

Suppose you wanted to know the value of $\binom{20}{3}$. You could calculate part of the Pascal triangle to get it, but that would be a lot of work.

Practice (7.1P) To calculate the value of $\binom{20}{3}$, what is the least portion of the Pascal triangle you would need?

What we do in our final result is find a formula for $\binom{n}{k}$.

Theorem (7.2) *The Binomial Theorem.* For each real number n, the binomial coefficient $\binom{n}{k}$ for each integer $k \geq 0$, defined by the relation $(1+x)^n = \sum_{0 \leq k} \binom{n}{k} x^k$, is given by the formula

$$\binom{n}{k} = \begin{cases} 1 & \text{if } k = 0 \\ \dfrac{n(n-1)\cdots(n-k+1)}{k!} & \text{if } k > 0. \blacksquare \end{cases}$$

That is, if $n \in \mathbf{Z}$, then $\binom{n}{k}$ is the product of k consecutive integers starting with n and going downward, divided by $k!$

Example (7.3) We know from working out the polynomial $(1+x)^4 = x^4 + 4x^3 + 6x^2 + 4x + 1$ that $\binom{4}{2} = 6$. Theorem (7.2) allows us to compute it directly: $\binom{4}{2} = 4 \cdot 3/2 \cdot 1 = 6$. We use (7.2) to find more values:

$$\binom{7}{3} = 7 \cdot 6 \cdot 5/3 \cdot 2 \cdot 1 = 7 \cdot 5 = 35.$$

$$\binom{20}{10} = \frac{20 \cdot 19 \cdot 18 \cdot 17 \cdot 16 \cdot 15 \cdot 14 \cdot 13 \cdot 12 \cdot 11}{10 \cdot 2 \cdot 9 \cdot 8 \cdot 5 \cdot 3 \cdot 7 \cdot 4 \cdot 6};$$

I wrote the factors of the denominator immediately below what they will divide when we simplify. Thus

$$\begin{aligned} \binom{20}{10} &= 19 \cdot 2 \cdot 17 \cdot 2 \cdot 13 \cdot 11 \\ &= 4 \cdot 11 \cdot 13 \cdot 17 \cdot 19 = 184{,}756. \end{aligned}$$

Proof *of (7.2)* The result follows immediately from the facts about Taylor series that you learned in calculus. If the function f has a Taylor series, that is, if there is a possibly infinite series $\sum a_k x^k$ equal to $f(x)$ for all real x in an interval around $x = 0$, then setting $x = 0$ shows us that $a_0 = f(0)$. Now differentiate, getting

$$f'(x) = \sum_{0 \leq k} k a_k x^{k-1},$$

and set $x = 0$ again. This yields $f'(0) = a_1$. Once more:

$$\begin{aligned} f''(x) &= \sum_{0 \leq k} k(k-1) a_k x^{k-2} \\ &= 2a_2 + 3 \cdot 2a_3 x + 4 \cdot 3a_4 x^2 + \cdots, \end{aligned}$$

and setting $x = 0$ gives us $f''(0) = 2a_2$, from which we can get a_2. Continuing in this way (really, doing an induction on k), we get

$$a_k = f^{(k)}(0)/k! \tag{7.4}$$

for all $k \geq 0$. Here $f^{(k)}$ denotes the kth derivative of f, and $f^{(0)}$ is by convention just f.

If we apply this general result (7.4) to the particular function defined by $f(x) := (1+x)^n$, we see that it is easy to calculate the derivatives:

$$(d/dx)^k(1+x)^n = n(n-1)\cdots(n-k+1)(1+x)^{n-k}.$$

Thus from (7.4) $a_k := \binom{n}{k}$ is given by the formula of the binomial theorem. QED

You have probably noticed that all of a sudden n became any real number instead of being restricted to the nonnegative integers as previously. The reason for the change is that the binomial theorem holds for all real n, and the proof just given is the simplest proof and works equally well for the general case.

(7.5) The definition of $\binom{n}{k}$ for real n is an extension of definition (2.1): $\binom{n}{k}$ is the coefficient of x^k in the *power series* for $(1+x)^n$ expanded around $x = 0$. That is,

$$(1+x)^n = \sum_{0 \le k} \binom{n}{k} x^k,$$

just as in (2.1) except now the series may be infinite. There is one slight gap, however: how do we know whether for general real n, the function $(1+x)^n$ is expressible as a power series? For that I have an exceedingly simple answer: see your calculus teacher. Rest assured that $(1+x)^n$ is so expressible, however.

Other proofs are possible when n is an integer; some of these are left to the Problems. Even if you don't know calculus you can reproduce all by yourself the proof we gave above if you confine yourself to polynomials! See Problem 7.8.

It doesn't take long to convince yourself, with (7.2), that if n is any real number other than a nonnegative integer, then the binomial coefficient $\binom{n}{k}$ is never zero no matter how large k is. In particular, the Taylor series for $(1+x)^n$ is not a polynomial. By the ratio test, it converges for all x in the range $-1 < x < 1$. If you wish to substitute numerical values for x in the equation $(1+x)^n = \sum \binom{n}{k} x^k$, you must confine them to that range $|x| < 1$ whenever $n \notin \{0,1,2,\ldots\}$.

Example (7.6) (*i*) From (7.2) we find

$$\frac{1}{1+x} = (1+x)^{-1} =: \sum_{0 \le k} \binom{-1}{k} x^k = 1 - x + x^2 - x^3 + - \cdots$$

That is, $\binom{-1}{k} = (-1)^k$, since from (7.2), $\binom{-1}{k} = [(-1)(-2)\cdots(-k)]/k!$.

(*ii*) We also may see that

$$\frac{1}{(1+x)^2} = (1+x)^{-2} =: \sum_{0 \le k} \binom{-2}{k} x^k$$

$$= 1 - 2x + \frac{(-2)(-3)}{2!}x^2 + \frac{(-2)(-3)(-4)}{3!}x^3 + \cdots$$

$$= 1 - 2x + 3x^2 - 4x^3 + - \cdots,$$

which we could also have obtained by differentiating the series for $1/(1+x)$. Thus $\binom{-2}{k} = (-1)^k(k+1)$.

(*iii*) The binomial theorem holds for any real exponent, so we may calculate, for example,

$$\begin{aligned}(1+x)^{1/3} :&= \sum_{0\le k}\binom{1/3}{k}x^k \\ &= 1 + \frac{1}{3}x + \frac{1}{3}\left(\frac{1}{3}-1\right)\frac{1}{2!}x^2 + \frac{1}{3}\left(\frac{1}{3}-1\right)\left(\frac{1}{3}-2\right)\frac{1}{3!}x^3 + \cdots \\ &= 1 + \frac{x}{3} - \frac{x^2}{9} + \frac{5x^3}{81} - + \cdots.\end{aligned}$$

In particular $\binom{1/3}{2} = -\frac{1}{9}$, and $\binom{1/3}{3} = \frac{5}{81}$.

But when n is a nonnegative integer, none of these complications arises, for then $(1+x)^n$ is simply a polynomial, a special case of Taylor series, so we can accept every bit of the proof with confidence, and we may freely substitute any real number for x as well. Doing so for appropriate choices of the value of x can lead to interesting identities on binomial coefficients, as you will see in the Problems.

Corollary (7.7) When n is a nonnegative integer,

(7.8)
$$\binom{n}{k} = \frac{n!}{k!(n-k)!}$$

The proof of this simple result is left to the problems.

This Corollary is more useful in theoretical than in numerical calculations. Those who like to calculate binomial coefficients using (7.8) and the factorial key on their calculator should try $\binom{100}{5}$ that way.

Notice that (7.8) does express the symmetry of binomial coefficients.

8. Identities Involving Binomial Coefficients

Equations involving binomial coefficients are often called *binomial coefficient identities* or simply *binomial identities*. The woods are full of them, and here we indicate just two, leaving several simple ones for the problems.

(8.1) *The knight's-move identity*. For all integers $n, k \ge 0$,

(8.2)
$$\sum_{0\le j\le k}\binom{n-j}{k-j} = \binom{n+1}{k}.$$

Proof To prove this identity we first look at what it says in terms of the Pascal triangle, which we represent as dots in the diagram below. We see that the left-hand side sums all the terms starting at the boundary point in row $n - k$ and "file" 0 and going into the triangle on a line parallel to the other boundary and ending at $\binom{n}{k}$. These terms are outlined in the box labeled L in the figure.

(8.2F)

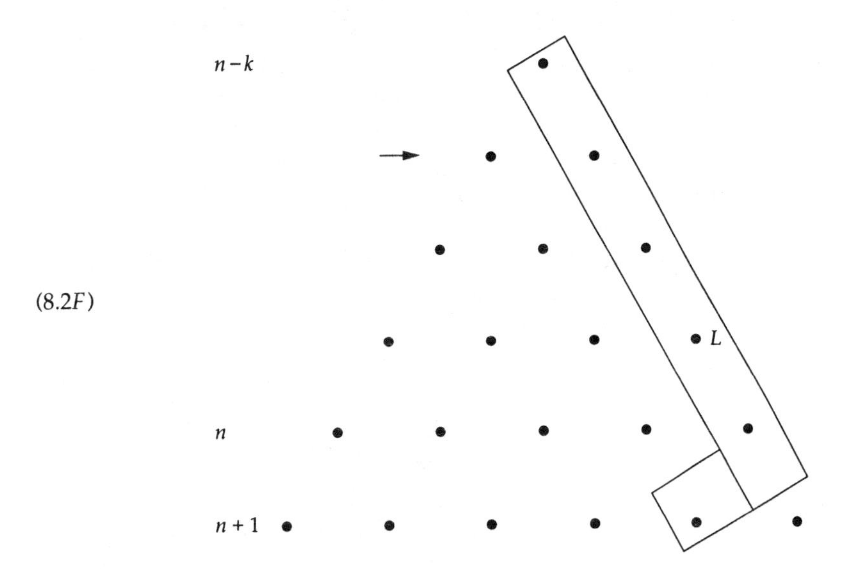

We chose the name because of this picture.

Here we need not assume that the top dot represents $\binom{0}{0}$, for the diagram represents the relative placements of the entries in any such triangular portion of the Pascal triangle. But we do take the left-hand boundary to be the $k = 0$ file, and in our picture k is 4. The dot in the small square represents $\binom{n+1}{k}$.

With this picture we can easily find the proof. We use the basic recursion (3.1): each binomial coefficient is the sum of the two above it and on either side in the Pascal triangle. Thus $\binom{n+1}{k} = \binom{n}{k-1} + \binom{n}{k}$, so we have expressed our target term as a sum of a term we want, $\binom{n}{k}$, and a term we don't want, $\binom{n}{k-1}$. We eliminate the latter by using the same recursion, (3.1), to express it as the sum $\binom{n-1}{k-1} + \binom{n-1}{k-2}$, which again has one wanted and one unwanted term. This process is best carried out directly on a diagram such as (8.2F), with nothing but check marks. After k steps one arrives at the boundary, where the arrow is; that term, equal to 1, is equal to the term above it in the box. And that's a way to prove it. Others, however, might prefer a proof by induction, which we leave to the problems.

(8.3) *The Vandermonde convolution.* This identity is useful and easy to prove. It is considered fundamental in more advanced treatments of binomial identities (see (10.4), page 169 and page 212).

Let $a, b \in \mathbf{R}$ and $m, n \in \mathbf{Z}$ (all independent of j). Then

(8.4)
$$\sum_{j \in \mathbf{Z}} \binom{a}{m+j}\binom{b}{n-j} = \binom{a+b}{m+n}. \ \blacksquare$$

Notice that we are now summing over all of **Z**, for the first time. You'll see in the proof how convenient this tactic is. Remember that for each real c,

$$\binom{c}{-1} = \binom{c}{-2} = \cdots = 0;$$

$\binom{c}{k}$ is always zero if k is negative, from our definition (7.5). Therefore, the sum in (8.4) is finite, running over the range $-m \le j \le n$. It's more convenient to sum over **Z** and not fuss with the limits.

Practice (8.5P) What if $-m > n$?

Proof of (8.4) This result merely recasts the definition (1.24) of polynomial multiplication. Consider the two power series

$$(1+x)^a = \sum_{j \in \mathbf{Z}} \binom{a}{m+j} x^{m+j}; \tag{8.6}$$

$$(1+x)^b = \sum_{j \in \mathbf{Z}} \binom{b}{n-j} x^{n-j}. \tag{8.7}$$

These equations are both true by definition (7.5), after a change of variable of summation.

Multiply (8.6) by (8.7) to get $(1+x)^{a+b}$ on the left. On the right let's single out only terms in x^{m+n}; those are obtained exactly by multiplying for each j the jth term of (8.6) by the jth term of (8.7). We add them all to get as the coefficient of x^{m+n} on the right

$$\sum_{j \in \mathbf{Z}} \binom{a}{m+j}\binom{b}{n-j}.$$

By definition (7.5), the coefficient of x^{m+n} on the left is $\binom{a+b}{m+n}$. QED

Practice (8.8P) Is (8.6) true if $m = -3$?

(8.9) *Application: Binomial inversion.* Allow negative n for once. Remember you can still use (7.2) to calculate, for example, $\binom{-10}{3} = (-10)(-11)(-12)/3! = -220$. Suppose we have the following system of $n+1$ simultaneous equations in the unknowns $v_0, v_1, \ldots, v_n$,

$$\sum_{0 \le i} \binom{i}{k} v_i = u_k \quad \text{for } k = 0, 1, \ldots, n, \tag{8.10}$$

where $u_0, u_1, \ldots, u_n$ are known. Then for all $k = 0, 1, \ldots, n$

$$v_k = \sum_{0 \le j} \binom{-k-1}{j-k} u_j \qquad \left(= \sum_{0 \le j} (-1)^{j-k} \binom{j}{k} u_j \right). \tag{8.11}$$

For convenience we set $u_j := v_j := 0$ for $j > n$.

To prove this claim, we use (8.10) to substitute for u_j in (8.11). We get

$$\sum_{0\le j}\binom{-k-1}{j-k}u_j = \sum_{0\le j}\binom{-k-1}{j-k}\sum_{0\le i}\binom{i}{j}v_i$$

$$= \sum_{0\le i} v_i \sum_{0\le j}\binom{-k-1}{j-k}\binom{i}{i-j}, \tag{8.12}$$

by interchanging the order of summation‡ and using symmetry (5.1). Notice that the lower indices $j-k$ and $i-j$ have sum $i-k$ independent of j, and that each upper index, $-k-1$ and i, is independent of j. So the Vandermonde convolution (8.4) applies to the inner sum of (8.12). The result is

$$\sum_{0\le i} v_i\binom{i-k-1}{i-k} = v_k\binom{-1}{0} + v_{k+1}\binom{0}{1} + v_{k+2}\binom{1}{2} + \cdots = v_k. \tag{8.13}$$

This is what we set out to verify.

Practice (8.14P) Show that the coefficients of $v_0, \ldots, v_{k-1}$ are zero in (8.13).

REMARK (8.15) The converse of the previous result also holds. The two can be more symmetrically stated as: Let $u_0, \ldots, u_n$ and $v_0, \ldots v_n$ be any sequences. Then (8.10) holds if and only if (8.11) holds. ■

The trickier proof of the converse (that (8.11) implies (8.10)) is omitted here; it is more easily done with matrices anyway.

A final comment: The parenthetical version of (8.11), which we did not use in the proof, comes from the following identity. For all real numbers n and all integers i,

$$\binom{-n}{i} = (-1)^i\binom{n+i-1}{i}, \tag{8.16}$$

which is easily proved from (7.2).

Some suggestions on the references listed in Section (10). Probably more binomial identities have been published than anyone wants to know. For example, Gould has organized more than 500 of them in (10.3), but with no exposition. An approach to these identities via hypergeometric functions provides real insight. Roy says in (10.8) that many published identities are special cases of just four identities on hypergeometric functions (hgf). Andrews in (10.1) says a table of 32 hgf identities includes almost all published identities on binomial coefficients. The hgf approach makes finding or proving binomial identities so routine that Andrews suggested in 1974 writing a computer program to do so. Now there is such a package in Macsyma.

Non-hgf introductions appear in Berge (10.2) and Knuth (10.6). Other more advanced approaches, including hgf, appear briefly in (10.5) (which, with (10.1),

‡Justification: We calculate some function a_{ij} at each point (i,j) in $I \times I$, where $I = \{0, 1, \ldots, n\}$. If we add the a_{ij} over each row and then add the row-sums, we get the same result as if we add each column and then add the column-sums. That is, we get the sum of all the values, $\sum a_{ij}$, where the sum is taken over $(i,j) \in I \times I$.

may be difficult for beginners). A nice introduction to binomial coefficients and hypergeometric functions appears in (10.4).

9. Applications to Computer Science

To convey a hint of how these ideas come up in computer science we shall present two applications. As you will see, one uses the counting property of binomial coefficients, the other the basic recursion.

(9.1) THE ZIPPER MERGE. Suppose we have two sequences of real numbers, already **sorted** (that is, arranged in increasing order):

$$a_1 < a_2 < \cdots < a_m$$
$$b_1 < b_2 < \cdots < b_n.$$

We assume that none of the a_i's equals any of the b_j's. To **merge** these two sequences is to put these $m + n$ numbers into a single sorted sequence, namely,

$$c_1 < c_2 < \cdots < c_{m+n},$$

where for each k, c_k is a_i or b_i for some i. A standard application of zipper merge is to merge the files on two tapes.

(9.2) **Zipper Merge**

Input: Two sorted sequences, as above.

Output: The merged sequence, as above.

Step 1. Set up an empty vector C (for the output).

Step 2. Set A = the input sequence of a's.

Step 3. Set B = the input sequence of b's.

Step 4. If A or B is empty, output the other sequence. Stop.

Step 5. If neither A nor B is empty, compare the first element of A with that of B, output the smaller, and remove it from A or B.

Step 6. Return to step 4.

Here by the verb *output* we mean, tack the number (or sequence of numbers) onto the right-hand end of C (In APL this tacking-on is called *catenation*, elsewhere often *concatenation*.) The algorithm works for any sorted sequences of real numbers as inputs, even when some are repeated.

Since all that matters in *zipper merge* is the relative sizes of these numbers and not their actual values, to analyze it we *assume that the inputs are the integers from 1 to $m + n$.*

Example (9.3) If we have

$$A = 7,\ 8.01,\ 10 =: a_1, a_2, a_3$$
$$B = 3.14,\ 7.5,\ 7.501 =: b_1, b_2, b_3,$$

then the sorted merged sequence is

$$\begin{aligned} C &= 3.14,\ 7,\ 7.5,\ 7.501,\ 8.01,\ 10 \\ &= b_1,\ a_1,\ b_2,\ b_3,\ a_2,\ a_3. \end{aligned}$$

Zipper merge produces this C by exactly the sequence of steps that it would use to merge

$$\begin{aligned} A' &= 2,5,6 \\ B' &= 1,3,4 \end{aligned}$$

to produce $C' = 1,2,3,4,5,6$. Why so? Because $b_1 = 3.14$ is the smallest number in C; $a_1 = 7$ is the second smallest, $b_2 = 7.5$ is the third smallest, and so on. To form A' and B' we replaced the smallest of A, B by 1, the second smallest by 2, and so on. Here is a nonlinear picture:

(9.3F)

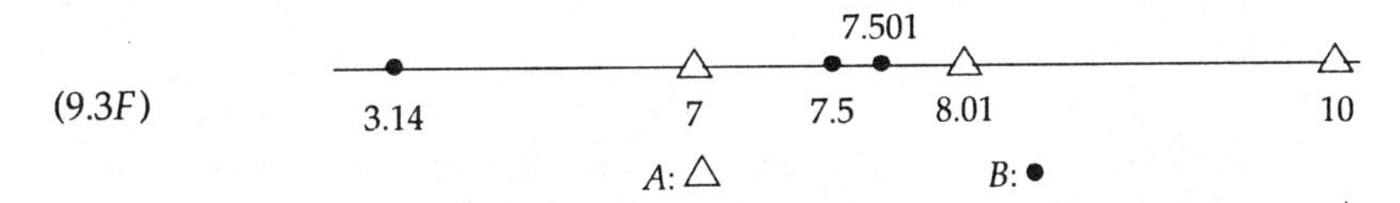

The "$x < y$?" comparison doesn't care how far apart x and y may be, only which one is to the left of the other. So we would find that *zipper merge* performs the same sequence of steps to output C', as the following figure shows:

(9.4F)

1 2 3 4 5 6

A: △ B: ●

In the latter case the outcome of the merging operation is always the same (what is it?), but for fixed m and n, there are $\binom{m+n}{n}$ different possible inputs, or choices of the two sequences of a's and b's. That is, if we choose a subset of n b's from the set $S = \{1, \ldots, m+n\}$, the rest must be the a's. The binomial coefficient is the number of n-subsets of S, or equivalently, by symmetry, the number of m-subsets, and so counts the number of inputs. Why do we care how many inputs there are? Because knowing that helps us understand the performance of the algorithm.

Example (9.4) We tabulate the working of the algorithm for the inputs $A = 2,5,6$ and $B = 1,3,4$. The table shows the state of A, B, and C from start (at left) to finish.

Step 4						Output A
Step 5	Compare	2 and 1	2 and 3	5 and 3	5 and 4	
	Result	1	2	3	4	
A	2,5,6	2,5,6	5,6	5,6	5,6	
B	1,3,4	3,4	3,4	4	$\varnothing$	
C	$\varnothing$	1	1,2	1,2,3	1,2,3,4	1,2,3,4,5,6

Note in particular the Result row, showing the outputs of the comparisons. The first comparison yields "1," because 1 is the first element in B, and of course 1 is smaller than the first element in A, since 1 is the smallest element of our entire input set. This is true in general, because 1 is the first element in A or B in every case, since those sequences are sorted to begin with.

After 1 is removed, 2 is the smallest element of the remaining set of input numbers, and so 2 is the first element of A or B. Thus the second comparison outputs 2. And so on.

(9.5) Analysis of *zipper merge*. The question we ask is this: In performing this algorithm, how many times, T, do we do Step 5? (T was 4 in Example (9.4).) We ask because T is proportional to the total effort used to perform the algorithm. The answer varies, of course, depending on what the input sequences are; it can be zero in the extreme case where one of the input sequences is empty; it can be $2n - 1$ when $m = n$ and the a_i's are the first n odd integers and the b_j's are the first n even integers. We, therefore, try to get an idea of the possible variation of this number T; in particular, we look at its minimum, average, and maximum values.

(9.6) Consider for example the case $m = n = 2$. We may diagram the workings of the algorithm for all $6 = \binom{4}{2}$ input sequences by means of a *tree*, a schematic representation of Steps 4 and 5 from start to finish. The input is $a_1, a_2;\ b_1, b_2$. The output, under our simplifying assumption, is $1, 2, 3, 4$.

To make our tree we let circles (*nodes*) stand for comparisons, and squares (*box nodes*) stand for final executions of Step 4. The algorithm begins with a comparison, since neither m nor n is zero. After each comparison, we draw a line (*edge*) down to the left to the next node if the smaller element was in the a sequence, and to the right if it was in the b sequence. The first comparison tells us which sequence 1 is in, the second which sequence 2 is in, and so on. For each $i \geq 1$, the output of the ith comparison is i.

(9.6F)

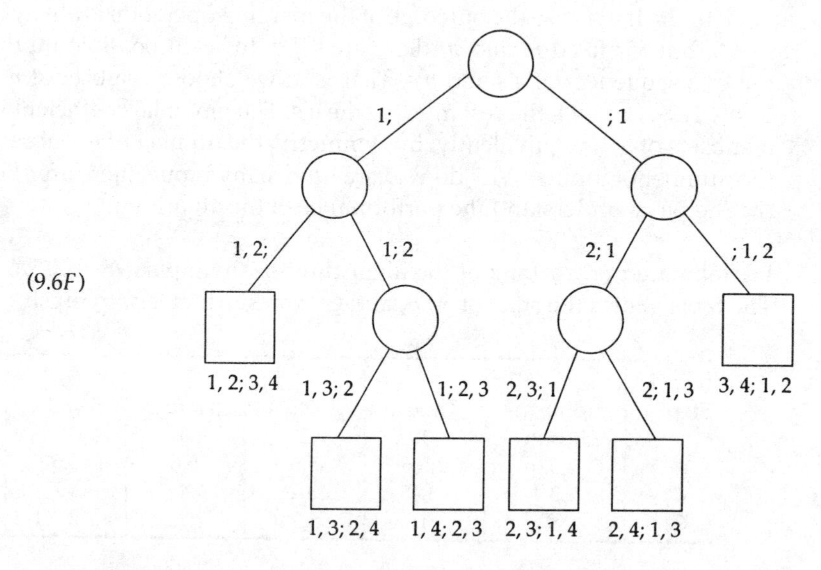

The numbers near each edge record what the algorithm has told us up to that point about the input. Thus "1; 2" means that so far we know that $a_1 = 1$ and $b_1 = 2$;

"2; 1, 3" means $a_1 = 2$ and $(b_1, b_2) = (1,3)$. And the numbers near the box nodes tell us what the whole input was. The number of comparisons needed to reach the box nodes, from left to right, is 2, 3, 3, 3, 3, 2. The average is $16/6 = 8/3$.

(9.7) We now investigate the average value of T in the general case; the minimum of T is clearly $\min\{m,n\}$ and the maximum is $m+n-1$ (unless $mn = 0$, in which case T is always 0). Our knight's-move identity, (8.2), will help us solve the problem.

Assume m and n are positive. Obviously, $m+n$ is the largest member of one of our two input sequences. The question is: what is the largest value M in the other input sequence?

(9.8)
$$M = \begin{cases} a_m & \text{if } b_n = m+n \\ b_n & \text{if } a_m = m+n \end{cases}.$$

We know M could range from $\min\{m,n\}$ to $m+n-1$. The key to this problem is the observation that the total number of comparisons made in the execution of this algorithm is exactly M. This is true because

> for each i, the output of the ith comparison is i, and the algorithm keeps making comparisons until it first empties a sequence.

And the sequence with M as its maximum will be emptied first, after exactly M comparisons. (See Example (9.4), where M was 4, the largest value in B, and T was also 4.)

Now we bring in the binomial coefficients. We need to count how many sequences there are with a given value of M as a maximum in one of the two input sequences. The answer is simple: If M is in the sequence of length m, then the preceding $m-1$ terms must be chosen from the integers $1, \ldots, M-1$. The number of such choices is the binomial coefficient $\binom{M-1}{m-1}$, by (6.2). The result is similar if M is instead in the other input sequence. Thus the total number of inputs having M as the maximum value in one of the two sorted sequences is

(9.9)
$$\binom{M-1}{m-1} + \binom{M-1}{n-1}.$$

Since each such input requires exactly M comparisons, we must multiply the quantity in (9.9) by M and add these numbers for all possible values of M in order to find the total number of comparisons made by the algorithm in zippering all inputs. And the average is this sum divided by the total number $\binom{m+n}{m}$ of inputs.

First we calculate the sum. Using Problem 7.4, we write it as

$$\sum_M \left\{ m\binom{M}{m} + n\binom{M}{n} \right\},$$

where the sum is over all M in the range $\min\{m,n\}$ to $m+n-1$. Now we apply the knight's-move identity (8.2), but in its symmetrized version (Problem 8.1), to evaluate this sum as

$$m\binom{m+n}{m+1} + n\binom{m+n}{n+1}.$$

Corollary (7.7) allows us to simplify this expression when we divide it by $\binom{m+n}{m}$, the total number of inputs. For the first of the two terms we get $mn/(m+1)$. By symmetry of argument, the other term is $mn/(n+1)$. The sum of these two is the average value that we seek:

$$\text{(9.10)} \qquad \begin{aligned} \text{Average value of } T &= \frac{mn}{(m+1)} + \frac{mn}{(n+1)} \\ &= (m+n) - \left[\frac{n}{(m+1)} + \frac{m}{(n+1)}\right] \end{aligned}$$

We write the average this way to show that it is not much less than the maximum value $m+n-1$. In fact, when m and n are large and close to each other, the average is about 1 less than the maximum. (Notice that (9.10) is correct even when $mn = 0$.)

For our example with $m = n = 2$, (9.10) yields $\frac{8}{3}$, which we got the hard way in Section (9.6).

We now examine the performance of another algorithm; again binomial identities will help us.

(9.11) THE k-SUBSETS OF AN n-SET IN LEXICOGRAPHICAL ORDER. Our object here is to design and analyze an algorithm which will produce the k-subsets in the stated order. We first define this order relation.

(9.12) *Lexicographic order.* Suppose X is a set on which there is a linear order defined. So as not to anticipate too much of Chapter 12, let us merely say that X is a set of letters from our alphabet or a set of positive integers. The order on X is alphabetical or numerical: $a < b$, or $1 < 2$. Two strings of the same length are ordered as follows:

$$x_1x_2\ldots x_n < y_1y_2\ldots y_n$$

if and only if $x_1 < y_1$, or if $x_1 = y_1$, then if and only if $x_2 < y_2$, and so on; the order between two strings is that between the leftmost symbols (with the same subscript) having a different value in one string than in the other. When X is a set of letters of the alphabet, lexicographic order is nothing but alphabetical order. Take, for example, $X = \{a,b,c,d,e\}$ and consider strings of length three. Then *aaa* precedes all other strings, *eee* follows all others, *abc* precedes *bbc* but follows *aae*, and *cab* precedes *cda*.

We represent a k-subset uniquely as the string of its elements, ordered by the given order on X. Thus $\{a,b,c\}$ is *abc*, and $\{d,a,e\}$ is *ade*. Now we list all the $10 = \binom{5}{3} = \binom{5}{2}$ 3-subsets of X in lexicographical order:

(9.12F)

		1	2	3	4	5
abc	123	o	o	o		
abd	124	o	o		o	
abe	125	o	o			o
acd	134	o		o	o	
ace	135	o		o		o
ade	145	o			o	o
bcd	234		o	o	o	
bce	235		o	o		o
bde	245		o		o	o
cde	345			o	o	o

where we have written the corresponding strings (3-subsets) for the set $Y = \{1,2,3,4,5\}$ as well, and even a matrix representation of the 3-subsets of Y.

Let us examine this matrix to see how we might create a general procedure for constructing the k-subsets of an n-set. We see that the first six rows are the linking together of 1 with the 2-subsets of $\{2,\ldots,5\}$, and that those 2-subsets are in lexicographic order. The next three rows are the same for 2 and $\{3,4,5\}$, and the last row is that for 3 and $\{4,5\}$. And, pressing on, we see that the first three rows are 1,2 linked with the 1-subsets of $\{3,4,5\}$, and so on. Perhaps you believe by now that we could produce all the k-subsets of an n-set by following some appropriately chosen procedure over and over again, on smaller and smaller sets each time. That is, we would produce them **recursively.** Thus we define the following algorithms **Lexord** and **Lex.**

(9.13.1) **Lexord**

Input: Integers $n \geq 0$, $k \geq 0$.

Output: The k-subsets of $\{1,\ldots,n\}$ in lexicographical order.

1. Lex$(k,1)$

End.

(9.13.2) **Lex**

This recursive algorithm has fixed integral paramaters n and k.

Input: Integers j and i with $0 \leq j \leq n$, $1 \leq i \leq n+1$, and $\{a_1,\ldots,a_{k-j}\}$ with $a_1 < \ldots < a_{k-j} < i$.

Output: The k-subsets of $\{1,\ldots,n\}$ that extend $\{a_1,\ldots,a_{k-j}\}$ with some j-subset of $\{1,\ldots,n\}$ in lexicographic order.

Lex (j,i) =

if $n-i+1<j$ **then** do nothing
  else if $j = 0$ **then** process $\{a_1,\ldots,a_k\}$
    else
      begin $a_{k-j+1} := i$;
        Lex $(j-1, i+1)$;
        Lex $(j, i+1)$
      end.

(9.14) *Explanation.* Lex(j,i) outputs the j-subsets of $\{i, i+1,\ldots,n\}$, in lexicographical order, by assigning the correct values to $a_{k-j+1},\ldots,a_k$. That is, $\{a_{k-j+1},\ldots,a_k\}$ is the j-subset of $\{i,\ldots,n\}$. The *process* instruction means that you do with the subset $\{a_1,\ldots,a_k\}$ what you generated it for. To run through all the k-subsets of $\{1,\ldots,n\}$ you would do Lex$(k,1)$. The "$x := y$" means the variable x is assigned the value y.

The meaning of the semicolons in the algorithm is this: the statement "$P;Q;R$" means "do P, and when finished, do Q, and when finished, do R."

The design of this algorithm follows closely the description we gave earlier. We first check whether j is too big; that step is the condition $|\{i,\ldots,n\}| = n-i+1 < j$. Then we see if we have just generated a k-subset, our goal. That this happens when j has been reset to zero will become apparent in a moment. Now with complete fidelity to our earlier description of the matrix of o's, we begin to produce the j-subsets by writing i as the first element of a j-subset ($a_{k-j+1} := i$); we then find, via Lex($j-1, i+1$), all the $j-1$ subsets of $\{i+1,\ldots,n\}$. They get produced as values of $a_{k-j+2},\ldots,a_k$. When those are all done, we do all the j-subsets of $\{i+1,\ldots,n\}$.

Why does $j = 0$ mean we have a full k-subset? If we started to do Lex(k, 1), our first "write" would be to set $a_1 = 1$. As we call Lex($j-1, i+1$) repeatedly, we reduce the j until we "write" a value for a_k; that happens exactly when j is 1. The next action is to do Lex($0, i+1$). But there we must *process* in accordance with the instruction after $j = 0$; and indeed we have just produced a k-subset.

(9.15) *Correctness of the algorithm Lex.* We prove correctness by induction—hardly a surprise. The proof will write itself once we state the predicate $S(n)$ correctly. We take for $S(n)$ the statement "Lex(j,i) produces in lexicographical order all the j-subsets of $\{i,\ldots,n\}$, for all i, j such that $1 \le i \le n$ and $0 \le j \le n$.

To verify the basis case, we need to show that Lex(1, 1) is correct. We march through the algorithm: The first two conditions are not satisfied, so we are told to write $a_1 = 1$, and do Lex(0, 2). Since the first coordinate is 0, we are to process $\{1\}$. Now we must do Lex(1, 2). Here the condition $n - i + 1 < j$ is satisfied, so we must do nothing, and then end. Thus $S(1)$ is true. (Notice that we made three calls of Lex in doing Lex(1, 1).)

Now let $n \ge 1$ be a fixed but arbitrary integer. Assume $S(m)$ is true for all $m \le n$. Consider $S(n+1)$. Since Lex(j,i) does not even deal with an $(n+1)$-set unless $i = 1$, we may confine our attention to that case. Then from our earlier description we see that the algorithm writes $a_1 = 1$, then calls Lex($j-1$, 2), where it is correct by the induction assumption, and then calls Lex(j, 2)—also correct by the induction assumption. QED

(9.16) *Analysis of the algorithm Lex.* Here the question is how many calls of Lex are made in the execution of Lex(j,i). We denote this number by $L(n-i+1, j)$.

Practice Work out the value of $L(n-i+1, j)$ for a few choices of parameters.

The following general properties of L leap at us.

(9.17) Boundary values: Lex is called just once to evaluate Lex(j,i) when $j = 0$ or when $i + j = n + 2$; these are the cases that satisfy the conditions on lines 2 and 3 of the algorithm.

Otherwise Lex(j,i) calls Lex($j-1, i+1$) and Lex($j, i+1$), so

$$L(n-i+1, j) = L(n-i, j-1) + L(n-i, j) + 1. \tag{9.18}$$

We may use the boundary values and the recursion (9.18) to find all the values of L, which are determined uniquely. (Why?) From scratch we calculated one value, $L(1,1) = 3$; it agrees with (9.18). Notice that our L-function, despite appearances, does not depend on n; i.e., for all d in $\mathbf{N}$, $L(n-i+1, j) = L(n+d-(i+d)+1, j)$. Why? Because both are just the number of calls of Lex in producing the j-subsets of an $(n-i+1)$-set. One set is $\{i,\ldots,n\}$ and the other is $\{i+d,\ldots,n+d\}$, but both have the same size.

Obviously L is independent of k. We may therefore tabulate L in an array shaped like the Pascal triangle. We get, using (9.18),

$$
\begin{array}{ccccccccccc}
 & & & & & 1 & & & & & \\
 & & & & 1 & & 1 & & & & \\
 & & & 1 & & 3 & & 1 & & & \\
 & & 1 & & 5 & & 5 & & 1 & & \\
 & 1 & & 7 & & 11 & & 7 & & 1 & \\
1 & & 9 & & 19 & & 19 & & 9 & & 1
\end{array}
$$

for the first few rows. The top entry is $L(-1,0) = 1$—slightly bogus, but no matter. The third row, 1 3 1, is $L(1,0), L(1,1), L(1,2)$. To push the analogy with the Pascal triangle, let us define $D(N,K) := L(N-1,K)$ for all integers $0 \le K \le N$. The recursion (9.18) becomes, for $0 < K < N$,

$$D(N+1,K) = D(N,K-1) + D(N,K) + 1, \tag{9.19}$$

and we may hope to find a simple expression for D, because its recursion is so close to that of the binomial coefficients (3.1). We take all the fun out of the search by observing that $D(N,K) = 2\binom{N}{K} - 1$ satisfies both (9.19) and the boundary conditions. Therefore,

$$L(N,K) = 2\binom{N+1}{K} - 1.$$

Since $L(N,K)$ is the number of calls of Lex in the execution of Lex$(K,1)$, it is the number of calls in the production of all the K-subsets of an N-set, of which there are a total of $\binom{N}{K}$. The average number of calls per K-subset is approximately $2\binom{N+1}{K}/\binom{N}{K}$, which is, by Problem 7.4,

$$2(N+1)/(N-K+1).$$

If $K \le N/2$, this average is at most 4. (For a different algorithm, see pages 26–38, (10.7).) ◀

10. Further Reading

(10.1) George Andrews, "Applications of basic hypergeometric functions," *SIAM Review*, Vol. 16 (1974), pages 441–484, Section 5.

(10.2) Claude Berge, *Principles of Combinatorics*, Academic Press, New York, 1971.

(10.3) H. W. Gould, *Combinatorial Identities*, unpublished manuscript, Morgantown, 1972.

(10.4) Ronald L. Graham, Donald E. Knuth, and Oren Patashnik, *Concrete Mathematics*, Addison-Wesley, Reading, 1989. (This book is a 600-page expansion of the first 100 pages of (10.6).)

(10.5) Daniel H. Greene and Donald E. Knuth, *Mathematics for the Analysis of Algorithms*, 2d ed., Birkhäuser, Boston, 1982.

(10.6) Donald E. Knuth, *Fundamental Algorithms*, Vol. 1 of *The Art of Computer Programming*, 2d ed., Addison-Wesley, Reading, 1973. (Cited also as (17.2) of Chapter 3.)

(10.7) A. Nijenhuis and H. S. Wilf, *Combinatorial Algorithms*, 2d ed., Academic Press, New York, 1978.

(10.8) Ranjan Roy, "Binomial identities and hypergeometric series," *Amer. Math. Monthly*, Vol. 94 (1987), pages 36–46.

11. Problems for Chapter 8

Unless stated otherwise, summations are taken over all $k \geq 0$.

Problems for Section 1

1.1 Multiply $(1 - x + x^2 - x^3)\,(1 + x + x^2 + x^3)$ by $(1 - x)\,(1 + x)$.

1.2 Prove by induction that $(1 + x)^n$ has degree n, for integral $n \geq 0$.

#1.3 Let $f(x)$ be a polynomial of degree n as in (1.2). Prove that $(x^n)f(1/x)$ is the polynomial $\sum a_{n-k}x^k$. What is the degree of the new polynomial?

Problems for Section 2

2.1 Calculate $\binom{5}{3}$ directly from the definition (2.1) or (2.3).

2.2[Ans] Prove by induction: $\binom{n}{n} = 1$ for all integers $n \geq 0$. Do not use the binomial theorem.

2.3 (*i*) Let x and y be variables and n any integer ≥ 0. Prove that $(x + y)^n = \sum \binom{n}{k} x^k y^{n-k}$. This two-variable version is often seen.

(*ii*) Find the base 8 representation of 100 (base 10) and of 1000 (base 9) without doing any division. Show how you solve these problems.

(*iii*) Use (*i*) to prove symmetry (5.1).

#2.4 Prove:

(*i*) $\sum \binom{n}{k} = 2^n$ for $n \geq 0$.

(*ii*) $\sum (-1)^k \binom{n}{k} = 0$ if n is a positive integer. What is the value of the sum if $n = 0$?

2.5 Prove that $\sum k\binom{n}{k} = n2^{n-1}$ for $n \geq 0$.

2.6 Prove that $\sum k(k-1)\binom{n}{k} = n(n-1)2^{n-2}$ for $n \geq 0$.

2.7 Prove that for integral $n \geq 0$, $\sum k^2\binom{n}{k} = (n+1)n2^{n-2}$.

2.8 Prove:

(*i*) $\sum_{0 \leq k} 2^{6-k}\binom{6}{k} = 3^6$.

(*ii*) $\sum_{0 \leq k} 2^{n-k}\binom{n}{k} = 3^n$ for any integer $n \geq 0$.

#2.9 For integers a, b with $0 \leq a < b$ prove that

$$\sum_{0 \leq k} (-1)^k k^a \binom{b}{k} = 0.$$

Problems for Section 3

3.1 Prove that $\binom{n}{n} = 1$ for all integers $n \geq 0$ by using the basic recursion (3.2).

3.2 Use induction (*not* the binomial theorem) to prove

$$\forall n \in \mathbf{Z},\ n \geq 0 \longrightarrow \binom{n}{1} = n.$$

3.3 Prove that for all $n \geq 2$ and all $k \geq 2$

$$\binom{n}{k} = \binom{n-2}{k-2} + 2\binom{n-2}{k-1} + \binom{n-2}{k}.$$

3.4 Prove: $\binom{n}{0} - \binom{n}{1} + \binom{n}{2} - \binom{n}{3} = -\binom{n-1}{3}$ for $n > 0$. Generalize.

3.5[Ans] Let $n \geq 0$. Prove: for all r and k such that $0 \leq r \leq n$ and $k \geq 0$,

$$\binom{n}{k} = \sum_{0 \leq j} \binom{r}{j}\binom{n-r}{k-j}.$$

[Hint: Try induction on r. Better hint: use (1.25).] Notice the result for $r = 0$ and for $r = 1$.

***3.6** For all integers $j \geq 0$ find integers a_k (which depend also on j) so that $\sum_{0 \leq k} a_k\binom{n}{k} = \binom{n+3}{j}$ holds for all $n \in \mathbf{N}$. [Hint: Try a few specific cases first.]

Problems for Section 4

4.1 Use the Pascal triangle to calculate $\binom{10}{2}$. Note: You don't need all of rows 0 through 10 for this problem.

4.2[Ans] Find integers a_1, a_2, a_3 such that $0 \leq a_1 < a_2 < a_3$ and $n = \binom{a_1}{1} + \binom{a_2}{2} + \binom{a_3}{3}$ for $n = 5,\ 10,\ 15$. Remember: $\binom{m}{m+1} = 0$ for all integers $m \geq 0$.

***4.3** Let j be a fixed positive integer. Prove that for every nonnegative integer n there is a unique sequence of integers $a_1, \ldots, a_j$ such that $0 \leq a_1 < a_2 < \cdots < a_j$ and

$$n = \binom{a_1}{1} + \binom{a_2}{2} + \cdots + \binom{a_j}{j}.$$

[Hint: Use the Pascal triangle. You may need to use $\binom{m}{m+1} = 0$ for various m.] Set $j = 3$ and write out the sequence a_1, a_2, a_3 for each of the integers $n = 0, 1, \ldots, 6$.

Problem for Section 5

5.1 Use induction to prove (5.1), the symmetry lemma.

Problems for Section 6

6.1 The five students in a seminar are each to choose a different topic from a list of eight for their final report. How many choices may the class make? A *choice* is the set of topics chosen.

6.2[Ans] How many committees of six members could one appoint from a set of eight men and five women? Explain your answer.

6.3 How many different committees consisting of two men and three women can be formed from a class of m men and w women?

6.4 Same question as in problem 6.2, but now every committee must contain at least two women. Explain your answer.

6.5 Derive $\binom{n}{n} = 1$ from (6.2), the counting theorem.

6.6 Derive symmetry (5.1) as a consequence of counting (6.2). [Hint: If X is a k-subset of the n-set T, how many points are in $T - X$? See also Chapter 5, (15.9).]

6.7 How many partitions are there of an 8-set

(*i*) into two cells, one of three elements, the other of five elements?

(*ii*) into three cells, one of two elements, and the other two of three elements each?

6.8 How many partitions are there of a 6-set into two cells of three elements each? Explain your answer. [No listing, please.]

6.9 Express your answer to the following questions (*i*) and (*ii*) in terms of binomial coefficients *and* as integers in base 10.

(*i*) How many partitions are there of a 7-set into two cells, one of four points and one of three points?

(*ii*) How many partitions of an 8-set are there into two cells of four points each? How many partitions are there of a 10-set into four cells, of, respectively, four, three, two, and one point(s)?

6.10 How many positive integers less than 1,024 are there which are the sum of exactly three distinct powers of two?

6.11[Ans] How many strings of n 0's and 1's are there having no 1's? having exactly one 1? having exactly two 1's?

6.12 How many strings of eight 0's and 1's are there having no more than two 1's? Express your answer in terms of binomial coefficients and explain.

6.13 Consider strings of length n in 0,1. How many strings differ in at most two positions from a fixed string?

6.14 How many strings of twenty-one 0's and 1's are there having exactly two 1's but also satisfying the condition that each 1 is immediately preceded by and immediately followed by a 0? Express the answer as a binomial coefficient.

6.15 Same question as in the preceding problem except that the word *immediately* is deleted from both places.

6.16 Let m and n be nonnegative integers. Consider strings of 0's and 1's having exactly m 1's and n 0's in all possible orderings. [Example: the set of all strings with two 0's and one 1 is $\{001, 010, 100\}$.]

(*i*) Prove that the total number of these strings having no two 0's adjacent is $\binom{m+1}{n}$. (This problem comes from reference (10.5) of Chapter 9.)

(*ii*) What is the total number of strings having no two 1's adjacent? Having no two 1's and no two 0's adjacent?

6.17[Ans] Let m and n be integers ≥ 0. Is the number of strings of $2m$ 0's and n 1's in which every 0 is next to another 0 less than, equal to, or more than $\binom{m+n}{n}$? Explain.

6.18 Prove the counting theorem (6.2) by induction.

6.19 Prove the basic recursion (3.1) by using the counting theorem (6.2).

6.20 [Recommended] You are at a corner on a rectangular grid of streets. You want to go to a corner m blocks east and n blocks north. You may travel only north or east. Prove that the total number of different routes you may take is $\binom{m+n}{m}$.

6.21 Let n and w be integers such that $0 \leq w \leq n$.

(*i*) Prove that

$$\binom{n}{2} = \binom{w}{2} + \binom{n-w}{2} + w(n-w).$$

$*$(*ii*) Find and prove an analogue of (*i*) for $\binom{n}{3}$.

Problems for Section 7

7.1[Ans] Calculate $a := \binom{11}{4}$, $b := \binom{11}{5}$, $b - a$, and b/a. Now find the value of $\binom{12}{6}$ in the simplest possible way; explain.

7.2 Find the prime factorization of $\binom{200}{6}$.

7.3 Prove (7.8).

#7.4 For integers $n \geq 0$ and $k \geq 1$,
$\binom{n}{k} = \binom{n}{k-1}(n-k+1)/k$, and $\binom{n}{k} = \frac{n}{k}\binom{n-1}{k-1}$.

7.5 Prove: for all integers $n, k \geq 0$, $n\binom{n}{k} = (k+1)\binom{n}{k+1} + k\binom{n}{k}$. [Hint: Work on the right-hand side.]

7.6 Prove the binomial theorem (7.2) for integers $n \geq 0$ by induction on n.

7.7 Prove: If $n \geq 2$ and $0 < k \leq n/2$, then $\binom{n}{k-1} < \binom{n}{k}$. Hence the middle binomial coefficient in each row is the largest.

***7.8** Take the following formula as defining the derivative $Df(x)$ of the polynomial $f(x)$ given by (1.2):

$$Df(x) := \sum k a_k x^{k-1}.$$

Prove that $D[f(x)g(x)] = [Df(x)]g(x) + f(x)[Dg(x)]$ for all polynomials $f(x)$ and $g(x)$. Conclude that $D[f(x)^m] = mf(x)^{m-1}[Df(x)]$ for $m \geq 0$. Prove Taylor's theorem for polynomials: $a_k = (D^k)f(x)/k!$ evaluated at $x = 0$.

***7.9** Prove that the result of problem 6.21*i* holds for all $n, w \in \mathbf{R}$.

Problems for Section 8

#8.1 For $k \geq 0$ and $m \geq 0$, prove that $\sum_{0 \leq n \leq m} \binom{n}{k} = \binom{m+1}{k+1}$.

8.2[Ans] Use Problem 8.1 to derive a formula for $\sum_{1 \leq n \leq m} n^2$.

Problems for Section 9

9.1 Let $m = 2$ and $n = 3$. Draw and label the tree analogous to that in (9.6F).

The next four problems deal with zipper merge. *The notation is the same as that in the text.*

9.2 List all inputs requiring exactly four comparisons when $m = 3$ and $n = 3$.

9.3 Again for $m = n = 3$, count without listing the total number of inputs requiring exactly M comparisons for each value of M.

9.4 Work out an example of zipper merge in the case $m = 5$, $n = 3$ by calculating the average number of comparisons made over all inputs as follows:

(*i*) For $M = 3, \ldots, 7$ calculate the number of inputs with $a_m = M$ ($b_n = M$). Don't list them, but use the ideas of the proof of (9.10) to count them.

(*ii*) Find the total number of comparisons used over all inputs.

(*iii*) Find the average and see that it agrees with (9.10). How close is the average to the maximum?

9.5 What are the inputs requiring the minimum number of comparisons? What is the value of that minimum?

9.6[Ans] Prove that if $n - 1 \leq m \leq n$ in the zipper merge algorithm, then a majority of input sequences require the maximum number of comparisons.

***9.7**[Ans] Prove the result of the previous problem for $2 \leq m = n - 2$. [Hint: Use (7.7).]

9.8 Let $E(n, k)$ be defined for all integers n and k satisfying $0 \leq k \leq n$. Suppose that for certain real numbers a and b, $E(n, 0) = E(n, n) = b$ for all n; and for all k with $0 < k < n$, $E(n, k) = a + E(n-1, k-1) + E(n-1, k)$. Find a simple formula for $E(n, k)$ in terms of binomial coefficients.

9.9 For $n = 7, k = 3$ write out the steps of the algorithm Lex(2, 5).

9.10 Prove that the values of L are determined uniquely by the boundary values and the recursion (9.18).

General Section

These problems are grouped by topics. They are ranked for difficulty within each group, but not overall.

G1. Find and prove a formula for $\sum \binom{n}{2k}$ for all $n \geq 1$.

———— o ————

G2. Express in terms of binomial coefficients the total number of strings (of length 9) having three 0's, three 1's, and three 2's. How many of these strings have 000 as a substring?

G3. Let n be any nonnegative integer. How many strings of length n in 0,1 are there beginning 000, ending 111, and having a total of seven 0's?

G4.[Ans] (*i*) How many bridge hands are there? The definition is: 13 cards from a 52-card deck. [A deck has four *suits* of 13 cards each. Each suit consists of cards marked 2, 3, ..., 10, J, Q, K, A.]

(*ii*) How many bridge hands containing all four aces are there?

G5. The point-count of a bridge hand is defined by the assignment to each ace (A) of four points; to each king (K), three points; to each queen (Q), two points; to each jack (J), one point; and zero to all other cards. The point-count of a hand is then the sum of all the points of the cards in it. How many bridge hands are there with a point-count of 36?

G6. Answer the question at the end of the World Series example in the Introduction, namely, "How many such strings can there be?" Explain your answer.

G7. Suppose you are in the Pascal triangle at $\binom{n}{k}$. You want to go to $\binom{m}{j}$, where $m > n$, and $k \le j \le k + m - n$. You may travel only along the paths suggested by the basic recursion (3.2), that is, from $\binom{a}{b}$ to $\binom{a+1}{b}$ or to $\binom{a+1}{b+1}$ for any integers a, b. How many different paths are there for your trip?

———— o ————

The problems in this group are based on the ideas of Chapter 7.

G8. Make a neat and regular Pascal triangle down to the row for $n = 15$, but reduce (to 0 or 1) the entries mod 2.

(*i*) Explain the resulting pattern.

(*ii*) Prove that for all integers $n \ge 0$, the total number of k for which $\binom{n}{k}$ is odd is a power of 2. (See also Problem G14.)

G9. Let p be prime. Prove that if $1 \le k \le p - 1$, then $\binom{p}{k} \equiv 0 \pmod{p}$.

G10. Use induction on a to prove Fermat's theorem (Chapter 7, (11.1)) in this form: If p is prime, then every integer a satisfies $a^p \equiv a \pmod{p}$.

G11. Let m be composite (that is, $m > 1$ and m is not prime). Prove that for some k such that $0 < k < m$, $\binom{m}{k} \not\equiv 0 \pmod{m}$.

G12. For all sequences $A = a_1, a_2, a_3, \ldots, a_{n-1}, a_n$ define an operator S by the rule

$$S(A) := a_1 + a_2,\ a_2 + a_3, \ldots, a_{n-1} + a_n.$$

S shortens the sequence by one term. Start with the sequence $1, a, b, c, d, 2$, choosing any integral values for a, b, c, d. Apply S until there is only one term left. What is that term congruent to mod 5? (After a puzzle in *Games* magazine). Generalize.

G13. Let p be prime and $q = p^i$ for $i \ge 0$. Prove that for all integers a, b, $(a + b)^q \equiv a^q + b^q \pmod{p}$. [This is a corollary of problem G9; use induction on i.]

$*$**G14.** For polynomials $f(x)$ and $g(x)$ with integral coefficients define $f(x) \equiv g(x) \pmod{m}$ iff all coefficients of $f(x) - g(x)$ are divisible by m. Use this idea to prove *Lucas's theorem*:

> Let p be prime. Let $n, k \ge 0$ have base p expressions $n = \sum n_i p^i$ and $k = \sum k_i p^i$, where $0 \le n_i, k_i < p$. Then
>
> $$\binom{n}{k} \equiv \binom{n_0}{k_0}\binom{n_1}{k_1}\binom{n_2}{k_2}\cdots \pmod{p}.$$

———— o ————

G15. Prove the multinomial theorem: Let $n \ge 0$ be an integer. For independent variables $x_1, \ldots, x_r$, with $r \ge 1$, define the multinomial coefficient

$$\binom{n}{e_1, \ldots, e_r},$$

where $e_1, \ldots, e_r$ are integers ≥ 0 with sum n, as the coefficient of $x_1^{e_1} \ldots x_r^{e_r}$ in the polynomial expression for $(x_1 + \cdots + x_r)^n$. Then

$$\binom{n}{e_1, \ldots, e_r} = \binom{n}{e_1}\binom{n - e_1}{e_2}\binom{n - e_1 - e_2}{e_3}\cdots\binom{e_r}{e_r},$$

where $\binom{a}{b}$ is the binomial coefficient (2.3). Compare problem 2.3(*i*). [Hint. Use induction on r.]

G16. Suppose $e_1, \ldots, e_r$ are positive integers with sum n. If no two of the e_i's are equal, then the multinomial coefficient

$$\binom{n}{e_1, \ldots, e_r}$$

is the number of partitions of an n-set into r cells in which some cell has size e_1, another size e_2, another e_3, and so on. Prove this claim.

G17. Prove: If $e_1, \ldots, e_r$ are integers ≥ 0 with sum n, then the multinomial coefficient

$$\binom{n}{e_1, \ldots, e_r}$$

is the number of ways of placing n labeled balls into r labeled cells in such a way that the ith cell holds exactly e_i balls. (For example, you play pool on a table with r labeled pockets. There are n balls numbered 1 to n. When you have sunk all the balls the outcome is that each pocket holds a certain subset of the balls. How many outcomes are possible when the size of the subset is specified for each pocket?)

G18. Prove: The multinomial coefficient in problem G17 is the number of strings of length n in r symbols $s_1, \ldots, s_r$ in which s_1 appears e_1 times, s_2 appears e_2 times, and so on.

G19. Prove the basic recursion for the trinomial coefficients:

$$\binom{n}{e-1,f,g} + \binom{n}{e,f-1,g} + \binom{n}{e,f,g-1} = \binom{n+1}{e,f,g},$$

where e, f, g, and n are integers ≥ 0 with $e+f+g = n+1$.

***G20.** State and prove an analogue of the basic recursion (3.2) for the multinomial theorem.

***G21.** (This problem is no. 215 in reference (10.5) of Chapter 9.) For integral $n \geq 0$ prove that $(n!)^{n+1}$ divides $(n^2)!$.

Some Test Problems

T1. How many juries of six members could be selected from a pool of 20 people? Express your answer in terms of binomial coefficients *and* as an integer in base 10. Show all steps to get the latter.

T2. Write your social security number (SSN). Add the leftmost digits of your SSN until you first reach a sum s greater than 10. Then, showing every step on your paper, calculate $\binom{s}{3}$.

T3. Prove that for all $n \geq 1$ $\sum_{0 \leq k} \binom{n}{2k} = 2^{n-1}$.

T4. Prove for all nonnegative integers n and k that

$$\binom{n}{k-2} \cdot \binom{n-k+2}{2} = \binom{n}{2} \cdot \binom{n-2}{k-2}.$$

T5. Find the prime factorization of

$$\binom{18}{4}.$$

T6. Consider zipper merge for inputs of size $m = 5$ and $n = 7$. Let M denote the total number of comparisons used to process an input. Over all possible inputs what is the

minimum value of M? (Explain.)

maximum value of M? (Explain.)

average value of M? (No explanation necessary.)

T7. Let n be any positive integer. Prove

$$\sum_{0 \leq k} \binom{n}{2k+1} = 2^{n-1}.$$

T8. How many partitions are there of a 7-set into two cells, one of which has three elements, the other cell four elements? Explain in terms of binomial coefficients.

T9. Consider zipper merge with $m = 2, n = 4$. Thus $A: a_1 < a_2$ and $B: b_1 < b_2 < b_3 < b_4$ are the inputs, where $\{a_1, a_2, b_1, \ldots, b_4\} = \{1, \ldots, 6\}$. For each $T = 1, 2, \ldots, 6$ give the total number of inputs requiring exactly T comparisons. Explain.

T10. Extend the basic recursion (3.1) by proving it for any real n.

12. Answers to Practice Problems

(1.11P) If $a_i = 0$ for all $i > N$, then the largest value of i for which $a_i \neq 0$ is certainly N or less. Therefore, by (1.8), (degree of $f(x)$) $\leq N$. But we could have $f(x) = 1 + x$ and take $N = 5$. Condition (1.3) holds, but the degree is 1.

(1.29P) Suppose $f(x)$ and $g(x)$ have degrees d and e, respectively. Thus

$$f(x) = ax^d + \text{(terms of lower degree)}$$

$$g(x) = bx^e + \text{(terms of lower degree)}$$

and $ab \neq 0$. It is easy to verify that the nonzero term of highest degree in the product $f(x)g(x)$ is

$$ax^d \cdot bx^e = abx^{d+e}.$$

(1.30P)(*i*) Let $f(x) = a$ be a constant polynomial. If $a \neq 0$, then degree of $f(x) = 0$. If $a = 0$, the answer is -1.

(*ii*) By (1.26):

$$\begin{array}{llllllll} 1 \cdot f(x) & : 1 & x & x^2 & \ldots & x^n & \\ -x \cdot f(x) : & & -x & -x^2 & \ldots & -x^n & -x^{n+1} \end{array}$$

The sum is $1 - x^{n+1}$.

By (1.27):

$$
\begin{array}{ccccccc|l}
0 & 1 & x & x^2 & \cdots & x^n & 0 & \textit{Result} \\
-x & 1 & & & & & & 1 = -x \cdot 0 + 1 \cdot 1 \\
 & -x & 1 & & & & & 0 = -x \cdot 1 + 1 \cdot x \\
 & & -x & 1 & & & & 0 = -x \cdot x + 1 \cdot x^2 \\
 & & & \ddots & & & & \vdots \\
 & & & & -x & 1 & & 0 = -x \cdot x^{n-1} + 1 \cdot x^n \\
 & & & & & -x & 1 - x^n & = -x \cdot x^n + 1 \cdot 0
\end{array}
$$

(*iii*)
$$
\begin{aligned}
(1 - x + x^2)(1 + x + x^2) &= (1 + x^2 - x)(1 + x^2 + x) \\
&= (1 + x^2)^2 - x^2 \\
&= 1 + 2x^2 + x^4 - x^2 \\
&= 1 + x^2 + x^4.
\end{aligned}
$$

To multiply by $(1 - x)(1 + x) = 1 - x^2$ we may use (*ii*) with $x := x^2$. We get $(1 - x^2)(1 + x^2 + x^4) = 1 - x^6$.

(2.4P) $(1 + x)^4 = (1 + x)(1 + x)^3$, so $\binom{4}{k}$ is the coefficient of x^k in $(1 + x)(1 + 3x + 3x^2 + x^3)$, which in the tabular form (1.26) is

$\binom{4}{0}$	$\binom{4}{1}$	$\binom{4}{2}$	$\binom{4}{3}$	$\binom{4}{4}$
1	3	3	1	
	1	3	3	1
1	4	6	4	1

(4.3P)

```
                         1
                       1   1
                     1   2   1
                   1   3   3   1
                 1   4   6   4   1
               1   5  10  10   5   1
             1   6  15  20  15   6   1
           1   7  21  35  35  21   7   1
         1   8  28  56  70  56  28   8   1
       1   9  36  84 126 126  84  36   9   1
     1  10  45 120 210 252 210 120  45  10   1
```

(4.4P) Since the binomial coefficients are defined for all $k \in \mathbf{Z}$, the Pascal triangle does not contain the values $\binom{n}{k} = 0$ when $n \in \mathbf{Z}$, $n \geq 0$, and $k < 0$ or $k > n$. The Pascal triangle is a triangular array of all the nonzero binomial coefficients $\binom{n}{k}$ for nonnegative integers n. (As you will see, binomial coefficients are defined for all real upper indices n. Although the basic recursion holds for all such n, it's no longer true that $\binom{n}{k} = 0$ if $k > n$ when $n \notin \{0, 1, 2, \dots\}$. That nonnullity kills triangularity.)

(6.9P) (*i*) $\binom{40}{2}$.

(*ii*) There are $\binom{10}{3}$ possible sets of three records for sale. Since no two of the n customers chose the same 3-subset of records, we know that $n \leq \binom{10}{3}$. (From above, $\binom{10}{3} = 120$.) In Section (7) we show how to calculate the value of $\binom{n}{k}$ without using the Pascal triangle.

(7.1P) You'd need the first four entries in each row down to row 20. For example,

```
          1
        1   1
      1   2   1
    1   3   3   1
  1   4   6   4
1   5  10  10
```

That's 74 terms.

(8.5P) If $-m > n$, then the set of all integers j satisfying $-m \leq j \leq n$ is empty. Therefore, the sum on the left should be zero and so should the term on the right of (8.4). Indeed, those are so: If $j \geq -m$ (which makes $\binom{a}{m+j}$ possibly nonzero), then $-j \leq m < -n$. Therefore, $n - j < 0$, so $\binom{b}{n-j} = 0$. On the right, $-m > n$ implies $0 > m + n$, making $\binom{a+b}{m+n} = 0$.

(8.8P) Of course (8.6) is true when $m := -3$. The summation over $j \in \mathbf{Z}$ makes it very easy to see that it's true, because $m + j$ runs over all of $\mathbf{Z}$, taking each value exactly once. Remember that m is fixed. (8.6) is true for all $m \in \mathbf{Z}$. If you want to three-putt it, you can say

$$\text{for } j = -m \text{ you get} \binom{a}{0} x^0,$$

$$\text{for } j = -m + 1 \text{ you get} \binom{a}{1} x,$$

$$\text{for } j = -m + 2 \text{ you get} \binom{a}{2} x^2,$$

and so on.

For $j < -m$ you get $\binom{a}{\;1}$ or $\binom{a}{\;2}$, and so forth, all of which are zero.

(8.14P) The coefficient of v_0 is $\binom{-k-1}{-k}$, which is zero if $k \geq 1$. If $k = 0$ it is $\binom{-1}{0}$, which appears on the right of (8.13) as the coefficient of v_0.

In general, if $i < k$, then $i - k$ is negative, making $\binom{*}{i-k} = 0$, whatever the value $*$ of the upper index.

CHAPTER

9

Counting

The most important principle in mathematics is that if you count the elements of a finite set in two ways, you get the same result both times.

Hermann Weyl

1. Introduction

Ten students were running in a campus-wide election for council. Five were to be elected. Each voter had five votes, which could be distributed in any way—all five could go to one candidate, or one vote could go to each of five different candidates, or anything in between (or even less, in that a voter could refuse to use some or all of the five votes).

(1.1) In the end 3100 people voted, and exactly 97 of them cast blank ballots. "Aha!" said John at this news, "That means at least two of the nonblank ballots were identical." How did he know?

(1.2) Even though the votes were to be tallied for individual candidates, the 10 candidates had formed into three groups (hoping to elect all members of their group so as to control the five-member council). They were

Group A: Alicia, Angela, Jim
Group B: Bill, Debbie, Sam
Group C: Christina, Kim, Michelle, Tom.

The outcome was, however, that no group won seats for more than two of its members. In a study of ticket-splitting, political science students found on examining the ballots that

2082 ballots had at least one vote for a candidate from Group A, and they made similar tallies for Group B and Group C. The results are summarized below.

Group A	2082
Group B	2306
Group C	2059
Groups A and B	1614
Groups A and C	1375
Groups B and C	1458

The last three tallies were of the total number of ballots with, in the first case, at least one vote for a candidate from Group A and at least one vote for a candidate from Group B (and similarly for the other two cases). They neglected, however, to tally the number of ballots with a vote for someone in each group (the most pronounced ticket-splitters); and before they could rectify things the ballots were sent off to be recycled. At this point, John, a student of discrete mathematics, told them that the number could be determined from the data they already had and that its value was, in fact, 1003. He also said there had been little straight-ticket voting, that there were only 96 ballots on which all votes were cast for candidates from Group A. How did John know?

In this chapter we'll develop methods for answering these questions. We take up only the most basic results in counting, except for the optional Section (8). Counting is the subject of hundreds of articles and dozens of books. It touches many other subjects in and out of mathematics, such as probability, physics, chemistry, information theory, and electrical engineering.

(1.3) To begin with we will make a more systematic study of how to count. The question is simple: if A is a set, what is the total number, denoted $|A|$, of elements in A? Fundamentally, the answer is that $|A| = n$, where n is a nonnegative integer, if and only if there is some bijection from A to the set of all positive integers from 1 to n, inclusive. (When $n = 0$ this set is $\varnothing$.) Often, however, we need not reduce the question to such basics. Indeed, the purpose of this chapter is to teach you how to count by various shortcuts (in which the bijections have been set up once and for all).

Example (1.4) The number of multiples of 3 between 1 and 50 is 16. Why? They are the elements of

$$A = \{3, 6, \ldots, 48\},$$

and if we divide each of these by 3 we have a bijection from A to $B := \{1, \ldots, 16\}$. That is, define $f : A \to B$ by the rule $f(x) := x/3$ for all $x \in A$. Then f is a bijection.

Practice (1.5P) What is the bijection from $\varnothing$ to $\varnothing$?

Recall that if A has exactly n elements we call A an n-set.

(1.6) A related, more basic result is that if A and B are sets, then $|A| = |B|$ if and only if there is some bijection from A to B. Someone has pointed out that children too young to know how to count still understand this principle. For example, they can determine that the number of hats equals the number of children at a party by putting a hat on each child and seeing that there are none left over. What's more, they do this without knowing the word "bijection."

We may use the result as follows: If we know the value of $|A|$ for some set A and if we can find a bijection from A to B, then we know that $|B| = |A|$.

Example (1.7) Suppose we ask how many right triangles have sides of integral lengths less than 16. Call B the set of all such triangles (one triangle for each congruence class—actually we are counting the cells of the (geometric) equivalence relation "congruence"). We know that B is in one-to-one correspondence with the set A of all triples of integers a, b, c such that

$$a < b < c \quad \text{and} \quad a^2 + b^2 = c^2.$$

After a bit of number theory we could find that

$$A = \{\,(3,4,5),\ (6,8,10),\ (9,12,15), (5,12,13)\,\}^{\dagger},$$

from which we see that $|A| = 4$, hence $|B| = 4$.

Practice (1.8P) Find a bijection between the set A of all relations on a set of three elements and the set

$$B := \{\, i\,;\, i \in \mathbf{Z},\ 0 \leq i < 512\,\}.$$

Hint: Consider the relation-matrix.

We look again at our earlier notations for sets (Ch. 1, (1.2) and Ch. 2, (19)) and now count them:

$$\begin{array}{ll} A = \{\,1,2,3,4\,\} & |A| = 4 \\ A = \{\,3,1,4,1,5,9,2,6,5\,\} & |A| = 7 \\ A = \{\,x\,;\, x \in \mathbf{Z}, 3 \leq x \leq 7\,\} & |A| = 5 \end{array}$$

Of course, the crucial fact is that we count each element of a set once and only once, no matter how many times it may appear in a list or other representation of the elements of the set. Thus if $A = \{\,x^2\,;\, -5 \leq x \leq 5, x \in \mathbf{Z}\,\}$, then $|A| = 6$, not 11.

2. Basic Results

So far it's been simple. And, in earlier chapters, we've already proved some general results, which we summarize below after one superobvious remark, **C1**, and the superbasic remark **C0**, which is (1.6).

(2.1) **C0. For sets A, B, $|A| = |B|$ iff there is a bijection from A to B.**

C1. If A and B are disjoint sets, then $|A \cup B| = |A| + |B|$.
For example, $|\{\,1,2,3\,\} \cup \{\,4,5\,\}| = 3 + 2 = 5$.

C2. For any finite sets A, B, the Cartesian product $A \times B$ has exactly $|A| \cdot |B|$ elements. More generally, the Cartesian product of m finite sets $A_1, \ldots, A_m$ has exactly $a_1 \cdots a_m$ elements, where $|A_i| = a_i$ for each i.

† See Stark, pages 151–152. (Cited in Chapter 7, (16.2).)

C3. For any integer $n \geq 0$, the total number of subsets of an n-set is 2^n.

C4. The total number of k-subsets of an n-set is $\binom{n}{k}$, for any nonnegative integer n and any integer k.

Again, $\binom{n}{k}$ stands for the binomial coefficient "n choose k" of Chapter 8, (2.1).

We proved **C3** by induction as Theorem (11.1) of Chapter 3. The result **C4** was the main point of Chapter 8, namely, Theorem (6.2). The other, **C2**, recalls explanations of multiplication given to children in schools:

Why 2 times 3 equals 6:

	•	•	•
•	•	•	•
•	•	•	•

Grade 3

	a	b	c
1	$(1,a)$	$(1,b)$	$(1,c)$
2	$(2,a)$	$(2,b)$	$(2,c)$

Grade 14

In any case, **C2** is easy to prove by induction on m in the general case. In addition, there is a special case of **C2** that occurs as problem 30 of Chapter 3, where you are asked to show that the total number of strings of length m over an "alphabet" of q letters is q^m. Recall that a "string" of length m over the alphabet A is just an element $(c_1, c_2, \ldots, c_m)$ of $A \times A \times \cdots \times A = A^m$ denoted without the parentheses or commas.

These five counting results are useful in many situations; indeed, **C0** and **C1** are basic. We have seen **C4** used in the analysis of algorithms in Chapter 8.

Example (2.2) Consider the familiar game "paper, scissors, stone" in which each of two players shows either an open hand ("paper"), two fingers ("scissors"), or a clenched fist ("stone"). How many outcomes are there of one play of this game? An outcome is a record of what player 1 "threw" and of what player 2 "threw." Answer: $9 = 3 \times 3 = |R \times R|$, where $R := \{\text{paper, scissors, stone}\}$. An outcome $(r_1, r_2) \in R \times R$ stands for player 1 throwing r_1 and player 2 throwing r_2.

Example (2.3) A tennis team has a choice of uniforms: white, blue, or yellow. Each member has two pairs of tennis shoes, one smooth-soled and one with treads. The coach decides how the team will dress for each match. (She makes the same choice for each member of the team.) How many choices are there? Answer: $6 = 2 \times 3$. Why? We use **C2**. A "choice" is any ordered pair (s, c), where s is one of the 2 shoe choices and c is one of the 3 choices of color. That is, the set of all possible choices is $S \times C$, where S is the 2-set of shoe choices, and C is the 3-set of colors.

		Color		
		white	blue	yellow
Shoes	smooth	○	○	○
	treaded	○	○	○

Each circle in (2.3F) represents an ordered pair (s, c).

(2.4) The key to using **C2** is that the choices are *independent*. That is, after any choice of shoe the coach is free to choose any of the three colors. That independence mirrors the definition of the Cartesian product $A \times B$ as the set of all ordered pairs (a, b), where a is *any* element of A and b is *any* element of B. In forming a Cartesian product you choose the a and b in all possible ways, independently of each other.

Why don't we add 2 and 3 to get 5 as our answer? That number represents $|S \cup C| = |S| + |C|$, since $S \cap C = \varnothing$, but it doesn't count what we want. It counts the total number of items each player has (where item:= uniform or pair of shoes). If players suited up for a match in a uniform with no shoes or in shoes with no uniform, $|S \cup C|$ would be the answer. Since the coach chooses one item from S and one from C independently, the answer is $|S \times C|$.

Notice that when we say a "choice" is an ordered pair (s, c), we are implicitly using **C0**. A choice is really a decision to use one of the 2 shoe types and one of the 3 colors. Without even thinking about the process, we made a bijection between the set of choices and $S \times C$.

Example (2.5) You have five programs to compute g and three to compute f. These programs are "compatible," i.e., range of $f \subseteq$ domain of g. Therefore you have 15 programs to compute the composition $g \circ f$. Say $f_1, \ldots, f_5$ compute f and g_1, g_2, g_3 compute g.

	f_1	f_2	f_3	f_4	f_5
g_1	$g_1 \circ f_1$	$\circ$	$\circ$	$\circ$	$\circ$
g_2	$\circ$	$\circ$	$\circ$	$\circ$	$\circ$
g_3	$\circ$	$\circ$	$\circ$	$\circ$	$\circ$

We display them using the "$\circ$" to stand for the composition of the row program with the column program.

Now here is a counting problem calling for all five of **C0**,..., **C4**.

(2.6) From a set $S = \{ p_1, p_2, \ldots, p_n \}$ of n people we are to appoint a committee with two cochairpersons. How many such appointments are possible?

Example (2.7) Let's do this first for $n = 4$, so $S = \{ p_1, \ldots, p_4 \}$. We can count the committees of size 2, then those of size 3, and finally those of size 4. There must be at least two members, so we ignore subsets with fewer than two elements.

Any subset of size 2 consists of just the two cochairs. There are $\binom{4}{2} = 6$ such committees.

Now suppose we have a committee of size 3, say $\{ p_1, p_2, p_3 \}$. We have to appoint two of its members as cochairs. We can do this in $\binom{3}{2}$ ways, indicated by underlining the cochairs:

$$\{ \underline{p_1}, \underline{p_2}, p_3 \}$$
$$\{ \underline{p_1}, p_2, \underline{p_3} \}$$
$$\{ p_1, \underline{p_2}, \underline{p_3} \}$$

Each of these three is a *different* committee with two cochairs appointed—because the subset of cochairs is different in each.

How many 3-subsets of S are there? Answer: $\binom{4}{3}$, of course. For each one there are $\binom{3}{2}$ committees with 2 cochairs appointed, so there are

$$\binom{4}{3}\binom{3}{2} = 4 \cdot 3 = 12$$

such committees of size 3. We used **C4** and **C2**.

Finally, for committees with four members, there is only one 4-subset, namely, S. There are $\binom{4}{2}$ ways to appoint two cochairs from S, so there are

$$\binom{4}{4} \cdot \binom{4}{2} = 6$$

committees of size 4 with two cochairs.

Since the subsets of such committees of sizes 2, 3, and 4, respectively, are mutually disjoint, we may use **C1** to get our answer for $n = 4$ as

$$6 + 12 + 6 = 24.$$

There is, however, a more elegant way to get this answer. We present it as the first part of the general solution to (2.6).

(2.8) There are two ways to approach this problem. One is the *"gestalt,"* the all-at-once, or big-picture approach. Using it, we say that there are $\binom{n}{2}$ choices for the two cochairs, and that for each such choice, there remain $n - 2$ people from whom the rest of the committee is to be chosen. Since there is no restriction on the size of the committee (except that it must have at least two members, the cochairs), the rest of the committee may be any subset of the $n - 2$ people. Thus there are $\binom{n}{2} 2^{n-2}$ such appointments. (Notice that this is the correct answer even if $n = 0$ or 1 and that it gives 24 when $n = 4$.)

If the argument just given does not convince you, we can make its dependence on **C0**,..., **C4** explicit. Think of one choice of the cochairs, say p_1, p_2. Then there are 2^{n-2} choices for the rest of the committee, as any subset of $\{ p_3, \ldots, p_m \}$. Thus there are exactly 2^{n-2} committees with p_1 and p_2 as cochairs. The same argument shows that there are exactly 2^{n-2} different committees with any particular 2-subset $\{ p_i, p_j \}$ as cochairs. Since committees with different chairpersons are different (in the terms of this problem), we have accounted for them all, and there are $2^{n-2} + \cdots + 2^{n-2}$ (with $\binom{n}{2}$ summands), which equals $\binom{n}{2} 2^{n-2}$.

A slightly more formal way to look at the preceding argument is this: the set of all committees with p_1, p_2 as cochairs is in one-to-one correspondence with

(2.9) $$\{ \{ p_1, p_2 \} \} \times \mathcal{P}(S - \{ p_1, p_2 \}).$$

The correspondence is that the element $(\{ p_1, p_2 \}, X)$ of the Cartesian product is mapped to the committee $\{ p_1, p_2 \} \cup X$ designated as having p_1, p_2 as cochairs. Here X is a subset of $S - \{ p_1, p_2 \}$. This correspondence is obviously a bijection. For example, if $X = \{ p_3, p_4, p_5 \}$, then the committee corresponding to $(\{ p_1, p_2 \}, \{ p_3, p_4, p_5 \})$ is $\{ p_1, p_2, p_3, p_4, p_5 \}$ with p_1, p_2 as cochairs. The ordered pair $(\{ p_2, p_3 \}, \{ p_1, p_5, p_4 \})$ corresponds to the same committee but with p_2, p_3 as cochairs. Thus we have set up a bijection between the things we want to count and the union (over all 2-subsets $\{i, j\}$) of the sets

$$A_{ij} := \{\{p_i, p_j\}\} \times \mathcal{P}(S - \{p_i, p_j\}).$$

We find $|A_{ij}|$ by use of **C2** and then **C3**. The first "factor" of the Cartesian product has only one element $\{p_i, p_j\}$. The second factor has 2^{n-2} elements. The next question is, how many sets A_{ij} are there? Answer: $\binom{n}{2} = n(n-1)/2$, from **C4**. Finally, since these sets A_{ij} are mutually disjoint (because "their" cochairs make different 2-subsets of S), the number we want,

$$\left|\bigcup A_{ij}\right|,$$

where the union is taken over all 2-subsets of $\{1, \ldots, n\}$, is equal by **C1** to $\sum |A_{ij}| = \binom{n}{2} 2^{n-2}$. This finishes the third version of the "*gestalt*" approach to the problem.

The point of these approaches to the solution is that by reducing a counting problem to questions about sets explicitly related to the problem, you can sometimes make the answer crystal clear. The more experience you get, the less you will need to go to the underlying sets this way, but doing so can be helpful in complicated situations, even for the expert.

(2.10) The second approach to the problem is the "one-brick-at-a-time" method, used in Example (2.7). We start with the committee and appoint the cochairs from it. If the committee has k members, then we can make $\binom{k}{2}$ different appointments of cochairs for it. Thus for all the $\binom{n}{k}$ committees of k members we find exactly $\binom{k}{2}\binom{n}{k}$ committees with their cochairs appointed. These are all different either because the committees are different or because we made their cochairs different. Using **C1** we may now add these numbers over all k to find that our answer is

$$\sum_{0 \le k} \binom{k}{2}\binom{n}{k}.$$

If our first answer is correct (and after all that work, it had better be), then we've proved an identity, that the above sum is $\binom{n}{2} 2^{n-2}$. Multiplying by 2 gives the form of it,

$$\sum_{0 \le k} k(k-1)\binom{n}{k} = n(n-1)2^{n-2}, \tag{2.11}$$

we saw in problem 2.6 of Chapter 8.

If we use the "one-brick-at-a-time" approach to the question of counting all the subsets of an n-set, we get another identity, $\sum \binom{n}{k} = 2^n$.

Practice (2.12P) Why not use $\{p_1, p_2\}$ as the first factor of the Cartesian product in (2.9)?

We next develop some new counting results of general usefulness.

3. Counting Functions

It is easy to count the functions from a finite set to a finite set.

(3.1) **C5. The total number of functions from an a-set to a b-set is b^a. Here a and b are any integers ≥ 0.**

Example (3.2) Recall problem 1.2 of Ch. 5: List all functions from $\{a,b,c\}$ to $\{1,2\}$. The way to do this is

	a	b	c		a	b	c
f_1 :	1	1	1	f_5 :	2	1	1
f_2 :	1	1	2	f_6 :	2	1	2
f_3 :	1	2	1	f_7 :	2	2	1
f_4 :	1	2	2	f_8 :	2	2	2

The meaning of the tables is, for example,

$$f_4(a) := 1, \qquad f_4(b) := 2, \quad \text{and} \quad f_4(c) := 2.$$

These eight functions are all the possible ones because such a function is defined by the row of the 3 values that it takes at a, b, c; the possible rows are precisely the strings of length 3 over $\{1,2\}$. There are q^n strings of length n over a set of size q (Chapter 3, Problem 30). For us, $q = 2$ and $n = 3$, so $2^3 = 8$ is the total number of functions.

This example anticipates the entire proof of **C5**.

Proof of **C5** (3.3) We take our a-set A as the set of all integers from 1 to a. The standard proof of **C5** sets up a table with a columns headed $1, \ldots, a$, with each row to be filled in in all possible ways with elements of the b-set. Each row stands, along with the column-headers, for the table of values defining a function from A to B. Thus

(3.4)

1	2	3	...	a
s	t	u	...	v
w	x	y	...	z

represents the headers and two rows, where the letters $s, t, \ldots, z$ stand for elements of B. The first row represents the function f such that $f(1) = s, f(2) = t$, etc.

We then argue that there are b unrestricted choices for each entry, and that there are a entries per row, hence b^a different ways to fill in rows, hence that number of functions from A to B.

We can justify the casual and vague "hence b^a" of the prior paragraph as follows: there is a 1–1 correspondence between the rows of the table indicated in (3.4) and the elements of $B^a := B \times \cdots \times B$ with a "factors." From **C2** $|B^a| = b^a$. QED

We used **C0** twice, first in setting up a correspondence between the functions and the rows of the table, and second between the latter and the elements of B^a.

Practice (3.5P) Give the proof for **C5** when a or b is 0.

(3.6) We say again that the functions from A to B are in one-to-one correspondence with the strings of length a over B.

To summarize our proof of **C5**: we used the following 1–1 correspondences, indicated by $\longleftrightarrow$.

$$\begin{aligned} \text{Function } f : A \longrightarrow B &\longleftrightarrow \text{table defining } f \\ &\longleftrightarrow \text{string over codomain } B \\ &\longleftrightarrow \text{element of } B^a \end{aligned}$$

Example (3.7) (*i*)Suppose a cafeteria offers five different complete meals. Each of 100 students chooses one complete meal for dinner. A workaholic dietitian records next to each student's name which dinner he or she chose. How many different such lists could there be? A "list" here means all 100 names with the choice next to each name. Answer: The list is the definition by a table of a function from the 100-set of students to the 5-set of complete meals. By **C5**, the answer is 5^{100}.

(*ii*) Now suppose the cafeteria offers four-course meals, with one choice to be made from the offerings for each course. There are two choices for the first course, three for the second, four for the third, and five for the fourth course. The poor dietitian records every choice on the same kind of list. Now how many lists could there be? We use **C2**. It tells us that there are a total of $2 \cdot 3 \cdot 4 \cdot 5 = 120$ different meals to be chosen, assuming that no student refuses any course. Now we use **C5** to see that there are a total of 120^{100} different lists possible. (If we allow students to refuse any but the vegetable course (the third), then there are $3 \cdot 4 \cdot 4 \cdot 6 = 288$ different meals and 288^{100} lists possible.)

(3.8) Here is another interpretation of **C5**: A function f from A to B is a placement of the elements of A into b bins labeled with the elements of B.

This is just a concrete way of saying that the function f is completely determined by the ordered pairs $(y, f^{-1}(y))$ as y ranges over B.† The subset $f^{-1}(y)$ of A is exactly the set of elements that are placed into the bin labeled y. Some texts use such language as "How many ways can a distinct things be placed in b labeled bins so that no bins are empty?" The question asks how many surjections there are from an a-set to a b-set. (We'll answer it in (5.11).)

Example (3.9) Suppose you are having a game of pocket billiards. There are balls numbered 1 to 15, which are to be sunk in 6 pockets called A, B, C, D, E, F, respectively. At the end of the game the contents of each pocket are recorded as the outcome. How many outcomes are possible? Answer: It is the number of functions from the 15-set of balls to the 6-set of pockets. By **C5** it is $6^{15} = 4701\ 8498\ 4576 \doteq 4.7 \times 10^{11}$.

(3.10) Further comments about **C5**: We see that **C5** is a special case of **C2** because of the 1–1 correspondences we used to prove **C5**.

C5 is also a generalization of **C3**. How so? Take $B = \{0, 1\}$; thus $b = 2$. Then any function f from A to B is the characteristic function of a subset of A, namely, of $f^{-1}(1)$.

†Here we use the customary notation f^{-1} instead of the "$\widehat{f}^{-1}$" of Ch. 5. We've doffed the hat.

Conversely, every subset of A has its characteristic function $g : A \longrightarrow \{0,1\}$. We saw this 1–1 correspondence between $\mathcal{P}(A)$ and the set of all functions from A to $\{0,1\}$, $\{f; f : A \longrightarrow \{0,1\}\}$, in Ch. 5, (19). Therefore the total number of subsets of A, which is $|\mathcal{P}(A)|$, satisfies

$$|\mathcal{P}(A)| = |\{f; f : A \longrightarrow \{0,1\}\}| = 2^a \text{ by } \mathbf{C5}.$$

We've gone pretty far with just multiplication and a simple idea taken from a children's party.

(3.11) *Notation*. Sometimes the set of all functions from A to B is denoted B^A. This notation is purely symbolic, of course, because A and B are sets. Using it we may restate **C5** as

$$|B^A| = |B|^{|A|}.$$

On the right we have an integer raised to an integral power, a standard usage of the superscript notation.

Examples (3.12) (*i*)Suppose 5 students in a seminar were to choose from a list of eight topics for a report. If no two students could choose the same topic, the total number of possible choices would be the number of injections from a 5-set to an 8-set (see section (5)). Without that restriction, the number would be, by **C5**, $8^5 = 2^{15} = 32,768$.

(*ii*) (*C&C*)† "I have six letters to be delivered in different parts of town, and two boys offer their services to deliver them; in how many different ways have I the choice of sending the letters?"

Each choice is a function from the 6-set of letters to the 2-set of boys, so there are $2^6 = 64$ choices. (The value of the function at a letter is the boy who is to deliver that letter.)

(3.13) *Zero to the power zero*. What is the value, if any, of 0^0? Answer: $0^0 = 1$, at least in a combinatorial context such as this chapter. Why? It is so because of **C5**, (3.1). If a^b is the total number of functions from a b-set to an a-set, then 0^0 should be the total number of functions from $\varnothing$ to $\varnothing$. There is exactly one such function, the empty set of ordered pairs. Therefore we define

$$0^0 := 1. \tag{3.14}$$

You will see from time to time how convenient (3.14) is.

(3.15) When limits are involved the situation can be different. The following functions of the real variable x all have the form 0^0 at $x = 0$, but their limits as $x \longrightarrow 0$ are different, as indicated (log stands for the natural logarithm):

$$x^x \longrightarrow 1; \qquad x^{\frac{1}{\log x}} \longrightarrow e \quad \text{as} \quad x \longrightarrow 0+;$$

$$\left(e^{-\frac{1}{x^2}}\right)^x \longrightarrow \begin{cases} 0 & \text{as} \quad x \longrightarrow 0+ \\ \infty & \text{as} \quad x \longrightarrow 0-. \end{cases}$$

The second example is a bit cooked, because $x^{\frac{1}{\log x}} = e$ for all $x > 0$.

† In this and later chapters "(C&C)" is a citation of (10.5), *Choice and Chance*.

4. The Method of Inclusion-Exclusion

We know how to calculate $|A \cup B|$ when A and B are disjoint; we use **C1**. We now present a method for finding it when they are not disjoint, and we promptly generalize it from two to any finite number of sets.

First, for $|A \cup B|$, consider the Venn diagram

(4.1*F*)

A B

It is clear that

(4.2) $$|A| + |B| = |A \cup B| + |A \cap B|,$$

since every point of $A \cap B$ is counted twice on the left—once in $|A|$ and once in $|B|$—whereas the other points of $A \cup B$ are counted just once on each side of (4.2). (Remember: $|A \cup B|$ counts each point in the union just once.) Therefore,

(4.3) $$|A \cup B| = |A| + |B| - |A \cap B|.$$

Notice that this generalizes **C1** to arbitrary sets. This result (4.3) for two sets is the key to a proof by induction of the general result, which we now state and prove.

(4.4) **Let $m \in \mathbf{N}$, and set $T := \{1, \ldots, m\}$. Then**

C6 $$\left|A_1 \cup \cdots \cup A_m\right| = -\sum_{\emptyset \neq S \subseteq T} \left|\bigcap_{i \in S} A_i\right| (-1)^{|S|}. \blacksquare$$

The notation is similar to that used in the proof of Theorem (6.2) of Chapter 8, the counting property of binomial coefficients. For $m = 3$, for instance, **C6** says

(4.5) $$|A_1 \cup A_2 \cup A_3| = \sum_{1 \le i \le 3} |A_i| - \left(\sum_{1 \le i < j \le 3} |A_i \cap A_j|\right) + |A_1 \cap A_2 \cap A_3|$$

$$= |A_1| + |A_2| + |A_3| - |A_1 \cap A_2| - |A_1 \cap A_3| - |A_2 \cap A_3| + |A_1 \cap A_2 \cap A_3|.$$

In particular, $\bigcap_{i \in S} A_i := A_j$ when S is a singleton $\{j\}$.

***Proof of* C6** By induction. Let $m = 1$. Then **C6** says simply $|A_1| = |A_1|$. Assuming **C6** for all unions of m sets for a fixed but arbitrary m, consider the union of $m + 1$ sets as $A \cup B$, with $A = A_1 \cup \cdots \cup A_m$ and $B = A_{m+1}$. Now the value of $|A|$ is given by the induction assumption. Using **C6** for $m = 2$, which we proved at the beginning, as (4.3), we see that $|A \cup B|$ is known except for the term $|A \cap B|$. By a distributive law, $A \cap B$ is the set

$$(A_1 \cap B) \cup \cdots \cup (A_m \cap B).$$

It is the union of m sets $B_i := A_i \cap B$, so we may apply the inductive assumption to it. When we do so, and notice that, for example, $B_1 \cap B_2 \cap B_3 = B \cap A_1 \cap A_2 \cap A_3$, we see that the case for $m = 2$ together with the inductive assumption yields **C6** for $m + 1$ sets. QED

You will need to fill in computational details omitted from the proof of **C6**. They are routine enough, but messy to write. The more advanced you get, the more you will be expected to fill in such gaps. (If necessary, start with $m = 3$.)

Example (4.6) A computerized sandwich machine makes sandwiches to order. Normally it keeps good records, but one day its totaling subprogram went bad. Since it did use 25 portions of salami, it must have made that many sandwiches with salami in them. Similarly, auditors were able to infer that it had made 20 with mustard, and 15 with pickles. It did record that it had made 12 sandwiches containing salami and mustard, seven containing salami and pickles, and four containing mustard and pickles. These numbers were vague, however, because some of the 12 with salami & mustard may have also had pickles. The case is similar for the other numbers: Some of the 25 with salami had other ingredients, and some did not, for example. Fortunately, the machine recorded two more numbers: There were exactly three sandwiches made with all three ingredients, and there were four "nothing" sandwiches (just two slices of bread) sold. Thus auditors could use inclusion-exclusion to find the total number T of sandwiches made by the machine.

They set $S :=$ the set of all sandwiches with salami, $M :=$ the set with mustard, and $P :=$ the set with pickles. Then they observed that

$$T - 4 = |S \cup M \cup P|.$$

The right-hand side equals $25 + 20 + 15 - (12 + 7 + 4) + 3 = 40$, by (4.5). Therefore $T = 44$.

Example (4.7) How many strings of length 9 over $\{0, 1\}$ are there that contain 10101 as a substring (in five consecutive places)?

Every such string is in one *or more* of the sets $A_1, \ldots, A_5$, where

$$\begin{array}{l}
A_1 = \{\, 1 \;\; 0 \;\; 1 \;\; 0 \;\; 1 \;\; \square \;\; \square \;\; \square \;\; \square \,\} \\
A_2 = \{\, \square \;\; 1 \;\; 0 \;\; 1 \;\; 0 \;\; 1 \;\; \square \;\; \square \;\; \square \,\} \\
A_3 = \{\, \square \;\; \square \;\; 1 \;\; 0 \;\; 1 \;\; 0 \;\; 1 \;\; \square \;\; \square \,\} \\
A_4 = \{\, \square \;\; \square \;\; \square \;\; 1 \;\; 0 \;\; 1 \;\; 0 \;\; 1 \;\; \square \,\} \\
A_5 = \{\, \square \;\; \square \;\; \square \;\; \square \;\; 1 \;\; 0 \;\; 1 \;\; 0 \;\; 1 \,\},
\end{array}$$

and the boxes are to be filled arbitrarily and independently with 0's or 1's.

The number we seek is $|A_1 \cup \cdots \cup A_5|$. In order to use inclusion-exclusion, **C6**, we need to calculate $|A_i|$, $|A_i \cap A_j|$, $|A_i \cap A_j \cap A_k|$, and so on. It is easy to see that

$$|A_i| = 16 \quad \text{for all } i = 1, \ldots, 5$$

since there are four places to be filled with 0's or 1's, and $2^4 = 16$. (To spell it all out, A_i is in 1–1 correspondence with the set of all strings of length 4, so by **C0** and **C2**, $|A_i| = 2^4 = 16$.)

For the next case, consider $A_1 \cap A_2$. This set is empty because all strings in A_1 have 0 as the second symbol, whereas all strings in A_2 have 1 as the second symbol. Similarly, we see that

$$A_i \cap A_{i+1} = \varnothing \quad \text{for} \quad i = 1,\ldots,4. \tag{4.8}$$

Now consider $A_1 \cap A_3$. This set is not empty because we may choose the first two symbols for a string in A_3 to be 10; we get

$$1\,0\,1\,0\,1\,0\,1\,\square\,\square, \tag{4.9}$$

which for any fill-in of the last two boxes is in $A_1 \cap A_3$. Conversely, any string in $A_1 \cap A_3$ must begin 10 since it's in A_1; and it must have 10101 for its next five symbols. Therefore, $A_1 \cap A_3$ is exactly the set of all strings of the form (4.9). Thus $|A_1 \cap A_3| = 4$.

Similarly we see that

$$|A_i \cap A_{i+2}| = 4 \quad \text{for} \quad i = 1,2,3. \tag{4.10}$$

Furthermore, it is obvious that

$$A_1 \cap A_5 = \{\, 101010101 \,\}, \tag{4.11}$$

so $|A_1 \cap A_5| = 1$.

It is easy to see that the remaining double intersections are empty:

$$A_1 \cap A_4 = A_2 \cap A_5 = \varnothing.$$

Now we will deal with the triple intersections. It is clear that

$$A_1 \cap A_3 \cap A_5 = A_1 \cap A_5 \tag{4.12}$$

and, from (4.8), that all other triple intersections are empty. That's because any 3-subset of $\{\,1,\ldots,5\,\}$ other than $\{\,1,3,5\,\}$ has two consecutive integers in it.

It follows that all quadruple intersections are empty and that $A_1 \cap A_2 \cap \cdots \cap A_5 = \varnothing$ also.

We can now find our answer. Using **C6** with $m = 5$ we get

$$|A_1 \cup \cdots \cup A_5| = 5 \cdot 16 - (4 + 4 + 4 + 1) + 1 = 68.$$

That is, the right-hand side is

$$\sum_{1 \le i \le 5} |A_i| - (|A_1 \cap A_3| + |A_2 \cap A_4| + |A_3 \cap A_5| + |A_1 \cap A_5|) + |A_1 \cap A_3 \cap A_5|;$$

we omitted the empty intersections.

(4.13) Now we can answer the question at the end of (1.2): How did John know there were 1003 ballots voted for a candidate from each group? John must have known inclusion-exclusion. If we define A as the set of all ballots with at least one vote for a candidate from Group A, and define similar sets for Groups B and C, then the data tallied by the political science students are the values of $|A|$, $|B|$, $|C|$, $|A \cap B|$, $|A \cap C|$, and $|B \cap C|$. Since we were also given the value of $|A \cup B \cup C|$ (as $3100 - 97$), we can solve for $|A \cap B \cap C|$ in (4.5).

To see that $|A \cap \overline{B} \cap \overline{C}| = 96$ we could use (4.3) on the two sets $A \cap B$ and $A \cap C$. That gives us

$$\begin{aligned} |(A \cap B) \cup (A \cap C)| &= |A \cap B| + |A \cap C| - |A \cap B \cap C| \\ &= 1614 + 1375 - 1003 \\ &= 1986. \end{aligned}$$

Since $(A \cap B) \cup (A \cap C) = A \cap (B \cup C)$, we see from **C1** that $A \cap \overline{B} \cap \overline{C} = |A| - 1986 = 96$.

Another way, which John used, is to fill in numbers in the Venn diagram of A, B, and C. It's almost the same as before. We start with 1003 in $A \cap B \cap C$.

(4.13*F*)

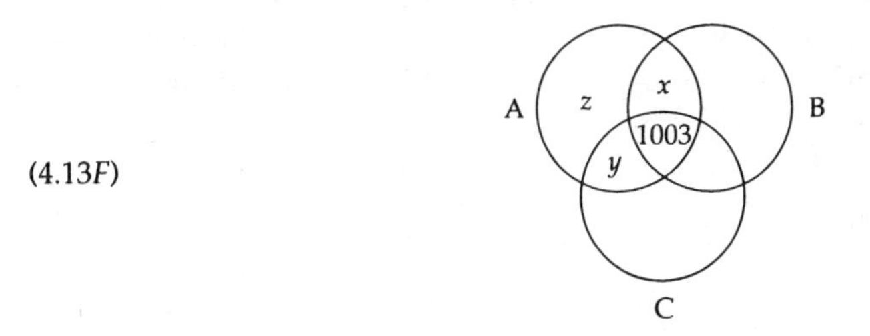

Since $|A \cap B| = 1614$, we know that $|A \cap B \cap \overline{C}| = 1614 - 1003 = 611 =: x$. Similarly, $|A \cap \overline{B} \cap C| = 1375 - 1003 = 372 =: y$. Therefore we know $z := |A \cap \overline{B} \cap \overline{C}| = |A| - x - y - 1003 = 96$.

Practice (4.14P) Find the sizes of the other 3 regions.

Later in this chapter, we'll see that inclusion-exclusion ("**in-ex**," as it's called for short) can be used to derive formulas as well as mere numbers.

5. Bijections, Injections, and Surjections

In this section we'll count the number of functions with the indicated property that map a k-set to an n-set.

(5.1) *Bijections.* If there is a bijection between two finite sets, then the sets have the same number of points. That is **C0**. So let $n \geq 0$ be an integer. Then

(5.2) **C7. The total number of bijections from an n-set to an n-set is $n!$**

Recall the definition of $n!$. It is

(5.3)
$$\begin{aligned} & 0! := 1 \\ \forall n \geq 1 \quad & n! := (n-1)! \cdot n. \end{aligned}$$

Proof of C7 We have to use induction. For $n = 0$ there is truly just one bijection, the empty set of ordered pairs; and 0! is indeed defined to be 1. Assuming the result for fixed but arbitrary $n \geq 0$, we consider a set A of $n+1$ points, say $1, \ldots, n+1$. We let B be any $(n+1)$-set; how many ways can we define a bijection $f : A \to B$? We define f at $n+1$ to be any of the $n+1$ values in B; but now the remaining assignment of f at $1, \ldots, n$ is restricted to the set $B - \{f(n+1)\}$. This has

n points. Therefore we may apply the induction assumption to see that there are $n!$ ways to define f for each choice of $f(n+1)$. We may now argue that the total number of bijections from A to B is $n! + \cdots + n!$ (with $n+1$ summands), which equals $n!(n+1)$, which is $(n+1)!$. Thus the result **C7** is true. QED

If we recall the definition of permutation as a bijection from a set to itself, we have an immediate corollary of **C7**, namely

(5.4) **C7′. The total number of permutations of an n-set is $n!$ for all integers $n \geq 0$.**

(5.5) We may also interpret the permutations of an n-set as the set of all strings of length n over the set with no "letter" (:= element of the set) repeated. That is because a permutation of the n-set X corresponds to an element of the Cartesian product X^n having no two coordinate-values equal.

Example (5.6) Suppose in a seminar of four students (A, B, C, and D) each one speaks once each session. There are $24 = 4!$ possible orders in which they could speak:

ABCD	*BACD*	*CABD*	*DABC*
ABDC	...	...	...
ACBD	...	...	...
ACDB	...	...	...
ADBC	...	...	...
ADCB	...	...	...

Practice (5.7P) Complete the above listing.

(5.8) *Injections*. Now we consider the problem of counting injections.

C8. The total number of injections from a k-set to an n-set is $n(n-1)\cdots(n-k+1)$. (This function is 1 by definition if $k = 0$.)

Proof (5.9) To prove **C8** we can use earlier results. Since an injection is a bijection between its domain and range, the range is a k-subset of the n-set, and we can choose it in $\binom{n}{k}$ ways.

(5.9*F*)

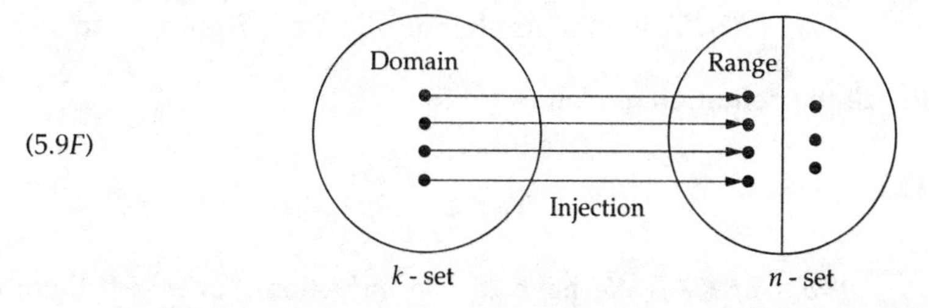

If we choose any one function f with a given range, we can get all functions having that range by composing f with all the $k!$ permutations of the range. Thus we get $k!\binom{n}{k}$ injections altogether. We used **C4**, **C7′**, and **C2**. Since we know $\binom{n}{k} = n(n-1)\cdots(n-k+1)/k!$ from the binomial theorem, Chapter 8, (7.2), we have proved **C8**. QED

The above paragraph is just a formal way of saying the usual proof, which counts the number of ways to fill in the table of values of the injection as n for the first entry, $n-1$ for the second (anything different from the first choice), $n-2$ for the third (anything different from the previous choices), and so on down to $n-k+1$ choices for the k-th and last entry. We then multiply the numbers of values for the entries together.

Notice that the formula in **C8** gives 0 when $k > n$. When $k = n$, **C8** becomes **C7** because an injection from an n-set to an n-set is automatically a surjection.

In other texts, an injection from $\{1, \ldots, k\}$ to an n-set is sometimes called a *permutation of n things taken k at a time, or a k-permutation of n things.* These names come from identifying the injection with its list of values on $1, \ldots, k$.

These results **C1** through **C8** are simple enough to prove. The tricky part can be to know which ones to apply in concrete problems. For example, consider the famous **birthday problem.**

(5.10) In a set of n people chosen at random, what is the probability that two or more of them have the same birthday?

We take "at random" to mean that all birthdays are equally likely, and we assume for simplicity that there is no February 29. Thus we consider a set B of 365 birthdays. A set of n people could have any one of 365^n "birthday functions"; i.e., for the n-set P of people, the function $f : P \to B$ defined by the rule

$$\text{For all } x \in P, \quad f(x) \text{ is the birthday of } x$$

is the birthday function for P. Our randomness assumption is that any one of these 365^n possible functions is as likely as any other. We want to know how many of these functions are NOT injections. The total number of injections from P to B is given by **C8**. It is $365(365-1)\cdots(365-n+1)$. Thus the total number of noninjections is

$$365^n - 365(365-1)\cdots(365-n+1).$$

Dividing this by 365^n gives us the probability we want:

$$1 - \frac{365(365-1)\cdots(365-n+1)}{365^n}. \tag{5.11}$$

The astonishing part of this result is that for n as small as 23, the probability that at least two of the 23 people have the same birthday is greater than 0.5. With 50 people the probability is 0.97+; with 100 it is 0.9999997−.

Practice (5.12P) Calculate (5.11) for $n = 23$.

(5.13) *Surjections.* How many surjections are there from a k-set to an n-set? The answer is

C9. The total number of surjections from a k-set to an n-set is

$$\sum_{0 \le j \le n} (-1)^j \binom{n}{j} (n-j)^k.$$

Proof We prove **C9** by using inclusion-exclusion, **C6**. Let X denote the k-set, the domain. Let the codomain Y be $\{1,2,\ldots,n\}$. Our strategy here will be similar to that in the birthday problem: Since we can easily count all the functions from X to Y by **C5**, we'll count up those we don't want (the nonsurjections) and subtract them.

Let's denote the set of all functions (from X to Y) *not* taking the value 1 by A_1. Similarly for $A_2,\ldots,A_n$. Since a nonsurjection fails to take one or more of the values $1,\ldots,n$, the set of all nonsurjections is

$$A_1 \cup A_2 \cup \cdots \cup A_n. \tag{5.14}$$

Our object is to count this set. To do so we introduce a more general notation.

If U is any subset of Y, let A_U denote the set of all functions f from X to Y such that f does not take any value in U, i.e.,

$$\begin{aligned} &\text{range of } f \subseteq Y - U; \\ A_U &:= \{f; f : X \longrightarrow Y, f(X) \cap U = \varnothing\}; \\ A_U &:= \{f; f : X \longrightarrow Y - U\}. \end{aligned} \tag{5.15}$$

Notice that the number of functions in A_U is given by **C5**:

$$|A_U| = |Y - U|^k = (n - |U|)^k. \tag{5.16}$$

To use "in-ex" we'll need to know the sizes of various intersections of the sets $A_1, A_2, \ldots, A_n$. It's easy to find the answer, because, for example,

$$A_1 \cap A_2 = A_{\{1,2\}}$$

in the notation of (5.15). To see this, notice that on the left we have the set of all functions not hitting 1 *and* not hitting 2—but that is the definition of $A_{\{1,2\}}$.

The same result holds for any number of the A_i's: to be fancy, let U be any subset of Y; then

$$\bigcap_{i \in U} A_i = A_U. \tag{5.17}$$

Practice (5.18P) Prove (5.17) by an epsilon-argument.

So now we can find the size of (5.14). We get

$$|A_1 \cup \cdots \cup A_n| = \sum_{1 \le i \le n} |A_i| - \sum_{1 \le i < j \le n} |A_i \cap A_j| + - \cdots$$

from **C6**, and from (5.17) and (5.16) we can calculate the value of every term on the right-hand side. We get

$$\begin{aligned} |A_1 \cup \cdots \cup A_n| &= \binom{n}{1}(n-1)^k - \binom{n}{2}(n-2)^k + - \cdots \\ &= \sum_{1 \le j \le n} (-1)^{j-1} \binom{n}{j} (n-j)^k. \end{aligned}$$

This is what we must subtract from the total number of functions from X to Y, namely, n^k. Thus, the total number of surjections from a k-set to an n-set is

$$n^k - \sum_{1 \le j \le n} (-1)^{j-1} \binom{n}{j} (n-j)^k,$$

which can be more elegantly written as in **C9**. QED

Here was a more serious use of in-ex.

REMARK (5.19) Our proof of the formula in **C9** assumes only that k and n are nonnegative integers, so it is correct even for $k < n$, when there are no surjections. Thus it has the value 0 when $k < n$. This fact might not be obvious if we look at the formula without knowing what it counts, but we could prove it just using simple properties of binomial coefficients. We'll mention it in Chapter 11.

Example (5.20) There are 10 people, all with driver's licenses and able to drive. They are to deliver five cars from point A to point B. How many ways may they group themselves into the cars? (A "way" is a list of who goes in what car. The cars are labeled, say, by their license plates. Who drives, if two or more are in a car, is not to be recorded on the list.)

Answer. Since every car must have a driver, the question asks for the number of surjections from a 10-set to a 5-set. The answer, from **C9** with $k = 10$ and $n = 5$, is

$$\sum_{0 \le j \le 5} (-1)^j \binom{5}{j} (5-j)^{10} = 5,103,000.$$

The total number of functions from a 10-set to a 5-set is $5^{10} = 9,765,625$, so we see that (for these numbers) well more than half are surjective.

Practice (5.21P) Show that the formula in **C9** gives 0 when $n = 0$ and $k > 0$. Show also that for $n = k = 0$ the formula has the value 1 and that 1 is correct.

6. Partitions

We can use the preceding result **C9** to get information about the number of partitions of a set. Our first result is

C10. Let k and n be integers ≥ 0. The total number of partitions of a k-set into n cells is

$$\frac{1}{n!} \sum_{0 \le j \le n} (-1)^j \binom{n}{j} (n-j)^k. \tag{6.1}$$

Proof This result is a simple corollary of **C9**. **C10** claims that the number of partitions we seek is $s/n!$, where s is the number of surjections from a k-set to an n-set. Here is the connection: let $f : X \longrightarrow Y$ be any such surjection, where $Y = \{1, \ldots, n\}$

and X is a k-set. Consider the sets $f^{-1}(1), f^{-1}(2), \ldots, f^{-1}(n)$. (Recall the definition from (9) of Chapter 5 and note that, as promised, we have doffed the "hat":

(6.2) $$f^{-1}(i) := \{x;\ x \in X,\ f(x) = i\}.)$$

Each of these n sets $f^{-1}(i)$ is nonempty, and they are the cells of a partition of X (see Chapter 5, (9.5) and problem 9.12 (i)).

Consider the Venn diagram for $n = 4$:

(6.2F)

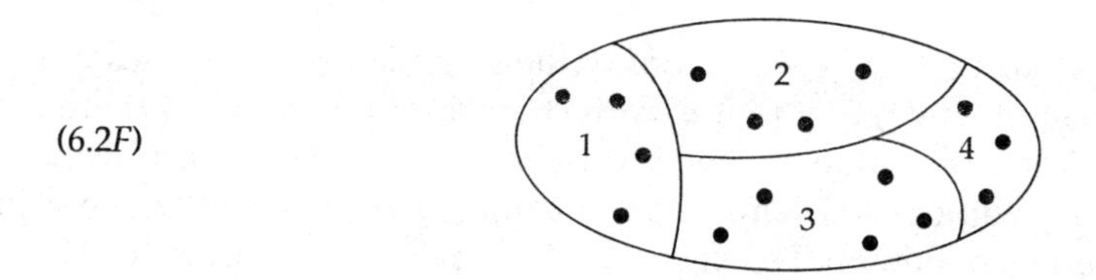

We have labeled the four cells according to the value f takes on them. There are many other surjective functions with the same set of cells, however. They arise when we permute the labels 1, 2, 3, 4 among the four cells. Thus, we could tabulate the functions as follows. Let the 4 cells be $C_1, \ldots, C_4$.

	C_1	C_2	C_3	C_4
f	1	2	3	4
g	1	3	4	2
h	2	4	1	3
⋮				

The table means that at each point of C_1, f takes the value 1, g takes the value 1, and h takes the value 2, and so on. You can see that each of the 4! permutations of $Y = \{1,2,3,4\}$ yields a function with the same partition $\{C_1, \ldots, C_4\}$ of X defined as in (6.2).

The same argument works for general n to show that the surjections from X to Y may be grouped into sets of $n!$ functions, each inducing the same partition (6.2) of X into n cells. Therefore the number of partitions of X into n cells is the number of surjections from X to Y, divided by $n!$ QED

Corollary (6.3) The number of surjections from a k-set to an n-set is divisible by $n!$

Example (6.4) In Example (5.20) we found that the number of surjections from a 10-set to a 5-set was 5,103,000. Since $5! = 120$, there are $5{,}103{,}000/120 = 42{,}525$ partitions of a 10-set into five cells. Thus there are 42,525 ways for the 10 people to split up before choosing which cell goes in which car.

(6.5) *Notation*. The number of partitions of a k-set into n cells is called a **Stirling number of the second kind**. It is denoted variously as

$$\left\{ {k \atop n} \right\} \quad \text{or} \quad S(k,n).$$

(6.6) The total number B_k of partitions of a k-set is

$$B_k = \sum_{0\le n} \left\{ {k \atop n} \right\} = \sum_{0\le n} \frac{1}{n!} \sum_{0\le j\le n} (-1)^j \binom{n}{j} (n-j)^k.$$

Although the **Bell number** B_k is the coefficient of $x^k/k!$ in e^{e^x-1}, as we said in Chapter 4, we do not prove that here. (See, for example, (10.1) for a proof.) In this sum, and in the sum in **C10**, (6.1), we could let the sum run over all $j \ge 0$, because $\binom{n}{j}$ is 0 for $j > n$. We could also limit the sum over n to $0 \le n \le k$, since, as we remarked in (5.19), the inner sum in (6.6) is 0 for $n > k$.

The Stirling numbers of the second kind satisfy a recursion that we state in the problems. ◀

7. Sets with Repetition

Now we consider the problem of counting the k-subsets of an n-set with repetitions allowed—whatever that means. After all, repetitions of elements "don't count" in sets, so what can this mean? It means, for example, that if our set is $\{a,b,c,d,e\}$, then $\{a,a,a,b\}$ is supposed to be regarded as a "4-subset with repetition." Of course, in our notation it stands for the 2-subset $\{a,b\}$, so we really need a different notation for it. Since there is no standard notation, let's invent one. We'll use brackets:

(7.1) $[a,a,a,b]$ is the set with repetition consisting of three a's and one b.

It's possible to define this new object in terms of previously defined ideas, however. Let X be a set. We define a **k-subset of X with repetition** in terms of functions f from X to $\mathbf{N}_0$, the set of all nonnegative integers. We restrict f as follows:

(7.2) $$\sum_{x\in X} f(x) = k.$$

Each such function f corresponds to the k-subset of X with repetition allowed in which for each $x \in X$, x occurs $f(x)$ times.

For $X = \{a,b,c,d,e\}$, $[a,a,a,b]$ corresponds to the function with table

x	a	b	c	d	e
$f(x)$	3	1	0	0	0

Practice (7.3P) List all 3-subsets with repetitions allowed of $\{a,b,c,d\}$.

If X is an n-set, then the total number of k-subsets of X with repetition is the number of solutions of the following equation:

(7.4) $$y_1 + y_2 + \cdots + y_n = k$$

with $y_1, y_2, \ldots, y_n \in \mathbf{N}_0$. The y_i's are the values of f at the points a_i of X. Next we find a formula for this number.

Instead of presenting the answer to the question: "How many k-subsets with repetition are there from an n-set?" and then verifying it, I prefer to take a straightforward "let x be the unknown" approach, from which you can learn some useful techniques. Thus, let $s(n,k)$ be the number we seek. We will get a recurrence relation for $s(n,k)$.

Suppose we focus on one element, a, of a k-subset with repetition (called a **multiset**, but we'll call it a **k-rep** for short from now on), and choose its frequency of occurrence, j, in the "k-rep" we are going to construct. Thus, j might be any integer between 0 and k — 0 if a does not appear in the k-rep at all. We fill out our k-rep by choosing a $(k-j)$-rep from the remaining $n-1$ points. If we do this for all possible j, we get all possible k-reps. Therefore, for all $n \geq 1$ and all $k \geq 0$

$$s(n,k) = \sum_{0 \leq j \leq k} s(n-1, k-j). \tag{7.5}$$

At this point we have no idea what the function s is, so let's tabulate it for small n and k and hope that lightning strikes. We can argue before we start that $s(n,0) = 1$ for all $n \geq 0$ because every set has exactly one empty subset. And, again directly from the definition, $s(0,k) = 0$ for all $k \geq 1$. Thus, we have reasoned out what the first row and column of the table must be. With that and the recursion (7.5), we can get the rest. The recursion says that each entry in the table is the sum of all the entries in the row just above starting from the left-hand edge and going over to the entry directly above the one being calculated. So we compute several values.

$n \backslash k$	0	1	2	3	4	5	6	...
0	1	0	0	0	0	0	0	
1	1	1	1	1	1	1	1	
2	1	2	3	4	5	6	7	
3	1	3	6	10	15	21	28	
4	1	4	10	20	35	56	84	
5	1	5	15	35	70	126	210	
6	1	6	21	56	126			
7	1	7	28	84				
8	1	8	36					
9	1	9						

$s(n,k)$

Well, well! We seem to have struck pay dirt. If we ignore the row $n = 0$ and tip the matrix 45° clockwise, we get the Pascal triangle! A modest bit of analytic geometry allows us to see that $s(n,k)$ is $\binom{N}{K}$ for $N = n+k-1$ and $K = k$. This is true because the "N-rows" of the Pascal triangle after the 45° rotation come from the lines $n + k = \textit{constant}$ in the table. The k-files are the columns.

So we have shown that $s(n,k) = \binom{n+k-1}{k}$ for the table-entries that we've calculated; we now prove it for all n, k by induction. Since $\binom{k-1}{k} = 1$ for $k = 0$ and 0 for $k \geq 1$, we've already verified our assertion for $n = 0$. Omitting the induction boilerplate for once, we need only show that for all $n \geq 0$

$$\binom{n+k}{k} = \sum_{0 \leq j \leq k} \binom{(n+k-1)-j}{k-j}, \tag{7.6}$$

But (7.6) is just (8.2) of Chapter 8, the knight's-move identity! That's obvious once you tip the table because the recursion (7.4) for s is "geometrically" just the knight's-move identity: It was (7.6.1F), and it becomes (7.6.2F).

(7.6.1F)

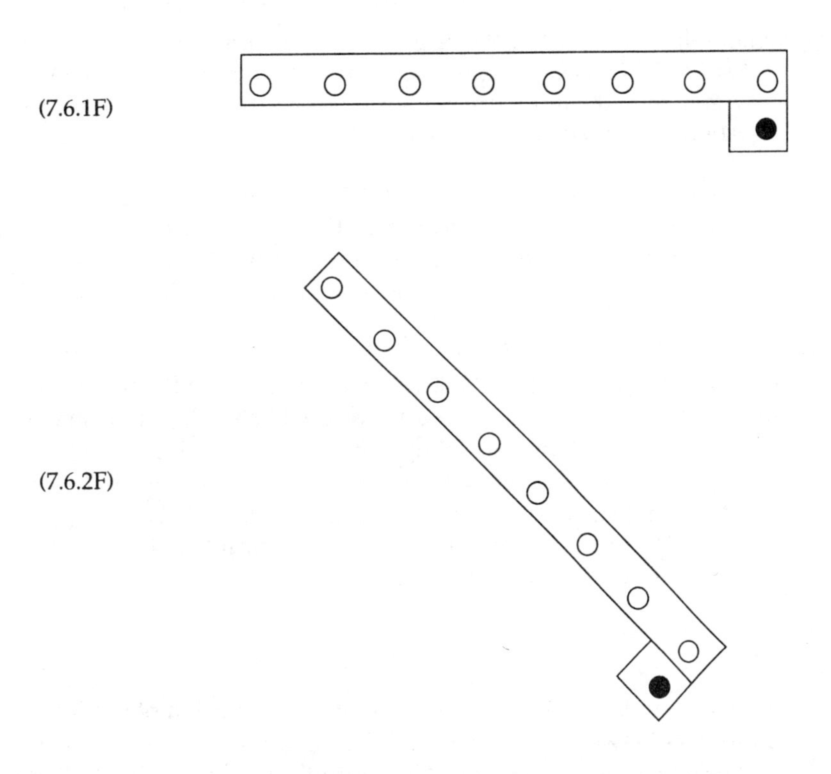

(7.6.2F)

Thus we have proved by induction that

C11. The total number of "k-subsets with repetition allowed" of an n-set is, for all $n, k \geq 0$,

(7.7)
$$s(n,k) = \binom{n+k-1}{k}.$$

Proof (7.8) *Second proof of* **C11**. Now that we know the answer we can give a direct, combinatorial proof, using **C0**. We set up a bijection between the k-reps of an n-set and the k-subsets of a certain $(n+k-1)$-set. For the n-set we take $\{1,\ldots,n\} = X$. Now let $R = \{R_1, R_2, \ldots, R_{k-1}\}$ be a set of $k-1$ different elements, and define $Y = X \cup R$. Elements of R will be our "repeaters." $|Y| = n+k-1$. We now set up a bijection β from the k-subsets of Y to the k-reps of X. Let K be any k-subset of Y. If $K \subseteq X$, then K is just a k-subset of X, and we set $\beta(K) = K$. But if $K = \{x_1, x_2, \ldots, x_j, R_{i_1}, \ldots, R_{i_{k-j}}\}$ with $j < k$, then we assume (as we may)

$$x_1 < x_2 < \cdots < x_j \quad \text{and} \quad i_1 < i_2 < \cdots < i_{k-j}.$$

Now we define $\beta(K)$ in a series of steps, one for each repeater in K, as follows:

$$R_{i_1} \quad \text{repeats} \quad x_{i_1}.$$

$$\text{Result:} \quad [y_1, y_2, \ldots, y_{j+1}, R_{i_2}, \ldots, R_{i_{k-j}}],$$

where the y's consist of the original x's but with x_{i_1} appearing twice. The process is repeated: R_{i_2} is used to repeat y_{i_2}, and so on until the R's have been all used up. At all times keep the numbers in their natural order. It is not hard to verify that β is a bijection, but we omit the details.

Example (7.9) $X = \{1, \ldots, 5\}$ and $k = 6$. Then $\beta(\{1, 3, 4, 5, R_2, R_4\}) = [1, 3, 3, 4, 4, 5]$, because our procedure calls for us first to repeat the second of the x's, namely 3, with result $[1, 3, 3, 4, 5, R_4]$. Now we are to repeat the fourth of the numbers in the string. Another example: Suppose we wanted to achieve $[1, 2, 2, 2, 5, 5]$. That is $\beta(\{1, 2, 5, R_2, R_3, R_5\})$.

Proof (7.10) *Third proof of* **C11**. Saving the best for last, we now give the traditional proof of **C11**. It relies on the symmetry of binomial coefficients. Consider the set $I := \{1, 2, \ldots, n + k - 1\}$. We establish a bijection between the set of all $(n-1)$-subsets of I and the set of all k-reps of $J := \{1, 2, \ldots, n\}$.

An example best conveys this idea. Suppose $n = 8$ and $k = 5$. Since $n - 1 = 7$, we choose a 7-subset of $I = \{1, \ldots, 12\}$, say $\{2, 3, 6, 7, 8, 10, 12\}$.

Number 12 places in a row and put vertical bars in the places numbered with elements of the chosen 7-set. Thus we get

$$\underset{1}{\underline{\ \ }}\quad \underset{2}{\underline{|}}\quad \underset{3}{\underline{|}}\quad \underset{4}{\underline{\ \ }}\quad \underset{5}{\underline{\ \ }}\quad \underset{6}{\underline{|}}\quad \underset{7}{\underline{|}}\quad \underset{8}{\underline{|}}\quad \underset{9}{\underline{\ \ }}\quad \underset{10}{\underline{|}}\quad \underset{11}{\underline{\ \ \ }}\quad \underset{12}{\underline{|}}$$

in this example. Put 1's on all the underlines to the left of the first divider, 2's on the underlines between the 1st and 2nd dividers, and so on, with 8's on all the underlines to the right of the last divider. This example gives us $[1, 3, 3, 6, 7]$. It has, for instance, no "2's" because there are no underlines between the first and second dividers. It should be reasonably clear that this process is the desired bijection. A more formal description and proof follow, however.

Let $\{x_1, \ldots, x_{n-1}\}$ be any $(n-1)$-subset of I with $x_1 < x_2 < \cdots < x_{n-1}$. We form a k-rep of $x_1 - 1$ "1's", $x_2 - x_1 - 1$ "2's", $x_3 - x_2 - 1$ "3's", ..., and finally $n + k - 1 - x_{n-1}$ "n's". The number of elements is k, so it is a k-rep from J.

Different $(n-1)$-subsets of I produce different strings, so the mapping is an injection.

Practice (7.11P)

(*i*) Write a formula for the number of occurrences of i for $1 \le i \le n$. Explain it. Prove that the length of the string is k.

(*ii*) Prove the injectivity.

To see that it is surjective, suppose our target k-rep has $a(j)$ j's for each $j \in J$. Set $x_1 = a(1) + 1$, and so on. In other words, reverse the procedure defining the mapping.

QED

Practice (7.12P) Prove the surjectivity in detail.

Example (7.13) At the cafeteria each student now chooses six items. There are still four courses, with the same two choices for the first, three for the second, four for the third, and five for the fourth; but the student now chooses any number of servings of any course or courses totaling six. How many different dinners are there in principle? Answer: Now there are just $2 + 3 + 4 + 5 = 14$ items from which to choose a 6-subset with repetition. By **C11** the answer is $\binom{19}{6} = 27{,}132$.

Example (7.14) Consider a function f of n variables. How many formally different partial derivatives of order k does it have? We assume that the order of differentiation does not matter, that $D_1D_2f = D_2D_1f$ for any two differential operators D_1 and D_2 (That is, D_1 is partial differentiation with respect to one variable, and D_2 is the same thing for another variable.) A derivative of f is "of order k" if it is the result of applying exactly k of these differential operators to f. Some may be repeated. For example, if $n = 3$ and $k = 7$ we might have

$$D_1D_2^4D_3^2f, \quad \text{or} \quad D_1^7f.$$

Clearly, the number of partial derivatives of order k is the number of k-reps of the n-set of differential operators. By **C11** it is $\binom{n+k-1}{k}$. In the small example just cited it is $\binom{9}{7} = \binom{9}{2} = 9 \cdot 8/2 = 36$.

Example (7.15) Here is an example that emphasizes the difference between functions and k-reps. Suppose five guests are offered any of three wines. Five identical glasses are available. Their order for the first round can be any 5-rep from the 3-set of wines, for a total of $\binom{5+3-1}{5} = \binom{7}{2} = 7 \cdot 6/2 = 21$ possible orders. For example, two people might order the red wine, one the white, and two the rosé. It doesn't matter what glasses the wines are poured into for the first round: Those who ordered red will take the glasses with red wine, and it doesn't matter which of the two they take. But for the second round, things change because we assume that they will use the same glasses again. They are free to change their order, too. Now an order is a function from the 5-set of people to the 3-set of wines! Why? Suppose the people are A, B, C, D, E; list their order for the second round as a set of ordered pairs:

$$\{(A,\ a),\ (B,\ b),\ (C,\ c),\ (D,\ d),\ (E,\ e)\}$$

Here a denotes the wine ordered by A, and so on. This function maps A to a, B to b, and so forth, and $\{a, b, \ldots, e\} \subseteq \{\text{red, white, rosé}\}$.

Thus, there are $3^5 = 243$ different possible orders for the second round. What's happening is that, in the interest of sanitation, each glass for the second round now belongs to a definite person. If the barkeep heard "One red, three whites, and one rosé" she couldn't pour that order until she learned which wine was to go into which glass.

Practice (7.16P) How many different second orders for two reds, one white, and two rosés could there be?

(7.17) Another way to state problems involving k-reps is in terms of balls in cells. Suppose you have k identical balls to be distributed into n labeled cells. The total number of ways to distribute the balls is the number of k-reps of an n-set. Why? Because if you label the cells $1, \ldots, n$, then a distribution corresponds to a set of pairs $(j, a(j))$, where j is the label on the cell and $a(j)$ is the number of balls in that cell. In this situation $a(j)$ lies between 0 and k, and $\sum a(j) = k$. This is the "model" introduced in (7.2). If the balls were labeled, i.e., distinguishable, then a distribution of the balls into the cells would be a function D.

Practice (7.18P) What are the domain and codomain of D?

(7.19) To sum up, we've seen three problems, all with the same solution $\binom{n+k-1}{k}$:

(*i*) How many k-reps are there from an n-set?

(*ii*) (See (7.4).) How many solutions in nonnegative integers y_i are there to the equation

$$y_1 + \cdots + y_n = k?$$

(*iii*) (See (7.17).) How many distributions are there of k identical balls into n labeled cells?

(7.20) Now we can answer the question of (1.10): How did John know there were two identical ballots? It's probably been obvious for most of this section that the answer is based on k-reps. The rules of this election allow the voter to choose any k-rep of the 10-set of candidates for $k = 0, 1, \ldots, 5$. The choice $k = 0$ results in a blank ballot; the choice $k = 2$ means three votes are not cast. Therefore the total number of different ballots possible is, by **C11**,

$$N = \sum_{0 \le k \le 5} \binom{10 + k - 1}{k} = \binom{9}{0} + \binom{10}{1} + \cdots + \binom{14}{5} = \binom{15}{5} = 3003.$$

The $\binom{15}{5}$ comes from the knight's-move identity, (8.2) of Chapter 8. And $N - 1 = 3002$ is the total number of nonblank ballots possible, because we counted the case $k = 0$ in the 3003. But there were exactly

$$3100 - 97 = 3003$$

nonblank ballots cast, so at least two of them must have been the same as each other.

The last piece of reasoning is the use of the pigeonhole principle (Chapter 5, (8)), which is not a counting technique but simply a useful tool. You'll meet it often.

(7.21) Notice that the function f of (7.2), which was used to define a k-rep from X, is a generalization of the characteristic function. To define an ordinary k-subset of X we could use the characteristic function χ:

$$\chi : X \longrightarrow \{0, 1\}$$

and

$$\sum_{x \in X} \chi(x) = k.$$

Any function χ with these two properties defines a k-subset of X (as $\chi^{-1}(1)$). For a k-rep from X we take a function

$$f : X \longrightarrow \{0,1,2,3,\dots\}$$

such that

$$\sum_{x \in X} f(x) = k.$$

The only difference is the codomain.

8. Flag Counts

(8.1) Now we consider what are called flag counts. These are better shown by example than given by a general definition. We can say that the flags are ordered pairs and that we count the set of flags in two ways, by the first coordinate and then by the second. For example, we prove (4.2), inclusion-exclusion for two sets A and B, by a flag count. Let $F = \{(X,a);\ X = A \text{ or } B,\ a \in X\}$. Counting by the first coordinate shows that $|F| = |A| + |B|$, by the second that $|F| = |A \cup B| + |A \cap B|$. We have counted F in two ways, so we've shown that

(8.2)
$$|A| + |B| = |F| = |A \cup B| + |A \cap B|,$$

which proves (4.2).

Example (8.3) If $A = \{1,2,3\}$ and $B = \{2,3,4,5\}$, then the set of "flags" is

$$F = \{(A,1),\ (A,2),\ (A,3),\ (B,2),\ (B,3),\ (B,4),\ (B,5)\}.$$

(*i*) Count F by the first coordinate: There are exactly $|A| = 3$ flags having A as first coordinate, and $|B| = 4$ with B as first coordinate, so the total number of flags is $|F| = |A| + |B| = 7$.

(*ii*) Count F by the second coordinate: Every point of $A \cup B$ is the second coordinate of at least one flag. If the point, say x, is in $A \cap B$, then it appears in two flags, (A,x) and (B,x). Otherwise it is in exactly one flag. Thus $|F| = |A \cup B| + |A \cap B| = 5 + 2$. The same argument works in general.

(8.4) For another example, let's go back to (2.6), the committees with two cochairs appointed. Suppose we consider only the committees of fixed size k. We consider the set F of "flags" (X,T), where X is a committee of k people chosen from a fixed n-set, and T is a 2-subset of X. T is the set of cochairs of the committee X. We count F by the first coordinate: There are $\binom{n}{k}$ first coordinates X; for each such X, there are $\binom{k}{2}$ choices of second coordinate T. Thus $|F| = \binom{n}{k}\binom{k}{2}$.

Now we count F by the second coordinate. There are $\binom{n}{2}$ possible second coordinates T; for each of these sets T, there are $\binom{n-2}{k-2}$ k-subsets X containing T. (That is, to fill out the committee, we choose the remaining $k-2$ members from the other $n-2$ people. We may choose them in $\binom{n-2}{k-2}$ ways; each choice gives us X as the $k-2$ members "plus" the two cochairs in T.) Therefore $|F| = \binom{n}{2}\binom{n-2}{k-2}$. We have proved an identity on binomial coefficients:

(8.5)
$$\binom{n}{k}\binom{k}{2} = \binom{n}{2}\binom{n-2}{k-2}.$$

Clearly this result can be generalized, and also proved directly, but more arduously, from the binomial theorem. The point is to illustrate the method of flag counting, which is good to know for other situations. The point of a flag count is often not to find the total number of flags but instead to get information about the size of the set of first (or second) coordinates.

Our *gestalt* approach (2.7) to this problem was to count all the flags (X,T) by the second coordinate when $|X|$ was unrestricted. Indeed, the original problem is easier to understand with the "flag" concept; it simply asks for the total number of flags (X,T), where X is any subset of the n people, and T is any 2-subset of X.

Example (8.6) Suppose we have a collection E of k-subsets of a given n-set Y, such that every 2-subset of Y is included in exactly m members of E. Flag counting gives us an easy way to determine the relation between the numbers m, n, and $|E|$. We define our set F of flags as all ordered pairs (X,T), where X is in E, and T is a 2-subset of X. (This is just like the preceding set F except that X is restricted to E.) Thus, the number of first coordinates is $|E|$, each of which has $\binom{k}{2}$ subsets of size 2. Therefore $|F| = |E|\binom{k}{2}$. We now count by the second coordinate: There are $\binom{n}{2}$ choices for T, and each of them is now included in exactly m members of E, so $|F| = m\binom{n}{2}$. Therefore $|E|k(k-1) = mn(n-1)$.

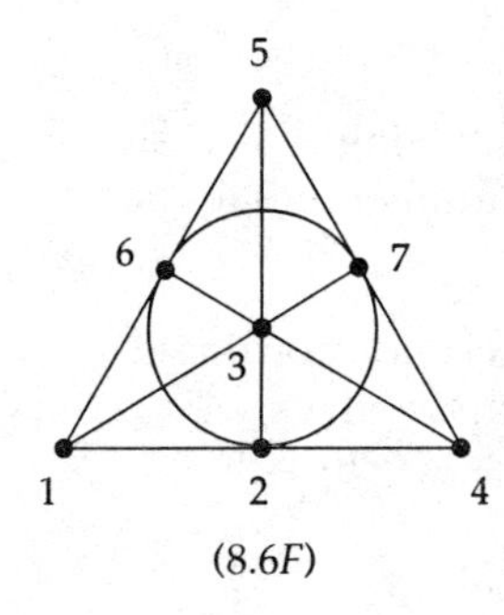

(8.6*F*)

For an example of this situation, consider E to be the seven 3-subsets of $\{1,\ldots,7\}$ indicated by the six lines and one circle in the diagram here.

Here $|E| = 7$, $k = 3$, and $m = 1$.

For a more general definition of flag counting, we can say that if we have a binary relation F from A to B, the flags are the ordered pairs in F. If we know enough about this relation to be able to count the number of pairs in F with any given first coordinate, and ditto for any given second coordinate, then we can count F in two ways and equate the results. We get from this procedure information about how many first coordinates there are (maybe second coordinates instead), or some other parameter of the situation. Usually we are not interested in the value of $|F|$. Strictly, a flag count gives us a relation between various parameters of the relation F. See, for example, (8.2), the flag-counting proof of in-ex for two sets.

If we consider the relation-matrix M of F, then a flag-count counts the total number of 1's in F first by counting those in the rows (counting by the first coordinate), and second by the columns (by the second coordinate).

(8.7) We now apply the technique of flag counting to prove a lemma of Kaplansky. We need first another of his results. Let us define $X(n)$ to be the set of integers from 1 to n, for any $n \geq 1$.

Lemma (8.8) The number of p-subsets P of $X(n)$ such that between any two consecutive points of P there are at least k points not in P is

$$\binom{n-(p-1)k}{p}.$$

Proof We define $S_1[S_2]$ as the collection of all p-subsets of $X(n-(p-1)k)$ [$X(n)$ having the gaps between successive points called for in Lemma]. We'll set up a bijection f from S_1 to S_2.

We choose any element of S_1, say $V = \{x_1, x_2, \ldots, x_p\}$. We expand the underlying set $X(n-(p-1)k)$ to $X(n)$ by inserting k "spots" just to the right of each of $x_1, x_2, \ldots, x_{p-1}$. Now renumber. Under the renumbering, the original p-subset V becomes

$$\{x_i + (i-1)k;\ i = 1, \ldots, p\} = \{x_1,\ x_2 + k, \ldots,\ x_{p-1} + (p-2)k,\ x_p + (p-1)k\};$$

and of course $f(v) \in S^2$.

Now define a map $g : S_2 \longrightarrow S_1$ as follows: For each $P \in S_2$, say $P = \{y_1, \ldots, y_p\}$, remove the k numbers $y_1 + 1, \ldots, y_1 + k$ from $X(n)$, and do the same for the k numbers to the right of $y_2, \ldots, y_{p-1}$. You have not removed any of $y_i, \ldots, y_p$. What remains are $n-(p-1)k$ of the numbers from $X(n)$. Renumber them from 1 to $n-(p-1)k$. The result is that the y_i's become renumbered as p elements of $X(n-(p-1)k)$; i.e., P has become an element, which we define as $g(P)$, of S_1.

Now notice that $g \circ f$ is the identity on S_1, and that $f \circ g$ is the identity on S_2. By Chapter 5, (15.6), f is bijective. QED

Illustration: Take $n = 10$, $k = 2$, and $p = 3$. Then $n-(p-1)k = 10 - 2 \cdot 2 = 6$. We choose a 3-subset from $X(6)$, say $\{1,3,4\}$. Represent it as

$$1\ \_\ 3\ 4\ \_\ \_\ ,$$

where we put underbars for 2, 5, and 6, the points of $X(6)$ not in the chosen subset. Now we insert $k = 2$ boxes just to the right of the 1 and the 3, getting

$$\begin{array}{cccccccccc} 1 & \square & \square & \_ & 3 & \square & \square & 4 & \_ & \_ . \\ 1 & 2 & 3 & 4 & 5 & 6 & 7 & 8 & 9 & 10 \end{array}$$

This linear array of $n = 10$ symbols is now numbered $1, \ldots, 10$. In the process, 1, 3, and 4 become $1, 5 = 3+2$, and $8 = 4 + 2 \cdot 2$; and $P = \{1,5,8\}$ is the desired 3-subset corresponding to our set $\{1,3,4\}$. QED

Lemma (8.9) The number b of p-subsets P of $Z(n)$, the integers from 1 to n in circular "order," such that between any two points of P (i.e., in each of the two arcs between the two points) there are at least k points of $Z(n)$ not in P is

$$b = \frac{n}{p}\binom{n-kp-1}{p-1} = \frac{n}{n-kp}\binom{n-kp}{p}.$$

Proof Consider the set F of all flags (P, x) where P is any of the p-subsets we are trying to count, and x is any point of P. Count by the first coordinate to see that

$|F| = bp$. Now focus on the second coordinate. Notice first that the set of all flags with a given second coordinate x is in a natural one-to-one correspondence with the set of all flags having any other second coordinate y: Just rotate the circle so that x goes to y. Thus, the number of flags with a chosen second coordinate is the same for all possible second coordinates. In other words, $|F| = n \times$ (the number of flags with $x = n - k$). A flag with $x = n - k$, however, has first coordinate P consisting of $n - k$ together with a $(p-1)$-subset of $X(n - 2k - 1)$ subject to the same spaced-apart-by-k condition. See the figure: We remove k points on either side of $n - k$ and are left with the linear problem on the remaining arc.

(8.9F)

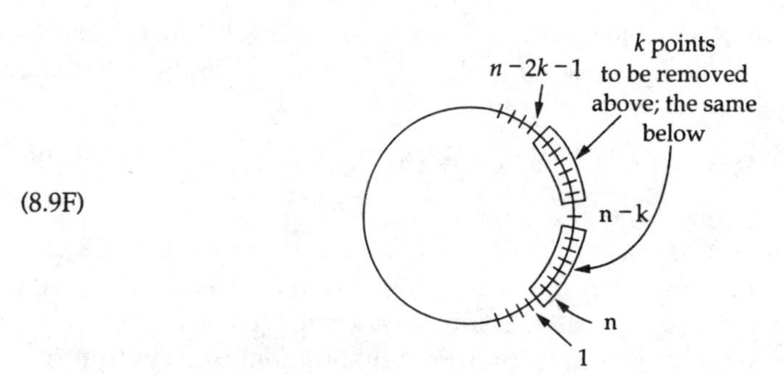

So the number of flags with a chosen second coordinate is given by Lemma (8.8); it is

$$\binom{n - 2k - 1 - (p-2)k}{p-1} = \binom{n - kp - 1}{p - 1} = f.$$

Thus $bp = fn$, and the result follows. QED

(8.10) *The "Problème des Ménages."*† How many ways $(=: M_n)$ are there to seat n couples at a round table with $2n$ places so that men and women alternate but no man sits next to his wife?

We answer this question by counting all forbidden seatings—those where one or more couples are side by side—and subtracting from the total number of seatings. As you will see, this approach uses inclusion-exclusion in a serious way. It also uses (8.9).

Let's use "seating" to mean any placement in which men and women alternate. Thus there are $2(n!)^2$ seatings altogether, because there are two choices for the set of n women's seats and then $n!$ ways to place the women in those seats, and $n!$ ways to place the men in the n men's seats.

In more detail: we could number the seats clockwise $1, 1', 2, 2', 3, \ldots, n, n'$ (and then be back at 1). Call S the set of seats numbered $1, 2, \ldots, n$, and S' those numbered

†This famous problem, discussed in many books and papers, is called by its French name. "Ménage" means "married couple." We follow a recent solution (Kenneth P. Bogart and Peter G. Doyle, "Non-sexist solution of the ménage problem," *American Mathematical Monthly* 93 (1986), pages 514–518.) For a history of this problem, see Jacques Dutka, "On the problème des ménages," *Mathematical Intelligencer*, vol. 8, no. 3 (1986), pages 18–25.

$1', 2', \ldots, n'$. A seating then requires that all the women be at seats S, or all at seats S': That's two choices.

(8.10F)

After that there are $n!$ ways to place the women [men] at their seats. By **C2** we multiply to get $2n!n!$.

We carefully dissect the set of all forbidden seatings by defining

For each subset W of *wives*

$$F(W) := \text{the set of all seatings in which each woman in } W \text{ sits next to her husband.}$$

Thus, for instance, $F(\emptyset)$ is the set of all seatings, and we know its size. There may be more couples together than specified by W, for example, if $n = 3$ and the couples are h, w; h', w'; and h'', w'', then $F(\{w\})$ is

$$\begin{array}{cccccc} w, & h, & w', & h'', & w'', & h' \\ w, & h, & w'', & h', & w', & h'' \\ w, & h, & w', & h', & w'', & h'' \\ w, & h, & w'', & h'', & w', & h' \end{array} \tag{8.11}$$

plus seatings obtained from these by rotation and reversal. (Understand that each row of (8.11) represents a circular placement with the last person next to the first.)

If a fourth couple h_0, w_0 joins the party then, for example,

$$w, h, w', h_0, w'', h', w_0, h''$$

is in $F(\{w\})$ but not in $F(W)$ for any other $W \neq \emptyset$.

We let L be the set of all n women. Then, letting $F(w) := F(\{w\})$,

$$F := \bigcup_{w \in L} F(w) \tag{8.12}$$

is the set of all forbidden seatings, because in any forbidden seating, at least one husband and wife are next to each other.

We'll use in-ex, **C6**, to find $|F|$. Before that, we need the following observations.

(8.13) For any two subsets W and W' of L, $F(W) \cap F(W') = F(W \cup W')$.

The proof of (8.13) is left to Problem (8.6).

Apply **C6** to (8.12) to find that

$$|F| = \sum_{\emptyset \neq W \subseteq L} (-1)^{|W|-1} |F(W)|, \tag{8.14}$$

since $F(w_1) \cap \cdots \cap F(w_j) = F(\{ w_1, \ldots, w_j \})$, by (8.13).

Now luckily we can simplify (8.14) greatly by showing that

$$\text{if } |W| = |W'|, \quad \text{then } |F(W)| = |F(W')|.$$

In fact, we calculate the value explicitly: Let $0 \leq j \leq n$, and let

$$W = \{ w_1, \ldots, w_j \}.$$

We ask how many $(=: d_j)$ ways there are to place j "dominos" around the table so they don't overlap, as in the sketch $(j = 3)$.

(8.15F)

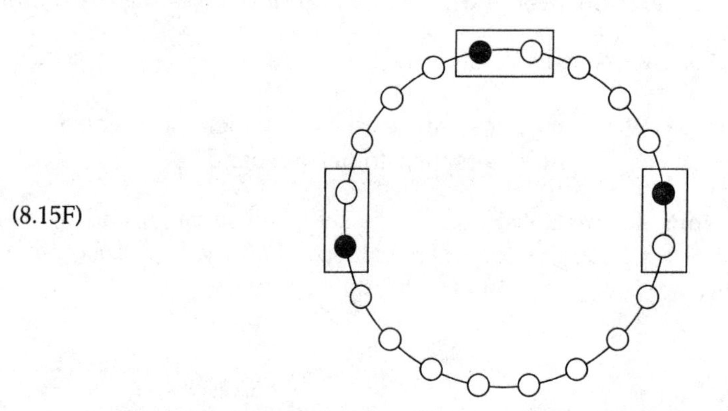

Each domino covers two adjacent places. Lemma (8.9) answers this question, if we

identify each domino with its point • reached first in a clockwise cycle of the table, and choose the "separation" k of Lemma (8.9) to be 1.

(Then the j-subset of points • is a subset P counted by (8.9)). Thus we find, setting $n := 2n, b := d_j, p := j$, and $k := 1$ in (8.9),

$$d_j = \frac{2n}{2n-j}\binom{2n-j}{j}. \tag{8.16}$$

The dominos are where we seat the women $w_1, \ldots, w_j$, either all in S or all in S' [two choices]. We seat each woman's husband on the other seat in "her" domino [one choice].

We can now count $|F(W)|$. There are two choices for the n women's seats, d_j choices for the placement of the dominos, $j!$ ways of seating the women from W in the j dominos, and $(n-j)!$ ways of seating the women of $L - W$ in the women's seats outside the dominos, and $(n-j)!$ ways to seat their husbands in the men's seats outside the dominos. Thus

$$|F(W)| = 2d_j \cdot j![(n-j)!]^2. \tag{8.17}$$

Since (8.17) depends on j but not on $\{w_1, \ldots, w_j\}$ it is the same for all j-subsets W of L. (Notice that it's right for $j = 0$, giving $2n!^2$.) Since there are a total of $\binom{n}{j}$ j-subsets of L, we find that

$$\begin{aligned} S_j := \sum_{W \subseteq L, |W|=j} |F(W)| &= 2d_j \binom{n}{j} j!(n-j)!^2 \\ &= 2 \cdot \frac{2n}{2n-j} \binom{2n-j}{j} \binom{n}{j} j!(n-j)!^2 \end{aligned}$$

Using $\binom{n}{j} = \frac{n!}{j!(n-j)!}$ from Chapter 8, (7.8), we simplify this to get

(8.18)
$$S_j = 2n! \frac{2n}{2n-j} \binom{2n-j}{j} (n-j)!.$$

Since the sum of any function $f : \mathcal{P}(L) \to \mathbf{R}$ over all nonempty $W \subseteq L$ may be broken up into the "sub sum" over the 1-subsets plus the "sub sum" over the 2-subsets, and so on, we may rewrite (8.14) as

(8.19)
$$|F| = \sum_{1 \le j \le n} (-1)^{j-1} S_j.$$

Now, since $S_0 = 2n!^2$ is the total number of seatings, our desired answer M_n is

(8.20)
$$\begin{aligned} M_n &= 2n!^2 - |F| \\ &= \sum_{0 \le j \le n} (-1)^j S_j \\ &= 2n! \sum_{0 \le j \le n} (-1)^j \frac{2n}{2n-j} \binom{2n-j}{j} (n-j)! \end{aligned}$$

from (8.19) and (8.18).

Some numerical values:†

(8.21)

n	M_n	$M_n/2n!$
3	12	1
4	96	2
5	3120	13
10	3191834419200	439792

(8.22) A final comment: You can see that

(8.23)
$$2n! \text{ divides } M_n,$$

at least for the values shown in (8.21). To prove (8.23) for all $n \ge 1$ is a nice little exercise in binomial coefficients, which we now work out.

It looks from (8.20) that $2n!$ divides M_n, but the denominator $2n - j$ raises doubts. So, going back to (8.18), we'll show that $2n!$ divides S_j. From (8.19) and (8.20), that will do it.

† From Bogart and Doyle, cited under (8.10).

From (8.18) we get

$$\frac{S_j}{2n!} = \frac{2n}{2n-j}\binom{2n-j}{j}(n-j)! \tag{8.24}$$

If $j = n$, this is obviously an integer. So consider $0 \le j < n$. Using the binomial theorem, Chapter 8, (7.2), we get (for $k \neq n$)

$$\binom{n}{k} = \frac{n}{n-k}\binom{n-1}{k}. \tag{8.25}$$

Now apply (8.25) to $\binom{2n-j}{j}$ in (8.24) to get

$$\frac{S_j}{2n!} = \frac{2n}{2n-j}\,\frac{2n-j}{2n-2j}\binom{2n-j-1}{j}(n-j)! = n\binom{2n-j-1}{j}(n-j-1)!,$$

an integer. That proves (8.23). QED

Another proof of (8.23) is possible. It doesn't require that we know M_n. It uses properties of permutations and equivalence relations. To save space we don't give it here, however.

9. On Terminology

The official term for the size of a set is **cardinality.** Thus $\{1,2,3\}$ has cardinality 3.

The official term for a set with repeated elements is **multiset**. That term doesn't follow the various usages of "set" and "subset," however. One can say "k-subset of X" but not "k-multisubset of X." Instead, one usage is "k-multiset on X" for what we call "k-rep from X," a term we introduced in this chapter for brevity and simplicity.

10. Further Reading

(10.1) Claude Berge, *Principles of Combinatorics*, Academic Press, New York, 1971.

(10.2) Kenneth P. Bogart, *Introductory Combinatorics*, Pitman, Boston, 1983.

(10.3) Richard Brualdi, *Introductory Combinatorics*, North-Holland, Amsterdam, 1977.

(10.4) Herbert John Ryser, *Combinatorial Mathematics*, Mathematical Association of America (distributed by Wiley), 1963. Carus Monograph No. 14.

(10.5) William Allen Whitworth, *Choice and Chance*, Hafner, New York, 1965 (reprinted from the fifth edition dated 1901).

Comments. Reference (10.1) is more advanced than the others. It's at a senior or first-year graduate level. Reference (10.4) is at about a senior level. Both are excellent; you will get a great deal from them if you can read them. Reference (10.5) is an old book famous for its 1,000 wonderful problems, many of which I have used in this and the next chapter.

11. Problems for Chapter 9

Problems for Section 2

2.1 How many strings are there of five 0's and five 1's that start with 101?

2.2 How many partitions are there of a 7-set into two cells, one of which has three elements and the other of which has four elements? Explain in terms of binomial coefficients.

2.3 How many strings are there of seven 0's, two 1's, and one 2? Explain, without listing them all and counting.

2.4 How many strings of eight 0's and 1's are there having no more than two 1's? Express in terms of binomial coefficients and explain.

2.5 You have one of each of the following coins: penny (1), nickel (5), dime (10), quarter (25), half-dollar (50), and dollar (100). The value of the coin in cents appears in parentheses after the name of the coin. How many different amounts of money can you make with these coins? Explain.

2.6 (C&C) You find yourself in a new country with a strange system of coinage. When you exchanged your money at the border you got $3n+1$ coins, n of which were all alike, each worth one "bog." The rest were all distinct: one was worth two "bigs," one was worth four bigs, and so on. In general you had n coins, all worth one bog, and for each $i = 1,2,\ldots,2n+1$ you had one coin worth 2^i bigs. How many different amounts of money could you make using exactly n of these coins? Prove that the answer is 2^{2n}. **Also:** 1 big $= n+1$ bogs.

2.7 (C&C) On an eight-by-eight checkerboard, how many subsets are there of two squares, one red and one black?

2.8[Ans] (C&C #169) Let n and r be nonnegative integers such that $r \geq n$. Consider the multinomial expansion of

$$(x_1 + \cdots + x_r)^n.$$

(*i*) Prove that the number of terms in this expansion in which none of the x_i's has an exponent of 2 or greater is $\binom{r}{n}$.

(*ii*) If you know a bit of Section (5), also prove that each such term has coefficient $n!$.

2.9 (C&C) Each side of a square is divided into 10 equal segments. How many triangles are there with vertices at the { points of division } ∪ { the four corners of the square }? Explain.

2.10 How many partitions are there of a 6-set into two cells of three points each?

Problems for Section 3

3.1 A class of 43 students votes on a date for an exam. Each votes for just one of the five possible days. How many different ways may they vote?

3.2 Each of 30 students is to choose one of five major fields of study. How many patterns of choice are possible if all fields are open to all students? A "pattern of choice" is a list of the students' names together with each student's choice.

3.3 Four (4) Cobol programs are entered in a beauty contest. Each member of the jury—Professors Able, Baker, Charlie, Duke, and Ezra—calls out his vote for the one program he thinks is the most beautiful. The next day the student newspaper reports the outcome, i.e., each prof's name and vote. To complete its story it lists all the outcomes that might have been but weren't. How many outcomes did it report?

3.4[Ans] Prove that the number of functions f from a k-set to an n-set $\{y_1,\ldots,y_n\}$ such that $\forall i\ |f^{-1}(y_i)| = a_i \geq 0$ (and, of course, $a_1 + \cdots + a_n = k$) is the multinomial coefficient

$$\binom{k}{a_1, a_2, \ldots, a_n}.$$

That is, if for each point of the codomain we prescribe the number of times that the point is taken as a value, we count the functions with a multinomial coefficient.

Problems for Section 4

4.1 A set of 60 people are surveyed. Twenty-six like pizza. Thirty-two like ice cream. Thirty like tofu. Fourteen like both pizza and ice cream. Seven like both pizza and tofu. Ten like ice cream and tofu. Two like all three. How many dislike all three?

4.2 Fifty-one students were polled. Twenty-one liked Fortran, 24 liked Basic, and 21 liked Cobol. Also, 10 liked Fortran and Basic, 12 liked Basic and Cobol, and 81 liked Fortran and Cobol. Finally, three students liked all three languages. How many students disliked all three languages?

4.3 There are four utility subroutines in a certain Cobol workspace. Various of them are shared by 21 different programs, each using at least one of the four, specifically:

The creative expense account subroutine is used in 11 programs.
Each of the other three subroutines is used in seven programs.
Each 2-subset of subroutines is used in n programs.
Each 3-subset of subroutines is used in two programs.
All four subroutines together are used in one program.
What is the value of n?

4.4 (*i*) Determine the exact number of sandwiches of each type in Example (4.6).

(*ii*) Suppose the machine recorded five sandwiches made with all three ingredients. What could you conclude?

4.5 Would you believe a market investigator who reports that of 1000 people surveyed, 816 like candy, 723 like ice cream, 645 like cake, while 562 like both candy and cake, 463 like both candy and ice cream, 470 like both ice cream and cake, and finally 310 like all three?

4.6[Ans] The 1989–1990 "Combined Membership List" of three organizations of mathematicians gives the following data on page *iv*:

Number of Names Listed

Total number of individuals	50,605
Total number of memberships	62,878
American Mathematical Society (AMS)	25,673
Mathematical Association of America (MAA)	30,735
Society of Industrial and Applied Mathematics (SIAM)	6,470
Joint memberships	
AMS and MAA	7,958
AMS and SIAM	1,214
MAA and SIAM	681
AMS, MAA, and SIAM	1,210

Check these numbers for consistency. Explain. [Hint: What is their definition of "joint membership"?]

4.7 A survey of 30 students reported that exactly 18 of them had received at least one grade of "A," 14 at least one "B," and 22 at least one "C." Furthermore, 12 had got at least one "A" and at least one "B," nine had got at least one "A" and at least one "C," and 10 had got at least one "B" and at least one "C." Finally, exactly seven had got at least one grade of "A" and of "B" and of "C." Would you believe this survey? Explain.

4.8 (*C&C*) In how many positive integers less than 1,000 (and represented in base 10) does the digit 9 occur?

4.9 How many strings of length 5 in the letters a, b, c, d, e have two or more consecutive a's? Explain.

4.10 How many strings of length 5 in the letters a, b, c, d, e have two or more consecutive vowels? Explain.

4.11 In how many integers between 1 and 10,000 does the digit 7 appear? (These integers are expressed in base 10.)

4.12 (*C&C*) In how many of the integers n such that $1 \leq n \leq 1,000$ do both digits 8 and 9 occur?

4.13 Explain the use of **C2** in (5.9).

4.14[Ans] Give a different proof of **C6** along the following lines: Let x be any point of $A_1 \cup \cdots \cup A_n$. Prove that the right-hand side of **C6** counts x exactly once (net). (Hint: First assume that x is in each of the sets A_i. Then do the general case similarly.)

Problems for Section 5

5.1 Each of five patients in a doctor's waiting room is reading one of the 10 magazines there. If no two read the same magazine, how many different choices of magazines could they make?

5.2 The five students in a seminar are each to choose a different one from a list of eight topics for their final report. How many choices may the class make?

5.3 Ten people are ordering dessert in a restaurant. The dessert trolley has one each of 15 different desserts, and there are no spares in the kitchen.

(*i*) How many different patterns of choice are possible if everyone has exactly one dessert?

(*ii*) How many if everyone has at most one dessert?

5.4 (*C&C*) On your shelf are 18 different books: two on Fortran, four on Jovial, five on Cobol, and seven on discrete math. How many ways could you shelve them keeping all those on a given subject grouped together? Explain.

5.5 Among all permutations f of $\{1,2,3,4,5\}$ how many have $f(1)$ odd? $f(2)$ odd? Both odd?

5.6 Imagine listing all permutations of $1,\ldots,n$ as rows of a function-table with columns headed by "1",...,"n". How many rows have the digits 1,2 adjacent with "1" to the left of "2"?

5.7 With the same setup as in Problem 5.6, how many rows have "1" to the left of "2"?

5.8 A social security number consists of nine decimal digits. Assume that every digit can be any of the numbers $0,1,\ldots,9$. Find the proportion of social security numbers with no repeated digits.

5.9 Five cars arrive at an empty parking lot with eight numbered spaces. How many ways are there for them to park? Explain.

5.10 (*C&C*) In how many ways can the letters of "*l'oiseau*" be arranged so that the vowels are in their natural order? [Ignore the apostrophe.] Explain.

5.11[Ans] How many cycles are there on an n-set? ["Cycle" is defined in Chapter 5, (16.3).] Explain.

5.12 (Stanley) How many permutations of a 6-set consist of exactly two cycles? Explain.

5.13 (*i*) How many surjections are there from a 5-set to a 5-set? Explain.

(*ii*) How many are there from a 5-set to a 3-set? (You may simply evaluate the formula in (5.13).)

5.14 (*C&C*) The 26 letters of the alphabet are written in a row (once each) so that no two vowels are together. How many ways may this be done? (Let the vowels be a, e, i, o, u.) Explain.

Problems for Section 6

6.1 Seven people divide into three teams to play Trivial Pursuit. If the only restrictions are that no team may be empty, everyone is on some team, and no two teams overlap, how many ways are there to choose the teams?

6.2 In the previous problem, suppose the three teams are intent on using markers colored red, blue, or green. The result is a "red" team, a "blue" team, and a "green" team. How many such results are possible?

6.3 With the same conditions as in the previous two problems, the three teams are now going to use any three of the six colors available. Now how many results are possible? [A "result" is three teams, each identified with a color.]

Problems for Section 7

7.1[Ans] (*i*) How many solutions in nonnegative integers are there to the equation

$$w + x + y + z = 10?$$

(*ii*) How many in positive integers? Explain.

7.2 (*C&C*) Show that the number of m-reps of an $(n+1)$-set equals the number of n-reps of an $(m+1)$-set.

7.3 (*C&C*) A bookbinder has 12 different books to bind in red, green, or black cloth.

(*i*) In how many ways can he bind them?

(*ii*) How many ways are there if he uses at least one of each color?

7.4 You have eight Hershey's kisses (identical pieces of candy), all of which you give to four people. How many ways are there to distribute the candy? Explain.

7.5 If students may take any number of items up to six in Example (7.13), how many meals could there be?

7.6 (*C&C*) How many fruit baskets can be made up from a large supply of apples, oranges, pomegranates, and plums if each basket is to have exactly eight pieces of fruit?

7.7 Each domino is marked by two numbers. The pieces are symmetrical, so the number-pairs are not ordered. How many different pieces can be made using the numbers $1,\ldots,n$?

7.8 [Recommended] Seven people enter an elevator in the basement. Each exits at floor 1, 2, 3, or 4. How many different ways can this happen? Explain.

7.9 [Recommended] Seven identical marbles, rolling around the floor of an elevator, disappear from it by rolling out the door at floors 1, 2, 3, or 4. Assume it stops once at each floor in the order given. How many ways can the seven marbles disappear? Explain.

7.10 (*C&C*, #182) You have three identical pennies and 10 identical nickels. How many ways can you distribute them to four labeled pockets if some pockets may be empty? Explain.

7.11 Seven of a total of 15 balls on a pool table are striped; each other ball is a solid color. The balls are sent to the six pockets labeled $A,\ldots,F$. In how many ways can this be done? Regard the seven "stripes" as identical to each other. Do the same for the eight "solids."

7.12[Ans] (*C&C*, #219) There are seven copies of one book, eight of a second book, and nine of a third book. How many ways can two people divide them if each takes 12 books? Explain.

7.13 Prove (7.11 P*ii*) using the direct form of the definition of "injective" in Chapter 5, (7.1). (That is, don't use (7.2), as is done in the answer to (7.11P).).

7.14 In the second proof of **C11**, prove that $i_1 \leq j$. See paragraph (7.8).

7.15 Consider the 5-set $X = \{a,b,c,d,e\}$. How many 12-reps are there from X in which a appears at least twice and d at least three times? Explain.

7.16 We defined in (7) $s(n,k) :=$ the total number of k-reps of an n-set.

(*i*) Prove directly from this definition that

$$(\star) \qquad s(n,k) = s(n-1,k) + s(n,k-1).$$

[Hint: See Ch. 3, Section (11).]

(*ii*) Give another proof of **C11** by using $(\star)$ instead of equation (7.5).

***7.17** Show that β, in the second proof of **C11**, is a bijection $\beta : \binom{Y}{k} \longrightarrow X_k$, where X_k is defined to be the set of all k-reps of X, and $\binom{Y}{k}$ stands for the set of k-subsets of Y.

***7.18** Define k-reps from a set X as equivalence classes of strings of length k over X.

***7.19** How many 5-reps of a 10-set are there in which no element appears more than twice? Express your answer in terms of binomial coefficients and as an integer in base 10. Explain.

Problems for Section 8

8.1 How many seatings (in the sense of (8.10)) of two married couples around a circular table with four places have no husband and wife next to each other? Answer in two ways: one, by reasoning it out from scratch; two, by calculating M_2 in (8.20). Show your steps.

8.2 Prove that $M_n/2(n!)$ is 1 for $n = 3$ and is 2 for $n = 4$ by

(*i*) evaluating (8.20)

(*ii*) using the definition in (8.10) and listing all possible arrangements for one fixed seating of wives. Explain why your list is correct, say by using a tree.

8.3 Suppose five couples are to sit around a table of 10 places as in (8.10), i.e., no man next to his wife. Let the wives be A,B,C,D,E and their husbands, respectively (and respectfully), a,b,c,d,e. Fix the positions of the wives in alternate seats, and fix the position of husband c as follows: A c B__ C __D __ E __ (A). The bars are places for the other husbands, and the "(A)" means that the last bar is between E and A. Show a convincing enumeration by tree or otherwise of all ways to fill in the bars.

8.4 Verify that $M_5/2(5!) = 13$ by assuming a fixed seating of the wives and making a tree to diagram the possible seatings of the husbands.

8.5 Prove (8.25).

8.6 Prove (8.13). [Reminder: To prove two sets equal, show that each is a subset of the other.]

8.7 State and prove a generalization of (8.5).

8.8 A sphere is exactly covered (i.e., with no gaps and no overlaps) with n regular pentagons, all of the same size. Find n. (You may think of the pentagons as bendable but not stretchable.) [Hint: You will need Euler's formula, Ch. 3, Problem 53.]

8.9 A set of 10 people are formed into committees of 4 members each in such a way that every 3-subset of people are members together of exactly two committees. How many committees are there altogether? (Hint: Count all flags (X,T), where X is a committee and T is a 3-subset of X.)

8.10 (*C&C* #154) A large floor is paved with square and triangular tiles, and all sides of the squares and triangles are equal. Every square is adjacent to four triangles, and every triangle to two squares and a triangle. Show that the number of vertices must be the same as the number of triangles, and that this must be double the number of squares. [Note. No such paving of a floor really exists. Why not? Consider the boundary. You can use Euler's formula (Chapter 3, Problem 53) to prove that no such covering of a ball is possible, either. But it is possible to paint lines on an inner tube (torus) to make such a pattern. Euler's formula for the torus is $v - e + f = 0$, holding for graphs that can be drawn on a torus with no edges crossing but that cannot be so drawn on a sphere.]

8.11 A soccer ball is covered with regular pentagons and hexagons, all of which have sides of the same length. Each pentagon is joined at its sides to five hexagons; each hexagon is joined to three pentagons and to three other hexagons. Deduce the number of each shape on the ball. [Hint: You will need Euler's formula, Ch. 3, Problem 53.]

8.12 Suppose a set of n students arrange themselves in several (possibly overlapping) study groups of three students each in such a way that each possible pair of students is together in exactly one study group.

(*i*) Prove that no two different study groups have two students in common.

(*ii*) How many study groups are there?

(*iii*) How many study groups is each student in?

*(*iv*) Prove n is congruent to 1 or 3 (mod 6).

8.13 (Speckman) Let $n, k \in \mathbf{N}$. Suppose a lottery drawing chooses a k-set from $\{1, \ldots, n\}$.

(*i*) How many choices of k-set have two consecutive integers? Express your answer in terms of binomial coefficients.

(*ii*) Find the ratio of your answer to the total number of possible choices. Express the ratio as a decimal number for $n = 54$ and $k = 6$.

General Problems

G1. How many ways are there to assign a total of 100 points to nine exam questions if each question gets at least six points?

G2. Let n be a positive integer. Consider the set S of all strings of 0's and 1's of length n. Define M as the subset of S of all strings with more 1's than 0's. Express $|M|$ as a function of n and explain your answer carefully.

G3.[Ans] Twenty boys and 20 girls spend the evening together at a party. There are n dances played. Each girl keeps a tally of her partners (of whom there are at most one per dance): Bill at dance 1, Bob at dance 2, Biff at dance 3, and so on. Afterwards, Sue looks over her card.

(*i*) How many ways might it have been marked?

(*ii*) If she recorded only how many times she had danced with each of her partners, and not also at which dances, how many ways are there?

Explain your answers in both cases.

G4. Consider four-digit numbers in the base 10, i.e., integers x such that $1000 \le x \le 9999$. How many of them have the sum of their digits equal to 8? No listing, please. Explain.

G5. The letters a, b, c, d, e are used twice each to make a string of 10 symbols. How many such strings are there? Explain, and express your answer as a multinomial coefficient.

G6. In how many ways can three nickels and 10 dimes be put into four pockets? Explain.

G7. How many different amounts of money can you make using some or all of the following coins?

3 pennies	3 dimes
1 quarter	2 half-dollars

G8. From a laser printer n wires lead. Each is to be connected to one of k computers. How many ways are there to make these connections

(*i*) if no two wires may be connected to the same computer?

(*ii*) in general?

Explain in both cases.

G9. How many strings in the letters a, b, c, d, e have three or more consecutive vowels? The strings all have length 5. Remember that letters in a string may be repeated. Explain your answer elegantly.

G10. There are three identical red balls, three identical white balls, and three identical green balls on a pool table with six labeled pockets. How many ways may they all be sent to the pockets?

G11. How many triangles are there with sides a, b, c such that

$$a,\ b,\ c \in \{4,5,6,7\}\text{? Explain carefully.}$$

G12. (*C&C*) How many arithmetic progressions of three terms are there among the numbers $1, 2, \ldots, 2n$ (where n is a nonnegative integer)? [An arithmetic progression is, in this case, a 3-subset $\{a, b, c\}$ of the set $\{1, \ldots, 2n\}$ such that $b - a = c - b$.]

G13. (*C&C*) Let n be a positive integer. How many 3-subsets of $\{1, 2, \ldots, 3n\}$ have a sum that is divisible by three?

G14. (*C&C*) How many ways are there to select three different integers that have an even sum from a set of 30 consecutive integers?

G15. (Modified from *C&C* #210). Two teachers—one in math and one in French—will examine each of four students, one on one, for 15 minutes, during four consecutive quarter-hour periods. How many ways can the exams be scheduled so that no student has both exams at the same time?

How would you approach this problem for five students (and five periods of 12 minutes each)?

G16. Arrange the letters of the word "combinatorial" in a diamond-shaped pattern with *c* at the top, two *o*'s in a horizontal row just below it, three *m*'s in a row just below the *o*'s, and so on. The middle row of the diamond should consist of seven *a*'s. Below it is a row of 6 *t*'s, then 5 *o*'s, and so on down finally to the bottom row, a single *l*. Here is the question: Start at the top and always go to an adjacent letter in the row below. How many such paths spell "combinatorial"? Explain your answer.

G17. Let S denote the set of all 3-subsets of $\{1, 2, \ldots, 10\}$.

What proportion of the elements of S contain two or more consecutive integers? Again, no listing, please. Explain.

G18. (*i*) What proportion of the 4-subsets of $J := \{1, \ldots, 10\}$ have two consecutive integers?

(*ii*) Optional. Use a random subset-generator, such as 4?10 in APL, to get 50 4-subsets of J. Find the proportion of these 50 with two consecutive integers. Compare your empirical proportion with the answer to part (*i*).

G19. You have strings of 15 "0's" and 10 "1's" arranged so that there are exactly seven transitions in each string. That is, "01" is a transition, and so is "10." How many such strings are there?

G20. Consider a predicate in three variables x, y, z. (*i*) How many different ways are there to quantify all three variables using $\forall$ or $\exists$ if we say that "different" means "typographically different"? For example, $\forall x \forall z \exists y$ is typographically different from $\forall z \forall x \exists y$. (*ii*) Now let's recognize that, for example, $\forall x \forall z$ has the same effect as $\forall z \forall x$. Extend that recognition to any two, or all three, variables, and also to $\exists$ in place of $\forall$. Now many different ways are there to quantify all three variables?

Test Questions

T1. A team of two investigators is to be appointed from a pool of 10 reporters for the *Daily Orange*. How many different teams could there be? Explain.

T2. How many different bridge hands are there with no aces, kings, queens, jacks, or tens? Explain.

T3. How many different strings of 80 "0's" and 20 "1's" are there? Explain.

T4. Fifty students were surveyed on their food preference. Thirty-one liked asparagus, 30 liked broccoli, and 30 liked cauliflower. Nineteen liked asparagus and broccoli, 17 liked asparagus and cauliflower, and 15 liked broccoli and cauliflower. Nine students liked all three vegetables. How many of these 50 students disliked all three vegetables?

T5. Each of a set S of students takes at least one of classes A, B, or C as follows: In A, B, and C, respectively, are 14, 17, and 12 students of S. Exactly nine students from S have A and B together. Exactly three students from S have A and C together. Exactly five students from S have B and C together. Exactly two students from S take all three classes. How many students are in S?

T6. Eight students (all of whom can drive) are to go from Syracuse to Colgate in three six-passenger cars. How many distributions of students to cars are possible if each car is to have at least one student? A formula is a good enough answer; you need not evaluate it. Explain your answer.

[A "distribution" is, for example,

Car 1 has Adam, Betty, and Chuck.
Car 2 has Ellen and Fred.
Car 3 has Dave, Gus, and Hope.

If Ellen and Dave exchange places, the result is a different distribution. If Ellen and Fred exchange places, the distribution is the same.]

T7. How many strings of "0's" and "1's" of length 7 contain 10101 somewhere in five consecutive places? [Listing the strings and counting them will get less credit than taking a more theoretical approach.] Explain your answer.

T8. A pool of 10 reporters is to be split up into five teams of two members each, and no teams are to have any members in common. How many such choices are there? Explain.

12. Answers to Practice Problems

(1.5P) $\varnothing$, the empty set of ordered pairs. It is a bijection because for any predicate $P(x)$, "$\forall x \in \varnothing\ P(x)$" is true. That is, in the definition of injectivity [surjectivity], the universal quantifier is applied over the domain [codomain] of the function.

(1.8P) Let g be the obvious bijection from A to the set S of all 3×3 relation-matrices. These matrices are

$$M = \begin{bmatrix} a_1 & a_2 & a_3 \\ a_4 & a_5 & a_6 \\ a_7 & a_8 & a_9 \end{bmatrix}$$

with $a_1, \ldots, a_9$ running freely over $\{0,1\}$. Now we find a bijection f from S to B.

For $M \in S$ we set

$$f(M) := \sum_{1 \le i \le 9} a_i 2^{9-i}.$$

Because base-2 representations are unique (Ch. 6, (43)), f is injective. Clearly f is surjective, because any number in B, being less than $512 = 2^9$, has a nine-bit base-2 expression. Then $f \circ g : A \longrightarrow B$ is a bijection.

(2.12P) Because $\{p_1, p_2\} \times \mathcal{P}(S - \{p_1, p_2\})$ is not what we want. In it are such things as (p_1, X) and (p_2, X) for $X \subseteq S - \{p_1, p_2\}$. We want instead ordered pairs like $(\{p_1, p_2\}, X)$.

(3.5P) If $a = 0$ we want the number of functions from $A = \varnothing$ to B. There is only the empty set of ordered pairs in this case (even if $b = 0$). So we get $b^0 = 1$ (even when $b = 0$. That is, as we explain in (3.13), $0^0 = 1$ by definition in a discrete context, and this result is the reason for that definition.)

If $b = 0$ and $a \ge 1$, then there are no functions from $A \neq \varnothing$ to $\varnothing$, so the answer is $0^a = 0$.

(4.14P)

$$\begin{aligned} \overline{A} \cap B \cap \overline{C} &= 237, \\ \overline{A} \cap B \cap C &= 455, \\ \overline{A} \cap \overline{B} \cap C &= 229. \end{aligned}$$

(5.7P)

BADC	*CADB*	*DACB*
BCAD	*CBAD*	*DBAC*
BCDA	*CBDA*	*DBCA*
BDAC	*CDAB*	*DCAB*
BDCA	*CDBA*	*DCBA*

(5.12P) In APL:

```
⎕IO ← 0
Z ← BIRTHDAY N;A;B;C
[1] A ← Nρ365
[2] B ← ιN
[3] C ← A − B
[4] C ← C ÷ 365
[5] C ← ×/C
[6] Z ← 1 − C
Z ← BIRTHDAY 23
Z = 0.507297+
```

(5.18P)

$$\begin{aligned} f \in \bigcap_{i \in U} A_i \quad &\text{iff } \forall i \in U \quad f \in A_i \\ &\text{iff no point of } U \text{ is a value of } f \\ &\text{iff (range of } f) \subseteq A - U \\ &\text{iff } f \in A_U. \end{aligned}$$

(5.21P) The answer rests on the term for $j = n$. It is $(-1)^n \cdot 0^k$. When k is positive, 0^k is 0. When $n = k = 0$, this term is $0^k = 0^0 = 1$. (See (3.13).) That value is correct, because the function from $\varnothing$ to $\varnothing$ is a surjection.

(7.3P)

$[a,a,a]$ $[b,b,b]$ $[c,c,c]$ $[d,d,d]$

$[a,a,b]$ $[a,a,c]$ $[a,a,d]$

$[b,b,a]$ $[b,b,c]$ $[b,b,d]$

(the same for $[c,c,x]$ and $[d,d,y]$),

and then there are the four 3-subsets, i.e., those with no repeated elements. The total number is $4 + 12 + 4 = 20$.

(7.11P) (*i*) Let $i \in \{1, \ldots, n\}$. Then i occurs $x_i - x_{i-1} - 1$ times in the k-rep corresponding to $\{x_1, \ldots, x_n\}$, where $x_0 := 0$ and $x_n := n + k$. This is true because i fills the spaces between the vertical dividers at x_{i-1} and x_i; these spaces are in number $x_i - x_{i-1} - 1$. The values of x_0 and x_n are where the bookends go. Hence the total number of occurrences is

$$\begin{aligned} \sum_{1 \le i \le n} (x_i - x_{i-1} - 1) &= x_n - x_0 - n \\ &= n + k - 0 - n = k \end{aligned}$$

because of telescoping.

(*ii*) Suppose we have two different $(n-1)$-subsets $x_1 < x_2 < \cdots < x_{n-1}$ and $y_1 < y_2 < \cdots < y_{n-1}$. Let j be the least subscript where they differ. That is $x_i = y_i$ for all $i < j$ but $x_j \neq y_j$. (Define $x_0 := y_0 := 0$.) Then j occurs $x_j - x_{j-1} - 1$ times in the k-rep for the "x"-set, but j occurs $y_j - y_{j-1} - 1 = y_j - x_{j-1} - 1 \neq x_j - x_{j-1} - 1$ times in the other k-rep.

(7.12P) Define the x_i's recursively. Set $x_0 := 0$ and for $1 \le i \le n$,

$$x_i := a(i) + x_{i-1} + 1.$$

Then $x_i - x_{i-1} - 1 = a(i)$ for all i. We need only show that $x_1, \ldots, x_{n-1}$ so defined are $n - 1$ distinct elements of I.

We know that $0 \le a(i)$. Therefore $x_i > x_{i-1}$ for all i. So we have $0 < x_1 < x_2 < \cdots < x_{n-1} < x_n$. What is the value of this superfluous x_n? Rewrite the defining equations as

$$x_i - x_{i-1} = 1 + a(i)$$

and sum over i. We get

$$\begin{aligned} x_n - x_0 &= n + \sum a(i) \\ &= n + k \end{aligned}$$

since $\sum a(i) = k$. Since $x_0 = 0$, we see that $x_n = n + k$, so $x_{n-1} \le n + k - 1$, making $x_1, \ldots, x_{n-1} \in I$.

(7.16P) There are $\binom{5}{2}$ ways in which two reds could be ordered, followed by $\binom{3}{1}$ ways for the white. These determine that the two remaining people order rosé. The answer is $\binom{5}{2}\binom{3}{1} = \binom{5}{2,1,2} = 30$.

(7.18P) The domain of D is the n-set of cells, and the codomain is the k-set of balls.

CHAPTER

10

Probability

...the leitmotif...was the search for the meaning of independence.
Mark Kac (1914–1984)

1. Introduction

(1.1) You and a friend decide to pass the time by playing the following game: You choose a number, say 3, from $\{1, \ldots, 6\}$. You pay your friend a dime and roll a die. If your number turns up, the game ends. If not, you repeat: Pay 10 cents and roll. This continues until your number turns up.

Your friend, however, should pay you for the privilege of receiving your dimes this way. How much should he or she pay if the game is to be fair? What does "fair" mean here, anyway?

One of the goals of this chapter is to teach you how to answer questions like these.

(1.2) Although it began with the study of gambling some centuries ago, the theory of probability is now a serious branch of mathematics with applications to many sciences. This chapter briefly introduces the most basic ideas of probability—sample space, event, independence, random variable, expected value, and variance. After that come several examples and applications, including derivation of a result in queueing theory.

2. Sample Spaces and Events

(2.1) A **sample space** is defined as a finite set S together with a function

$$\mathbf{Pr} : S \longrightarrow U := \{t;\ t \in \mathbf{R}, 0 \leq t \leq 1\}$$

such that

$$\sum_{a \in S} \mathbf{Pr}(a) = 1. \qquad (2.2)$$

Strictly, we denote this sample space by $(S, \mathbf{Pr})$.

Examples (2.3) (*i*) $S_1 := \{1, \ldots, n\}, \mathbf{Pr}_1(a) := 1/n$ for all $a \in S_1$.

(*ii*) Make S_2 the set of the 26 letters of the alphabet and for all α in S_2, define $\mathbf{Pr}_2(\alpha)$ as the proportion of letters equal to α in Lincoln's Gettysburg Address. Thus, if the total number of letters in the words of this speech were 5,000 and the letter a appeared 200 times, then $\mathbf{Pr}_2(a)$ would be $200/5000 = 0.04$.

We extend $\mathbf{Pr}$ to $\mathcal{P}(S)$ by the rule: for all $A \subseteq S$,

(2.4)
$$\mathbf{Pr}(A) := \sum_{a \in A} \mathbf{Pr}(a).$$

Thus in particular $\mathbf{Pr}(\{a\}) = \mathbf{Pr}(a)$ for any $a \in S$. $\mathbf{Pr}$ is called the **probability function** of the sample space.

Example (2.5) For $(S_1, \mathbf{Pr}_1)$ as defined above in (2.3), $\mathbf{Pr}_1(\{1,2\}) = 1/n + 1/n = 2/n$. For $(S_2, \mathbf{Pr}_2)$, supposing that the letter z does not appear in the Gettysburg Address we find $\mathbf{Pr}_2(\{a, z\}) = \mathbf{Pr}_2(a) + \mathbf{Pr}_2(z) = \mathbf{Pr}_2(a)$.

Practice (2.6P) For any two subsets A, B, prove

$$\mathbf{Pr}(A \cup B) = \mathbf{Pr}(A) + \mathbf{Pr}(B) - \mathbf{Pr}(A \cap B).$$

(2.7) An **event** in a sample space S is by definition a subspace of S. The **probability of the event** A is by definition $\mathbf{Pr}(A) := \sum_{a \in A} \mathbf{Pr}(a)$.

Two events A, B in the sample space S are called **independent** if and only if

(2.8)
$$\mathbf{Pr}(A \cap B) = \mathbf{Pr}(A)\mathbf{Pr}(B).$$

Examples (2.9) (*i*) For S_1 as in (2.3) we may take $A := \{1, 3, 5, \ldots\}$ and $B := \{2, 4, 6, \ldots\}$, i.e., $A := \{x;\ x \in S_1, x \text{ is odd}\}$ and $B := \{x;\ x \in S_1, x \text{ is even}\}$. Now $A \cap B = \varnothing$ and $\mathbf{Pr}_1(A) = 1/2 = \mathbf{Pr}_1(B)$ (if n is even), but $\mathbf{Pr}_1(A \cap B) = 0$, so A and B are not independent events, since $\mathbf{Pr}_1(A)\mathbf{Pr}_1(B) = (1/2)^2 = 1/4$. Thus independence may at first appear a nonintuitive concept.

(*ii*) But consider this example: We take $n = 6m$ and define $A' := \{x;\ x \in S_1,\ 2 \mid x\}$ and $B' := \{x;\ x \in S_1, 3 \mid x\}$. That is, the events are "being a multiple of 2" and "being a multiple of 3." Now $A' \cap B'$ is the event "being a multiple of 6," i.e., $A' \cap B' = \{x;\ x \in S_1, 6 \mid x\}$. Now $\mathbf{Pr}_1(A') = 1/2$, $\mathbf{Pr}_1(B') = 1/3$ and $\mathbf{Pr}_1(A' \cap B') = 1/6 = \mathbf{Pr}_1(A')\mathbf{Pr}_1(B')$, so A' and B' are independent events.

The idea of independence is that even if you know that the element x of S has one property (belongs to the event A), you still have no more information than you'd have without this knowledge about whether x has the other property (belongs to the event B). With this understanding, it becomes clear why disjoint events cannot be independent (unless one has probability 0).

(2.10) To continue with the previous example, (2.9), take $n = 12$ and suppose an oracle, selecting a point x from S_1, tells us only that x is in $A' = \{2, 4, 6, 8, 10, 12\}$. Assume that we want to know, however, if x is in B'. In other words, is x in

$A' \cap B' = \{6, 12\}$? Since x is one of the six numbers in A', the probability that $x = 6$ or $x = 12$ is $\frac{1}{6} + \frac{1}{6} = \frac{1}{3}$. (Since we *know* that x is in A', the probability of each individual point in A' is $\frac{1}{6}$ in this situation because we are temporarily in a new sample space A' instead of S_1.)

But now start from S_1, choose a point at random, and calculate the probability that it is in B'. The answer is

$$\mathbf{Pr}_1(B') = \mathbf{Pr}_1(\{3, 6, 9, 12\}) = \tfrac{4}{12} = \tfrac{1}{3}.$$

In other words, x in A' has probability $\frac{1}{3}$ of being in B', and x in S_1 has the same probability of being in B'. The oracle was of no help. Being told x was in A' gave us no information about whether x was in B'. That's why we call the two events independent.

(2.11) To carry it further, keep $n = 12$ and define $C' := \{2, 3, 5, 7, 11\}$. Now $B' \cap C' = \{3\}$ and $\mathbf{Pr}_1(B' \cap C') = \frac{1}{12}$. But $\mathbf{Pr}_1(B') = \frac{1}{3}$ and $\mathbf{Pr}_1(C') = \frac{5}{12}$, so $\mathbf{Pr}_1(B')\mathbf{Pr}_1(C') = \frac{1}{3}\frac{5}{12} = \frac{5}{36} \neq \frac{1}{12}$. Thus B' and C' are not independent events. If the oracle chose an x from S_1 and told us that it is in C', we'd know that its probability of being in B' was only $\frac{1}{5}$. That's more information than we'd have had without the oracle's help. Why? Because C' and B' are not independent events.

A collection of n events $A_1, \ldots, A_n$ ($n \geq 2$) in S is called **independent** if and only if for *every* subcollection $A_{j_1}, \ldots, A_{j_k}$

$$\mathbf{Pr}(A_{j_1} \cap \cdots \cap A_{j_k}) = \mathbf{Pr}(A_{j_1}) \cdots \mathbf{Pr}(A_{j_k}). \quad (2.12)$$

(2.13) **Letters in a Word.** Problems concerning choice of letters from a given word are like this:

Example

Suppose the word is *beet*. You always choose each letter in the word with equal probability, but the repetition of letters affects the resulting probability function on the sample space. Suppose you ask, "What is the probability that a letter chosen from *beet* is a vowel?" The sample space is $\{b(\frac{1}{4}), e(\frac{1}{4}), e'(\frac{1}{4}), t(\frac{1}{4})\}$, which is represented this way temporarily for clarity. The word is *bee't*, to distinguish the first and second *e*'s from each other; and the value of the probability function is in the parentheses after the letter. Now the sample space proper is $\{b(\frac{1}{4}), e(\frac{1}{2}), t(\frac{1}{4})\}$.

The event "is a vowel" is $\{e\}$ and its probability is $\mathbf{Pr}(e) = \frac{1}{2}$. The event "is a consonant" is $\{b, t\}$ and $\mathbf{Pr}(\{b, t\}) = \frac{1}{4} + \frac{1}{4} = \frac{1}{2}$.

Suppose you ask, "What is the probability that two letters chosen from *beet* are different from each other?" The sample space is now (in the first way)

$$S_2 = \{be, be', bt, ee', et, e't\};$$

each point has probability $1/6$. Note that *eb* is considered the same as *be* because you are working with 2-subsets of letters from *beet*. Note also that *bb* and *tt* are not in S_2. Thus, S_2 is properly

$$S_2 = \{be(1/3),\ bt(1/6),\ ee(1/6),\ et(1/3)\}.$$

The event "letters are different" is

$$\{ be, bt, et \}$$

with probability $1/3 + 1/6 + 1/3 = 5/6$.

With longer words, these lists become impractical and unappealing, and you must use your counting techniques to get the values of **Pr**. For the first question it's easy: *Beet* has two vowels and two consonants, so the probability that one letter is a vowel is $2/(2+2) = 1/2$. For the second question there are $\binom{4}{2} = 6$ 2-"subsets" of letters, of which only *ee* is not in the defined event, so the probability is $1 - \mathbf{Pr}(ee) = 1 - 1/6 = 5/6$. How do we know $\mathbf{Pr}(ee) = 1/6$? Because as a choice of two letters from *beet* it arises in only one way.

Suppose the question were, "What is the probability that two letters chosen from *beet* cannot be used to spell an English word?" Since *be* is the only such word, our answer is $1 - \mathbf{Pr}(be)$. But *be* arises from two 2-subsets, so it has probability $1/6 + 1/6 = 1/3$.

(2.14) **Odds.** We sometimes speak of the **odds** of an event E; the odds of, or for, E are by definition

$$\text{(2.15)} \qquad \text{Odds of E} := \frac{\mathbf{Pr}(E)}{\mathbf{Pr}(\overline{E})} = \frac{\mathbf{Pr}(E)}{1 - \mathbf{Pr}(E)}.$$

Odds are often expressed in such language as "x to y" for positive x and y, which means that the odds are x/y.

If you know the odds r of an event, you know its probability p; i.e.,

(2.16) For each r satisfying $0 \leq r < \infty$, there is a unique p such that

$$\frac{p}{1-p} = r,$$

$$\text{and} \quad 0 \leq p < 1.$$

Practice (2.17P) Prove the foregoing statement (2.16).

Example (2.18) You might hear that the odds of winning something in a lottery are 1 to 5. This means that the probability of winning is $\frac{1}{1+5} = \frac{1}{6}$.

Practice (2.19P) If the odds of E are x to y, i.e., x/y (for positive x and y), prove that $\mathbf{Pr}(E) = \frac{x}{x+y}$.

(2.20) *Another use of "odds."* Two people, A and B, bet against each other on an athletic contest. Suppose a tie is impossible, and A gives odds to B, say, 2 to 1. These odds represent merely an agreement between the bettors: If A's team wins, B gives A 1 cent; if not, A gives B 2 cents. The probability (if it can even be defined) that the stronger team wins may be different than the 2/3 suggested by these "odds."

3. Random Variables

A **random variable** is by definition a function

$$\text{(3.1)} \qquad f : S \longrightarrow \mathbf{R},$$

where S is a sample space (and $\mathbf{R}$ is, of course, the set of all real numbers). Any such function is called a random variable.

Example (3.2) We could define random variables f, g on S_1 by the rules

$$f(x) := \begin{cases} 0 & \text{if } x \text{ is even} \\ 1 & \text{if } x \text{ is odd} \end{cases}$$

$$g(x) := \begin{cases} 0 & \text{if } 3 \mid x \\ 1 & \text{if } 3 \mid (x-1) \\ 2 & \text{if } 3 \mid (x-2) \end{cases}$$

for all x in S_1. But there need not be a "nice" rule for defining a random variable; it may be *any* real-valued function defined on a sample space. We could take

$$h(i) := \text{the } i\text{th decimal digit of } \pi$$

as a rule defining a random variable h on S_1.

To the random variable $f : S \longrightarrow \mathbf{R}$ we associate the **probability distribution** of f, also denoted **Pr**, defined by the rule: For all $y \in \mathbf{R}$

$$\mathbf{Pr}(f = y) := \mathbf{Pr}(\{a; f(a) = y\}) = \mathbf{Pr}(f^{-1}(y)). \tag{3.3}$$

On the right-hand side of $:=$, **Pr** is the probability function defined for S in (2.4). That is, if $y \in \mathbf{R}$ then $f^{-1}(y)$ is an event in S and so **Pr** is defined for that event. $\mathbf{Pr}(f = y)$ is 0 if y is not in the range of f.

Example (3.4) For the functions of (3.2) just above, taking $n = 6m$ we find $\mathbf{Pr}_1(f = 0) = \mathbf{Pr}_1(f^{-1}(0)) = \sum_{2i \in S_1} \mathbf{Pr}_1(2i) = 1/2$. Also $\mathbf{Pr}_1(f = 1) = \mathbf{Pr}_1(f^{-1}(1)) = \sum_{2i+1 \in S_1} \mathbf{Pr}_1(2i+1) = 1/2$. Finally, $\mathbf{Pr}(f = y) = 0$ if $y \neq 0, 1$. Similarly $\mathbf{Pr}_1(g = 0) = \mathbf{Pr}_1(g = 1) = \mathbf{Pr}_1(g = 2) = 1/3$, and $\mathbf{Pr}_1(g = y) = 0$ if $y \notin \{0, 1, 2\}$.

(3.5) If we denote the range of the random variable f by T (thus $T \subseteq \mathbf{R}$) and define a function $\mathbf{Q} : T \longrightarrow U$ by the rule

$$\forall y \in T \quad \mathbf{Q}(y) := \mathbf{Pr}(f = y),$$

then $(T, \mathbf{Q})$ is a sample space.

(3.5F)

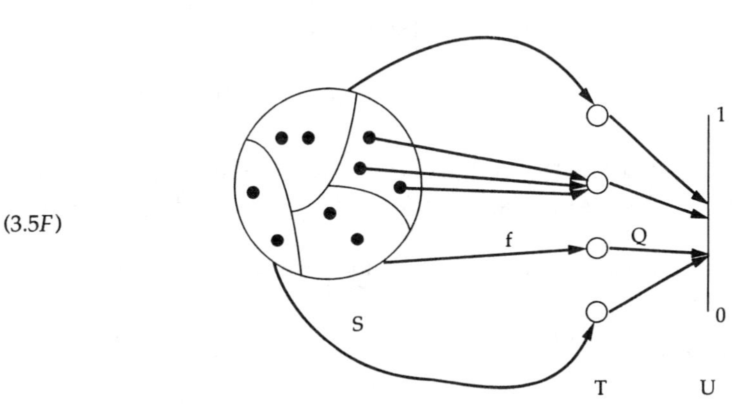

Practice (3.6P) Prove (3.5).

Example (3.7) For the f of (3.2) we get $T = \{0, 1\}$ and $\mathbf{Q}(0) = \mathbf{Q}(1) = 1/2$ if n is even. If $3 \mid n$ and we take g, we get $T = \{0, 1, 2\}$ and $\mathbf{Q}(0) = \mathbf{Q}(1) = \mathbf{Q}(2) = 1/3$.

$(T, \mathbf{Q})$ is an abstract version of $(S, \mathbf{Pr})$ in which the only events are $\varnothing$ and the $f^{-1}(y)$ for $y \in T$, and unions of these.

4. Expected Value

Let f be a random variable on the sample space $(S, \mathbf{Pr})$. The **expected value**, or **expectation**, of f is defined as

(4.1)
$$Ef := \sum_{a \in S} f(a)\mathbf{Pr}(a).$$

Sometimes Ef is called the **mean** of f.

Example (4.2) If $S = \{1, \ldots, n\}$ and $\mathbf{Pr}(a) = 1/n$ for each $a \in S$, then $Ef := (1/n)\sum f(a)$ is just the arithmetic mean of $f(1), f(2), \ldots, f(n)$.

Let f and g be random variables on the sample space $(S, \mathbf{Pr})$. Then the functions $f + g$ and fg defined by the rules: $\forall a \in S$

(4.3)
$$\begin{aligned}(f + g)(a) &:= f(a) + g(a)\\ (fg)(a) &:= f(a)g(a)\end{aligned}$$

are also random variables on S. For the first of these we easily see the following result.

Proposition (4.4) $E(f + g) = Ef + Eg$. For all constants c, $E(cf) = cEf$, and $E(c) = c$.

Proof By definition (4.1),

$$E(f + g) = \sum_{a \in S} [(f + g)(a)]\mathbf{Pr}(a).$$

By definition (4.3), however, this equals

$$\sum_{a \in S} (f(a) + g(a))\mathbf{Pr}(a) = \sum [f(a)\mathbf{Pr}(a) + g(a)\mathbf{Pr}(a)].$$

This is $Ef + Eg$. The second part follows immediately from the distributive law. QED

(4.5) The result (4.4) extends to any nonempty finite set of random variables.

Practice (4.6P) Prove $Ec = c$ for constant c. First define the random variable precisely.

Practice (4.7P) Prove (4.5).

5. Variance

The **variance** of the random variable f is defined to be

(5.1) $$\mathrm{Var}\, f := E((f - Ef)^2).$$

Proposition (5.2) $\mathrm{Var}\, f = E(f^2) - (Ef)^2$.

Proof The proof is simple. Notice that $Ef =: u$ is just some real number and that $(f - u)^2$ and f^2 are squares (see (4.3) for definition) of random variables. Then

$$\begin{aligned}\mathrm{Var}\, f : &= E(f - u)^2 = E(f^2 - 2uf + u^2)\\ &= E(f^2) - 2uEf + Eu^2,\end{aligned}$$

by (4.4). But $Ef = u$ and $Eu^2 = u^2$ since u is constant. Therefore

$$\mathrm{Var}\, f = E(f^2) - 2u^2 + u^2 = E(f^2) - u^2.$$ QED

In working with the variance we usually use (5.2) instead of (5.1).

Example (5.3) Let's find the expected value and variance of the f of (3.2). As we remarked, Ef is the arithmetic mean because **Pr** is a constant function. Thus $Ef = 1/2$ if n is even (which we assume). Now, by (5.2), $\mathrm{var}\, f = E(f^2) - (Ef)^2$. But since $f^2 = f$ for this function, $Ef^2 = Ef = 1/2$. Thus $\mathrm{var}\, f = 1/2 - 1/4 = 1/4$.

Practice (5.4P) Find $\mathrm{Var}\, g$ for the g of (3.2). Assume $3 \mid n$.

6. Independence

We have defined independence for events (subsets of the sample space S). We now extend the definition to random variables.

Let $f : S \longrightarrow \mathbf{R}$ and $g : S \longrightarrow \mathbf{R}$ be random variables on the sample space $(S, \mathbf{Pr})$.

DEFINITION (6.1) The random variables f and g are called **independent** if and only if $\forall y \in \mathrm{range}\, f$ and $\forall z \in \mathrm{range}\, g$

$$\begin{aligned}&\mathbf{Pr}(\{\, a;\ a \in S,\ f(a) = y,\ \text{and } g(a) = z \,\})\\ &\qquad = \mathbf{Pr}(\{\, a;\ a \in S,\ f(a) = y \,\})\mathbf{Pr}(\{\, a;\ a \in S,\ g(a) = z \,\}).\end{aligned}$$

More succinctly: $\mathbf{Pr}(f^{-1}(y) \cap g^{-1}(z)) = \mathbf{Pr}(f^{-1}(y))\mathbf{Pr}(g^{-1}(z))$.

(6.2) It follows immediately from Definition (2.8) that f and g are independent random variables iff $\forall y, z \in \mathbf{R}, f^{-1}(y)$ and $g^{-1}(z)$ are independent events. Because $\mathbf{Pr}(\emptyset) = 0$ we need not require y and z to be in the ranges of f and g, respectively.

Example (6.3) Consider the random variables f and g of (3.2). We'll check that they are independent for the case $n = 6m$. First notice that

$$
\begin{array}{ll}
 & \textbf{Pr} \\
f^{-1}(0) = \{2,4,6,\ldots,6m\} & \mathbf{Pr}_1(f = 0) = 1/2 \\
f^{-1}(1) = \{1,3,5,\ldots,6m-1\} & \mathbf{Pr}_1(f = 1) = 1/2 \\
g^{-1}(0) = \{3,6,\ldots,6m\} & \mathbf{Pr}_1(g = 0) = 1/3 \\
g^{-1}(1) = \{1,4,7,\ldots,6m-2\} & \mathbf{Pr}_1(g = 1) = 1/3 \\
g^{-1}(2) = \{2,5,8,\ldots,6m-1\} & \mathbf{Pr}_1(g = 2) = 1/3
\end{array}
$$

There are six intersections $f^{-1}(y) \cap g^{-1}(z)$ for $y = 0,1$ and $z = 0,1,2$. It is easy to see directly that each of the six has probability $1/6 = (1/2)(1/3) = \mathbf{Pr}_1 f^{-1}(y)\mathbf{Pr}_1(g^{-1}(z))$. We could even use the Chinese Remainder Theorem (Ch. 7, (7.5)): The set of all solutions x to the simultaneous congruences $x \equiv y \pmod 2$ and $x \equiv z \pmod 3$ is a congruence class mod 6.

7. Mean and Variance of Independent Random Variables

There are two results. The first is

Theorem (7.1) If f and g are independent variables on the sample space $(S, \mathbf{Pr})$, then

$$E(fg) = (Ef)(Eg). \blacksquare$$

To prove Theorem (7.1) we'll use the lemma below on characteristic functions. Recall from Chapter 5, (19), that if $A \subseteq S$, the characteristic function χ_A of A takes the value 1 at each point of A and 0 on $\overline{A}$.

Lemma (7.2) The characteristic functions χ_A, χ_B of independent events A, B are independent random variables, and

(7.3)
$$E(\chi_A\chi_B) = (E\chi_A)(E\chi_B).$$

Proof of (7.2) We leave the proof of the first part to the problems. For the second part, (7.3), we see that

$$
\begin{aligned}
E(\chi_A\chi_B) :&= \sum_{a\in S} \mathbf{Pr}(a)\chi_A(a)\chi_B(a) \\
&= \sum_{a\in A\cap B} \mathbf{Pr}(a) \\
&= \mathbf{Pr}(A \cap B) = \mathbf{Pr}(A)\mathbf{Pr}(B).
\end{aligned}
$$

The last equality holds because A and B are independent. But

$$
\begin{aligned}
\mathbf{Pr}(A) :&= \sum_{a\in A} \mathbf{Pr}(a) = \sum_{a\in S} \mathbf{Pr}(a)\chi_A(a) \\
&:= E\chi_A.
\end{aligned}
$$

Hence $E(\chi_A\chi_B) = (E\chi_A)(E\chi_B)$. QED

Proof of Theorem (7.1) We express f and g in terms of the characteristic functions of the cells of the partitions of S induced by f and g. We use Chapter 5, (9.5). Thus, let Rf $[Rg]$ be the range of f $[g]$; then

(7.4) $$f = \sum_{y \in Rf} y\chi_y, \qquad g = \sum_{z \in Rg} z\theta_z,$$

where $\chi_y : S \longrightarrow \{0, 1\}$ is the characteristic function of $f^{-1}(y) \subseteq S$ and θ_z is that of $g^{-1}(z) \subseteq S$. We now explain (7.4).

The meaning of (7.4) is this: For any a in S,

$$f(a) = \sum_{y \in Rf} y \cdot \chi_y(a),$$

and the dot stands for multiplication of the two real numbers y and $\chi_y(a)$.

Why is (7.4) true? It is true because for all $a \in S$, $\chi_y(a) = 0$ if $y \neq f(a)$; and $\chi_{f(a)}(a) = 1$. In other words, every term in the sum is 0 except when $y = f(a)$; then we get $f(a)\chi_{f(a)}(a) = f(a) \cdot 1 = f(a)$. Now using (7.4) we see that

(7.5) $$fg = \sum_{\substack{y \in Rf \\ z \in Rg}} yz\chi_y\theta_z.$$

Apply E to (7.5) and get, using (4.5),

(7.6) $$E(fg) = \sum_{\substack{y \in Rf \\ z \in Rg}} yzE(\chi_y\theta_z).$$

Now we use the independence of f and g, which tells us that $f^{-1}(y)$ and $g^{-1}(z)$ are independent events for all $y \in Rf, z \in Rg$. That is, χ_y and θ_z are the characteristic functions of independent events. Therefore, by Lemma (7.2),

$$E(\chi_y\theta_z) = (E\chi_y)(E\theta_z).$$

Using this in (7.6) we get

$$\begin{aligned} E(fg) &= \sum yzE\chi_y E\theta_z \\ &= \sum_{y \in Rf} yE\chi_y \sum_{z \in Rg} zE\theta_z \\ &= (Ef)(Eg). \end{aligned}$$

QED

The second result is

Theorem (7.7) If f and g are independent random variables on the sample space $(S, \mathbf{Pr})$, then

$$\text{Var}(f + g) = (\text{Var} f) + (\text{Var} g).$$

Proof We use (5.2), then (4.4).

$$\begin{aligned}\text{Var}(f+g) &= E((f+g)^2) - (E(f+g))^2 \\ &= E(f^2+2fg+g^2) - (Ef+Eg)^2 \\ &= E(f^2) + 2E(fg) + E(g^2) - (Ef)^2 - 2(Ef)(Eg) - (Eg)^2 \\ &= E(f^2) - (Ef)^2 + E(g^2) - (Eg)^2 + 2[E(fg) - (Ef)(Eg)].\end{aligned}$$

The term in brackets is 0 because f and g are independent, by Theorem (7.1). Since by (5.2) $E(f^2) - (Ef)^2 = \text{Var} f$, we are done. QED

8. Real Life

(8.1) So what? Well might you ask that question of all that precedes.

The idea of sample space is meant to be an abstract mathematical model of real situations in which "chance" or "randomness" (undefined terms!) plays a role. For example, let the sample space S_1 be the six faces $\{D_1, \ldots, D_6\}$ of a die, and define $\mathbf{Pr}_1(D_i) = 1/6$ for each face. Define a random variable f_1 on S_1 by the rule

$$f_1(D_i) = i \text{ (the number on that face).}$$

Thus the range of f_1 is $\{1, \ldots, 6\}$. It is our earlier S_1 with $n = 6$ if we use (3.5), the transfer of the probability function on the domain S to the range of the random variable.

If we roll an honest die we expect that each face is equally likely to come up. We model the honest die with $(S_1, \mathbf{Pr}_1)$ and f_1. Rolling the die corresponds in the model to "selecting a face D_i with probability $\mathbf{Pr}_1(D_i)$" and evaluating $f_1(D_i)$.

In general the value of the probability function $\mathbf{Pr}$ at $a \in S$ is supposed to equal the likelihood of the "selection" of a. The term "sample" in "sample space" has the sense of the verb "sample": "Pick one and see what you get." In real life, we would let the roll of the die be our selection or sampling process in $(S_1, \mathbf{Pr}_1)$. Thus, if the die were loaded so that D_6 came up $\frac{1}{4}$ of the time and D_1 $\frac{1}{12}$ of the time (with the others $\frac{1}{6}$ as before), then we would model it with $(S_1, \mathbf{Pr}_1')$, where

$$\mathbf{Pr}_1'(D_1) = \frac{1}{12} \qquad \mathbf{Pr}_1'(D_6) = \frac{1}{4}$$

$$\mathbf{Pr}_1'(D_i) = \frac{1}{6}, \qquad 1 < i < 6.$$

An *event* in S_1 is a subset of S_1. So $\{D_1, D_2, D_3\} = F_1$ is an event and $\mathbf{Pr}_1(\{D_1, D_2, D_3\}) = \frac{1}{6} + \frac{1}{6} + \frac{1}{6} = \frac{1}{2}$. Interpretation: When we roll the die, the probability (of the event) that a 1, 2, or 3 shows is $\frac{1}{2}$. The model fits our intuition reasonably well. The event $F_2 = \{D_2, D_3, D_4, D_5\}$ has probability $\frac{4}{6} = \frac{2}{3}$; and F_1 and F_2 are independent events, because

$$\mathbf{Pr}_1(F_1 \cap F_2) := \mathbf{Pr}_1(\{D_2, D_3\}) := \frac{1}{6} + \frac{1}{6} = \frac{1}{3}$$

$$= \mathbf{Pr}_1(F_1)\mathbf{Pr}_1(F_2) := \frac{1}{2} \cdot \frac{2}{3} = \frac{1}{3}.$$

Practice (8.2P) Find more examples of independent sets in $(S_1, \mathbf{Pr}_1)$ and $(S_1, \mathbf{Pr}_1')$.

(8.3) *Random variables*. The term "random variable" is just a name. We defined it to mean a function on a sample space. Notice we have not defined the term "random." A random variable need not be "random" at all; a constant function is, after all, a random variable. The use of random variables is to measure attributes of the elements of a sample space. The f_1 that we defined before tells us the number of dots on the face of the die. We could take a set of n people for S_2, with $\mathbf{Pr}_2(x) = 1/n$ for each $x \in S_2$, and define several different random variables on S_2:

$$\begin{aligned} f_2(x) :&= \text{the height of } x, \\ f_2'(x) :&= \text{the number of years of schooling } x \text{ has had}, \\ f_2''(x) :&= \text{the number of siblings of } x. \end{aligned}$$

These exemplify measures of attributes.

We might use the range of a random variable as a sample space (as in (3.5)). For the die with $(S_1, \mathbf{Pr}_1)$ and f_1 we would substitute $\{1, \ldots, 6\}$ for S_1 and define the associated probability function as $\mathbf{Q}_1(i) = \mathbf{Pr}_1(f_1^{-1}(i)) = \mathbf{Pr}_1(D_i) = \frac{1}{6}$. If we used $\mathbf{Pr}_1'$ (for the loaded die) we would have $\mathbf{Q}_1'(1) = \frac{1}{12}$, $\mathbf{Q}'(6) = \frac{1}{4}$, and $\mathbf{Q}_1'(i) = \frac{1}{6}$ for $1 < i < 6$.

The expected value of a random variable is a number that tells something about the random variable. There is an analogy with physics: If $\mathbf{Pr}(a)$ is the distance from the origin of the point a on the line at which a mass $f(a)$ is placed, then Ef is the center of gravity of the system of such masses.

(8.4) For the die the expected value of f_1 is

$$\begin{aligned} Ef_1 &= \frac{1}{6} \cdot 1 + \frac{1}{6} \cdot 2 + \cdots + \frac{1}{6} \cdot 6 \\ &= \frac{1}{6}(1 + 2 + \cdots + 6) = \frac{1}{6} \cdot \frac{6 \cdot 7}{2} = \frac{7}{2}. \end{aligned}$$

Thus the expected value of a random variable need not be an actual value of it.

Example (8.5) Let $S_3 = \{2, 3, \ldots, 101\}$ and for all a in S_3, set $P(a) = \frac{1}{100}$ and $f_3(a) :=$ the smallest prime dividing a. Then $f_3^{-1}(2)$ is the set of 50 even integers $\{2, 4, 6, \ldots, 100\}$. $f_3^{-1}(3)$ is the 17-set of odd multiples of 3: $\{3 \cdot 1, 3 \cdot 3, 3 \cdot 5, 3 \cdot 7, \ldots, 3 \cdot 33\}$.

$$\begin{aligned} f_3^{-1}(5) &= \{5, 25, 35, 55, 65, 85, 95\} \\ &= \{5k;\ 1 \le k \le 20, 2 \nmid k, 3 \nmid k\}. \\ f_3^{-1}(7) &= \{7, 49, 77, 91\} \\ f_3^{-1}(11) &= \{11\}. \end{aligned}$$

(This is closely related to the sieve of Eratosthenes, found in Chapter 6, (12).) For all primes q in S_3 larger than $\sqrt{101}$, $f_3^{-1}(q) = \{q\}$. (Why?) The 21 remaining primes in S_3 are 13, 17, 19, 23, 29, 31, 37, 41, 43, 47, 53, 59, 61, 67, 71, 73, 79, 83, 89, 97, 101. Thus, $Ef_3 = \frac{1}{100}(50 \cdot 2 + 17 \cdot 3 + 7 \cdot 5 + 4 \cdot 7 + 11 + 13 + 17 + 19 + 23 + \cdots + 101)$, where the sum is over the primes just listed. Thus, $Ef_3 = 13.58$.

If we regarded the range of f_3 as the sample space, our probability function would be

$$\begin{array}{cccccccc} 2 & 3 & 5 & 7 & 11 & 13 & \cdots & 101 \\ \frac{1}{2} & \frac{17}{100} & \frac{7}{100} & \frac{4}{100} & \frac{1}{100} & \frac{1}{100} & \cdots & \frac{1}{100} \end{array}.$$

The variance of a random variable tells us roughly how much the variable deviates from its expected value. A large variance means a large deviation. The variance of f_1 is, from (5.2),

$$\begin{aligned} \operatorname{Var} f_1 &= E(f_1^2) - (Ef_1)^2 \\ &= \frac{1}{6}(1^2 + 2^2 + \cdots + 6^2) - \left(\frac{7}{2}\right)^2 \\ &= \frac{1}{6}\frac{6 \cdot 7 \cdot 13}{6} - \frac{49}{4} = 3 - \frac{1}{12}; \end{aligned}$$

we used Chapter 3, Problem 6. The variance of f_3 is

$$\frac{1}{100}(50 \cdot 2^2 + 17 \cdot 3^2 + 7 \cdot 5^2 + 4 \cdot 7^2 + 11^2 + 13^2 + \cdots + 101^2) - (13.58)^2 = 581.9236,$$

where again the indicated sum is over the primes between 13 and 101.

(The square root of the variance, called the *standard deviation* of the random variable, is more clearly related to the values of the random variable. We touch on it under Chebyshev's inequality below in (13).)

The standard deviation of f_1 is $(35/12)^{\frac{1}{2}} = 1.71-$; the standard deviation of f_3 is $(581.9236)^{\frac{1}{2}} = 24.12+$. The latter is quite large—almost twice the mean.

Practice (8.6P) Define a random variable g on $S = \{1, \ldots, 10\}$ with $\mathbf{Pr}(a) = \frac{1}{10}$ for all $a \in S$ by: $g(a) = 0$ if $a < 10$, and $g(10) = y$. Calculate Eg, $\operatorname{Var} g$, and $(\operatorname{Var} g)^{\frac{1}{2}}$.

To find examples of independent random variables we could follow the tame procedure of constructing independent sets and taking their characteristic functions as random variables. They are independent by Lemma (7.2).

(8.7) An example of nonindependent random variables is f_3 (above) and f_4, where $f_4(a) :=$ the second smallest prime divisor of a, unless a is a power of a prime, in which case $f_4(a) := 1$. Now $f_3^{-1}(29) = \{29\}$. Notice that $f_4^{-1}(7)$ is nonempty and disjoint from $\{29\}$ because it consists of certain multiples of 7, for example 14. Disjoint sets A, B are never independent unless $\mathbf{Pr}(A) = 0$ or $\mathbf{Pr}(B) = 0$.

(8.8) A "better" example of independent random variables arises from the sample space $(\{0,1\}^n, \mathbf{Pr}_5)$ where $\mathbf{Pr}_5(x) := 2^{-n}$ for all $x \in \{0,1\}^n$. We may regard the elements of $\{0,1\}^n$ as the base-2 representations of the integers $0, 1, \ldots, 2^n - 1$. We define n random variables $g_1, \ldots, g_n$ as follows: For $i = 1, \ldots, n$ and $x = x_1x_2 \ldots x_n \in \{0,1\}^n$, $g_i(x) := x_i$.

It is easy to see that g_1 and g_2 are independent, for example. Suppose we list the sample space and indicate the four subsets $g_1^{-1}(0)$, $g_1^{-1}(1)$, $g_2^{-1}(0)$, and $g_2^{-1}(1)$.

$$
\begin{array}{cccccccc}
 & 0 & 0 & \cdot & \cdot & \cdot & 0 & \\
 & 0 & 0 & \cdot & \cdot & \cdot & 1 & g_2^{-1}(0) \\
 & & & \vdots & & & & \\
g_1^{-1}(0) & 0 & 0 & 1 & 1 & \cdots & 1 & \\
 & 0 & 1 & 0 & \cdot & \cdot & 0 & \\
 & 0 & 1 & 0 & \cdot & \cdot & 1 & g_2^{-1}(0) \\
 & & & \vdots & & & & \\
 & 0 & 1 & 1 & \cdot & \cdot & 1 & \\
 & 1 & 0 & 0 & \cdot & \cdot & 0 & \\
 & & \vdots & & & & & g_2^{-1}(0) \\
g_1^{-1}(1) & 1 & 0 & 1 & 1 & \cdots & 1 & \\
 & 1 & 1 & 0 & 0 & \cdots & 0 & \\
 & & \vdots & & & & & g_2^{-1}(1) \\
 & 1 & 1 & 1 & 1 & \cdots & 1 &
\end{array} \tag{8.9}
$$

Now $g_1^{-1}(0) \cap g_2^{-1}(0)$ is the first 2^{n-2} vectors in the list, a set on which $\mathbf{Pr}_5$ has the value $\frac{1}{4} = \mathbf{Pr}_5(g_1^{-1}(0))\mathbf{Pr}_5(g_2^{-1}(0)) = (\frac{1}{2}) \cdot (\frac{1}{2})$. The result is similar for the other three cases that must be verified to show that g_1 and g_2 are independent. Thus they *are* independent, and, likewise, so are g_i and g_j for any $i \neq j$.

Furthermore, define $h = g_1 + g_2$. Then h and g_3 are independent. To prove this claim, note that the range of h is $\{0, 1, 2\}$ and that

$$
\begin{aligned}
h^{-1}(0) &= g_1^{-1}(0) \cap g_2^{-1}(0) \\
h^{-1}(1) &= [g_1^{-1}(0) \cap g_2^{-1}(1)] \cup [g_1^{-1}(1) \cap g_2^{-1}0)] \\
h^{-1}(2) &= g_1^{-1}(1) \cap g_2^{-1}(1).
\end{aligned} \tag{8.10}
$$

We can see by looking at the vectors displayed in (8.9) that g_3 is 0 on half the vectors of each of these three subspaces in (8.10) and that g_3 is 1 on the other half. Therefore h and g_3 are independent random variables.

Practice (8.11P) Show that $g_1 + g_2$ and $g_2 + g_3$ are not independent. (After you solve it, please read the answer.)

9. Repeated Trials

Consider rolling a die twice. Each roll is a sampling to see what face comes up. We may also call the roll a "trial."

The model here is $S_1 \times S_1$. The outcome of the sampling process is the ordered pair (a,b), where $a,b \in \{D_1,\ldots,D_6\} = S_1$.†

> Since the die has no memory, the outcome of the second roll is governed by the same probability function that governed the first roll.

The probability that our first roll "selects" $a \in S_1 = \{D_1,\ldots,D_6\}$ is $\mathbf{Pr}_1(a)$. The probability that our second roll "selects" $b \in S_1$ is $\mathbf{Pr}_1(b)$, as we just remarked. Therefore the probability that on two successive rolls we "select" (a,b) is $\mathbf{Pr}_1(a)\mathbf{Pr}_1(b)$. Thus we define our probability function $\mathbf{Pr}$ on $S_1 \times S_1$ as

$$\mathbf{Pr}(a,b) = \mathbf{Pr}_1(a)\mathbf{Pr}_1(b), \tag{9.1}$$

and this definition makes $(S_1 \times S_1, \mathbf{Pr})$ a sample space. The definition of Cartesian product captures the intuitive idea of independence, for to get all $(x,y) \in A \times B$ we fill in each coordinate in all possible ways independently of the value in the other coordinate. The definition (9.1) of the probability function $\mathbf{Pr}$ on $S_1 \times S_1$ seems to be just a natural recognition of the underlying nature of Cartesian products.

Practice (9.2P) Prove that $(S_1 \times S_1, \mathbf{Pr})$ is sample space.

Recall that the random variable f_1 maps D_i to i for each $i = 1,\ldots,6$. We use f_1 to define two random variables on $S_1 \times S_1$. For all $(a,b) \in S_1 \times S_1$

$$\begin{aligned} h_1(a,b) &:= f_1(a) \\ h_2(a,b) &:= f_1(b). \end{aligned} \tag{9.3}$$

Now h_1 and h_2 are independent random variables on $S_1 \times S_1$! Why? Consider, for example, $h_1^{-1}(3)$ and $h_2^{-1}(4)$.

$$\begin{aligned} h_1^{-1}(3) &= \{(D_3,D_j);\ j = 1,\ldots,6\}. \\ h_2^{-1}(4) &= \{(D_i,D_4);\ i = 1,\ldots,6\}. \end{aligned}$$

By our definitions (9.1) and (2.4),

$$\mathbf{Pr}(h_1^{-1}(3)) = \frac{1}{36} + \cdots + \frac{1}{36} = \frac{1}{6} = \mathbf{Pr}(h_2^{-1}(4)).$$

Also $h_1^{-1}(3) \cap h_2^{-1}(4) = \{(D_3,D_4)\}$, for which $\mathbf{Pr}$ has the value $\frac{1}{36} = \frac{1}{6}\cdot\frac{1}{6} =: \mathbf{Pr}(h^{-1}(3))\mathbf{Pr}(h^{-1}(4))$.

There is nothing special about 3 and 4 in this example; we can show in the same way that for all $k,l \in \{1,\ldots,6\}$

$$\mathbf{Pr}(h_1^{-1}(k) \cap h_2^{-1}(l)) = \mathbf{Pr}(h_1^{-1}(k))\mathbf{Pr}(h_2^{-1}(l)). \tag{9.4}$$

Therefore h_1 and h_2 are independent random variables.

We generalize the foregoing example as follows.

Theorem (9.5) Let $(S,\mathbf{Pr})$ and $(T,\mathbf{Q})$ be any two sample spaces. Let f and g be random variables on S and T, respectively.

†This sample space is so close to our earlier S_1 with $n = 6$ that we use the same notation for it.

Define a probability function π on $S \times T$ as

(9.6) $$\forall (a,b) \in S \times T \qquad \pi(a,b) := \mathbf{Pr}(a)\mathbf{Q}(b).$$

Then $(S \times T, \pi)$ is a sample space; and the random variables h_1 and h_2 are independent, where, for all $a \in S, b \in T$,

$$h_1(a,b) := f(a)$$
$$h_2(a,b) := g(b).$$

Proof This proof mimics the one for the example just above. The key is the definition of π: We select b with probability $\mathbf{Q}(b)$ independently (in the colloquial sense) of which a we selected. Thus if $y \in \text{range} f$ and $z \in \text{range}\, g$, then

$$h_1^{-1}(y) \cap h_2^{-1}(z)$$

is a "rectangle" in $S \times T$, sketched in (9.7F) below, namely

$$(f^{-1}(y) \times T) \cap (S \times g^{-1}(z))$$
$$= f^{-1}(y) \times g^{-1}(z) = A$$

Evaluating π at each point of this "rectangle" A and summing yields

$$\mathbf{Pr}(f^{-1}(y))\mathbf{Q}(g^{-1}(z)).$$

(Check this claim.) But $\mathbf{Pr}(f^{-1}(y)) = \pi(h_1^{-1}(z))$, and $\mathbf{Q}(g^{-1}(z)) = \pi(h_2^{-1}(z))$. Therefore $\pi(h_1^{-1}(y) \cap h_2^{-1}(z)) = \pi(h_1^{-1}(y))\pi(h_2^{-1}(z))$. QED

(9.7F)

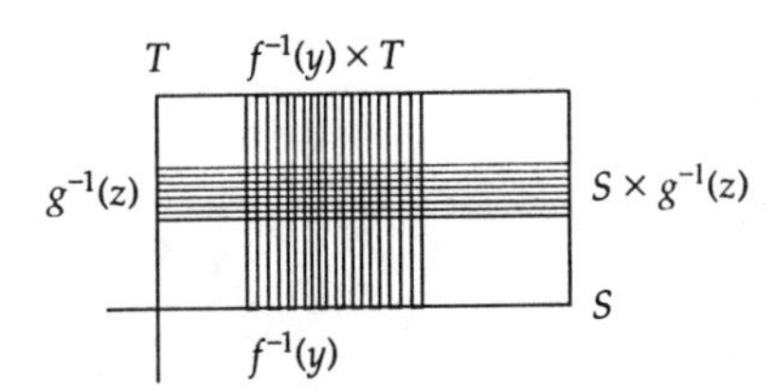

This model $S \times T$ and Theorem (9.5) are expressed only to make our intuitive idea that f and g should be independent under successive trials fit our definition (6.1) of independent random variables.

Now that we have seen how to model two successive trials we can extend the idea to any number of successive trials. In subsequent sections we do that for two important random variables. But first...

(9.8) We can now begin to answer the questions raised in (1.1) about the gambling game played with a single die. The game is fair if your friend pays you as a "starting fee" the amount that you would pay out on an average during a game. And what is your "average" payout? It is the expected value of the amount you pay. In more detail, let's define the random variable

$$f(G) := \text{the amount you pay to the completion of the game } G.$$

G consists of n rolls of the die in which your chosen number ("3" in the example of (1.1)) appears for the first time at the nth roll. Thus we said $f(G) = n$ dimes.

We can classify the games (partition them) by their length n.

A game of length 1 consists of your rolling a "3" first thing. That happens with probability $\frac{1}{6}$, so the probability that your payout is only one dime is $\frac{1}{6}$.

A game of length 2 consists of a roll of a non-"3" followed by a roll of a "3." That has probability $\frac{5}{6} \cdot \frac{1}{6}$, as in (9.1), since the event is $\{1,2,4,5,6\} \times \{3\}$. In any game of length 2 you pay out two dimes.

In general, the probability that the game ends at roll n (has length n) is

$$\left(\frac{5}{6}\right)^{n-1} \frac{1}{6}. \tag{9.9}$$

Therefore, we have this infinite set of games partitioned according to length, and we know the probability of each cell of the partition; it's (9.9). Your payout (in dimes) equals the length of the game.

What, then, would you expect to pay if you played this game? Shouldn't it be the average of the payouts of the cells with the payout of each cell weighted according to the probability that you roll a game in that cell? Namely, in dimes,

$$\frac{1}{6} \cdot 1 + \frac{5}{6} \cdot \frac{1}{6} \cdot 2 + \left(\frac{5}{6}\right)^2 \frac{1}{6} \cdot 3 + \left(\frac{5}{6}\right)^3 \frac{1}{6} \cdot 4 + \cdots = \sum_{1 \le n} \left(\frac{5}{6}\right)^{n-1} \frac{1}{6} \cdot n. \tag{9.10}$$

Formula (9.10) "says" you pay just one dime with probability $\frac{1}{6}$, just two dimes with probability $\frac{5}{6} \cdot \frac{1}{6}$, and so on. You see also that (9.10) is the expected value of f as defined in (4). We discuss this further in (14).

Example (9.11) To evaluate (9.10): It is

$$\frac{1}{6} \sum_{1 \le n} \left(\frac{5}{6}\right)^{n-1} n. \tag{9.12}$$

We differentiate the geometric series

$$\frac{1}{1-x} = \sum_{0 \le n} x^n$$

to get

$$\frac{1}{(1-x)^2} = \sum_{1 \le n} n x^{n-1}. \tag{9.13}$$

Setting $x = \frac{5}{6}$ gives 36 for the value of (9.13). Therefore (9.12), hence (9.10), has the value 6, so Your friend should pay 60 cents as an entry fee.

This example stems from a problem in *Choice and Chance*, (17.4).

10. The Binomial Distribution

Suppose we consider a two-valued random variable, on the sample space $(V, \mathbf{Pr})$, such as the outcome of heads or tails on the flip of a coin, or whether or not the computer breaks down while someone is using it, or whether a student's room number is divisible by 13. We define one of the two values as "success" and the other as "failure" (of the "experiment," which is "to evaluate the random variable at a point of the sample space"). Thus, we could call the outcome "heads" of the coin flip a "success."

Any random variable f with at least two values can give rise to a two-valued random variable: We partition the range of f into two cells and call any outcome in one cell a success and any outcome in the other a failure. Thus, we could call the outcome of rolling a die success if it is in $\{2,3,5\}$ and failure if it is in $\{1,4,6\}$.

There are many applications for the idea of repeating trials of such a random variable. We make the trials independent by using the natural definition (9.6), extended to n factors, of the probability function on the sample space $V^n = V \times \cdots \times V$.

Customarily we denote the probability of success in one trial by p, of failure by q. Thus $p + q = 1$. Also we ignore the sample space V and represent it by the values of the random variable as in (3.5). We call these values S (for success) and F (for failure). (Any two distinct symbols would do instead: T, F, or 1, 0, or "hot," "cold.") The sample space for n repeated trials is then the set C of all 2^n strings of length n over $\{S, F\}$. The probability function **Pr** on C is given by the formula

(10.1) $$\mathbf{Pr}(x) := p^k q^{n-k}$$

for all strings x in C consisting of k S's and $n-k$ F's in any order. That (10.1) is true follows from our extension of (9.6): For example, if $x = SSFSFFS$, then $\mathbf{Pr}(x) = p \cdot p \cdot q \cdot p \cdot q \cdot q \cdot p = p^4 q^3$. (Remember that strings are just shorthand for elements of Cartesian products.)

DEFINITION (10.2) Consider the random variable $b : C \longrightarrow \{0, 1, \ldots, n\}$ defined by the rule $\forall X \in C, b(x) :=$ the number of successes in x. The probability distribution D_b of b is called the **binomial distribution.**

Of course D_b depends on n and p, so we sometimes denote it as $D_b(n, p;\ )$. From the definition in (3.3) we see that

(10.3) $$D_b(k) := \mathbf{Pr}(b^{-1}(k)) = \binom{n}{k} p^k q^{n-k} =: D_b(n, p;\ k)$$

for all integers k, because $b^{-1}(k)$ is the set of all strings with exactly k "S's."From Chapter 8, (6.2), the counting property of binomial coefficients, we know that there are $\binom{n}{k}$ such strings.

We plot here the values of D_b for $n = 15$ and $p = 0.5$, 0.3, and 0.1.

(10.3F)

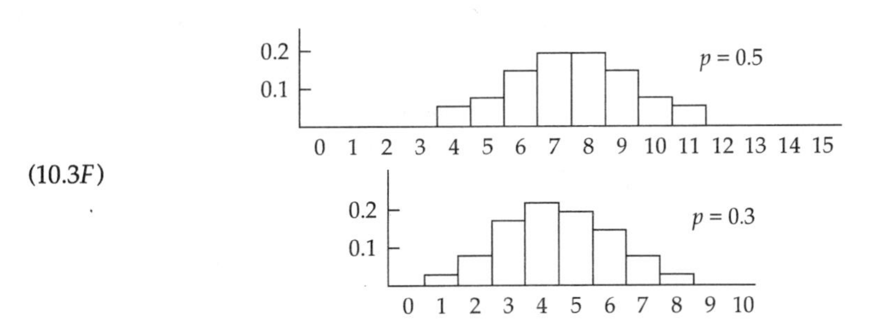

(10.3F *continued*)

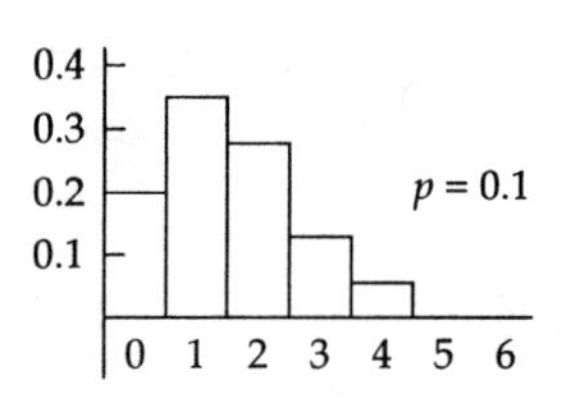

(10.4) The binomial distribution (10.2) arises when we study the number of misprints per page in books—see (17.1), page 170. It also comes up in countless other places. For example, flipping an honest coin n times in succession has probability distribution $D_b(n, \frac{1}{2};\ )$. In a cafeteria, the selection (success) or refusal (failure) of a dish of Brussels sprouts by customers has the distribution $D_b(n, p;\ )$, where n is the number of customers and p is the probability that a customer chooses the dish.

Example (10.5) An airline finds that on the average 20 percent of people with reservations do not show up for their flights. What is the least number of reservations that the airline should book for a 300-seat flight so that there is a probability of 0.9 or more that no more than 10 seats will be vacant?

Solution. We take failure as a "no-show." Thus $p = 0.8$. We want 290 or more successes. The probability of that event X is

(10.6)
$$\mathbf{Pr}_1(X) = \sum_{290 \le k} D_b(n, 0.8;\ k)$$

and we want the smallest n making $\mathbf{Pr}_1(X) \ge 0.9$. The n we seek is approximately 374. I found this not by laboriously working out (10.6) for various n, but by using the *normal approximation* to the binomial distribution. If you're interested, you may learn about that in a course on probability. If you can't wait, you'll find it in (17.1) Feller, Chapter VII.

(10.7) The expected value and variance of the "binomial" random variable b are easy to calculate. To find them we'll use the more general form of the binomial theorem given in Chapter 8, Problem 2.3:

(10.8)
$$(x + y)^n = \sum_{0 \le k} \binom{n}{k} x^k y^{n-k};$$

here x and y are independent (in the sense of calculus) variables. We shall prove

Theorem (10.9) The expected value and variance of the binomial random variable b are

$$Eb = np, \ \mathrm{Var}\, b = npq.$$

Proof We seek

$$Eb(n, p;\ ) := \sum_{0 \le k} k \binom{n}{k} p^k q^{n-k}.$$

To find it we differentiate (10.8) with respect to x, getting

$$n(x+y)^{n-1} = \sum_{0 \le k} k\binom{n}{k} x^{k-1} y^{n-k}.$$

We multiply by x:

$$nx(x+y)^{n-1} = \sum_{0 \le k} k\binom{n}{k} x^{k} y^{n-k}. \tag{10.10}$$

Now set $x = p$ and $y = q$:

$$\begin{aligned} np(p+q)^{n-1} &= \sum_{0 \le k} k\binom{n}{k} p^{k} q^{n-k} =: Eb \\ &= np. \end{aligned}$$

To find the variance we use (5.2): $\operatorname{Var} f = Ef^2 - (Ef)^2$. Thus

$$Eb^2 = \sum_{0 \le k} k^2\binom{n}{k} p^{k} q^{n-k}.$$

We differentiate (10.10), again with respect to x, to find

$$n(x+y)^{n-1} + n(n-1)x(x+y)^{n-2} = \sum_{0 \le k} k^2\binom{n}{k} x^{k-1} y^{n-k}.$$

Multiply by x and set $x = p, y = q$:

$$np + n(n-1)p^2 = \sum_{0 \le k} k^2\binom{n}{k} p^{k} q^{n-k} =: Eb^2.$$

From this quantity, subtract $(Eb)^2 = n^2p^2$ to get

$$\operatorname{Var} b = np + n^2p^2 - np^2 - n^2p^2 = np(1-p) = npq. \qquad \text{QED}$$

Thus the binomial distribution with probability of success p has mean np and variance npq on n repeated trials.

There is a tremendous literature on the binomial distribution; if you are interested, the references at the end of this chapter, or any book on probability, can guide you to it.

11. The Poisson Distribution

Suppose items arrive at a counter from time to time so that

(11.1) the probability of one or more arrivals during any time interval is proportional to the length of that interval.

Let the constant of proportionality be λ. The assumption (11.1) above makes sense only for interval lengths λ^{-1} or less; as you will see, we apply it only to small intervals.

We also assume that

(11.2) the probability of an arrival in any time-interval is not affected by an arrival or nonarrival in any disjoint time interval. Thus if I and J are disjoint time intervals of respective lengths a and b, then the probability of an arrival during I and another arrival during J is $\lambda^2 ab$ (if a and b are small).

The assumptions above lead to the so-called **Poisson distribution,** which agrees with many observed distributions. For example, alpha-particle emissions, which introduce errors into computer memories, and the arrival of sign-on requests from users of time-sharing systems fit Poisson distributions (for appropriate λ).

Call the variable with this distribution φ. Let's find the probability $\mathbf{Pr}(\varphi = k)$ that there are exactly k arrivals in a *unit* time interval. We will use limits and the simple ideas of the previous section.

Divide the interval into n mutually disjoint subintervals of equal lengths $1/n$. The probability of an arrival in such a subinterval is λ/n. We take n large enough to make λ/n small, and so that the probability that two or more items arrive in one subinterval is negligibly small.

From (10.3) we find that for any $k \geq 0$ the probability that there are exactly k arrivals in our unit interval is

$$\binom{n}{k}\left(\frac{\lambda}{n}\right)^k\left(1 - \frac{\lambda}{n}\right)^{n-k}, \tag{11.3}$$

This (11.3) is an approximation because we neglected the possibility of two arrivals in an interval of length $1/n$. We recall from calculus (without proof) the result that for any real x

$$\lim_{n\to\infty}\left(1 + \frac{x}{n}\right)^n = e^x. \tag{11.4}$$

We rewrite (11.3) as

$$\frac{\lambda^k}{k!}\left[\frac{n}{n}\frac{(n-1)}{n}\cdots\frac{(n-k+1)}{n}\right]\left[\left(1 - \frac{\lambda}{n}\right)^{-k}\right]\left(1 - \frac{\lambda}{n}\right)^n$$

The term in the first pair of brackets is

$$1\left(1 - \frac{1}{n}\right)\left(1 - \frac{2}{n}\right)\cdots\left(1 - \frac{k-1}{n}\right).$$

Remember: k is fixed and n gets arbitrarily large. This is a product of a *fixed* number of terms each of which tends to 1 as $n \longrightarrow \infty$. Therefore the product tends to 1. For the same reason, the term in the second pair of brackets tends to 1. The final term tends to $e^{-\lambda}$, by (11.4). Therefore

$$\mathbf{Pr}(\varphi = k) = e^{-\lambda}\frac{\lambda^k}{k!}; \tag{11.5}$$

it is the probability of exactly k arrivals during a time interval of length 1.

If the interval I has length t, then the probability of exactly k arrivals during I is

$$e^{-\lambda t}\frac{(\lambda t)^k}{k!},$$

since λt is the "λ" for I rescaled. That is, t is now the unit of length.

You can see, incidentally, that (11.5) is a probability distribution because

$$\sum_{0\le k}\mathbf{Pr}(\varphi = k) = e^{-\lambda}\sum_{0\le k}\frac{\lambda^k}{k!} = 1$$

(from the well-known Taylor series $e^{\lambda} = \sum \lambda^k/k!$).

Here are graphs of three Poisson distributions, in which the values of $\mathbf{Pr}(\varphi = k)$ are rounded to the nearest 0.02.

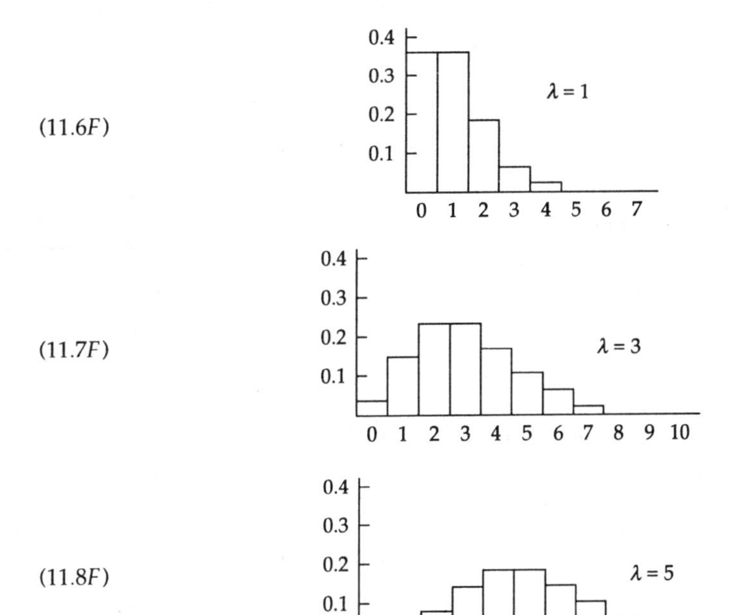

12. Mean and Variance

(12.1) The mean of this Poisson random variable φ defined via

$$\mathbf{Pr}(\varphi = k) = e^{-\lambda}\frac{\lambda^k}{k!}$$

is λ. The proof is easy:

$$E\varphi := e^{-\lambda}\sum_{0\le k} k\frac{\lambda^k}{k!} = e^{-\lambda}\cdot\lambda\sum_{1\le k}\frac{\lambda^{k-1}}{(k-1)!} = \lambda. \tag{12.2}$$

(12.3) The variance of φ is also λ:

(12.4)
$$\begin{aligned} \operatorname{Var} \varphi := E(\varphi^2) - (E\varphi)^2 &= e^{-\lambda} \sum_{0 \le k} k^2 \frac{\lambda^k}{k!} - \lambda^2 \\ &= e^{-\lambda} \sum_{1 \le k} k \frac{\lambda^k}{(k-1)!} - \lambda^2. \end{aligned}$$

We can evaluate the series as follows: Change the variable of summation to get

$$\sum_{1 \le k} k \frac{\lambda^k}{(k-1)!} = \sum_{0 \le k} (k+1) \frac{\lambda^{k+1}}{k!}.$$

Now $xe^x = \sum x^{k+1}/k!$ so $(d/dx)xe^x = \sum (k+1)x^k/k!$. Multiply by x to get

$$x\left(\frac{d}{dx}\right)xe^x = \sum \frac{(k+1)x^{k+1}}{k!}.$$

But $x(d/dx)xe^x = xe^x + x^2e^x$, too. Put $x = \lambda$ and get

$$\operatorname{Var} \varphi = e^{-\lambda} \cdot \lambda e^{\lambda}(1 + \lambda) - \lambda^2 = \lambda.$$

13. Chebyshev's Inequality†

This result is simple but useful. We've remarked that a random variable with large variance has greater probability of taking values far from its mean than if it has a small variance. Chebyshev's inequality makes that description more precise. It also leads us immediately to the "law of large numbers," which comes up all the time.

Theorem Let f be any random variable. Then for any $y > 0$

(13.1) $$\mathbf{Pr}(|f| \ge y) \le (Ef^2)/y^2$$

If $m = Ef$, then

(13.2) $$\mathbf{Pr}(|f - m| \ge y) \le (\operatorname{Var} f)/y^2.$$

Proof If we replace f by $f - m$ in (13.1), we get (13.2) because $\operatorname{Var} f$ is defined in (5.1) to be $E(f - m)^2$. So we need prove only (13.1).

By definition

$$\mathbf{Pr}(|f| \ge y) := \sum_{x:|f(x)| \ge y} \mathbf{Pr}(x).$$

But if $x \in S$ is such that $|f(x)| \ge y$ then, since $y > 0$, $1 \le |f(x)|/y$ and $1 \le f(x)^2/y^2$. Thus we may say

(13.3) $$\sum \mathbf{Pr}(x) \le \frac{1}{y^2} \sum \mathbf{Pr}(x) f(x)^2,$$

with both sums taken over $\{x; |f(x)| \ge y\}$. The second of the sums in (13.3) is at most Ef^2.

QED

† Also known as the Bienamé-Chebyshev inequality, recognizing a less well known codiscoverer.

Practice (13.4P)

(*i*) Suppose f takes the tabulated values $f(x)$ with the indicated probabilities. Calculate $\mathbf{Pr}(|f| \geq y)$ and $(Ef^2)/y^2$ for $y = 1$ and for $y = 2$.

$f(x)$	-3	-2	-1	0	1	2	3
$\mathbf{Pr}(x)$	0.1	0.2	0.2	0.2	0.1	0.1	0.1

(*ii*) Calculate (13.2) for f_3, defined in (8.5), and get the true value, for $y = 2\sigma$.

Notice that (13.2) tells us nothing about f if $y^2 \leq \operatorname{Var} f$, i.e., if y is at most the standard deviation $\sigma = (\operatorname{Var} f)^{\frac{1}{2}}$ of f. But if $y = 2\sigma$, for example, then (13.2) says that the probability that f is 2σ or more away from its mean is at most $\frac{1}{4}$.

Remark (13.5) The only reason for using $f(x)^2/y^2$ in (13.3) was to get rid of the absolute value. If we use $|f(x)|/y$ instead we get

(13.6) $$\mathbf{Pr}(|f| \geq y) \leq (E|f|)/y$$

as a "linear" version of (13.1). The random variables f and $|f|$ are different in general; the expectation of one may be easier to calculate than the other.

14. The Law of Large Numbers

This theorem allows many statistical observations to be conveniently summarized in terms of expected values. It is more complicated to state than to prove.

Theorem (14.1) *Let $f_1, f_2, \ldots$ be a sequence of mutually independent random variables all having the same probability distribution. (See (9.5)). Let $m = Ef_i$ be their (common) expected value. Then, for every $\epsilon > 0$,*

$$\lim_{n\to\infty} \mathbf{Pr}\left(\left|\frac{f_1 + \cdots + f_n}{n} - m\right| > \epsilon\right) = 0. \blacksquare$$

This result can be restated as

(14.2) The probability that the difference between the mean of the sample and the mean of the whole population is less than ϵ approaches 1 as the number of observations tends to infinity.

Proof of (14.1) We use (4.4): $E(f + g) = Ef + Eg$. Set $g_n = (f_1 + \cdots + f_n)/n$. Then $Eg_n = m$. Thus from (13.1) we see that

(14.3) $$\mathbf{Pr}(|g_n - m| > \epsilon) \leq (\operatorname{Var} g_n)/\epsilon^2.$$

Though ϵ may be small, once we choose it, it is fixed; and we are then to let n become arbitrarily large. From (7.7) we get $\operatorname{Var} g_n = \frac{1}{n^2}\operatorname{Var}(f_1 + \cdots + f_n) = \sigma^2/n$, where $\sigma^2 = \operatorname{Var} f_1 = \operatorname{Var} f_2 = \cdots$. Therefore (14.3) becomes

$$\mathbf{Pr}(|g_n - m| > \epsilon) \leq \frac{\sigma^2}{\epsilon^2}\frac{1}{n},$$

which tends to 0 as $n \longrightarrow \infty$. QED

Example (14.4) Suppose we roll a die n times. We know from (8.4) that its expected value is 3.5. Theorem (14.1) says that the probability that the average of the first n outcomes is between 3.4 and 3.6 (if we take $\epsilon = 0.1$) approaches 1 as n approaches ∞.

It might be instructive to model this example in terms of Cartesian products. We have, in terms of (3.5), $S = \{1,\ldots,6\}$. We set $S^n = S \times \cdots \times S$ (n factors). The outcome of n rolls of the die is a point in S^n. Each point in S^n has probability $1/6^n$. We define the *mean* of a point $a = (a_1,\ldots,a_n) \in S^n$ to be $m(a) := \frac{1}{n}(a_1 + \cdots + a_n)$. The theorem says that if the set L_n of all points in S^n with mean less than 3.4 has exactly u_n elements, then its probability, which is

$$u_n/6^n,$$

tends to 0 as $n \longrightarrow \infty$. Similarly $v_n/6^n \longrightarrow 0$, where v_n is the number of points with mean greater than 3.6.

The numbers u_n and v_n are not so easy to calculate, but we can easily calculate the size of a certain subset of L_n. Consider the set L_1 of all points with no coordinate greater than three. For example, $(1, 2, 3, 3, 1, 2, 2, 3) \in L_1$ (for $n = 8$). Every point in L_1 has mean 3 or less, so $L_1 \subseteq L_n$. Now $|L_1| = 3^n$. And $3^n/6^n = 2^{-n} \longrightarrow 0$ as $n \longrightarrow \infty$.

The theorem says that, for large n, an overwhelmingly large proportion of the points in S^n have mean very close to the mean of S^n, defined as

$$\frac{1}{6^n}\sum_{a\in S^n}(a_1 + \cdots + a_n)/n \tag{14.5}$$

where $a = (a_1,\ldots,a_n)$. This quantity is 3.5. It is the expected value of the random variable m defined on S^n. It is a nice exercise to verify without using probabilistic arguments that 3.5 is the value of (14.5).

To prove Theorem (14.1) in this situation by counting L_n, say, would be quite a task. The beauty of the probability approach is that it gives such a simple proof of (14.2). It does not tell us, however, how large n must be to put the probability greater than .9, or .99, or whatever. The counting approach would achieve that.

We make several applications of (14.1) in the next sections.

(14.6) The law of large numbers justifies our use of expected value in answering the question about the game of (1.1). If you played the game many times, the arithmetic mean of your payout would with high probability be close to the expected value 6 which we calculated in (9.11).

To see the connection, let the $f_1,\ldots,f_n$ in Theorem (14.1) be the f defined on the sample space of all games as we discussed that space in (9.8).

15. Little's Formula

This basic formula applies to **queues**, which in applications are ordered sets of customers waiting for service at some place. They might be jobs in a time-sharing computer system waiting their turn for processing or people waiting in line for service at a bank or gas station.

In a queue, the rule for deciding who gets served next (the *service rule*) could be first-come, first-served (*FCFS* or *FIFO*), or priority for service could be made a function of the length of the job, or it could be the so-called *round-robin*, in which the $FCFS$ rule is used but with a limited time for service; uncompleted jobs then go to the end of the queue. Even last-in, first-out ($LIFO$) is a possible rule.

Little's formula is a simple relation between the rate at which customers arrive at the queue and the number of customers in the queue and amount of time they wait. Since all these quantities vary, the formula is expressed in terms of expected values of them considered as random variables. It holds for any service rule.

We define the situation starting from some time called 0 when the queue is empty. We assume first of all that customers arrive with probability given by a Poisson distribution with mean λ.† That is, by (12.2), λ is the expected number of arrivals in any unit time interval.

We define

(15.1) $$A(t) := \text{the total number of arrivals at the queue between time 0 and time } t.$$

$A(t)$ is a nonnegative integer for each $t \geq 0$.
A is a random variable that we regard for convenience as depending on the continuous variable t. Its expectation is λ. But the situation is really discrete, since everything about it is determined by the time a_i of arrival at, and the time d_i of departure from, the queue of the ith customer, for $i = 1, 2, \ldots$. Assuming no two customers arrive together we have

$$0 \leq a_1 < a_2 < \cdots;$$

the order of the d_i's, however, depends on the rule of priority for service. For $FCFS$ we have $d_1 < d_2 < \cdots$.

We also are straying out of the finiteness we have always before assumed. Although we allow time to go on forever, we always deal with a finite interval of time (from 0 to t) in which only a finite number of customers has arrived. We will assume that certain limits exist as time runs on.

We also define the **length** of the queue at time t to be

(15.2) $$L(t) := \text{the total number of customers in the queue at time } t.$$

$L(t)$ is a nonnegative integer for each $t \geq 0$; it increases by 1 *just after* each time a_i and decreases by 1 *just after* each time d_j (unless $a_i = d_j$ for some i and j; at such times L does not change.)

Let us now graph the functions A and L for a small example.

(15.2F)

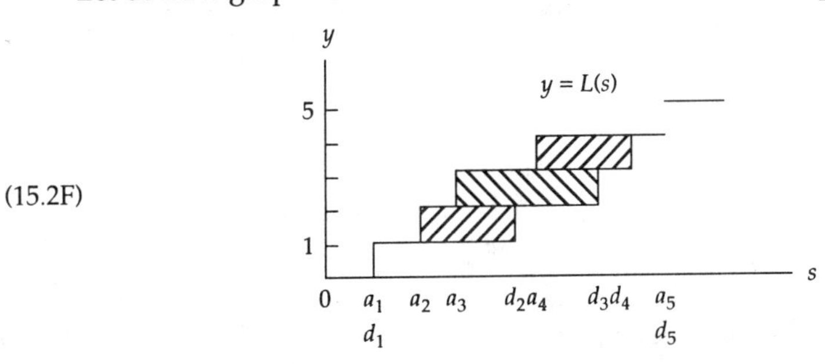

†Poisson distributions are accurate models of many such queues, according to actual observations.

Here the upper envelope is A, which jumps up one step at each arrival-time a_i. The height of the shaded portion at time t is $L(t)$. The lower boundary of the shaded portion jumps up one step at the time of a departure d_i.

We have assumed that a customer arriving at an empty queue departs immediately for service. Thus $d_1 = a_1$ and $d_5 = a_5$ in the example.

We also assume in the example a first-come, first-served service rule. A different rule, say one that interchanged the departure-times of the second and third customers, would yield the graph

(15.3*F*)

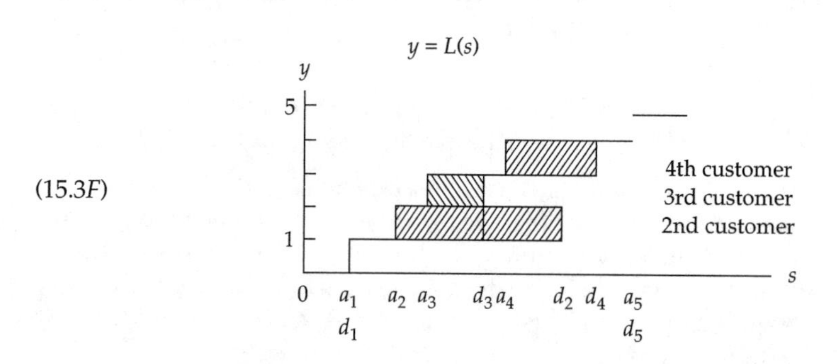

The length of the queue is still the height of the shaded portion!

(15.4) In the figures, each shaded rectangle has horizontal length equal to the amount of time that its customer has spent in the queue. Since its vertical length is 1, its *area* is the **customer's waiting time.** The waiting time of the ith customer is

$$w_i = d_i - a_i \tag{15.5}$$

up to time t if $d_i \leq t$. If $d_i > t$, then the ith customer is still in the queue at time t and has waited $w_i = t - a_i$ time units.

The observation (15.4) yields a basic equation leading to Little's formula: For any queue

(15.6) the total of the waiting times of all the customers up to time t equals the area under the curve $y = L(s)$ for $0 \leq s \leq t$.

In (15.2*F*) the "curve" for L is the step function

(15.7*F*)

y $\quad y = L(s)$

1

0 $\;a_1\;$ $a_2\; a_3$ $\;d_2 a_4\;$ $d_3\; d_4\; a_5$ $\;s$

d_1 $\quad d_5$

It is just the shaded portion of the graph in (15.2*F*) "bottom justified."

In (15.3*F*) the curve for L is the same! It counts how many are in the queue, not which ones they are.

Practice (15.8P) Draw the graph analogous to that of (15.2F) for $(a_1, \ldots, a_5) = (1,3,4,6,8), d_1 = a_1$, and $d_i = a_i + 2$ ($FCFS$) for $i > 1$. Graph L for this queue. Now do the same for the following permutation of the departure times:

$$d_1 = a_1 = 1,\ d_2 = 6,\ d_3 = 5,\ d_4 = 10, \text{ and } d_5 = 8.$$

We now define the sample mean of the random variable L as

$$\overline{L} = \frac{1}{t}\int_0^t L(s)ds. \tag{15.9}$$

In words: Up to time t, the average length of the queue for the observed arrival and departure times is the height $\overline{L}$ of the rectangle with the same base (= line-segment from 0 to t) and the same area as the area under the curve $y = L(s)$ over $0 \leq s \leq t$.

By the law of large numbers, suitably extended to this infinite situation, $\overline{L}$ approaches the expected value EL as t gets larger.

In the example above, (15.2F), if $t = a_5$, then $t\overline{L} = (a_3 - a_2) + (a_4 - d_2) + (d_4 - d_3) + 2(d_2 - a_3) + 2(d_3 - a_4)$.

We also define the sample mean $\overline{w}$ of the waiting times as the average of the times waited by all the customers up to time t:

$$\overline{w} = \frac{1}{A(t)}\sum_{i:a_i \leq t} w_i. \tag{15.10}$$

Again the law of large numbers tells us that $\overline{w}$ approaches the expected value of the random variable w. (We assume that these expected values EL and Ew exist. They are the limits we mentioned earlier.) In particular $A(t)/t \longrightarrow \lambda$. We need not assume a Poisson distribution of arrival times but only that $A(t)/t$ has a limit. The random variable w is defined on the space of all customers; its probability distribution depends on the distributions of arrival and departure times. The beauty of (14.1), the law of large numbers, is that we need not know the probability distribution of w, because by (14.1) we approximate Ew by taking the arithmetic mean of samples of w. Now we can state and prove Little's formula:

Theorem (15.11) Under the above definitions $EL = \lambda Ew$.

Proof In (15.6) we saw that

$$\int_0^t L(s)ds = \sum_{a_i \leq t} w_i \tag{15.12}$$

We want to relate the *time* averages of L and A to the *per person* average of the w_i. We divide (15.12) by t and multiply and divide on the right by $A(t)$, getting

$$\frac{1}{t}\int_0^t L(s)ds = \left(\frac{A(t)}{t}\right)\left(\frac{1}{A(t)}\sum w_i\right). \tag{15.13}$$

As we have said, the law of large numbers applied to each of the three averages here tells us that the limiting form of this equation is $EL = \lambda Ew$. QED

One way to use Little's formula is to estimate waiting times by observing the length of the queue and measuring λ (if it is unknown). You could try it the next time you're in a long line at a bank.

Little's formula is true no matter what rule for priority of service holds in the queue. In (15.3*F*), where we interchanged the departure times of the second and third customers, we saw that the function L was unchanged. But any permutation of the departure times can be achieved as a succession of interchanges of two, even of two adjacent, departure times (Chapter 5, Problems 18.3 and 18.4). Therefore L is the same function for given arrival and departure times, no matter how the latter are permuted. Under our assumptions on the existence of limits, the proof of Little's formula (15.11) depends only on the equality in (15.12), which we've just shown holds under all service rules.

16. More About Queues

Now we apply Little's formula to get more information about the queue with its server. We have ignored the latter but now assume for simplicity that

> the server completes every job in constant time s and does only one job at a time.

We will show that

$$\overline{L} = \frac{1}{2}[(\lambda s)^2 + (\lambda s)^3 + \cdots] = \frac{1}{2}\cdot\frac{(\lambda s)^2}{1-\lambda s} = \frac{1}{2}\cdot\frac{\rho^2}{1-\rho}. \tag{16.1}$$

and

$$\overline{N} = \rho + \overline{L} = \rho + \frac{1}{2}\frac{\rho^2}{1-\rho} \tag{16.2}$$

where $\rho = \lambda s$, and $\overline{N}$ is the expected total number of customers in the system, i.e., in the queue Q or in the server S.

We assume $\rho = \lambda s < 1$.

The key to our method of proof is to treat the whole system Q, S as *another* queue. We then play off the two queues against each other by applying Little's formula to both.

As before, we denote the times at which customers arrive as $a_1, a_2, \ldots$. Remember, we count the queue Q by adding 1 just *after* the arrival-time a_i for each i. Thus, if there is no departure near time a_i, the function L looks like

(16.3*F*)

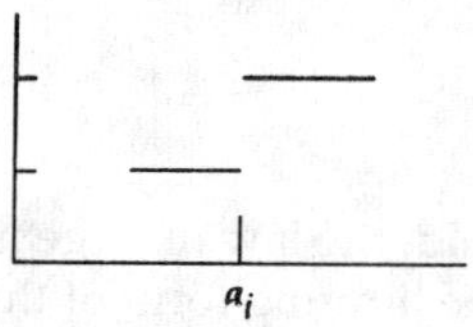

Define w_i as the waiting-time in the queue Q of the ith customer. Then

$$w_i = sL(a_i) + r(a_i), \tag{16.4}$$

where

$$r(t) := \text{the remaining service-time of the job in the server.} \tag{16.5}$$

Formula (16.4) is obvious: There are $L(a_i)$ prior customers in Q awaiting service, and they have to wait $r(a_i) := r_i$ time-units before the first of them enters service.

First we sum (16.4) up to time t and divide by $A(t)$, the number of arrivals up to time t. We get

$$A(t)^{-1} \sum_{a_i \le t} w_i = sA(t)^{-1} \sum_{a_i \le t} L(a_i) + A(t)^{-1} \sum_{a_i \le t} r(a_i). \tag{16.6}$$

Using the law of large numbers (14.1) we say that as $t \longrightarrow \infty$ (16.6) tends to

$$\overline{w} = s\overline{L} + \overline{r}, \tag{16.7}$$

where $\overline{w}$ and $\overline{L}$ are the expected values defined earlier in (15.9) and (15.10), and $\overline{r} = Er$ is the expected value of r. The assumption that $\lambda s < 1$ allows $\overline{L}$ to exist! If the service rate $1/s$ were greater than the arrival rate λ the queue would grow beyond bound as time increased.

Since $\overline{L} = \lambda\overline{w}$ by (15.11), Little's formula, we may multiply (16.7) by λ to get

$$\overline{L} = \lambda\overline{w} = \lambda s\overline{L} + \lambda\overline{r}.$$

Thus

$$\overline{L}(1 - \lambda s) = \overline{r}\lambda, \tag{16.8}$$

and

$$\overline{L} = \frac{\overline{r}\lambda}{1 - \lambda s}. \tag{16.9}$$

Thus, to proceed we must find the expected value $\overline{r}$ of r. We do it indirectly by looking at the whole system Q, S as a queue. Let $N(t)$ be the number of customers in Q, S at time t. Thus

$$N(t) = L(t) + x(t) \tag{16.10}$$

where

$$x(t) := \begin{cases} 1 & \text{if } r(t) > 0 \\ 0 & \text{if } r(t) = 0. \end{cases}$$

That is, $x(t)$ is 1 if and only if there is a customer being served (in S) at time t.

Since $N(t)$ is the length of the queue for the whole system Q, S, and since its mean arrival rate is the same λ, Little's formula applies again. It yields

$$\overline{N} := EN = \lambda(EW_i) = \lambda\overline{W} \tag{16.11}$$

where W_i is the waiting time of the ith customer for $i = 1, 2, \ldots$. It is immediate that $W_i = w_i + s$, hence

$$\overline{W} = \overline{w} + s. \tag{16.12}$$

Since, from (16.10),

$$\overline{N} = \overline{L} + Ex = \overline{L} + \overline{x}, \tag{16.13}$$

we find from (16.11) that

$$\begin{aligned} \overline{x} &= \lambda\overline{W} - \overline{L} \\ &= \lambda\overline{w} + \lambda s - \lambda\overline{w} = \lambda s, \end{aligned} \tag{16.14}$$

where we used (16.12) and Little's formula (15.11) applied to Q.

Little algebra and less thought have brought us the value of $\overline{x}$. From it we can easily find $\overline{r}$. Look at the graph of a portion of the curve $y = r(t)$:

(16.15*F*)

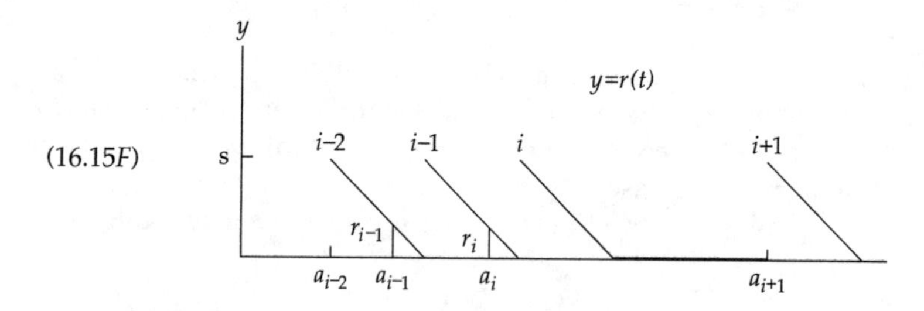

Explanation: Customer $i-2$ arrives at time a_{i-2} and is served immediately. The number at the top each slanting line is the number of the customer being served. Customer $i-1$ arrives before the service of $i-2$ is finished. By r_j we denote $r(a_j)$. In this example, r_{i-1} is the amount of time that customer $i-1$ must wait for service. Notice also that $r_{i+1} = 0$ because there are no customers in the queue at time a_{i+1}.

Let's calculate $\overline{r}$ using the law of large numbers (14.1). It tells us that

$$\overline{r} = \lim_{t\to\infty} \frac{1}{t}\int_0^t r(u)du.$$

This leads to an independent calculation of $\overline{r}$ *and* to the result that $L(t)/t \longrightarrow 0$. That is, $\overline{r}$ is the (limit of the) average height of the curve $y = r(u)$ (over $0 \le u \le t$ as $t \longrightarrow \infty$). Over the portion of the t-axis where $r(t) \ne 0$ the average height is $s/2$. The *probability* that $r \ne 0$ is exactly $\overline{x}$, by definition (16.10). That is, $\overline{x}$ is the proportion of the t-axis where $r(t) \ne 0$. Therefore

$$\overline{r} = \overline{x}\frac{s}{2} = \lambda s^2/2. \tag{16.16}$$

Thus we've proved (16.1) and (16.2).

Notice that for ρ near 1 the expected queue-length $\overline{L}$ of (16.9) becomes very large. As we mentioned earlier, ρ is the ratio of the arrival rate λ to the *service* rate $1/s$. Thus $\rho = \lambda s$. If $\rho = 0.9$ then $\overline{L} = 4.05$; if $\rho = 0.99$ then $\overline{L} = 49.005$. For many types of queues the averages of the queue lengths and waiting times take the form $A + B/(1 - \rho)$ for constants A and B. The systems do not work well as ρ gets close to 1.

17. Further Reading

(17.1) William Feller, *An Introduction to Probability Theory and Its Applications*, vol. I, third edition, John Wiley & Sons, Inc., New York, 1968.

(17.2) Mark Kac, *Statistical Independence in Probability, Analysis, and Number Theory*, Mathematical Assoc. of America, Washington, 1959. (Carus Monograph No. 12).

(17.3) Leonard Kleinrock, *Queueing Systems*, John Wiley & Sons, New York, vol. I, 1975; vol. II, 1976.

(17.4) William Alan Whitworth, cited in Chapter 9, (10.5).

18. Problems for Chapter 10

Problems for Section 2

2.1[Ans] Let the word be "calculator." (See (2.13).) What is the probability that

(*i*) a letter from the word is a consonant?

(*ii*) two letters from the word are different? Explain.

(*iii*) List the sample space and probability function for part (*i*).

2.2 Three letters are chosen at random from "calculator." What is the probability that

(*i*) one of them is a t?

(*ii*) one or more of them is a c?

2.3 In bridge, a 52-card deck is dealt out in four hands of 13 cards each. What is the probability that your hand has no ace? [Assume that all 52! shuffles are equally likely.]

2.4 Give an example of three events A, B, C in a sample space that are mutually independent (i.e., any two are independent) but for which

$$\Pr(A \cap B \cap C) \neq \Pr(A)\Pr(B)\Pr(C).$$

You may present your answer as a Venn diagram with the probabilities proportional to the areas of the regions.

2.5 Getting low on gasoline, you decide to buy some—but at the lowest possible price. Once you pass a station, however, you cannot turn back. Each station posts its price, and let's assume no two stations have the same price. Suppose you adopt the following strategy: You buy gas at the first station having a lower price than the one before it. (If there is no such station, buy at the last one. You know how many stations there are.) That means, in particular, that you pass the first station. Assume there are four stations; instead of assuming prices for them, rank them as 1, 2, 3, and 4, with 1 standing for the lowest price and 4 the highest. Apply the strategy to this situation and find the expected value of the rank. Assume each of the 4! possible distributions of the order in which you pass the stations.

2.6 Again you want to buy gasoline. With the same situation as in the preceding problem, find the expected rank of this second strategy: Reject the first station you come to. After that, accept the first one cheaper than all those before it. If none is, choose the last one.

2.7 Now there are five stations. Show that the expectation of the second strategy improves on that of the first strategy by $3/40$. [Hint. Don't list all 120 permutations, but argue on the five placements of the number 5. Show that both strategies yield the same result for all $x5xxx$ and all $xx5xx$ permutations. Find the net changes in result for each strategy for the $5xxxx$, $xxx5x$, and $xxxx5$ permutations.]

2.8 Ten students, A, B, $C\ldots$, are randomly arranged in a line. What is the chance that A and B are next to each other? Express your answer in base 10 after explaining how you got it. The sample space is the set of 10! permutations, each of which has equal probability. What is the event?

2.9[Ans] We modify the old New York Lotto game. To play you chose a 6-subset T from $\{1,\ldots,48\}$. To choose winners, the organizers drew a 6-subset W of such integers at random. They also drew a seventh "supplemental" number s at random. The winners were those for which

$$s \in T \text{ and } |T \cap W| = 3 \text{ or for which } |T \cap W| \geq 4.$$

[Winnings depended on the number of tickets sold and number of winners in each category.] Find the probability of being a winner. Explain your calculations in detail.

2.10 Suppose the probability of an event is x/y, where x and y are positive real numbers. Express the odds of the event in terms of x and y. Explain.

Problems for Section 3

3.1 Consider rolling two dice, the first honest and the second loaded as in $(S_1, \mathbf{Pr}_1')$ in (8.1). Define the random variable f to be the sum of the numbers showing on the tops of the dice. Find the probability distribution $\mathbf{Pr}(f = k)$ for $k = 2,\ldots,12$. Explain.

Problems for Section 4

4.1 Define $S := \{0,1,\ldots,n\}$ and $\mathbf{Pr}(k) := \binom{n}{k}2^{-n}$; also define, for $x \in S, f(x) := x, g(x) := 1$.

(i) Verify that $(S, \mathbf{Pr})$ is a sample space.
(ii) Calculate the expected values of the random variables f and g.

4.2[Ans] It costs \$1 to play the New York instant lottery, which advertises the following schedule of prizes and states odds for each.

Prize	*Odds*
\$2	1:14.29
\$5	1:100
\$10	1:300
\$50	1:1,200
\$100	1:4,445
\$1,000	1:120,000
\$10,000	1:2,520,000.

(i) Find the probability, correct to three significant figures, of a winning ticket.
(ii) What is the expected return (to the nearest cent)? Show your calculations or program.

4.3 [*C&C*, #818] You throw a common die. If an even number turns up you get that number of dollars. If an odd number turns up you pay that number of dollars. What is the value of your expectation? Explain.

4.4 [*C&C*] You throw two dice. If the sum of the faces turned up is even, you win that much; if odd, you lose that amount. What is your expectation? Explain.

4.5 [*C&C*, #817] You draw a card from a 52-card deck. If it is a King, Queen, or Jack you pay 14 cents. Otherwise you get a number of cents equal to the number of the card. What is your expectation? Explain.

4.6 [*C&C*, #565] A bag has

one counter marked 1
two counters marked 4
three counters marked 9

and so on up to n counters marked n^2 (where n is a positive integer). You draw one counter and are paid the amount shown on it. Show that your expectation is equal to the number of counters.

4.7[Ans] [C&C, #278] You are to toss a coin at most 10 times. If the first toss comes up heads you win a penny. If the second toss is also heads you win two more pennies. In that case you toss again, and if you get another heads you win four more pennies, and so on. The amount doubles each time you toss another heads, but as soon as you toss tails you are out of the game. What is your expected winning?

4.8 Suppose in Los Primos there were a machine that at the press of a button returned an integer from 2 to 101, all with equal probability.

(i) You and a friend agree to gamble as follows: you will pay him d dollars, he pushes the button, and pays you p dollars, where p is the smallest prime dividing the integer returned by the machine. What should d be so the game will be "fair" (:= each player has zero expected winning.)
(ii) Assume you played ten times at the fee of d = \$10 per play. Give a careful explanation of what your expected net winning would be at the end.

4.9 [*C&C*, #395] In a bag there are ten counters marked with numbers. A person is allowed to draw two of them. If the sum of the numbers drawn is an odd number, he receives that number of dollars; if it is an even number, he pays that number of dollars. Find the expected return if the counters are marked

(i) 0 to 9
(ii) 1 to 10.

Problems for Section 6

6.1[Ans] What does it mean to say "x and y are independent variables in the sense of calculus"?

6.2 Let A and B be disjoint events in a sample space. Prove that A and B are not independent unless $\mathbf{Pr}(A)$ or $\mathbf{Pr}(B)$ is 0.

6.3 Let $S = \{00, 01, 10, 11\}$ and define $\mathbf{Pr}(00) = 1/2$, $\mathbf{Pr}(01) = 1/4$, $\mathbf{Pr}(10) = 1/8$, and $\mathbf{Pr}(11) = 1/8$. Prove or disprove that the random variables f and g are independent, where, for each x in $S, f(x) :=$ the first coordinate of x, and $g(x) :=$ the second coordinate of x.

6.4 Suppose S is a sample space of exactly two points: $S = \{a, b\}$ and $\mathbf{Pr}(a)\mathbf{Pr}(b) \neq 0$. Prove that no two random variables f and g on S are independent unless one is constant.

Problems for Section 7

7.1 Prove the first part of (7.2), that χ_A and χ_B are independent random variables.

Problems for Section 9

9.1 [C&C] You throw two dice. What is the probability that neither shows a 1 on the top face nor are the two top faces alike?

9.2[Ans] [C&C, #282] Out of a set of 21 dominoes, numbered from double one to double six, one is drawn at random. At the same time a pair of common dice are thrown.

(*i*) What is the chance that the numbers turned up on the dice will be the same as those on the domino?

(*ii*) What is the chance that they will have at least one number in common?

(*iii*) Identify explicitly the sample space and probability function for this problem.

9.3 [C&C, #808] Throwing two dice you win $|x - y|$, where x and y are the numbers showing on the top faces. Show that your expected win is $\frac{35}{18}$.

9.4 [C&C] You are to throw three dice and win the difference between the maximum and minimum numbers turned up. Show that your expectation is $\frac{35}{12}$.

9.5 [C&C, #819] An honest die is thrown six times. Define m as the minimum of the six values that come up. Prove that the expected value of m is approximately 1.44.

Problems for Section 10

10.1 Over a large set of inputs a program runs twice as often as it aborts. What is the probability that of the next six attempts, four or more will run?

10.2 You win a set at tennis once you have won six games *and* are ahead by at least two games. If you and your opponent are evenly matched, find the probability that you win the set with a score of

(*i*) 6 to 3

(*ii*) 7 to 5

(*iii*) 6 to 4

(*iv*) 8 to 6

Express your answers as rational numbers and as decimals to three significant figures.

10.3 What is the probability that after six games of tennis you lead your opponent four games to two if your probability of winning each game is

(*i*) 0.5?

(*ii*) 0.6?

(*iii*) 0.4?

Express your answers as rational numbers and as decimals to three significant figures.

10.4[Ans] A carnival pitchman sets up his game: The player flips a coin three times. If no heads come up the player gets no payoff. If only one head comes up the player wins one dollar. If exactly two heads come up in any order the player wins four dollars. If all three flips come up heads the player wins nine dollars.

(*i*) With an honest coin in use, what is the expected amount won by a player in one game? Explain.

(*ii*) Charging $4 per game the pitchman got too few customers, so he dropped the price to $3 but substituted a dishonest coin. What threshold of dishonesty gives him the same expected net profit per game as with the honest coin at $4 per play? Explain.

10.5 A class of 15 students has three computer terminals available in the lab for their course. Each student uses a terminal with probability $1/10$. What is the probability that one or more students from this class will have to wait to use a terminal? Calculate accurate to two significant figures.

10.6 For masochists only: calculate $\mathbf{Pr}_1(X)$ in (10.6) for $n = 374$ to two significant figures.

10.7[Ans] [C&C] A boy tries to jump a ditch: In jumping from the upper bank to the lower he succeeds five times out of six; in jumping from the lower to the upper he succeeds three times out of five. What is the chance that after four trials he leaves off on the same side on which he began?
Assume he starts from the upper bank.

Problems for Section 11

11.1 Errors in statements of homework problems in a certain course occur with a Poisson distribution having an average of two errors per weekly assignment.

(*i*) What is the probability that there are no errors in a weekly assignment?

(*ii*) What is the probability that there are three or more errors in a weekly assignment?

11.2 Counts of wrong numbers obtained from a set of telephones over a given time period were found to obey a Poisson distribution with $\lambda = 8$. Find the probability of no more than five wrong numbers.

11.3 On the average, 15 out-of-state cars per hour pass a certain point on a road. What is the probability that exactly four out-of-state cars pass that point in a 12-minute period?

11.4 Suppose 5 percent of cars here are of European manufacture. You observe 100 cars chosen at random. Concerning the probability that exactly 10 of them are of European manufacture:

(*i*) Write and *carefully* evaluate numerically the binomial formula for this probability.

(*ii*) Evaluate the Poisson approximation to your binomial formula.

Be careful! Your answers should be within 10 percent of each other. Which was easier to evaluate, the exact or the approximate formula?

Problems for Section 13

13.1 Assume (13.1) and prove (13.2).

Problems for Section 14

14.1 Assuming (14.1), justify (14.2), the restatement of the law of large numbers.

14.2 Verify (14.5) without using probabilistic arguments. Then give a probabilistic proof. Which method do you prefer: algebraic or probabilistic?

14.3 You pay \$11 to play the following game: An arrow spins and comes to rest on an integer $k, 1 \le k \le 20$. Each number is equally likely to be the one the arrow points at. After 200 spins you are paid the arithmetic mean (in dollars) of the 200 numbers it pointed at. Show that your probability of being paid more than the \$11 entry fee is less than $1/3$.

Problems for Section 15

15.1 Students arrive randomly to see a professor during office hours at an average rate of one every 10 minutes. Each waits his or her turn. The professor sees each student alone for eight minutes. What is the expected number of students waiting to see this professor?

15.2 Students call the computer center to get their account numbers and passwords. George gives them the information, which takes 15 seconds each time. Anyone who calls while George is talking to a student is put on hold and waits his turn. Find the expected number of students on hold if the average number of calls is

(*i*) two per minute;

(*ii*) three per minute. Explain.

(*iii*) If the "arrivals" (of calls) are Poisson, what is the probability that there will be four calls in a minute in case (*ii*)?

Problems for Section 16

***16.1** Prove that as $t \to \infty$, $L(t)/t \to 0$.

General Problems

G0. In bridge a "Yarborough" is a hand of 13 cards with no Ace, King, Queen, Jack, or 10. Find the probability of being dealt a Yarborough.

G1. The World Series ends as soon as one team wins four games. There are no ties, and there are only two teams. What is the probability that the series lasts for seven games if

(*i*) the teams are evenly matched;

(*ii*) one team is a 3-to-2 favorite (the true odds) over the other for each game.

G2. A deep-space probe sends back information as bits (0's and 1's). At the receiver, each bit, independently of all other bits, has an error probability 10^{-4}. Find the probability of that exactly one of 10,000 bits is wrong.

G3. Suppose that on the average 10 percent of the cars passing a certain point are from out of state.

(*i*) What is the probability that the next two cars that pass that point are from out of state?

(*ii*) What is the probability that exactly one of the next 10 cars that pass the point is from out of state?

G4.[Ans] What is the probability that a social security number (*ssn*) is composed of precisely three digits? Give your answer correct to five significant figures. (Assume that each nine-digit number is equally likely to be an *ssn*.)

***G5.** What is the most likely number of distinct digits in a social security number?

***G6.** What is the expected number of distinct digits in a social security number? [Hint: If you've done the preceding two problems, you won't have much trouble with this one.]

G7. From a set of q different elements, you repeatedly choose an element at random, record its name, and replace it. After n choices what is the probability $g(q,n)$ that you have chosen every element at least once? (Assume that each element is chosen with probability q^{-1}.)
Example. $q = 4$: You have the set of four aces from a deck of cards. What amounts to the same thing: a deck of 52 cards in which your goal is to draw a card from each of the four suits.

G8. Evaluate the formula you found for the preceding problem for $q = 4$ and $n = 1, 2, \ldots, 12$. (You should use a computer.)

G9. In the formula of problem G7 for $g(q,n)$, let q be large. Assume the known result that as $m \to \infty$

$$(\star) \qquad \forall x \in \mathbf{R} \qquad \left(1 + \frac{x}{m}\right)^m \to e^x.$$

Example. $\left(\frac{99}{100}\right)^{100} = 0.366+$, and $e^{-1} = 0.368-$.

(*i*) Take $n \geq q$. Use ($\star$) to derive the approximation

$$g(q,n) \doteq (1 - e^{-n/q})^q.$$

(*ii*) *The converse of the birthday problem* (See Chapter 9, (10)). Use a computer to evaluate this approximation for $q = 365$ and $n = 1500 + 100k$ for $k = 0, 1, \ldots, 15$.

(*iii*) Find the least value of n such that $g(365, n) > 1/2$, according to this approximation. (The true value is only 1 less, so the approximation is pretty good.) This is the smallest number of randomly chosen people such that the probability that their birthdays fall on every day of the year is greater than $1/2$.

G10. A professor reports grades in his course by filling in small ovals with a soft pencil on the registrar's form. [The technical term for this activity is "bubbling."] With probability p he bubbles an adjacent grade instead of the correct one, for each student and independently from one student to another for grades B, C, and D, (shown schematically in ($\star\star$).) The grades on the form are in the order shown here, and the correct number of students in his course with each grade is also shown:

($\star$)

A	B	C	D	F
5	10	20	5	0

$$(\star\star) \qquad A \ \underset{p}{\longleftarrow} \ B \ \underset{p}{\overset{p}{\longleftrightarrow}} \ C \ \underset{p}{\overset{p}{\longleftrightarrow}} \ D \ \underset{p}{\longrightarrow} \ F$$

He never misbubbles (i.e., bobbles) intended grades of A or F.
The registrar reports back to the professor the grades as bubbled and also reports the total number of students with each grade. If the professor made just one error—say he bubbled a C instead of a B—then the registrar's tally would show

A	B	C	D	F
5	9	21	5	0

Comparing this tally to his own would tell the professor he had made a mistake. If he made two errors, say the above and a D instead of a C, the tally would be

A	B	C	D	F
5	9	20	6	0,

and again he'd know there was at least one error. If, however, the second error were a B instead of a C, then the registrar would report the same tally as the professor ($\star$); and he would therefore not know that he had made two errors that canceled each other.

(*i*) Find formulas for the probability of an error (pe), of exactly one error ($pe1$), and of exactly two errors ($pe2$).

(*ii*) Find a formula for the probability of exactly two errors that cancel ($cpe2$).

(*iii*) Use a computer to tabulate

$$pe \qquad pe1 \qquad pe2 \qquad cpe2 \qquad ncpe2$$

for $2p = 0.01$ and 0.001, where $ncpe2 :=$ the probability of exactly 2 noncanceling errors.

(*iv*) Assume that exactly two errors occur; prove that the odds against their canceling are 104 to 15 (almost 7 to 1).

(*v*) (For discussion) What do you think of this method of reporting grades? Do you think the bubble sort of takes the human element out of it?

G11. In 1990 the New York State Lottery advertised the game "Joker's Wild" which cost \$1 to play with the following schedule of prizes and odds:

Prize	*Odds*	*Prize*	*Odds*
\$1	1:8.93	\$ 10	1:162.50
2	1:26.32	20	1:250
4	1:38.46	5000	1:170,909.09

They state that the overall odds are 1:5.1.

(*i*) Check their statements for consistency.

(*ii*) Calculate the expected winnings for one play of this game.

G12. In 1989 the New York State Lottery advertised a game called "Winner Take All" with the following schedule of prizes and odds:

Prize	*Odds*	*Prize*	*Odds*
\$1	1:10.87	\$ 10	1:100
2	1:41.67	15	1:500
2	1:50	25	1:500
4	1:100	100	1:18,461.54
5	1:55.56	500	1:120,000

The ad stated that overall odds were 1:5.62. (The double occurrence of the \$2 prize represents two different tickets, each with that prize.) The game cost \$1 per ticket.

(*i*) See if the overall odds are as advertised.

(*ii*) Find the expected value of a ticket.

G13. (Feller) You park your car in a row of 15 cars, not at either end. While you are gone, five of the other cars leave. What is the probability that when you return the two spaces on either side of your car are empty?

19. Answers to Practice Problems

(2.6P) This is inclusion-exclusion, but generalized in an obvious way:
If f is a function $f : X \longrightarrow \mathbf{R}$, and A and B are subsets of X, then

$$\sum_{x \in A \cup B} f(x) = \sum_{x \in A} f(x) + \sum_{x \in B} f(x) - \sum_{x \in A \cap B} f(x).$$

Like in-ex, this generalizes to any finite number of sets. The proof is the same as for in-ex (Chapter 9, (4)). In-ex has $f(x) := 1$.

(2.17P) Solve for p to get $p = r/(1 + r)$. Since $0 \leq r < 1 + r$, we see on dividing by $1 + r$ that $0 \leq p < 1$.

(2.19P) From (2.15) and (2.16) we see that

$$\mathbf{Pr}(E) = \frac{r}{1+r} = \frac{x/y}{1+x/y} = \frac{x}{x+y}.$$

(3.6P) T is a finite set because it is the range of a function defined on the finite domain S. Since $\mathbf{Q}(y)$ is defined as the probability of an event in S, we see that $0 \leq \mathbf{Q}(y) \leq 1$ for all $y \in T$. Thus $\mathbf{Q}$ satisfies the first part of Definition (2.1), that $\mathbf{Q}$ map T to the unit interval U. The second part of the definition calls for $\mathbf{Q}$ to satisfy (2.2), namely,

$$\sum_{y \in T} \mathbf{Q}(y) = 1.$$

This follows from the definition of $\mathbf{Q}$ and from the same property of $\mathbf{Pr}$. Since $\mathbf{Q}(y) := \mathbf{Pr}(f = y) = \mathbf{Pr}(f^{-1}(y))$, by definition

$$\mathbf{Q}(y) = \sum_{a \in f^{-1}(y)} \mathbf{Pr}(a).$$

But the events $f^{-1}(y)$ for $y \in T$ are the cells of a partition of S (see Chapter 5, (9.5)). Therefore

$$\sum_{y \in T} \mathbf{Q}(y) = \sum_{a \in S} \mathbf{Pr}(a) = 1,$$

the last equality holding by definition (2.2).

(4.6P) The constant c is the random variable $f : S \longrightarrow \mathbf{R}$ which takes the value c at each point of S. It is the constant function. Its mean is by definition $\sum \mathbf{Pr}(a)c = c \sum \mathbf{Pr}(a) = c$.

(4.7P) We prove $E(f_1+\cdots+f_n) = Ef_1+\cdots+Ef_n$ by a simple induction on n. Basis: we know it's true for $n = 2$. In general, then, $E(f_1+\cdots+f_n) = E(f_1+\cdots+f_{n-1}) + Ef_n$ (assuming $n \geq 2$) by the basis step. By the inductive assumption $E(f_1+\cdots+f_{n-1}) = Ef_1+\cdots+Ef_{n-1}$.

(5.4P) g takes each of the values 0, 1, 2 on one-third of the points of the domain S_1. The probability function for S_1 is defined in (2.3) as $\mathbf{Pr}_1(x) = 1/n$ for all x in S_1. Thus the mean of g is

$$Eg = \frac{1}{n}\left(\frac{n}{3}\cdot 0 + \frac{n}{3}\cdot 1 + \frac{n}{3}\cdot 2\right) = 1.$$

So $\text{Var } g = Eg^2 - 1^2 = \frac{1}{n}\left(\frac{n}{3}\cdot 0^2 + \frac{n}{3}\cdot 1^2 + \frac{n}{3}\cdot 2^2\right) - 1 = \frac{1}{3} + \frac{4}{3} - 1 = 2/3.$

(8.2P) For $\mathbf{Pr}_1$ the events A and S_1 are trivially independent, for any event A. Suppose we set $A := \{D_1, D_2\}$. What events $B \neq \emptyset$ are independent of A? We know B cannot be disjoint from A. Can B include A? If so, then $\mathbf{Pr}(A \cap B) = \mathbf{Pr}(A) = 1/3$; and for this to equal $\mathbf{Pr}(A)\mathbf{Pr}(B)$ we'd have to have $B = S_1$, so take $B \cap A = \{D_1\}$. Then $\mathbf{Pr}(A \cap B) = 1/6$. If this is to equal $\mathbf{Pr}(A)\mathbf{Pr}(B)$, then

$$\frac{1}{3}\cdot \mathbf{Pr}(B) = \frac{1}{6},$$

so $\mathbf{Pr}(B) = 1/2$. Thus $B = \{D_1, D_i, D_j\}$, for $2 < i < j \leq 6$ is independent of A. (The same is true for D_2 in place of D_1.) For $\mathbf{Pr}_1'$ let's ask the same question: What events B' are independent of $A' := \{D_1, D_2\}$? Again, to avoid trivial cases, we must have intersection without inclusion. If we take $A' \cap B' = \{D_1\}$, then

$$\mathbf{Pr}(A' \cap B') = \frac{1}{12} = \mathbf{Pr}'(A')\,\mathbf{Pr}'(B') = \frac{1}{3}\,\mathbf{Pr}'(B')$$

implies $\mathbf{Pr}'(B') = 1/4 = \mathbf{Pr}'(D_1) + \mathbf{Pr}'(B' - \{D_1\})$. Thus

$$\mathbf{Pr}'(B' - \{D_1\}) = \frac{1}{4} - \frac{1}{12} = \frac{1}{6}.$$

Therefore we must take $B' = \{D_1, D_i\}$ for some i satisfying $1 < i < 6$.

(8.6P) $Eg = \frac{1}{10}y$. $\text{Var } g = \frac{1}{10}y^2 - \left(\frac{y}{10}\right)^2 = 0.09y^2$. $(\text{Var } g)^{1/2} = 0.3|y|$.

(8.11P) We partition the sample space into 8 cells $A_1, \ldots, A_8$. A_1 is the first one-eighth of the points; in fact, $A_1 = g_1^{-1}(0) \cap g_2^{-1}(0) \cap g_3^{-1}(0)$. A_2 is the next one-eighth, and so on. Then we can tabulate the values of the functions as

	A_1	A_2	A_3	A_4	A_5	A_6	A_7	A_8
$F := g_1 + g_2$	0	0	1	1	1	1	2	2
$G := g_2 + g_3$	0	1	1	2	0	1	1	2

First proof: We work directly from Definition (6.1) to find y and z such that $\mathbf{Pr}(F^{-1}(y) \cap G^{-1}(z)) \neq \mathbf{Pr}(F^{-1}(y))\mathbf{Pr}(G^{-1}(z))$. For example, $y = z = 0$ works:

$$\begin{aligned} F^{-1}(0) &= A_1 \cup A_2 \\ G^{-1}(0) &= A_1 \cup A_5 \\ \therefore F^{-1}(0) \cap G^{-1}(0) &= A_1, \text{ and} \\ \mathbf{Pr}(F^{-1}(0) \cap G^{-1}(0)) &= \mathbf{Pr}(A_1) = \frac{1}{8} \\ \text{But } \mathbf{Pr}(F^{-1}(0)) &= \mathbf{Pr}(A_1) + \mathbf{Pr}(A_5) \\ &= \frac{1}{4} \\ &= \mathbf{Pr}(G^{-1}(0)). \end{aligned}$$

Thus $\mathbf{Pr}(F^{-1}(0))\mathbf{Pr}(G^{-1}(0)) = \frac{1}{4}\cdot\frac{1}{4} \neq \frac{1}{8}$ so F and G are not independent.

Second proof: We use Theorem (7.1). Obviously g_i has mean 1/2 (for each i) because it takes the values 0 and 1 equally often. Therefore $g_1 + g_2$ and $g_2 + g_3$ each have mean 1, by Proposition (4.4). So if they are independent, then their product $(g_1 + g_2)(g_2 + g_3) = FG$ has mean equal to $1 \cdot 1 = 1$. We use the table to find $E(FG)$ as follows:

$$E(FG) = \frac{3}{8}\cdot 0 + \frac{2}{8}\cdot 1 + \frac{2}{8}\cdot 2 + \frac{1}{8}\cdot 4 = \frac{5}{4}.$$

Therefore the random variables $g_1 + g_2$ and $g_2 + g_3$ are not independent. This second method of proof is important. The logic of it is this: You know a result of the form

Theorem. For all x, if x has property P, then x has property Q.

You then show that a certain a does not have property Q. It follows from the contrapositive (Chapter 2, (11.1)) that this a does not have property P.

(9.2P) $\displaystyle\sum_{a,b\in S_1} \mathbf{Pr}(a,b) = \sum_{a\in S_1} \mathbf{Pr}(a) \sum_{b\in S_1} \mathbf{Pr}(b) = \sum_{a\in S_1} \mathbf{Pr}(a)\cdot 1 = 1.$

(13.4P) (*i*) $\mathbf{Pr}(|f| \geq 1) = \mathbf{Pr}(f \neq 0) = 1 - 0.2 = 0.8.$

$$\begin{aligned} (Ef^2)/1^2 = Ef^2 &= (0.2 + 0.1)1^2 + (0.2 + 0.1)2^2 \\ &\quad + (0.1 + 0.1)3^2 \\ &= 0.3 + 1.2 + 1.8 = 3.3 \end{aligned}$$

Because 3.3 is greater than 1, it is a poor upper bound for a probability, so $y = 1$ is not a large enough number to make Chebyshev's theorem useful.

$$\mathbf{Pr}(|f| \geq 2) = 0.1 + 0.1 + 0.2 + 0.1 = 0.5.$$

$(Ef^2)/2^2 = Ef^2/4 = 3.3/4 = 0.825$—much better.

(*ii*) We get $\mathbf{Pr}(|f_3 - 13.58| \geq 2\sigma) \leq \frac{1}{4}$. Thus it says that $\mathbf{Pr}(f_3 \geq 13.58 + 2\sigma = 61.8+) \leq \frac{1}{4}$, since f_3 is positive. (σ was calculated at the end of (8.5).) The true value for $\mathbf{Pr}(f_3 \geq 61.8)$ is $\frac{8}{100} = 0.08$ since there are exactly eight primes greater than 61 in S_3. The approximation is less good in this case.

CHAPTER

11

Recurrence Relations

1. Introduction

You've probably heard of the **Fibonacci sequence**

(1.1) $$0, 1, 1, 2, 3, 5, 8, 13, 21, \ldots,$$

in which every number after the first two is the sum of the preceding two numbers. That is,

(1.2) $$F_0 := 0, \quad F_1 := 1, \quad \text{and}$$

(1.3) $$\forall n \geq 2, \quad F_n := F_{n-1} + F_{n-2}.$$

This sequence, dating from 1202 A.D., is still of such interest that there is a journal, called *The Fibonacci Quarterly*, just for papers on it.

The equation (1.3) is called a **recurrence relation**, or **recursion**, for the sequence $\langle F_n \rangle$, and (1.2) states the **initial values** for the sequence. Using (1.2) and the recurrence relation we could get the value of F_n for any value of n within reason, but questions like "How fast does F_n grow as n tends to infinity?" seem hard to answer from what we know so far. The purpose of this chapter is to develop a general method for getting closed-form formulas for $\langle F_n \rangle$ and for other sequences defined by recurrence relations. The formula for F_n will yield an instant answer to the question $F_n = \Theta(?)$.

Examples (1.4) (*i*) The basic recursion

$$\binom{n}{k} = \binom{n-1}{k-1} + \binom{n-1}{k}$$

for the binomial coefficients is a recurrence relation in *two* variables. See Chapter 8, (3.2). The initial values are either

$$\binom{0}{0} = 1 \quad \text{and} \quad \binom{0}{k} = 0 \quad \text{for all } k \in \mathbf{Z}, k \neq 0$$

or they are replaced by *boundary values*

$$\binom{n}{0} = \binom{n}{n} = 1 \quad \text{for all } n \in \mathbf{Z}, n \geq 0.$$

We found a formula for $\binom{n}{k}$ in the binomial theorem, Chapter 8, (7.2). This two-variable recurrence is special, however. In this chapter we'll treat only one-variable recurrences, like (1.3).

(*ii*) In Chapter 3, problems 29 and 43 asked you to verify by induction that certain recursions had the solutions that were given. Methods of this chapter allow you to find those solutions.

(1.5) *Notation*. We'll denote any sequence $a_0, a_1, a_2, \ldots$ by $\langle a_n \rangle$.

Where do recurrences come from? Answer: from lots of counting problems, to name one source.

Example (1.6) Let $n \geq 0$ and find the number s_n of strings over $\{0, 1\}$ of length n *not* containing 101 as a substring. (By "substring" here we mean three consecutive symbols in the string. Thus 1 1 0 1 0 is a forbidden string of length 5 because the three middle symbols are 101. But 11001 is allowed because no three *consecutive* symbols are 101.)

We can see immediately that $s_0 = 1$, $s_1 = 2$, $s_2 = 4$, and $s_3 = 7$, so we have our initial conditions (and more, as you'll see).

Practice (1.7P) Explain the values for $s_0, \ldots, s_3$ in (1.6).

To find a recursion for s_n, it's easiest to break the set S_n of the desired strings into two subsets—those starting with 0 and those starting with 1. Thus let

$z_n :=$ total number of strings of length n starting with 0 and not containing 101

$u_n :=$ total number of strings of length n starting with 1 and not containing 101.

Now of course

(1.8) $$s_n = z_n + u_n.$$

Our first observation is that

(1.9) $$z_{n+1} = s_n.$$

Practice (1.10P) Prove (1.9).

Now we want to determine u_{n+1} in terms of u_j, z_j, or s_j for $j \leq n$. We could almost get away with the same recurrence for u_{n+1}, namely, $u_{n+1} = s_n$, but it's wrong. The trouble is that a string

$$t := 1\, x_1\, x_2\, \ldots\, x_n \qquad (x_i = 0 \text{ or } 1)$$

with $x := x_1 \dots x_n \in S_n$ is not in S_{n+1} if $x_1x_2 = 01$, because t starts with 101. But any other x in S_n does make $t \in S_{n+1}$. So we define

$(zu)_n :=$ total number of strings of length n starting with 01 and not containing 101

and, since we're going to need it,

$(zz)_n :=$ total number of strings of length n starting with 00 and not containing 101.

We see that, obviously,

$$z_n = (zu)_n + (zz)_n, \tag{1.11}$$

and (now we're cooking)

$$u_{n+1} = s_n - (zu)_n. \tag{1.12}$$

We just explained (1.12): $1x$, for $x \in S_n$, is in S_{n+1} iff x does not start with 01.

Practice (1.13P) Let x be any string of length n.
(i) Prove that $0x \in S_{n+1}$ iff $x \in S_n$.
(ii) Prove that if $1x \in S_{n+1}$, then $x \in S_n$.

Notice also that

$$(zz)_n = s_{n-2}, \tag{1.14}$$

similarly to (1.9).

Using (1.14) and (1.11) we get

$$\begin{aligned}(zu)_n &= z_n - (zz)_n \\ &= z_n - s_{n-2} \\ &= s_{n-1} - s_{n-2}.\end{aligned}$$

The last comes from (1.9). Now we're in the home stretch: For all $n \geq 2$,

$$\begin{aligned}s_{n+1} &= z_{n+1} + u_{n+1} \\ &= s_n + s_n - (zu)_n \\ &= 2s_n - s_{n-1} + s_{n-2}.\end{aligned} \tag{1.15}$$

We restate (1.15) as

$$s_{n+3} - 2s_{n+2} + s_{n+1} - s_n = 0 \qquad (n \geq 0). \tag{1.16}$$

This is our desired recurrence for s_n. Notice that we need the three values s_0, s_1, s_2 as initial values and that they give, from (1.6), $s_3 = 7$ when substituted in (1.15) with $n = 2$.

We solved a similar problem for a specific value of n by inclusion-exclusion in Chapter 9, (4.7). We could (in principle) use in-ex on this problem for general n, but it would be cumbersome to express and count all the intersections of the various sets, and the formula might have n or n^2 or some large number of terms. If so, it would not be in closed form.

You'll see eventually that the method of recurrence relations can be more powerful than in-ex, because the former yields a reasonable closed-form expression for s_n.

We have to postpone solving our problem (1.16), the recursion for s_n, until we develop the general theory.

2. The Simplest Recurrences

The recurrences treated in this section are **linear, homogeneous, and constant-coefficient with no repeated roots.**

Example (2.1) Consider

(2.2) $$a_{n+2} + 2a_{n+1} - 3a_n = 0$$

for all $n \geq 0$, with initial values a_0 and a_1. It is *linear* because each term of the unknown sequence $\langle a_n \rangle$ appears linearly, i.e., it is not multiplied by any term of the sequence. The recurrences $b_{n+1}^2 + b_{n-1}b_n = 0$ and $b_{n-1}b_{n+1} = b_{n+2}$ are nonlinear. (We solve a nonlinear recurrence in Chapter 13, (11).)

It is *homogeneous* because when expressed in the form

(2.3) $$\sum_{0 \leq i \leq k} d_i \, a_{n+k-i} = V_n$$

it has $V_n := 0$.
It has *constant coefficients* because the coefficients d_i in (2.3) are constants.

The recurrence $b_{n+2} - nb_{n+1} + b_n = 0$ is linear and homogeneous, but one of the coefficients, $-n$, is not constant.

We'll explain "no repeated roots" in the example below.

Here is the procedure for solving the recurrence

(2.4) $$d_0 \, a_{n+k} + d_1 \, a_{n+k-1} + \cdots + d_k \, a_n = 0$$

with initial values $a_0, \ldots, a_{k-1}$, where the d_i are constant and the **characteristic polynomial**

(2.5) $$p(x) := d_0 \, X^k + d_1 \, X^{k-1} + \cdots + d_k$$

has no repeated roots.

(2.6) To find the solution of (2.4), do

Step 1. Identify the characteristic polynomial $p(x)$ and find its roots. Call them $r_1, \ldots, r_k$.

Step 2. The general solution of (2.4) is then

$$\forall n \geq 0 \qquad a_n = c_1 r_1^n + \cdots + c_k r_k^n$$

for constants $c_1, \ldots, c_k$.

Step 3. Determine the constants from the initial values.

Here is a summary of this procedure in the form of a theorem.

Theorem (2.7) Suppose the characteristic polynomial of the recurrence relation (2.4) has no repeated roots. Say its roots are $r_1, \ldots, r_k$.

Then there are unique constants $c_1, \ldots, c_k$ depending on the initial values $a_0, \ldots, a_{k-1}$ such that for all integers $n \geq 0$

$$a_n = c_1 r_1^n + \cdots + c_k r_k^n. \blacksquare$$

We'll give the proof of Theorem (2.7) in section (4).

Now we show the procedure for solving the recursion (2.2) in order to illustrate the general procedure.

(2.8) To solve recursion (2.2):

Step 1. Identify the characteristic polynomial $p(x)$ and find its roots. Call them $r_1, \ldots, r_k$. In this case

$$p(x) := x^2 + 2x - 3$$

corresponding to

$$a_{n+2} + 2a_{n+1} - 3a_n.$$

To remember how to find $p(x)$, just blindly replace a_{n+i} in the recurrence by x^i. Since $p(x) = x^2 + 2x - 3 = (x-1)(x+3)$, the roots are 1 and -3. (Since they are not equal to each other we say $p(x)$ *has no repeated roots*.)

Step 2. The general solution of (2.2) is then

$$\forall n \geq 0 \qquad a_n = c_1 \cdot 1^n + c_2(-3)^n = c_1 + c_2(-3)^n$$

for constants $c_1, \ldots, c_2$. That is, we take a linear combination of the nth powers of the roots of $p(x)$.

Step 3. Determine the constants from the initial values.

If we have initial values $a_0 = 1$ and $a_1 = -1$ we get

$$n = 0: \; a_0 = 1 = c_1 + c_2$$
$$n = 1: \; a_1 = -1 = c_1 - 3c_2$$

The solution is $c_1 = c_2 = \frac{1}{2}$. Thus

$$a_n = \frac{1}{2} + \frac{1}{2}(-3)^n \qquad (n \geq 0)$$

is the unique solution to

$$\begin{cases} a_0 = 1, \quad a_1 = -1 \\ a_{n+2} + 2a_{n+1} - 3a_n = 0 \quad (n \geq 0). \end{cases}$$

Practice (2.9P) For the same recurrence but with initial values $a_0 = 1$, $a_1 = -3$, find a_n.

3. Repeated Roots

Here we show how to write the solution to a recurrence in which the characteristic polynomial has repeated roots. As we said in (2.5)

(3.1) The **characteristic polynomial** of the recursion

$$\sum_{0 \leq i \leq k} d_i \, a_{n+k-i} = V_n$$

is the polynomial

$$p(x) := \sum_{0 \le i \le k} d_i \, x^{k-i}.$$

We assume here that the d_i's are constants.

Assume now that $p(x)$ has a root r that appears exactly m times (r is called an ***m*-fold root** of $p(x)$). We also say "r is a root of **multiplicity** m." This means that $p(x)$ can be factored as

$$p(x) = (x - r)^m g(x)$$

for some *polynomial* $g(x)$ and that r is not a root of $g(x)$, i.e., $g(r) \neq 0$.

Example (3.2) $p_1(x) := x^2 - 2x + 1 = (x - 1)^2$, so 1 is a double, or 2-fold, root of $p_1(x)$. $p_2(x) := (x^3 - 1)(x^5 - 1)$ has 1 as a double root, because

$$p_2(x) = (x - 1)^2(x^2 + x + 1)(x^4 + x^3 + x^2 + x + 1);$$

obviously, substituting $x := 1$ in the latter two factors does not produce 0. (If factorizations like those of $x^3 - 1$ and $x^5 - 1$ here are unfamiliar, see Chapter 3, (16.1).)

$$\begin{aligned} p_3(x) :&= x^4 - 4x^3 + 5x^2 - 4x + 4 \\ &= (x - 2)^2(x^2 + 1) \end{aligned}$$

has 2 as a double root, and the two square roots of -1, i and $-i$, are single roots (1-fold, or nonrepeated, roots).

$$\begin{aligned} p_4(x) :&= x^6 - 3x^5 - 6x^4 + 24x^3 - 48x + 32 \\ &= (x - 2)^3(x + 2)^2(x - 1), \end{aligned}$$

so 2 is a triple root (3-fold root), -2 has multiplicity 2, and 1 is a single (sometimes called *simple*) root.

Relax. You won't have to find roots of polynomials of degree greater than 3. Not for my problems, anyway.

I found these examples by starting with the factors, and I used Macsyma to expand them. The point of the examples is simply to illustrate the idea of a multiple, or repeated, root. But you can find roots numerically with such packages as Macsyma or Maple.

Here now is the solution for the homogeneous, constant-coefficient case for any characteristic polynomial.

Theorem (3.3) The recurrence relation

$$\sum_{0 \le i \le k} d_i \, a_{n+k-i} = 0$$

has as its general solution

$$a_n = \sum_{r \in R} r^n \left(\sum_{1 \le j \le m_r} c_{r,j} \, n^{j-1} \right),$$

where R is the set of (distinct) roots of the characteristic polynomial

$$p(x) := \sum_{0 \le i \le k} d_i \, x^{k-i}$$

and m_r is the multiplicity of the root $r \in R$; the $c_{r,j}$ are constants that may be determined from the initial values $a_0, a_1, \ldots, a_{k-1}$. ■

(3.4) Notice that this result (3.3) includes the previous solution, in Theorem (2.7), as the special case $m_r = 1$ for all $r \in R$. The proof of Theorem (3.3) is in Section (5).

Example (3.5) Let's do problem 29 of Chapter 3 from scratch. The recurrence is

$$b_{n+2} + 2b_{n+1} + b_n = 0,$$

and two sets of initial values were given:

(i) $b_0 = b_1 = 1$;

(ii) $b_0 = 1$, $b_1 = -3$.

Theorem (3.3) allows us to find a formula for b_n easily.

The characteristic polynomial is $x^2 + 2x + 1 = (x+1)^2$, so -1 is a double root. By (3.3) we see that

$$b_n = (-1)^n \, (c_1 + c_2 n).$$

To find c_1 and c_2 for case (i), we set

$$b_0 = c_1 = 1$$
$$b_1 = -c_1 - c_2 = 1,$$

which imply $c_1 = 1$ and $c_2 = -2$. Thus for (i), $b_n = (-1)^n \, (1 - 2n)$, the formula you were given in Chapter 3, problem 29.

For (ii) we get

$$b_0 = c_1 = 1$$
$$b_1 = -c_1 - c_2 = -3.$$

Therefore $c_1 = 1$ and $c_2 = 2$. Thus $b_n = (-1)^n \, (1 + 2n)$, as advertised.

Recurrence relations like (2.3) are sometimes also called **difference equations**. If you've studied differential equations you'll see the analogy between them and difference equations.

Example (3.6) The differential equation, for $t_0 \le t \le t_1$,

$$\frac{d^2y}{dt^2} + 2\frac{dy}{dt} + y = 0$$

has the characteristic polynomial $x^2 + 2x + 1$ with -1 as a double root. The general solution is

$$y = c_1 e^{-t} + c_2 t e^{-t},$$

as you can check even if you have not studied differential equations. In the continuous case we don't use r^t but $(e^r)^t$ instead. The constants c_1 and c_2 are determined by the boundary values $y(t_0)$ and $y(t_1)$ (or the initial values $y(t_0)$ and $\frac{dy}{dt}(t_0)$).

4. Proof for the Simplest Case

We have the recursion, holding for all $n \geq 0$,

$$d_0 a_{n+k} + d_1 a_{n+k-1} + \cdots + d_k a_n = 0 \tag{4.1}$$

with $d_0 d_k \neq 0$ and initial values $a_0, a_1, \ldots, a_{k-1}$. For the moment we don't assume anything about the roots of the characteristic polynomial

$$p(x) = d_0 x^k + d_1 x^{k-1} + \cdots + d_k. \tag{4.2}$$

As we remarked in (1), there is obviously a unique sequence $\langle a_n \rangle$ satisfying (4.1) and having the given initial values $a_0, \ldots, a_{k-1}$, because we may solve (4.1) with $n = 0$ for a_k, then with $n = 1$ for a_{k+1}, and so on. But suppose we didn't know anything stated in Sections (2) and (3). How might we proceed to find a formula for a_n? One way, a kind of elaborate version of the "Let X be the unknown" approach, is to define an unknown function

$$\begin{aligned} g(x) :&= a_0 + a_1 x + a_2 x^2 + \cdots + a_n x^n + \cdots \\ &= \sum_{0 \leq n} a_n\, x^n. \end{aligned} \tag{4.3}$$

We've made the unknowns $a_0, a_1, \ldots$ the coefficients in a Taylor series. Then we'd try to use the recursion (4.1) to get more information about the infinite series $g(x)$, called the **generating function** of $\langle a_n \rangle$.

Example (4.4) Suppose we got lucky and found a closed form for $g(x)$, say

$$g(x) = \frac{1}{(1+x)^2} = (1+x)^{-2}.$$

Then from the binomial theorem (Chapter 8, (7.2)) we'd conclude that

$$\begin{aligned} a_n &= \binom{-2}{n} = \frac{(-2)(-3)\cdots(-2-n+1)}{n!} \\ &= (-1)^n \frac{(n+1)!}{n!} = (-1)^n (n+1). \end{aligned} \tag{4.5}$$

Practice (4.6P) Derive (4.5) more easily by first justifying

$$\frac{1}{(1+x)} = 1 - x + x^2 - x^3 + \cdots$$

without using the binomial theorem, and then differentiating.

OK, but how would we get such a closed form for $g(x)$? Our only hope would be the recursion (4.1) and the initial conditions. So here goes.

Notice that if we multiply $g(x)$ by x we shift the coefficients one step to the right:

$$\begin{aligned} xg(x) &= a_0x + a_1x^2 + a_2x^3 + \cdots + a_nx^{n+1} + \cdots \\ &= \sum_{0\le n} a_n\, x^{n+1}. \end{aligned}$$

Multiply by x^2 and shift them two steps to the right, and in general

$$\begin{aligned} x^jg(x) &= a_0x^j + a_1x^{j+1} + \cdots + a_nx^{n+j} + \cdots \\ &= \sum_{0\le n} a_n\, x^{n+j}. \end{aligned} \tag{4.7}$$

Suppose we make an array by writing (4.7) for $j = 0, 1, \ldots, k$, but to save space let's put in the powers of x only once, as headers, and omit the + signs:

(4.8)

	x^0	x^1	x^2	$\ldots$	x^{k-1}	x^k	$\ldots$	x^{n+k}	$\ldots$
$g(x) =$	a_0	a_1	a_2	$\ldots$	a_{k-1}	a_k	$\ldots$	a_{n+k}	$\ldots$
$xg(x) =$	0	a_0	a_1	$\ldots$	a_{k-2}	a_{k-1}	$\ldots$	a_{n+k-1}	$\ldots$
$x^2g(x) =$	0	0	a_0	$\ldots$	a_{k-3}	a_{k-2}	$\ldots$	a_{n+k-2}	$\ldots$
$\vdots$	$\vdots$	$\vdots$	$\vdots$		$\vdots$	$\vdots$		$\vdots$	
$x^{k-1}g(x) =$	0	0	0	$\ldots$	a_0	a_1	$\ldots$	a_{n+1}	$\ldots$
$x^kg(x) =$	0	0	0	$\ldots$	0	a_0	$\ldots$	a_n	$\ldots$

The header x^{n+k} is meant to stand for any column headed by x^i for any $i \ge k$. So here n can be *any nonnegative integer*, though we recognize that if n were 0, the column for x^{n+k} would duplicate that for x^k.

Now we multiply the rows of (4.8) by the coefficients d_j of our recursion (4.1). We multiply the first row by d_0, the second by d_1, and so on. Forget about the columns at the left, up to x^{k-1}. Look at what the column for x^{n+k} becomes.

$$\begin{array}{c} x^{n+k} \\ \hline d_0a_{n+k} \\ d_1a_{n+k-1} \\ d_2a_{n+k-2} \\ \vdots \\ d_{k-1}a_{n+1} \\ d_ka_n \end{array} \tag{4.9}$$

Still forgetting for now the left-hand columns, see what happens if we add the rows: The coefficient of x^{n+k} is the sum of the entries in the column (4.9),

$$d_0a_{n+k} + d_1a_{n+k-1} + \cdots + d_ka_n. \tag{4.10}$$

But this quantity is the left-hand side of (4.1)! It is 0. It is 0 for *every* $n \ge 0$.

We got to (4.10) by multiplying the row $x^jg(x)$ by d_j, and therefore the sum of all the rows after such multiplication by the d_j's is

$$d_0g(x) + d_1xg(x) + \cdots + d_kx^kg(x) = (d_0 + d_1x + \cdots + d_kx^k)g(x). \tag{4.11}$$

Now, finally, we "remember" the left-hand columns of (4.8), those up to x^{k-1}. Notice that every nonzero entry in those columns is one of the initial values $a_0, a_1, \ldots, a_{k-1}$. They are given to us in advance. The d_j's are also known, since we know the recursion. Therefore multiplying the rows by the d_j's and summing produces some known value in each of those columns up to x^{k-1}. The result is

$$(4.12)\qquad \begin{aligned}(d_0a_0)x^0 + (d_0a_1 + d_1a_1)x^1 + (d_0a_2 + d_1a_1 + d_2a_0)x^2 + \cdots \\ + (d_0a_{k-1} + d_1a_{k-2} + d_2a_{k-3} + \cdots + d_{k-1}a_0)x^{k-1} =: N(x).\end{aligned}$$

The $N(x)$ of (4.12) is the entire right-hand side since the same process (multiply by d_j and sum) produces 0 in every column x^k and beyond. Therefore (4.11) and (4.12) are equal, so

$$(4.13)\qquad (d_0 + d_1x + \cdots + d_kx^k)g(x) = N(x),$$

our polynomial of degree less than k defined in (4.12). Again, the coefficients of $N(x)$ are determined by the coefficients d_j of the recursion and by the initial values $a_0, \ldots, a_{k-1}$ (which are given in any particular problem).

Equation (4.13) is just about it. It determines $g(x)$ as the quotient of two polynomials, $N(x)$ in the numerator and

$$(4.14)\qquad D(x) := d_0 + d_1x + \cdots + d_kx^k$$

in the denominator:

$$(4.15)\qquad g(x) = \frac{N(x)}{D(x)}.$$

This is a closed form for $g(x)$. We can go further, though, and prove both (2.7) and (3.3), if we look more closely at $D(x)$.

$D(x)$ is the reverse of the characteristic polynomial $p(x)$. From (3.1):

$$p(x) = d_0x^k + d_1x^{k-1} + \cdots + d_k.$$

Perhaps you recall our use of reversed polynomials in proving the symmetry of binomial coefficients in Chapter 8, (5.1). If not, no matter; just consider $x^kp\left(\frac{1}{x}\right)$. It is

$$(4.16)\qquad \begin{aligned}x^kp\left(\frac{1}{x}\right) &= x^k\left(d_0\left(\frac{1}{x}\right)^k + d_1\left(\frac{1}{x}\right)^{k-1} + \cdots + d_k\right) \\ &= d_0 + d_1x + \cdots + d_kx^k = D(x).\end{aligned}$$

We'll exploit this fact (4.16) neatly to solve our problem.

You've probably heard that every polynomial of degree k with real coefficients has k roots, say $r_1, r_2, \ldots, r_k$, (which may be complex and may be repeated). Please accept that fact without proof. From it you can easily see that the polynomial $p(x)$ factors as

$$(4.17)\qquad p(x) = d_0(x - r_1)(x - r_2)\cdots(x - r_k).$$

The roots are the values of x for which $p(x)$ becomes 0. We saw several examples of this factorization in (3.2).

What does the factorization (4.17) tell us about $D(x)$? We do the reversal (4.16) on (4.17) and get

$$
\begin{aligned}
x^k p\left(\frac{1}{x}\right) &= x^k d_0\left(\frac{1}{x} - r_1\right)\left(\frac{1}{x} - r_2\right)\cdots\left(\frac{1}{x} - r_k\right) \\
&= d_0\,(1 - r_1x)\,(1 - r_2x)\cdots(1 - r_kx) \\
&= D(x).
\end{aligned}
\tag{4.18}
$$

Let's simplify by setting $d_0 = 1$. There's no loss, because we could divide (4.1) by d_0 from the start and not change the solution $\langle a_n \rangle$ at all. Thus from now on

$$d_0 = 1.$$

We may now restate (4.15) as

$$
g(x) = \frac{N(x)}{(1 - r_1x)(1 - r_2x)\cdots(1 - r_kx)}, \tag{4.19}
$$

where $r_1, \ldots, r_k$ are all the roots of the characteristic polynomial $p(x)$, and we still allow them to be repeated. Equation (4.19) is what we'll use to prove both (2.7) and (3.3).

(4.20) ***Proof of Theorem*** (2.7), the simplest case. We're now assuming that the k roots are all different from each other. Some or all might be nonreal, but no two are equal. Our proof will use induction on k.

Basis. For $k = 1$ we have a geometric series,

$$
\begin{aligned}
g(x) &= \frac{a_0}{1 - r_1x} = a_0(1 + r_1x + (r_1x)^2 + \cdots) \\
&= a_0 \sum_{0 \le n} r_1^n x^n.
\end{aligned}
\tag{4.21}
$$

Thus

$$g(x) := \sum_{0 \le n} a_n x^n = \sum_{0 \le n} a_0 r_1^n x^n.$$

Therefore $a_n = a_0 r_1^n$. This is exactly what Theorem (2.7) says when $k = 1$; the constant must be a_0, as it is here, to make things right when we set $n := 0$ in $a_n = cr_1^n$.

Notice that we don't need all the labor of (4.7) through (4.19) to settle the case when $k = 1$, because we could do it instantly from the recursion:

$$a_{n+1} + d_1 a_n = 0 \quad (\text{for } n \ge 0)$$

immediately implies

$$
\begin{aligned}
a_n &= -d_1 a_{n-1} = (-d_1)^2 a_{n-2} \\
&= (-d_1)^3 a_{n-3} = \cdots = (-d_1)^n a_0.
\end{aligned}
$$

And $-d_1$ is the root r_1 of the characteristic polynomial $x + d_1$.

(4.22) Now consider $k = 2$. This is the interesting case. We'll use the same partial fractions you may have used in calculus to integrate functions like $g(x)$. We confront

$$g(x) = \frac{n_0 + n_1 x}{(1 - r_1 x)(1 - r_2 x)}$$

and we claim we can express it as a sum of two terms like (4.21), namely,

$$\frac{n_0 + n_1 x}{(1 - r_1 x)(1 - r_2 x)} = \frac{A}{1 - r_1 x} + \frac{B}{1 - r_2 x}, \tag{4.23}$$

where A and B are constants. (By the way, let's represent the coefficients of $N(x)$ from (4.12) simply as $n_0, n_1, \ldots$.) Why is (4.23) possible? Because if we combine the terms of the right-hand side of (4.23), we get

$$\frac{A(1 - r_2 x) + B(1 - r_1 x)}{(1 - r_1 x)(1 - r_2 x)} = \frac{(A + B) - (r_2 A + r_1 B)x}{(1 - r_1 x)(1 - r_2 x)}, \tag{4.24}$$

and setting the numerator equal to $n_0 + n_1 x$ gives us two equations to solve:

$$\begin{aligned} A + B &= n_0 \\ -r_2 A - r_1 B &= n_1. \end{aligned} \tag{4.25}$$

Multiply the first equation by r_1, and then add the equations. You get

$$\begin{aligned} r_1 A + r_1 B &= r_1 n_0 \\ -r_2 A - r_1 B &= n_1 \\ \hline (r_1 - r_2)A &= r_1 n_0 + n_1. \end{aligned}$$

Since $r_1 - r_2 \neq 0$, we may solve for A and then for B as $n_0 - A$. So (4.23) is possible, and it depends crucially on the distinctness of the roots of $p(x)$.

If we apply the result (4.21) for $k = 1$ to $A/(1 - r_1 x)$ and to $B/(1 - r_2 x)$ we see that Theorem (2.7) also holds for $k = 2$.

(4.26) *A quick way to find A and B*: For A, multiply (4.23) by $1 - r_1 x$ to get

$$\frac{n_0 + n_1 x}{1 - r_2 x} = A + \frac{B(1 - r_1 x)}{1 - r_2 x}. \tag{4.27}$$

Now set $x = r_1^{-1}$, making $1 - r_1 x = 0$. The "B-part" disappears.

Thus

$$A = \frac{n_0 + n_1 r_1^{-1}}{1 - r_2 r_1^{-1}}. \tag{4.28}$$

You may get B the same way; the result is obtained by interchanging r_1 and r_2 in (4.28). Alternatively, you may use $B = n_0 - A$.

At this point we need to identify carefully just what we'll prove by induction. It's really a lemma, to be used in proving Theorem (2.7).

Lemma (4.29) Let $k \in \mathbf{N}$, and suppose $r_1, \ldots, r_k$ are distinct (real or) complex numbers. If $N(x)$ is any polynomial of degree less than k, then there exist complex constants $C_1, \ldots, C_k$ such that

$$\text{(4.30)} \qquad \frac{N(x)}{(1-r_1x)\cdots(1-r_kx)} = \frac{C_1}{1-r_1x} + \cdots + \frac{C_k}{1-r_kx}. \blacksquare$$

(4.31) *Comment*. Once we prove (4.29) all we need do to finish the proof of (2.7) is apply the geometric-series argument (4.21) of the case $k = 1$ to each term on the right of (4.30).

(4.32) ***Proof of*** (4.29) We've already done the cases $k = 1$ and $k = 2$, so we take k to be a fixed but arbitrary integer with $k \geq 2$. We assume (4.29) for this value of k and try to prove it for $k+1$. Thus we have $k+1$ distinct integers $r_1, \ldots, r_k, r_{k+1}$ and a polynomial $N(x)$ of degree less than $k+1$. We simply split off one factor from the denominator as follows:

$$\frac{N(x)}{(1-r_1x)\cdots(1-r_{k+1}x)} = \frac{1}{(1-r_{k+1}x)} \cdot \frac{N(x)}{(1-r_1x)\cdots(1-r_kx)}.$$

Our approach is to express

$$\frac{N(x)}{(1-r_1x)\cdots(1-r_kx)}$$

as a sum $\sum C_j/(1-r_jx)$ by using the inductive assumption. Then we multiply by $1/(1-r_{k+1}x)$ and break up the resulting products of two factors on the right by calling in the result for $k = 2$.

(4.33) There's only one hitch: $N(x)$ might have degree k. That would prevent us from using the inductive assumption. We can easily get around that, as follows. We divide $N(x)$ by the denominator polynomial $D(x)$ and get a constant C as (partial) quotient with remainder $N_0(x)$ *of degree less than* k.

Example (4.34) Consider

$$\frac{x^3}{(1-x)(1+x)(1-2x)}.$$

The denominator $D(x) = 1 - 2x - x^2 + 2x^3$. Thus $x^3 = \frac{1}{2}D(x) + \frac{1}{2}x^2 + x - \frac{1}{2}$. (Check this.) Therefore

$$\frac{x^3}{D(x)} = \frac{1}{2} + \frac{1}{2} \cdot \frac{x^2 + 2x - 1}{D(x)}.$$

Now the quotient has smaller degree on top than on the bottom.

The general situation is simpler than the example. We have two polynomials $N(x)$ and $D(x)$, both of degree k. Say they are

$$N(x) = ax^k + N_0(x)$$
$$D(x) = bx^k + D_0(x),$$

where $N_0(x)$ and $D_0(x)$ are polynomials of degree less than k. In other words, a and b are the leading coefficients of the two polynomials, and neither one is 0. Then

$$N(x) - \frac{a}{b}D(x) = N_0(x) - \frac{a}{b}D_0(x) =: R(x),$$

a polynomial of degree less than k. Therefore

$$N(x) = \frac{a}{b} D(x) + R(x),$$

where (degree of $R(x)$) $< k$. Set $C := \frac{a}{b}$; we see that

$$\frac{N(x)}{D(x)} = C + \frac{R(x)}{D(x)}. \tag{4.35}$$

Now we are entitled to apply the inductive assumption to $R(x)/D(x)$.

(4.36) All right—back to the proof. We carry out the strategy mentioned in (4.32). Our target is

$$\begin{aligned} \frac{N(x)}{\prod\limits_{1 \le j \le k+1} (1 - r_j k x)} &= \frac{1}{(1 - r_{k+1}x)} \frac{N(x)}{\prod\limits_{1 \le j \le k} (1 - r_j x)} \\ &= \frac{1}{(1 - r_{k+1}x)} \left[C + \frac{R(x)}{\prod\limits_{1 \le j \le k} (1 - r_j x)} \right], \end{aligned} \tag{4.37}$$

where we've used (4.35). Understand that if (degree of $N(x)$) $< k$, then $C = 0$ and $R(x) = N(x)$. But degree of $R(x)$ *is* less than k, so we apply the inductive assumption to say there are constants $C_1, \dots, C_k$ such that

$$\frac{R(x)}{\prod\limits_{1 \le j \le k} (1 - r_j x)} = \frac{C_1}{1 - r_1 x} + \cdots + \frac{C_k}{1 - r_k x}.$$

Now

$$\frac{N(x)}{\prod\limits_{1 \le j \le k+1} (1 - r_j x)} = \frac{C}{1 - r_{k+1}x} + \sum_{1 \le j \le k} \frac{C_j}{(1 - r_j x)(1 - r_{k+1}x)}. \tag{4.38}$$

Finally, each of the last k terms has a split-up according to the case for $k = 2$, equation (4.23). Thus for $1 \le j \le k$ there are constants C_j' and C_j'' such that

$$\frac{C_j}{(1 - r_j x)(1 - r_{k+1}x)} = C_j \left(\frac{C_j'}{1 - r_j x} + \frac{C_j''}{1 - r_{k+1}x} \right). \tag{4.39}$$

Combining terms with denominator $1 - r_{k+1}x$ yields our existence proof:

$$\frac{N(x)}{\prod\limits_{1 \le j \le k+1} (1 - r_j x)} = \left(\sum_{1 \le j \le k} \frac{C_j C_j'}{1 - r_j x} \right) + \frac{C + \sum C_j C_j''}{1 - r_{k+1}x}. \tag{4.40}$$

That is, the case for k implies that for $k + 1$, so by induction we've proved Lemma (4.29).

And, as we remarked in (4.31), that completes our proof of Theorem (2.7). QED

Comments (4.41)

(*i*) That proof was long, but I insist it's easy. The whole thing depends on nothing but the multiplication of polynomials. And that is nothing but the distributive law! Even the existence of roots for a polynomial is founded on polynomial multiplication.

(*ii*) This was an existence proof. We would hardly ever solve a recursion by carrying out the partial fraction decomposition of $g(x) = N(x)/D(x)$ and then turning each term into its geometric series. What we did was show in the case of nonrepeated roots that the partial-fraction decomposition exists. From that it's easy to get the procedure (2.6) for the solution of the recursion.

(*iii*) In the proof the initial conditions are built in. They determine the numerator polynomial $N(x)$. That in turn determines the numerator constants C_j in (4.30) and gives the particular solution to the recursion that satisfies the given initial conditions. Procedure (2.6), on the other hand, lets the C_j's be k undetermined constants and then imposes the k initial conditions at the end to find the actual values of those constants.

(*iv*) If r is any root of the characteristic polynomial $p(x)$, then $a_n := r^n$ is a solution to the recurrence (4.1). We've given a lengthy proof in the nonrepeated-roots case, but it's true in general and easy to verify. Just substitute:

$$\begin{aligned} \sum_{0 \le j \le k} d_j\, a_{n+k-j} &= \sum d_j\, r^{n+k-j} \\ &= r^n \sum d_j\, r^{k-j} \\ &= r^n\, p(r) = 0. \end{aligned}$$

The virtue of our long proof is that it shows exactly what the solution is, namely, a certain linear combination of the r_j^n. If we merely observed that for each j, r_j^n is a solution to the recurrence, we wouldn't know that we could satisfy arbitrary initial conditions by appropriate choice of the constants c_j in $\sum c_j r_j^n$—unless, of course, we developed some theory of linear equations and, in particular, showed the Vandermonde determinant† is nonzero. (That is a corollary of our result, incidentally.)

Other virtues of our proof: It shows the existence and form of partial-fractions decompositions of rational functions (those are quotients of polynomials) and a use of the generating function of a sequence.

Practice (4.42P)

(*i*) In the proof of (4.29), where did we use the hypothesis that all the roots are different?

(*ii*) Why in (4.1) did we put in the condition $d_0 d_k \neq 0$, and where did we use it?

(*iii*) Find a partial-fractions expression for $\frac{2}{1+x^2}$. Do the same for $\frac{2+2x}{1+x^2}$.

†If you want to know what that is, your teacher will be delighted to tell you. Or see Problem G6.

Example (4.43) Suppose you have the recurrence $a_{n+2} - 5a_{n+1} + 6a_n = 0$ for $n \geq 0$ and $a_0 = 1$, $a_1 = 3$. Find a formula for a_n good for all integers $n \geq 0$.

We follow (2.6). $p(x) = x^2 - 5x + 6$. It factors as $p(x) = (x-2)(x-3)$. Thus

$$a_n = c_1 2^n + c_2 3^n$$

for appropriate constants c_1, c_2. We have

$$\begin{aligned} a_0 &= c_1 + c_2 = 1 \\ a_1 &= 2c_1 + 3c_2 = 3. \end{aligned}$$

These give $c_2 = 2$, $c_1 = -1$. Thus for all $n \geq 0$

$$a_n = 2 \cdot 3^n - 2^n.$$

(4.44) Let's solve the recursion of our Example (1.6), where we said s_n is the number of strings in 0, 1 not having 101 as a substring. We found the recursion (1.16) for s_n:

$$s_{n+3} - 2s_{n+2} + s_{n+1} - s_n = 0,$$

and we had initial conditions $s_0 = 1$, $s_1 = 2$, and $s_2 = 4$.

The characteristic polynomial is $p(x) = x^3 - 2x^2 + x - 1$. Cubics can be solved exactly in terms of square and cube roots, but it can be tedious to do them that way. Using Macsyma instead, I got the following numerical approximations to the roots (with the command *allroots* $(p(x))$):

(4.45)
$$\begin{aligned} r_1 &= 1.75488 \\ r_2, r_3 &= 0.12256 \pm 0.74482i, \end{aligned}$$

where $i^2 = -1$. The Macsyma command *solve* $(p(x), x)$ even gives the roots exactly, as

(4.46)
$$\begin{aligned} r_1 &= \frac{2}{3} + A + \frac{1}{9A} \\ r_2 &= \frac{2}{3} + A\omega + \frac{1}{9A\omega} \\ r_3 &= \frac{2}{3} + A\omega^2 + \frac{1}{9A\omega^2} \end{aligned}$$

where

$$A := \left(\frac{\sqrt{23}}{6\sqrt{3}} + \frac{25}{54} \right)^{1/3};$$

the real cube root is intended. Also, ω is a primitive cube root of 1; that is, $\omega^3 = 1$ but $\omega \neq 1$. Thus $\omega^2 + \omega + 1 = 0$; and $\omega = -\frac{1}{2} + (\sqrt{3}/2)i$, and $\omega^2 = -\frac{1}{2} - (\sqrt{3}/2)i$.

From (2.6) we see that

(4.47)
$$s_n = c_1 r_1^n + c_2 r_2^n + c_3 r_3^n$$

for appropriate constants c_1, c_2, c_3. We can determine them from the initial conditions.

Thus

$$\begin{aligned} c_1 + c_2 + c_3 &= 1 \qquad (n = 0) \\ r_1c_1 + r_2c_2 + r_3c_3 &= 2 \qquad (n = 1) \\ r_1^2c_1 + r_2^2c_2 + r_3^2c_3 &= 4 \qquad (n = 2) \end{aligned} \tag{4.48}$$

But before we find the values, notice one point: r_2 and r_3 each have absolute value less than 1.

$$|r_2| = |r_3| \doteq \sqrt{(.12)^2 + (.74)^2} \doteq 0.75$$

Since $r_2 \doteq 0.75e^{i\theta}$ and $r_3 \doteq 0.75e^{-i\theta}$ for appropriate real θ, it follows that r_2^n and r_3^n approach 0 as n tends to infinity. Therefore for all large enough n (no matter what c_2 and c_3 turn out to be),

$$s_n = \text{the nearest integer to } c_1r_1^n. \tag{4.49}$$

Certainly, then, $s_n \sim c_1r_1^n$ (i.e., s_n is asymptotic to $c_1r_1^n$).

Notice that we can now conclude, from (4.49), that the proportion of all strings of length n not containing 101 as a substring tends to 0 exponentially, as we might, or might not, have expected. That is,

$$\frac{s_n}{2^n} \sim c_1\left(\frac{r_1}{2}\right)^n \doteq c_1(0.88)^n.$$

Solving (4.48) for c_1, c_2, c_3 in APL (where you represent the complex number $a + ib$ $(a, b \in \mathbf{R})$ as "aJb") we get

$$\begin{aligned} c_1 &\doteq 1.26724 \\ c_2 &\doteq -0.13362 + i(0.12828) \\ c_3 &\doteq -0.13362 - i(0.12828). \end{aligned} \tag{4.50}$$

There is no point carrying the calculations further, because Macsyma gave us the roots to only 6 decimal places.

Let's check how accurate this all is. Using the approximations (4.45) and (4.50) in (4.47) yields the following values

n	3	4	5	6
s_n	7.000006	12.00002	21.00002	37.00004

n	10	15	20
s_n	351.0007	5842.02	97229.37

,

so already at $n = 20$ the formula obtained from the approximate values becomes suspect (although the true value is indeed 97229, as calculated from the recursion).

If we wanted to improve matters numerically we could calculate the roots more accurately from (4.46) and then c_1, c_2, and c_3 more accurately from (4.48).

We could even do it exactly by using (4.46) directly in (4.48). It's easier than it looks. After all, c_1 and r_1 are the only ones that matter; we know that the others are negligible for large n, so our current approximations to them are good enough.

A mere page of hand calculation shows that

$$c_1 = \frac{r_2r_3 + 4 - 2(r_2 + r_3)}{(r_1 - r_2)(r_1 - r_3)}. \tag{4.51}$$

Getting a little fancy, we can say that $r_2 r_3 = 1/r_1$ and $r_2 + r_3 = 2 - r_1$. That's because $p(x) = (x - r_1)(x - r_2)(x - r_3) = x^3 - 2x^2 + x - 1$. So $r_1 r_2 r_3 = 1$ and $r_1 + r_2 + r_3 = 2$. Then the denominator is

$$\begin{aligned} r_1^2 - (r_2 + r_3) r_1 + r_2 r_3 &= r_1^2 - (2 - r_1) r_1 + 1/r_1 \\ &= 2r_1^2 - 2r_1 + 1/r_1. \end{aligned}$$

Now that we've expressed c_1 in terms of r_1 we see that:

$$\begin{aligned} c_1 &= \frac{\frac{1}{r_1} + 4 - 2(2 - r_1)}{2r_1^2 - 2r_1 + 1/r_1} \\ &= \frac{2r_1^2 + 1}{2r_1^3 - 2r_1^2 + 1} . \end{aligned}$$

One more simplification: Since r_1 is a root of $p(x)$, $p(r_1) = 0$. Thus

$$r_1^3 - 2r_1^2 + r_1 - 1 = 0,$$

that is,

$$2r_1^3 = 4r_1^2 - 2r_1 + 2;$$

therefore

$$c_1 = \frac{2r_1^2 + 1}{2r_1^2 - 2r_1 + 3} = 1 + \frac{2(r_1 - 1)}{2r_1(r_1 - 1) + 3} . \tag{4.52}$$

Numerical evaluation of this formula with r_1 as in (4.45) gives the value of c_1 in (4.50).

Now we could evaluate A in (4.46) to as many places as we wish and thus get r_1 and c_1 to about that number of places. ◁

5. Proof When Roots Are Repeated

Again we seek the solution $\langle a_n \rangle$ of (4.1). We set $g(x) := \sum a_n x^n$, and from (4.15) we know

$$g(x) = \frac{N(x)}{D(x)} = \frac{N(x)}{\prod_{1 \le j \le k} (1 - r_j x)} , \tag{5.1}$$

where $N(x)$ is a known polynomial of degree less than k; and now we allow $D(x)$ to have multiple roots. Again the whole burden is the proof that $N(x)/D(x)$ has an expansion into partial fractions.

First let's look at the result for a_n in the simplest such case.

Example (5.2) Let $D(x) = (1 - rx)^m$, where $m \ge 1$. Take $N(x) := 1$. Then

$$g(x) = \frac{N(x)}{D(x)} = \frac{1}{(1 - rx)^m} = \sum_{0 \le n} \binom{-m}{n} (-r)^n x^n \tag{5.3}$$

by Definition (2.1) of Chapter 8, extended to negative upper index. It isn't hard to show (see, for example, (4.5)) that for any positive integer m

$$(5.4)\qquad \binom{-m}{n} = (-1)^n\binom{n+m-1}{n} = (-1)^n\binom{n+m-1}{m-1}.$$

Thus for all $n \geq 0$

$$(5.5)\qquad a_n = \binom{-m}{n}(-r)^n = \binom{n+m-1}{m-1}r^n.$$

Notice that the coefficient of r^n in (5.5) is a polynomial in n of degree $m-1$. That's because

$$(5.6)\qquad \binom{n+m-1}{m-1} = \frac{(n+m-1)(n+m-2)\cdots(n+1)}{(m-1)!}$$

Remember that m is fixed and n runs over all nonnegative integers.

Practice (5.7P) Prove (5.4).

The preceding example is actually generic for this whole section, so we'll refer to it later.

Let's change notation to allow for multiple roots. We take

$$(5.8)\qquad \begin{aligned} D(x) &= (1-r_1x)^{m_1}\cdots(1-r_\ell x)^{m_\ell} \\ &= \prod_{1\le j\le \ell}(1-r_jx)^{m_j}. \end{aligned}$$

Thus we take $p(x)$ to have ℓ *distinct* roots $r_1,\ldots,r_\ell$ with the indicated multiplicities $m_1,\ldots,m_\ell$, respectively. Since the degree of $D(x)$ is k, $m_1+\cdots+m_\ell = k$.

When no roots are repeated, the partial-fractions expansion is simple; there's just one term, of the form $c/(1-rx)$, for each root r. But when a root r is repeated m times, it contributes m terms of the form

$$\frac{c_1}{1-rx}, \quad \frac{c_2}{(1-rx)^2}, \ldots, \quad \frac{c_m}{(1-rx)^m}.$$

To see why this is so easy if we piggyback on our result, Lemma (4.29), for the nonrepeated case, and adapt a nice idea of M. R. Spiegel.[†]

Theorem (5.9) Let $N(x)$ be a polynomial of degree less than k. Let $D(x) = \prod_{1\le j\le \ell}(1-r_jx)^{m_j}$ be as described in (5.8). Then the rational function $N(x)/D(x)$ has an expansion in partial fractions of the form

$$(5.10)\qquad \sum_{1\le j\le \ell}\left(\frac{A_{j,1}}{1-r_jx} + \frac{A_{j,2}}{(1-r_jx)^2} + \cdots + \frac{A_{j,m_j}}{(1-r_jx)^{m_j}}\right). \blacksquare$$

That is, there exist complex constants $A_{j,1}, A_{j,2}$, etc., that make $N(x)/D(x)$ equal to the expression in (5.10).

[†] "Partial fractions with repeated linear or quadratic factors," *American Mathematical Monthly*, Vol. 57 (1950), pages 180–181.

Proof We'll just prove the existence of, and not worry about how to find, the A's. After all, our method is to impose the initial conditions at the end and determine the constants that way.

(5.11) Consider the nonrepeated denominator $D_0(x) = (1-r_1x)\cdots(1-r_\ell x)$ having the same roots but only once each.

Introduce new variables $z_1, \ldots, z_\ell$ by replacing the factor $1 - r_jx$ by $z_j - r_jx$. Call the result

$$D_1(x) = (z_1 - r_1x)(z_2 - r_2x) \cdots (z_\ell - r_\ell x). \tag{5.12}$$

Consider now $N(x)/D_1(x)$. The denominator has degree ℓ as a polynomial in x, so if (degree of $N(x)$) $\geq \ell$, divide $N(x)$ by $D_1(x)$ to get

$$N(x) = D_1(x)\,q(x) + N_1(x), \tag{5.13}$$

where $q(x)$ and $N_1(x)$ are polynomials in x with coefficients that may depend on the z's, and the degree of $N_1(x)$ is less than ℓ.

▶COMMENT (5.14) Actually (5.13) is the division algorithm for polynomials in one variable. We'll prove it just as we proved our basic result ((7.5) of Chapter 6) on the greatest common divisor.

(5.15) *Division algorithm for polynomials.* Let $a(x)$ and $b(x) \neq 0$ be polynomials in x with complex coefficients. Then there are unique polynomials $q(x)$ and $r(x)$ such that

$$a(x) = b(x)\,q(x) + r(x)$$

and such that (degree of $r(x)$) $<$ (degree of $b(x)$).

Proof Consider the set S

$$S := \{\, a(x) - b(x)\,u(x);\ u(x) = \text{ any polynomial}\,\}.$$

If the zero polynomial is in S, then $b(x)$ divides $a(x)$. If not, there is (by well ordering) a polynomial of least degree in S; call it $s(x)$. By the simplest of algebraic maneuvers (see, for example, (4.34), where we did it in a special case), you can show that (degree of $s(x)$) $<$ (degree of $b(x)$). Therefore from $a(x) - b(x)\,u(x) = s(x)$ you get the result: $q(x) = u(x)$ and $r(x) = s(x)$. ◀

The result is

$$\frac{N(x)}{D_1(x)} = q(x) + \frac{N_1(x)}{D_1(x)}. \tag{5.16}$$

At this point we begin a running example of the procedure we are describing. Don't be dismayed by the calculations, because we'll never *do* them. We'll just talk knowingly about them. Remember—we're only in this for the existence.

Example (5.17) Take $D(x) := (1-x)^3(1+x)$ and $N(x) := 8x^3$. Then $D_0(x) = (1-x)(1+x)$, and $D_1(x) = (z-x)(1+x)$. We don't need z_2, as you'll see. Since (degree of $N(x)$) $= 3 > 2 =$ (degree of $D_1(x)$), we divide by $D_1(x)$ (and I'm using Macsyma for this):

$$\begin{aligned} N(x) := 8x^3 &= (z-x)(1+x)[-8x - 8z + 8] \\ &\quad + 8[(z^2 - z + 1)x + z^2 - z]. \end{aligned}$$

Thus

$$\frac{8x^3}{(z-x)(1+x)} = -8(x+z-1) + 8\,\frac{(z^2-z+1)x + z^2 - z}{(z-x)(1+x)} \tag{5.18}$$

—To be continued.

Now the rational function $N_1(x)/D_1(x)$ satisfies the conditions of Section (4) for the existence of a partial fraction decomposition, so put the decomposition of $N_1(x)/D_1(x)$ into (5.16) to get

$$\frac{N(x)}{D_1(x)} = q(x,z) + \frac{A_1(z)}{z_1 - r_1x} + \frac{A_2(z)}{z_2 - r_2x} + \cdots + \frac{A_\ell(z)}{z_\ell - r_\ell x}, \tag{5.19}$$

where we emphasize that $q(x)$ and the constants A_j are functions of z, which we use to stand for $z_1,\ldots,z_\ell$. That is, $q(x,z)$ is a polynomial in x with coefficients that are rational functions of $z_1,\ldots,z_\ell$, and $A(z)$ is a rational function of $z_1,\ldots,z_\ell$.

You can see from (5.18) that they have to depend on z in the example. It's really clearer not to pursue the example now, so we postpone the rest of it to the end of the proof.

Now if $m_1 > 1$, differentiate (5.19) *with respect to* z_1. Look what happens. On the left you get

$$-\frac{N(x)}{(z_1-r_1x)^2(z_2-r_2x)\cdots(z_\ell-r_\ell x)}; \tag{5.20}$$

on the right you get

$$\frac{\partial}{\partial z_1}q(x,z) + \frac{\frac{\partial}{\partial z_1}A_1(z)}{z_1-r_1x} - \frac{A_1(z)}{(z_1-r_1x)^2} + \sum_{2\le j\le l}\frac{\frac{\partial}{\partial z_1}A_j(z)}{z_j-r_jx}. \tag{5.21}$$

Continue to differentiate with respect to z_1 a total of $m_1 - 1$ times. On the left we'll have

$$\frac{(-1)^{m_1-1}(m_1-1)!\,N(x)}{(z_1-r_1x)^{m_1}(z_2-r_2x)\cdots(z_\ell-r_\ell x)}. \tag{5.22}$$

You see where this is going. On the right we have

$$\left(\frac{\partial}{\partial z_1}\right)^{m_1-1} q(x_1z) + \sum_{2\le j\le \ell}\frac{\left(\frac{\partial}{\partial z_1}\right)^{m_1-1}A_j(z)}{z_j-r_jx} + \left(\frac{\partial}{\partial z_1}\right)^{m_1-1}\frac{A_1(z)}{z_1-r_1x}. \tag{5.23}$$

What is this last term? If you look at (5.21) a moment, you'll believe that it has the form

$$\frac{B_1(z)}{z_1-r_1x} + \frac{B_2(z)}{(z_1-r_1x)^2} + \cdots + \frac{B_{m_1}(z)}{(z_1-r_1x)^{m_1}}, \tag{5.24}$$

where the B's are rational functions of z (not involving x, since $A_1(z)$ is constant with respect to x). To prove it has the form (5.24) is a simple induction on m_1, which we leave to the problems. For our lofty purpose we don't care what the various functions $B(z)$ are explicitly; we care only that they exist.

At this point we set $z_1 := 1$. We have from (5.22), (5.23), and (5.24)

$$\frac{(-1)^{m_1-1}(m_1-1)!\,N(x)}{(1-r_1x)^{m_1}\prod_{2\le j\le \ell}(z_j-r_jx)} = \left(\frac{\partial}{\partial z_1}\right)^{m_1-1} q(x,z)\Big|_{z_1=1}$$

(5.25)

$$+ \sum_{2\le j\le \ell} \frac{\left(\frac{\partial}{\partial z_1}\right)^{m_1-1} A_j^*(z)}{z_j - r_jx}$$

$$+ \frac{B_1^*(z)}{1-r_1x} + \frac{B_2^*(z)}{(1-r_1x)^2} + \cdots + \frac{B_{m_1}^*(z)}{(1-r_1x)^{m_1}},$$

where by the asterisk (*) I merely mean that z_1 has been set equal to 1.

I know the foregoing looks bad, but it really isn't. It's just the simple idea that by differentiating on the left side you can go from the non-repeated- to the repeated-roots case. And so you just chase it down on the right side.

To finish the proof, you do the same thing for each other repeated root in turn. If a root r_j appears only once, this procedure simply sets $z_j = 1$, so you could omit z_j from the start if you wished. The result is

$$\text{(5.26)} \qquad (\text{constant}) \times \frac{N(x)}{D(x)} = \sum_{1\le j\le \ell}\ \sum_{1\le s\le m_j} \frac{B_{j,s}}{(1-r_jx)^s}$$

where the $B_{j,s}$ are now constants, since all the z_j's have finally been set equal to 1. The $A_{j,s}$ of (5.10), our targets, arise when we divide by the constant on the left.

(5.27) One little gap: How do we know that $q(x,z)$ has gone to 0 under this procedure?

The only term that could survive the hammer blows of all those differentiations would have to be

$$\text{(5.28)} \qquad z_1^{m_1-1}\, z_2^{m_2-1} \cdots z_\ell^{m_\ell-1} \cdot x^i.$$

(It might have larger exponents.) But the total degree of the monomial (5.28) is

$$i - \ell + \sum_j m_j = k - \ell + i,$$

and even if $i = 0$ it's too big. No term in $q(x,z)$ can have total degree greater than $k - \ell - 1$.[†] Therefore q has become 0 by the end of the process.

Example (5.29) *Example (5.17), continued.* We now find the partial-fraction expansion of the rational function on the right-hand side of (5.18). (Actually, the Macsyma command *Partfrac*

[†]This part is a little technical—sorry. It's explained at the end, in (5.43). Try a few cases in Macsyma to get a feel for this situation.

$(N(x)/D_1(x), x)$ gives us $q(x,z)$ and that expansion at the same time.) The result is

$$\frac{8x^3}{(z-x)(1+x)} = -8(x+z-1) + \frac{8z^3/(1+z)}{z-x} - \frac{8/(1+z)}{1+x} =: R_1 . \tag{5.30}$$

(About (5.27): Notice that $k = 4$ in our problem, and that $N(x)$ has degree 3. Also $\ell = 2$, and the largest degree of any term in $q(x)$ is $3 - 2 = 1$ in either variable x or z.)

Now we differentiate (5.30):

$$LHS: \frac{d}{dz} \frac{8x^3}{(z-x)(1+x)} = \frac{-8x^3}{(z-x)^2(1+x)} =: L_2$$

$$RHS: \frac{d}{dz} R_1 = -8 + \frac{24z^2/(1+z)}{(z-x)} - \frac{8z^3/(1+z)^2}{z-x} - \frac{8z^3/(1+z)}{(z-x)^2} + \frac{8/(1+z)^2}{1+x} =: R_2 .$$

We differentiate again:

$$\begin{aligned} \frac{d}{dz} L_2 = \frac{2 \cdot 8x^3}{(z-x)^3(1+x)} &= \frac{d}{dz} R_2 \\ &= \frac{48z/(1+z)}{z-x} - \frac{48z^2/(1+z)^2}{z-x} \\ &\quad + \frac{16z^3(1+z)^3}{z-x} - \frac{48z^2/(1+z)}{(z-x)^2} + \frac{16z^3/(1+z)^2}{(z-x)^2} \\ &\quad + \frac{16z^3/(1+z)}{(z-x)^3} - \frac{16/(1+z)^3}{1+x} =: R_3 . \end{aligned}$$

And there it is. Now we divide by 2 and set $z := 1$ in R_3 with Macsyma's *subst* $(z = 1, R_3)$. The result is

$$\frac{8x^3}{(1-x)^3(1+x)} = \frac{7}{1-x} - \frac{10}{(1-x)^2} + \frac{4}{(1-x)^3} - \frac{1}{1+x} . \tag{5.31}$$

(5.32) *The Home Stretch.* It's now easy to wrap it all up. We've proved that

$$g(x) := \sum_{0 \le n} a_n x^n \tag{5.33}$$

has the partial-fraction expression (5.10). Every term of (5.10) has the form

$$\frac{A}{(1-rx)^m} , \tag{5.34}$$

where A is a constant, r is a root of the characteristic polynomial $p(x)$, and m is no greater than the multiplicity of that root r. We saw in Example (5.2) that the function (5.34) has the power series

$$\frac{A}{(1-rx)^m} = A \sum_{0 \le n} \binom{n+m-1}{m-1} r^n x^n , \tag{5.35}$$

and we remarked that the binomial coefficient here is a polynomial in n of degree $m - 1$. We see in (5.10) that we must add terms (5.34) for all $m = 1, \ldots, m_j$ and for

$r = r_1, \ldots, r_\ell$ where m_j is the multiplicity of r_j. Let's focus on one root, say r_1. It comes into m_1 terms like (5.35). The coefficients of x^n in them are

(5.36)
$$
\begin{aligned}
A_1\binom{n}{0} r_1^n &= A_1 r_1^n \\
A_2\binom{n+1}{1} r_1^n &= A_2(1+n) r_1^n \\
A_3\binom{n+2}{2} r_1^n &= \frac{A_3}{2}(2+3n+n^2) r_1^n \\
&\vdots \\
A_{m_1}\binom{n+m_1-1}{m_1-1} r_1^n &= \frac{A_{m_1}}{(m_1-1)!}\left((m_1-1)! + \cdots + n^{m_1-1}\right) r_1^n .
\end{aligned}
$$

We don't have to work out exactly these polynomials in n. We just need to know their degrees.

What do we get if we add all the terms in (5.36)? We get some polynomial in n (coefficients determined by the values of the constants $A_1, \ldots, A_{m_1}$) multiplied by r_1^n. This polynomial has degree less than m_1, since its maximum degree, reached when $A_{m_1} \neq 0$, is $m_1 - 1$. In other words, the contribution of r_1 to the solution is

(5.37)
$$\left(c_{1,1} + c_{1,2}\, n + \cdots + c_{1,m_1}\, n^{m_1-1}\right) r_1^n$$

where the constants $c_{1,1}, \ldots, c_{1,m_1}$ are to be determined.

The same holds for all the roots, and we add the expressions like (5.37) to get, finally,

(5.38)
$$a_n = \sum_{1 \le j \le \ell} \left(c_{j,1} + c_{j,2}\, n + \cdots + c_{j,m_j}\, n^{m_j-1}\right) r_j^n .$$

We determine the k constants as follows:

(5.39) *Option 1*. Since we are given k initial values, $a_0, a_1, \ldots, a_{k-1}$, we may set up k equations (for $n = 0, 1, \ldots, k-1$) from (5.38) and solve them for the k unknowns $c_{j,—}$.

Practice (5.40P) Why is k the total number of constants?

(5.41) *Option 2*. If we wish we may calculate the numerator polynomial $N(x)$, which incorporates the initial values, and actually work out the partial-fraction expansion of $N(x)/D(x)$. Then we'd use the binomial theorem, or Example (5.2), on each term. That option amounts to turning the proof from an existence proof into a constructive proof.

Example (5.42) *Example (5.17), concluded*. We use the second option (5.41) since we were given $N(x)$ to start with. From (5.31) and Example (5.2) we get

$$\frac{8x^3}{(1-x)^3(1+x)} = 7\sum x^n - 10\sum (n+1)x^n + 4\sum \binom{n+2}{2} x^n - \sum (-1)^n x^n,$$

all sums being over $0 \le n$. It follows that

$$\begin{aligned} a_n &= 7 - 10 - 10n + 2\,(n+2)\,(n+1) - (-1)^n \\ &= 1 - 4n + 2n^2 \; - \; (-1)^n. \end{aligned}$$

With this example we never saw a recursion. Just for the fun of it, can we find one that $\langle a_n \rangle$ satisfies? We know the initial values from the formula for a_n:

$$a_0 = 0,\; a_1 = 0,\; a_2 = 0,\; a_3 = 8.$$

But of course we know the recursion. It has the same coefficients as $D(x)$, but reversed. We saw that before, in (4.16). Therefore, since

$$\begin{aligned} D(x) &= (1-x)^3(1+x) = (1-x^2)\,(1-x)^2 \\ &= 1 - 2x + x^2 - x^2 + 2x^3 - x^4 \\ &= 1 - 2x + 2x^3 - x^4, \end{aligned}$$

the recursion satisfied by $\langle a_n \rangle$ is

$$a_{n+4} - 2a_{n+3} + 2a_{n+1} - a_n = 0.$$

(5.43) *The gap of (5.27).* Here we fix the "little gap" mentioned in (5.27).

(*i*) The **total degree** of a monomial in several variables is defined as the sum of the degrees of the individual variables. Thus

$$x_1\, x_2^3\, x_3^2 \quad \text{has total degree} \quad 6 = 1 + 3 + 2.$$

A polynomial in several variables may have more than one term of the same total degree, for example,

$$f(x,y) := x^2 + xy + y^2.$$

(*ii*) A polynomial in several variables in which every term has the same total degree s is called **homogeneous of degree** s. *Example:* $f(x,y)$ above is homogeneous of degree 2.

If $g(x_1,\ldots,x_m)$ is homogeneous of degree s, then

$$g(tx_1,\ldots,tx_n) = t^s g(x_1,\ldots,x_n). \tag{5.44}$$

where t is another variable.

Practice (5.45P) Prove (5.44).

Notice that $D_1(x) = (z_1 - r_1x)\cdots(z_\ell - r_\ell x)$ is homogeneous of degree ℓ in the variables $x, z_1, \ldots, z_\ell$ (the r_j's are constants).

(*iii*) We had the polynomial in one variable $N(x)$ of degree less than k and were considering the partial quotient $q(x, z_1, \ldots, z_\ell)$ in

$$N(x) = D_1(x,z)\,q(x,z) + r(x,z), \tag{5.46}$$

where we write z for $z_1, \ldots, z_\ell$. We treated the z_j's temporarily as constants in regarding $D_1(x,z)$ as a polynomial in x of degree ℓ, and we concluded from (5.15), the division algorithm for polynomials, that the degree in x of $r(x,z)$ is less than ℓ.

Practice (5.47P) Let $f(x,z)$ and $g(x,z) \neq 0$ be polynomials in x and $z = z_1, \ldots, z_\ell$ with real coefficients. Suppose also that when $g(x,z)$ is regarded as a polynomial in x, its leading coefficient is a (nonzero) real number. Consider the partial quotient $q(x,z)$ and the remainder $r(x,z)$ in

$$f(x,z) = g(x,z)\, q(x,z) + r(x,z),$$

where (degree in x of $r(x,z)$) $<$ (degree in x of $g(x,z)$).

Prove that $q(x,z)$ and $r(x,z)$ are polynomials in $z_1, \ldots, z_\ell$, as well as in x. (That is, no denominators that are polynomials in z arise.)

Our target is to show that no term of the form (5.28) appears in the $q(x,z)$ of (5.46). It will be easier to prove this for the special case $N(x) = x^n$ with $n < k$, because the general case is a linear combination of these cases.

Practice (5.48P) Prove the assertion just made.

Thus consider

$$x^n = D_1(x,z)\, q(x,z) + r(x,z). \tag{5.49}$$

Since $D_1(x,z)$ has degree ℓ in x, $q(x,z)$ and $r(x,z)$ are polynomials in x and z, according to (5.47). If $n < \ell$, then $q(x,z) = 0$ (and $r(x,z) = x^n$), so there's nothing to prove, since no term (5.28) can be in $q(x,z)$ in this case.

Therefore assume $\ell \leq n < k$. We use the homogeneity of x^n and $D_1(x,z)$. In fact we'll prove that $q(x,z)$ and $r(x,z)$ are homogeneous of degrees $n - \ell$ and n, respectively. Introducing the new variable t as above we find

$$t^n x^n = D_1(tx,tz)\, q(tx,tz) + r(tx,tz).$$

Since D_1 is homogeneous of degree ℓ, this equation becomes

$$t^n x^n = t^\ell D_1(x,z)\, q(tx,tz) + r(tx,tz).$$

We divide by t^n:

$$x^n = D_1(x,z)\left[t^{\ell-n} q(tx,tz)\right] + t^{-n} r(tx,tz).$$

Subtract this equation from (5.49):

$$0 = D_1(x,z)\left[q(x,z) - t^{\ell-n} q(tx,tz)\right] + r(x,z) - t^{-n} r(tx,tz). \tag{5.50}$$

If the quantity in brackets is not identically 0, then the product of it and $D_1(x,z)$ is a polynomial in x of degree at least ℓ. (The coefficients are polynomials in z and t possibly divided by powers of t.) But the rest of (5.50) is a polynomial in x of degree *less* than ℓ, and it is the negative of the first part. That's impossible unless both are 0, so we've proved the homogeneity (with degrees as advertised) of $q(x,z)$ and $r(x,z)$.

So, in particular, every term in $q(x,z)$ has total degree $n - \ell$. But (5.28) is

$$z_1^{m_1 - 1} z_2^{m_2 - 1} \cdots z_\ell^{m_\ell - 1}.$$

and it has total degree

$$-\ell + \sum_{1 \le j \le \ell} m_j = k - \ell > n - \ell.$$

So $q(x,z)$ is totally wiped out by our differentiations. QED

6. The Nonhomogeneous Case

Suppose that the right-hand side of the recursion is not 0. Thus the problem is to find a sequence $\langle a_n \rangle$ satisfying, for all $n \ge 0$,

(6.1) $$d_0\, a_{n+k} + \cdots + d_k\, a_n = V(n)$$

where $d_0 d_k \ne 0$, initial values $a_0, \ldots, a_{k-1}$ are given, and $V(n)$ is a given nonzero function of n.

(6.2) The strategy for solving such a problem is this:

Step 1. Find the general solution to the homogeneous problem. That is, you set $V(n) = 0$ and solve the problem

$$\sum_{0 \le j \le k} d_j b_{n+k-j} = 0.$$

Use the methods of Sections (4) or (5). Say the answer is $\langle b_n \rangle$. Don't solve yet for the coefficients $c_1, \ldots, c_k$.
Step 2. Find any solution whatever to (6.1); call it $\langle p_n \rangle$. Then

(6.3) $$a_n = b_n + p_n.$$

$\langle p_n \rangle$ is sometimes called a **particular solution** to (6.1).
Step 3. Set $n = 0, 1, \ldots, k-1$ in (6.3) and solve for $c_1, \ldots, c_k$.

Practice (6.4P) Verify that the result of Step 3 is a solution to the problem.

We won't go much into the various methods for finding the particular solutions. There is quite a bagful of tricks developed to do it. They are analogous to those used to find particular solutions of differential equations. A good source for them is (9.1).

We give just one. Suppose $V(n) := r^n v(n)$, where r is a fixed real number, and $v(n)$ is a polynomial in n of degree e with real coefficients. The result we'll prove is

Theorem (6.5) Let r be a real number and $v(x)$ a polynomial of degree e. The solution of the nonhomogeneous recursion

(6.6) $$d_0\, a_{n+k} + d_1\, a_{n+k-1} + \cdots + d_k\, a_n = r^n v(n),$$

with given initial values $a_0, \ldots, a_{k-1}$, is the solution of the homogeneous recursion with characteristic polynomial

(6.7) $$p(x)\,(x - r)^{e+1},$$

where $p(x)$ is the characteristic polynomial of the original recursion:

$$p(x) := d_0\, x^k \;+\; d_1\, x^{k-1} \;+\cdots+\; d_k.$$

The extra $e + 1$ initial values are the values of $a_k, a_{k+1}, \ldots, a_{k+e}$, which are determined from (6.6) and the given initial values $a_0, \ldots, a_{k-1}$. ■

To prove Theorem (6.5) we'll need a simple item from Chapter 8, namely, problem 2.9:

If $a, b \in \mathbf{Z}$ and $0 \le a < b$, then

$$\sum_j (-1)^j\, j^a \binom{b}{j} = 0\,. \tag{6.8}$$

From (6.8) it follows that for any polynomial $f(x)$ of degree less than b,

$$\sum_j (-1)^j f(j) \binom{b}{j} = 0\,, \tag{6.9}$$

since $f(j)$ is a linear combination of powers j^a for various $a < b$.

The next corollary of (6.9) is that for any integer n (not varying with j)

$$\sum_j (-1)^j\, (j+n)^a \binom{b}{j} = 0\,,$$

still with $0 \le a < b$, of course. That's because $(j + n)^a$ is a polynomial in j of degree less than b. And, finally, for $f(x)$ as above

$$\sum_j (-1)^j f(j+n) \binom{b}{j} = 0\,, \tag{6.10}$$

for the same reason we used to show (6.9).

Equation (6.10) will serve as a lemma in the proof of Theorem (6.5).

Proof of Theorem (6.5) (*i*) First we write (6.6) for $e + 2$ successive values of n, namely, for $n, n+1, \ldots, n+e+1$. I'll represent those equations in a kind of Venn diagram with a shaded rectangle standing for the left-hand side of (6.6):

(6.11F)

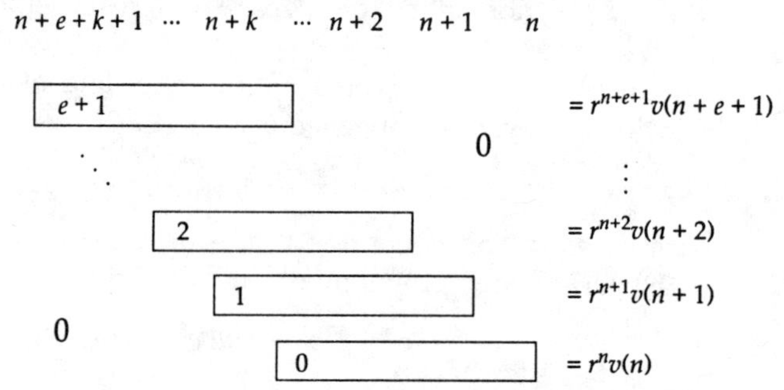

$n+e+k+1 \;\cdots\; n+k \;\cdots\; n+2 \quad n+1 \quad n$

[$e+1$] $= r^{n+e+1}v(n+e+1)$

0

$\ddots$ $\quad\vdots$

[2] $= r^{n+2}v(n+2)$

[1] $= r^{n+1}v(n+1)$

0

[0] $= r^{n}v(n)$

The rectangle with j in it, for $j = 0, 1, \ldots, e+1$, stands for

$$d_0\, a_{n+k+j} + d_1\, a_{n+k-1+j} + \cdots + d_k\, a_{n+j}.$$

(*ii*) Second, for each j we multiply the jth equation of (6.11) by r^{e+1-j}, producing the same power of r on the right in each equation.

(*iii*) Third, for each j we multiply the jth equation by $(-1)^j\binom{e+1}{j}$.

We then add the equations. On the right we get

$$r^{n+e+1} \sum_{0 \le j \le e+1} (-1)^j \binom{e+1}{j} v\,(n+j)$$

which is 0 by (6.10), since the degree of the polynomial $v(x)$ is e.

On the left we get some recursion spanning terms from $a_{n+e+k+1}$ to a_n. We don't have to kill ourselves to find out what it is, however, because we can easily find its characteristic polynomial directly. We know that in any recursion the characteristic polynomial arises when we substitute (for each j) x^j for a_{n+j} in the recursion. We do that in (6.11) as modified by our steps (*i*), (*ii*), and (*iii*)—we'll now write down the results of these steps for the first time, but with $a_{n+j} := x^j$.

Under that substitution, box number 0 holds $p(x)$, box number 1 holds $xp(x)$; and in general box j holds $x^j p(x)$. Multiplication by a power of x merely shifts the polynomial (Chapter 8, (1.26)). Thus the characteristic polynomial of the new, "big" recursion is

$$\sum_{0 \le j \le e+1} (-1)^j \binom{e+1}{j} r^{e+1-j}\, x^j\, p(x)\,. \tag{6.12}$$

But this is the product of $p(x)$ by

$$\sum_j (-1)^j \binom{e+1}{j} r^{e+1-j}\, x^j\,,$$

which we can simplify drastically. It is

$$r^{e+1} \sum_j \binom{e+1}{j} \left(-\frac{x}{r}\right)^j = r^{e+1} \left(1 - \frac{x}{r}\right)^{e+1},$$

by Chapter 8, Definition (2.1). We multiply through by the r^{e+1} to get

$$(r - x)^{e+1}.$$

Thus the characteristic polynomial of the new homogeneous recursion that we found a_n to satisfy is

$$(x - r)^{e+1} p(x), \tag{6.13}$$

where we may have multiplied by -1. (If so, it doesn't change the roots of the polynomial.)

Notice that this holds irrespective of whether r is a root of $p(x)$; if it is, then it merely has multiplicity in (6.13) greater than $e+1$. And the particular solution for the original recursion is now seen to be

$$r^n \times (\text{polynomial in } n \text{ of degree } e) \times r^m,$$

where m is the multiplicity of r as a root of $p(x)$ ($m = 0$ if $p(r) \neq 0$). QED

Practice (6.14P) Since (6.13) takes no account of the original polynomial $v(x)$ on the right-hand side of (6.6), how can (6.13) be correct?

Example (6.15) Consider the recursion

(6.16) $$a_{n+2} - a_{n+1} - 2a_n = 2$$

with initial conditions

(6.17) $$a_0 = 3, \ a_1 = -2.$$

The characteristic polynomial is $p(x) = x^2 - x + 2 = (x+1)(x-2)$, and the right-hand side of (6.16) is

(6.18) $$V(n) = 1^n \cdot 2,$$

i.e., it is a polynomial in n of degree $e = 0$. By Theorem (6.5) the solution of (6.16) and (6.17) is the solution of the homogeneous recurrence with characteristic polynomial

(6.19) $$p(x)\,(x-1)^{e+1} = (x+1)\,(x-2)\,(x-1),$$

subject to the further initial condition derived from (6.16) and (6.17), namely,

(6.20) $$\begin{aligned} a_2 &= a_1 + 2a_0 + 2 \\ &= -2 + 6 + 2 = 6. \end{aligned}$$

From (6.19) and Theorem (2.7) we know that the solution to (6.16) and (6.17) has the form

(6.21) $$a_n = c_1(-1)^n + c_2 \cdot 2^n + c_3 \cdot 1^n$$

for appropriate constants c_1, c_2, c_3. We can solve for the c's using the values for a_0, a_1, and a_2 in (6.17) and (6.20). But it is possible to simplify the procedure if we recognize that $c_1(-1)^n + c_2 2^n$ produces 0 when substituted for a_n in the recurrence in (6.16). That is true because of Theorem (2.7) for the homogeneous case. Therefore, in the language of (6.3), the rest (c_3 in this case) must be the particular solution. We substitute $a_n := c_3$ in (6.16) to get

(6.22) $$\begin{aligned} c_3 - c_3 - 2c_3 &= 2 \\ c_3 &= -1. \end{aligned}$$

We now recast (6.21) using (6.22):

$$a_n = c_1(-1)^n + c_2 2^n - 1.$$

From (6.17) we get equations to determine c_1 and c_2:

$$\begin{aligned} a_0 &= c_1 + c_2 - 1 = 3 \\ a_1 &= -c_1 + 2c_2 - 1 = -2. \end{aligned}$$

We add these to find

$$\begin{aligned} 3c_2 - 2 &= 1 \\ 3c_2 &= 3 \\ c_2 &= 1. \end{aligned}$$

It follows that $c_1 = 3$. Therefore

$$a_n = 3(-1)^n + 2^n - 1 \qquad (6.23)$$

is the solution to (6.16) and (6.17). We didn't need to use (6.20) after all, because we imposed the condition in (6.22).

Another illustration of the shortcut for finding the particular solution: Suppose the right-hand side of (6.16) were $n2^n$. Then $e = 1$ and $r = 2$ in the terms of Theorem (6.5), so the new characteristic polynomial would be

$$p(x)\,(x-2)^2 = (x+1)\,(x-2)^3.$$

Now the solution would be

$$a_n = c_1(-1)^n + c_2 2^n + c_3 n \cdot 2^n + c_4 \cdot n^2 2^n, \qquad (6.24)$$

according to Theorem (3.3) for the repeated-roots case. Again the particular solution must be the last two terms of (6.24) for appropriate choice of c_3 and c_4, and we can determine those constants by substitution. We set

$$a_n := c_3 n 2^n + c_4 n^2 2^n$$

in (6.16), with our new right-hand side $n2^n$, to find

$$\begin{aligned} c_3\big[(n+2)\,2^{n+2} - (n+1)\,2^{n+1} - n2^{n+1}\big] \\ + c_4\big[(n+2)^2\,2^{n+2} - (n+1)^2\,2^{n+1} - n^2\,2^{n+1}\big] = n2^n. \end{aligned}$$

We divide by 2^n and begin to work it out:

$$c_3[4n + 8 - 2n - 2 - 2n] + c_4\,[4n^2 + 16n + 16 - 2n^2 - 4n - 2 - 2n^2] = n,$$

which is the same as

$$(6c_3 + 14c_4) + 12c_4 n = n.$$

Thus we'd have $c_4 = 1/12$ and $6c_3 + 14/12 = 0$, or $c_3 = -7/36$. We'd put these values in (6.24) and then use the initial conditions for a_0 and a_1 to solve for c_1 and c_2.

(6.25) Sometimes a simple solution to (6.1) presents itself. Suppose, for example, we had asked in Example (1.6) to find the number t_n of strings of length n that do contain 101. Then $t_n + s_n = 2^n$, and we'd have reached the recursion

$$t_{n+3} - 2t_{n+2} + t_{n+1} - t_n = 2^n$$

with initial conditions $t_0 = t_1 = t_2 = 0$. This nonhomogeneous recurrence, of course, becomes homogeneous for s_n if we set $s_n := 2^n - t_n$; it becomes (1.16).

7. Summary, and a Look Outward

Summary: You might have found the algebra in this chapter forbidding. But the chapter isn't heavily algebraic, it's just relentlessly algebraic.

The structure of the arguments is pretty simple, however, and the techniques hardly go beyond multiplication of polynomials.

We set up the generating function $g(x) := \sum a_n x^n$ for the unknown sequence $\langle a_n \rangle$.

We then used the given recurrence relation (4.1) to express $g(x)$ as a quotient of two polynomials

$$g(x) = \frac{N(x)}{D(x)}, \tag{7.1}$$

in which the numerator depends on the initial values $a_0, \ldots, a_{k-1}$ and the denominator is the reverse of the characteristic polynomial of the recurrence. It has degree k.

The whole question was then to express $N(x)/D(x)$ in some more useful form. Partial fractions are that form. The rest of the proof in each case is to show that we can express $N(x)/D(x)$ in partial fractions.

We saw an immediate solution in the simplest case, degree 1, namely the geometric series (4.21):

$$g(x) = \frac{a_0}{1 - r_1 x} = a_0 \sum_{0 \le n} r_1^n x^n. \tag{7.2}$$

(7.3) If the general case (7.1) is a sum of terms like those in (7.2), then we've solved it.

Partial fractions allow us to conclude just that in the nonrepeated-roots case. That proof, starting at (4.22), rested on the simple observation that (7.3) holds for degree 2: There exist constants A and B such that

$$\frac{n_0 + n_1 x}{(1 - r_1 x)(1 - r_2 x)} = \frac{A}{1 - r_1 x} + \frac{B}{1 - r_2 x}. \tag{7.4}$$

You can't say that proving (7.4) was any kind of algebraic big deal, but it led promptly to a proof of the general inductive step (in a typical reduction of an inductive proof to the case for the variable equal to 2).

Now if we had one repeated root we could go, by differentiating with respect to x, from (7.2) to

$$\frac{a_0 r_1^{m-1}}{(1 - r_1 x)^m} = a_0 \sum_{0 \le n} r_1^n n(n-1) \cdots (n-m+1)\, x^{n-m},$$

which is a solution for that case.

If there are two or more distinct roots, some of which are repeated, then we can't differentiate, say (7.4), with respect to x because it would affect all terms. So we follow Spiegel's idea and introduce a new variable z_j into the term that contains r_j. If we then differentiate with respect to z_j we don't alter the denominators of the other terms as we increase the exponent on $(z_j - r_j x)^e$ to the value m_j, which is our target. The formulas get messy, but the idea behind the procedure is simple.

Finally, we reduced one nonhomogeneous case to a homogeneous problem of higher degree. All we used was polynomial multiplication, including the definition of binomial coefficients.

A look outward. It would be nice to have ideas of linear algebra available to discuss some of this chapter. The set of all solutions to a homogeneous linear recurrence is a vector space of dimension equal to the degree, say k, of the characteristic polynomial. For the nonhomogeneous case, the solution set is a coset of that space; the "particular solution" is any element of that coset. In the homogeneous case, the map f from the solution space to the space of the k initial values,

$$f(\langle\langle a_n \rangle\rangle) := (a_0, \ldots, a_{k-1}),$$

is a nonsingular linear transformation onto $\mathbf{R}^k$. The abstract can illuminate the concrete.

8. Chebyshev and Legendre Polynomials

There is an important use of recurrences that we will go into only briefly. That is to define sequences of functions (instead of the sequences of numbers we've focused on).

The sequence of **Chebyshev polynomials** $T_n(x)$ may be defined by the recurrence

$$T_{n+2}(x) = 2x\, T_{n+1}(x) - T_n(x) \tag{8.1}$$

for $n \geq 0$, with the initial values

$$T_0(x) = 1, \quad T_1(x) = x. \tag{8.2}$$

The Chebyshev polynomials are used in numerical analysis to approximate other polynomials over the range $-1 \leq x \leq 1$.

The **Legendre polynomials** $L_n(x)$ are defined by the recurrence

$$(n+2)L_{n+2}(x) - (2n+3)xL_{m+1}(x) + (n+1)L_n(x) = 0 \tag{8.3}$$

and the initial conditions

$$L_0(x) = 1, \quad L_1(x) = x. \tag{8.4}$$

We define generating functions for these sequences of polynomials just as we did for sequences of numbers, except that we need a new variable, say z. For the Chebyshev polynomials we have

$$g_T(x,z) := \sum_{0 \leq n} T_n(x) z^n. \tag{8.5}$$

As before, we multiply (8.5) by the coefficient of the recursion for $T_n(x)$ and by a suitable power of z to get

$$\begin{aligned} z^2 g_T(x,z) &= T_0(x)z^2 + T_1(x)z^3 + \cdots + T_n(x)z^{n+2} + \cdots \\ -2xz g_T(x,z) &= -2xT_0(x)z - 2xT_1(x)z^2 - \cdots - 2xT_n(x)z^{n+1} - \cdots \\ g_T(x,z) &= T_0(x) + T_1(x)z + T_2(x)z^2 + \cdots + T_n(x)z^n + \cdots \end{aligned} \tag{8.6}$$

In these three equations we see that the coefficients of z^{n+2} are, respectively,

$$T_n(x),\ -2xT_{n+1}(x),\ T_{n+2}(x). \tag{8.7}$$

The sum of these functions is 0, according to (8.1). Therefore the sum of the three equations in (8.6) is

$$\begin{aligned}(z^2 - 2xz + 1)g_T(x,z) &= -2xz + 1 + xz \\ &= 1 - xz,\end{aligned}$$

where we used (8.2) for $T_0 = 1$, $T_1 = x$. Therefore the generating function of the Chebyshev polynomials $T_n(x)$ is

$$g_T(x,z) = \frac{1 - xz}{1 - 2xz + z^2}. \tag{8.8}$$

The analogue for the Legendre polynomials is

$$g_L(x,z) = \frac{1}{(1 - 2xz + z^2)^{1/2}}, \tag{8.9}$$

the tricky proof of which is left to you in the problems.

9. Further Reading

(9.1) Charles (Karoly) Jordan, *Calculus of Finite Differences*, Chelsea, New York, 1965. (See Chapter XI.)

(9.2) G. S. Lueker, "Some techniques for solving recurrences," *Computing Surveys*, Vol. 12, no. 4 (1980), pages 419–436.

(9.3) Ronald E. Mickens, *Difference Equations*, Van Nostrand Reinhold, New York, 1987.

10. Problems for Chapter 11

In this set of problems F_n always denotes the nth Fibonacci number. See (1.2) and (1.3) for the definition.

Problems for Section 2

2.1 Suppose a sequence $\langle a_n \rangle$ of integers satisfies the recurrence

$$a_{n+2} = a_{n+1} + 2a_n$$

for all $n \geq 0$, and $a_0 = 2$ and $a_1 = 1$. Find a closed form for a_n.

2.2 Suppose the sequence $\langle a_n \rangle$ satisfies

$$a_{n+2} - a_{n+1} - 6a_n = 0$$

for all $n \geq 0$. Find a closed form for a_n if

(*i*) $a_0 = 0$ and $a_1 = 1$

(*ii*) $a_0 = 2$ and $a_1 = 1$.

2.3 Suppose a sequence $\langle a_n \rangle$ satisfies $a_{n+2} = -a_n$ for all $n \geq 0$, and $a_0 = a_1 = 2$. Find a closed form for a_n.

2.4 Find a formula for the nth term of the Fibonacci sequence $\langle F_n \rangle$ of (1.2) and (1.3). Show how you followed the procedure of (2.6). Also find a simple function to which F_n is asymptotic.

2.5 Prove that

$$\lim_{n\to\infty} \frac{F_{n+1}}{F_n} = \frac{1+\sqrt{5}}{2}.$$

This number is known as the *golden mean*.

2.6[Ans] Suppose you have the recursion of the Fibonacci sequence,

$$\forall n \geq 0 \qquad a_{n+2} = a_{n+1} + a_n,$$

but the initial conditions $a_0 = 1 = a_1$. Find a formula for a_n.

2.7 Same problem as 2.6, but now the initial conditions are $a_0 = 1, a_1 = 3$.

2.8 Consider the recurrence $a_{n+2} = 2a_{n+1} - 2a_n$ with $a_0 = a_1 = 2$.

(*i*) Find a closed-form expression for a_n.
(*ii*) Tabulate a_n for $n = 0, 1, \ldots, 10$.
(*iii*) Prove that $a_n = 0$ for all n of the form $n = 4N + 2$.

***2.9** Find a formula for $t_n :=$ the number of strings over 0,1 of length n not containing 11 as a substring. You should

(*i*) Develop a recursion for t_n.
(*ii*) Solve the recursion.

***2.10**[Ans] Prove the following properties of Fibonacci numbers:

(*i*) $F_{n+1}F_{n-1} - F_n^2 = (-1)^n$, for $n \geq 1$
(*ii*) F_{n+1} and F_n are relatively prime, for $n \geq 0$.
(*iii*) $F_n = F_m F_{n-m+1} + F_{m-1}F_{n-m}$, for $1 \leq m \leq n$
(*iv*) Prove: $\forall m, n \geq 0$ if $m|n$, then $F_m|F_n$.
(*v*) $\gcd(F_m, F_n) = F_{\gcd(m,n)}$, for $m, n \geq 0$

Problems for Section 3

3.1 Suppose a sequence $\langle a_n \rangle$ satisfies the recurrence

$$a_{n+2} = 4a_{n+1} - 4a_n$$

and the initial conditions $a_0 = 1, a_1 = 4$. Find a formula for a_n.

3.2 Suppose a sequence $\langle b_n \rangle$ satisfies the recurrence

$$b_{n+3} - 4b_{n+2} + 5b_{n+1} - 2b_n = 0$$

and the initial conditions $b_0 = 4$, $b_1 = 7$, $b_2 = 17$. Find a formula for b_n.

Problems for Section 4

4.1 Find the partial fraction expansion of

$$\frac{18}{x^3 - 6x^2 + 3x + 10}.$$

Show your steps and explain your procedure. [Hint. If you can find a root r of the polynomial $f(x)$ in the denominator, then you can divide $f(x)$ by $x - r$. That will reduce the degree to 2; then you can factor it.]

4.2 Find the partial fraction expansion of

$$\frac{x^3 + 1}{x^3 + 6x^2 + 11x + 6}.$$

Show your steps and explain your procedure. [Hint. Same as previous problem.]

4.3 (*i*) Calculate A to several decimal places of accuracy in (4.46), and then use your A in (4.46) and (4.52) to calculate r_1 and c_1.
(*ii*) Write a program to calculate s_n using your values of c_1 and r_1 in (4.49), or in (4.47) if you prefer. Tabulate your answers for $n = 20, 30, 40, 50$, and 60.
(*iii*) Write a program to calculate s_n exactly from the recursion (1.16) (repeated under (4.44)). Find a value of n where your formula under (*ii*) differs from the exact value.

***4.4**[Ans] Prove that for each $n \geq 0$ the Fibonacci number F_n satisfies

$$F_n = \sum_{0 \leq k \leq n-1} \binom{n-k-1}{k}$$

[Hint: Use the generating function but not the partial-fraction expansion. Another way: look at the Pascal triangle.]

Problems for Section 5

5.1 Find the partial fraction expansion of

$$\frac{4}{(x-1)^2(x+1)^2}.$$

Show your steps and explain your method.

5.2 Solve the recurrence $a_{n+3} - 6a_{n+2} + 12a_{n+1} - 8a_n = 0$, with initial conditions $(a_0, a_1, a_2) = (1, 0, 4)$.

5.3[Ans] Prove that if $m \geq 1$ and $A(z)$ is a rational function of the single variable z which does not depend on x, then

$$(\star) \qquad \left(\frac{d}{dz}\right)^{m-1} \frac{A(z)}{z - rx}$$

has the form (5.24) with $r_1 := r$ and $m_1 := m$.

5.4 Write out a complete proof of (5.15).

Problems for Section 6

6.1 A sequence $\langle a_n \rangle$ satisfies the recurrence (for $n \geq 0$)

$$a_{n+2} = 3a_{n+1} - 2a_n + 2 \cdot 3^n$$

and the initial conditions

$$a_0 = 1, \qquad a_1 = 2.$$

Find a formula for a_n.

6.2 A sequence $\langle b_n \rangle$ satisfies the recurrence (for $n \geq 0$)

$$b_{n+3} - 6b_{n+2} + 11b_{n+1} - 6b_n = 12n3^n$$

and has

$$b_0 = 2, \qquad b_1 = -25, \qquad b_2 = -140$$

as initial values. Find a formula for b_n.

6.3 (*i*) A sequence $\langle a_n \rangle$ satisfies the recurrence

$$a_{n+2} - 5a_{n+1} + 6a_n = 4n^2 - 2n - 3.$$

If $a_0 = 6$ and $a_1 = 14$, find a formula for a_n.

(*ii*) Identify the dominant term in the formula for a_n. (It is the term to which a_n is asymptotic—see Chapter 5, (27).)

(*iii*) If you have a computer available, find the least n for which the ratio r of a_n to this dominant term satisfies $|r - 1| < 0.01$.

Problems for Section 8

***8.1** Show that the generating function

$$g_L(x,z) := \sum_{0 \leq n} L_n(x) z^n$$

for the Legendre polynomials (8.3), (8.4) is

$$g_L(x,z) = \frac{1}{(1 - 2xz + z^2)^{1/2}}.$$

[Hint. Differentiate the generating function with respect to z. To solve this problem is a nice application of a simple differential equation in a discrete setting.]

8.2 Define $a_n := T_n\left(\frac{1}{2}\right)$ for all $n \geq 0$.

(*i*) Find a formula for a_n.

*(*ii*) What simple property holds for the sequence $\langle a_n \rangle$?

General Problems

G1.[Ans] Find a closed-form formula for $a_n := F_0 + \cdots + F_n$.

[There are at least three ways to solve this problem: find a recursion for a_n; use certain properties of F_n developed in other problems; another way.]

G2. Define $a_n := \lfloor \frac{n}{2} \rfloor$. Find a recurrence relation for $\langle a_n \rangle$, and solve it by the methods of this chapter.

G3. Define $b_n := \lfloor \frac{n}{3} \rfloor$. Find a recurrence relation for $\langle b_n \rangle$, and solve it by the methods of this chapter.

G4. Define $c_n := \lfloor \frac{n}{2} \rfloor + \lfloor \frac{n}{3} \rfloor$. Find a recurrence relation for $\langle c_n \rangle$ and solve it by the methods of this chapter.

G5.[Ans] (This problem does not involve difference equations, but it is in the algebraic spirit of this chapter.) Prove the counting theorem for binomial coefficients, Theorem (6.2) of Chapter 8, as follows. Define $B := \{0,1\}$. Let $n \in \mathbf{N}$ and consider

$$(\star) \qquad \sum_{(v_1,\ldots,v_n) \in B^n} X^{v_1 + \cdots + v_n} = \sum_{(v_1,\ldots,v_n) \in B^n} \prod_{i=1}^{n} X^{v_i}$$

[Hint: Interchange the order of sum and product. Justify.]

G6. (To solve this problem, you need to know some linear algebra.) Let $r_1, \ldots, r_k$ be $k \geq 1$ distinct real numbers. For $0 \leq i < k$ and $1 \leq j \leq k$, define $a_{ij} = (r_j)^i$.

(*i*) Prove that the matrix (a_{ij}) (a *Vandermonde* matrix) is nonsingular.

(*ii*) Hence prove (1.17) of Chapter 8: Any polynomial of degree $k - 1$ is determined by its values at any k distinct points.

11. Answers to Practice Problems

(1.7P) When $n < 3$ the forbidden substring 101 cannot exist, so we get all strings of length n in S_n. When $n = 3$ only 101 is forbidden, so we have $s_3 = 2^3 - 1 = 7$.

(1.10P) $z_{n+1} = s_n$ because there is a bijection between the two sets. That is, $Z_{n+1} :=$ the subset of all strings in S_{n+1} starting with 0 is in 1–1 correspondence with S_n. In fact, the correspondence is very simple: $Z_{n+1} = 0S_n$, the latter meaning the set of all strings in S_n with an extra 0 placed in front. More precisely,

$$0S_n := \{\, 0x;\; x \in S_n \,\}$$

where $0x$ denotes the catenation of the string "0" with the string x. (Chapter 1, (19.9)). To prove this claim, let $x \in S_n$. Then $0x \in S_{n+1}$ because 101 is not a substring of x, hence not of $0x$. Thus $0x \in Z_{n+1}$. Now let $y \in Z_{n+1}$. Then $y \in S_{n+1}$, so 101 is not a substring of y. But $y = 0x$ for some string x of length n, and therefore 101 is not a substring of x. So $x \in S_n$.

(1.13P) (*i*) is answered by our proof for (1.10). (*ii*) is best proved as a contrapositive. If $x \notin S_n$, then 101 is a substring of x, hence also of $1x$, making $1x \notin S_{n+1}$.

Note: The converse is false if $n \geq 2$, because x could begin with 01 and be in S_n, but $1x$ would then have 101 as a substring.

(2.9P) From Step 3 we see that

$$c_1 + c_2 = 1$$
$$c_1 - 3c_2 = -3$$

for which the solution is $c_2 = 1$ and $c_1 = 0$. Thus $a_n = (-3)^n$.

(4.6P) $(1+x)^{-1} = 1 - x + x^2 - x^3 + - \cdots$ because it is a geometric series $(1-(-x))^{-1}$:

$$\frac{1}{1-z} = 1 + z + z^2 + z^3 + \cdots,$$

and the series converges for all z with $|z| < 1$. This can be verified by methods of calculus, or simply by using (16.2) of Chapter 3:

$$\text{If } T := 1 + z + \cdots + z^n$$
$$\text{then } T = \frac{1 - z^{n+1}}{1-z}.$$

Let $n \longrightarrow \infty$. Since $|z| < 1$, $T \longrightarrow \frac{1}{1-z}$, and T is the partial sum of the infinite series in question. If we differentiate the equation at the beginning, we get

$$\frac{-1}{(1+x)^2} = -1 + 2x - 3x^2 + 4x^3 - + \cdots,$$

from which (4.5) follows.

(4.42P) (*i*) When we applied the inductive assumption, obviously, to know that $r_1, \ldots, r_k$ were all distinct. And crucially when we used the $k = 2$ case; we needed to know in particular that $r_j \neq r_{k+1}$ for all $j = 1, \ldots, k$.

(*ii*) If $d_0 d_k = 0$, then one of d_0 and d_k is 0. In that case we could express the same recursion (4.1) with a smaller value of k. Suppose, for example, that d_0 and d_1 were 0, and $d_2 \neq 0$, and $d_k = 0$ but $d_{k-1} \neq 0$. Then we'd have the same recursion with $k-3$ in place of k. (4.1) would be

$$\forall n \geq 0,\ d_2 a_{n+k-2} + \cdots + d_{k-1} a_{n+1} = 0$$

which is the same as

$$\forall n \geq 1, e_0 a_{n+k-3} + e_i a_{n+k-4} + \cdots + e_{k-3} a_n = 0,$$

where $e_j := d_{j+2}$ and $e_0 e_{k-3} \neq 0$. The only difference is that we've "lost" a_0; we can't include it in the solution (because $d_k = 0$).

If we had specified k initial conditions there might be no solution, because we are free to specify no more than $k-3$ initial values $a_1, \ldots, a_{k-3}$. Another way to put it is to say that the degree of $D(x)$ is equal to $k - (i + t)$, where i is the number of initial 0's in $d_0, d_1, \ldots, d_k$, and t is the number of 0's at the right-hand end of that sequence. The reason is that $p(x)$ has degree $k - i$ and has 0 as a t-fold root, so

$$p(x) = \left(d_i x^{k-(i+t)} + \cdots + d_{k-t}\right) x^t$$

where $d_i d_{k-t} \neq 0$. The reverse of $p(x)$, namely,

$$D(x) := x^{k-i} p\left(\frac{1}{x}\right),$$

is

$$x^{k-i-t}\left(d_i\left(\frac{1}{x}\right)^{k-i-t} + \cdots + d_{k-t}\right) x^t \left(\frac{1}{x}\right)^t$$
$$= d_{k-t} x^{k-i-t} + \cdots + d_i.$$

The x^t factor in $p(x)$ disappears. We use the assumption $d_0 d_k \neq 0$ in saying that $D(x)$ has degree k. We need to know the degree of $D(x)$ for definiteness in the entire discussion, because it equals the number of "effective" roots of $p(x)$. (A zero root contributes nothing to the solution a_n.)

(*iii*) $1 + x^2 = (1 + ix)(1 - ix)$, where $i^2 := -1$. Thus to find A, B in

$$\frac{2}{(1+ix)(1-ix)} = \frac{A}{1+ix} + \frac{B}{i - ix}$$

we multiply by $1 + ix$ to get

$$\frac{2}{1-ix} = A + \frac{B(1+ix)}{1-ix},$$

set $1 + ix = 0$ (i.e., $x := i$), and find $1 = 2/(1-i^2) = A$. It follows that $B = 1$ (see (4.26)).

$$\frac{2+2x}{(1+ix)(1-ix)} = \frac{A'}{1+ix} + \frac{B'}{1-ix}.$$

By the same procedure we find

$$A' = \frac{2+2i}{1-i^2} = 1 + i.$$

We get B' by interchanging roots in the formula for A'. The roots are i and $-i$, so $B' = 1 - i$.

(5.7P) By (7.2) of Chapter 8, which holds for any real upper index,

$$\binom{-m}{n} := \frac{(-m)(-m-1)\cdots(-m-n+1)}{n}$$
$$=(-1)^n\frac{(m)(m+1)\cdots(m+n-1)}{n!}$$
$$=(-1)^n\binom{m+n-1}{n}.$$

Symmetry (Chapter 8, (5)) holds whenever the upper index is a nonnegative integer. So if $n \geq 0, m+n-1 \geq 0$. Thus

$$\binom{-m}{n} = (-1)^n\binom{m+n-1}{m-1}.$$

(5.40P) For each j with $1 \leq j \leq \ell$ the root r_j has multiplicity m_j. There are m_j constants for r_j, as (5.38) shows. The total number of constants is therefore

$$\sum_{1\leq j\leq \ell} = \text{degree of } D(x) = k.$$

(5.45P) $g(x_1,\ldots,x_n)$ is a sum of terms of the form (constant) $x_1^{e_1}x_2^{e_2}\cdots x_n^{e_n}$ where $e_1+e_2+\cdots+e_n = s$. It's the same s for every term. Thus $g(tx_1,\ldots,tx_n)$ is a sum of terms of the form

$$(\text{constant})\,(tx_1)^{e_1}(tx_2)^{e_2}\cdots(tx_n)^{e_n}$$
$$=(\text{constant})\,t^{e_1+\cdots+e_n}x_1^{e_1}\cdots x_n^{e_n}$$
$$=t^s\,(\text{constant})\,x_1^{e_1}\cdots x_n^{e_n}.$$

So t^s factors out of every term, leaving it as it was originally.

(5.47P) We're given $f(x,z) = a(z)x^m + \sum_i a_i(z)x^{m-i}$ for some nonzero polynomial $a(z)$ in $z_1,\ldots,z_l$. We also have $g(x,z) = bx^d + \sum_i b_i(z)x^{d-i}$ for a nonzero real number b. The degrees of $f := f(x,z)$ and $g := g(x,z)$ are m and d, respectively. We prove the desired result with the following procedure.

Step 0. Set $q(x,z) := 0$.
Step 1. If $m < d$, then $r(x,z) := f$. Stop.
Step 2. $M := x^{m-d}a(z)/b; f := f - Mg$.
Step 3. $m :=$ degree of f as polynomial in x;
$a(z) :=$ leading coefficient of f as polynomial in x ($:= 0$ if $f = 0$);
$q(x,z) := M + q(x,z)$.
Step 4. Go to Step 1.

This procedure is merely the familiar division of f by g. The only actual division occurs at Step 2, where we divide by the real number b. Thus M is always a polynomial in x and z, and so is $q(x,z)$. The subtraction in Step 2 wipes out the leading term, so the degree drops by 1 or more, ensuring that the procedure ends.

(5.48P) Prove: The general case is a linear combination of cases of the form $N(x) = x^n$. The general case is:

Nothing of the form (5.28) appears in the $q(x,z)$ of (5.46). "The form (5.28)" is any term

$$z_1^{m_1-1}z_2^{m_2-1}\ldots z_\ell^{m_\ell-1}$$

multiplied by any polynomial in x,z. If we can't get a term of such degree in each of the z_j's from $x^n/D(x)$ for any n, then we can't get it from a linear combination of such things. But $N(x)/D(x)$ is such a linear combination.

(6.4P) We find $b_n = \sum_i c_iu_i(n)$, where we know $u_i(n)$ from Theorem (2.7) or (3.3). We've also found somehow a specific sequence $\langle p_n\rangle$ that satisfies (6.1). It's already obvious for any choice of constants c_i that $a_n := b_n + p_n$ satisfies the recurrence (6.1) because the left-hand side (lhs) of (6.1) is linear in the a's. That is, substituting $a_n := b_n + p_n$ in (6.1) gives lhs(b_n), which is 0, plus lhs(p_n), which is $V(n)$. So the only question is whether we can find constants $c_1,\ldots,c_k$ that make $\langle a_n\rangle$ satisfy the initial conditions. We set up the equations and see. They are

$$u_1(0)c_1 + \cdots + u_k(0)c_k = a_0 - p_0$$
$$u_1(1)c_1 + \cdots + u_k(1)c_k = a_1 - p_1$$
$$\cdots$$
$$u_1(k-1)c_1 + \cdots + u_k(k-1)c_k = a_{k-1} - p_{k-1}.$$

Why can we solve these simultaneous equations? Because they are the same equations we'd have to solve for a homogeneous problem with the initial conditions $a_i - p_i$ for $0 \leq i < k$. And we already know that a homogeneous problem has a solution for any initial conditions.

(6.14P) The answer is suggested in the last sentence of Theorem (6.5). The new characteristic polynomial (6.13) has degree $k+e+1$, where k is the degree of the original one, $p(x)$. To solve the new problem we need the first $k+e+1$ values of the sequence as initial values. We're given $a_0,\ldots,a_{k-1}$, and we get $a_k,\ldots,a_{k+e}$ by solving (6.6) successively with $n = 0,1,\ldots,e$. Doing so calls in the values $r^0v(0)$, $r\ v(1),\ldots,r^ev(e)$ on the right-hand side. As big experts on Vandermonde determinants know, a polynomial of degree e is determined by its values at any $e+1$ points. See problem G6.

CHAPTER

12

Matrices and Order Relations

"After you, Alphonse."
"After you, Gaston."

1. Introduction

Relations come up frequently in computer science and mathematics. You have probably heard of "relational data bases," for example. They are collections of data organized as relations. Thus for each student at a school there might be recorded in a computer a vector $(x_1, x_2 \ldots, x_n)$, where x_1 is the student's name, x_2 his or her identifying number, x_3 his or her home address, and so on. If there are n coordinates in each vector, the set of all such vectors, one per student, is called an **n-ary relational data base.**

We have already studied equivalence relations (Chapter 4), and we have used without much comment such relations as divisibility, being less than, and (set) inclusion. The latter are instances of order relations, which will be the focus of this chapter. We begin, however, with general relations. Although this chapter repeats some of the definitions first made in Chapter 4, you will need to know that chapter fairly well in order to profit from this one.

DEFINITION (1.1) A **binary relation** R from the set A to the set B is a subset of $A \times B$. Thus

$$R \subseteq A \times B.$$

If $A = B$, we say that R is a binary relation *on* A.

We will drop the term *binary* from now on because we won't be considering n-ary relations for $n > 2$.

We often use an infix notation, as introduced in Chapter 4, (4). Thus we say "$a\ R\ b$" for "$(a, b) \in R$."

Examples (1.2) A is the set of students at a school, B is the set of car manufacturers, R_1 is the relation $R_1 = \{(x,c);\ x \in A,\ c \in B,\ x$ owns a car made by $c\}$. Another: same $A, B := \{0,1\}$, $R_2 = \{(x,b);\ b = 0$ if x lives on campus, $b = 1$ if x lives off campus$\}$. A third: same A, now $B = A$, $R_3 = \{(a,b);\ a,b \in A,\ a$ and b have had a class together$\}$.

Since most students do not own cars, most students are not the first coordinate of any point in R_1.

DEFINITION (1.3) For any relation $R \subseteq A \times B$, the subset of A

$$\{a;\ a \in A,\ \exists b \in B,\ (a,b) \in R\}$$

is called the **domain** of R. The subset of B

$$\{b;\ b \in B,\ \exists a \in A,\ (a,b) \in R\}$$

is called the **range** of R.

Thus the domain is the set of first coordinates of all the ordered pairs of the relation; the range is the set of all second coordinates.

(1.4) Consider the examples in (1.2). In most schools, the domain of R_1 is a small proportion of the student body. The domain of R_2 is the whole student body, and usually the range of R_2 is B. (It would be an unusual school with range $R_2 = \{0\}$ or $= \{1\}$.) The domain and range of R_3 are both the set of all students A, (unless we are dealing with freshmen who have never enrolled in any class). The relation R_3 is not, however, $A \times A$. It might even be as small as the *diagonal* of $A \times A$, namely $\{(a,a);\ a \in A\}$; this would hold for a school in which every class is a private lesson. Such a situation still makes range $R_3 =$ domain $R_3 = A$, and, in the general case, if every student has enrolled in a class, then R_3 must contain the diagonal. Therefore $A =$ range $=$ domain.

2. Comparison with Functions

(2.1) Every function (defined as a set of ordered pairs) is a relation, but some relations are not functions. You recall from Chapter 5, (5), that a function is a set f of ordered pairs subject to a certain restriction, namely, that

$$\forall (a,b),(x,y) \in f, \qquad a = x \quad \text{implies} \quad b = y.$$

(2.2) R_2 above is a function (if every student has just one address), and so is R_1 if no student owns two cars. But R_3 is not a function unless R_3 is merely the diagonal, in which case it is the identity function on A.

If the function f is defined instead as a rule mapping each point x of the domain X to some point $f(x)$ of the codomain Y, then the relation naturally associated with f is

$$R_f = \{(x, f(x));\ x \in X\}.$$

R_f is a relation from X to Y. Under either definition of function, the domain and range of the function as defined in Chapter 5 coincide with the present definition of domain and range of the relation R_f.

The general relation from A to B is a "multiple valued" partial function from A to B.

3. The Converse of a Relation

DEFINITION (3.1) If R is a relation from A to B, the **converse** of R is the relation R^c from B to A defined as

$$R^c := \{ (b,a) ; \ (a,b) \in R \}.$$

Thus the domain of R^c is the range of R, and the range of R^c is the domain of R.

(3.2) If the relation R is a function f, then the converse of R is closely related to f^{-1}, the function we defined (as $\widehat{f}^{-1}$) from $\mathcal{P}(Y)$ to $\mathcal{P}(X)$, the power sets of the codomain and domain of f, respectively. (See Chapter 5, (9).) Since we are starting with the ordered-pair definition of f, let us take Y to be the range of f. Then from our definition of f^{-1} we have, for all $b \in Y$,

$$f^{-1}(b) = \{ a; \ a \in X, \ (a,b) \in f \}.$$

In terms of the converse relation we see that

$$f^{-1}(b) = \{ a; \ a \in X, \ (b,a) \in f^c \}. \tag{3.3}$$

The set $f^{-1}(b)$ equals $f^c(b)$—if we temporarily think of f^c as a function from Y to $\mathcal{P}(X)$.

Examples (3.4) If R is the relation (on a set of people) "is a child of," then R^c is the relation "is a parent of." If R is the relation of divisibility on $\mathbf{N}$, thus $a \, R \, b$ iff $a \mid b$ iff a divides b, then R^c is the relation "is a multiple of." That is, from our definition of "divides" in Chapter 6, (1.1), a divides b iff b is a multiple of a—which expresses the relation divisibility, $a \, R \, b$, in terms of its converse $b \, R^c \, a$.

4. The Composition of Two Relations

Suppose we have sets A, B, and C and relations $R \subseteq A \times B$ and $S \subseteq B \times C$. Thus R is a relation from A to B, and S is a relation from B to C. We define

(4.1) The **composition** $R \circ S$ of R and S is the relation from A to C consisting of

$$\{ (a,c); \ a \in A, \ c \in C, \ \exists b \in B \ (a,b) \in R \text{ and } (b,c) \in S. \}$$

Example (4.2) If $A = B = C = \mathbf{N}$, suppose R, S are the relations $a \, R \, b$ iff a and b are both even, and $b \, S \, c$ iff b and c are both divisible by 3. Then *some* pairs from R are

$$(2,2), \ (2,4), \ (2,6), \ (4,6), \ (4,8), \ (10,12);$$

some from S are

$$(3,3),\ (3,6),\ (3,9),\ (6,3),\ (12,9).$$

Thus $(2,3) \in R \circ S$ because $(2,6) \in R$ and $(6,3) \in S$.

R	S	$R \circ S$
(2,6)	(6,3)	(2,3)
(4,6)	(6,3)	(4,3)
(10,12)	(12,9)	(10,9)

Practice (4.3P) Prove that $R \circ S = D_2 \times D_3$, where $D_i := \{x;\ x \in \mathbf{N},\ i \mid x\}$.

The graphs show that composition of relations generalizes composition of functions.

(4.4) Consider the example

(4.4F)

a_1 a_2 a_3 b_1 b_2 b_3 b_4 c_1 c_2

R S

A B C.

If we follow an arrow from a_1 in A to b_1 in B and thence to C (since there *is* an arrow out from b_1), we arrive at c_1 in C such that $(a_1, c_1) \in R \circ S$. Doing this for all arrows from A yields $R \circ S = \{(a_1,c_1),\ (a_3,c_1),\ (a_3,c_2)\}$.

Notice that in some cases $R \circ S \neq S \circ R$, even if both compositions make sense. Also notice that $R \circ S$ can be $\varnothing$, even if neither R nor S is $\varnothing$.

Practice (4.5P) Give an example where $R \circ S = \varnothing$ but neither R nor S is empty.

There is an important result connecting the ideas of converse and composition. It is

Theorem (4.6) Let R be a relation from A to B, and S a relation from B to C. Then

$$(R \circ S)^c = S^c \circ R^c. \tag{4.7}$$

In words: the converse of a composition is the composition of the converses in reverse order.

Proof Let's first use a Venn diagram to see if (4.7) makes sense.

(4.8F)

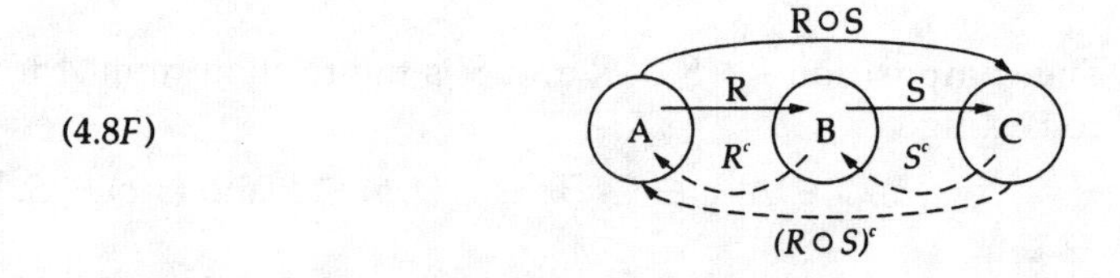

$(R \circ S)^c$ is a relation from C to A. S^c is a relation from C to B, and R^c from B to A. So $S^c \circ R^c$ is a relation from C to A. Thus (4.7) is at least possibly true.

The proof is a nice exercise in epsilon-arguments. Let's first prove that

$$(4.9) \qquad (R \circ S)^c \subseteq S^c \circ R^c.$$

Let $(x,y) \in (R \circ S)^c$. Then $(y,x) \in R \circ S$ by definition (3.1) of converse. There is an element $b \in B$ such that

$$(y,b) \in R \quad \text{and} \quad (b,x) \in S$$

by definition (4.1) of composition. Now

$$(b,y) \in R^c \quad \text{and} \quad (x,b) \in S^c,$$

so again by definition of composition, $(x,y) \in S^c \circ R^c$. Thus (4.9) is true.

It only remains to prove the converse (yes),

$$(4.10) \qquad S^c \circ R^c \subseteq (R \circ S)^c.$$

We leave that to you in the Problems. QED

This result (4.6) is surprisingly useful, as you'll see later in the chapter. We proved a special case of it long ago: See Chapter 5, (10.6), on composition of functions and their inverses. Also see the comment after (3.3) in this chapter.

Example (4.11) Consider the relations R, S of (4.4). We already calculated that

$$R \circ S = \{ (a_1,c_1), (a_3,c_1), (a_3,c_2) \}.$$

Thus we reverse the pairs to get

$$(R \circ S)^c = \{ (c_1,a_1), (c_1,a_3), (c_2,a_3) \}.$$

Now

$$S^c = \{ (c_1,b_1), (c_1,b_4), (c_2,b_4) \},$$

and

$$R^c = \{ (b_1,a_1), (b_2,a_1), (b_3,a_2), (b_3,a_3), (b_4,a_3) \}.$$

The first pair in each list gives $(c_1,a_1) \in S^c \circ R^c$. The second and fourth pairs, respectively, give $(c_1,a_3) \in S^c \circ R^c$. The last pair from each gives $(c_2,a_3) \in S^c \circ R^c$. Therefore

$$S^c \circ R^c = \{ (c_1,a_1), (c_1,a_3), (c_2,a_3) \},$$

and this is exactly $(R \circ S)^c$.

5. Graphs of Relations

As we said in Chapter 4, a graph of a relation R on A is a drawing of dots for the points of A and an arrow from a to b for each ordered pair $(a,b) \in R$. It is easy to see that the graph of the converse of a relation is the same as the graph of the relation except that all the arrows are reversed. (We usually say *the* graph of R even though the drawing is not unique.)

(5.1) Thus if $R = \{(1,2),(1,3),(1,4),(4,4)\}$, then the graph of R is

(5.1*F*)

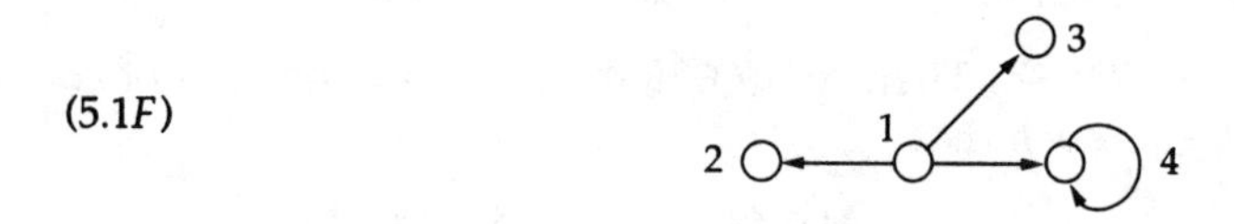

R^c is $\{(2,1),(3,1),(4,1),(4,4)\}$ and its graph is

(5.2*F*)

Notice that reversing the arrow in the loop at 4 was unnecessary (because a loop reversed is just itself: $(4,4) = (4,4)$).

(5.3) Any graph on A in which no point has more than one arrow coming out from it (an **out-arrow**) is the graph of a partial function from A to A. If every point has exactly one out-arrow then the graph is that of a function with domain A. If every point has exactly one out-arrow and exactly one **in-arrow** (defined analogously), then the graph represents a permutation of A (a bijection from A to A).

Example (5.4) Let $A = \{a,b,c\}$ and consider:

(5.4*F*)

R is a partial function, but R^c is not.

(5.5*F*)

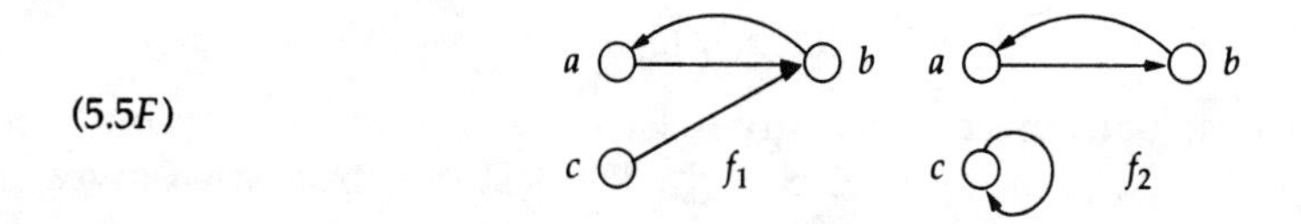

f_1 is a function on A to A; f_2 is a permutation of A.

Practice (5.6P) Characterize the properties injectivity and surjectivity in terms of in-arrows or out-arrows in the graph of the function $f : X \longrightarrow Y$ viewed as a relation from X to Y.

6. Relation-Matrices $M(R)$

(6.1) Instead of specifying a relation R from A to B as a set of ordered pairs, we sometimes use the **relation-matrix** $M = M(R)$ of R. M is a matrix of m rows and n columns, where $m = |A|$ and $n = |B|$. Each row is labeled by a different element of

A; the same is true for the columns. The entries in M are 0's and 1's; the entry in row a and column b (for any $a \in A$ and $b \in B$) is 1 if and only if the ordered pair (a, b) is in R.

Example (6.2) For $R_1 = \{(1,2), (1,3), (2,2), (2,1), (3,2)\}$ the relation-matrix M_1 is (if $A = B = \{1,2,3\}$)

$$M_1 = \begin{array}{c|ccc} & 1 & 2 & 3 \\ \hline 1 & 0 & 1 & 1 \\ 2 & 1 & 1 & 0 \\ 3 & 0 & 1 & 0. \end{array}$$

The first coordinate of the ordered pair determines the row, the second coordinate the column.

Practice (6.3P) Prove that a relation from A to B is a function from A to B if and only if the relation-matrix has exactly one 1 in each row.

Representing a relation as its relation-matrix is usually the best way to compute properties of the relation.

7. General Matrices

We will spend some time now on basic ways to think about matrices in general while we explore the matrices belonging to relations.

The entries of a general matrix might be from $\mathbf{R}$, or $\mathbf{Z}$, or $\mathbf{Z}_m$. They might be polynomials, or other matrices! The entries in a relation-matrix are from $\{0,1\}$.

(7.1) We may call a matrix of m rows and n columns an "$m \times n$" matrix (read as "m by n"). We sometimes denote an $m \times n$ matrix M as $M = (a_{i,j})$ for $1 \le i \le m$ and $1 \le j \le n$. This means that $a_{i,j}$ is the entry in row i and column j. Usually we omit the comma in the subscript.

(7.2) Thus for $m = 2, n = 3$, and $a_{ij} = i - j$, we would have the matrix

$$(a_{ij}) = (i - j) = \boxed{\begin{array}{ccc} 0 & -1 & -2 \\ 1 & 0 & -1 \end{array}}$$

DEFINITION (7.3) The **main diagonal** of the $m \times n$ matrix $M = (a_{ij})$ is the diagonal line or band through the entries

$$a_{11}, a_{22}, \ldots, a_{rr}, \quad \text{where} \quad r := \min\{m, n\}$$

Examples (7.4)

$$M_1 = \left(\binom{j}{i}\right) = \begin{pmatrix} 1 & 2 & 3 & 4 \\ 0 & 1 & 3 & 6 \\ 0 & 0 & 1 & 4 \\ 0 & 0 & 0 & 1 \end{pmatrix} \qquad M_2 = \begin{pmatrix} 1 & 2 & 3 & 4 & 5 \\ 6 & 7 & 8 & 9 & 10 \\ 11 & 12 & 13 & 14 & 15 \end{pmatrix}.$$

In M_1 the four 1's are on the main diagonal. In M_2 the main diagonal is occupied by $1, 7, 13$.

The concept of the main diagonal is geometric; it often helps us focus on problems and properties of matrices.

DEFINITION (7.5) The **transpose** ${}^tM$ of any matrix M is defined to be the reflection of M in its main diagonal.

(7.6) Thus if

$$M = \begin{pmatrix} 1 & 2 & 3 & 4 & 5 \\ 6 & 7 & 8 & 9 & 10 \end{pmatrix}$$

then

$${}^tM = \begin{pmatrix} 1 & 6 \\ 2 & 7 \\ 3 & 8 \\ 4 & 9 \\ 5 & 10 \end{pmatrix};$$

the main diagonals are shaded. The first row of M becomes the first column of ${}^tM$, the second row becomes the second column. The remaining columns become rows in like manner. The main diagonals in M and ${}^tM$ are the same.

(7.7) In terms of entries, we may express the transpose of the $m \times n$ matrix $M = (a_{ij})(1 \le i \le m;\ 1 \le j \le n)$ as the $n \times m$ matrix ${}^tM = (b_{ji})$, where

$$b_{ji} = a_{ij} \qquad 1 \le i \le m, \quad 1 \le j \le n.$$

(7.8) The 2×5 matrix M in (7.6) is $M = (a_{ij})$ for $1 \le i \le 2$ and $1 \le j \le 5$. Notice that the "1, 4" entry, namely 4, in M is the "4, 1" entry in ${}^tM$. The "2, 3" entry 8 of M is the "3, 2" entry of ${}^tM$. That is, $a_{1,4} = 4 = b_{4,1}$, and $a_{2,3} = 8 = b_{3,2}$, where the transpose of M is ${}^tM = (b_{j,i})$.

If we express the entries of M in a formula, namely,

$$a_{ij} = j + 5(i-1),$$

then we may simply say that

$$b_{ji} = j + 5(i - 1).$$

The point is that j is the *column*-index for M but is the *row*-index for ${}^tM$. (Vice-versa for i, of course.) The first index always counts the rows.

(7.9) In all cases the notation $M = (a_{ij})$ means that the entry a_{ij} is in row i and column j of the matrix M, for all i in $\{1,\ldots,m\}$ and all j in $\{1,\ldots,n\}$.

(7.10) *Rows and Columns.* Thus the transposed matrix (b_{ji}) has, by definition, the entry b_{ji} in row j and column i. Since we defined $b_{ji} = a_{ij}$† for all i,j (in their respective domains), we see that the transpose has n rows and m columns. It will soon be clear—if it isn't already so—that transposing interchanges rows and columns.

If $(a_{ij}) = M$ is any $m \times n$ matrix, then the first row of M is the vector

$$(a_{11}, a_{12}, \ldots, a_{1n});$$

and, in general, the ith row is the vector

$$(a_{i1}, a_{i2}, \ldots, a_{in}).$$

The first column of M is the vector

$$(a_{11}, a_{21}, \ldots, a_{m1});$$

the jth column of M is the vector

$$(a_{1j}, a_{2j}, \ldots, a_{mj}).$$

We *could* write these column-vectors as vertical rather than horizontal arrays, but a vector is a vector (here an element of $\{0,1\}^m$ if M is a relation-matrix), so it really doesn't matter. Sometimes it seems necessary to write vectors as columns to avoid confusion, but that is usually because the vectors are really $m \times 1$ matrices being used in matrix products (defined later).

At any rate, here is M—notice its jth column is precisely the vector just above:

$$(7.11) \qquad M = \begin{bmatrix} a_{11} & a_{12} & \cdots & a_{1j} & \cdots & a_{1n} \\ a_{21} & a_{22} & \cdots & a_{2j} & \cdots & a_{2n} \\ \vdots & & & & & \cdot \\ a_{i1} & a_{i2} & \cdots & a_{ij} & \cdots & a_{in} \\ \vdots & & & & & \\ a_{m1} & a_{m2} & \cdots & a_{mj} & \cdots & a_{mn} \end{bmatrix}$$

Rather than approach the matrix M entry by entry, as (7.11) does—and as our definition of transpose as (b_{ji}) with $b_{ji} = a_{ij}$ does—it often saves effort to define an

†Paradoxically, although $b_{ji} = a_{ij}$ for all i,j, it is not true in general that as matrices $(b_{ji}) = (a_{ij})$! As matrices the two are transposes of each other.

$m \times n$ matrix as the ordered set of its m rows (row-vectors) or of its n columns (column-vectors). Thus we could perfectly well say that the M of (7.11) is the $m \times 1$ matrix (R_i) where R_i is the vector consisting of the ith row of M (for $i = 1, \ldots, m$). We could equally well say that M is the $1 \times n$ matrix (C_j), where C_j is the vector consisting of the jth column (for $j = 1, \ldots, n$). The reason for this alternate point of view is that sometimes it is more natural or convenient to deal with the row- or column-vectors as things in their own right than it is to break them up into their individual entries. For example

DEFINITION (7.12) If M is any $m \times n$ matrix, then the transpose ${}^tM$ of M is the $n \times m$ matrix in which the rows are the columns of M (in the same order).
Equivalently, ${}^tM$ is the $n \times m$ matrix in which the columns are the rows of M in the same order.

This equivalent definition (7.12) is simpler than the earlier one, (7.7), in terms of the entries. Strictly speaking we should prove that the four definitions of *transpose* in (7.5), (7.7), and (7.12) are equivalent, but we leave that as a problem for you.

The transpose has a simple but useful role to play in the study of binary relations. We state the result as a theorem.

Theorem (7.13) Let R be a relation from A to B and let M be the relation-matrix $M(R)$ of R. Then the converse R^c of R has as relation-matrix the transpose ${}^tM$:

$$M(R^c) = {}^tM(R).$$

Proof The theorem follows immediately from the definitions (3.1) of converse and (7.7) of transpose. QED

8. Matrix Multiplication

In order to do calculations with relations we usually use the relation-matrix. Often then we need to do matrix multiplication. We define that operation now in case you haven't seen it before.

(8.1) Our matrices are assumed to have their entries from a set A which has *two* associative binary operations, one denoted $+$ and called addition, the other denoted $\cdot$ and called multiplication. (See Chapter 5, (28.2).) For example, A might be the set $\mathbf{R}$ of all real numbers and $+$ and $\cdot$ the usual addition and multiplication on $\mathbf{R}$. Or, A might be $\{0, 1\}$, but with $+$ equal to **Boolean addition**, defined by this table

(8.2)

+	0	1
0	0	1
1	1	1

$+$ is Boolean addition;

we define $\cdot$ on $\{0,1\}$ by the table

(8.3)
$$\begin{array}{c|cc} \cdot & 0 & 1 \\ \hline 0 & 0 & 0 \\ 1 & 0 & 1 \end{array} .$$

Notice that this **Boolean multiplication** $\cdot$ is the same as multiplication in $\mathbf{Z}$ restricted to $\{0,1\}$. Notice also that if we interpret 0 as "false" and 1 as "true", then Boolean addition + is "or," and $\cdot$ is "and."

We are now using + and $\cdot$ to stand for abstract, or general, binary operations. That's because we've run out of abstract symbols, *and* we want to suggest similarities with the usual addition and multiplication. We assume that both + and $\cdot$ are commutative.

(8.4) There are other properties required of these two operations + and $\cdot$ on A. They must satisfy the distributive law:

$$\forall a,b,c \in A \qquad a \cdot (b+c) = (a \cdot b) + (a \cdot c),$$

and there must be two elements, denoted 0 and 1, in A such that for all a in A,

$$0 + a = a + 0 = a;$$
$$1 \cdot a = a \cdot 1 = a.$$

The two examples above satisfy the distributive law. We know this already for the first; if $A = \{0,1\}$, and + and $\cdot$ are defined as in (8.2) and (8.3), then since $0 \cdot a = 0$ and $1 \cdot a = a$ for all a in $\{0,1\}$,

$$a \cdot (b+c) = \begin{cases} 0 & \text{if } a = 0 \\ b+c & \text{if } a = 1 \end{cases}$$

for all $b,c \in \{0,1\}$. But also

$$0 \cdot b + 0 \cdot c = 0 + 0 = 0$$
$$1 \cdot b + 1 \cdot c = b + c;$$

thus (8.4) holds for $\{0,1\}$ with Boolean addition (8.2) and multiplication (8.3) as the two operations.

Let A be a set with such operations + and $\cdot$ defined. Let $\mathbf{x}$ and $\mathbf{y}$ be any two "r-vectors over" A. This means $\mathbf{x}$ and $\mathbf{y}$ are both in $A^r = A \times \cdots \times A$, the Cartesian product of r copies of A. We define the **dot product** of the vectors $\mathbf{x}$ and $\mathbf{y}$ as

(8.5)
$$\mathbf{x} \bullet \mathbf{y} := x_1 \cdot y_1 + x_2 \cdot y_2 + \cdots + x_r \cdot y_r$$

where $\mathbf{x} = (x_1, \ldots, x_r)$ and $\mathbf{y} = (y_1, \ldots, y_r)$. Thus $\mathbf{x} \bullet \mathbf{y}$ is an element of A.

For example, with the usual addition and multiplication in $\mathbf{Z}$, the dot-product of $\mathbf{x} = (1,0,1,1)$ and $\mathbf{y} = (0,1,1,1)$ is $1 \cdot 0 + 0 \cdot 1 + 1 \cdot 1 + 1 \cdot 1 = 0 + 0 + 1 + 1 = 2$. If we use Boolean addition for the same vectors, the dot product is 1.

We define $\mathbf{x} + \mathbf{y}$ as the r-vector

(8.6)
$$\mathbf{x} + \mathbf{y} := (x_1 + y_1, \ldots, x_r + y_r);$$

we simply add the corresponding coordinates. We use the same symbol + for addition of vectors as for addition of elements in the underlying set A.

For the same two vectors the sum over **Z** is (1,1,2,2), and with Boolean addition it is $\mathbf{x} + \mathbf{y} = (1,1,1,1)$.

We also define the **scalar** product $a \cdot \mathbf{x}$ if $a \in A$ and $\mathbf{x}$ is a *vector*, as above:

(8.7)
$$a \cdot \mathbf{x} := (a \cdot x_1, a \cdot x_2, \ldots, a \cdot x_r).$$

Notice that for $a, b \in A$

(8.8)
$$\begin{aligned} a \cdot (\mathbf{x} + \mathbf{y}) &= a \cdot \mathbf{x} + a \cdot \mathbf{y} \\ (a + b) \cdot \mathbf{x} &= a \cdot \mathbf{x} + b \cdot \mathbf{x}. \end{aligned}$$

Practice (8.9P) Prove both formulas of (8.8).

Because of the distributive law (8.4) for $\cdot$ and +, there is also such a law for $\bullet$ and +: For all r-vectors $\mathbf{x}, \mathbf{y}, \mathbf{z}$

(8.10)
$$(\mathbf{x} + \mathbf{y}) \bullet \mathbf{z} = \mathbf{x} \bullet \mathbf{z} + \mathbf{y} \bullet \mathbf{z}.$$

Finally $\mathbf{x} \bullet \mathbf{y} = \mathbf{y} \bullet \mathbf{x}$ for any two r-vectors, because $\cdot$ is commutative.

Example (8.11) In $\{0,1\}$ with + as Boolean addition (8.2) and $\cdot$ defined by (8.3),

$$\begin{aligned} (1,0,1) \bullet (1,1,1) &= 1 \cdot 1 + 0 \cdot 1 + 1 \cdot 1 \\ &= 1 + 0 + 1 = 1 \\ (1,0,0) \bullet (0,1,0) &= 1 \cdot 0 + 0 \cdot 1 + 0 \cdot 0 \\ &= 0 + 0 + 0 = 0 \end{aligned}$$

Our definition of matrix multiplication uses the dot product. Let L and M be two matrices, with L of size $m \times r$ and M of size $r \times n$, for positive integers m, r, n. The **product** LM of these matrices is defined as the $m \times n$ matrix

(8.12)
$$LM := (a_{ij}),$$

where $a_{ij} :=$ (Row i of L) $\bullet$ (Column j of M) for $i = 1, \ldots, m$ and $j = 1, \ldots, n$.

This definition may be familiar to you when the matrices are over the real numbers; we have merely extended it to any set A having appropriate operations + and $\cdot$.

(8.13) For example, suppose we multiply a 2×3 matrix L by a 3×3 matrix M. Suppose the *rows* of L are L_1 and L_2 and the *columns* of M are M_1, M_2, M_3. Then

$$\underset{L}{\boxed{\begin{matrix} L_1 \\ L_2 \end{matrix}}} \; \underset{M}{\boxed{\begin{matrix} M_1 & M_2 & M_3 \end{matrix}}} = \underset{LM}{\boxed{\begin{matrix} L_1 \bullet M_1 & L_1 \bullet M_2 & L_1 \bullet M_3 \\ L_2 \bullet M_1 & L_2 \bullet M_2 & L_2 \bullet M_3 \end{matrix}}}$$

If again A is $\{0,1\}$, + is Boolean addition (8.2), and $\cdot$ is given by (8.3), then

(8.14)
$$\boxed{\begin{matrix} 1 & 0 & 1 \\ 0 & 1 & 1 \end{matrix}} \; \boxed{\begin{matrix} 0 & 1 & 0 \\ 1 & 0 & 1 \\ 0 & 1 & 1 \end{matrix}} = \boxed{\begin{matrix} 0 & 1 & 1 \\ 1 & 1 & 1 \end{matrix}}$$

as you can easily check: $L_1 = 101$ and $M_1 = 010$, so $L_1 \bullet M_1 = 1 \cdot 0 + 0 \cdot 1 + 1 \cdot 0 = 0$. $L_2 = 011$ and $M_3 = 011$, so $L_2 \bullet M_3 = 0 \cdot +1 \cdot 1 + 1 \cdot 1 = 0 + 1 + 1 = 1$. The other entries are computed similarly.

You have probably seen all of this before, except perhaps the name "dot product" and the use of Boolean addition. Now let us take another view of matrix multiplication, the insider's approach. We take L and M as above. Then

(8.15) For $1 \le i \le m$, row i of the product LM of matrices is the linear combination of the rows of M with coefficients taken from row i of L. That is, *row* i of LM is obtained as follows: First multiply each row of M by the corresponding entry in *row* i of L, using the "scalar" product (8.7). Then add the resulting row-vectors.

(8.16) *A schematic example for* $r = 3$. Suppose row i of L is the vector (a, b, c). Then row i of LM is obtained as follows:

1. Set up a, b, c opposite rows 1, 2, and 3 of M.

a	row 1 of M
b	row 2 of M
c	row 3 of M

2. Scalar multiply row 1 of M by a, row 2 of M by b, and row 3 of M by c.
3. Add the resulting row-vectors. The sum is row i of the matrix product LM.

If we look at (8.14) from this viewpoint we see that row 1 of LM is the sum of rows 1 and 3 of M, because the first row of L is 101. The second row of LM is the sum of rows 2 and 3 of M because row 2 of L is 011. *Sum* is used in the sense of Boolean addition, because that is how we defined "+" in this situation.

We easily see why this algorithm (8.15) is correct. From the definition (8.12) of matrix multiplication, row i of LM is the n-vector

$$(L_i \bullet M_1, L_i \bullet M_2, \ldots, L_i \bullet M_n),$$

where M_k is the kth column of M, for $1 \le k \le n$. If $L = (a_{ij})$ for $1 \le i \le m$ and $M = (y_{jk})$ for $1 \le j \le r$, and $1 \le k \le n$, then $L_i = (a_{i1}, a_{i2}, \ldots, a_{ir})$, and row i of LM is the sum of the n-vectors

$$\begin{array}{llll}
(a_{i1} \cdot y_{11}, & a_{i1} \cdot y_{12}, & \ldots, & a_{i1} \cdot y_{1n}) \\
(a_{i2} \cdot y_{21}, & a_{i2} \cdot y_{22}, & \ldots, & a_{i2} \cdot y_{2n}) \\
\vdots & \vdots & & \vdots \\
(a_{ir} \cdot y_{r1}, & a_{ir} \cdot y_{r2}, & \ldots, & a_{ir} \cdot y_{rn})
\end{array}$$

because, for example, the sum of the r first coordinates is $L_i \bullet M_1$. If R_j stands for *row* j of M, these vectors are just

$$\begin{array}{c}
a_{i1} \cdot R_1 \\
a_{i2} \cdot R_2 \\
\vdots \\
a_{ir} \cdot R_r,
\end{array}$$

when we factor out the common a_{ij} from row j and use our definition (8.7) of "scalar" multiplication. We have shown that

(8.17) Row i of LM is the sum of each row R_j of M multiplied by the corresponding entry a_{ij} of row i of L.

Rule (8.17) is a shortened definition of "linear combination of the rows of M with coefficients taken from row i of L." The full definition appears in (8.15).

This viewpoint is useful when we think about matrices and when we calculate small products by hand. It is much easier to work with a whole row than to focus on the individual entries.

There is an analogous viewpoint "coming from the other side." We state it, leaving the explanation to you.

(8.18) The matrix product LM has its kth column equal to the linear combination of the columns of L with coefficients taken from the kth column of M.

To summarize: To multiply M on the left by L is to act on the *rows* of M according to the entries in the *rows* of L by the algorithm (8.15). To multiply L on the right by M is to act on the *columns* of L according to the entries in the *columns* of M by the algorithm (8.18).

It is far better to view matrix multiplication via the algorithms (8.15) and (8.18) than by the definition (8.12).

Now let's do examples of matrix multiplication using (8.15) and (8.18). We'll confine ourselves to 0-1 matrices (i.e., all entries are 0 or 1) for simplicity and because relation-matrices are 0-1 matrices.

Example (8.19) Calculate LM, where

$$L = \begin{matrix} 1 & 0 & 1 & 0 \\ 1 & 1 & 0 & 0 \\ 0 & 0 & 0 & 1 \\ 1 & 0 & 1 & 1 \end{matrix} \quad \text{and} \quad M = \begin{matrix} 1 & 0 & 0 & 1 \\ 0 & 1 & 0 & 1 \\ 1 & 0 & 1 & 0 \\ 1 & 1 & 0 & 1 \end{matrix}$$

(*i*) Let's do it first as real matrices; i.e., take $0, 1 \in \mathbf{Z}$ and $+$ and $\cdot$ as the usual addition and multiplication in $\mathbf{Z}$.

$$LM = \begin{matrix} 1 & 0 & 1 & 0 \\ 1 & 1 & 0 & 0 \\ 0 & 0 & 0 & 1 \\ 1 & 0 & 1 & 1 \end{matrix} \quad \begin{matrix} \surd \\ \\ \surd \\ \\ \end{matrix} \begin{matrix} 1 & 0 & 0 & 1 \\ 0 & 1 & 0 & 1 \\ 1 & 0 & 1 & 0 \\ 1 & 1 & 0 & 1 \end{matrix} = \begin{matrix} 2 & 0 & 1 & 1 \\ 1 & 1 & 0 & 2 \\ 1 & 1 & 0 & 1 \\ 3 & 1 & 1 & 2 \end{matrix}$$

How did we do this? Easy: the first row of LM is the sum of row 1 and row 3 of M because row 1 of L is 1010; we use (8.15) or (8.17). The second row of LM is the sum of rows 1 and 2 of M, because row 2 of L is 1100. The third row of

LM is row 4 of M because row 3 of L is 0001. Row 4 of LM is the sum of rows 1, 2, and 4 of M because row 4 of L is 1101. With small matrices like these you can read off the row-sums by looking, or perhaps by checking off which ones to add each time, as with the $\surd$ marks in (8.20), which were used to get row 1 of LM.

(*ii*) If we look at the product in (8.20) from the point of view of the columns, then column 1 of LM is the sum of columns 1, 3, and 4 of L because column 1 of M is 1011 (as a vector). Column 3 of LM equals column 3 of L because column 3 of M is 0010 (as a vector).

(*iii*) If we want to view L and M as matrices over $\mathbf{Z}_2$ with $\oplus$ as $+$ and $\otimes$ as $\cdot$, then all we do is reduce the product LM in (8.20) mod 2:
We get

$$LM = \begin{matrix} 0 & 0 & 1 & 1 \\ 1 & 1 & 0 & 0 \\ 1 & 1 & 0 & 1 \\ 1 & 1 & 1 & 0 \end{matrix} \quad \text{as mod-2 matrices.}$$

If we want to use Boolean addition for $+$ then we replace each positive integer in (8.20) by 1. We get

$$LM = \begin{matrix} 1 & 0 & 1 & 1 \\ 1 & 1 & 0 & 1 \\ 1 & 1 & 0 & 1 \\ 1 & 1 & 1 & 1 \end{matrix} \quad \text{as Boolean matrices.}$$

You should construct more such examples until matrix multiplication becomes second nature. (In APL, the matrix product LM is $L + . \times M$ in $\mathbf{Z}$; the Boolean product is $L \vee . \wedge M$.)

9. Compositions, Products, and Transposes

We start this section with a result that connects composition of relations and matrix products. We next use that result to get a result about the converse of a composition. We'll then generalize that to the product of any two matrices. The whole process is reasonably easy because it piggybacks on previous results. The only work we have to do is in proving the first theorem.

Theorem (9.1) Let R_1 and R_2 be binary relations, from A to B and from B to C, respectively. Suppose M_1 and M_2 are the respective relation-matrices: $M_1 = M(R_1)$, $M_2 = M(R_2)$.

Then the relation-matrix $M(R_1 \circ R_2)$ of the composition of R_1 and R_2 is the Boolean matrix product of M_1 and M_2:

$$M(R_1 \circ R_2) = M_1 M_2 = M(R_1)\, M(R_2). \blacksquare$$

Example (9.2) Setting $L = M_1$ and $M = M_2$, look at Example (8.19), *(iii)*. If we take $A := \{1,2,3,4\}, B := \{a,b,c,d\}$, and $C := \{5,6,7,8\}$, then from row 1 of M_1 we see that $(1,a)$ and $(1,c)$ are in R_1. From M_2 we get

$$\text{Row } a : (a,5) \quad \text{and} \quad (a,8) \in R_2;$$
$$\text{Row } c : (c,5) \quad \text{and} \quad (c,7) \in R_2.$$

It follows from the definition of composition (4.1) that since $(1,a) \in R_1$,

$$(1,5) \quad \text{and} \quad (1,8) \in R_1 \circ R_2;$$

and since $(1,c) \in R_1$,

$$(1,5) \quad \text{and} \quad (1,7) \in R_1 \circ R_2.$$

Thus row 1 of $M(R_1 \circ R_2)$ must be 1011. (The "1's" stand for (1,5), (1,7), and (1,8).) That is exactly the first row of the Boolean matrix product calculated in (8.19), *(iii)*.

Proof of Theorem *(9.1).*

Step 1. We get a "1" in the "a,c" position of $M(R_1 \circ R_2)$, that is, in row a, column c, if and only if $(a,c) \in R_1 \circ R_2$.

Source: definition of relation-matrix.

Step 2. This happens iff for some $b \in B$, both $(a,b) \in R_1$ and $(b,c) \in R_2$.

Source: definition (4.1) of composition.

Step 3. $(a,b) \in R_1$ iff M_1 has a "1" in the "a,b" position; $(b,c) \in R_2$ iff M_2 has a "1" in the "b,c" position.

Source: definition of relation-matrix.

Let $M_1 =: (x_{i,j})$ and $M_2 =: (y_{j,k})$, with $i \in A$, $j \in B$, and $k \in C$, since we are indexing the rows and columns with the names of the elements of the sets. Also set

$$M(R_1 \circ R_2) =: (z_{i,k}).$$

Step 4. The conditions in Step 3 say

$$x_{a,b} = 1 \quad \text{and} \quad y_{b,c} = 1. \tag{9.3}$$

These imply that the "a,c" element of $M(R_1 \circ R_2)$, namely $z_{a,c}$, is 1. The reason is that

$$z_{a,c} := (\text{ row } a \text{ of } M_1) \bullet (\text{col } C \text{ of } M_2)$$
$$= \sum_j x_{a,j} \cdot y_{j,c}.$$

In this sum the term for $j := b$ is $1 \cdot 1 = 1$, by (9.3). Since the sum is Boolean, the value is 1 as soon as one term is 1.

Source: definition of matrix multiplication (8.12).

Since a and c were arbitrary elements of A and C, this completes the proof.QED

Comments.

(*i*) You can see that the proof is just a general version of the example in (9.2). Everything was a simple matter of definition until Step 4, where matrix multiplication first came in—and that step was pretty simple, too. I spelled things out in detail in this proof.

(*ii*) The set B indexes the columns of M_1 and the rows of M_2. It is important that the elements of B be in the same order for both matrices.

Example (9.4) Suppose the relations are defined by their relation-matrices as

$$M_1 := \begin{matrix} 0 & 0 & 1 \\ 1 & 0 & 1 \\ 0 & 1 & 1 \end{matrix}, \qquad M_2 := \begin{matrix} 0 & 1 & 0 & 1 \\ 1 & 1 & 0 & 0 \\ 0 & 0 & 1 & 1 \end{matrix}.$$

The Boolean product M_1M_2 is

$$M_1M_2 := \boxed{\begin{matrix} 0 & 0 & 1 \\ 1 & 0 & 1 \\ 0 & 1 & 1 \end{matrix}} \boxed{\begin{matrix} 0 & 1 & 0 & 1 \\ 1 & 1 & 0 & 0 \\ 0 & 1 & 0 & 0 \end{matrix}} = \boxed{\begin{matrix} 0 & 1 & 0 & 0 \\ 0 & 1 & 0 & 1 \\ 1 & 1 & 0 & 0 \end{matrix}}$$

as you can easily work out with (8.15).

Let's draw the graphs of the three relations to verify that M_1M_2 is the matrix of the composition. Taking $A := \{1,2,3\}$, $B := \{a,b,c\}$, and $C := \{4,5,6,7\}$ we find the graphs are

(9.5*F*)

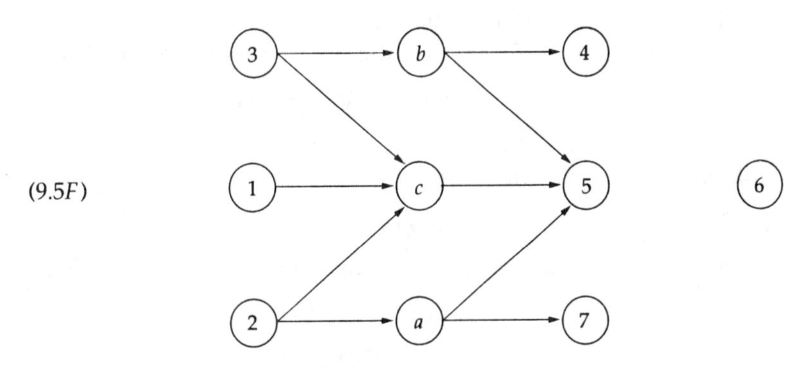

You can easily see from (9.5*F*) that in the composition we have

$$(1,5)$$
$$(2,5) \quad \text{and} \quad (2,7)$$
$$(3,4) \quad \text{and} \quad (3,5),$$

which correspond precisely to the five 1's in M_1M_2.

Now we can easily prove the next result, which expresses the relation-matrix of the converse of a composition in terms of the original relation-matrices.

Theorem (9.6) Let R_1 and R_2 be relations as in (9.1), with relation-matrices M_1 and M_2, respectively. Then

(9.7) $$M((R_1 \circ R_2)^c) = M(R_2^c) \cdot M(R_1^c);$$

(9.8) $$^t(M_1M_2) = {}^tM_2\,{}^tM_1.$$

In words, (9.8) says this: The transpose of a product of relation-matrices is the product of the transposes in reverse order.

Proof The result is easy to prove. We know from Theorem (4.6) that

$$(R_1 \circ R_2)^c = R_2^c \circ R_1^c.$$

We take the relation-matrices of each side and use Theorem (9.1) on the right-hand side. That yields (9.7). Theorem (7.13), which relates converses and transposes, turns (9.7) into (9.8). QED

Actually, the result (9.8) holds for any two matrices and under any definition of the $+$ and $\cdot$ used to define the matrix product. The proof is even simpler than the one we just gave.

Theorem (9.9) Let M_1 be an $m \times n$ matrix and M_2 an $n \times s$ matrix, with any binary operations $+$ and $\cdot$ used to define matrix multiplication (satisfying (8.1) and (8.4)). Then

(9.10) $$^t(M_1M_2) = {}^tM_2\,{}^tM_1.$$

Proof Let $M_1 = (a_{ij})$ and $M_2 = (b_{jk})$ (with $1 \le i \le m$, $1 \le j \le n$, and $1 \le k \le s$). We'll prove the equality in (9.10) by showing that entries in the same position are the same on the left as on the right.

By definition (8.12), the entry in the i,j position of M_1M_2 is

(9.11) $$(\text{row } i \text{ of } M_1) \bullet (\text{col } j \text{ of } M_2).$$

By definition (7.12) of transpose, the entry in row j, column i of $^t(M_1M_2)$ is this same quantity (9.11), which we may write as

(9.12) $$(\text{col } j \text{ of } M_2) \bullet (\text{row } i \text{ of } M_1).$$

if we use the commutativity of multiplication. But, again by (7.12),

$$\text{col } j \text{ of } M_2 = \text{row } j \text{ of } {}^tM_2;$$
$$\text{row } i \text{ of } M_1 = \text{col } i \text{ of } {}^tM_1.$$

Therefore (9.12) is

$$(\text{row } j \text{ of } {}^tM_2) \bullet (\text{col } i \text{ of } {}^tM_1),$$

which is the entry in row j, column i of ${}^tM_2\,{}^tM_1$, by definition (8.12) again. QED

Comments (9.13)

(*i*) We proved that for general matrices the transpose of a product is the product of the transposes in reverse order. That was Theorem (9.9). The proof was purely algebraic (easy, in fact) and did not depend on any of the previous results about relations and relation-matrices. Therefore we could find different proofs for two of our earlier results on compositions of relations, provided the sets are finite. Here are all those results in brief:

$$(R \circ S)^c = S^c \circ R^c \qquad [(4.6)]$$
$$M(R^c) = {}^t(M(R)) \qquad [(7.13)]$$
$$M(R_1 \circ R_2) = M(R_1) \cdot M(R_2) \qquad [(9.1)]$$
$$M((R_1 \circ R_2)^c) = M(R_2^c) \cdot M(R_1^c) \qquad [(9.6)]$$

We would still need to prove (7.13) and (9.1) as we did before. They establish the connections between converse and transpose and between composition and matrix product. Once we had those we could prove (9.6) and then (4.6) as corollaries of (9.9), (9.1), and (7.13). We leave the details to you in the problems.

(*ii*) No matter how we prove it, the result (4.6) allows an easier proof of (10.6) of Chapter 5. That concerned composition of functions. It read: If $f : X \longrightarrow Y$ and $g : Y \longrightarrow Z$, then

(9.14) $$(g \circ f)^{-1} = f^{-1} \circ g^{-1};$$

we've dropped the ˆ. Instead of proving it by viewing f^{-1} as a function from $\mathcal{P}(Y)$ to $\mathcal{P}(X)$ (and doing likewise for g^{-1}), regard f and g as relations. Then f^{-1} is f^c and g^{-1} is g^c, the converses (as we remarked after (3.3)), so (9.14) is a special case of (4.6).

(*iii*) ▶We could even carry out the alternative proofs under (*i*) if the sets were infinite. Then we'd replace the relation-matrix by the characteristic function of the relation (see Chapter 5, ((19.6)). We'd imitate the definition of Boolean matrix multiplication for these functions. There's no problem with infinite sums under Boolean addition.◀

10. Order Relations

You should review the properties reflexivity, symmetry, and transitivity discussed in Chapter 4. Although we repeat definitions here as we need them, we refer you to the earlier chapter for examples.

We begin with a simplification. Suppose R is a relation from A to B. Then R is a relation *on* $A \cup B$ (i.e., from $A \cup B$ to $A \cup B$). That is, the same set R of ordered pairs is a subset of $(A \cup B) \times (A \cup B)$:

(10.1) $$R \subseteq A \times B \subseteq (A \cup B) \times (A \cup B).$$

Therefore, simply redefining the sets on which the relation is defined allows us to associate just one set instead of two sets to the relation. Thus from now on, without loss of generality, we will speak of a general relation as a relation on the set A.

A major focus of this chapter is the idea of *order relation*. Such relations come up frequently in computer science and mathematics. The common relations mentioned at the beginning of this chapter are all order relations:

DEFINITION (10.2) A **partial order relation** on the set A is any relation R on A that is reflexive, antisymmetric, and transitive.

The new property here is antisymmetry. It is defined as follows:

DEFINITION (10.3) The relation R is called **antisymmetric** iff for all $a, b \in A$, $(a,b) \in R$ and $(b,a) \in R$ imply $a = b$.

A relation is antisymmetric if and only if its graph does not contain any "back arrows." That is, the graph must not have any two distinct points a, b with arrows in both directions between them:

(10.3*F*) a ⇄ b is forbidden.

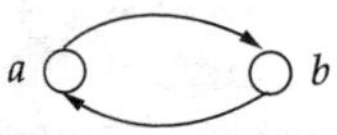

Only the trivial back arrow, the loop from a point to itself, may appear (but it need not be there).

As we said in Chapter 4, a relation R on A is called *reflexive* if and only if for all $a \in A$, $(a,a) \in R$. R is called *transitive* if and only if for all $a, b, c \in A$, (a,b) and $(b,c) \in R$ imply $(a,c) \in R$. We always define these properties, and symmetry, for relations on a set. It makes no sense to call a relation from A to B reflexive and little sense to call it symmetric unless $A = B$. Transitivity can be, but usually isn't, defined when $A \neq B$.

Example (10.4) Here are relations with the indicated properties:

(*i*) Symmetric

10.4*F* (*ii*) Nonsymmetric and not antisymmetric

(*iii*) Antisymmetric

Notice the difference in the definitions:

(*i*) $\forall a, b \in A \quad a\,R\,b \longrightarrow b\,R\,a$ (symmetry)

(*ii*) $\exists a, b \in A \quad a\,R\,b \wedge b\,\not R\,a$ (nonsymmetry)

$\exists a, b \in A \quad a\,R\,b \wedge b\,R\,a \wedge (a \neq b)$ (nonantisymmetry)

(*iii*) $\forall a, b \in A \quad a\,R\,b \wedge b\,R\,a \longrightarrow a = b$ (antisymmetry)

11. Comparison with Equivalence Relations

Before we look at examples of partial order relations, we compare them with equivalence relations. The common properties of both types of relations are reflexivity and transitivity. Look at the example below of a reflexive, transitive graph (in which all loops and some arrows implied by transitivity, as from a to b and from b to c, are omitted for clarity).

(11.1F)

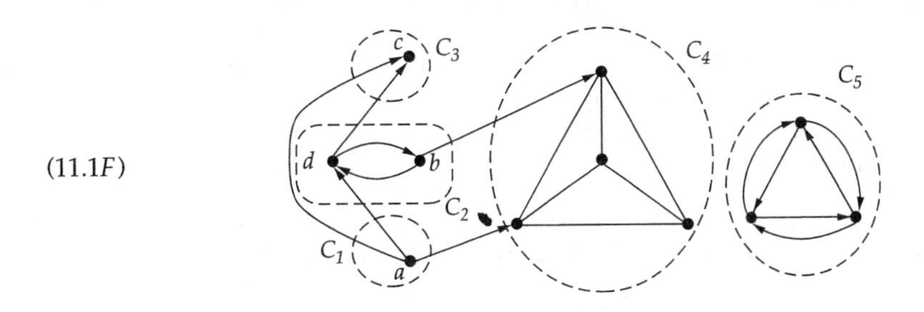

Here each line between points in C_4 stands for two arrows, one in each direction. The other omitted arrows are (a, b) and all but one of those from b and d and a to each point of C_4.

The subsets $C_1, \ldots, C_5$ are called the *strongly connected components* of the graph. In general,

DEFINITION (11.2) For the reflexive transitive relation R on A, the **strongly connected component**, *scc*, of $a \in A$ is

$$scc(a) := \{x;\ x \in A,\ (a,x) \in R \text{ and } (x,a) \in R\}.$$

Proposition (11.3) The set of all *scc*'s of a reflexive, transitive relation on A is a partition of A.

Proof We see that the set of all *scc*'s is a partition of A by noting first that $a \in scc(a)$ for all $a \in A$, by reflexivity of R. Second, we will show that for all $a, b \in A$, $scc(a)$ and $scc(b)$ are disjoint or equal. By Chapter 4, (5.1), that will prove the *scc*'s are the cells of a partition of A.

Suppose $scc(a) \cap scc(b) \neq \emptyset$. Then $\exists x \in scc(a)$ such that $x \in scc(b)$. Thus $(a,x) \in R$ and $(x,b) \in R$, by definition of *scc*. By transitivity, $(a,b) \in R$.

We'll now show $scc(a) \subseteq scc(b)$. Let u be any element of $scc(a)$. Then

$$(u,a) \in R.$$

Since we just saw that $(a,b) \in R$, transitivity tells us that

$$(u,b) \in R.$$

We also know that $(x,a) \in R$, $(b,x) \in R$, and $(a,u) \in R$, since x and u belong to *scc*'s. As before, we arrive at

$$(b,u) \in R.$$

Thus $(u,b) \in R$ and $(b,u) \in R$, so

$$u \in scc(b).$$

Therefore $scc(a) \subseteq scc(b)$. The reverse inclusion holds by symmetry of argument, so $scc(a) = scc(b)$.
QED

Practice (11.4*P*) Justify the invocation of "symmetry of argument" just above.

You can see that the *scc*'s are just the equivalence classes when the relation R is also symmetric. In general they help us see more clearly the difference between equivalence relations and partial order relations among reflexive, transitive relations R:

(11.5) R is an equivalence relation iff there is no arrow between any two distinct *scc*'s. R induces a partial order relation *on the set of its scc's.*

By *induces* we mean: If one or more arrows go between *scc*'s, replace them by one "big" arrow. Note that if there is an arrow from $scc(a)$ to $scc(b)$—i.e., if $(a, b) \in R$—then the back arrow does not exist if the two *scc*'s are distinct, for that would make $b \in scc(a)$. In other words, the *induced* relation on the *scc*'s is antisymmetric. Here is its graph for the example above (loops omitted):

(11.5*F*)

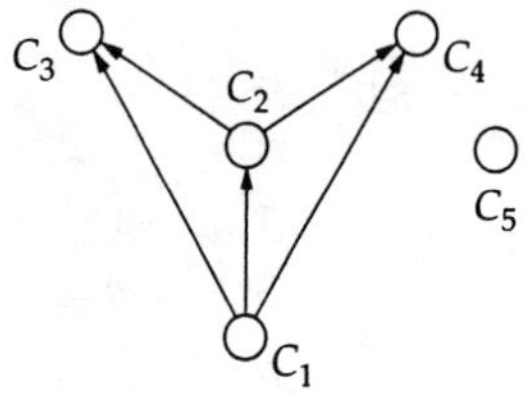

An equivalence relation is the trivial partial order on its set of *scc*'s, the order in which all *scc*'s are isolated, as C_5 is isolated, above.

(11.6) Another example of a reflexive, transitive relation F: for all $a, b \in \mathbf{N}$, aFb iff every prime dividing a divides b.

Practice (11.7P) Restrict F to $A = \{2^i 3^j 5^k;\ 0 \leq i, j, k \leq 10\}$; determine the *scc*'s, and graph the induced partial order.

Example (11.8) Students' homework papers are returned in "partial" alphabetical order as follows: All names beginning with "A" are together in any order, then all "B's" are together in any order, and so on. The rule for arranging the papers in this way arises from a reflexive-transitive relation in which the strongly connected components are the set of "A's," the set of "B's," and so on.

We make it reflexive by fiat. We don't care what order the "A's" are in, so each two "A's" are joined by arrows in each direction.

If A stands for the student named "Able" and A' For "Abbott" and A'' for "Archer," B for "Bravo," and so on, then

(11.8*F*)

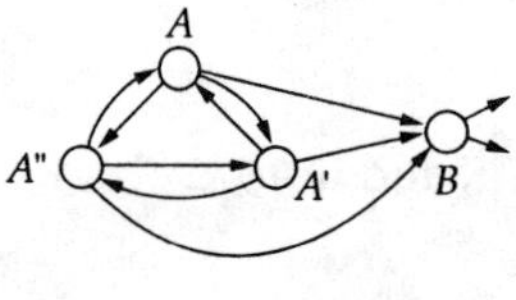

the underlying relation is the one shown in (11.8*F*) (where loops are omitted). The only order imposed by this relation is that all the "A's" come first, then all the "B's," and so on.

12. Examples of Partial Order Relations

Example (12.1) *Less than or equal to*, on **N**. We use the infix notation and define, for all $a, b \in \mathbf{N}$, $a \le b$ iff $a - b \notin \mathbf{N}$. Of course, all we need do to show that $\le$ is a partial order on **N** is to show that it satisfies the definition (7.11). Thus $\le$ is antisymmetric, because

$$a \le b \quad \text{and} \quad b \le a \quad \text{imply}$$
$$a - b \notin \mathbf{N} \quad \text{and} \quad b - a \notin \mathbf{N}.$$

Thus neither $a - b$ nor its negative $b - a$ is positive, yet both are integers. Therefore they must both be 0. We conclude $a = b$.

The reflexivity and transitivity of $\le$ are easily verified: $a \le a$ because $a - a = 0 \notin \mathbf{N}$; $a \le b$ and $b \le c$ imply

$$a - b \notin \mathbf{N} \quad \text{and} \quad b - c \notin \mathbf{N}.$$

Since $\mathbf{Z} - \mathbf{N}$ is closed under addition (!), $(a - b) + (b - c) = a - c \notin \mathbf{N}$. Therefore $\le$ is transitive.

Example 2 (12.2) *Divisibility*, on **N**. We have already defined divisibility of integers: a divides b iff $\exists x \in \mathbf{Z}$ such that $ax = b$. We use an infix notation, the *vertical* bar of Chapter 6:

$$a \mid b \quad \text{if and only if} \quad a \text{ divides } b.$$

We emphasize that the bar is *vertical*, that it stands for the relationship and *not* the quotient b/a. The negation of this relation is denoted $a \nmid b$; it means "a does not divide b." Thus

$$2 \mid 4,\ 2 \nmid 3,\ 7 \mid 28,\ 4 \nmid 2,\ 10 \nmid 5,$$
$$1 \mid a \quad \text{for all } a \in \mathbf{N},$$
$$23 \mid (2^{11} - 1), \quad \text{and}$$
$$a \equiv 0 \pmod{m} \quad \text{iff} \quad m \mid a.$$

To prove that divisibility is a partial order on **N**, we see that for all $a \in \mathbf{N}$, $a \mid a$; thus divisibility is reflexive. (Referring to this relation by its infix notation alone, as in "| is reflexive," just isn't done.) For antisymmetry, suppose $a, b \in \mathbf{N}$ and $a \mid b$ and $b \mid a$. Then both

$$\frac{b}{a} \quad \text{and} \quad \frac{a}{b}$$

are positive integers. Since their product is 1, they must each be 1, so $a = b$.

For transitivity, if a, b, c are positive integers such that $a \mid b$ and $b \mid c$, then

$$\frac{b}{a} \quad \text{and} \quad \frac{c}{b}$$

are both positive integers. Therefore so is their product, c/a; this means $a \mid c$.

Example 3 (12.3) *Inclusion*, on a collection X of sets. We defined inclusion of sets in Chapter 1. We remarked there that every set is a subset of itself,

$$A \subseteq A;$$

therefore inclusion is reflexive. Is it antisymmetric? Yes, because if

$$A \subseteq B \quad \text{and} \quad B \subseteq A$$

then $A = B$; that is our definition of equality between sets, and transitivity is equally obvious (we posed it as an exercise in Chapter 1, (10.7)).

(12.4) Let us graph some of the points in these three examples. Letting G_i denote the graph of Example i ($i = 1, 2, 3$), we see that

(12.5F) G_1 : 1 → 2 → 3 → 4 → ···

The arrows drawn stand for $1 \le 2, 1 \le 3, 1 \le 4, 2 \le 3, 2 \le 4$, and $3 \le 4$.

(12.6F) G_2 :

2 → 4 → 8 → ···

6

3 → 9 → 18 → ···

⋮

Here the arrows drawn stand for $2 \mid 4$, $4 \mid 8$, $2 \mid 8$, $2 \mid 6$, $3 \mid 6$, $3 \mid 9$, $9 \mid 18$, $6 \mid 18$, $3 \mid 18$, and $2 \mid 18$. Only a few of the points of the infinite set **N** on which the relation is defined are pictured in G_2. In particular, 1 is missing; there is an arrow from 1 to every other integer.

(12.7) For G_3 let us take X to be the collection of all subsets of the 3-set $\{a, b, c\}$. Then we see that

(12.7F)

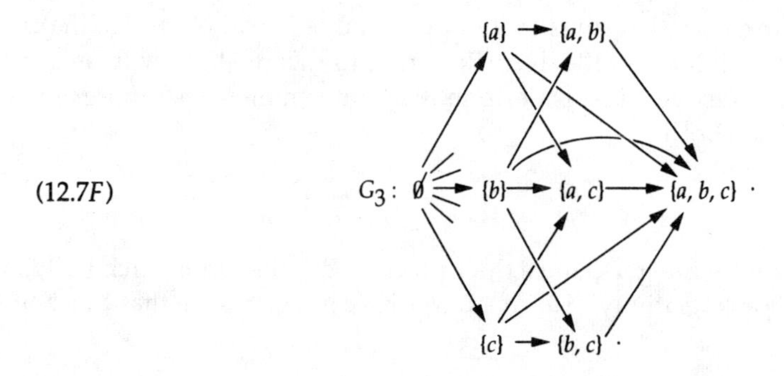

Here we started but did not complete four arrows from $\varnothing$ to the sets with two and three elements.

In every one of these graphs we did not draw the arrows of reflexivity, the loops. The reason is simple: They would clutter the page. We knew they were there, so we omitted them from the picture. In fact it is customary to omit more arrows from the graphs of order relations and to draw them according to a fixed convention, the Hasse diagram.

13. The Hasse Diagram

The **Hasse diagram** of a partial order on the set A is a drawing of the points of A and of some of the arrows of the graph of the order relation. There are rules to determine which arrows are drawn and which are omitted, namely

(13.1) Rules for drawing the Hasse diagram of a partial order:

- omit all arrows that can be inferred from transitivity
- omit all loops
- draw "arrows" without heads
- understand that all arrows point upwards.

Here are the Hasse† diagrams of the three examples in Section (12).

(13.1*F*)

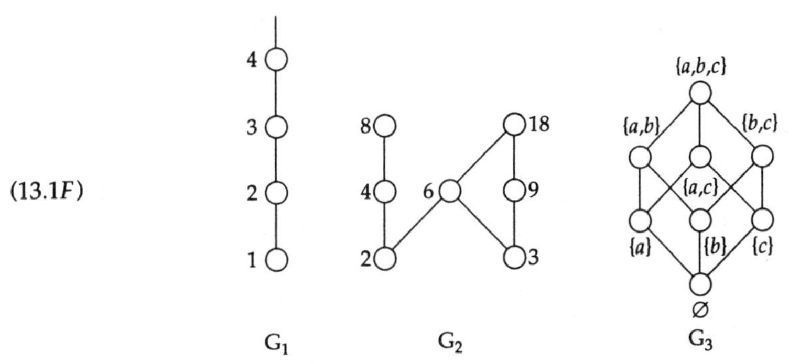

Hasse Diagrams for Examples from (12).

It is much easier to grasp the relation from its Hasse diagram than from its graph.

But how do we know that we can always draw the Hasse diagram? The problem

(13.2*F*)

is in confidently assuming that the "arrows all point up" convention never leads to a contradiction (as it would, for example, if we tried to graph (13.2*F*) by the rules (13.1)). In fact the Hasse diagram can always be drawn, but we put the proof in the Appendix.

†Pronunciation: *Hasse* rhymes with *casa*.

14. The "Idea" of Order Relations

Now you can see why we call a partial order relation R by that name: It imposes some kind of "order" on the underlying set A. It does so by telling us that some elements of A *precede* others.

(14.1) If $(a, b) \in R$, then a **precedes** b.

Because of antisymmetry we never find that b also precedes a (unless $b = a$). So this definition of precedence fits with our intuitive understanding of the term.

We express the precedence $(a, b) \in R$ graphically in the Hasse diagram of R by drawing a below b *and* by connecting a to b by one or more lines, *always moving upward*.

In general there may be elements a, b in A such that neither precedes the other in the given partial order.

(14.2) For example, consider divisibility on $\{2, 3, \ldots, 10\}$. The Hasse diagram is

(14.2*F*)

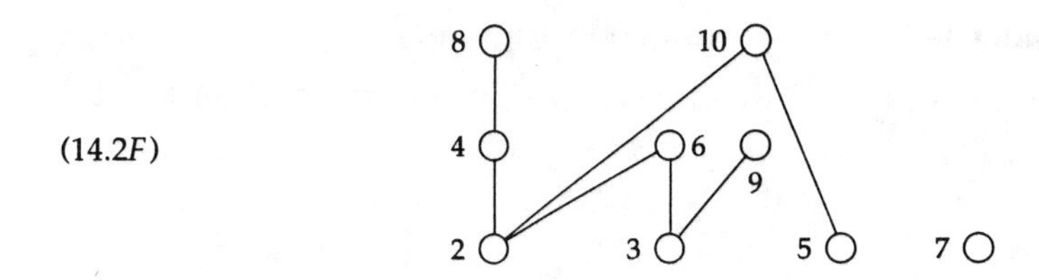

No two primes are related by divisibility, but also 5 and 6 are not related in either direction: 5 does not divide 6, and 6 does not divide 5. (There is a path from 5 to 6 but it is "disallowed" because it has a downward leg, from 10 to 2.) In fact 5 precedes only itself and 10.

(14.3) Another term sometimes used to express a partial order is *comparable*. If $(a, b) \in R$, we might say that

a and b are **comparable,** and a precedes b.

If neither $(a, b) \in R$ nor $(b, a) \in R$ we would say

a and b are **incomparable,** or **not comparable.**

(14.4) But if we define the relation $\leq$ on $\{2, \ldots, 10\}$, namely $a \leq b$ iff $b - a$ is positive, then *any* two elements are comparable; the Hasse diagram, tipped, is

(14.4*F*)

2 3 4 5 6 7 8 9 10

The latter is an example of a *linear,* or *total* order.

DEFINITION (14.5) A **linear**, or **total**, order R on a set A is a partial order on A in which every two elements of A are comparable:

$$\forall a, b \in A, \text{either}$$
$$(a, b) \in R \text{ or } (b, a) \in R.$$

The name "linear" is justified by the following result.

Proposition (14.6) The Hasse diagram of a linear order R on a finite set A may be drawn as a chain of vertical line-segments joining the points of A. ▪

Example (14.7) Suppose we had $A = \{1, \ldots, 5\}$ and R were the usual $\leq$. Then the Hasse diagram would be

(14.7F)

We postpone the proof of (14.6) for a bit, and now make two obvious definitions. Let R be a partial order on the set A.

DEFINITION (14.8) A **maximal** element for R on A is any element t of A such that

$$\forall x \in A \qquad t\, R\, x \quad \text{implies} \quad x = t.$$

A **minimal** element is any element b of A such that

$$\forall y \in A \qquad y\, R\, b \quad \text{implies} \quad y = b.$$

Thus in the Hasse diagram there is no element above a maximal element and none below a minimal element.

Examples (14.9) (*i*) Consider the examples of Section (12). In (12.1), $(\mathbf{N}, \leq)$ has a unique minimal element, 1, and no maximal element. In (12.2), $(\mathbf{N}, |)$, i.e., divisibility on $\mathbf{N}$, there is again a unique minimal element and no maximal element. In (12.3), inclusion on a collection of sets, there may or may not be minimal or maximal elements. If the collection were $\mathcal{P}(X)$ for some set X, there would be a unique minimal element $\varnothing$ and a unique maximal element X.

(*ii*) Consider again $A := \{2, \ldots, 10\}$ with R as divisibility. Then each prime is a minimal element, and the maximal elements are 6, 7, 8, 9, and 10. The Hasse diagram is shown in (14.2*F*)

Practice *(14.10P) Define a collection of sets such that under inclusion there are no minimal elements and no maximal elements.

Lemma (14.11) In any partial order R on a nonempty finite set A there is at least one maximal element and at least one minimal element. If the order is linear these elements are unique.

Proof Let $x \in A$. If there is no $y \in A$ (except $y = x$) such that $x\ R\ y$, then x is maximal. If not, then let $y \in A$ satisfy $x\ R\ y$ and $y \neq x$.

Now repeat the process until either you arrive at a maximal element or at an element you have already chosen. The finiteness of the set guarantees one of these outcomes. The latter outcome, however, is impossible, because it would produce a "circle" in the graph of the relation. That is, suppose our choices in order were $x_1, x_2, \ldots$ and $x_i = x_j$ for $i < j$.

(14.11*F*)

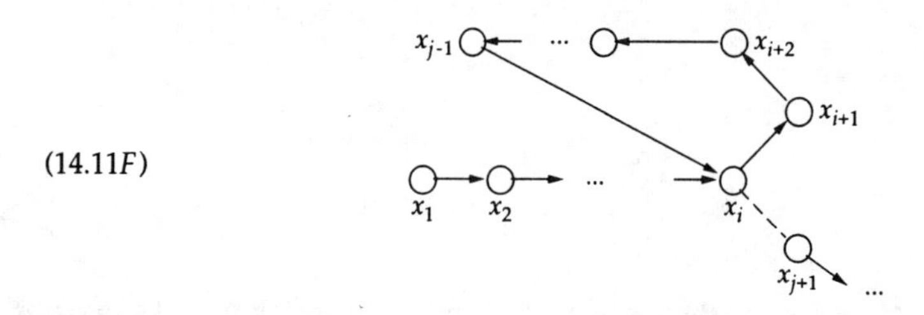

By transitivity, any two points in a circle are related to each other by R, violating antisymmetry. Thus a maximal element exists.

By symmetry of argument a minimal element exists.

Finally, if R is linear, we prove uniqueness of the maximal element. Let x be a maximal element. Let y be any other element of A. Thus $y \neq x$, but because R is linear, $x\ R\ y$ or $y\ R\ x$ holds. It can't be $x\ R\ y$, because x would not be maximal. Thus $y\ R\ x$. Therefore y is not maximal. Thus there is only one maximal element.

Arguing symmetrically, we could show there is a unique minimal element.QED

Now we can prove Proposition (14.6).

(14.12) Proof of (14.6) By induction on the size of A. Let b and t denote the unique minimal and maximal elements of A guaranteed by Lemma (14.11). If there is another point, say x, in A, then consider the two subsets of A of all points above x and all points below x:

$$B := \{y;\ y \in A,\ x\ R\ y\},$$
$$C := \{y;\ y \in A,\ y\ R\ x\},$$

Obviously $t \in B$ and $b \in C$.

(14.12F)

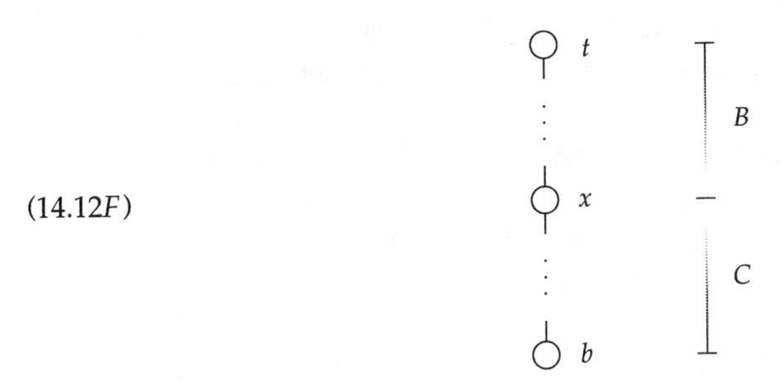

Let R_B be the restriction of R to B. Then (B, R_B) is a linearly ordered set of fewer points than A. By induction (B, R_B) has a Hasse diagram satisfying (14.6). The same is true of C. If we join these two Hasse diagrams at x, we get a Hasse diagram for (A, R) satisfying (14.6). QED

(That was a lot of work for an obvious result. Sorry.)

15. Transfinite Induction

There is a generalization of the second theorem of mathematical induction (Chapter 3, (5.1)) that holds for certain partially ordered sets. (Throughout this section we use the notation A for our set and $<$ for a partial order on A.) These are partially ordered sets satisfying the **minimum condition:**

(15.1) Every nonempty subset has a minimal element.

The condition (15.1) means that in any nonempty subset B of A

$$\exists c \in B \; \forall b \in B \qquad b < c \;\text{ implies }\; b = c. \tag{15.2}$$

We speak of c as a **minimal** element of B because we think of the abstract partial order $< R$ as $\leq$ on **R** or perhaps $\subseteq$ on a collection of sets. (We use the term "least" element for linear orders; then condition (15.1) becomes the well-ordering principle.)

In a partially ordered set there may be more than one minimal element, as we saw with divisibility on $\{2, 3, \ldots, 10\}$.

(15.3) The minimum condition is equivalent to the following property:

(15.4) There are no infinite descending chains in A.

DEFINITION (15.5) An **infinite descending chain** in A is a sequence of mutually distinct points a_n (for $n \in \mathbf{N}$) such that

$$a_1 > a_2 > a_3 > \cdots .$$

By "$x > y$" we mean "$y < x$ but $y \neq x$."

In other words, the function $f : \mathbf{N} \longrightarrow A$ defined by the rule

$$\forall n \in \mathbf{N} \qquad f(n) := a_n$$

is an injection, and $\forall n \in \mathbf{N}\, f(n+1) < f(n)$.

We leave to you the problem of proving the equivalence of the minimum condition and nonexistence of infinite descending chains.

Example (15.6) **N**, $\leq$ has no infinite descending chains. **Z**, $\leq$ does have such chains. **N** with divisibility ($|$) as the order relation has no infinite descending chains. The converse of this last order does have infinite descending chains.

Here is the theorem mentioned earlier.

Theorem. (15.7) (Transfinite induction). Suppose A is a set with a partial order $<$ satisfying the minimum condition. If $S(x)$ is a predicate with $D_x = A$ satisfying

(15.8) $$\forall a \in A \quad (\forall b \in A,\ b < a \text{ and } b \neq a \text{ imply } S(b)) \quad \text{implies } S(a),$$

then $\forall a \in A\ S(a)$ is true.

Proof If the theorem is false, then "$\forall a \in A\ S(a)$" is false. That is, there is an element $c \in A$ such that $S(c)$ is false. This means that the set

$$G := \{x;\ x \in A,\ \sim S(x)\}$$

is nonempty. By the minimum condition, G has a minimal element; call it z. Since (15.8) is true, setting $a := z$ there tells us that the inner proposition in (15.8)

(15.9) $$\forall b \in A,\ b < z \text{ and } b \neq z \quad \text{imply } S(b)$$

is false (since it implies the false $S(z)$). We now derive a contradiction by showing that (15.9) is true.

Since z is a minimal element in G there is no element $b \in G$ such that $b < z$ and $b \neq z$. Therefore the antecedent in (15.9),

(15.10) $$b < z \quad \text{and} \quad b \neq z,$$

is true only for points $b \in A - G$. But $S(b)$ is true for all such points, since $b \notin G$.

Therefore the implication in (15.9) is true (as $T \longrightarrow T$) for all $b \in A - G$ satisfying (15.10), and it is true as ($F \longrightarrow ?$) for all $b \in A$ that make (15.10) false. This contradiction establishes the theorem as true. QED

Comments

(*i*) We stated (15.8) for **N**, $\leq$ in Chapter 3, Section (10).

(*ii*) In proving (15.8) for a given situation it is necessary to do a basis step for every minimal point in the Hasse diagram. For example, if we have **N**, $\leq$ there is only one lowest point, 1. Then (15.8) says (for $a := 1$)

$$(\forall b \in \mathbf{N}, b < 1 \text{ implies } S(1)) \quad \text{implies } S(1),$$

which is

$$T \to S(1).$$

This is true iff $S(1)$ is true; thus structural induction generalizes the second form of induction of Chapter 3, (5.1).

(15.11) *An example of structural induction.* (This example depends on Chapter 8.) To construct an example we first need an infinite set with a nontrivial partial order. We'll use *lexicographic order* (which you should meet for its own sake anyway). Suppose we had a predicate $S(m,n)$ of two variables m,n with domains $D_m = D_n = D = \mathbf{N} \cup \{0\}$. We could define a partial order $<$ on $D \times D$ as **lexicographic order**:

$$\forall a,b,x,y \in D \quad (a,b) < (x,y) \text{ iff } a < x \text{ or } (a = x \text{ and } b \leq y). \tag{15.12}$$

In general we may define with (15.12) lexicographic order on $A \times A$ whenever A has a linear order $\leq$. In our example A is $\mathbf{N} \cup \{0\}$ and $\leq$ is the usual magnitude relation.

Practice (15.13*P*) Verify that $<$ is a partial order. Draw the Hasse diagram for the subset of points (a,b) with $a \leq 2, b \leq 2$.

Now (15.8) would require us to prove $S(0,0)$ as the basis step, since (0,0) is the only least element. For the inductive step we would have to prove that for any $(a,b) \in D \times D$,

If $S(x,y)$ is true for all (x,y) such that
$x < a$ or $(x = a$ and $y < b)$
then $S(a,b)$ is true.

In other words our inductive assumption would be that S is true on the (infinite, if $a > 0$) set of points (x,y) with $x < a$ and on $(a,0),(a,1),\ldots,(a,b-1)$ if $b \geq 1$. (Notice that $D \times D$ with this order has no infinite descending chains even though each element (a,b) with $a > 0$ has infinitely many descendants! In fact it is a linear order (so is any lexicographic order). It "is" $A_0 < A_1 < A_2 < \cdots$, where A_i is $(i,0) < (i,1) < (i,2) < \cdots$. Thus it is the same as infinitely many copies of the integers $\{0,1,\ldots\}$ ordered by magnitude, with the copies ordered as just described.

Example (15.14) Let us prove for all integers $m,n \geq 0$ that $P(m,n)$ is true, where $P(m,n)$ is the predicate

$$f(m,n) := \sum_{0 \leq k} (-1)^k \binom{m}{k}\binom{k}{n} = \begin{cases} 0 & m \neq n \\ (-1)^n & m = n. \end{cases} \tag{15.15}$$

Thus we have a function f defined at each point with integral coordinates in the first quadrant, and we are to prove that it is 0 except on the line $n = m$, where it alternates between 1 and -1.

(15.15*F*)

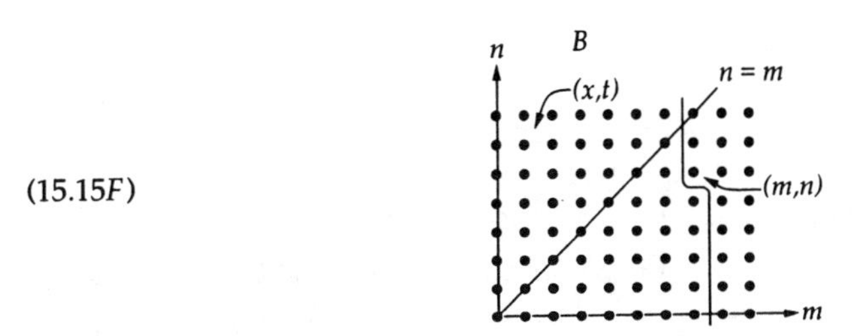

It will prove convenient to expand our basis step to the whole half-line $n = 0$. Thus

$$f(m,0) := \sum_{0\le k}(-1)^k\binom{m}{k} = \begin{cases} 0 & \text{if } m \ge 1 \\ 1 & \text{if } m = 0, \end{cases} \tag{15.16}$$

by Problem 2.4 of Chapter 8. Therefore (15.15) is true for $n = 0$ for all m.

We need to prove $P(0,n)$ for all n, another kind of basis step. It's very simple. $P(0,n)$ is the statement

$$\sum_{0\le k}(-1)^k\binom{0}{k}\binom{k}{n} = \begin{cases} 0 & \text{if } n \ne 0 \\ 1 & \text{if } n = 0, \end{cases}$$

but the left-hand side is simply $\binom{0}{n}$, because $\binom{0}{k} = 1$ for $k = 0$ and is 0 for $k \ge 1$. That says it for $\binom{0}{n}$ as well, so $P(0,n)$ is true for all n. Thus we know that $P(m,n)$ is true on all the boundary points of $D \times D$.

Now we do the inductive step. We let (m,n) be any point of $D \times D$ with $m \ge 1$ and $n \ge 1$. Using (15.8), we assume that $P(x,t)$ is true for all points (x,t) of $D \times D$ that precede (m,n). They form the set

$$B := \{(x,t);\ x < m, \text{or } x = m \text{ and } t < n\},$$

outlined in the strip in the figure above. And we try to prove that $P(m,n)$ must then be true.

Letting (x,t) be any point of B with $t \ge 1$, we first prove that

$$\sum_{0\le k}(-1)^k\binom{1+x}{1+x}\binom{k}{t} = \begin{cases} 0 & \text{if } x < t \\ (-1)^t & \text{if } x \ge t \end{cases}. \tag{15.17}$$

You will see that (15.17) is our connection between the points of B and the "target point" (m,n). Although $(x,t) \in B$, the point $(x+1,t)$ may be outside of B. Since by assumption $f(u,t)$ is given by (15.15) for all $u \le x$, we add (15.15) over all such u and conclude that

$$\sum_{0\le u\le x}\ \sum_{0\le k}(-1)^k\binom{u}{k}\binom{k}{t} = \sum_{0\le u\le x} f(u,t) = \begin{cases} 0 & \text{if } x < t \\ (-1)^t & \text{if } x \ge t \end{cases}$$

since $f(u,t)$ is 0 except at (t,t). But if we sum over u first we get

$$\sum_{0\le k}(-1)^k\binom{k}{t}\sum_{0\le u\le x}\binom{u}{k}.$$

The inner sum, that over u, from the symmetrized knight's-move identity (Chapter 8, (8.2) or problem 8.1), equals $\binom{x+1}{k+1}$. This proves (15.17).

Now we use the basic recursion of Chapter 8, (3.1). Since $\binom{k}{t} + \binom{k}{t-1} = \binom{k+1}{t}$, and since $(x,t-1) \in B$, $P(x,t-1)$ is also true. Therefore (15.17) holds for $t := t-1$. Adding

(15.17) as shown above to (15.17) for t replaced by $t-1$ gives us

$$\sum(-1)^k\binom{1+x}{1+k}\binom{1+k}{t} = \begin{cases} 0 & \text{if } x < t \\ (-1)^t & \text{if } x \geq t \end{cases} + \begin{cases} 0 & \text{if } x < t-1 \\ (-1)^{t-1} & \text{if } x \geq t-1 \end{cases}$$

(15.18)

$$= \begin{cases} 0+0 & \text{if } x < t-1 \\ 0+(-1)^{t-1} & \text{if } x = t-1 \\ (-1)^t + (-1)^{t-1} = 0 & \text{if } x > t-1 \end{cases}$$

If we multiply (15.18) by -1 we get

(15.19)
$$\sum_{0\leq k}(-1)^{k+1}\binom{1+x}{1+k}\binom{1+k}{t} = \begin{cases} 0 & \text{if } x+1 \neq t \\ (-1)^t & \text{if } x+1 = t, \end{cases}$$

which on the left lacks only the first term (set $k = -1$),

$$\binom{1+x}{0}\binom{0}{t} = \binom{0}{t},$$

of (15.15) for $(1+x, t)$. But $\binom{0}{t} = 0$ if $t \geq 1$, as we have assumed. Since $(m-1, n) \in B$ we may set $x := m-1$ and $t = n$ in (15.19). It follows that $P(m,n)$ must be true. Therefore $P(m,n)$ is true for all $m, n \geq 0$, by (15.7). QED

Comments

(*i*) On which points of B did we actually use the inductive assumption? Since at the end we set $x := m-1$ and $t := n$, we used it on all the points of B with $t = n$, the horizontal line-segment to the left of (m,n), to prove (15.17).

Then we used it on $(m, n-1)$ to prove (15.18).

(*ii*) Notice that the above proof is correct for all m and n in D. Compare this to Chapter 3, Example (7.2).

(*iii*) Even though the order is linear and it may be easier to prove (15.15) some other way, the proof just concluded is a good example of the use of structural induction.

16. More on Composition

Now we'll take a closer look at compositions. The purpose is to develop your intuition and facility with both this idea and the associated matrices. First we consider some simple examples.

Examples (16.1) Let $R = \{(1,1),(1,2),(2,3),(1,3),(3,4)\}$ Then $R \circ R = \{(1,2),(1,3),(1,4),(2,4),(1,1)\}$. The fact that all four ordered pairs listed in $R \circ R$ truly belong to $R \circ R$ is easy to see: $(1,2)$ in $R \circ R$ comes from $(1,1)$ and $(1,2)$ in R; $(1,3)$ in $R \circ R$ from $(1,1)$ and $(1,3)$ in R; $(1,1)$ in $R \circ R$ from $(1,1)$ and $(1,1)$ in R; $(1,4)$ in $R \circ R$ from $(1,3)$ and $(3,4)$ in R; and $(2,4)$ in $R \circ R$ from $(2,3)$ and $(3,4)$ in R. The fact that there are no more ordered pairs in $R \circ R$ is tedious to verify directly from the list of ordered pairs in R. We'll do it more pleasantly in a moment.

Let's also look at the graph of R. Remember we saw in Chapter 4 that there are three ways to view relations: as sets of ordered pairs, as directed graphs, and as matrices of 0's and 1's. Each way has something to contribute to our understanding. Each is logically independent of the others, in that you can determine the properties of a relation (at least in principle) entirely from any one of the three approaches alone. But some properties are easier to study with, say, matrices than with graphs, and some are better with ordered pairs. At any rate, the graph of R is

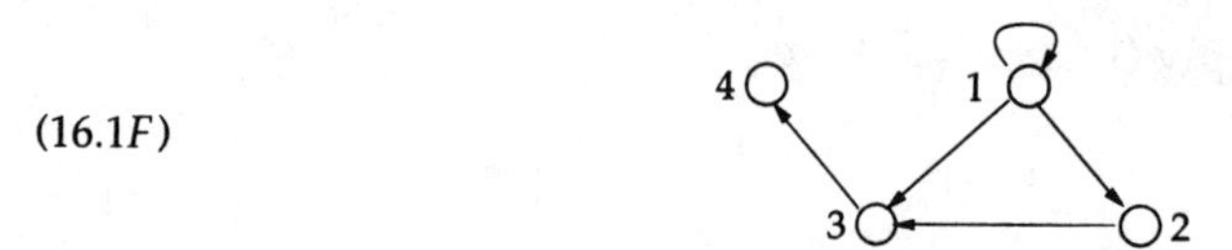

(16.1*F*)

(16.2) We now notice that definition (4.1) tells us to form $R \circ R$ by looking for every path of length 2 in $G(R)$ and replacing its two arrows by one new arrow from start to finish. That is, wherever we have in $G(R)$

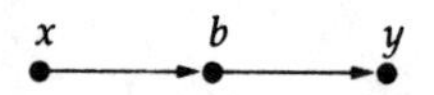

we make the arrow from x to y in $G(R \circ R)$:

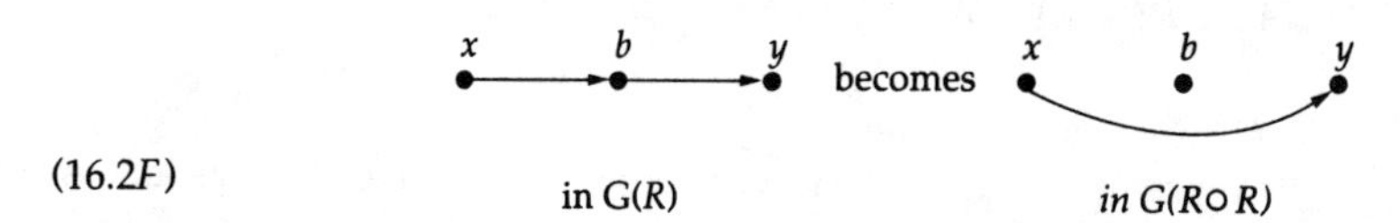

(16.2*F*)

Let us redraw $G(R)$ for the example and superimpose on it the dashed arrows of $G(R \circ R)$.

(16.3*F*)

Now we see that $G(R \circ R)$ is

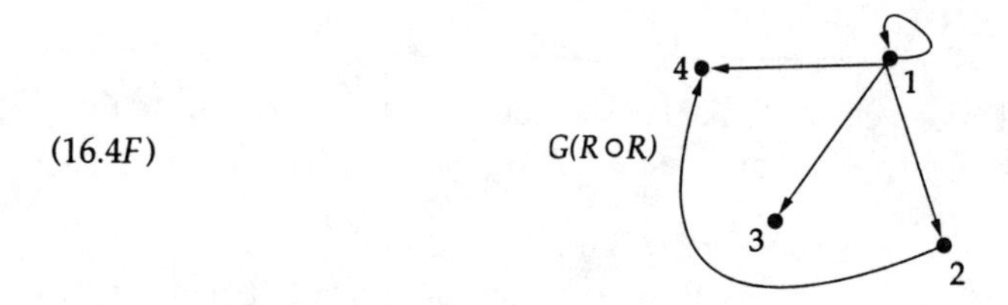

(16.4*F*)

which we could better draw as

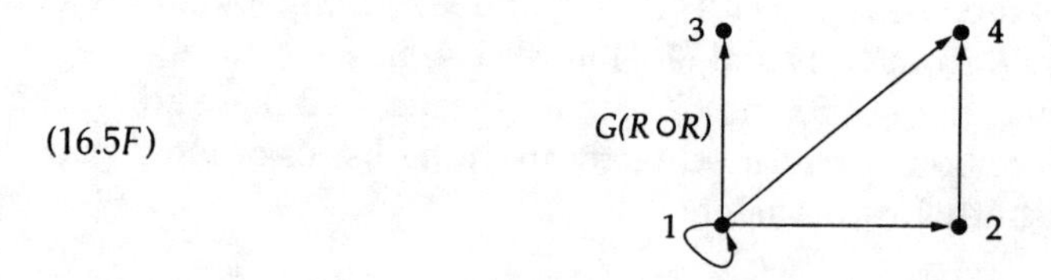

(16.5*F*)

Even with the graph it is not so easy to make sure that we have spotted all the paths of length 2 in $G(R)$. We'll see that using the matrix $M(R)$ offers the most reliable procedure for calculating $R \circ R$.

One fact we can see is generally true:

(16.6) $$\forall a, b \in A \quad (a,a) \in R \text{ and } (a,b) \in R \text{ imply } (a,b) \in R \circ R.$$

Now we show how to use the relation-matrix. As we know from Theorem (9.1), $M(R)^2$ is the relation-matrix of $R \circ R$. By $M(R)^2$ we mean, of course, the matrix product of $M(R)$ with itself, using Boolean addition and multiplication. Let's work it out for Example (16.1).

$$M(R) = \begin{array}{c|cccc|} & 1 & 2 & 3 & 4 \\ \hline 1 & 1 & 1 & 1 & 0 \\ 2 & & & 1 & 0 \\ 3 & & & & 1 \\ 4 & & & & 0 \\ \hline \end{array}$$

(16.7) $$M(R) \cdot M(R) = \begin{bmatrix} 1 & 1 & 1 & 0 \\ 0 & 0 & 1 & 0 \\ 0 & 0 & 0 & 1 \\ 0 & 0 & 0 & 0 \end{bmatrix} \begin{bmatrix} 1 & 1 & 1 & 0 \\ 0 & 0 & 1 & 0 \\ 0 & 0 & 0 & 1 \\ 0 & 0 & 0 & 0 \end{bmatrix}$$

$$= \begin{bmatrix} 1 & 1 & 1 & 1 \\ 0 & 0 & 0 & 1 \\ 0 & 0 & 0 & 0 \\ 0 & 0 & 0 & 0 \end{bmatrix}$$

which we calculated using (8.15) or (8.18). (The latter is easier to use on this matrix.) This is indeed the relation matrix of what we claimed was $R \circ R$ or $G(R \circ R)$.

We show now how one of the 1's in $M(R) \cdot M(R)$ represents a path of length 2 in $G(R)$: Look at the 1 in the upper right corner, for the ordered pair $(1,4)$. How did it get there? It came from the dot product of row 1 of $M(R)$ with column 4 of $M(R)$, i.e.

$$\begin{aligned} (1\,1\,1\,0) \bullet (0\,0\,1\,0) &= 1 \cdot 0 + 1 \cdot 0 + 1 \cdot 1 + 0 \cdot 0 \\ &= 1. \end{aligned}$$

The "1" is there because of the 1 in position $(1,3)$ in row 1 of $M(R)$, and because of the 1 in position $(3,4)$ in column 4 of $M(R)$. But of course those 1's in $M(R)$ are there just because the ordered pairs $(1,3)$ and $(3,4)$ are in R. So the matrix product has nicely captured the idea of the definition of $R \circ R$.

Theorem (9.1) makes computation of compositions feasible.

(16.8) Another example of the key idea why $M(R)M(S)$ represents $M(R \circ S)$: Suppose row i of $M(R)$ is 101010 and column j of $M(S)$ is, as a vector, 010110. Then the dot-product is

$$\begin{aligned} (1\,0\,1\,0\,1\,0) \bullet (0\,1\,0\,1\,1\,0) &= 1 \cdot 0 + 0 \cdot 1 + 1 \cdot 0 + 0 \cdot 1 + 1 \cdot 1 + 0 \cdot 0 \\ &= 0 + 0 + 0 + 0 + 1 + 0 \\ &= 1. \end{aligned}$$

If the points of B are $1, 2, \ldots, 6$, then there is an arrow from i to 5 in R and one from 5 to j in S.

(16.8F)

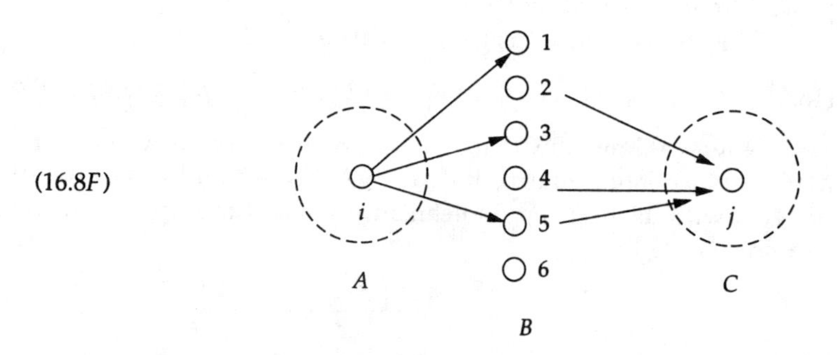

There is a path from i to 5 to j using first an arrow from R and then one from S. This path becomes a single arrow in the graph of $R \circ S$, from i to j. It is picked up precisely in the dot-product.

Another Example (16.9) Suppose row i of $M(R)$ is 101010 and column j of $M(S)$ is 111000. Then the dot-product is

$$\begin{aligned}(1\,0\,1\,0\,1\,0) \bullet (1\,1\,1\,0\,0\,0) &= 1\cdot 1 + 0\cdot 1 + 1\cdot 1 + 0\cdot 0 + 1\cdot 0 + 0\cdot 0\\ &= 1 + 0 + 1 + 0 + 0 + 0\\ &= 1,\end{aligned}$$

still 1, even though now there are two different paths of length 2 from i to j as the sketch shows:

(16.9F)

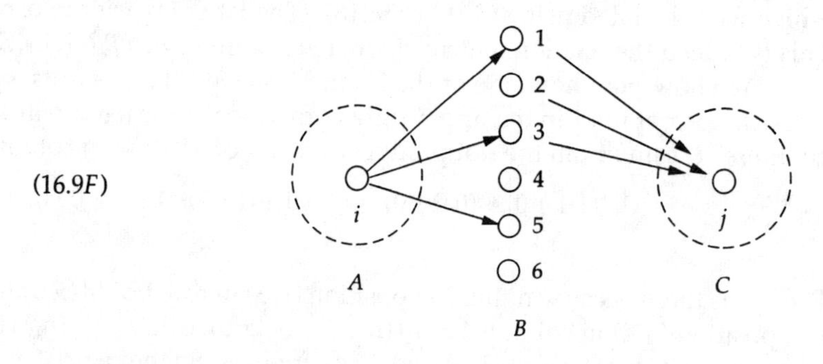

If we used addition in **Z** the sum would be 2.

(16.10) In general, if we multiply relation-matrices with $+$ and $\cdot$ from **Z**, then the entry in the product at (i, j) is the total number of paths from i to j in the composition.

17. Transitivity

We can use the foregoing ideas of matrices and matrix multiplication to study transitivity. Let's first state a criterion for transitivity in terms of composition and then express the same criterion in terms of matrices.

Lemma (17.1) The relation R on A is transitive if and only if $R^2 := R \circ R \subseteq R$. ▪

The proof of this result is a nice exercise for you. In terms of the relation-matrices it says (with + as Boolean sum)

(17.2) R is transitive if and only if $M(R)^2 + M(R) = M(R)$.

Consider the logical "and" defined on $\{0, 1\}$ as

(17.3) $$0 \wedge 0 = 0 \wedge 1 = 1 \wedge 0 = 0, \qquad 1 \wedge 1 = 1.$$

The operation $\wedge$ is just the Boolean product (8.3). We extend it to two 0,1 matrices (a_{ij}) and (b_{ij}) of the same size:

(17.4) $$(a_{ij}) \wedge (b_{ij}) := (a_{ij} \wedge b_{ij});$$

that is, we "and" together each pair of corresponding entries.

Practice (17.5P)
(*i*) Prove that (17.2) is equivalent to (17.1).
(*ii*) Prove R is transitive iff $M(R)^2 \wedge M(R) = M(R)^2$.

We have already seen (Chapter 4) that R is reflexive iff the main diagonal of $M(R)$ consists entirely of 1's. (Remember that the main diagonal is defined as the entries in positions (x, x) for all x in A.) Furthermore, it is obvious that R is symmetric if and only if $M(R) = {}^tM(R) = M(R^c)$. So the relation-matrix can be used to discuss more than just transitivity.

(17.6) Let us try to express the property of antisymmetry in matrix terms. Remember that antisymmetry is more than just nonsymmetry, for the latter occurs if there is even one pair (a, b) in R with (b, a) not in R. For antisymmetry our relation-matrix must have a 0 opposite every 1 not on the main diagonal. (By *opposite* I mean "placed symmetrically with respect to the main diagonal to.") That is, the matrix must look like

(17.6*F*)

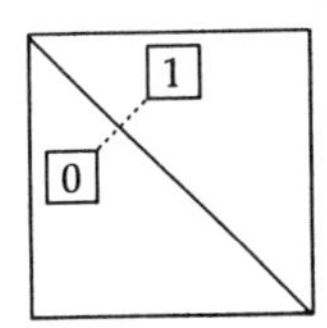

for every 1 not on the main diagonal. The reflected position must have a 0. We can easily express this condition in matrix terms, as

(17.7) R is antisymmetric if and only if $M(R) \wedge {}^tM(R)$ has nothing but 0's off the main diagonal.

Condition (17.7) is easier to express, but perhaps harder to compute, in terms of the relation itself:

(17.8) R is antisymmetric if and only if $R \circ R^c \subseteq I_A$,

where I_A is the diagonal of $A \times A$:

(17.9) $$I_A := \{ (x, x);\ x \in A \}.$$

The problem with expressing (17.8) in matrices is that we have no inclusion relation for matrices. We could define one, of course, for 0, 1 matrices, although it's rare (and maybe unique to this book):

(17.10) The $m \times n$ 0,1 matrices M_1 and M_2 satisfy the relation $\subseteq$, as $M_1 \subseteq M_2$, if and only if $M_1 \wedge M_2 = M_1$.

The definition (17.10) is merely a restatement in matrix terms of the usual inclusion relation $R_1 \subseteq R_2$ for the relations.

(17.11) Union and intersection of relations are expressed in terms of the relation-matrices by Boolean sum and term-by-term Boolean product (*and*, i.e., $\wedge$). Thus if R_1 and R_2 are relations then

$$M(R_1 \cup R_2) = M(R_1) + M(R_2),$$

where + is Boolean sum;

$$M(R_1 \cap R_2) = M(R_1) \wedge M(R_2).$$

We defined the second of these operations on matrices in (17.4); it is the entry-by-entry *and* of the two matrices. The Boolean sum of matrices is defined analogously; it is the entry-by-entry *or* of the two matrices.

(17.12) For example, set

$$M(R_1) = \begin{matrix} 1 & 1 & 1 & 1 \\ 0 & 0 & 0 & 0 \\ 0 & 1 & 0 & 1 \\ 0 & 0 & 0 & 0 \end{matrix} \qquad M(R_2) = \begin{matrix} 1 & 0 & 0 & 0 \\ 1 & 0 & 0 & 1 \\ 1 & 0 & 0 & 1 \\ 1 & 0 & 0 & 1 \end{matrix}$$

Then

$$M(R_1) + M(R_2) = \begin{matrix} 1 & 1 & 1 & 1 \\ 1 & 0 & 0 & 1 \\ 1 & 1 & 0 & 1 \\ 1 & 0 & 0 & 1 \end{matrix}$$

and

$$M(R_1) \wedge M(R_2) = \begin{matrix} 1 & 0 & 0 & 0 \\ 0 & 0 & 0 & 0 \\ 0 & 0 & 0 & 1 \\ 0 & 0 & 0 & 0 \end{matrix}$$

And, to sum up, we showed above that composition of relations is mirrored by the more complicated matrix product:

$$M(R_1 \circ R_2) = M(R_1)\, M(R_2).$$

We form the matrix product using Boolean addition.

18. Closures

Suppose R is a relation and P is a property of some relations. For example, P might be reflexivity, or symmetry or transitivity, or antisymmetry. For *some* properties P we define the P-**closure** of R as the smallest relation that both includes R (as a set of ordered pairs) and has property P. The properties P for which we define the P-closure are precisely those properties P such that

the complete relation $A \times A$ has property P,

and

P is preserved under intersection.

The latter means that if R_1 and R_2 have property P, then $R_1 \cap R_2$ has property P. Then we can prove that every relation has a P-closure.

Lemma (18.1) Suppose P is a property of relations preserved under intersection. Suppose also that $A \times A$ has property P. Let R be *any* relation on A. Then R has a unique P-closure, namely the intersection of all relations including R and having property P.

Proof Since $A \times A$ has property P, the set $\mathcal{S}$ of *all* relations including R and having property P is nonempty. The intersection R_0 of *all* the relations in $\mathcal{S}$ includes R and has property P, by hypothesis. Finally, for any $R' \in \mathcal{S}$, $R_0 \subseteq R'$ by the definition of R_0 as the intersection of all the relations in $\mathcal{S}$. Therefore R_0 is the smallest relation including R and having property P. QED

(18.2) As usual, we use *smallest* in the sense of inclusion; to say a set is smallest among all sets with some property means it is a subset of each of them. In the case of finite sets, then, *smallest* always implies the minimum number of elements, too.

Examples (18.3)

(*i*) Reflexivity is the simplest example. The intersection of reflexive relations is reflexive, and $A \times A$ is reflexive. Therefore there exists a reflexive closure for any relation R, by Lemma (18.1). But we need not look far to see what it is; it is just

$$R \cup I_A$$

where $I_A = \{(x,x);\ x \in A\}$, as defined already in (17.9). The matrix is $M(R) + M(I_A)$. In other words, we just turn the main diagonal of $M(R)$ into all 1's to get the matrix of the reflexive closure of R.

(*ii*) Symmetry is almost as easy. Again, symmetry is preserved under intersection (check this), and $A \times A$ is symmetric. Therefore, by Lemma (18.1), every relation R on A has a symmetric closure. Again we can find the symmetric closure of R by inspection: it is

$$R \cup R^c. \tag{18.4}$$

The matrix of the symmetric closure of R is

$$M(R) + {}^tM(R) \qquad \text{(Boolean+)}.$$

You should verify these claims.

(*iii*) Transitivity: $A \times A$ is transitive, and the intersection of transitive relations is transitive, so every relation on A has a unique transitive closure, by (18.1).

The above proof of Lemma (18.1) is a typical "existence proof" from mathematics. It is elegant and simple, but it doesn't offer a practical way to find the P-closure of a given relation. We saw immediately how to construct the reflexive and symmetric closures of relations; those were easy. To find the transitive closure is less obvious, however. We'll present two *constructive* proofs of the existence of the transitive closure in a moment. These proofs will also show how to calculate the transitive closure. In fact they are *algorithms* for finding the transitive closure. But before we go on to that...

There is no antisymmetric closure for the relation R unless R is already antisymmetric, in which case it is its own antisymmetric closure. So the concept of antisymmetric closure is pointless. (P-closures exist for relations *not* having property P as long as the hypotheses of (18.1) hold.) It's true that antisymmetry is preserved under intersection, but $A \times A$ is *not* antisymmetric (unless A is a singleton), so the Lemma (18.1) does not apply.

19. Algorithms for Finding Transitive Closures

So how do we find the transitive closure $R^\star$ of a relation R on the set A? One way is to construct

(19.1)
$$\begin{aligned} & R \cup R^2 \cup R^3 \cup \cdots \\ &= \bigcup_{m \geq 1} R^m, \end{aligned}$$

where R^m is defined as $R \circ R^{m-1}$ for all $m \geq 1$ and $R^0 := I_A = \{ (x,x);\ x \in A \}$.

Practice (19.2P) Prove that $R^1 = R$ for any relation R.

Proposition (19.3) Let R be any relation on the set A. Then the transitive closure $R^\star$ of R is

$$\begin{aligned} R^\star &= R \cup R^2 \cup R^3 \cup \cdots \\ &= \bigcup_{m \geq 1} R^m. \end{aligned}$$

Proof Almost all we have to show, by Lemma (17.1), is that $(R^\star)^2 \subseteq R^\star$. That will show $R^\star$ is transitive. We'll then show it's the smallest such relation including R. By the distributivity of composition over union of relations, $R^{\star 2}$ is just

(19.4)
$$R^{\star 2} = R^2 \cup R^3 \cup \cdots,$$

which is, of course, a subset of $R^\star$. (I.e., if R, R', and R'' are relations on A, then

(19.5)
$$R \circ (R' \cup R'') = (R \circ R') \cup (R \circ R'').)$$

Practice (19.6P) Prove (19.5) and show how it leads to (19.4).

Therefore $R^\star$ is transitive. We'll now show that $R^\star$ is a subset of any transitive relation S that includes R.

We explained in (16.2) that R^2 replaces every path of length 2 in R by a shortcut path of length 1:

(19.6*F*)

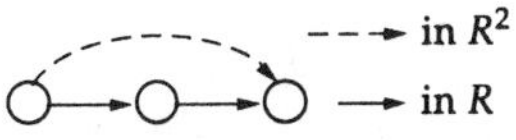

- - → in R^2
→ in R

In general, for $m \geq 1$, R^m replaces every path of length m in R by a one-arrow shortcut, the dashed arrow in the sketch:

(19.7*F*)

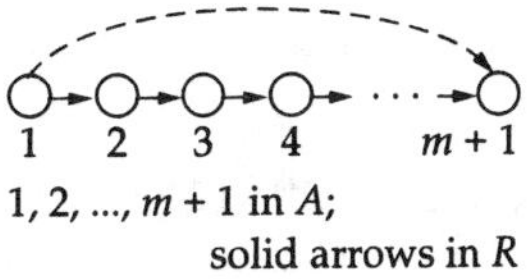

1 2 3 4 $m+1$
1, 2, ..., $m+1$ in A;
solid arrows in R

We could (but we won't) do a simple proof by induction on m to show that such dashed arrows must be in any transitive relation including R. Therefore $R^m \subseteq S$ (for all $m \geq 1$), so $R^\star \subseteq S$. QED

NOTE We've used the term *length* of a path in a graph to mean the total number of arrows in the path. We take the definition of path to be self-evident. ■

If A has exactly n points, then we may state Proposition (19.3) as

(19.8) $$R^\star = R \cup R^2 \cup \ldots \cup R^n.$$

It is easy to construct an example showing the need to go as high as n in (19.8). (You should do that.)

To prove (19.8), which says that to form $R^\star$ we may ignore paths of length greater than n in an n-point set A, we use, not for the first time, the famous pigeonhole principle. (See Chapter 5, (8).)

Consider a path of length $q > n$ composed of arrows from R. Thus

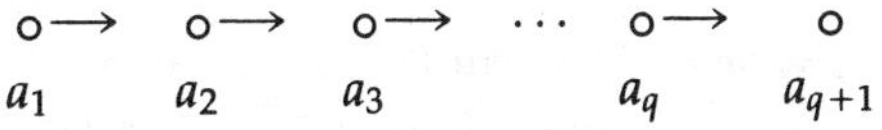

$$\begin{array}{cccccc} \circ\!\longrightarrow & \circ\!\longrightarrow & \circ\!\longrightarrow & \cdots & \circ\!\longrightarrow & \circ \\ a_1 & a_2 & a_3 & & a_q & a_{q+1} \end{array}$$

is such a path, with a_i in A for all i. Now A has only n points, so at least two of the first q of these a_i's must equal each other: $\exists i < j$ such that $a_i = a_j$ and $1 \leq i < j \leq q$. That is, the portion of the path from a_i to a_j is just a circle of length $j - i$.

(19.9*F*)

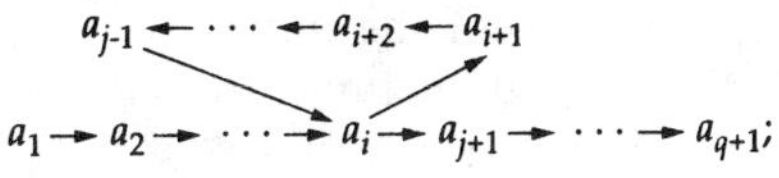

$a_{j-1} \leftarrow \cdots \leftarrow a_{i+2} \leftarrow a_{i+1}$

$a_1 \rightarrow a_2 \rightarrow \cdots \rightarrow a_i \rightarrow a_{j+1} \rightarrow \cdots \rightarrow a_{q+1};$

Remove the arrows $(a_i, a_{i+1}), \ldots, (a_{j-1}, a_j)$ and the points $a_{i+1}, \ldots, a_j$. The remaining path has length $q - (j - i)$; its one-arrow shortcut appears in $R^{q-(j-i)}$.

Practice (19.10P) What points and arrows do you remove if $j = i + 1$?

Since $j > i$ we've shown that for any two points $a, b \in A$, if there is a path in R of length greater than n from a to b, then there is a shorter path in R from a to b.

Repeating this process if the length is still greater than n (wringing out *all* the circles) brings the length down to n or less. Therefore (19.8) is true. QED

This process (19.8) can be mechanized with the relation-matrix:

(19.11)
$$\begin{aligned} M(R^\star) &= M(R) + M(R)^2 + M(R)^3 + \cdots \\ &= \sum_{1 \le m \le n} M(R)^m \qquad \text{(Boolean sum).} \end{aligned}$$

In practice we stop the calculation as soon as $M(R)^m$ adds no new 1's.

(19.12) *Warshall's Algorithm.* There is a faster procedure than that of (19.3) for finding $M(R^\star)$. It is called **Warshall's algorithm.**

Theorem (19.13) Let R be any relation on the set $A = \{1, \ldots, n\}$ with relation-matrix $M = (m_{ij})$. Then the relation-matrix $M^\star$ of the transitive closure of R is $M^{(n)}$, where the sequence $M^{(0)}, M^{(1)}, \ldots, M^{(t)} = (m_{ij}^{(t)}), \ldots$ of relation-matrices is defined as follows. We define

$$M^{(0)} := M, \quad \text{i.e., } m_{ij}^{(0)} = m_{ij}$$

and

(19.14)
$$\forall t \ge 1, \quad m_{ij}^{(t)} := m_{ij}^{(t-1)} + m_{i,t}^{(t-1)} \cdot m_{t,j}^{(t-1)}$$

where the + stands for Boolean addition. ▪

Notice that (19.14) defines the entire matrix $M^{(t)}$ in terms of the entries in the preceding matrix $M^{(t-1)}$. Rule (19.14) says: if the i,j entry is 1 at stage $t - 1$, it remains 1 from then on. If it is 0, it will become 1 in the next matrix iff $M^{(t-1)}$ has a 1 in the i,t place *and* in the t,j place. The instructions are precise; we change an entry from 0 to 1 only under a certain condition; we never change a 1 to a 0. This gradual filling out of M to $M^\star$ means that (19.13) is more efficient than (19.8).

Proof of Theorem (19.13) First we remark that $m_{ij}^\star = 1$ iff there is a path in R from i to j. This remark follows immediately from Proposition (19.3). Now we prove by induction that for all $t \ge 0$

(19.15) $\forall i,j$ in $A, m_{ij}^{(t)} = 1$ iff there is a path in R from i to j going through *only* $\{1, 2, \ldots, t\}$ as intermediate points.

For the basis step of the induction we must show that $m_{ij}^{(0)} = 1$ iff there is a path from i to j in R with *no* intermediate points, since by $\{1, \ldots, t\}$ we understand, more precisely, the set $\{x;\ x \in \mathbf{N}, 1 \le x \le t\}$. This set is empty if $t = 0$. But $m_{ij}^{(0)}$ is defined to be m_{ij}, so the basis step is okay. Now let t be a fixed but arbitrary integer ≥ 0, and assume (19.15), which is $S(t)$, a predicate. We shall try to prove $S(t+1)$, so consider any i,j in A, and look at $m_{ij}^{(t+1)}$. If $m_{ij}^{(t)} = 1$ then (as we remarked before starting the proof) $m_{ij}^{(t+1)}$ is also 1, and by (19.15), the induction assumption, there is a path from i to j through $\{1,\ldots,t\}$, *a fortiori* through $\{1,\ldots,t+1\}$. The only other way $m_{ij}^{(t+1)}$ can be 1 is for both $m_{i,t+1}^{(t)}$ and $m_{t+1,j}^{(t)}$ to be 1, by the definition (19.14). By (19.15), which holds for *all* subscripts, there are paths in R from i to $t+1$ and from $t+1$ to j both having all their intermediate points in $\{1,\ldots,t\}$. Pasting them together produces a path in R from i to j with all its intermediate points in the set $\{1,\ldots,t+1\}$. This completes the proof by induction of (19.15) for all $t \geq 0$.

To finish the proof of the Theorem all we need to remark is that every path from i to j has its intermediate points in $\{1,\ldots,n\} = A$, so $M^{(n)} = M^{\star}$. QED

Notice what (19.14) says in terms of rows:

(19.16) Row i of $M^{(t)}$ is the Boolean sum of row i of $M^{(t-1)}$ and the scalar product of one entry of $M^{(t-1)}$, $m_{i,t}^{(t-1)}$, with row t of $M^{(t-1)}$.

If we push this further we see that the only new 1's in $M^{(t)}$ can come from whatever 1's there may be in column t and row t of $M^{(t-1)}$, because the new contribution out of (19.14) is the product of an entry from column t with one from row t.

Look more closely at the 1's this procedure produces. At i,j we get a 1 in $M^{(t)}$, by definition (19.14),

$$\text{if } \alpha := m_{t,j}^{(t-1)} = 1$$

$$\textit{and} \text{ if } \beta := m_{i,t}^{(t-1)} = 1.$$

If there's a 1 in $M^{(t-1)}$ in this i,j place, then we don't care about α and β; but we get a new 1 if $\alpha = \beta = 1$ and there was a 0 there.

The sketch will illuminate all:

(19.17F) $M^{(t-1)} =$ [sketch: matrix with row t and column j meeting at α; row i and column t meeting at β; box at row i, column j]

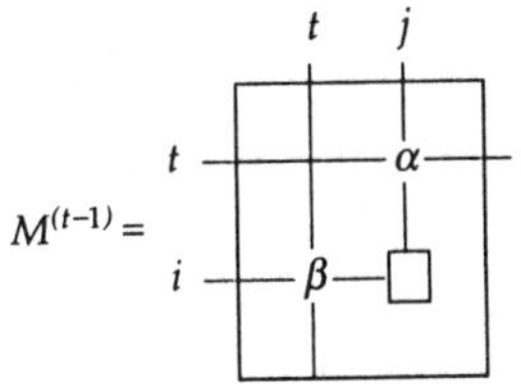

This sketch shows that we set to 1 every entry in such a "box" position in $M^{(t-1)}$ if the corresponding entries α and β are both 1. In other words, we

(19.18) pick out all the 1's in row t and column t of $M^{(t-1)}$ and "complete the rectangle" by setting the "box" to 1 for *each* such pair of 1's α and β.

The procedure is illustrated in the following example.

(19.19) Consider the relation-matrix M, given by

$$
M = \begin{array}{c|cccccc}
 & \square & \downarrow & & & \downarrow & \\
\square & 0 & 1 & 0 & 0 & 1 & 0 \\
\rightarrow & 1 & 0_1 & 1 & 0 & 0_1 & 1 \\
\rightarrow & 1 & 0_1 & 0 & 1 & 0_1 & 0 \\
 & 0 & 0 & 0 & 1 & 0 & 0 \\
\rightarrow & 1 & 0_1 & 1 & 1 & 1 & 0 \\
 & 0 & 0 & 1 & 0 & 0 & 1
\end{array} = M^{(0)}
$$

To apply Warshall's algorithm (by hand) we set $t = 1$, and mark row t and column t with a $\square$ as done above. We now make a mark, say an arrow, at each column where row 1 has a 1, and at each row where column 1 has a 1, as done above. The positions in the matrix where *both* row *and* column are marked with an arrow are all set to 1. This requires the five changes of 0 to 1 indicated by the 0's with subscript 1 in M (0_1). The result is

$$
M^{(1)} = \begin{array}{c|cccccc}
 & & \downarrow & & & & \\
 & \downarrow & \square & \downarrow & & \downarrow & \downarrow \\
\rightarrow & 0_2 & 1 & 0_2 & 0 & 1 & 0_1 \\
\rightarrow\ \square & 1 & 1 & 1 & 0 & 1 & 1 \\
\rightarrow & 1 & 1 & 0_2 & 1 & 1 & 0_2 \\
 & 0 & 0 & 0 & 1 & 0 & 1 \\
\rightarrow & 1 & 1 & 1 & 1 & 1 & 0_2 \\
 & 0 & 0 & 1 & 0 & 0 & 1
\end{array}
$$

Now we set $t = 2$, mark row 2 and column 2 with $\square$, and "arrow" the rows and columns where there are 1's in column 2 and row 2, respectively, as done above. We then set to 1 all entries for which both row and column have an arrow. This changes six 0's, marked 0_2, to 1's. Notice that we changed the 0 in position 1,1 this time.

$$
M^{(2)} = \begin{array}{c|cccccc}
 & & & \square & & & \\
 & 1 & 1 & 1 & 0 & 1 & 1 \\
 & 1 & 1 & 1 & 0 & 0 & 1 \\
\square & 1 & 1 & 1 & 1 & 1 & 1, \\
 & 0 & 0 & 0 & 1 & 0 & 0 \\
 & 1 & 1 & 1 & 1 & 1 & 1 \\
 & 0 & 0 & 1 & 0 & 0 & 1
\end{array}
$$

to which we now apply the same procedure for $t = 3$. But since row 3 is all 1's we must "arrow" every column; since column 3 is all 1's except at row 4, we "arrow" rows 1, 2, 3, 5, 6. The result is to change every 0 except those in row 4 to a 1.

$$
M^{(3)} = \begin{array}{cccccc}
 & & & \square & & \\
1 & 1 & 1 & 1 & 1 & 1 \\
1 & 1 & 1 & 1 & 1 & 1 \\
1 & 1 & 1 & 1 & 1 & 1 \\
0 & 0 & 0 & 1 & 0 & 0 \\
1 & 1 & 1 & 1 & 1 & 1 \\
1 & 1 & 1 & 1 & 1 & 1
\end{array}
$$

$M^{(3)}$ is the matrix of the *total* relation on $\{1,2,3,5,6\} = A'$ (i.e., $A' \times A'$) together with all arrows from A' to 4 (to mix the metaphor). Clearly this relation is transitive, so we need go no further. But if we did, we would arrow all rows, since column 4 is all 1's. Only column 4 is all 1's, since row 4 is 000100. No new 1's come into $M^{(4)}$. Thus $M^{(4)} = M^{(3)}$. To compute $M^{(5)}$ and $M^{(6)}$ we would never "arrow" row 4 since neither column 5 nor column 6 has a 1 in row 4. Therefore we would never change any 0 in row 4. Thus

$$M^{(3)} = M^{(4)} = M^{(5)} = M^{(6)} = M^*.$$

Nice procedure, isn't it? It's almost this fast on a computer.

It is not true in general, though, that if $M^{(t)} = M^{(t+1)}$ then $M^{(t)} = M^*$. Try

(19.20)
$$M = \begin{matrix} 0 & 0 & 0 & 1 \\ 0 & 1 & 1 & 0 \\ 0 & 1 & 1 & 0 \\ 1 & 0 & 0 & 0 \end{matrix}$$

and see! $M^{(1)} = M^{(2)} = M^{(3)} \neq M^{(4)}$!

You can see how much faster this algorithm is than (19.11) if you also do this example using (19.11).

(19.21) *Warshall's Algorithm Mechanized*. Perhaps the best way to mechanize (19.13) is to use the following procedure. Assume M is of size n (i.e., $n \times n$), and for $i = 1, \ldots, n$ denote the ith row of M as $R_i^{(0)}$.

(19.22)

Input: $n \times n$ relation-matrix M
Output: M^*
$M^{(0)} := M$
for $t = 1, \ldots, n$ **do**
  for $i = 1, \ldots, n$ **do**
    $R_i^{(t)} := (R_i^{(t-1)}) \vee m_{i,t}^{(t-1)} \cdot (R_t^{(t-1)})$
    Row i of matrix $M^{(t)} := R_i^{(t)}$
  end
Output $M^{(n)}$
end

This procedure is taken directly from (19.14) and (19.16).

20. Complexity Analysis

We analyze our two procedures for finding transitive closure—Equation (19.11) and Warshall's algorithm (19.13)—by counting the number of operations they use. By an "operation" we mean here either of the Boolean operations + or ·,

(20.1)
$$x + y \quad \text{or} \quad x \cdot y,$$

an *or* or an *and*.

To analyze (19.11) we need to count the number of bit operations (20.1) used to multiply two $n \times n$ matrices, say A and B. Each entry in AB is the dot product (8.12) of a row of A with a column of B. Thus n *and*'s and $n-1$ *or*'s are used to compute each entry, so each entry in AB requires $2n-1$ operations. To calculate the entry in a fixed position in each of $M^2, \ldots, M^n$ thus takes $(n-1)(2n-1)$ operations. To add these n entries together takes $n-1$ *or*'s. So far, per entry we have

$$(n-1)(2n-1) + (n-1)$$

operations. There are n^2 entries, so the overall complexity of (19.11) is

$$n^2(n-1)(2n),$$

which is $O(n^4)$.

For Warshall's algorithm we use (19.14) to see that to find an entry in $M^{(t)}$ requires one *and* and one *or*. Since there are n^2 entries the complexity of getting $M^{(t)}$ is $2n^2$. Now there are n $M^{(t)}$'s to calculate, so the complexity of Warshall's algorithm is $2n^3$, which is $O(n^3)$, significantly less than for (19.11).

Notice that our analysis works equally well for any operations $+$ and $\cdot$, not merely for the Boolean.

21. The Equivalence Closure

How could we find the smallest equivalence relation including a given relation? Obviously there *is* one, because $A \times A$ is an equivalence relation on the set A, and the intersection of equivalence relations is an equivalence relation. By (18.1) any relation R on A has a unique equivalence closure. Let's call it R_{eq}.

We know R_{eq} must be reflexive, symmetric, and transitive. It must also satisfy

(21.1) $$R \subseteq R_{eq}.$$

Therefore, clearly R_{eq} must include each of the reflexive, symmetric, and transitive closures of R:

(21.2) $$I_A \cup R^c \cup R^\star \subseteq R_{eq}.$$

But the union of these closures is not necessarily symmetric or transitive.

Practice (21.3P) Find a relation R such that $I_A \cup R^c \cup R^\star$ is
(*i*) not symmetric
(*ii*) not transitive.

So the simpleminded approach above doesn't always give us R_{eq}. What would work is this:

(21.4) $$R_{eq} = (I_A \cup R \cup R^c)^\star.$$

To prove (21.4) consider first that

(21.5) $$R_{eq} \supseteq I_A \cup R \cup R^c,$$

the reflexive-symmetric closure of R. Since R_{eq} is in particular transitive, it must include the transitive closure of any subrelation:

(21.6) $$R_{eq} \supseteq (I_A \cup R \cup R^c)^\star.$$

But now:

(21.7) The transitive closure of any reflexive, symmetric relation is an equivalence relation.

It follows from (21.7) that

(21.8) $$R_{eq} \subseteq (I_A \cup R \cup R^c)^\star,$$

since the latter is an equivalence relation including R, and R_{eq} is by definition the intersection of all such equivalence relations. Therefore, by (21.6) and (21.8), (21.4) is true. (We let you prove (21.7) as a problem.)

Practice (21.9P)

(*i*) If R is transitive, is $R \cup R^c$ transitive?

(*ii*) Prove that if R is transitive, then $R \cup I_A$ is transitive.

Finally, using (21.4) is **not** the way to find R_{eq}. We'll see in Chapter 13, on trees, a good algorithm for finding R_{eq}. The trouble with (21.4) is that it treats the three attributes of an equivalence relation, namely, reflexivity, symmetry, and transitivity, too separately. In particular, the transitive closure is more complicated to compute than the equivalence closure. You'll see that the better algorithm pulls the three attributes together by recognizing the structure of an equivalence relation.

22. Appendix. Proof that Hasse Diagrams Exist

Let us use induction to prove that for any partial order on a finite set a Hasse diagram can always be drawn. First we recall the earlier result, Lemma (14.11), on the existence of minimal and maximal elements.

We'll use it to give an easy proof that Hasse diagrams exist and are unique.

Proposition (22.1) For any partial order R on a finite set A it is possible to draw the Hasse diagram (13.1) of R. Moreover, the Hasse diagram of R is unique. ■

What do we mean by "uniqueness" here? After all, a drawing can be done in many ways. We mean that the set of ordered pairs corresponding to the edges in the Hasse diagram is uniquely determined by the partial order. The same is true of the set of isolated points.

Example (22.2) If we have $A = \{2, 3, 4, 5, 6\}$ with divisibility as the partial order, we know the Hasse diagram is

(22.2F)

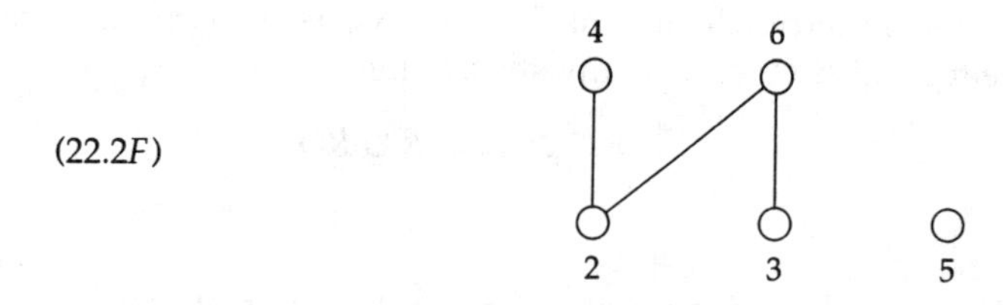

The set of ordered pairs corresponding to this Hasse diagram is $\{ (2,4), (2,6), (3,6) \}$. The set of isolated points is the singleton $\{ 5 \}$.

The ordered pair corresponding to the edge with endpoints a and b in the Hasse diagram is (a,b) if a is lower than b.

Proof of Proposition (22.1) We use induction on the number n of points in A. If $n = 1$ we draw $\circ$. This proves existence and uniqueness for $n = 1$. Let n be a fixed but arbitrary positive integer, and assume that Proposition (22.1) is true for any partial order on any set of n points. Consider now any partial order R on a set A of $n+1$ points. (So far, nothing but triviality and boilerplate.)

Use Lemma (14.11) to find a minimal element x in A. Remove x from A, leaving a set of n points. Consider the relation $R' = R \cap (A - \{ x \}) \times (A - \{ x \})$ on $A - \{ x \} = A'$. It is all the ordered pairs of R except those with x as a coordinate. R' is a partial order on A'. (You should verify this claim.) By the induction assumption, there is a unique Hasse diagram D for R' on A'.

If x is isolated, i.e., (x,x) is the only ordered pair in R with x as a coordinate, there's obviously no problem: We just add an isolated dot for x to D. For the uniqueness, we add x to the set of isolated points of R'. Since x must be in that set, any Hasse diagram for R on A must consist of the isolated dot for x and a Hasse diagram for $A' = A - \{ x \}$ and R'. By the inductive assumption the latter is unique. Hence the Hasse diagram for A is unique.

If x is not isolated, we add a dot for x below D and consider the nonempty set

$$B := \{ a;\ a \in A',\ x \mathrel{R} a \}$$

There is an induced partial order on B, namely, $(B \times B) \cap R$. Since B has no more than n points, our inductive hypothesis tells us that B has a unique Hasse diagram. Thus B has a uniquely determined set $B_{\min}$ of minimal elements. We add to D a line from x to each element of $B_{\min}$. Call the resulting diagram D^+. We claim that D^+ is the unique Hasse diagram for A.

Practice (22.3P) Prove that every pair $(c,c') \in R$ that is not represented by a line in D^+ is implied by transitivity from lines in D^+.

We leave the rest of the proof to the problems. QED

23. Further Reading

(23.1) Andrew M. Gleason, cited in Chapter 1, (20.1). Chapters 3, 5, and 6.

(23.2) Paul R. Halmos, cited in Chapter 1, (20.2).

24. Problems for Chapter 12

Problems for Section 3

3.1 Let R and S be relations on a set A.

(*i*) Prove that if $R \subseteq S$ and S is symmetric, then

$$R \cup R^c \subseteq S$$

(*ii*) Also prove that $R \cup R^c$ is symmetric.

Problems for Section 4

4.1 Consider sets A, B, and C and relations R from A to B and S from B to C. The definitions are

$$A := \{1,2,3,4,5\} \qquad B := \{a,b,c,d\} \qquad C := \{x,y,z\}.$$
$$R := \{(1,a),\ (1,c),\ (2,b),\ (3,a),\ (4,b),\ (4,c),\ (5,d)\}.$$
$$S := \{(a,x),\ (c,x),\ (d,y),\ (d,z)\}.$$

(*i*) Graph these relations with one representation of B common to both graphs.

(*ii*) On the same graph draw the composition $R \circ S$ of R and S.

(*iii*) Express $R \circ S$ as a set of ordered pairs.

4.2 Give an example of two relations R, S on a nonempty set A such that $R \circ S \neq S \circ R$.

4.3 Let A and B be sets. For which sets is it true that

$$(A \times B) \circ (B \times A) = A \times A?$$

For which sets is it false? Explain.

4.4[Ans] Let R be a reflexive relation on the set A which is not necessarily symmetric or transitive. Let R^c denote the converse of R.

(*i*) Show R^c is reflexive.

(*ii*) Show $R \circ R^c$ is symmetric ($\circ$ denotes composition).

(*iii*) Prove or disprove: $R \circ R^c$ is transitive.

So: Is $R \circ R^c$ necessarily an equivalence relation?

4.5 Let R be a reflexive relation. Prove that

$$R \cup R^c \subseteq R \circ R^c.$$

4.6[Ans] Give a concise description of the domain and range of $R \circ S$ in terms of those of R and S. Assume that R is a relation from A to B and S is a relation from B to C.

4.7 Prove (4.10).

Problems for Section 6

6.1 Consider the relation $R = \{(1,2),(1,3),(1,1),(2,1),(3,2)\}$. Write the relation-matrix of R

(*i*) if R is defined on $\{1,2,3\}$.

(*ii*) if R is defined on $\{1,2,3,4,5\}$.

6.2 If R is a linear order relation on $A := \{1,\ldots,n\}$ and if $k\ R\ (k+1)$ for all $k = 1,2,\ldots,n-1$, prove that $M(R)$ is a matrix with nothing but 0's below the main diagonal (when the rows and columns are numbered in the order $1,2,\ldots,n$). Determine all entries in $M(R)$.

Problems for Section 7

7.1 Prove that definitions (7.5), (7.7), and the two in (7.12) are mutually equivalent.

7.2 Write the transposes of M_1 and M_2 of Example (7.4).

Problems for Section 8

8.1 Find the matrix product M_1M_2 using first addition in $\mathbf{Z}$ and then Boolean addition. Show your calculation of row 3 in detail.

$$M_1 := \begin{matrix} 1 & 0 & 1 & 0 & 1 \\ 0 & 1 & 0 & 1 & 0 \\ 0 & 1 & 0 & 1 & 1 \\ 1 & 0 & 1 & 1 & 0 \\ 1 & 1 & 1 & 0 & 1 \end{matrix} \qquad M_2 := \begin{matrix} 0 & 0 & 1 & 1 & 0 \\ 1 & 0 & 1 & 1 & 1 \\ 1 & 1 & 1 & 0 & 1 \\ 1 & 1 & 0 & 1 & 0 \\ 1 & 0 & 0 & 1 & 1 \end{matrix}.$$

8.2 Calculate the matrix product LM, where and $+$ and $\cdot$ are defined as follows:

$$L = \begin{matrix} 1 & 1 & 0 & 1 \\ 0 & 0 & 1 & 1 \\ 0 & 1 & 0 & 1 \\ 1 & 0 & 1 & 0 \end{matrix} \qquad M = \begin{matrix} 1 & 0 & 0 & 1 \\ 0 & 1 & 0 & 1 \\ 1 & 0 & 1 & 0 \\ 1 & 1 & 0 & 1 \end{matrix},$$

(*i*) the usual operations in $\mathbf{Z}$

(*ii*) $\oplus$ and $\otimes$, resp., in $\mathbf{Z}_2$

(*iii*) Boolean sum and product.

8.3 Prove that the transpose of a sum of matrices (of the same size) is the sum of the individual transposes. "Sum of matrices" is defined in problem 8.5.

8.4 Define the Boolean sum of relations on the same set as follows:

$$R_1 + R_2 := R_1 \cup R_2.$$

Thus $M(R_1 + R_2) = M(R_1) + M(R_2) := (a_{ij} + b_{ij})$, where $M(R_1) = (a_{ij})$ and $M(R_2) = (b_{ij})$ and the $+$ stands for Boolean sum. Prove that composition distributes over Boolean sum:

$$R \circ (R_1 + R_2) = (R \circ R_1) + (R \circ R_2).$$

8.5 Prove that matrix multiplication distributes over matrix addition in general. Matrix addition is defined as

$$(a_{ij}) + (b_{ij}) := (a_{ij} + b_{ij}),$$

where the two matrices are the same size and the + stands for the addition given for the underlying set.

Problems for Section 9

9.1 Calculate the relation-matrix of the composition $R \circ S$ where

$$M(R) = \begin{matrix} 1 & 0 & 0 & 0 & 1 \\ 0 & 1 & 0 & 0 & 1 \\ 0 & 1 & 1 & 0 & 1 \\ 1 & 0 & 1 & 0 & 0 \end{matrix} \quad \text{and} \quad M(S) = \begin{matrix} 0 & 1 & 0 \\ 1 & 1 & 0 \\ 0 & 0 & 1 \\ 1 & 0 & 0 \\ 1 & 0 & 1 \end{matrix}.$$

9.2 Use matrix multiplication to find the relation-matrix of the composition $R \circ S$ if

$$M(R) = \begin{matrix} 1 & 0 & 1 & 0 & 1 \\ 0 & 1 & 1 & 1 & 0 \\ 0 & 1 & 0 & 0 & 1 \end{matrix} \qquad M(S) = \begin{matrix} 0 & 1 & 1 & 0 \\ 1 & 1 & 0 & 0 \\ 0 & 0 & 0 & 1 \\ 1 & 0 & 1 & 0 \\ 0 & 1 & 0 & 0 \end{matrix}.$$

Here we take R as a relation from some 3-set A to a 5-set B, and S as a relation *from* B to some 4-set C.

9.3 Calculate the matrix product of the two matrices in Problem 9.2 using real-number addition and multiplication (instead of Boolean).

9.4 Prove Theorems (9.6) and (4.6) as corollaries of Theorems (9.9), (7.13), and (9.1). You may assume that the sets on which the relations are defined are finite.

***9.5** Do (9.13 *iii*).

Problems for Section 10

10.1 Prove that the converse of a partial order A is a partial order.

10.2 (*i*) Find all relations that are symmetric and antisymmetric.

(*ii*) How many are there on an n-set?

10.3 (*i*) Give an example of an antisymmetric relation having no loops.

(*ii*) How many such relations are there on a set of n points?

10.4 [Ans] The complement R' of a relation R on A is $R' := A \times A - R$.

(*i*) Determine whether the complement of a reflexive relation is reflexive.

(*ii*) Same question for symmetry.

(*iii*) Same for transitivity.

(*iv*) Same for antisymmetry.

10.5 Consider the set A of all strings in 0,1. τ is the empty string. Order them lexicographically; that is,

$$\tau \preccurlyeq \alpha \quad \text{for all strings } \alpha \text{ in } A$$
$$\alpha \preccurlyeq \beta \quad \text{iff } \alpha = \gamma\sigma \text{ and } \beta = \gamma t$$

where γ is the longest common prefix of both α and $\beta, and (\sigma = \lambda)$ or (σ starts with a 0 and t starts with a 1).

(*i*) Prove that $\preccurlyeq$ is a partial order on this set A.

(*ii*) Prove that $\preccurlyeq$ is a linear order on A.

10.6 Let R be a partial order on the set A. Also prove $\preccurlyeq$ is a linear order on A. Let S be a partial order on the set B. Define a relation T on $A \times B$ as follows:

$$(a,b)\ T\ (x,y) \quad \text{iff} \quad a\ R\ x \text{ and } b\ S\ y.$$

Prove that T is a partial order on $A \times B$.

Problems for Section 11

11.1 Consider the set X of all strings of 0's and 1's of length 4. Define a function wt on X as follows:

$$\forall a \in X,\ wt(a) := \text{the total number of 1's in } a.$$

Define a relation R on X via this rule: For all a, b in X, $a\ R\ b$ iff $wt(a)$ divides $wt(b)$.

(*i*) Prove that R is reflexive and transitive.

(*ii*) Find the *scc*'s of R.

(*iii*) Draw the graph of the partial order induced by R on the set of *scc*'s.

11.2 With the same set X as in the previous problem, define a relation S as follows: Letting $a = a_1a_2a_3a_4$ and $b = b_1b_2b_3b_4$ be in X with a_i, b_i in $\{0, 1\}$ for all i, define

$$aSb \text{ iff } n(a_1a_2a_3) \le n(b_1b_2b_3),$$

where $n(000) := 0, n(001) := 1$ in $\mathbf{Z}$, $n(010) := 2$ in $\mathbf{Z}$, etc. Thus the function n maps a string s of 0's and 1's to the integer of which s is the base-2 representation.

(*i*) Prove that S is reflexive and transitive.

(*ii*) Find the *scc*'s of S.

(*iii*) Draw the partial order induced by S on the set of *scc*'s, but omit the arrows implied by transitivity.

11.3 Let R be the relation on $B = \{ 2^i 3^j 5^k;\ 0 \le i,j,k \le 2 \}$ defined by the rule: for all a, b in B,

$a\ R\ b$ iff every prime dividing a divides b.

(*i*) Prove that R is reflexive and transitive.

(*ii*) Determine the strongly connected components (*scc*'s) of R.

(*iii*) Draw the graph of the partial order on the set of *scc*'s induced by R.

11.4 Prove that every *scc* C of a reflexive transitive relation R is a complete directed graph. This means: $\forall a, b \in C, a\ R\ b$ and $b\ R\ a$.

Problems for Section 13

13.1 On $\mathbf{N}$ define the relation $R := \{ (a,b);\ a, b \in \mathbf{N}$ and b divides $a \}$.

(*i*) Prove that R is a partial order.

(*ii*) Draw the Hasse diagram for R restricted to the points $1, \ldots, 12$.

For the following two problems prove or disprove that the relation defined by the given relation-matrices is a partial order. Draw the graph of the relation; if it is a partial order, draw the Hasse diagram instead.

13.2

	a	b	c	d
a	1	1	0	1
b	0	1	0	1
c	1	0	1	0
d	0	0	1	1

13.3

	a	b	c	d	e
a	1	0	1	1	1
b	0	1	0	1	1
c	0	0	1	0	1
d	0	0	0	1	1
e	0	0	0	0	1

Problems for Section 14

14.1 Write and explain the relation-matrix of a linear order on a five-point set.

14.2 Suppose L_1 and L_2 are two partial orders on the same set A. Prove that $L_1 \cap L_2$ is a partial order on A.

14.3 Let L_1 and L_2 be linear orders on A. Define R on $A \times A$ as follows: For all $(a,b), (x,y)$ in $A \times A$

$$(a,b)\ R\ (x,y) \quad \text{if } a L_1 x, \qquad \text{or} \quad \text{if } a = x, \quad \text{then } b\ L_2\ y.$$

Prove that R is a linear order on $A \times A$.
This is called **lexicographic order**. If $L_1 = L_2$ is alphabetical order and A is the alphabet, then R is alphabetical order on all strings of length 2 over A. Extended to strings of any length this is the usual alphabetic order. If $L_1 = L_2$ is the magnitude relation $\le$ on $A = \{0, 1, \ldots, 9\}$, then this order is the magnitude relation on 2-digit numbers expressed in base 10. (Compare problem 10.5.)

14.4 [Ans] Let L_1 and L_2 be linear orders on A.

(*i*) Prove: If $L_1 \subseteq L_2$, then $L_1 = L_2$.

(*ii*) Prove that the intersection of two different linear orders on the same set is never a linear order.

14.5 (*i*) Prove that every finite partially ordered set A can be **topologically sorted**; this means there is a linear order L on A such that

$$\forall a, b \in A \qquad a\ R\ b \text{ implies } a\ L\ b,$$

where R is the given partial order on A.

(*ii*) With the same notation, prove that if the rows and columns are ordered by L, then $M(R)$ is 0 below the main diagonal.

14.6 Suppose R is a partial order on the finite set A. Let L be a topological sort of R (see Problem 14.5). Prove that

(*i*) The Hasse diagram of L^c is the upside-down version of that for L;

(*ii*) L^c is a topological sort of R^c.

Problems for Section 15

15.1 Prove that the ordered set $D \times D$ of (15.12) has no infinite descending chains. [Suggestion: Prove that any descending chain is finite.]

15.2 Prove that the minimum condition (15.1) is equivalent to the nonexistence of infinite descending chains (15.5).

Problems for Section 16

16.1 Determine whether the converse of (16.6) is true.

16.2 Prove (16.10).

Problems for Section 17

17.1 Let R be transitive. Prove that R^c is transitive. Use the matrix condition (17.2).

17.2 Prove that a relation R on A is

(*i*) transitive iff $(R \circ R) \subseteq R$

(*ii*) antisymmetric iff $R \cap R^c \subseteq I_A$. [I_A is the diagonal of $A \times A$. See (17.9).]

17.3[Ans] Consider the relations on a 2-set.

(*i*) How many are there?

(*ii*) How many are transitive?

Find a systematic way to answer this question; do not use exhaustion.

17.4 Prove that if the relation R on A is reflexive and transitive, then $R = R^2$. Show that the converse is false even if R has no isolated points (i.e., even if

$$(\text{domain of } R) \cup (\text{range of } R) = A).$$

17.5 Use the relation-matrix M to find all the nontransitive relations on a 2-set. [Hint. Let

$$M = \begin{bmatrix} a & b \\ c & d \end{bmatrix}.$$

Calculate M^2 with Boolean operations. Find all cases where a 0 in M becomes a 1 in M^2. These are the nontransitives.] Explain.

Problems for Section 18

18.1 Define the function $f : D \longrightarrow \mathbf{Z}$ by the rule, for all x in $D, f(x) = x^2 + 1$, where $D = \{0, \pm 1 \pm 2, \ldots, \pm 5\}$. Consider f as a relation.

(*i*) Express the converse of f as a set of ordered pairs.

(*ii*) What is the transitive closure of f? Explain.

18.2 Suppose the graph of the relation R on A has a circuit of length m. This means there are m ordered pairs in R of the form $(a_1, a_2), (a_2, a_3) \ldots, (a_m, a_1)$. The points $a_1, \ldots, a_m$ need not be all different. Prove that the transitive closure of R must include the complete relation on $\{a_1, \ldots, a_m\}$.

18.3 Prove the uniqueness of the P-closure in Lemma (18.1).

Problems for Section 19

19.1 Give an example of a relation R on an n-point set for which the transitive closure R^* of R is $R^* = R \cup \cdots \cup R^n$ and $R^* \neq R \cup \cdots \cup R^{n-1}$. Thus it is necessary in some cases to compute R^n to get the transitive closure of R.

19.2[Ans] Prove that in computing the transitive closure R^* of a relation R on the set A as

$$R^* = R \cup R^2 \cup \cdots = \bigcup_{1 \le i} R^i$$

you can stop as soon as you get no new 1's. That is, if $R^j \subseteq R \cup \cdots \cup R^{j-1}$ and $j \geq 2$, show that $R^* = R \cup \cdots \cup R^{j-1}$. (See the remark after (19.11).)

19.3 (*i*) Find the transitive closure of the relation with relation-matrix M. Use the method indicated in Problems 19.1 and 19.2.

$$M = \begin{matrix} 1 & 0 & 1 & 0 & 1 & 0 & 0 \\ 0 & 1 & 0 & 0 & 0 & 0 & 0 \\ 0 & 0 & 1 & 1 & 1 & 0 & 0 \\ 0 & 1 & 0 & 0 & 0 & 0 & 0 \\ 1 & 0 & 0 & 0 & 1 & 0 & 0 \\ 0 & 0 & 0 & 0 & 0 & 1 & 0 \\ 1 & 0 & 0 & 0 & 0 & 0 & 1 \end{matrix}$$

(*ii*) Draw the graphs of the original relation and of the transitive closure.

(*iii*) Find the strongly connected components of the transitive closure and draw the Hasse diagram of the partial order on the *scc*'s.

19.4 Use Warshall's algorithm to find the matrix of the transitive closure of the relation with matrix M in Problem 19.3. Show your steps.

19.5 Use Warshall's algorithm to find the matrix of the transitive closure for the relation with matrix M_1.

$$M_1 = \begin{matrix} 1 & 0 & 0 & 0 & 1 \\ 0 & 1 & 1 & 1 & 0 \\ 1 & 0 & 0 & 0 & 0 \\ 0 & 0 & 0 & 1 & 0 \\ 0 & 0 & 1 & 0 & 0 \end{matrix}$$

19.6 Let R be a relation with relation-matrix M, where

$$M = \begin{matrix} 0 & 1 & 0 & 0 & 0 \\ 0 & 0 & 1 & 0 & 0 \\ 1 & 0 & 0 & 0 & 0 \\ 0 & 1 & 0 & 0 & 0 \\ 0 & 0 & 0 & 1 & 0 \end{matrix}.$$

(*i*) Use Warshall's algorithm to find M^*, the relation-matrix of the transitive closure of R.

(*ii*) Verify that the M^* you found is the matrix of a transitive relation including R.

(*iii*) Draw the graph of R with solid arrows. Then use dashed lines to fill in the extra arrows from R^*.

19.7 Let R be a reflexive relation on an n-point set. Prove that the transitive closure of R is

$$R \cup \cdots \cup R^{n-1}.$$

Problems for Section 21

21.1 If R is transitive, then $I_A \cup R \cup R^c$ is transitive. True or false? Explain.

21.2 Prove (21.7).

Problem for Section 22

22.1 Complete the proof of Proposition (22.1) by proving the following:

(*i*) The placement of the new lines from x into D^+ does not make any line of D redundant.
(*ii*) None of the lines from x in D^+ is redundant.
(*iii*) D^+ is a Hasse diagram for A and is unique.

Test Questions

T1. Consider the relation-matrix M of the relation R on $A = \{a, b, c, d, e\}$ defined as

$$M = \begin{matrix} 1 & 0 & 1 & 0 & 1 \\ 0 & 1 & 0 & 1 & 0 \\ 0 & 0 & 1 & 0 & 1 \\ 0 & 0 & 0 & 1 & 1 \\ 0 & 0 & 1 & 0 & 1 \end{matrix}$$

Is the relation R a partial order on A? Explain.

T2. With R defined as in the previous problem, let S be the relation on A with relation-matrix N defined below. Find the relation-matrix of the composition $S \circ R$.

$$N = \begin{matrix} 1 & 0 & 1 & 0 & 0 \\ 0 & 1 & 0 & 1 & 0 \\ 0 & 0 & 1 & 0 & 1 \\ 1 & 0 & 0 & 1 & 0 \\ 0 & 1 & 0 & 0 & 1 \end{matrix}$$

Show your calculation for two rows.

T3. Prove that the converse of a partial order is a partial order. [Recall that the converse R^c of a relation R is defined as $R^c := \{(b, a);\ (a, b) \text{ in } R\}$.]

T4. Calculate the relation-matrix of the composition $R \circ S$, where

$$M(R) = \begin{matrix} 1 & 0 & 1 & 0 & 0 \\ 0 & 1 & 0 & 1 & 0 \\ 0 & 0 & 1 & 0 & 1 \\ 1 & 0 & 0 & 1 & 0 \\ 0 & 1 & 0 & 0 & 1 \end{matrix} \qquad M(S) = \begin{matrix} 1 & 0 & 1 & 1 & 1 \\ 1 & 1 & 0 & 0 & 1 \\ 1 & 0 & 0 & 1 & 0 \\ 1 & 0 & 1 & 1 & 0 \\ 0 & 0 & 0 & 0 & 1 \end{matrix}$$

T5. Consider the relation R with relation-matrix

$$M(R) = \begin{matrix} 1 & 1 & 0 & 1 & 1 \\ 0 & 1 & 0 & 0 & 0 \\ 0 & 1 & 1 & 0 & 0 \\ 0 & 0 & 0 & 1 & 1 \\ 0 & 0 & 0 & 0 & 1 \end{matrix}.$$

Prove that R is a partial order relation. Draw the Hasse diagram of R.

T6. Let R be a reflexive relation. Prove that $R \subseteq R^2$. (Note. R^2 is by definition the composition $R \circ R$.)

25. Answers to Practice Problems

(4.3P) As usual, to prove two sets equal we show that each is a subset of the other.
So let $(a, c) \in R \circ S$. Then $\exists b \in B = \mathbf{N}$ such that

$$(a, b) \in R \quad \text{and} \quad (b, c) \in S.$$

These are defined to mean

$$2 \mid a \quad \text{and} \quad 2 \mid b; \quad \text{and} \quad 3 \mid b \quad \text{and} \quad 3 \mid c,$$

respectively. In particular, $2 \mid a$ and $3 \mid c$, so $(a, c) \in D_2 \times D_3$. Conversely, suppose $(x, y) \in D_2 \times D_3$. Then $2 \mid x$ and $3 \mid y$. Let $u \in \mathbf{N}$ satisfy $6 \mid u$. Then u is even, so $(x, u) \in R$; also $3 \mid u$, so $(u, y) \in S$. Thus $(x, y) \in R \circ S$.

(4.5P) This happens iff

$$(\text{range of } R) \cap (\text{domain of } S) = \varnothing,$$

so we may find many examples. We could take $R = \{(1, 1)\}$ and $S = \{(2, 2)\}$.

(5.6P) A function is injective if in its graph no point of the codomain has more than one in-arrow. It is surjective if every point of the codomain has at least one in-arrow.

(6.3P) For each $a \in A$ there has to be a 1 in row a in order that the function be defined at a. But if there are two or more 1's you violate the "singlevalued" condition for a function, (5.2) of Chapter 5.

If the matrix has at most one 1 in each row, then it represents a partial function from A to B; its domain is a subset of A.

(8.9P) Since $\mathbf{x} + \mathbf{y} = (x_1 + y_1, \ldots, x_r + y_r), a \cdot (\mathbf{x} + \mathbf{y}) = (a \cdot (x_1 + y_1), \ldots, a \cdot (x_r + y_r))$ by (8.7). By the distributive law (8.4), this is $(a \cdot x_1 + a \cdot y_1, \ldots, a \cdot x_r + a \cdot y_r)$, which equals

$$(a \cdot x_1, \ldots, a \cdot x_r) + (a \cdot y_1, \ldots, a \cdot y_r)$$

by (8.6). Using (8.7) again we see that this is $a \cdot \mathbf{x} + a \cdot \mathbf{y}$.

The other formula is proved similarly.

(11.4P) Symmetry of argument results from interchanging a and b throughout. To justify that here:

(*i*) Our initial hypothesis, $scc(a) \cap scc(b) \neq \varnothing$, is unaltered by the interchange of a and b. From that hypothesis, using the definition of scc, of course, we proved $scc(a) \subseteq scc(b)$. We showed

$$(\star)\ \forall a, b \in A \quad scc(a) \cap scc(b) \neq \varnothing \longrightarrow scc(a) \subseteq scc(b).$$

It follows immediately from $(\star)$ that $\forall a, b \in A$, $scc(a) \cap scc(b) \neq \varnothing \longrightarrow scc(b) \subseteq scc(a)$.

We don't even have to imagine going through the whole proof with a and b interchanged.

(11.7P) There are 8 scc's, one for each choice of the triple $(i,j,k) \in \{0,1\}^3$. For example, $(i,j,k) = (1,1,0)$ corresponds to the scc

$$S_{110} := \{2^i 3^j;\ 1 \leq i,j \leq 10\};$$

the exponent on 5 is 0 in all of these and the i and j are both positive. Thus 2 and 3 divide each element of S_{110}. If we identify each scc by its smallest element we get $S_{000} = scc(1) = \{1\}$, $S_{001} = scc(5) = \{5^k;\ 1 \leq k \leq 10\}$, and so on. The graph, with loops left out, is

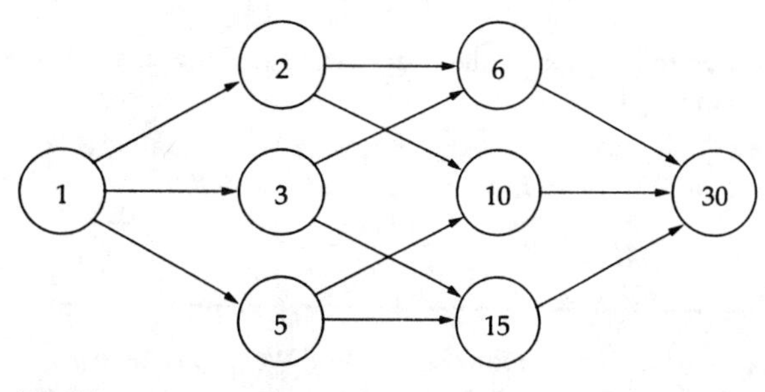

(25.1*F*)

So the induced partial order is just divisibility on the least elements of the scc's.

***(14.10P)** We can't take the sets to be subsets of a finite set because there would be minimal and maximal elements. We might try the collection of all finite subsets of **N**; it has no maximal elements, but it has $\varnothing$, a minimal element. One solution (you should verify it) is the collection of all infinite subsets X of **N** such that $\mathbf{N} - \mathrm{X}$ is also infinite. For instance, $\mathrm{X} := \{2n - 1;\ n \in \mathbf{N}\}$.

(15.13P) Reflexivity is easy: $(a,b) \prec (a,b)$, because the two first coordinates are equal and the second coordinates, being equal, satisfy $b \leq b$, the second alternative of the definition (15.12).

Antisymmetry: $(a,b) \prec (x,y)$ and $(x,y) \prec (a,b)$ imply $a = x$ and $b = y$ because:

Since $a, b, x, y \in \mathbf{N}$ and $\leq$ is a total order on **N**, only one of the conditions

$$a < x,\ a = x,\ a > x$$

holds. Of these only $a = x$ allows both

(*i*) $(a,b) \prec (x,y)$ and
(*ii*) $(x,y) \prec (a,b)$

to hold. So $a = x$.

Now use the second alternative of (15.12): $b \leq y$ from (*i*), and $y \leq b$ from (*ii*). Therefore, by the antisymmetry of $\leq$, $b = y$.

Transitivity: Suppose $(a,b) \prec (x,y) \prec (u,v)$. Target: Show $(a,b) \prec (u,v)$.

Since (15.12) implies $a \leq x$ and $x \leq y$, we see that $a \leq u$. If $a < u$, then $(a,b) \prec (u,v)$ without further discussion.

If $a = u$, then $x = a = u$ from antisymmetry of $\leq$, since $a = u$ implies $a \leq x$ and $x \leq a$. Now the second alternative of (15.12) gives us $b \leq y$ and $y \leq v$, hence $b \leq v$ by transitivity of $\leq$. Therefore $(a,b) \prec (u,v)$ in this case also. The same kind of reasoning shows that $\prec$ is a linear order.

The Hasse diagram for $a, b \in \{0,1,2\}$ is

o (2,2)
|
o (2,1)
|
o (2,0)
|
o (1,2)
|
o (1,1)
|
o (1,0)
|
o (0,2)
|
o (0,1)
|
o (0,0)

(17.5P) (*i*) Suppose $R^2 \subseteq R$. Then every "1" in $M(R^2)$ appears in $M(R)$. But $M(R^2) = M(R)^2$ by Theorem (9.1), so $M(R)^2 + M(R) = M(R)$. Thus $R^2 \subseteq R$ implies $M(R^2) + M(R) = M(R)$. The converse is equally obvious. Therefore $R^2 \subseteq R$ iff $M(R)^2 + M(R) = M(R)$. It follows that (17.1), which says (in brief)

$$\text{Transitivity} \quad \text{iff} \quad R^2 \subseteq R$$

is equivalent to (17.2),

$$\text{Transitivity} \quad \text{iff} \quad M(R)^2 + M(R) = M(R).$$

(*ii*) $M(R)^2 \wedge M(R) = M(R)^2$ is simply another way of saying that every "1" in $M(R)^2$ is in $M(R)$. We just showed that that condition is equivalent to "$R^2 \subseteq R$."

(19.2P) $R^1 := R \circ I_A$. Therefore

$$(a,b) \in R \text{ implies } (a,b) \circ (b,b) = (a,b) \in R^1$$

if we may temporarily abuse the notation. Thus $R \subseteq R^1$.

Conversely, $(a,b) \in R^1$ implies

$$\exists x \in A \quad (a,x) \in R \quad \text{and} \quad (x,b) \in I_A$$

but $(x,b) \in I_A$ forces $x = b$. Therefore $(a,b) \in R$, so $R^1 \subseteq R$. That is, $R = R \circ I_A = I_A \circ R$. QED

(19.6P) We'll do (19.5) in terms of the relation-matrices. Set $M := M(R)$, $M' := M(R')$, and $M'' := M(R'')$. Then

$$M(R \circ (R' \cup R'')) = M(M' + M'') = MM' + MM''$$

by the distributivity of matrix multiplication over matrix addition.

This yields (19.4) as follows:

$$\begin{aligned}(R^\star)^2 &:= R^\star \circ (R \cup R^2 \cup \cdots) \\ &= (R^\star \circ R) \cup (R^\star \circ R^2) \cup \cdots \\ &= (R^2 \cup R^3 \cup \cdots) \cup (R^3 \cup R^4 \cup \cdots) \cup \cdots,\end{aligned}$$

where we have just used the turned-around version of (19.5),

$$(R' \cup R'') \circ R = (R' \circ R) \cup (R'' \circ R).$$

The above winds up as

$$R^{\star 2} = R^2 \cup R^3 \cup \cdots.$$

(19.10P) If $j = i+1$, there is a loop at a_i, since $a_i = a_j = a_{i+1}$. You remove only the loop, shortening the length by 1.

(21.3P) (*i*) Take $R = \{(a,b), (b,c)\}$. Then $I_A \cup R^c \cup R^\star$ is not symmetric, because it has (a,c) but not (c,a). (Draw the graph.) It is also not transitive.

(*ii*) Take $R = \{(a,b), (c,b)\}$. Then $S := I_A \cup R^c \cup R^\star$ is not transitive, because $R^\star = R$. Therefore $S = I_A \cup R \cup R^c$, and $(a,b), (b,c) \in S$ but $(a,c) \notin S$. (Again it helps to draw the graph.)

(21.9P) (*i*) Not necessarily. Consider the relation $R = \{(a,b)\}$. It is vacuously transitive, but $R \cup R^c = \{(a,b), (b,a)\}$ is not transitive because it lacks the loops (a,a) and (b,b).

(*ii*) We use Lemma (17.1), in its matrix form (17.2). Let $M = M(R)$, and let I denote $M(I_A)$; thus I is the identity matrix. Then we want to show that $(M+I)^2 \wedge (M+I) = (M+I)^2$, i.e., that every "1" in $(M+I)^2$ is already in $M+I$. Since matrix multiplication obeys the distributive law,

$$\begin{aligned}(M+I)^2 &= M^2 + MI + IM + I^2 \\ &= M^2 + M + I\end{aligned}$$

since $MI = IM = M$ and since the $+$ stands for the Boolean sum. But since $R^2 \subseteq R$ (from: R is transitive), we know that $M^2 + M = M$. Therefore

$$(M+I)^2 = M + I.$$

So $(M+I)^2 + (M+I) = M+I$, as we wanted to show.

(22.3P) We are given $c\ R\ c'$. If both c and c' are in A', then the conclusion is true because D is a Hasse diagram for A', by the inductive hypothesis. The only remaining possibility is $x = c$ or $x = c'$.

If $x = c$, then $c' \notin B_{\text{min}}$. There is thus a b in B_{min} such that $b\ R\ c'$. (Explain the last statement.) Thus $x\ R\ b$ is a line in D^+, and $b\ R\ c'$ is a path in D, hence in D^+. So $c\ R\ c'$ is implied by transitivity from lines in D^+.

If $x = c'$, then also $x = c$, since x is minimal.

CHAPTER

13

Trees

1. Binary Search

(1.1) You play the game Twenty Questions,[†] trying to guess an unknown integer n satisfying $1 \leq n \leq N$. How large may N be so that you can still find n?

Answer: If $N = 2^{20}$ or less, you can be sure to win. Here's how (you probably know this already): First ask, "Is $n \leq 2^{19}$?" If the answer is "yes," then ask "Is $n \leq 2^{18}$?" If the answer to the first question is "no," then ask, "Is $n \leq 2^{19} + 2^{18}$?" And so on.

The strategy is to cut the unknown set S in half with each question. If you have an unknown set of size 2^q, then a best yes-or-no question reduces it to size 2^{q-1}. One question for the set $S = \{i+1, \ldots, i+2j\}$ is this: Is $n \leq i+j$? That is, is n in the first half of the set S? If "yes," you replace S by its first half; if "no," you replace S by its second half. And then you ask the same question for the new S. At the end you've reduced the set S containing n to a singleton, so you know n.

It is easy to see by induction that if you are allowed q questions, then you may take N as large as 2^q and be assured of finding n. No larger value of N allows you to be certain of winning every time. This strategy for playing Twenty Questions is an example of *binary search*. You seek an element x of a set, and you find it efficiently by dividing the set in two pieces as nearly equal in size as possible and discarding the piece not containing x. You keep repeating the process, each time on a set half the size of the one before.

Example Suppose you played Three Questions: $N = 8$ and you are to find a number n, $1 \leq n \leq 8$, with three queries. Here is a diagram showing what question to ask at each stage.

[†]You may ask up to 20 "yes or no" questions.

(1.1F)

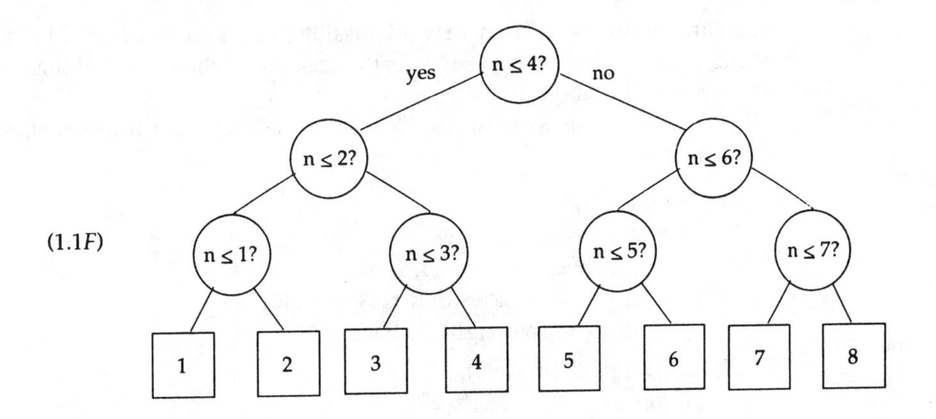

If the answer is "yes," go left. If "no," go right. The value of n is in the box. Your first question is, "Is $n \leq 4$?" Suppose the answer is "no." Your second question is then, "Is $n \leq 6$?" Suppose the answer is "yes." You then ask, "Is $n \leq 5$?" If "yes," then $n = 5$; if "no," then $n = 6$.

Your first question reduces the set containing n to $\{5,6,7,8\}$, your second to $\{5,6\}$.

(1.2) The same procedure would work in the following situation: Some universities and other large institutions control long-distance calls by assigning to each authorized user his or her own secret number. The user must then dial that number to get a long-distance line. There could easily be 1,000 such secret numbers. What is an efficient way for the system to check someone's input number to see if it's genuine? *Binary search* is a good answer.

The system plays the same game we described at the beginning, except from the opposite viewpoint; knowing the number n (your input), it proceeds to find out whether n is in its set S of assigned secret numbers. S is now a set of, say, 1,000 integers less than 1,000,000. It is no longer a set of consecutive integers, but no matter.

Example (1.3) Consider the small example $S = \{7,13,22,35,41,59,64,77,82,95,106,111\}$. We organize the system according to the following diagram, called T_S:

(1.2F)

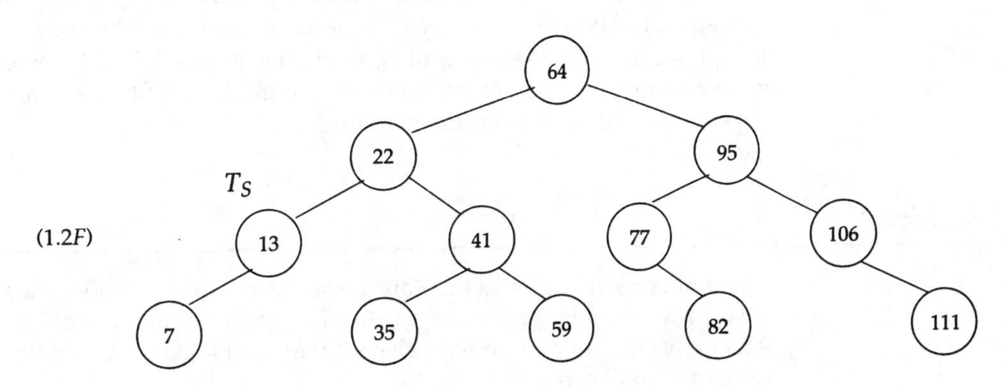

You can see that 64 is (as nearly as possible) the median of S; 22 is (ditto) the median of the subset of S less than 64; 95 is the median of the subset of S greater than 64, and so on down the diagram.

How does the system use T_S? Someone dials in a number n, supposedly in S; the system

1. Checks if $n < 64$:
   If yes, it goes left, to 22.
   If no, it checks if $n = 64$.
      If yes, it decides that $n \in S$; it stops.
      If no, it goes right, to 95.
2. Checks if $n < 22$ [$n < 95$].
   If yes, it goes to 13 [to 77].
   If no, it checks if $n = 22$ [$n = 95$].
      If yes, it decides that $n \in S$; it stops.
      If no, it goes to 41 [to 106].

Each time it goes to a number s in T_S, the system tests if $n < s$. If so, it goes to the number below *and left*; if not, it tests if $n = s$. If n is s, it lets the person have a long-distance line, and then stops. If $n \neq s$, then it goes to the number below *at the right*. The process repeats until n is identified as an element of S or is exposed as a fraud.

What's actually in the computer closely resembles the diagram (1.2*F*). Each number in T_S, for example, 64, is represented as a word with three *fields*.

(1.3*F*)

address of 22	64	address of 95

The first field contains the *address* of the word for 22; the second contains the number, here 64; the third, the address of the word for 95. These addresses, called *pointers*, tell the computer where to go next when n is not equal to the number in the middle field. Both pointers with each of the numbers at the bottom, for example, 7, would point to the instruction "Deny service." So would the right pointer for 13 and the left pointers for 77 and 106.

The diagram in (1.2*F*) is called a **binary search tree**. It grew naturally from our strategy for playing the Twenty Questions game of (1.1). It's just a way of organizing the questions to be asked. A similar diagram (Chapter 8, (9.6*F*)) helped us understand zipper merge. Another situation where you might want to use a binary search tree is playing Go Fish with a million-card deck.

2. Trees

The diagrams (1.1*F*) and (1.2*F*) are instances of *trees*, probably the most commonly used idea in computer science for the storage and manipulation of data. Trees have a recursive definition that is so elegant that at first glance you'd think the foregoing examples don't come under it.

DEFINITION (2.1) A **tree** is an ordered triple

$$T = (N(T),\ r(T),\ L(T)),$$

where
$N(T)$ is a nonempty finite set;
$r(T) \in N(T)$; and
$L(T)$ is a collection of trees such that $\{ N(X); X \in L(T) \}$ is a partition of $N(T) - \{ r(T) \}$.

$N(T)$ is the set of **nodes** of the tree T; $r(T)$ is the **root** of T; $L(T)$ is the collection of **immediate subtrees** of T.†

This forbiddingly abstract definition is surprisingly easy to use in making inductive proofs about trees. Even for such proofs, however, a picture is worth, say, 256 words. We develop the interplay between the definition and the pictures in the examples below.

(2.2) *Criterion for equality.* Two trees T_1 and T_2 are **equal** iff their triples are equal: $T_1 = T_2$ iff $(N(T_1),\ r(T_1),\ L(T_1)) = (N(T_2),\ r(T_2),\ L(T_2))$. This condition means, of course, $N(T_1) = N(T_2)$, $r(T_1) = r(T_2)$, and $L(T_1) = L(T_2)$. This criterion is a logical consequence of the definition of a tree as a triple.

The definition (2.1) defines *tree* in terms of *tree,* but it works because any trees in $L(T)$ are smaller than T; i.e., their sets of nodes are proper subsets of $N(T)$. To make this claim believable we show that in the ultimate case, when $N(T)$ is a singleton, say $\{ a \}$, then (2.1) defines T clearly: We find $T = (\{ a \}, a, \varnothing)$. Why? Since the root $r(T) \in N(T)$, $r(T)$ must be a. Then $L(T)$ is necessarily empty, because if X were a tree in $L(T)$ then the root $r(X)$ would have to be in $N(T) - \{ r(T) \} = \varnothing$. We diagram this tree as ⓐ, despite the fact that the definition says nothing about drawings.

Examples (2.3) There are two trees on the set of nodes $\{ a, b \}$. One is $T_1 = (\{ a, b \}, a, \{ X \})$, where $X = (\{ b \}, b, \varnothing)$. The other is $T_2 = (\{ a, b \}, b, \{ Y \})$, where $Y = (\{ a \}, a, \varnothing)$. We diagram them as

(2.4F)

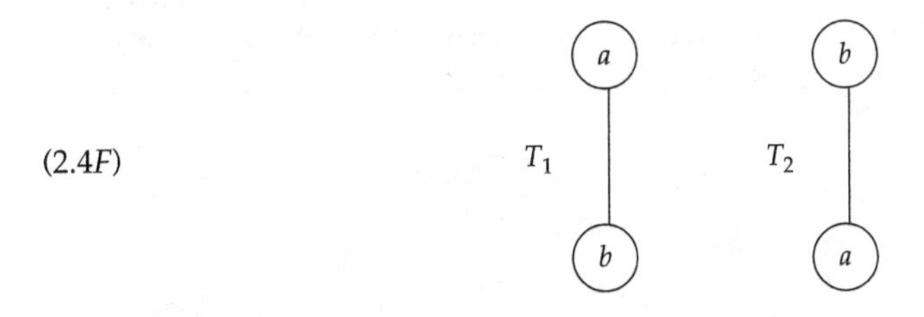

(2.5) The root is at the top, and there is a line from the root to the root of each of the immediate subtrees. That is the rule for drawing trees.

†Strictly speaking, (2.1) defines a *rooted, unordered* tree. This definition is due to Frank J. Oles (private communication, 1983).

Of the several trees on the set of nodes $\{a, b, c\}$ we draw three:

(2.6F)

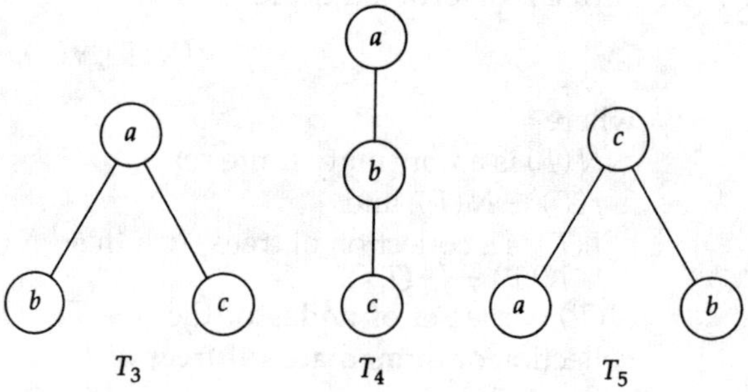

There is one immediate subtree for T_4, namely, the tree T_6 drawn as

(2.7F)

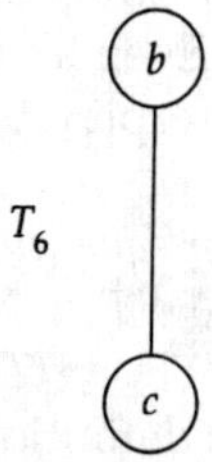

T_6 itself has an immediate subtree, ⓒ. T_3 has two immediate subtrees, ⓑ and ⓒ. In the terms of the abstract definition (2.1),

(2.8)
$$T_3 = (\{a,b,c\}, a, \{(\{b\}, b, \varnothing), (\{c\}, c, \varnothing)\}),$$
$$T_4 = (\{a,b,c\}, a, \{(\{b,c\}, b, \{(\{c\}, c, \varnothing)\})\}).$$

You can see from these examples that the drawing is far easier to understand than the formal description as an ordered triple. The drawing tells you right away what is $N(T)$, what is the root $r(T)$, and what are the subtrees. For this reason we use the drawing and the idea of the drawing of the tree instead of (2.1) in almost every encounter with trees. The tree T_S of (1.2F), for example, has two immediate subtrees, each of which has two immediate subtrees. Three of those four trees (i.e., those with roots at 13, 41, 77, 106) have only one subtree on a single node, and the other has two such subtrees. Expressing it in the form (2.1) would be messy. We draw T_S again, putting tents over all subtrees except those with only one node.

(2.9F)

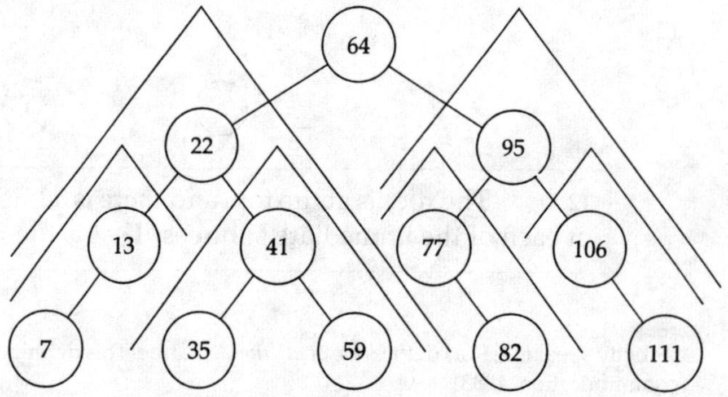

(2.10) **Subtrees.** We defined the term *immediate subtree* in (2.1), but we just used the word *subtree* by itself. Informally, a **subtree** of a tree is any immediate subtree of that tree, any immediate subtree of an immediate subtree, and so on. Thus the tree of (2.9*F*) has 12 subtrees, one for each node.

To be formal we define **subtree** as follows: The idea of immediate subtree defines a relation $\mathcal{IS}$ on a set of trees. T_1 is, or is not, an immediate subtree of T_2. Let $\mathcal{S}$ denote the reflexive transitive closure of $\mathcal{IS}$. Then we call $\mathcal{S}$ the subtree relation. A moment's thought will convince you that the transitivity in this formal definition captures the preceding informal idea. The reflexivity adds a new note: every tree is a subtree (but not an immediate subtree!) of itself.

We may as well admit that these drawings or diagrams such as (1.2*F*) and (2.6*F*) are called the **graphs** of the trees they represent. To avoid confusing trees with ordinary graphs, however, we follow the convention of (2.5) to distinguish the root. You will see that the recursive definition (2.1) leads to many elegant inductive proofs, which are convincing even though most of them appeal to the formal definition (2.1) only through the graphs of the trees.

Example (2.11) We give a simple example. From our drawings it's "obvious" that every node in a tree T is a root of a subtree of T, as just defined in (2.10). Let's prove it formally, by induction on the number of nodes n in $N(T)$. Thus our predicate $P(n)$ is this: For any tree on n nodes, every node of the tree is the root of a subtree of that tree.

We always use the second form of induction (Chapter 3, (5)) for trees. We can prove the basis and inductive steps on one foot with the observations that in $T = (N(T), r(T), L(T))$,

(*i*) $r(T)$, as the root of T, is the root of the subtree T of T; and

(*ii*) each tree in $L(T)$ has fewer nodes than T has.

That is, for any tree T as above, its root is the root of a subtree, by (*i*); and by induction, from (*ii*), each tree X in $L(T)$ satisfies the predicate. Of course, X is a subtree of T; thus each node of X is the root of a unique subtree of T by the transitivity of $\mathcal{S}$. (This proof recalls the compact version of induction found in Chapter 3, (5.3).) QED

The proof can be done more easily with diagrams and handwaving than with the definition (2.1) of a tree as an ordered triple. That is, we'd draw the basis step as ⓐ, and say $P(1)$ is obviously true. Then we'd draw the inductive step as, say,

(2.12*F*)

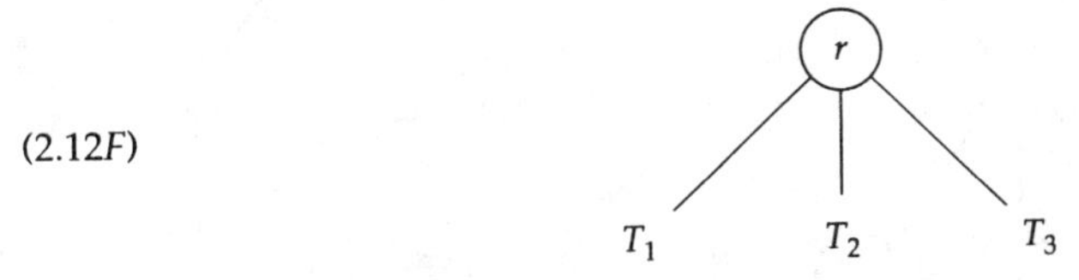

remark that T_1 has fewer nodes than T, and conclude by the inductive assumption that the desired result holds for T_1. It's clear from the picture (2.12*F*) that every subtree of T_1 is a subtree of T, and that these conclusions are valid for all the immediate subtrees of T.

(2.13) *Ordered trees.* We now make a further assumption about trees, that the set $L(T)$ in (2.1) is a *linearly ordered set of ordered trees.* For example, if T has exactly three immediate subtrees, then one is *first*, another is *second*, and another is *third*. Such a tree is called an **ordered tree**; *we assume from now on that all our trees are ordered.* We draw the first immediate subtree at the left, the second next, and so on.

Example The ordered trees

(2.13F)

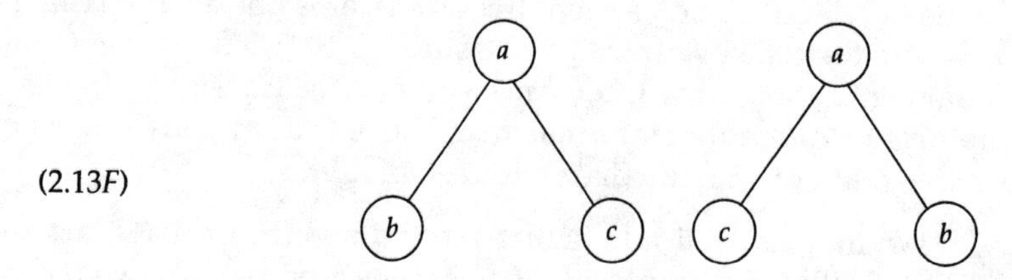

are different; as unordered trees under the definition (2.1), they would be the same.

(2.14) *Terminology.* In any tree T the roots of the immediate subtrees are called the **sons** or **daughters** of the root of T. The root of T is called the **father** of its sons. The **descendants** of $r(T)$ are the elements of $N(T) - \{r(T)\}$. The **ancestors** of any a in $N(T)$ are the father of a, the father of the father of a, and so on (up to $r(T)$ if $a \neq r(T)$). The *descendants* of the node a are all the nodes different from a in the subtree of which a is the root.

The root $r(T)$ of the tree T is at *level* 0 in T; each node $a \neq r(T)$ of the tree is at *level* $1+$ (level of father of a). In other words, the **level** of a node is the total number of ancestors of that node.

The **height** of the tree T is $\max\{\text{level of } a; a \in N(T)\}$.
A **leaf** of T is any node with no sons.
We sometimes say that T is a tree **on** the set of nodes $N(T)$.

Example (2.15)

(2.15F)

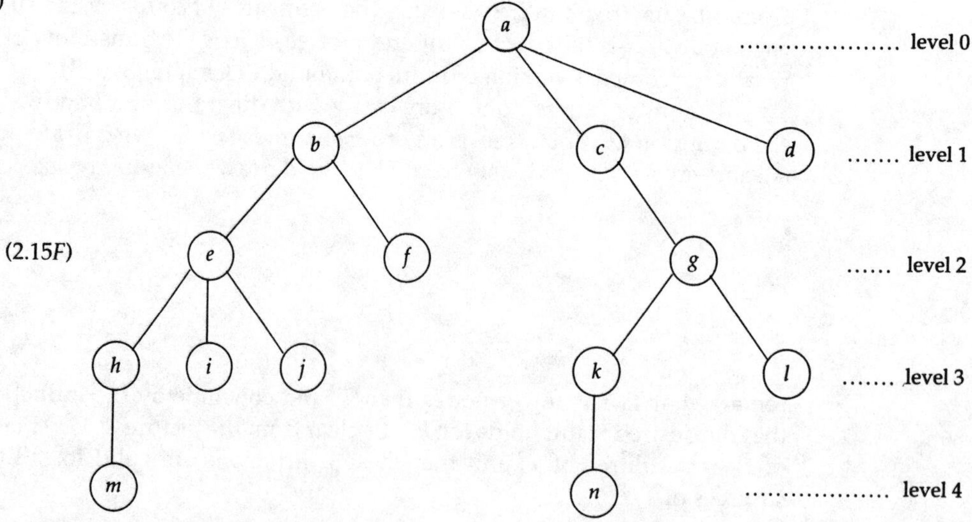

This tree has height 4. The sons of a are b, c, and d. The father of h is e. The ancestors of g are c and a. The descendants of c are g, k, n, and l. The leaves are m, i, j, f, n, l, and d.

3. Forests

(3.1) A **forest** is defined as a finite set of trees in which different trees have no nodes in common. A forest may be empty. An **ordered forest** is by definition a forest of which the set of trees is a finite linearly ordered set of *ordered* trees. We consider only ordered forests from now on, and we refer to them simply as forests. We sketch forests as we do trees, putting the first tree at the left.

Example (3.2) Here is a graph of a forest.

(3.3F)

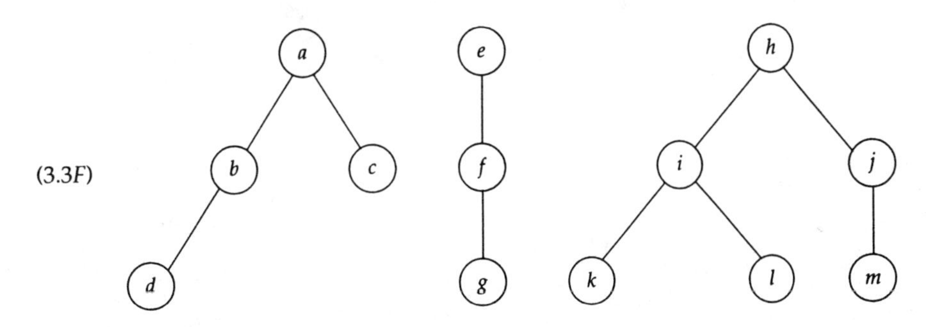

Note that a forest may be the empty set of trees, but there are no empty trees in a forest. There are many situations naturally representable by forests but not by trees.

Example (3.4) Biologists have conceived a huge forest (probably not ordered) to classify living organisms. It has five trees, one for each of plants, animals, fungi, bacteria, and protozoans. Each tree has height six, except that some scientists insert more levels. They have named each level in the forest: level 0 is Kingdom; 1, Phylum, sometimes Division; 2, Class; 3, Order; 4, Family; 5, Genus; and 6, species. (A mnemonic for these names is "King Philip came over from Geneva, Switzerland.") For instance, the house cat is in the kingdom *Animalia* under phylum *Chordata*, class *Mammalia*, order *Carnivora*, family *Felidae*, genus *Felis*, species *catus*.

Some nodes in these trees have many sons. There are some 30 orders of insects, covering to date more than 750,000 known species. One of these orders, *Coleoptera* (beetles), has more than 275,000 species. In the subtree of height 3 with root *Coleoptera*, some parent has at least $\lceil (275,000)^{1/3} \rceil = 66$ sons.

Practice (3.5P) Explain the last sentence.

For us, the natural question is: Where are the trees in this forest? The answer is kingdom *Plantae*, in which they are widely distributed. One division, *Ginkgophyta*, contains only one nonextinct species (the ginkgo tree). Pines and hemlocks are in division *Coniferophyta*, others are in divisions *Cycadophyta* and *Gnetophyta*, and most of the trees we know in Europe and the Americas (e.g., oaks) are in the division *Anthophyta* (Angiosperms). The lower levels of oaks are class *Dicotyledoneae*, order *Fagales*, family *Fagaceae*, genus *Quercus*, species *alba* (white oak), or *rubra* (northern red oak), among others.

Our last biological example may be called the tree of trees. (No, it is not an instance of Russell's paradox.) It is a tree structure showing the probable evolutionary development of certain orders of trees and other plants. As a concession to the biologists we put the root at the bottom. The idea in (3.6*F*) is that each son is an evolutionary descendant of its father.†

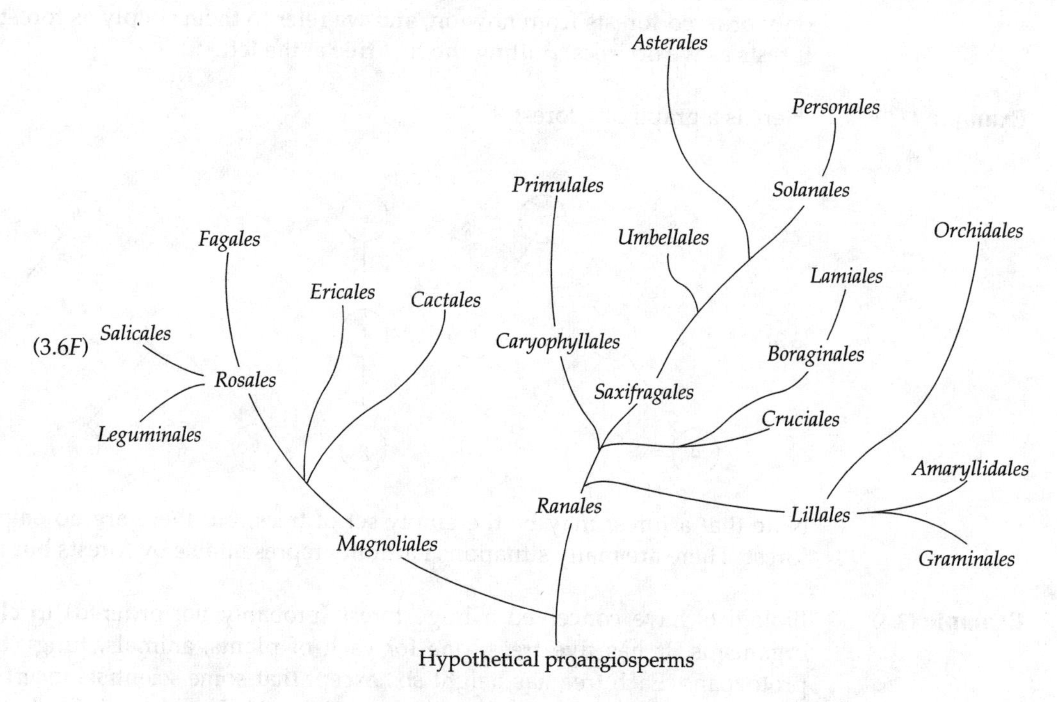

Biologists are quick to say that the classifications are not carved in stone, even that everything above genus is subjective. One of them told me, "If you want to make Gymnosperms a phylum [over five divisions of plants], be my guest."

A mathematician once wrote that the variety of all mathematical topics, results, and methods rivals in size and interrelationships the biologists' forest that we've just mentioned.

4. Binary Trees

An idea close to that of tree is *binary tree*.

DEFINITION (4.1) A **binary tree** B is either empty or is an ordered quadruple

$$(N(B), r(B), B_L, B_R),$$

†Reprinted with permission from Robert F. Seagel, et al., *An Evolutionary Survey of the Plant Kingdom*. Copyright© 1965 by Wadsworth Publishing Co.

where $N(B)$ is a nonempty set, $r(B) \in N(B)$, and B_L and B_R are binary trees such that

$$N(B_L) \cup N(B_R) = N(B) - \{\, r(B) \,\}.$$

If B is empty, we denote it by a square $\square$, and we define

$$N(\square) := \varnothing.$$

We say that $N(B)$ is the set of **nodes** of B, that $r(B)$ is the **root** of B, and that B_L and B_R are, respectively, the **left** and **right immediate subtrees** of B.

Subtree is defined for binary trees as it is for trees in (2.10). We usually omit the term *immediate* when it is implied by the context.

Example (4.2) Here is an example of a binary tree.

(4.3*F*)

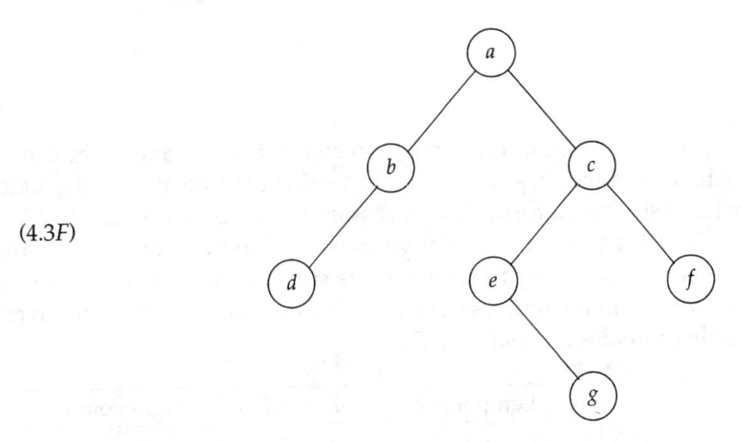

In drawings of binary trees, the convention is that the left subtree, here $(\{\, b,d \,\}, b, (\{\, d \,\}, d, \square, \square), \square)$, goes to the left of the root and below. Similarly for the right subtree, here $(\{\, c,e,f,g \,\}, c, (\{\, e,g \,\}, e, \square, (\{\, g \,\}, g, \square, \square)), (\{\, f \,\}, f, \square, \square))$. As with trees, the drawings are far easier to understand than the definition as a quadruple.

The terms *level, height, son, daughter, father, ancestor, descendant,* and *leaf* are defined for binary trees just as they are for trees in (2.14). (Exception: The height of the empty binary tree is not defined.) The term **leftson [rightson]** means the root of the left [right] immediate subtree.

What are the differences between *tree* and *binary tree*? First, a tree must be nonempty, but a binary tree may be empty. Second, two different binary trees may be the same if viewed as trees. For example, the binary trees

(4.4*F*)

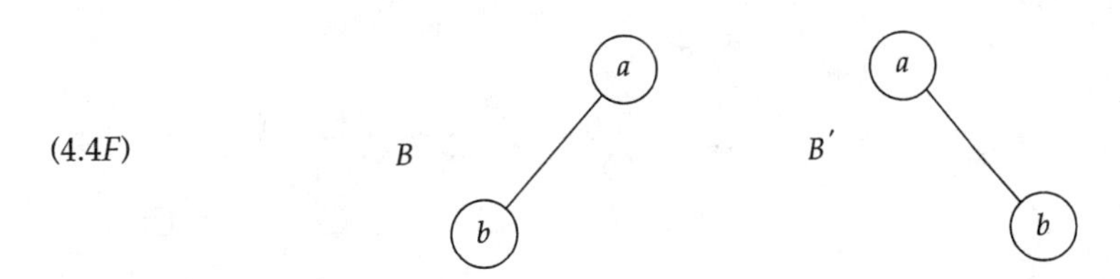

are different; as quadruples they are

$$B = (\{a,b\},\ a,\ (\{b\},b,\square,\square),\ \square)$$

and

$$B' = (\{a,b\},\ a,\ \square, (\{b\},b,\square,\square)).$$

In other words, the left subtree of B is nonempty but that of B' is empty, so they are different binary trees. But if we regarded them as trees they would be the same, namely, $(\{a,b\},a,\{(\{b\},b,\varnothing)\})$, i.e.,

(4.5F)

The point is that in a tree there's no concept of left and right but only first, second, third,.... If a tree has just one subtree there's no way to tell whether it's the left or the right subtree, information necessary if it is to be a binary tree.

Third, a tree may have any number of immediate subtrees; but a binary tree, if nonempty, has exactly two immediate subtrees. Binary trees are more attractive than trees for use in computers because of this uniformity: Each node can be represented as a three-field word as in (1.3*F*).

Left pointer	Node info	Right pointer

The left [right] pointer is the address of the leftson [rightson]. A special pointer is used for all empty subtrees. To represent a tree we might need words of many fields for some nodes because the number of sons of a node is not restricted. As we shall see, however, every tree can be represented as a binary tree, so binary trees are particularly desirable.

Here are all the binary trees on n nodes for $n \leq 3$.

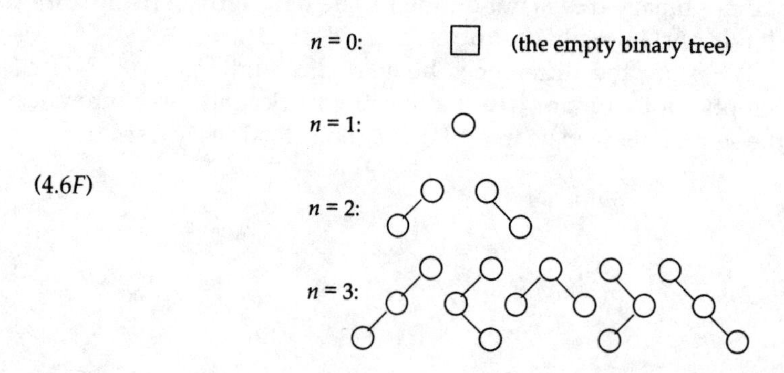

In Section 11 we'll develop a formula for the total number of binary trees on n nodes, good for all $n \geq 0$.

There is a simple relationship holding for any binary tree. We state it as a theorem, leaving the proof for the problems.

Theorem (4.6) Denote for any nonempty binary tree $n_i :=$ the total number of nodes with exactly i sons. Then $n_0 = 1 + n_2$. ■

In drawing binary trees we sometimes show the empty subtrees. Thus the second binary tree on three nodes in (4.6*F*) could be drawn as

(4.7*F*)

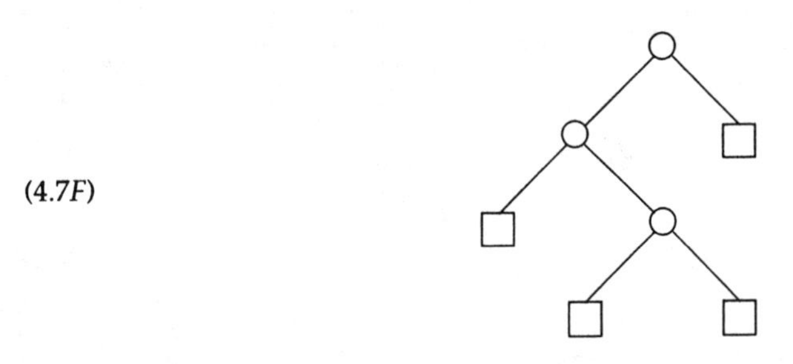

Practice (4.8P) Without drawing any trees use (4.6*F*) to find out how many binary trees there are on four nodes.

The representation of the secret telephone codes in (1.2) is a binary tree; the "go left—go right" rule for constructing it was essential. The left [right] subtree of the root has all the numbers smaller [larger] than the root, a property holding recursively for every subtree.

5. Forests and Binary Trees

Surprisingly, there is a simple bijection (one-to-one correspondence) between the set of ordered forests and the set of binary trees. This fact allows the seemingly more general types of data representable with forests to be represented as binary trees, which we've seen are much easier to store in computers than general trees. This bijection is best understood through an example. Consider the ordered forest in (5.1*F*). Ignore the dashed lines for now.

(5.1*F*)

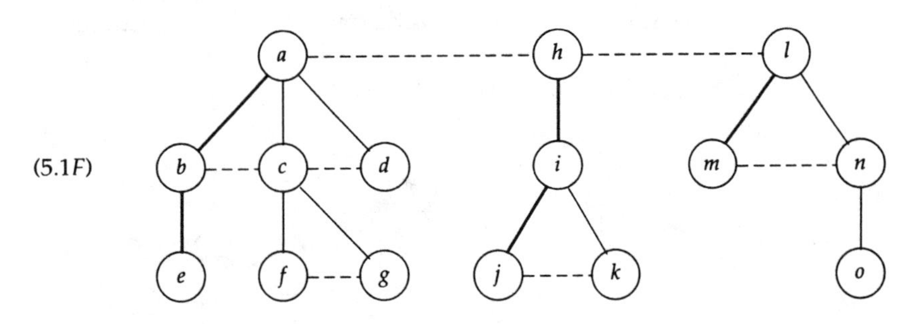

Convert this forest to a binary tree as follows:

(*i*) Erase all edges except those to first sons. In (5.1*F*) we keep only the bold edges.

(*ii*) For each node introduce a horizontal line from the first son to the second son, from the second son to the third son, and so on.

These are the dashed horizontal lines in (5.1*F*). The roots too are joined by such lines as shown. The binary tree corresponding to this ordered forest has a graph obtained by tipping the graph in (5.1*F*) 45° clockwise. Figure (5.2*F*) shows the result.

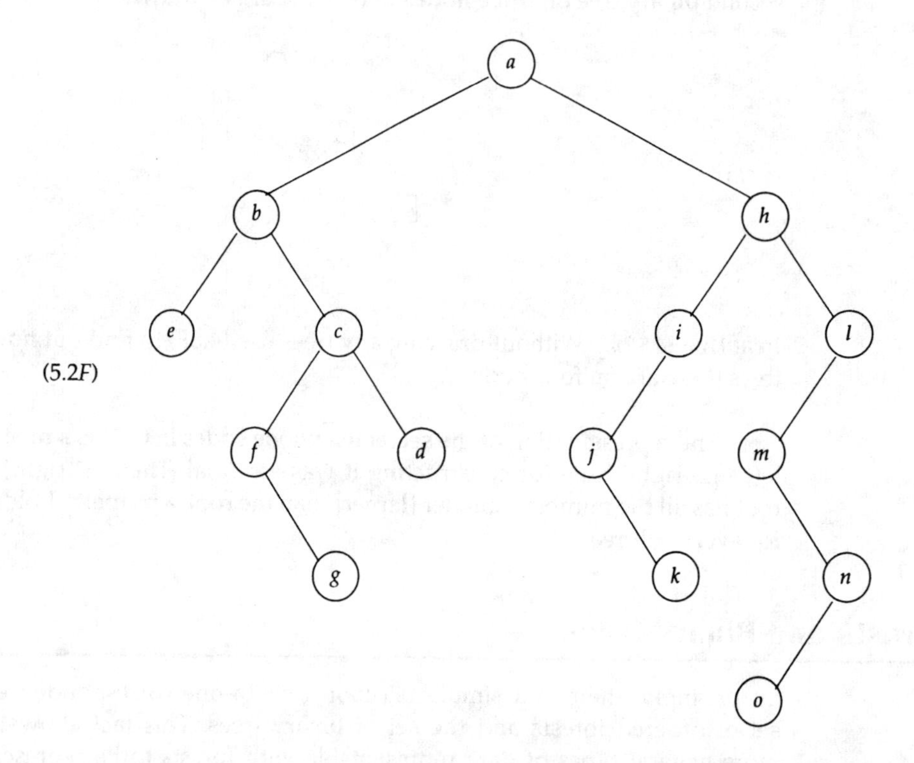

(5.2*F*)

(5.3) Here is the general definition of the correspondence. The mapping β that takes an ordered forest (of ordered trees) to a binary tree is this: with the roots of the trees viewed as the first, second, and so on, sons of a fictitious node, the first son of each node becomes the left son of that node (the first root becomes the root of the binary tree); for each $n > 1$ the nth son of that node becomes the right son of the $(n-1)$st son of that node.

Going backward, the two binary trees

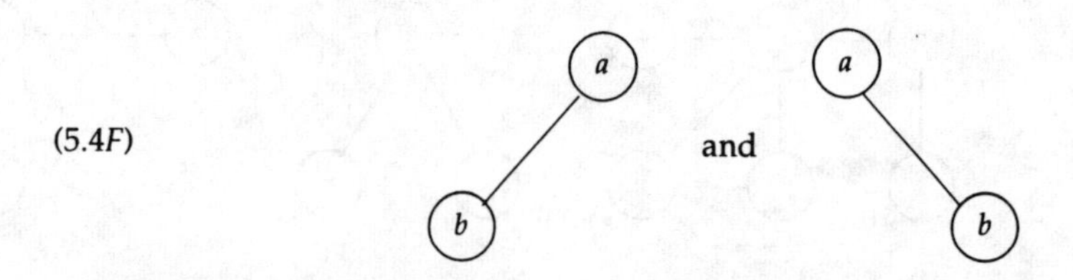

(5.4*F*)

come from the forests

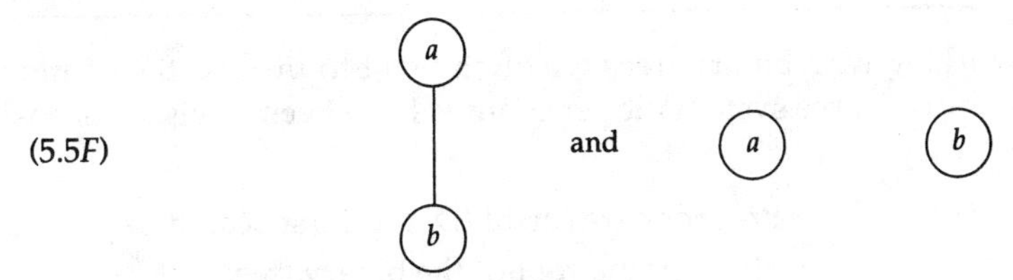

respectively (under this correspondence β). The binary tree B of (5.6F) comes from the forest F in (5.7F): $\beta(F) = B$.

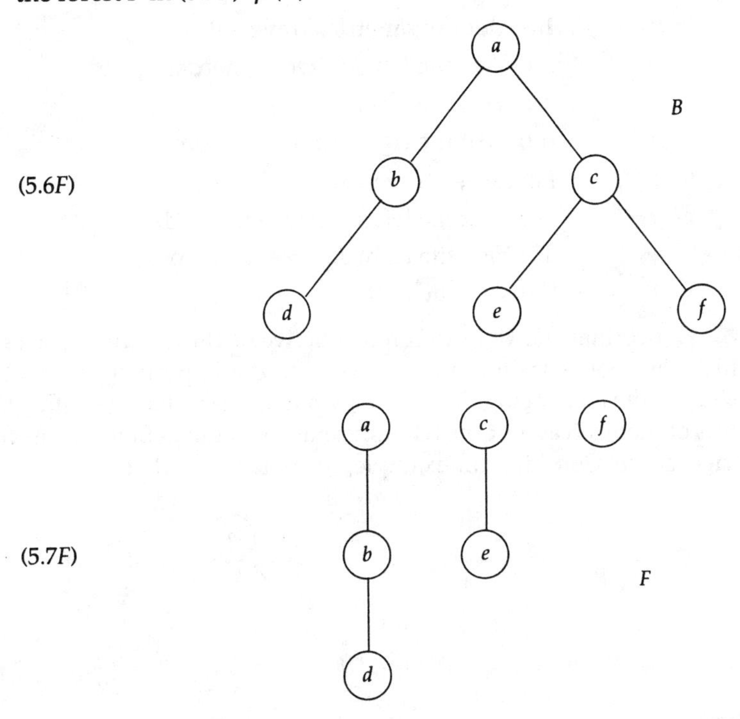

Practice (5.8P) Draw the binary tree $B := \beta(F)$ for the forest F in (5.8F). Then let T be the *ordered* tree with the same graph as B, and draw the binary tree $B_1 = \beta(F')$, where F' is the forest consisting of the one tree T.

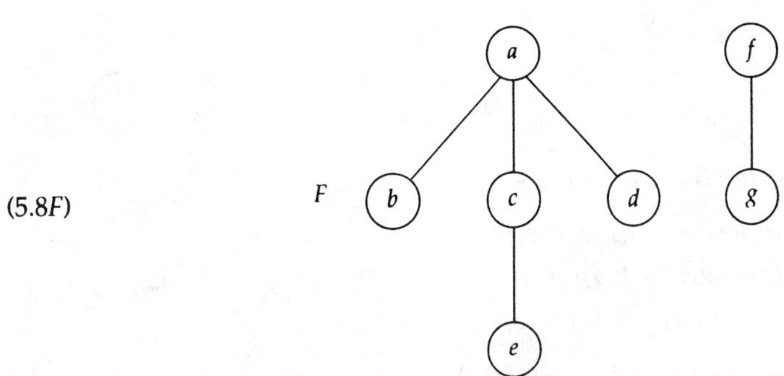

6. Traversal of Binary Trees

In working with binary trees we often want to process the contents of each node in some way. Three systematic procedures have been devised for visiting all the nodes of a binary tree.

(6.1) **Preorder** Traversal (Depth-First Search)†

(*i*) Visit the root of the binary tree;
(*ii*) Visit the left subtree in preorder;
(*iii*) Visit the right subtree in preorder.

(6.2) **Inorder** (Symmetric) Traversal

(*i*) Visit the left subtree in inorder;
(*ii*) Visit the root;
(*iii*) Visit the right subtree in inorder.

(6.3) **Postorder** Traversal

(*i*) Visit the left subtree in postorder;
(*ii*) Visit the right subtree in postorder;
(*iii*) Visit the root.

Notice that the only difference between these traversal rules is the order in which the root is visited, plus, of course, the important fact that the subtrees are visited in the *order* being defined. Again it appears that we define these traversals in terms of themselves, but the recursive nature of the definitions means that they make perfect sense. Consider, for example, the binary tree B of (6.4*F*).

(6.4*F*)

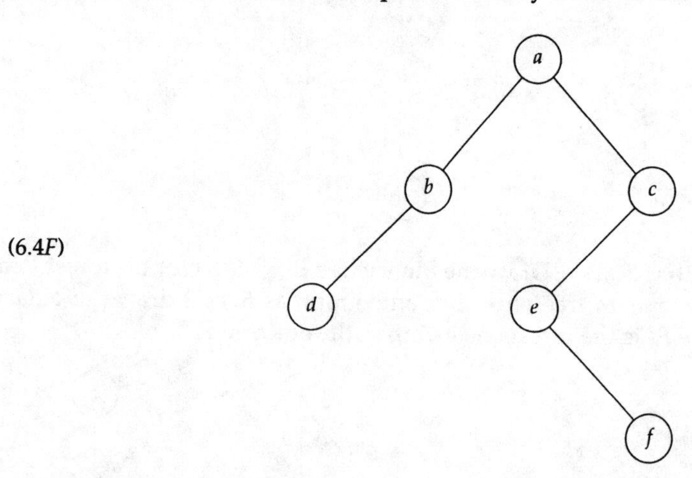

We list the three traversals of B:

Preorder: a, b, d, c, e, f
Inorder: d, b, a, e, f, c
Postorder: d, b, f, e, c, a.

†In view of the terminology of (2.14), this would better be called height-first search, but it isn't.

Partial explanation. For preorder, we must first visit the root a. Then in Step (*ii*) we must visit the left subtree B_L in preorder. That means we start preorder again, now on the tree B_L. It has root b, so that's our next stop. Now we must apply Step (*ii*) to B_L, thus visit the left subtree of B_L in preorder. This tree has only one node d. Since □ is the right subtree of B_L, there's nothing to record at Step (*iii*). At this point we've completed our visit of B_L in preorder, so we go to Step (*iii*) of our preorder visit of B. That means we visit B_R in preorder. You can easily check the result claimed for B_R, and, in the same spirit, the others.

(6.5) Perhaps you have seen algebraic formulas represented as binary trees; for example, $(u + a \times v) \div c$ is "parsed" as the tree

(6.5*F*)

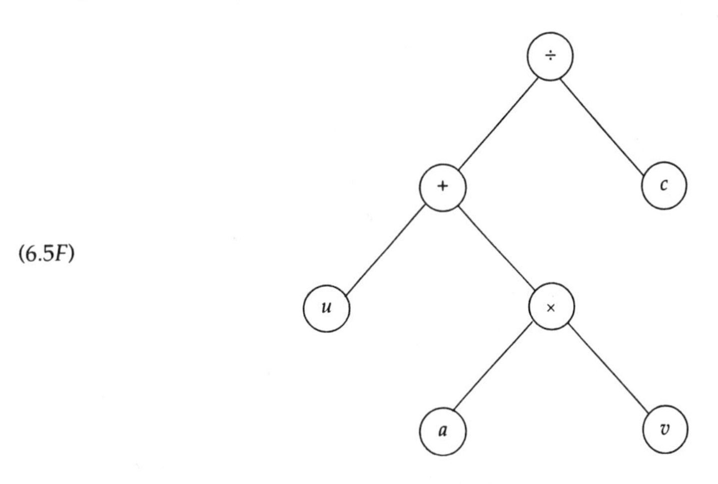

The symmetric traversal of this tree is

$$u, +, a, \times, v, \div, c.$$

This order is the same as that in which the terms are written in the formula. But if we were actually calculating this quantity we would want to find $a \times v$ first, then add u, and then divide by c. That sounds like postorder traversal, which for this tree is:

$$u, a, v, \times, +, c, \div. \quad (6.6)$$

This order corresponds to our actual order of calculation if we read from left to right and interpret each binary operation as acting on the two quantities immediately to its left:

$$\alpha, \beta, \gamma, \delta, *, \cdots$$

means

$$\alpha, \beta, (\gamma * \delta), \cdots$$

if $\alpha, \beta, \gamma, \delta$ are numbers and $*$ is a binary operation. Under this interpretation the sequence (6.6) becomes, successively,

$$\begin{gathered} u,\ a \times v,\ +,\ c,\ \div \\ u + a \times v,\ c,\ \div \\ (u + a \times v) \div c. \end{gathered} \quad (6.7)$$

This order of expressing formulas in variables joined by binary operations is called *Polish postfix notation* or *reverse Polish notation*. No book on discrete mathematics would be complete without some mention of it.

For the record, the preorder traversal of the same tree produces

(6.8) $$\div, +, u, \times, a, v, c.$$

If we read from right to left and make the same convention about the binary operations but interchange the roles of left and right, this sequence becomes

$$\div,\ +,\ u,\ a \times v,\ c$$

$$\div,\ u + a \times v,\ c$$

$$(u + a \times v) \div c.$$

The string (6.8) is known as *Polish prefix notation*. Notice that the variables u, a, v, and c, at the leaves of the tree, appear in the same order in all three traversals. The other nodes separate them in various ways, of course.

(6.9) "Formula trees" like that in (6.5F) have an elegant recursive function defined on them to yield their "value." The function defines a value for each subtree in this way:

> *Basis:* each leaf, which contains a variable or a number, has that entry as its value.
>
> (6.10) *Inductive step:* each subtree T' with root having a binary operation $*$ as entry has value $L * R$, where L [R] is the value of the left [right] subtree of T'.

If you think about this definition a moment, you'll see it's just postorder or preorder traversal with the processing (= assigning a value to each node) built in. In fact, the steps in (6.7) exactly follow the procedure (6.10) if we evaluate the right subtree before the left.

7. A Sorting Algorithm

Now we present an application of binary trees to the problem of sorting numbers (i.e., arranging them in numerical order). After that we analyze this sorting procedure. Let us go back to the tree-building procedure of (1.2) where, from an arbitrary sequence of distinct integers, we made those integers the labels on nodes of a binary tree. The idea of that procedure was this:

> To insert an integer x into the tree, we would chase down the tree constructed to date, going to the left or right subtree after each comparison. As soon as we came to an empty subtree T' we would replace T' by x, (strictly, by $(\{x\}, x, \square, \square)$).

We may use the following procedure as a formal description. We generalize the procedure described earlier by allowing an arbitrary choice of element from the set S, not necessarily the first one from S in list form. As before, $\square$ stands for the empty binary tree. Descriptions are in braces { }.

(7.1) **Procedure Kilmer**

S is a set of numbers, no two equal,
to be made into a binary search tree B

1. $B := \square$.
2. **If** $S \neq \varnothing$
   Pick $x \in S$
   $S := S - \{x\}$
   $B :=$ **Insert** (x, B)
   Return to Step 2.
3. **If** $S = \varnothing$, **end.**

The procedure **Insert** is defined under the assumption that x is not one of the labels of B.

(7.2) **Insert** $(x, \square) :=$

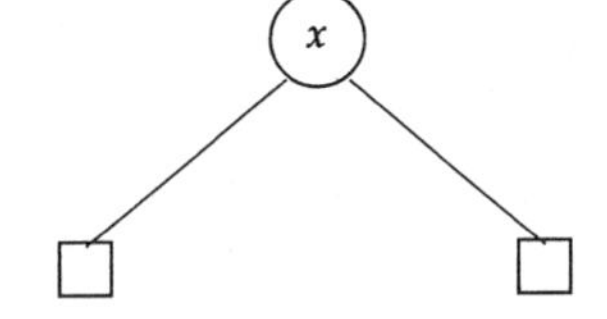

If $B \neq \square$, **let** $B = (N(B),\ r,\ B_L,\ B_R)$.

Insert $(x, B) :=$ **if** $x < r$ **then**

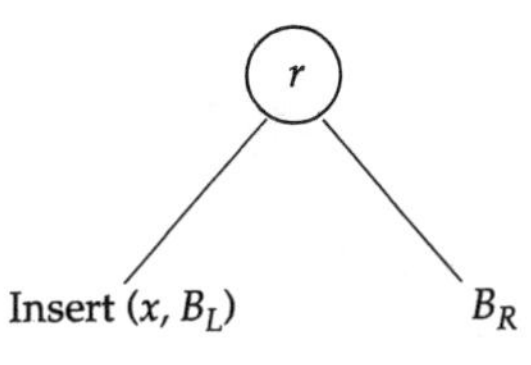

otherwise

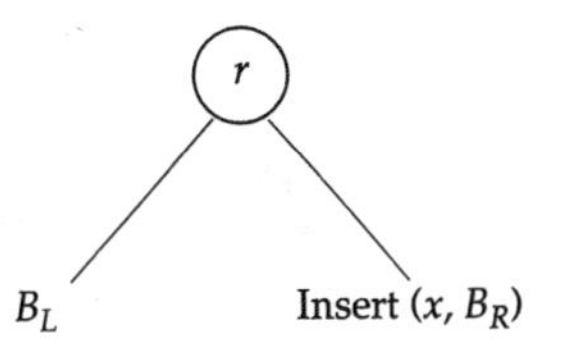

This yields a binary search tree, the set of labels of which is $\{x\} \cup \{\text{labels of } B\}$.

We can then list the numbers of S in increasing order by simply traversing the tree in symmetric order. The example below, at least, bears out this claim. We'll prove it in a moment. This procedure is called **Treesort**. We first define for all binary trees B,

DEFINITION (7.3) $Sym(B) :=$ the nodes of B listed in the order visited in a symmetric traversal of B.

Now we define **Treesort.**

DEFINITION (7.4) Procedure **Treesort.**
Input S, a list of distinct numbers;
Output S, sorted.

$B :=$ **Kilmer**(S)
Treesort$(S) := Sym(B)$.

Example (7.5) Suppose S is 47, 16, 30, 38, 15, 92, 99, 35, 52, 40, 59, 71. Instead of choosing $x \in S$ at random, let's always pick the first element. Then **Kilmer** produces the binary tree

(7.5F)

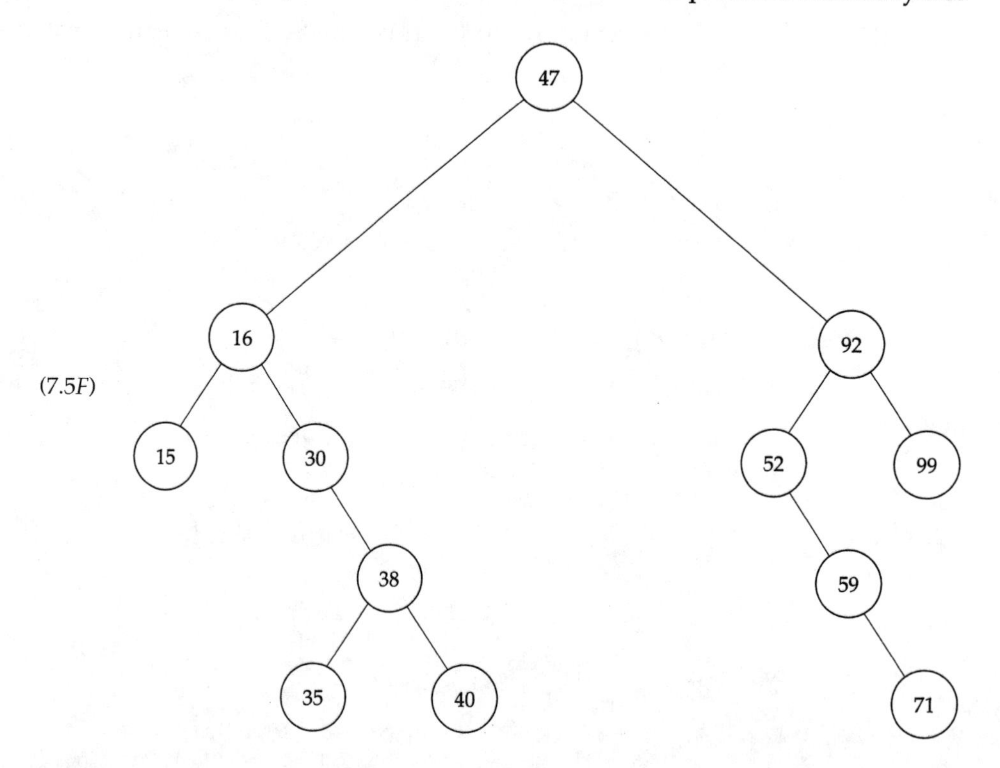

The symmetric traversal of this tree indeed sorts S: Once we visit 15 we have traversed the left subtree of 16 in symmetric order; we then visit its root 16. The root of its right subtree is 30; since 30 has no left, we next visit 30. The rightson of 30 has a leftson, 35, with no descendants. So we visit 35, then its father 38, and then the only right descendant of 38, namely 40. Having completed our visit of the left subtree of **Kilmer** (S), we visit the root 47 and then the right subtree in symmetric order.

We now prove that **Treesort** sorts S. We use the property of the tree B already mentioned: for every x in the set S, the left [right] subtree of the node x consists of all elements of S that are less [greater] than x. We also use the recursive definition (6.2) of symmetric order traversal.

Proposition (7.6) For any finite set S of integers, **Treesort** sorts S. That is, it lists the elements of S in increasing order.

Proof We use induction on the number of points n in S. If $n = 0$ there is nothing to prove. Let n be a fixed but arbitrary integer ≥ 0 and assume the Proposition is true for all sets of n or fewer integers. Let S be a set of $n+1$ integers. Choose an element, say v, of S and let B be any binary tree formed by Kilmer acting on S with v as root. Let B_L be the left subtree, and B_R the right subtree, of B. Now B_L was formed by the same procedure as B. Therefore its nodes, a set of n or fewer integers, are sorted by symmetric traversal of B_L, according to the induction hypothesis. Furthermore, symmetric traversal of B begins with symmetric traversal of B_L, by the definition (6.2).

We constructed B so that every node of B_L is less than the root v of B. So symmetric traversal of B produces the list

$$(\text{sorted nodes of } B_L), v, (\text{sorted nodes of } B_R),$$

which is S, sorted. Therefore the inductive step is proved, and the Proposition holds for all finite sets S. QED

This proof illustrates the key step in proofs by induction of results about trees.

(7.7) *It is almost always better to mimic the inductive definition of trees by going from a tree to its subtrees at the inductive step, rather than to remove a leaf.*

Practice (7.8P) Try to prove (7.6) by an inductive proof in which you remove a leaf from the tree on $n+1$ vertices and apply the inductive assumption to the resulting tree on n vertices.

(7.9) *Analysis of* **Treesort**. The most time-consuming part of **Treesort** is doing the comparisons needed to set up the tree B. The less costly symmetric traversal of that tree requires no comparisons. We now analyze **Treesort** by estimating the total number of comparisons it uses. Remember, the *level* of a node, defined in (2.14), is the total number of ancestors of that node. The number of comparisons needed to insert each node x into B is the level of x, denoted as $\ell(x)$. Thus the total number N_c of comparisons used to form B is

$$\sum_{x \in N(B)} \ell(x) =: N_c. \tag{7.10}$$

For example, the quantity N_c is computed for each of the following binary trees on five nodes. The level of each node is written next to it.

(7.10F)

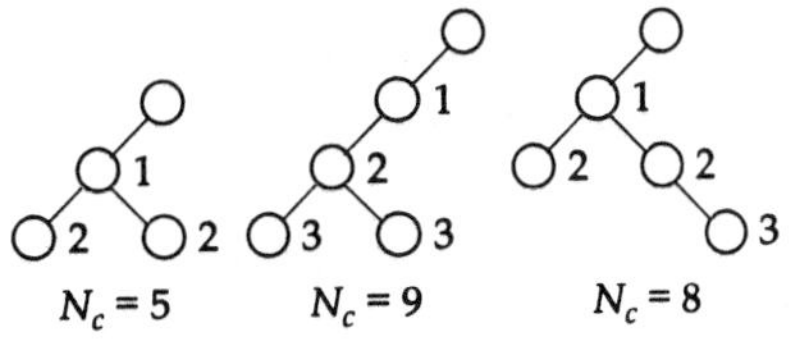

$N_c = 5$ $N_c = 9$ $N_c = 8$

It is not so tricky after all to estimate the *average* value of N_c over all inputs S of size n. We assume for definiteness that S is any permutation of the integers $1, \ldots, n$ and that we always choose the first entry of S for the root of the tree. We'll find an upper bound on the average number of comparisons used in **Kilmer** to set up the tree.

Theorem (7.11) The average number $A(n)$ of comparisons performed in the execution of **Treesort** is approximately $1.39n \log_2 n$. (The minimum and maximum of N_c are approximately $n \log_2 n$ and $\frac{1}{2}n^2$, respectively.)

Proof Notice that $A(0) = 0$. Let n be a fixed positive integer. There are the same number of permutations starting with 1 as with 2 as with 3... as with n.

We claim that

(7.12) $$A(n) = n - 1 + \frac{1}{n}\left\{ \sum_{1 \le j \le n} A(j-1) + \sum_{1 \le j \le n} A(n-j) \right\}$$

because

- the first element of S could be any of $1, \ldots, n$ with equal likelihood. Say it is j. Then j is the root of B.
- Each of the remaining $n - 1$ entries in S is compared with j and then assigned to B_L or B_R.
- The numbers $1, \ldots, j-1$ are the nodes of B_L; since they may be presented in any order in S, they need on average $A(j-1)$ comparisons to be formed into B_L by the algorithm.
- The same is true for the $n - j$ numbers $j + 1, \ldots, n$ and B_R; they need on average $A(n-j)$ comparisons.
- We total these three quantities to find

  $$n - 1 + A(j-1) + A(n-j)$$

  comparisons needed when j is the first element of S. Add this quantity over all $j = 1, \ldots, n$ and divide by n to get (7.12).

Notice that the sums in (7.12) are equal. Thus (7.12) becomes

(7.13) $$A(n) = n - 1 + \frac{2}{n} \sum_{1 \le i \le n-1} A(i).$$

We now prove by induction that

(7.14) $$A(n) \le 2n \log n$$

for all $n \ge 1$, where log denotes the natural logarithm. For our predicate $P(n)$ we take (7.14). Since $A(1) = 0$ and $\log 1 = 0$, $P(1)$ is true. Let n be a fixed but arbitrary integer, $n \ge 2$, and assume $P(1), \ldots, P(n-1)$ are all true. We try to prove $P(n)$. Thus from (7.13) and (7.14) we see that

(7.15) $$A(n) \le n - 1 + \frac{4}{n} \sum_{2 \le i < n} i \log i.$$

We find an upper bound on the sum in (7.15) as follows: Define $f(x) = x \log x$; then the sum in (7.15) is $\sum f(i)$. Consider the graph of the curve $y = f(x)$ over $2 \le x \le n$, and the "bar graph" of $f(i)$ for $i = 2, \ldots, n-1$.

(7.16F)

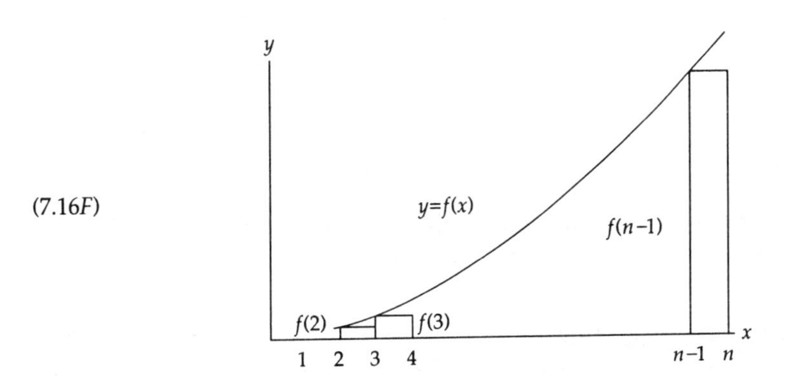

The area under the bar graph is precisely the sum in (7.15), that is, $f(2) + f(3) + \cdots + f(n-1)$. Simply because $f(x)$ is an increasing function (since $f'(x) = 1 + \log x > 0$), the area under the curve $y = f(x)$ over $2 \le x \le n$ is greater than that under the bar graph. Therefore

$$A(n) \le n - 1 + \frac{4}{n} \int_2^n x \log x \, dx. \tag{7.17}$$

A simple integration by parts yields $\frac{1}{2}(x^2 \log x - x^2/2)$ as the indefinite integral. Now (7.17) gives us

$$A(n) \le n - 1 + 2n \log n - n - (4/n)(2 \log 2 - 1) \le 2n \log n. \tag{7.18}$$

Thus we've proved that $P(1) \wedge \cdots \wedge P(n-1)$ imply $P(n)$; by the second theorem of induction (Chapter 3, (5)), $P(n)$, i.e., (7.14), is true for all integers $n \ge 1$. QED

Since $\log_e x = (\log_e 2) \log_2 x$, and $\log_e 2 = 0.693+$, we may state (7.14) as

$$A(n) \le 1.39 n \log_2 n.$$

The upshot is that **Treesort** requires, on the average, at most $1.39 n \log_2 n$ comparisons for large n.

We could derive (7.14) from (7.13) by discrete methods,[†] except that we'd need calculus for the approximation

$$1 + \frac{1}{2} + \cdots + \frac{1}{n} \sim \log_e n.$$

There are input sequences S that take $1 + 2 + \cdots + n - 1 = n(n-1)/2$ comparisons, namely, $S = (1, 2, \ldots, n)$. If we always insert the first element of the input, we get (7.19F) for **Kilmer**(S). This input S maximizes N_c, as we showed in Chapter 3, (12.13). You can figure out how that result applies here.

[†]See Donald E. Knuth, *The Art of Computer Programming*, Vol. 3, *Sorting and Searching*, (Reading: Addison-Wesley, 1973), pages 120–121.

(7.19*F*)

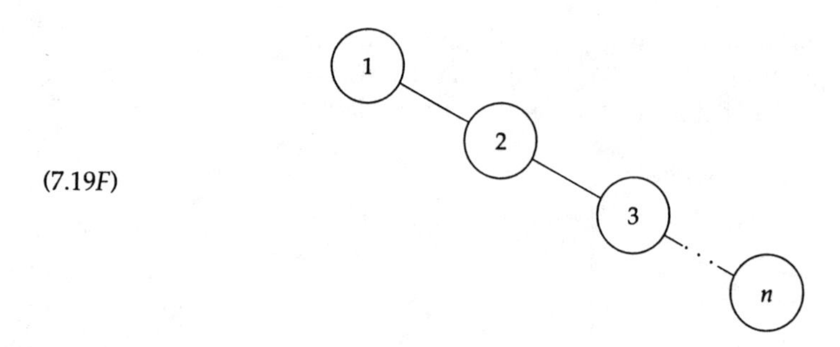

And we now show that the minimum number of comparisons occurs when the tree B is "compressed" as much as possible.

8. The Complete Binary Tree

To complete the proof of Theorem (7.11) we now discuss a special binary tree for which the sum (7.10), $\sum \ell(x)$, is a minimum. Let n be the total number of nodes in the binary tree B, and let n_i be the total number of nodes at level i, for $i \geq 0$. Then $n = \sum_i n_i$, and the sum of the levels of all the nodes is

$$L(B) := \sum_{x \in N(B)} \ell(x) = \sum i n_i. \tag{8.1}$$

For each $i, n_i \leq 2^i$. (You can easily prove that by induction, for all binary trees.)

Examples For B_1, $n_0 = 1, n_1 = 2, n_2 = 3$, and $\sum n_i = 6$; thus $\sum \ell(x) = 0+1+1+2+2+2 = 0 \cdot 1 + 1 \cdot 2 + 2 \cdot 3 = 8$.

(8.1*F*)

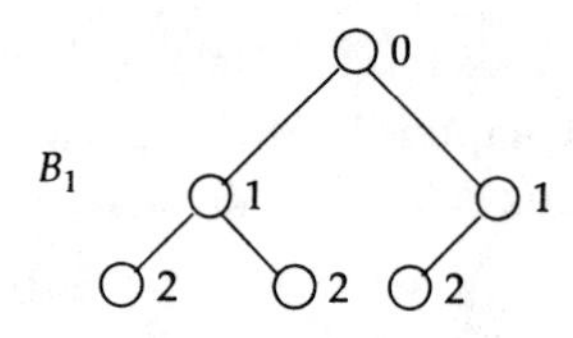

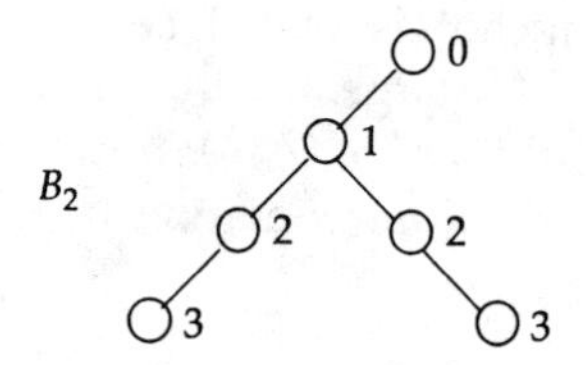

For $B_2, n_0 = 1, n_1 = 1, n_2 = 2, n_3 = 2$, and $\sum n_i = 6$; thus $\sum \ell(x) = 0+1+2+2+3+3 = 0 \cdot 1 + 1 \cdot 1 + 2 \cdot 2 + 3 \cdot 2 = 11$.

(8.2) Now we define $B(n)$, **the complete binary tree on n nodes.** The idea is simple: we fill in all levels with the maximum number of nodes (2^i for level i) until we get to the last level; then we left-justify the remaining nodes. Several examples are shown in figure (8.2*F*).

(8.2F)

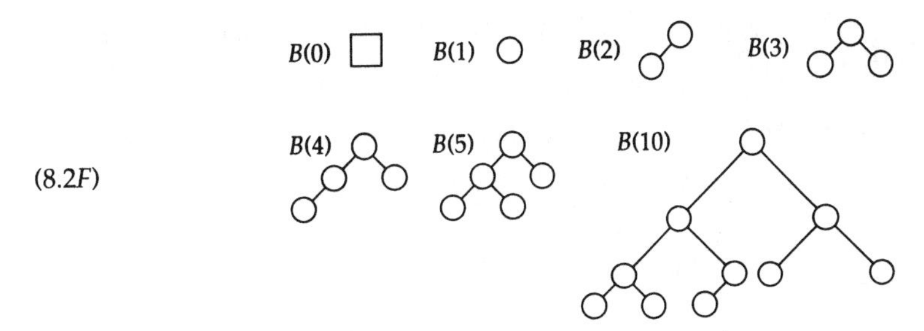

The left-justification is an arbitrary convention; the nodes at level h ($:=$ height of $B(n)$) could be sons of any of the nodes at level $h-1$ without changing the value of $\sum \ell(x)$.

The tree $B(n)$ is sometimes called a *completely balanced* binary tree.

Now we show:

(8.3) Among all binary trees B on n nodes, $B(n)$ has the least value of $\sum \ell(x)$.

Proof To prove (8.3), let B be any binary tree on n nodes. Suppose h denotes the height of B; thus $h = \max\{i;\ n_i \neq 0\}$.

Case 1. If $n_i = 2^i$ for all $i < h$, then we know that B is the same as $B(n)$ except that the nodes at level h might not be left-justified. Thus $L(B) = L(B(n))$.

Case 2. If for some $j < h$, $n_j < 2^j$, find the smallest such j. Then remove a node at level h and introduce a node at such level j. The result is a binary tree B' with the same number of nodes such that

$$L(B') = L(B) - (h-j),$$

since n_h has been decreased by 1, and n_j increased by 1. Thus $L(B') < L(B)$, since $j < h$.† (Recall that $L(B) := \sum_x \ell(x) = \sum_i in_i$.)

Example (8.4) Example of Case 2.

(8.4F)

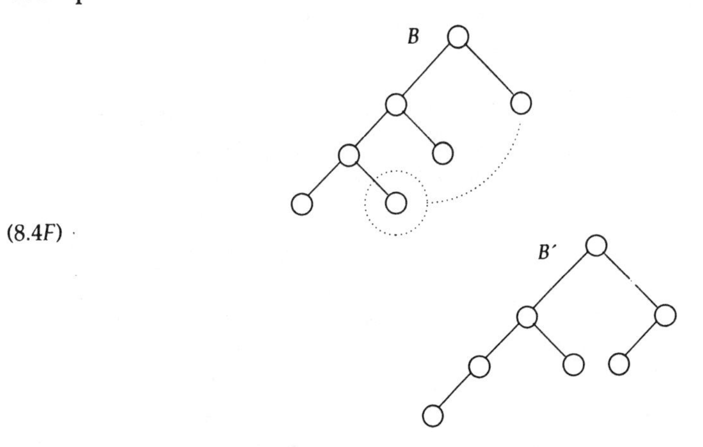

In B, $n_2 = 2 < 2^2$, so we take one node from level $h = 3$ and move it to level 2.

†The proof could stop here, because we've shown that no tree B under Case 2 can have minimum $L(B)$. But see the answer to (8.5P).

Now replace B by B' and start this process again. It stops with a binary tree B_0 satisfying $L(B_0) = L(B(n)) \leq L(B)$. We have proved (8.3) algorithmically. QED

Practice (8.5P) Find an example for which the procedure of Case 2 in the proof of (8.3) cannot be carried out if j is not the minimum.

Note that we proved that $B(n)$ minimizes $\sum \ell(x)$ without knowing the value of $\sum \ell(x)$ for $B(n)$. We'll now find it explicitly. It is

(8.6) $$L(B(n)) := \sum_{x \in N(B(n))} \ell(x) = h(n+1) - 2^{h+1} + 2 \sim n \log_2 n,$$

where $h :=$ height of $B(n) = \lfloor \log_2 n \rfloor$. Proving (8.6) will complete the proof of Theorem (7.11).

Proof To prove (8.6), we first verify that $h = \lfloor \log_2 n \rfloor$. We count the nodes in $B(n)$ as follows: For each i there are exactly 2^i nodes at level i if $0 \leq i < h$. Thus $B(n)$ has

$$1 + 2 + 2^2 + \cdots + 2^{h-1} = 2^h - 1$$

nodes at levels less than h, and

(8.7) $$n_h = n - (2^h - 1)$$

nodes at level h. Since $1 \leq n_h \leq 2^h$ it follows from (8.7) that

(8.8) $$2^h \leq n < 2^{h+1}.$$

Taking logarithms, we get

$$h \leq \log_2 n < h + 1.$$

Therefore $h = \lfloor \log_2 n \rfloor$.

We need to find the value of

(8.9) $$L(B(n)) = \sum_{0 \leq i < h} i2^i + hn_h.$$

Since $(x-1)\sum_{0 \leq i < h} x^i = (x^h - 1)$ (see Chapter 3, (16.1), for a proof), we evaluate the sum in (8.9) as follows: We differentiate the equation

$$\sum_{0 \leq i < h} x^i = \frac{x^h - 1}{x - 1}$$

to get

$$\sum_{0 \leq i < h} ix^{i-1} = \frac{hx^{h-1}(x-1) - (x^h - 1)}{(x-1)^2}.$$

Multiply by x and then set $x = 2$. The result is

$$\sum_{0 \leq i < h} i2^i = 2(h2^{h-1} - (2^h - 1)).$$

Thus, from (8.9) and (8.7),

(8.10) $$L(B(n)) = h2^h - 2^{h+1} + 2 + hn - h2^h + h = h(n+1) - 2^{h+1} + 2;$$

this is (8.6). Notice that the dominant term in (8.6) is hn, which is asymptotic to $n \log_2 n$. The term 2^{h+1} is of the same order as n, because from (8.8)

$$1/2 \leq n/2^{h+1} < 1.$$

This completes the proof of (8.6), and thus of Theorem (7.11). QED

Practice (8.11P) Prove $hn \sim n \log_2 n$.

Example (8.12) If $n = 100$, then $h = 6$. The exact formula (8.10) gives

$$L(B(100)) = 6 \cdot 101 - 2^7 + 2 = 606 - 128 + 2 = 480,$$

versus the approximation $6 \cdot 100 = 600$.
If $n = 1000$, then $h = 9$ and

$$L(B(1000)) = 9 \cdot 1001 - 2^{10} + 2 = 7987 \text{ (versus 9000)}.$$

If $n = 10^6$, then $h = 19$, and

$$L(B(10^6)) = 17,951,445 \text{ (versus 19,000,000)}.$$

The ratio of the exact values to the approximations rises slowly toward 1.

$$\frac{480}{600} = 0.80, \qquad \frac{7987}{9000} = 0.89-, \qquad \frac{17951445}{19 \times 10^6} = 0.94+.$$

After all, we did toss out $2^{h+1} = O(n)$.

(8.13) *The complete tree $B(n)$ as an array*. The tree $B(n)$ is so regular in structure that it can be represented as a vector (one-dimensional array). We illustrate by numbering the nodes in a natural order for the case $n = 10$.

(8.14*F*)

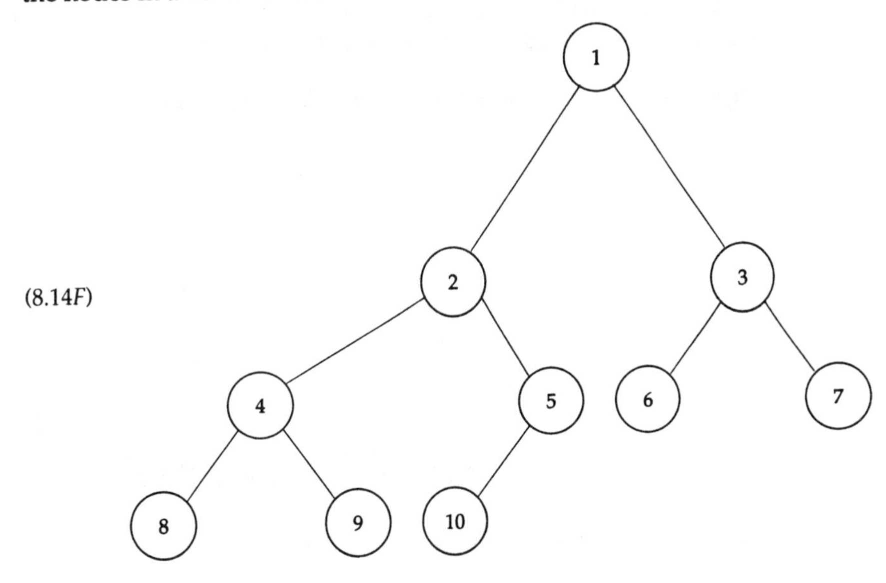

You can see that for each node i its sons are $2i$ and $2i + 1$. This fact, easily proved in general, is left to the problems. It is easier to represent and process the array than the tree. The idea is that instead of setting up a tree $B(10)$ with some numbers $a_1, \ldots, a_{10}$ at the nodes (a_i at node i), you keep the array $a_1, \ldots, a_{10}$ and say: For each i, the sons of i are $2i$ and $2i + 1$ (if those numbers are less than 11). Algorithms are written in these terms, but we won't go into them.

9. A Lower Bound

Any sorting method that is based entirely on binary comparisons must use (approximately) at least $n \log_2 n$ comparisons to sort some input of n numbers. We explain why this is so.

(9.1) *The decision tree.* To any sorting algorithm based only on binary comparisons (comparisons of the relative sizes of two numbers) there corresponds the binary **decision tree** of the algorithm. The first comparison ("Is $a < b$?") has two possible outcomes, $a < b$ or $b < a$.[†] For the second step, if $a < b$, then the algorithm tells you to compare a certain two things, and if $b < a$ it tells you to compare two possibly different things. In a diagram:

(9.1F)

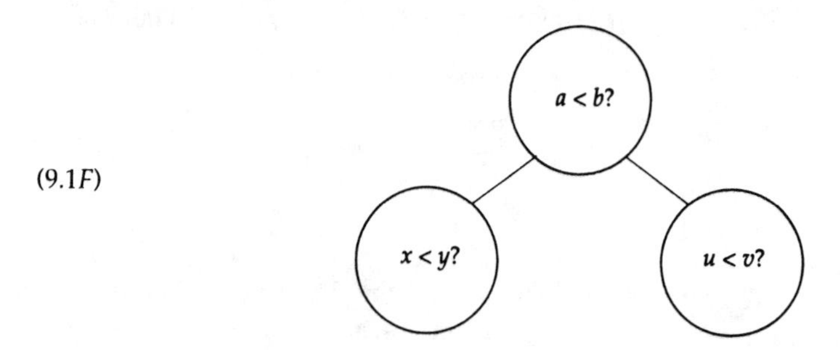

It has to look like this at every comparison, except at a last comparison, when it determines the sorted order of the input.

Example (9.2) Suppose we have three numbers a, b, c. A decision tree for sorting them is

(9.3F)

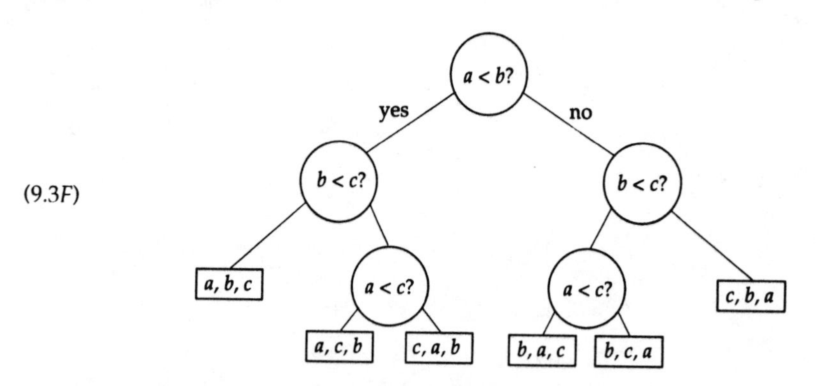

[†]We assume for simplicity that all input numbers are distinct.

Some inputs require three comparisons, others two. Notice that there are $6 = 3!$ leaves, the "box" nodes, where we find the input sorted.

The decision tree of an algorithm is not necessarily constructed or drawn, but it must exist.

If an algorithm is to sort each of the $n!$ possible permutations of n numbers input to it, its decision tree must have at least $n!$ leaves. If the height of this tree is h, then it has at most 2^h leaves (see Problem 14.5). Thus

(9.4) $$n! \leq \text{number of leaves} \leq 2^h,$$

so $h \geq \log_2(n!)$. But h is the number of comparisons used to sort each input having its box node at level h. Therefore

(9.5) For any sorting algorithm using only binary comparisons and for all $n \geq 1$, there is a least one input of n numbers the sorting of which requires $\log_2 n!$ or more comparisons.

(9.6) By Stirling's approximation (Chapter 5, (27.6)) we see that

$$\log_2 n! \sim \log_2 \left(\frac{n}{e}\right)^n \sqrt{2\pi n} = n \log_2 n - n \log_2 e + \frac{1}{2} \log_2 n + \log_2 \sqrt{2\pi};$$

in this relationship the dominant term is $n \log_2 n$. Thus the number of comparisons required for all large n is at least about $n \log_2 n$, as we set out to show.

10. Extended Binary Trees

(10.1) An *extended binary tree* is, by definition, a binary tree in which every node has zero or two sons.

Examples (□ := empty binary tree.)

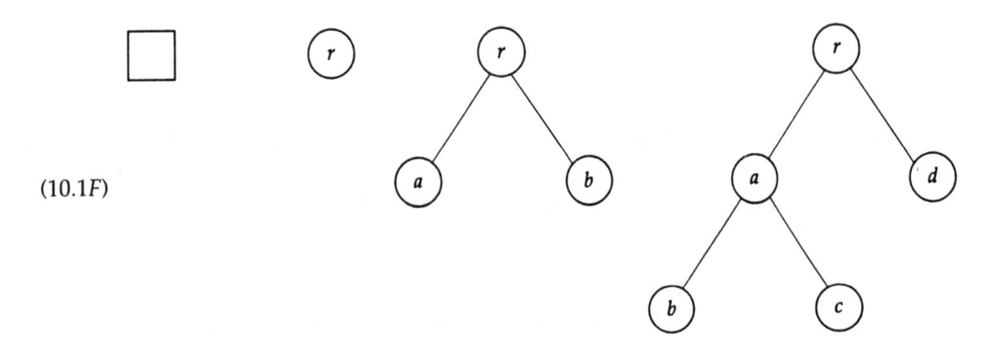

(10.1*F*)

We saw another such tree in Chapter 8, (9.6*F*), when we analyzed zipper merge. They also arise in algorithms for sorting and maximum-finding. We can even regard *any* binary tree as an extended binary tree if we use □ as a node. For example, the binary tree T

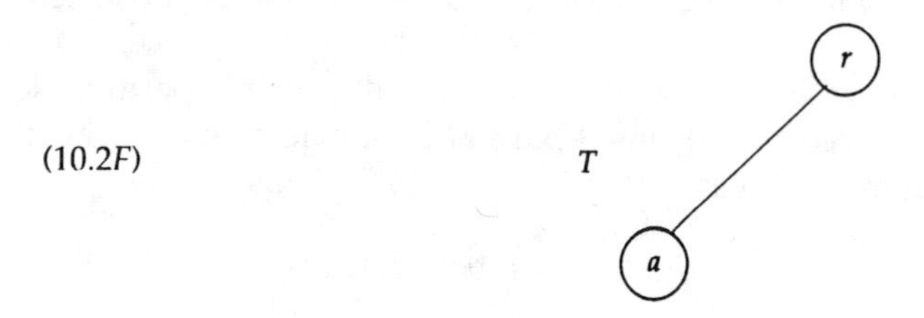

is not an extended binary tree, but we may regard it as the extended binary tree T_1:

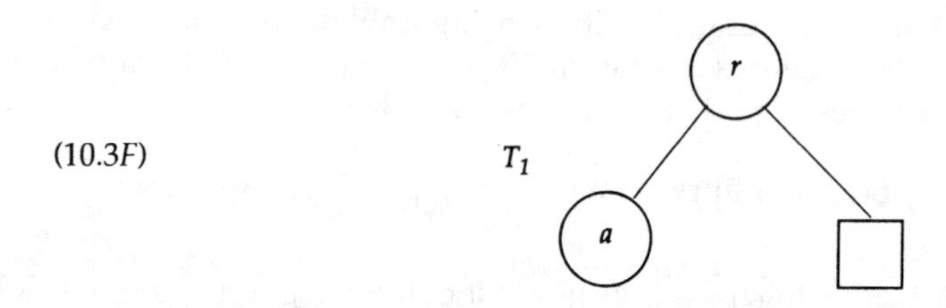

That is, T has an empty right subtree, which we denote explicitly when we draw □ in the graph. If we think of □ as a node, then T_1 is an extended binary tree.

(10.4) We now define, for any binary tree T, a certain extended binary tree $ex(T)$ by the following rule: We replace each empty subtree of T by a leaf. This is exactly the process of considering empty subtrees as nodes that we just discussed. For example, T might be as shown,

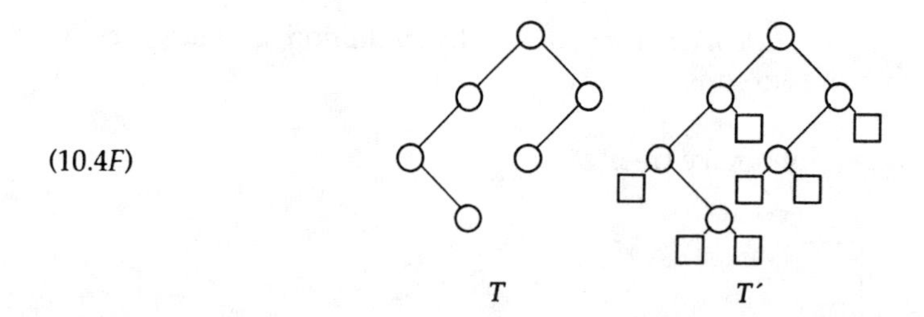

and T' is T with all the empty subtrees drawn. If we make each □ a leaf, we get $ex(T)$. In the rest of this section we'll use □ to stand for a leaf or for the empty binary tree; the meaning will be clear from the context. This function ex maps the set of all binary trees into itself.

Notice that

- $ex(T) \neq T$ for any binary tree T.
- $ex(T)$ includes T in a sense easy to grasp from a drawing and tedious to define in terms of ordered quadruples (4.1).

Example 10.5) Suppose our tree is B, sketched as

(10.5*F*) B

Then the following four extended binary trees all include B, but only the last is $ex(B)$.

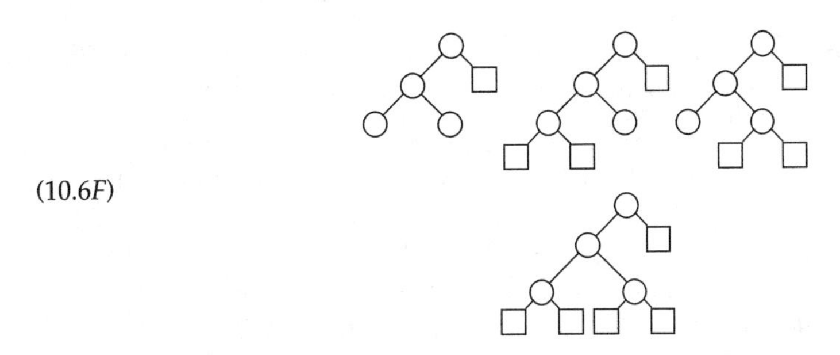

(10.6*F*)

Our results on extended binary trees apply to all such trees, not just to those obtained by the process (10.4). But see Problem 11.2.

Practice (10.7P) Find an extended binary tree B such that $B \neq ex(T)$ for any binary tree T.

DEFINITION (10.8) In an extended binary tree, any node with two sons is called an **internal** node; any leaf is called an **external** node.

Lemma (10.9) Let $n \in \mathbf{N}$ and let T be an extended binary tree with a total of n internal nodes. Then T has a total of $n + 1$ external nodes.

Proof We use induction on n. For $n = 1$ there is only the tree with a root and two leaves, for which the result is apparent.

Now assume the Lemma for all extended binary trees with n or fewer internal nodes, where n is a fixed but arbitrary positive integer. (Thus, as is usual with trees, we use the second form of induction.) Let T be any extended binary tree with $n + 1$ internal nodes. We prove the Lemma for T. Look at the sketch:

(10.10*F*) T r T_L T_R

The left and right subtrees of T are nonempty extended binary trees (why?), each on n or fewer internal nodes (because the root of T is an internal node). Say T_L has n' internal nodes, and T_R n'' of them. Then $n' + n'' = n$, and

$$\text{(10.11)} \qquad \begin{aligned} &T_L \text{ has } n' + 1 \text{ external nodes;} \\ &T_R \text{ has } n'' + 1 \text{ external nodes.} \end{aligned}$$

The result (10.11) comes from the induction assumption. But the set of external nodes of T is the union of those sets for T' and T'', which are disjoint, of course. Thus T has exactly $n' + n'' + 2 = (n + 1) + 1$ external nodes. QED

Practice (10.12P) At the inductive step, why are T_L and T_R nonempty extended binary trees, and why is the root of T an internal node?

Again, we emphasize that the way to do inductive proofs on trees is to remove the root, not a leaf. Don't use pruning shears; use a big ax. Trying to do the preceding proof by removing a leaf from T would not even change the number of internal nodes. Maybe we could lop off an internal node x having two external nodes as sons and replace x by a new external node, but first we'd have to prove that such x exists. It's *much* simpler to go from the top down, as in the proof above.

There is another remarkably simple relationship between internal and external nodes in any extended binary tree. It is between the internal and external path lengths.

We have already seen in analyzing **Treesort** that the levels of the nodes of a tree can give useful information about the process represented by the tree. We wanted to know the sum of the levels of all the nodes of the tree T. This quantity is the internal path length of the extended tree T' of T. We define that term and its counterpart:

DEFINITION (10.13) The **internal path length** $I(T)$ of an extended binary tree T is the sum of the levels of all the internal nodes; the **external path length** $E(T)$ is the sum of the levels of all the external nodes.

We proved in (8.3) that the minimum value of $L(B) := \sum \ell(x)$ for all binary trees B of n nodes was attained by the complete binary tree $B(n)$. If we set $T = B'$ we see that $L(B) = I(T)$, so we know that $B'(n)$ minimizes $I(T)$.

Theorem (10.14) For any extended binary tree T on n internal nodes,

$$E(T) - I(T) = 2n.$$

Proof Again we use induction on n. Clearly the Theorem is true for $n = 0$, for the tree T is then either the empty tree or the tree of just one node. In either case $E(T) = 0 = I(T)$, by (10.14) of Chapter 1 if necessary.

For the inductive step, we let n be a fixed but arbitrary nonnegative integer, and assume the Theorem for all extended binary trees on n or fewer internal nodes. Now we shall prove that any extended binary tree T on $n + 1$ internal nodes satisfies the Theorem. [This paragraph is boilerplate for tree proofs; it's plagiarized from the previous proof.]

Again the subtrees T_L and T_R of T are extended binary trees, each on n or fewer internal nodes. Therefore $E(T_L) - I(T_L) = 2n'$, and $E(T_R) - I(T_R) = 2n''$, by the inductive assumption, where n' and n'' are the numbers of internal nodes in T_L and T_R, respectively. The level of node x used in computing these quantities is its level in the subtree. In T its level is 1 greater.

Example (10.15) On the left is the level in T_L; on the right, in T.

(10.16*F*)

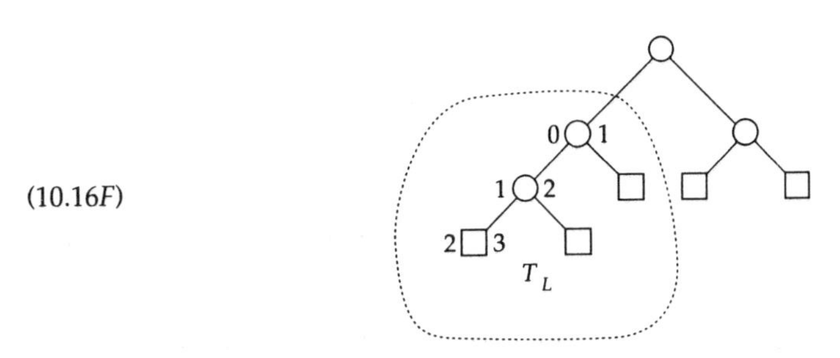

Thus

(10.17)
$$E(T) = E(T_L) + E(T_R) + n + 2,$$

since T has $n + 2$ external nodes by Lemma (10.9), and we must add 1 to the level of each of them in passing from either subtree to the full tree T.

Similarly

(10.18)
$$I(T) = I(T_L) + I(T_R) + n,$$

since T has n internal nodes other than its root. We combine (10.17) and (10.18) to get

$$E(T) - I(T) = E(T_L) + E(T_R) - I(T_L) - I(T_R) + 2,$$

and using the inductive assumption we get

$$\begin{aligned} E(T) - I(T) &= 2(n' + n'') + 2 \\ &= 2(n + 1). \end{aligned}$$

QED

Example (10.19) We calculate $E(T)$ for the extended binary tree T shown in (10.20*F*).

(10.20*F*)

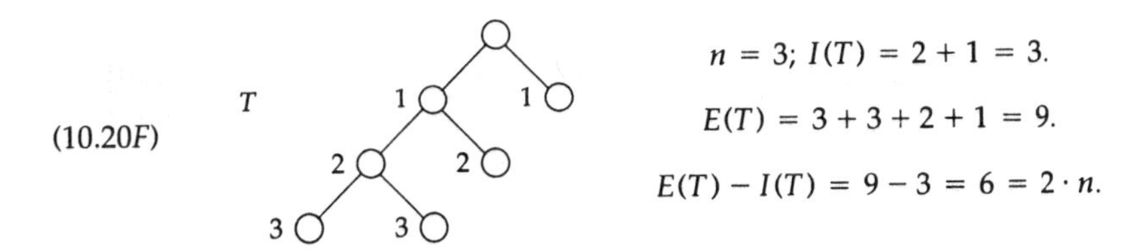

11. Counting Trees

(11.1) Here is a problem. How many binary trees are there on n nodes? For this count, two trees are the same if they have the same "shape"; we ignore any labels there may be on the nodes. Thus there is a total of two binary trees on two nodes, namely,

(11.1*F*)

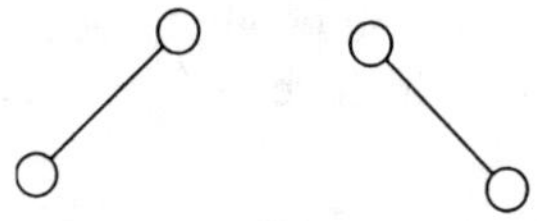

and as we saw in (4.6*F*) there are five on three nodes. Let us call the number we seek b_n. What we know so far is

(11.2)

n	0	1	2	3
b_n	1	1	2	5.

We'll now develop a formula for b_n, using mathematical recursion and polynomial multiplication (!)—see Chapter 8, (1.24) ff. The start is easy: Every binary tree on n nodes has two subtrees, in which the total number of nodes is $n - 1$. Thus if B_L [B_R] is the left [right] subtree of B, let B_L have exactly i nodes and B_R exactly j. Then $i + j = n - 1$, and $0 \leq j \leq n - 1$.

(11.3*F*)

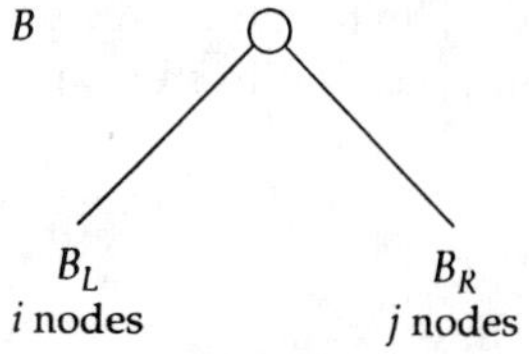

B

B_L
i nodes

B_R
j nodes

> For any fixed i, j with $i + j = n - 1$ there are exactly $b_i b_j$ binary trees on n nodes with i nodes in the left subtree and j nodes in the right.

This is the first key to the problem. (It follows from counting principles **C0** and **C2** of Chapter 9: B_L and B_R may be chosen independently of each other in b_i and b_j ways, respectively.) Therefore

(11.4)
$$b_n = \sum_{i+j=n-1} b_i b_j.$$

But (11.4) reminds us of the definition of polynomial multiplication (Chapter 8, (1.24)). And that is the second key. Let us define an infinite series

(11.5)
$$y := \sum_{0 \leq n} b_n x^n.$$

Then, by definition,

(11.6)
$$y^2 := \left(\sum_{0 \leq i} b_i x^i\right)\left(\sum_{0 \leq j} b_j x^j\right).$$

Thus

$$y^2 = \sum_{0\le k}\left(\sum_{i+j=k} b_i b_j\right)x^k \tag{11.7}$$
$$= \sum_{0\le k} b_{k+1}x^k,$$

from (11.4).

Now we see that on the right of (11.7) we almost have y. To improve the resemblance to y we multiply (11.7) by x:

$$xy^2 = \sum_{0\le k} b_{k+1}x^{k+1} \tag{11.8}$$
$$= b_1x + b_2x^2 + \cdots$$
$$= y - b_0 = y - 1.$$

Therefore

$$xy^2 - y + 1 = 0. \tag{11.9}$$

To find y we simply use the beloved quadratic formula:

$$y = \frac{1 \pm \sqrt{1-4x}}{2x}. \tag{11.10}$$

But since we want to know the coefficients of the power series for y about zero instead of the closed form for y in (11.10), we find the power series for $(1-4x)^{1/2}$. We use the binomial theorem (Chapter 8, (7.2)) for exponent $\frac{1}{2}$. We get

$$(1-4x)^{1/2} = \sum_{0\le k}\binom{1/2}{k}(-4x)^k \tag{11.11}$$
$$= 1 - \left(\frac{1}{2}\right)\cdot 4x + \frac{\left(\frac{1}{2}\right)\left(-\frac{1}{2}\right)}{2}4^2x^2 - \frac{\left(\frac{1}{2}\right)\left(-\frac{1}{2}\right)\left(-\frac{3}{2}\right)}{3!}4^3x^3 + - \cdots$$
$$= 1 - 2x - 2x^2 - 4x^3 - \cdots.$$

The nth term is

$$\frac{x^n\cdot\left(\frac{1}{2}\right)\left(\frac{-1}{2}\right)\left(\frac{-3}{2}\right)\cdots\left(\frac{1}{2}-n+1\right)\cdot(-4)^n}{n!} = \frac{-x^n\cdot 1\cdot 1\cdot 3\cdot 5\cdots(2n-3)2^{-n}\cdot 4^n}{n!}. \tag{11.12}$$

The sign is always negative (for $n \ge 1$) because there are $n-1$ negative factors in $\binom{1/2}{n}$ and n more in $(-4)^n$. We now multiply top and bottom by $(n-1)!$ and use $n-1$ of the 2's to make $2^{n-1}(n-1)! = 2\cdot 4\cdot 6\cdots(2n-2)$. We get for (11.12)

$$\frac{-x^n\cdot 1\cdot 2\cdot 3\cdot 4\cdot 5\cdots(2n-3)(2n-2)\cdot 2}{n!(n-1)!}.$$

We may write this quantity as

$$-x^n\frac{(2n-2)!}{(n-1)!(n-1)!}\frac{2}{n} = -x^n\cdot\frac{2}{n}\binom{2n-2}{n-1}. \tag{11.13}$$

Now we solve for y. From (11.11) we see that we must choose the minus sign in (11.10); otherwise y would have a term $1/x$, contradicting our definition (11.5) of y as a power series. With the minus sign we get

$$y = \sum_{1\le n} \frac{1}{n}\binom{2n-2}{n-1}x^{n-1};$$

(11.14)
$$y = \sum_{0\le k} \frac{1}{k+1}\binom{2k}{k}x^{k}.$$

We got the latter simply by changing the index via $n =: k+1$. But now look: (11.14) is in the same form as the definition of y in (11.5). Therefore, since power series are equal if and only if their coefficients are equal,

(11.15)
$$b_k = \frac{1}{k+1}\binom{2k}{k}, \quad \text{for } k = 0,1,2,\ldots.$$

This completes our solution of (11.1).

Practice (11.16P) Check that (11.15) agrees with (11.2) and with the further values $b_4 = 14,\ b_5 = 42,\ b_6 = 132$.

The numbers $1,1,2,5,14,\ldots,\frac{1}{(n+1)}\binom{2n}{n},\ldots$ are called *Catalan numbers*. They count lots of things besides binary trees; for example, b_n is the number of ways $2n$ people seated around a circular table can shake hands in n pairs without any arms crossing.† It is also the number of ways to parenthesize a product of $n+1$ terms $a_0a_1\ldots a_n$, for example, $a_0(a_1a_2)$ and $(a_0a_1)a_2$; $a_0(a_1(a_2a_3))$, $a_0((a_1a_2)a_3)$, $(a_0a_1)(a_2a_3)$, $(a_0(a_1a_2))a_3$, and $((a_0a_1)a_2)a_3$.

The method we used to find b_n was fairly standard. First, we found a recursion (11.4) for b_n. Second, we set up the generating function y for b_n in (11.5) and used the recursion to find an equation (11.9) that y satisfied. The recursion suggested the equation; since the recursion was nonlinear, so was the equation. Third, we solved the equation for y in the form of a power series.

We now estimate b_n by means of Stirling's approximation (Chapter 5, (27.6)), which says that

$$n! \sim (n/e)^n\sqrt{2\pi n}.$$

By (7.7) of Chapter 8,

(11.17)
$$\binom{2n}{n} = \frac{(2n!)}{(n!)^2} \sim \frac{\left(\frac{2n}{e}\right)^{2n}\sqrt{2\pi\cdot 2n}}{\left(\frac{n}{e}\right)^{2n}\cdot 2\pi n} = \frac{2^{2n}}{\sqrt{\pi n}}.\ddagger$$

Thus the Catalan number for large n is approximately

(11.18)
$$\frac{2^{2n}}{(n+1)\sqrt{\pi n}}.$$

†H. W. Gould, *Bell and Catalan Numbers*, (Research Bibliography of Two Special Number Sequences) (Morgantown, Unpublished manuscript 1976).

‡From (11.17) we see, incidentally, how small is the probability that if you flip an honest coin 100 times, you get exactly 50 heads: It is only about $1/\sqrt{50\pi} = 0.08-$.

The error in using (11.18) is less than 1 percent if $n \geq 13$. It overestimates b_n; for example, for $n = 4$ it gives 14.44 (versus $b_4 = 14$), for $n = 7$ it gives 436.7 (versus $b_7 = 429$). For $n = 13$ it gives 750,075.95 (versus $b_{13} = 742,900$).

12. The Equivalence Closure

To find the equivalence closure of a given relation we use the basic fact of an equivalence relation, namely, that it determines a unique partition of the underlying set, and conversely. That result is explained in Chapter 4, Section (6).

Example Suppose that a relation R is given by the graph G_R:

(12.1F)

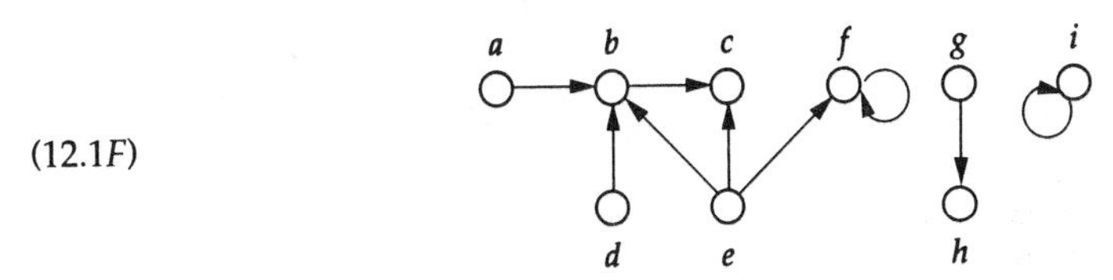

We can see instantly that its equivalence closure R_{eq} has to be

$$R_{eq} = A_1 \times A_1 \cup A_2 \times A_2 \cup A_3 \times A_3,$$

$$\text{where } A_1 = \{a, b, c, d, e, f\}$$

$$A_2 = \{g, h\}$$

$$A_3 = \{i\}.$$

That is, since in the graph G_R any two points of A_1 are "connected," there has to be an arrow between those points in R_{eq}.

A computer, however, can't draw a picture and take in the "connections" at a glance—and maybe we couldn't either if R were a big relation. So computer scientists created the following algorithm to find the cells (A_1, A_2, A_3 in the preceding example) of the partition.

(12.2) **Algorithm for the Equivalence Closure**

Input: a relation R

Output: the cells of the partition of $A :=$ (domain of R) $\cup$ (range of R) corresponding to the smallest equivalence relation including R.

Step 1. $P := \{\{a\}; a \in A\}$
Step 2. **while** $R \neq \varnothing$
Step 3. Choose $(v, w) \in R$
Step 4. Delete (v, w) from R
Step 5. Find $v \in W_1 \in P$ and $w \in W_2 \in P$
Step 6. If the cells W_1 and W_2 are different,
Step 7. Replace them with the union $W_1 \cup W_2$.
End

Notice that at Step 1 P is the partition of A into singleton cells, and at the end P is the partition of A corresponding to the equivalence closure R_{eq} of A.

In practice this algorithm would be considered too vague. It does not tell you *how* to find which cell v is in at Step 5, and it doesn't tell you how to make the union efficiently at Step 7, either. But those two operations are a little tricky and technical, so I've omitted the detailed **union-find** algorithm (as it's called) from this book. It's enough to say that the cells of P are represented as trees. These trees grow from single-node trees at Step 1 to trees of as many nodes as there are points in an equivalence class at the end.

I adapted this algorithm from (13.1, page 110).

Practice (12.3P) Represent the example of (12.1F) as a set R of ordered pairs. Then carry out Algorithm (12.2) on R.

13. Further Reading

(13.1) Alfred V. Aho, John E. Hopcroft, and Jeffrey D. Ullman, *The Design and Analysis of Computer Algorithms*, Addison-Wesley, Reading, Mass., 1974.

(13.2) Donald E. Knuth, cited in Chapter 3, (17.2).

14. Problems for Chapter 13

Problems for Section 1

These four problems concern Twenty Questions (1.1). The only question you may ask is: "Is $n \le a$?" for any integer a that you choose.

1.1 Prove by induction that if $N > 2^k$, then there are at least two numbers $n, n' \in \{1, \ldots, N\}$ that you cannot determine with k questions.

1.2 Hence prove that any deviation from the "cut in half" strategy described in (1.1) will lead to failure for at least two values of n. (Here $N = 2^{20}$.)

1.3 Suppose $N = 40$. How many questions would you need to be sure to determine any $n \in \{1, \ldots, 40\}$? Explain.

1.4 Adapt the game for any $N \in \mathbf{N}$ so that the answers give the coefficients in the base 2 expression for n.

Problems for Section 2

2.1 We may consider a tree T as a relation on the set $N(T)$ of nodes of the tree. For example, we define R on $N(T)$ as

$$a \; R \; b \text{ iff } b \text{ is a son of } a.$$

What can you say about R and the converse of R? What are the domain and range of R? Is R reflexive? symmetric? transitive?

Problems for Section 4

4.1 Using (4.6F) explain, without drawing, how many unlabeled binary trees there must be on four nodes if one immediate subtree is empty.

4.2 Draw the binary trees on four nodes in seven pairs so that the trees in a pair are symmetric to each other on reflection about a vertical line through the root. Use the result of the previous problem to help verify that the total number of trees here is indeed 14.

4.3 Prove Theorem (4.6): Let B be any binary tree on n nodes. If $n \ge 1$, set $n_i :=$ the total number of nodes of B having exactly i sons, for $i = 0, 1, 2$. Prove that $n_0 = 1 + n_2$.

4.4[Ans] Define $\max_n :=$ the maximum number (and $\min_n :=$ the least number) of leaves on any binary tree of n nodes.

(*i*) What is the value of $\min_n$? Explain.

*(*ii*) What is the value of $\max_n$? Explain.

(*iii*) Prove that for any integers $j, n \ge 1$ satisfying $\min_n \le j \le \max_n$, there is a binary tree on n nodes with exactly j leaves.

4.5 In the notation of Problems 4.3 and 4.4, describe and count the number of binary trees on n nodes for which $n_0 = \min_n$.

$*$**4.6**[Ans] In the notation of Problem 4.4, what are the values taken by the height of all binary trees on n nodes with $\max_n$ leaves?

Problems for Section 5

5.1 Draw the binary tree $\beta(F)$, where F is the forest:

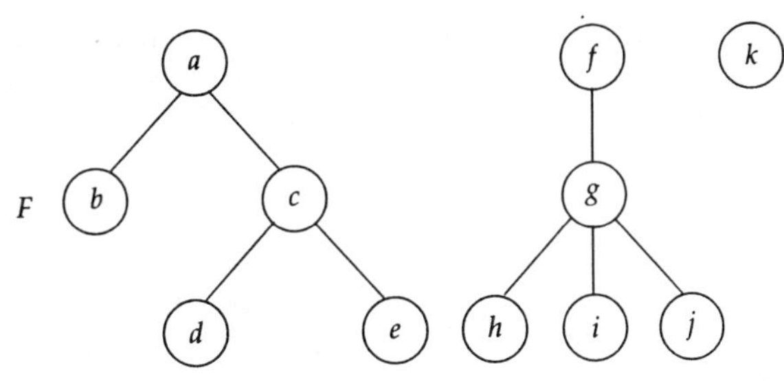

(14.1F)

5.2 Using the extended definition of β found in Problem 5.5, draw $\beta(F)$ and $\beta^2(F)$, where F is the ordered forest:

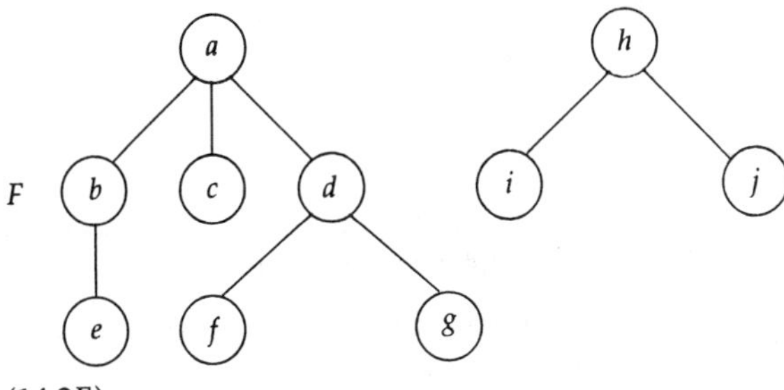

(14.2F)

5.3 What forest F satisfies $\beta(F) = B$ for the following binary tree B? Explain by reversing the process defined in (5.3).

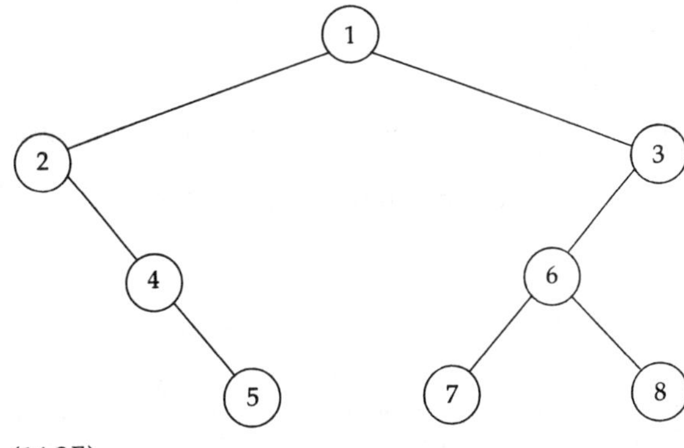

(14.3F)

5.4[Ans] Let X be a set. Let $\mathcal{F}$ be the set of all ordered forests F such that the nodes of F are elements of X. Let $\mathcal{B}$ be the set of all binary trees B such that the nodes of B are elements of X. Prove that $\beta : \mathcal{F} \longrightarrow \mathcal{B}$ is a bijection.

5.5 For the purposes of this problem let's extend the domain of the function β defined in (5.3) to include the set of all binary trees: If B is a binary tree, let T be the ordered tree with the same graph as B, and define $\beta(B) := \beta((T))$, where (T) is an ordered 1-tuple, the list of one item T in parentheses. Let F be any binary tree or forest on $n \geq 1$ nodes. Consider the sequence of binary trees $\beta(F), \beta(\beta(F)) =: \beta^2(F), \beta^3(F), \ldots$.

(*i*) Prove that for some m, $\beta^m(F) = \beta^{m+1}(F)$ and that if $k \geq m$, then $\beta^k(F) = \beta^m(F)$.

(*ii*) Can you find a lower bound for m?

(*iii*) What is the tree $\beta^m(F)$? Explain.

Problems for Section 6

6.1 The following tree is to be traversed in symmetric order. Write the nodes in the order traversed.

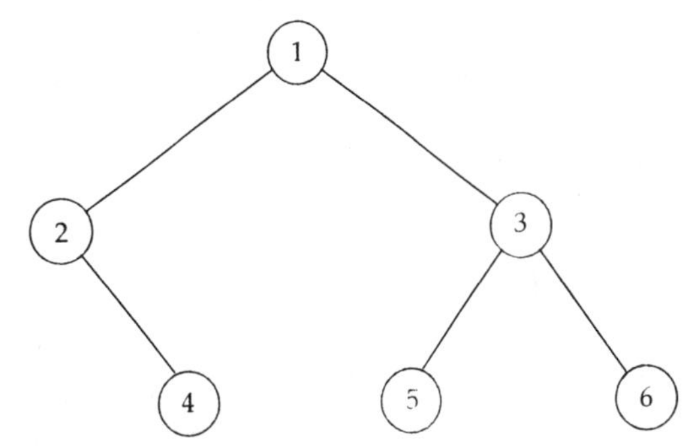

(14.4F)

6.2 List the nodes of the following binary tree in preorder and in inorder (symmetric order).

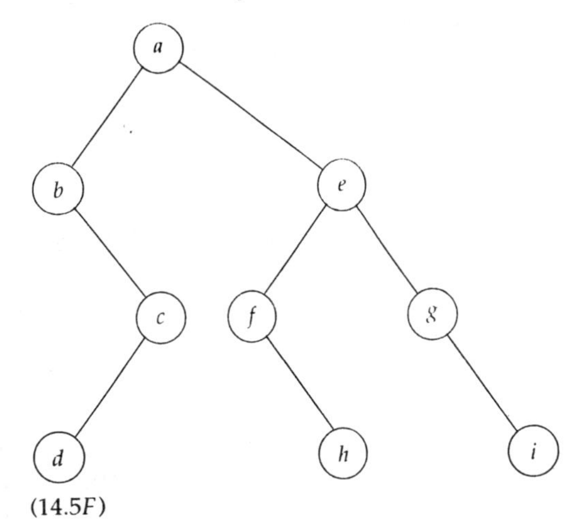

(14.5F)

6.3 The binary tree T is traversed in preorder. List the nodes in the order in which they are "visited." Do the same for postorder.

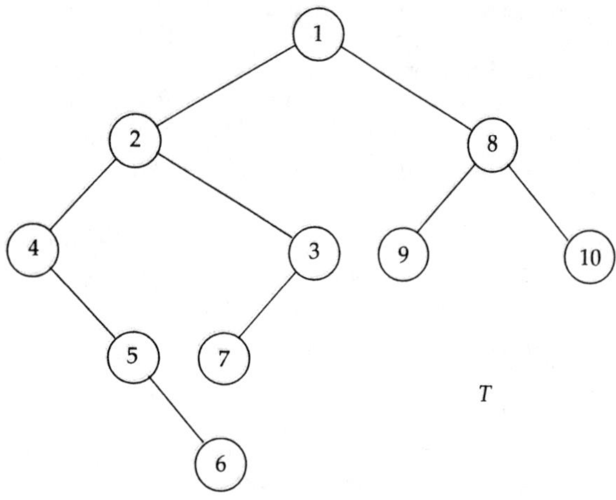

(14.6*F*)

6.4 Draw two different binary trees on the same set of eight nodes that are the same tree when viewed as ordinary trees. Then list the nodes for each of your binary trees in the order visited under the preorder and inorder traversals.

6.5 Find all binary trees the nodes of which are listed in the same order in both preorder and symmetric order. Explain.

6.6[Ans] Prove that the leaves of any binary tree appear in the same order (relative to each other) in preorder, symmetric order, and postorder traversals.

6.7 The following sequences of nodes are the orders of visit indicated for a binary tree:

preorder: A, B, F, C, D, G, E
inorder: B, F, A, D, G, C, E.

Draw the tree.

6.8 (*i*) Prove that a binary tree is uniquely determined by the preorder and inorder sequences of its nodes.

(*ii*) Prove the same, with *postorder* replacing *preorder*.

(*iii*) Show that (*i*) is false if *postorder* replaces *inorder*.

6.9 Prove by induction on the number of nodes that a binary tree has preorder traversal equal to the backward version of its postorder traversal if every node of the tree has no more than one son (e.g, if the postorder traversal is a, b, c, d, then the backward version of it is d, c, b, a.)

6.10 How many binary trees on four nodes labeled (somehow) a, b, c, d are there such that the inorder traversal is a, b, c, d? Explain.

***6.11**[Ans] Let a and b be nodes of a binary tree. Prove that b is a descendant of a if and only if a precedes b in preorder and a follows b in postorder.

Problems for Section 7

7.1 Assume that **Treesort** always inserts the first element of the sequence into the tree. How many comparisons does **Treesort** make in sorting the sequence

(*i*) $5, 6, 1, 3, 4, 2, 8, 7$?
(*ii*) $7, 3, 1, 8, 2, 6, 4, 5$?
(*iii*) $4, 6, 2, 5, 3, 7, 1$?

Explain your answer in each case.

Problems for Section 8

8.1 Prove $\lceil n_h/2 \rceil$ is the number of fathers of the n_h leaves on level h of the complete binary tree of height h.

8.2 Let $n \geq 1$. Suppose the complete binary tree $B(n)$ has nodes labeled $1, \ldots, n$, as in (8.13), by the rule

> For each $i, 0 \leq i < h$, the nodes on level i are labeled $2^i, 1 + 2^i, \ldots, 2^{i+1} - 1$ from left to right. The nodes on level h are labeled $2^h, 1 + 2^h, \ldots, n$.

Prove that for each $j \in \{1, \ldots, n\}$,

(*i*) j is the label of a leaf iff $j > n/2$.
(*ii*) If $j = n/2$, then j has exactly one son and that son is labeled $2j = n$.
(*iii*) If $j < n/2$, then j has two sons and they are labeled $2j$ and $2j + 1$.

8.3 Find a simple expression for the number of leaves on the complete binary tree $B(n)$. Prove your answer.

8.4 Let B be any binary tree of height h on n nodes. Prove that $n < 2^{h+1}$. Also fill in the gaps leading to (8.8): Prove that if B is the complete binary tree, then $2^h \leq n$.

8.5 Prove (8.8).

Problems for Section 10

10.1 (*i*) Prove by induction directly from the definition that the total number of nodes of an extended nonempty binary tree is odd.

(*ii*) Prove the same result more briefly.

10.2 Draw all the different extended binary trees on seven nodes. Calculate the internal and external path lengths for each. (It should be clear that you need make the calculations for only two of the trees.)

10.3 Use (4.6) to prove (10.9) differently than in the text.

10.4 Consider an extended binary tree B as an ordered tree by setting leftson = first son, rightson = second son. Map the forest $\{ B \}$ to a binary tree B_1 by the map β of (5.3). Prove that B_1 is not an extended binary tree if B has more than one node.

10.5 Prove that the mapping *ex* applied to any binary tree on $n \geq 0$ nodes introduces exactly $n+1$ new nodes.

10.6 (*i*) Prove that the mapping *ex* defined in (10.4) is an injection from the set of all binary trees to the set of all extended binary trees.

(*ii*) Prove that *ex* is a bijection between the set of all binary trees on n nodes and the set of all extended binary trees on $2n+1$ nodes, for $n \geq 0$.

10.7 Let n be odd. Prove that the extended binary trees on n nodes all have the same number L of leaves, and that the other binary trees on n nodes all have fewer than L leaves. (Hence the total number of binary trees on an odd number n of nodes with $\max_n$ leaves is the total number of extended binary trees on n nodes. This number is to be determined in Problem 11.1.)

Problems for Section 11

11.1 How many extended binary trees are there on $2n+1$ nodes, for each $n \geq 0$? [See Problem 10.6.]

11.2 For $n = 3$ and 4, count how many binary trees there are on n nodes with k leaves, for $\min_n \leq k \leq \max_n$. Try to do the same for $n = 5$ without simply drawing all the trees.

11.3[Ans] (*i*) Prove that b_n is even if n is even and positive.

*(*ii*) Prove that b_n is odd iff $n+1 = 2^k$ for $k = 0, 1, 2, \ldots$. [Hint. Look at reflections.]

In Problems 11.4 through 11.7, always input the leftmost remaining number from the input list to ***Kilmer****.*

11.4 List all the permutations S of 1,2,3,4 that yield the same tree as T (below) when we input S to **Kilmer**. Explain your answer.

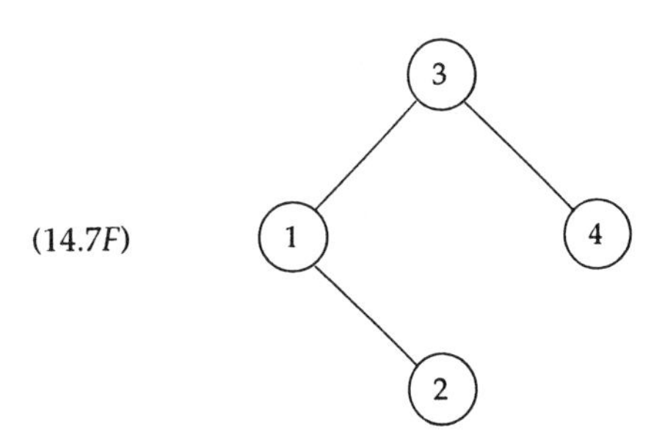

(14.7*F*)

11.5 Find the number of permutations of $1, \ldots, 8$ yielding the following tree when imput to **Kilmer**.

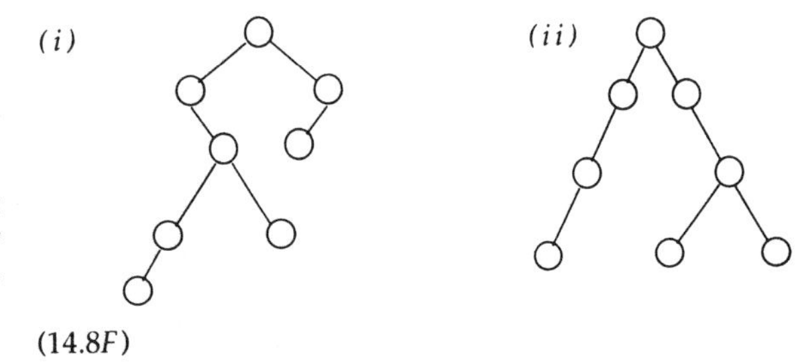

(14.8*F*)

11.6 Refer to Problem 11.5. How many permutations of $1, \ldots, n$ produce the complete binary tree $B(n)$

(*i*) for $n = 6$?

(*ii*) for $n = 8$?

(*iii*) for $n = 10$?

11.7 Find a formula for the total number T_m of permutations of $1, 2, \ldots, 2^m - 1$ yielding $B(2^m - 1)$ when input to **Kilmer**. For $m = 3$ and 4, evaluate your formula for T_m.

***11.8**[Ans] Let e_n denote the total number of binary trees on n nodes having the minimum height and the maximum number of leaves. Find a formula for e_n.

***11.9**[Ans] How many binary trees on n nodes are there having exactly two leaves?

General Problems

Some of these problems are weaker versions of prior problems.

G1. Prove that, in any nonempty binary tree on an even number of nodes, there is at least one node having just one son.

G2. Prove that for all odd n there is a binary tree on n nodes in which no node has exactly one son.

G3. Suppose you have a sequence of integers $n_0, n_1, \ldots, n_h$. Find necessary and sufficient conditions for the existence of a binary tree having exactly n_i nodes at level i, for $i = 0, 1, \ldots, h$. Prove your answer.

G4. Prove: If $\exists i \geq 1, n_i = 2^i$, then $n_{i-1} = 2^{i-1}$. Here n_i denotes the number of nodes in a binary tree at level i.

G5. Prove that any binary tree of height h has at most 2^h leaves.

G6. For all $n \geq 0$, prove that the number of ordered (unlabeled) trees on $n + 1$ nodes is equal to the number of binary (unlabeled) trees on n nodes. [Hint: You do not need to know the value of b_n of Section (11) to solve this problem.]

G7[Ans] How many binary trees on n nodes have the maximun number of leaves?

15. Answers to Practice Problems

(3.5P) The subtree with root *Coleoptera* has height 3, and the levels are named Order, Family, Genus, Species.

Order

Coleoptera

Family

Genus

Species

(15.1*F*)

If all nodes in the subtree with root *Coleoptera* had at most s sons, then the number of nodes at the Family level would be at most s, at the Genus level at most s^2, and at the species level at most s^3. If $s = 65$, then $s^3 = 274,625$, which is less than 275,000. So $s \geq 66$.

(4.8P) The left and right subtrees must have a total of three nodes. They can be distributed in four ways: three left and zero right, denoted (3,0). And also (2,1), (1,2), and (0,3). Thus from (4.6*F*) we see there are five possible left subtrees in the (3,0) case. In the (2,1) case there are two possible left subtrees and one possible right subtree, making $2 \cdot 1 = 2$ in this case. The other two cases are symmetric to these, so the answer is $2(5 + 2) = 14$.

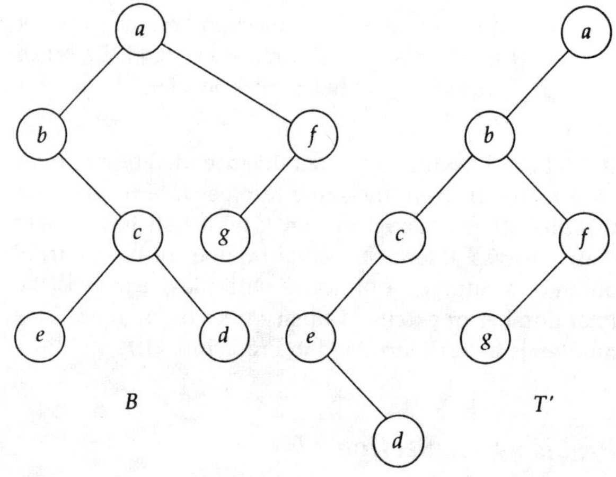

(15.2*F*)

(7.8P) B is a binary tree formed by **Kilmer** acting on a set S of $n + 1$ integers. We remove a leaf from B, getting a tree B'. Say the leaf holds $v \in S$. Then B' is the result of applying **Kilmer** to $S' := S - \{v\}$ if we insert the elements of S' in the same order as before. By the inductive assumption, **Treesort** $(S') = \text{Sym}(B')$ is S' sorted. That wasn't so bad, but now what? How do you show that v gets put in the right place at the $n + 1$ stage?

Maybe you can do it, but it's quite a drag, isn't it? You'd let x be the father of v, and consider two cases: $v < x$ or $v > x$. If $v < x$, you'd have to show that any $y \in S$ such that $y < x$ and $y \neq v$ satisfies $y < v$, because v immediately precedes x in Sym(B). And so on—if you want to do it this way, enjoy yourself.

(8.5P) Consider B and B' here:

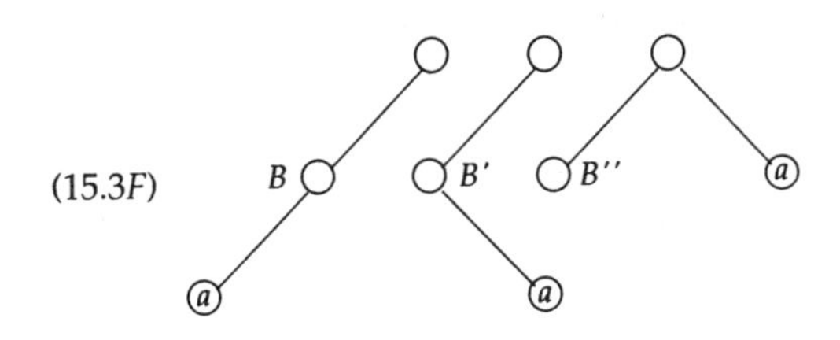

We should have turned B into B'', since $\min j = 1$ for B. Instead we patched the leaf into level 2, where it was to start with. The result is $L(B) = L(B')$—no decrease occurs.

Here is another example:

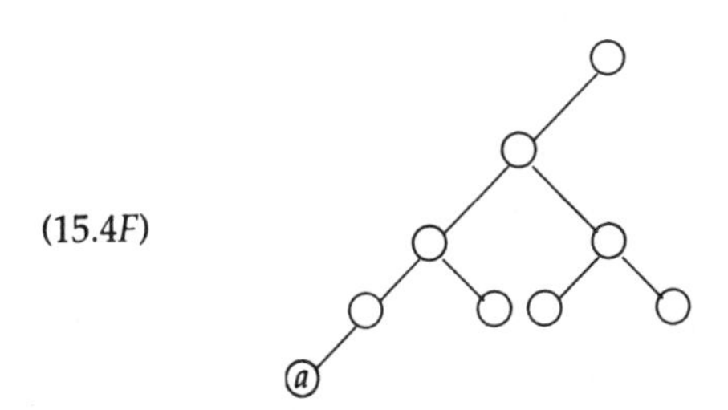

Here $n_i < 2^i$ for all $i \geq 1$. We should put the leaf a at level 1; if instead we try to put it at level 2 or 3, we find it's impossible.

In other words, there's a gap in the proof. If $n_i < 2^i$, then i is positive, because $n_0 = 1 = 2^0$. If j is the least such value of i, then $n_{j-1} = 2^{j-1}$. Therefore there is a node x at level $j-1$ without two sons. We may attach the leaf to x.

Moral: When you see "choose this" or "do that," make sure it's possible. In the proof we said "introduce a node at such level j," without showing it was possible to do so.

(8.11P) We proved after (8.8) that $h = \lfloor \log_2 n \rfloor$. Therefore

$$h = \log_2 n - \alpha_n$$

for some α_n with $0 \leq \alpha_n < 1$. Thus

$$\frac{h}{\log_2 n} = 1 - \frac{\alpha_n}{\log_2 n},$$

which tends to 1 as $n \longrightarrow \infty$. That is, $h \sim \log_2 n$, which is equivalent to the target result.

(10.7P) The answer is the empty binary tree. All extended binary trees on $n \geq 1$ nodes are in the range of *ex* (see Problem 10.6 *ii*).

(10.12P) We know $n > 0$, and T has $n+1$ internal nodes. Therefore the root of T has at least one son; and since T is extended, it has two sons. Thus the root of T is internal. Likewise, T_L and T_R are nonempty. They are extended because T is so: every node of T (hence of T_L and T_R) has exactly zero or two sons.

(11.16P) This is just a routine use of the binomial theorem (Chapter 8, (7.2)) for numerical calculation. Feller (Chapter 10, (17.1)) wouldn't even stoop to calling it "trite."

(12.3P) $R = \{ (a,b), (d,b), (b,c), (e,b), (e,c), (e,f), (f,f), (g,h), (i,i) \}$

The procedure (12.2) is this:

Step 1. $P := a|b|c|d|e|f|g|h|i$ in the partition notation of Chapter 4, (5.4).
Step 2. $R \neq \varnothing$
Step 3. (a,b) is chosen.
Step 4. $R := R - \{ (a,b) \}$.
Steps 5 and 6: $P := a,b|c|d|e|f|g|h|i$
Steps 2, 3, 4: $R \neq \varnothing$, (d,b) is chosen,
$R := R - \{ (d,b) \}$.
Steps 5, 6: $P := a,b,d|c|e|f|g|h|i$
Steps 2, 3, 4: $R \neq \varnothing$, choose (b,c), $R := R - \{ (b,c) \}$.
Steps 5, 6: $P := a,b,c,d|e|f|g|h|i$
Steps 2, 3, 4: $R \neq \varnothing$, choose (e,b), $R := R - \{ (e,b) \}$.
Steps 5, 6: $P := a,b,c,d,e|f|g|h|i$

You can complete it.

CHAPTER

14

Graphs

1. Introduction

(1.1) Try to draw the figure here without lifting your pencil and without drawing any line-segment twice. You may go through the points of intersection as many times

(1.1*F*)

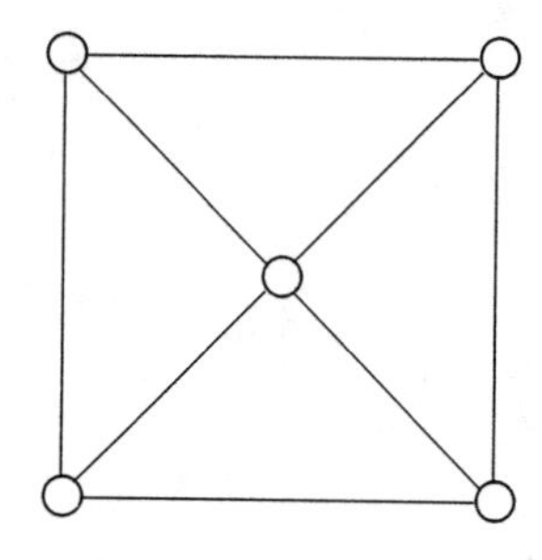

as you wish. Can't do it, can you? It's an old puzzle; maybe you've seen it.

(1.2) Now put a roof on it, as in (1.2*F*), and try to draw it.

(1.2*F*)

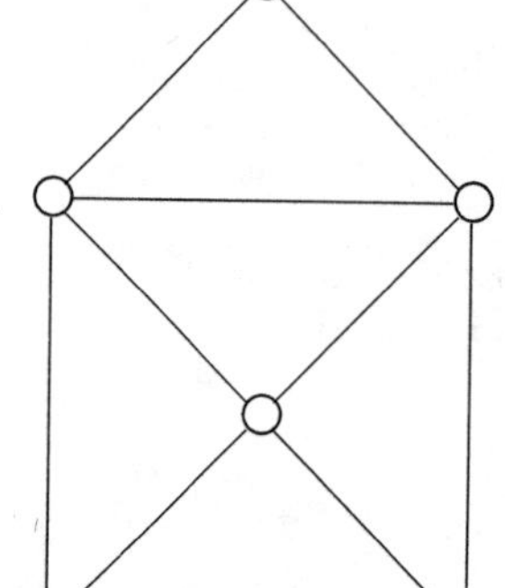

Miraculously, now you can do it.

Why can you draw the second but not the first? The surprisingly simple answer is the first result on graph theory in this chapter.

(1.3) The puzzles presented in (1.1) and (1.2) are simplifications of the first problem ever posed in graph theory. That was the famous problem of the seven bridges of Königsberg. In that Prussian city there is an island in the River Pregel flowing through the town. There are five bridges to the island and two more between mainland points. The artist's rendition shows all:

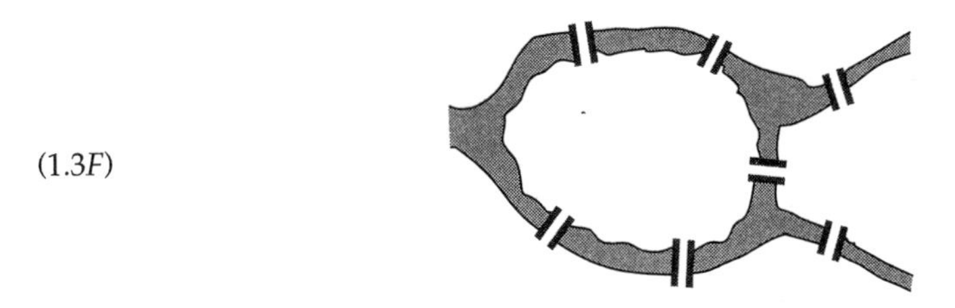

(1.3F)

People wondered if one could walk through the town in such a way as to cross each bridge exactly once. In answering the question in 1736, Euler founded the subject of graph theory, which, after long gestation, has grown enormously in the last fifty years. We now present his answer.

(1.4) First, we turn the problem into the same type of puzzle as those in (1.1) and (1.2). We name the mainland areas A, B, C and the island D. We represent those areas as points ○ and the bridges as lines joining two points. We get

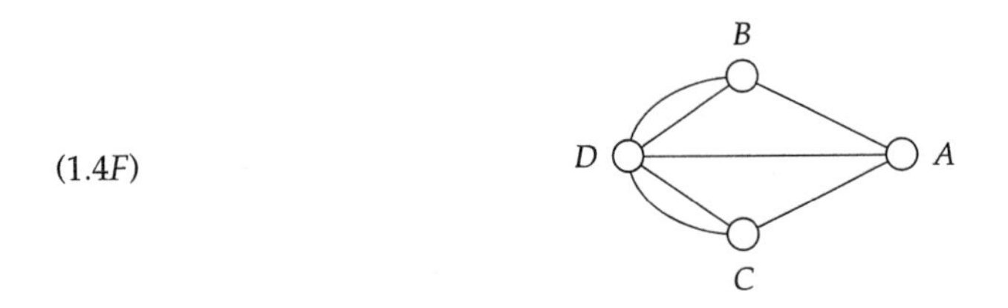

(1.4F)

Since the problem rules out swimming, it's the same as before: draw the figure without lifting the pencil or going over any line twice. If you try it you'll see that you can cross six bridges this way, but not all seven.

To present Euler's proof of the impossibility of this walk we'll need to introduce a few terms. We present those in Section 2 and solve the problem in Section 3.

2. Basic Concepts

DEFINITIONS (2.1) By the term **graph** we mean in this chapter an **undirected graph**, that is, a nonempty set V of points called **vertices** (singular: **vertex**) and a set E

of **edges** joining some pairs of vertices. We call the graph G, and our notation is

$$G = (V, E).$$

Examples (2.2) It's much easier to think of small graphs in terms of drawings than as sets. In these examples the vertices are the small circles ○ and the lines are the edges.

(2.2F)

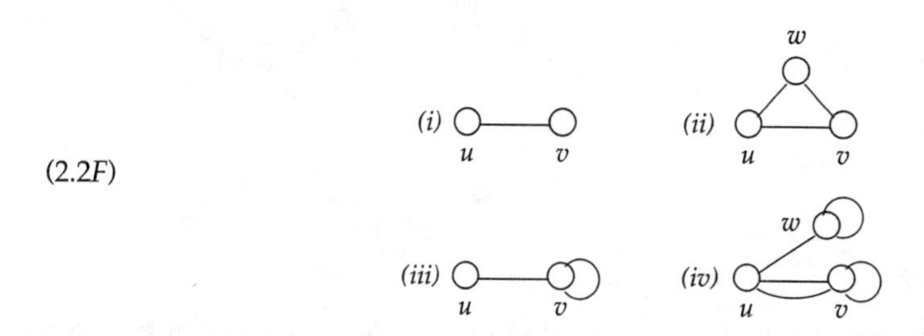

(2.3) You see that

- loops are allowed, and
- edges may be repeated.

(2.4) Loops were defined in Chapter 4: a **loop** is an edge joining a vertex to itself.

We've used the term *join* intuitively, motivated by the drawings (1.1F), (1.2F), (1.4F), and (2.2F). There are ways to make it precise (i.e., define it in terms of sets like everything else in mathematics). If you're interested, see references (16.2) and (16.3) for two different ways to do this. Sometimes we'll specify the edge between distinct vertices u and v as the 2-set $\{u, v\}$. If the edge is a loop at u, we might denote it as the 2-rep $[u, u]$.

We stay with our intuitive sense of *join*. An edge joins its endpoints. An edge is a line drawn between two (possibly equal) vertices, and those vertices are the **endpoints** of that edge.

Now we introduce the idea on which the solution of the Königsberg bridge problem depends:

(2.5) The **degree** of a vertex in a graph G is the total number of occurrences of that vertex as an endpoint of an edge of G.

NOTATION: $d(v)$

If there are no loops at v, then $d(v)$ is the number of edges with v as an endpoint. If there are ℓ loops at v and e nonloop edges with v as endpoint, then

$$d(v) = 2\ell + e.$$

The reason a loop contributes two to the degree of its vertex is that a loop is represented as a 2-rep, say $[v, v]$; and v occurs twice in $[v, v] =: e$ as an endpoint of e.

Examples (2.6) Here we've put the degree of each vertex next to it in the drawings.

(2.6F)

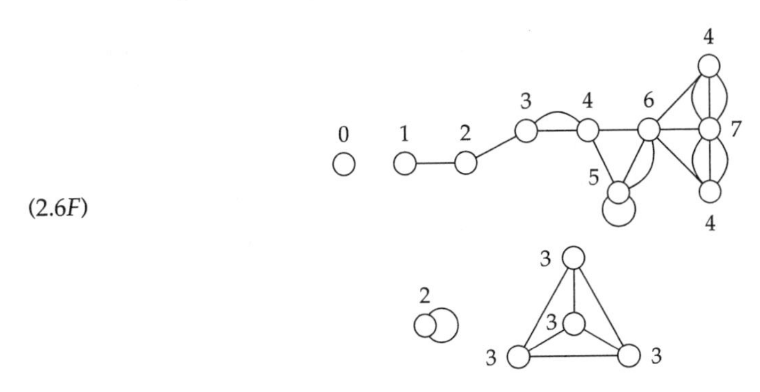

We also need to put three simple ideas on the table: *walk*, *cycle*, and *connectedness*.

DEFINITION (2.7) A **walk** in a graph is a sequence, alternating between vertices and edges, $v_0, e_1, v_1, e_2, \ldots, e_n, v_n$ (with $n \geq 0$) in which $v_0, \ldots, v_n \in V$ and $e_1, \ldots, e_n \in E$; and for each $i = 1, \ldots, n$,

$$\text{the endpoints of } e_i \text{ are } v_{i-1} \text{ and } v_i.$$

We say that this walk is **from** v_0 **to** v_n. The **length** of a walk is the number of edges in it (n in the preceding notation).

NOTE: We allow n to be zero in the definition of a walk. That is, for any vertex v, the sequence v denotes a walk of length zero from v to v.

In other words, imagine the graph as a map of a city showing only its bridges (as the edges) and the land areas (vertices) which the bridges join. Then our new term *walk* is a record of the bridges crossed in the order they were crossed in an ordinary walk around town; the land areas are also recorded to reduce ambiguity.

Examples (2.8) (*i*) In the graph

(2.8F)

e_1

v_0 v_1

e_2

there are walks

$$v_0, e_1, v_1 \text{ of length } 1,$$
$$v_0, e_1, v_1, e_2, v_0 \text{ of length } 2,$$
$$v_1, e_1, v_0, e_2, v_1 \text{ of length } 2,$$
$$v_0, e_1, v_1, e_1, v_0 \text{ of length } 2,$$
$$v_0, e_1, v_1, e_2, v_0, e_1, v_1 \text{ of length } 3.$$

The last two walks use edge e_1 twice. Because of the multiple edges you could not tell the starting vertex if you recorded only the edges, as e_1, e_2; both the second and third walks in this example have e_1, e_2 as the sub-sequence of edges. Similarly, listing only the vertices would fail to tell the story, as the second and fourth walks show.

(*ii*) Here there are walks of any length $n \geq 0$ consisting of n appearances of e in the sequence. We have v (length 0), v, e, v (length 1), v, e, v, e, v (length 2), and so on.

(2.9*F*) v e

(*iii*) Some possible walks in the following figure include $x,\ a,\ u,\ d,\ y;\ \ x,\ b,\ v,\ e,\ y,\ f,\ w$; and $x,\ a,\ u,\ d,\ y,\ f,\ w,\ c,\ x$.

The second walk is shown in bold lines.

(2.10*F*)

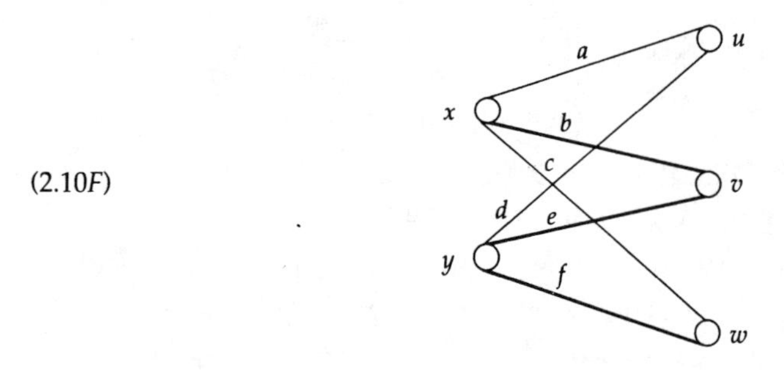

If the graph has no repeated edges, then we sometimes specify a walk by listing only its vertex-sequence. There is no ambiguity.

Example (2.11) In (2.10*F*) the three walks listed would in vertex-sequence notation be specified as

$$x,\ u,\ y$$
$$x,\ v,\ y,\ w$$
$$x,\ u,\ y,\ w,\ x.$$

We may also specify a walk by listing the first vertex and then the sequence of edges, as $x; b, e, f$, for example, in (2.10*F*).

(2.12) Instead of introducing new terms for special kinds of walks, we'll usually spell out what we want with extra words. For example, graph theorists call a walk that does not use any edge twice a *trail*, but we won't use that term. We'll just say "walk with no repeated edges." Otherwise this chapter might sink under the weight of the new vocabulary. Graph theory teems with terms.

The following definition is one of our few exceptions.

DEFINITION (2.13) A **cycle** in the graph G is a walk in G that begins and ends at the same vertex v, has at least one edge, and uses no edge twice nor any vertex except v twice.

Example (2.14) Consider the bow-tie graph in (2.14*F*). It has two cycles.

(2.14*F*)

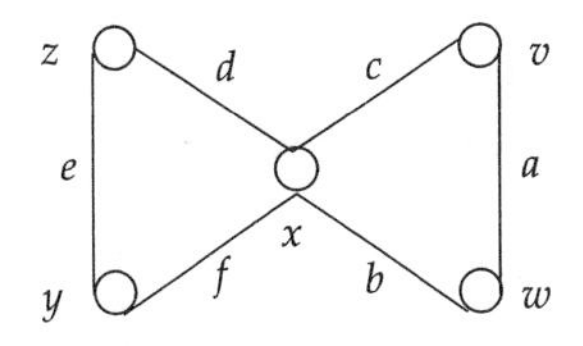

One is

$$v,\ a,\ w,\ b,\ x,\ c,\ v.$$

Another is

$$x,\ d,\ z,\ e,\ y,\ f,\ x.$$

It has a walk that is not a cycle, but that starts and ends at the same vertex, namely,

$$v,\ c,\ x,\ f,\ y,\ e,\ z,\ d,\ x,\ b,\ w,\ a,\ v.$$

The vertex x appears twice in the list.

(2.15) Now we define *connectedness*. It is the most intuitive of all ideas about graphs. A graph is *connected* if it is all in one piece, as for example in (2.16*F*):

(2.16*F*)

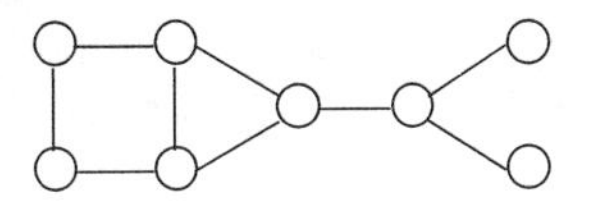

It is *disconnected* if it has two or more pieces; the graph in (2.17*F*) has three pieces.

(2.17*F*)

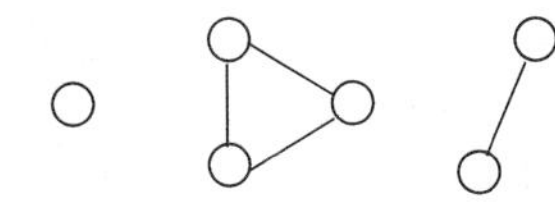

We make the definition precise using the idea of walk.

DEFINITION (2.18) A graph is **connected** iff for any two vertices u and v there is a walk from u to v.

We can even use the idea of walk to define the "pieces" so casually mentioned before. We also use the idea of equivalence relation (Chapter 4). Consider the vertex-set V of a graph G. Define a relation W on V by the rule

(2.19) $\quad \forall u, v \in V, \quad uWv$ iff there is a walk in G from u to v.

It's obvious that W is an equivalence relation. (It is reflexive since n can be zero in the definition (2.7) of walk.) The equivalence classes are the pieces mentioned earlier; in a graph we call them the *components*. Thus a graph is connected iff it has only one component; the graph of (2.17F) has three components.

The term *component* not only means the subset of vertices of G making up an equivalence class of W, but it also includes all the edges of G that join those vertices. Thus a component of G is a graph in its own right.

DEFINITION (2.20) A **component** of the graph $G = (V, E)$ is a graph $G_1 = (V_1, E_1)$ with vertex-set V_1 equal to an equivalence class of the relation W defined in (2.19) and with edge-set E_1 equal to the set of all edges of G having both endpoints in V_1.

(2.21) *The edge-degree lemma.* As a coda to this section we present an absurdly simple but useful result relating the number of edges of a graph and the degree of its vertices. It's a natural remark as soon as the word *degree* has appeared. As in (2.5), we denote by $d(v)$ the degree of the vertex v.

Lemma (2.22) In any graph $G = (V, E)$,

$$\sum_{v \in V} d(v) = 2|E|.$$

In words: the sum of the degrees of the vertices is twice the number of edges.

Proof If we sum $d(v)$ we have counted at each vertex the number of edges having that vertex as an endpoint. As we run through all the vertices, we encounter every edge exactly twice, because it has two endpoints. That completes our first proof of (2.22).

Actually we have here a flag count, however (Chapter 9, (8)). Assume for the moment that G has no loops. Define a set F of *flags* as

$$F := \{ (v, e);\ v \in V,\ e \in E,\ v \text{ is an endpoint of } e \}.$$

We count F by the first coordinates to find

(2.23)
$$|F| = \sum_{v \in V} d(v),$$

since for each v there are exactly $d(v)$ edges having v as an endpoint.

Now count F by the second coordinate. Since every edge has two endpoints, there are exactly two ordered pairs (v, e) in F for each e in E. Therefore

$$|F| = 2|E|.$$

With (2.23) this equation yields the lemma in the loopless case.

If now we introduce a loop at v, we add one edge and increase the degree of v by two. Thus the equation

$$\sum_{v \in V} d(v) = 2|E| \tag{2.24}$$

still holds. We may introduce any number of loops and still preserve equation (2.24). So the lemma holds for any graph; that's our second proof.

There is a more intuitive way to express this flag-count argument. Notice that F is a binary relation from V to E, i.e., a subset of $V \times E$. Consider the relation-matrix of F, called the **incidence matrix** of the graph G. The rows are indexed by the vertices and the columns by the edges of G.

Example (2.25) The graph G_1

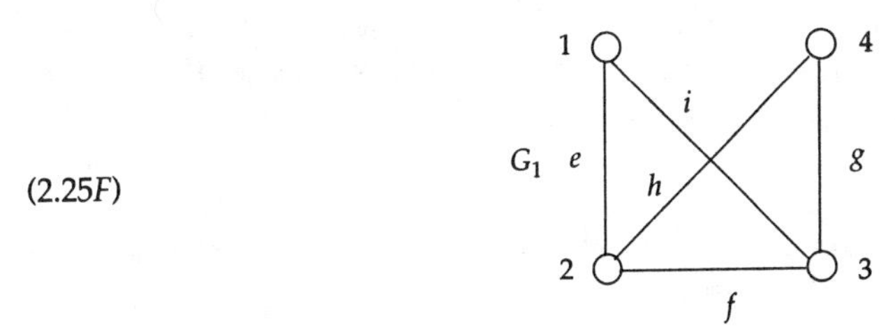

has incidence matrix

	e	f	g	h	i
1	1	0	0	0	1
2	1	1	0	1	0
3	0	1	1	0	1
4	0	0	1	1	0

The incidence matrix can handle repeated edges and even loops if we allow the entry "2." Here is an example.

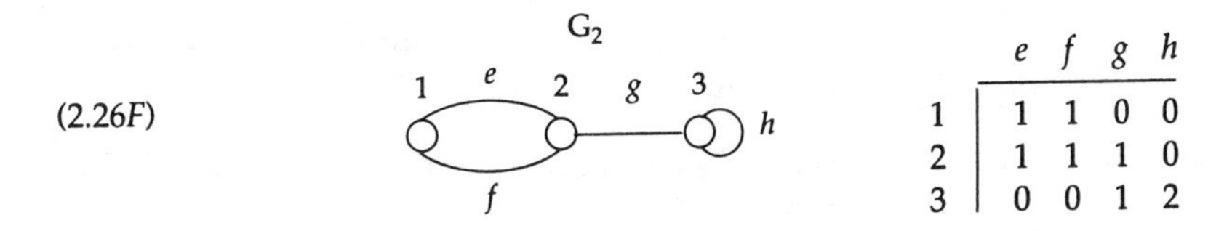

	e	f	g	h
1	1	1	0	0
2	1	1	1	0
3	0	0	1	2

We prove the edge-degree lemma as follows. For each vertex v, $d(v)$ is the sum of the entries in row v of the incidence matrix. So $\Sigma_v d(v)$ is the sum of all the entries in the matrix.

Now sum a column. The sum is two, because each edge has two endpoints. Therefore the sum of all the entries in the matrix is twice the number of edges.

That ends the third proof. QED

(2.27) We call a vertex v

even if $d(v)$ is even,

odd if $d(v)$ is odd.

The obvious edge-degree lemma (2.22) has a less obvious consequence:

Corollary (2.28) In any graph the total number of vertices having odd degree is even.

Proof On the left-hand side of

$$\sum_{v \in V} d(v) = 2|E| \tag{2.29}$$

the total number of odd integers must be even, or the sum would not be even. QED

Alternatively, we could reduce equation (2.29) mod 2, using ideas of Chapter 7. This reduction yields zero for each even vertex on the left, and one for each odd vertex. The result would be

$$n_{\text{odd}} \equiv 0 \pmod 2,$$

where n_{odd} denotes the total number of vertices of odd degree in G. That's a restatement of the Corollary.

3. Euler Tours

Now we can state and prove Euler's theorem.

Theorem (3.1) (Euler, 1736). Let $G = (V, E)$ be a connected graph.

(*i*) If G has a walk that uses each edge of G exactly once, then the total number (n_{odd}) of vertices of odd degree is zero or two.
Conversely,

(*ii*) if $n_{\text{odd}} = 0$, then such a walk exists, and it may start at any vertex. It must end at the same vertex.

(*iii*) If $n_{\text{odd}} = 2$, then such a walk exists. It must start at a vertex of odd degree and finish at the other one. ∎

Before proving (3.1) we present some examples.

Examples (3.2) (*i*) We traverse the graph of (3.2F)

(3.2F) v ○———○ w $\quad d(v) = d(w) = 1,$

by going from v to w along the only edge, or from w to v. This is a case where $n_{\text{odd}} = 2$.

(*ii*) In the following triangle,

(3.3*F*)

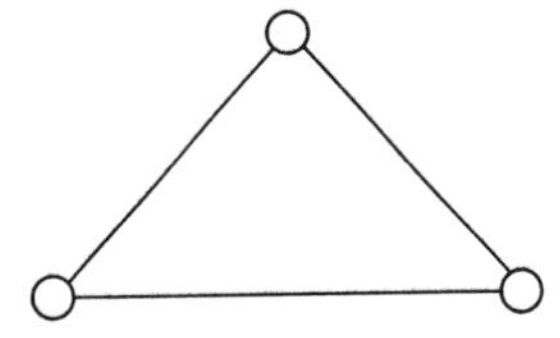

each vertex has degree 2, so $n_{\text{odd}} = 0$. Obviously we may start anywhere and must then finish at the starting vertex.

(*iii*) In Figure (1.1*F*) there are four vertices of odd degree:

(3.4*F*)

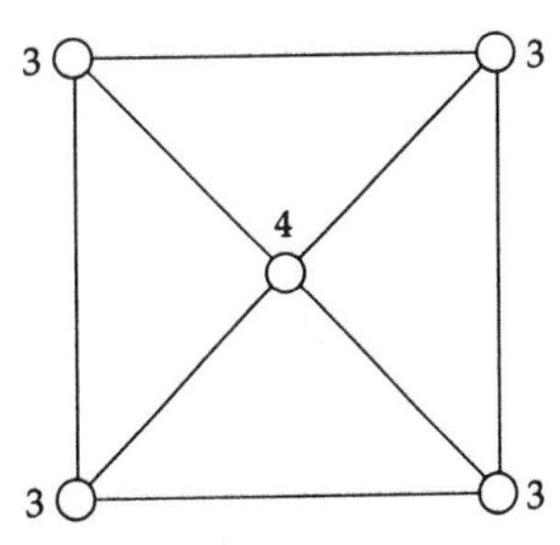

According to Theorem (3.1) no walk using each edge exactly once exists in this graph.

(*iv*) In Figure (1.2*F*), however, there are only two vertices of odd degree:

(3.5*F*)

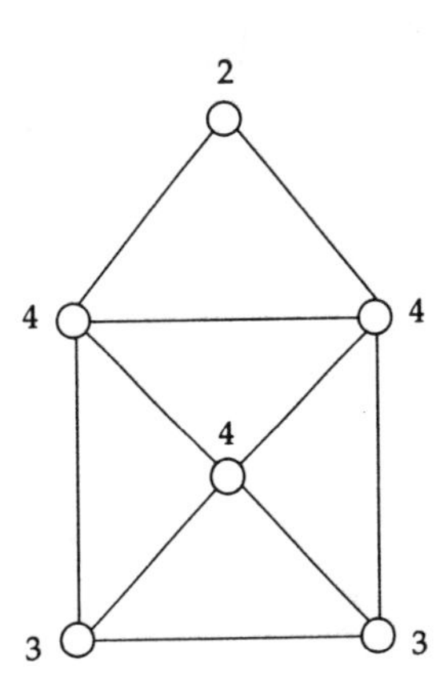

According to the theorem there is a walk starting at either vertex of odd degree, ending at the other, and using each edge exactly once. In fact, you will see from the proof of Euler's theorem (3.1) that there must be many such walks.

At this point it will be convenient to introduce a new term, *Euler tour.*

DEFINITION (3.6) An **Euler tour** is a walk in a graph that uses each edge of the graph exactly once.

You may also see *trail, path, cycle,* or *circuit* in place of *tour* in other books. The latter two terms are for tours that end at the starting point. Bringing in the term *Euler tour* goes against (2.12), but it will save lots of writing; and the *Euler* in it makes it easy to remember what it means.

Proof (3.7) The whole proof of Euler's Theorem (3.1) rests on one simple idea: In any walk not using any edge twice, we use up *two* of the edges ending at v as we "pass through" a vertex v.

(3.7*F*)

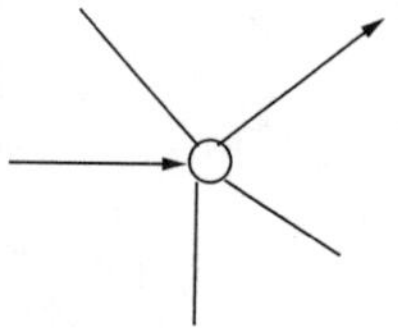

We're trying to make an Euler tour, so once we use two edges ending at v, we can't use them again. We can use the other edges ending at v, however, and they are $d(v) - 2$ in number. As we use two edges and remove them from consideration, the number of remaining edges ending at v has the same *parity* as before; that is,

$$d(v) - 2 \text{ is } \begin{cases} \text{even if } d(v) \text{ is even} \\ \text{odd if } d(v) \text{ is odd.} \end{cases} \tag{3.8}$$

In particular, as we try to construct an Euler tour,

(3.9) if $d(v)$ is even, then any time we can enter v on an edge not used yet, we can leave v on another edge not previously used.

Practice (3.10P) Explain (3.9).

Now for the proof. To prove (3.1*i*), assume G has an Euler tour. For this Euler tour define for each vertex v

$$\begin{aligned} t_{\text{in}}(v) &:= \text{the number of entrances into } v \text{ in this tour;} \\ t_{\text{out}}(v) &:= \text{the number of exits from } v \text{ in this tour.} \end{aligned} \tag{3.11}$$

Since an Euler tour uses each edge exactly once, counting entrances into and exits from a vertex is to count the degree of that vertex:

$$d(v) = t_{\text{in}}(v) + t_{\text{out}}(v). \tag{3.12}$$

Either the tour ends at the starting vertex or not.

Case 1. The tour ends where it begins. Then the tour enters each vertex exactly as many times as it leaves that vertex. That is, $t_{\text{in}}(v) = t_{\text{out}}(v)$, so by (3.12) $d(v)$ is even for all vertices v. Therefore $n_{\text{odd}} = 0$.

Case 2. The tour starts at u and ends at $v \neq u$. Let w be any vertex different from u and v. The argument of Case 1 applies to w, showing $d(w)$ is even.

The tour may visit u (or v) more than once, but every time it passes through u after the start it uses up two edges ending at u. The tour enters on one of those edges and exits on the other one. Therefore the tour exits u ($t_{\text{out}}(u)$ times) once more than it enters u ($t_{\text{in}}(u)$ times). Thus $t_{\text{out}}(u) = 1 + t_{\text{in}}(u)$. As before,

$$d(u) = t_{\text{in}}(u) + t_{\text{out}}(u) = 1 + 2t_{\text{in}}(u),$$

so $d(u)$ is odd. Similarly, $d(v)$ is odd. Thus $n_{\text{odd}} = 2$.

We've proved that if there is an Euler tour, then $n_{\text{odd}} = 0$ and the tour ends where it starts, or $n_{\text{odd}} = 2$ and the ending and starting vertices differ but are the two vertices of odd degree.

The converse has two parts, (3.1*ii* and *iii*). We first dispose of the presence of loops in G. It's obvious that G has an Euler tour if and only if G' has one, where G' is the same graph as G except that all loops have been removed. And removing loops at a vertex preserves the parity of the degree of that vertex. Therefore we assume that G has no loops.

Practice (3.13P) Explain why G has an Euler tour iff G' has one.

To prove (3.1*ii*), assume $n_{\text{odd}} = 0$. Let v be any vertex. We now construct a walk W. Choose any edge a ending at v. Since a is not a loop, the other endpoint w of a is not v. Since $d(w)$ is even, there must be at least one edge other than a ending at w. Choose, for the second edge of the walk, any such edge. Now there are $d(w) - 2$ unused edges ending at w, still an even number. Remember (3.9): if we can enter a vertex of even degree, we can exit from it.

Only at v is the number of unused edges odd (so far). We continue from w in this way until we can go no farther; that point must be v.

Practice (3.14P) Explain why the process ends at v.

At this point we may have included all the edges of the graph in our walk W. If so, we're finished; we have an Euler tour. If not, there remains by (3.8) an *even* number of unused edges at each vertex, so we'll complete the proof of (*ii*) by induction on the number of edges.

We interrupt the proof for some examples.

Example (3.14) Suppose we chose edges a, b, c as the first three of our tour in this graph:

(3.14*F*)

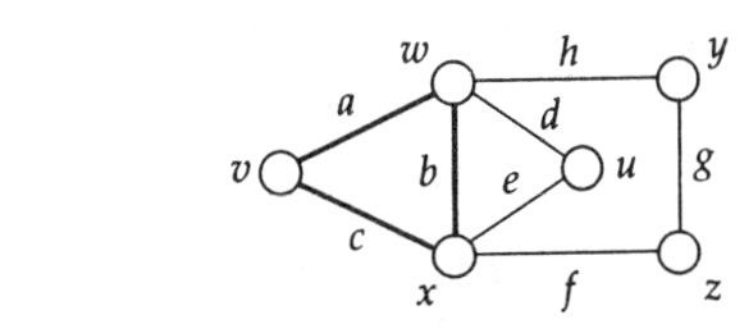

We'd end at v without using all edges. But the number of unused edges ending at each vertex is still even:

v	w	x	y	z	u
0	2	2	2	2	2

Thus we make an Euler tour of this *remaining graph* starting at x or w and patch it into the first one. For example,

$$w,\ d,\ u,\ e,\ x,\ f,\ z,\ g,\ y,\ h,\ w. \tag{3.15}$$

This tour ends where it starts, as it must, at w. The first walk was

$$v,\ a,\ w,\ b,\ x,\ c,\ v\ .$$

We simply replace w in the latter list by the whole of (3.15) to get an Euler tour of the graph. The sequence of edges is then

$$a,\ d,\ e,\ f,\ g,\ h,\ b,\ c.$$

Example (3.16) Sometimes the unused edges remaining after the first partial tour is constructed form a disconnected graph. Consider the graph in (3.16*F*). Suppose our first tour, starting at v, were a, b, c, d. Then we'd have two components left to tour, but each one, of course, satisfies the condition $n_{\text{odd}} = 0$, so we find Euler tours for each and paste them into our first tour.

(3.16*F*)

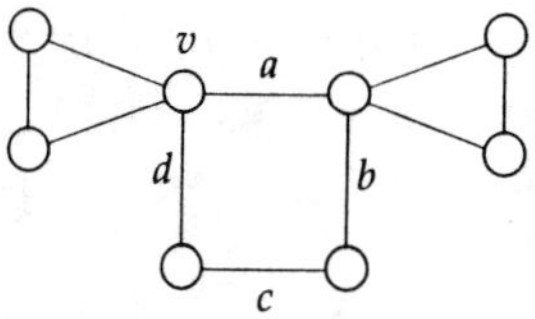

Back to the proof: It will take some work to justify the intuitively obvious pasting procedure of these examples. Here goes. Let $E_0 \subseteq E$ be the set of all edges of G not used in the walk W. Let V_0 be the set of all endpoints of edges in E_0. Set $G_0 := (V_0, E_0)$. The *remaining graph* G_0 may not be connected, but in each component of G_0, every vertex is even.

To show that we can really do the pasting, we now need to prove that

(⋆) each component of G_0 has a vertex in common with W.

So let u be any vertex in any component C of G_0. Since G is connected, there is a walk W_1 in G from v to u.

(⋆⋆) If an edge e of W_1 has its endpoints in different components of G_0, then e is in W.

Why is (⋆⋆) true? Its contrapositive is obvious from the definition of *component*. Let W_1 be

$$W_1 : v = u_0, e_1, u_1, e_2, \ldots, e_n, u_n = u.$$

Let u_i be the first vertex of W_1 in C. (Since u is a vertex of C, u_i exists.)

If $u_i = v$, then we've proved ($\star$) for C. But if $u_i \neq v$, then we'll show that e_i is an edge of W; that implies that u_i is in W as well as in C. We thus consider the preceding vertex u_{i-1} in W_1. There are two possibilities:

Case 1. If u_{i-1} is in a component C' of G_0, then $C' \neq C$, so by ($\star\star$), e_i is in W.

Case 2. If u_{i-1} is not in any component of G_0, then W has used all the edges through u_{i-1}, so e_i in particular is in W. Thus we've proved ($\star$).

By the inductive assumption ("by induction," we often say), each component of G_0 has an Euler tour starting and ending at any of its vertices; we paste the appropriate tour into W. We've completed the proof of (*ii*).

Now to prove (*iii*). First we restate it:

To prove: If $n_{\text{odd}} = 2$, then an Euler tour exists. It must start at a vertex of odd degree and finish at the other one.

So now we have two vertices of odd degree, say v_1 and v_2. To prove (*iii*) we first appeal to (*i*), which says that if a tour exists when $n_{\text{odd}} = 2$, then it must start at one of v_1, v_2 and finish at the other one. So we know our only hope to prove existence is to start at v_1 or v_2.

Now it's just like the previous case. Choose an edge e ending at v_1 (say) to start the walk. That leaves an even number of unused edges with v_1 as an endpoint (since e is not a loop). We choose edges one after the other as before, never using any edge twice, always continuing if possible. Our walk—call it W—must end at v_2.

If W hasn't used all the edges, the graph G_0 of the remaining (unused) edges has even degree at every vertex. Therefore each component of G_0 has an Euler tour (again by induction—or by (*ii*)). The same proof as that for (*ii*) shows that W goes through some vertex of each component of G_0. Therefore it's possible to paste the Euler tours of the components into the walk W to satisfy (*iii*). QED

Example (3.17) Consider the graph of (3.17*F*).

(3.17*F*)

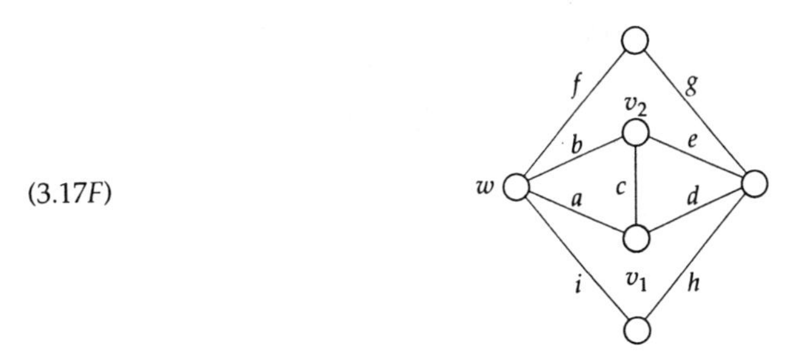

The only odd vertices are v_1 and v_2, of degree 3. If we start at v_1 and take the edges labeled a, b, c, d, e, then we end at v_2 and can go no farther. What remains is a square, which we may obviously traverse by an Euler tour, say f, g, h, i, starting at w. We paste that into the first tour as

$$v_1 : a,\ \mathbf{f},\ \mathbf{g},\ \mathbf{h},\ \mathbf{i},\ b,\ c,\ d,\ e.$$

4. Free Trees

DEFINITION (4.1) We define a **free tree** as a special kind of graph, namely, a connected graph with no cycles.

Three free trees are drawn in (4.1*F*).

(4.1*F*)

Free trees come up in many theoretical and applied studies of graphs or things modeled by graphs.

Our first result on free trees is an alternate definition, a characterization of them. You can see in the free trees of (4.1*F*) that

(4.2) For all vertices x, y there is exactly one walk from x to y that uses no vertex or edge twice.

Let's call (4.2) the *unique walk property*. It turns out to hold for all free trees; and, conversely, any graph with the unique walk property (4.2) is a free tree. We summarize these facts in the following statement.

Theorem (4.3) A graph is a free tree if and only if it has the unique walk property (4.2).

Proof We want to show that the set $\mathcal{F}$ of all free trees is the same as the set $\mathcal{W}$ of all graphs with the unique walk property (4.2).

We first show $\mathcal{W} \subseteq \mathcal{F}$. Let G be any graph with the unique walk property. Thus G is any element of $\mathcal{W}$. Then G is connected, since there is a walk joining any two vertices. We now show

(4.4) If a graph has the unique walk property, then it has no cycles.

We prove the contrapositive of (4.4): Assume the graph has a cycle through the vertex v, say

$$v, e, \ldots, v,$$

where e is an edge. (There must be at least one edge in a cycle, by Definition (2.13).) Then the sequence v is a walk (of length 0) from v to v, but the cycle is another walk (by definition (2.13), having no repeated edges or vertices except v) from v to v. Thus if a graph has a cycle, then it does not have the unique walk property. That proves (4.4), and so $\mathcal{W} \subseteq \mathcal{F}$.

Now we prove $\mathcal{F} \subseteq \mathcal{W}$. This statement is the converse of (4.4), but for *connected* graphs:

(4.5) Let G be a connected graph. If G is acyclic, then G has the unique walk property.

(**Acyclic** means *having no cycles*.)

Intuitive proof of (4.5): We prove the contrapositive. If there are different walks W and W' from x to y, then there must be a cycle in the graph. See the sketch in (4.4F) to get the idea.

(4.4F)

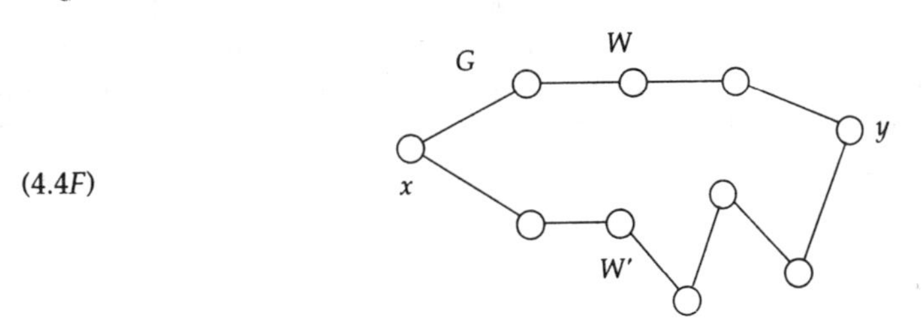

Proof of (4.5) We first remark that since G is connected, for all vertices x, y of G there is at least one walk from x to y. By cutting cycles from it we may assume the walk uses no vertices or edges twice. The burden now is to show that if G is acyclic, then for each x, y there is *only* one such walk. We do so by proving the contrapositive:

(4.6) Let G be connected. If G does not have the unique walk property, then G has a cycle.

Thus we assume that for some vertices x, y, there are two walks from x to y. If either walk has a cycle in it, we are finished; so we assume that in each walk no edges or vertices appear twice.

To be precise, let the walks be

$$W : x = u_0, e_1, u_1, e_2, \ldots, e_m, u_m = y$$
$$W' : x = v_0, f_1, v_1, f_2, \ldots, f_n, v_n = y.$$

The u's and v's are vertices, and the e's and f's are edges.

If our graph has different edges joining the same endpoints, then it has a cycle; so from now on we treat the case in which G has no repeated edges.

The walks W and W' may coincide at first, as

(4.7F)

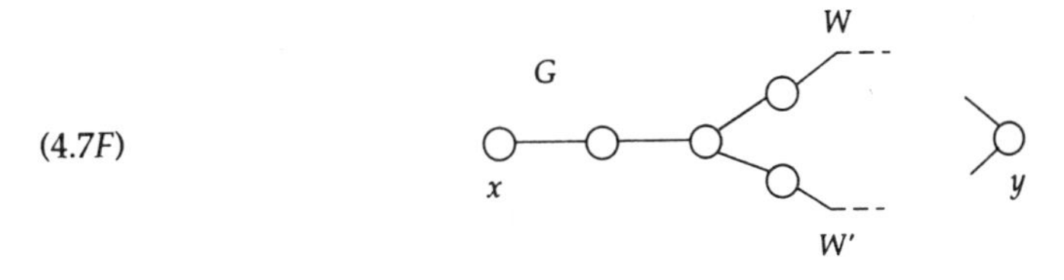

We now define i as the least subscript for which $e_{i+1} \neq f_{i+1}$. There must be such a value of i because we assume $W \neq W'$. (In (4.4F) $i = 0$; in (4.7F) $i = 2$. This is the first vertex at which the walks diverge.)

In other words, $(e_1,\ldots,e_i) = (f_1,\ldots,f_i)$, and $e_{i+1} \neq f_{i+1}$.

Practice (4.8P)

(*i*) Prove that if $e_i = f_i$ for $i = 1,\ldots,m$, then $n = m$ and $W = W'$.
(*ii*) Prove $i < m$.

The two walks have, after $u_i = v_i$, another vertex in common, namely, $y = u_m = v_n$. So we consider the vertices $u_{i+1},\ldots,u_m$; from these we choose the one of least subscript that is also in the walk W'. Call it u_j. Of course $j \leq m$.

Now we consider the part of the walk W from u_i to u_j, and the part of the walk W' from u_i to u_j. (Remember, both u_i and u_j are vertices in the walk W'.) None of the vertices $u_{i+1},\ldots,u_{j-1}$ is in W', by definition of j. It follows that none of the edges between them in W is in W' either. Figure (4.8*F*) may help you think about the situation.

(4.8*F*)

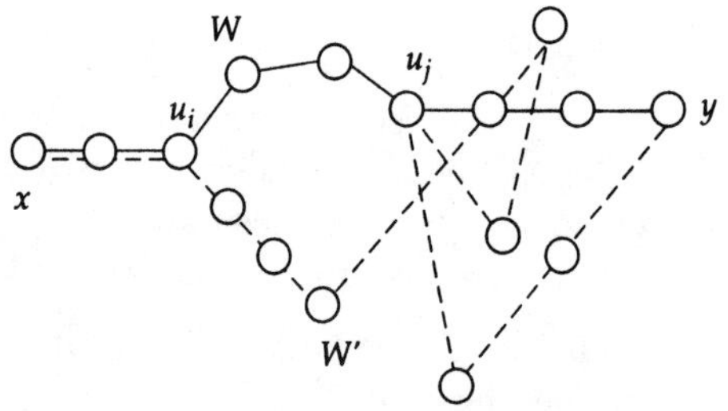

The walk W' may overlap with W after u_j, but by definition not between u_i and u_j. It follows that W from u_i to u_j, followed by W' backward from u_j to u_i, is a cycle. By our construction we know it does not use any vertex or edge twice (except its starting vertex), as required in Definition (2.13). QED

(4.9) *Comment:* You see how fussy a proof in graph theory can be, even though it is intuitively obvious.

(4.10) *Terminology:* We defined *tree* in Chapter 13 differently than here, by recursion, for one thing. We said there that, strictly speaking, we were defining a rooted tree, i.e., one with a root. We now point out that if we specify a vertex of a free tree as its root, we get a tree in the sense of Chapter 13.

Example (4.11) Consider the free tree T of (4.11*F*). Designate vertex r as its root.

(4.11*F*)

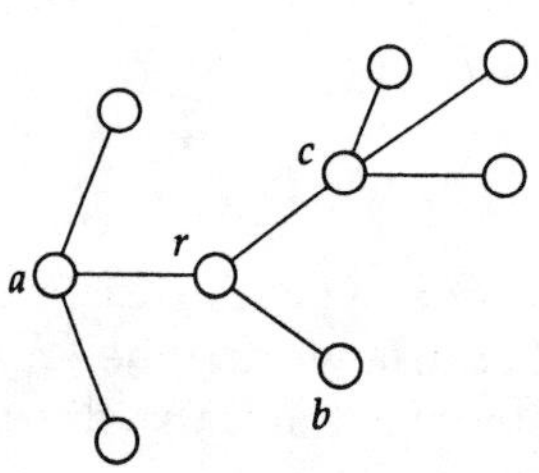

Then redraw the same tree in the style of Chapter 13:

(4.12*F*)

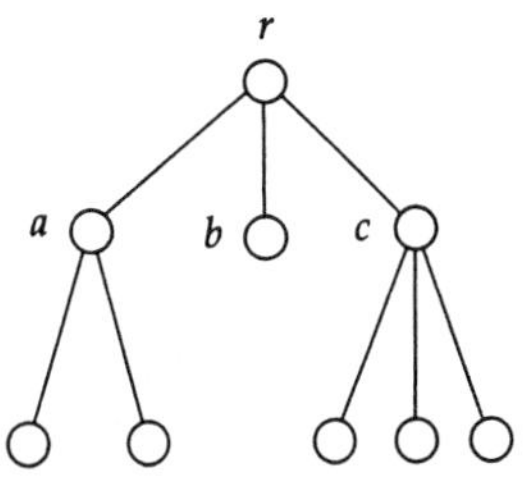

You see it now looks just like the tree of Chapter 13, (2).

(4.13) *Further properties of free trees.* There is a simple relationship between the numbers of edges and vertices in a free tree. Here are all the free trees on five vertices, for example.

(4.13*F*)

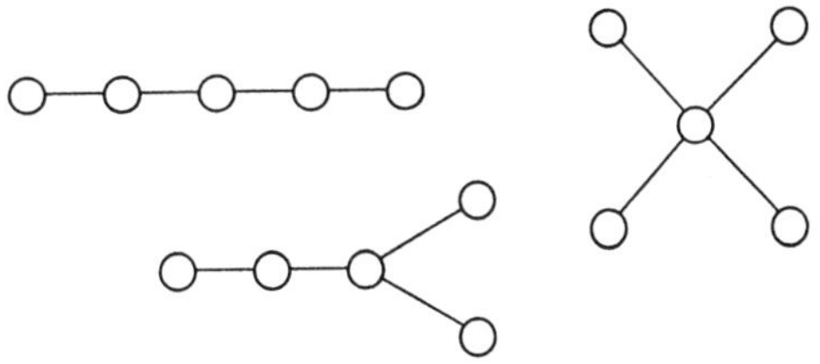

Notice that each one has exactly four edges.

Practice (4.14P) Explain why the three trees in (4.13*F*) are all the free trees on five vertices.

That observation generalizes easily.

Theorem (4.15) Let $n \in \mathbf{N}$. Any free tree $T = (V, E)$ on n vertices has exactly $n - 1$ edges.

Proof We use the unique walk property (4.2) of T. We also use induction on n. A free tree on one vertex has no edges, so the theorem is true for $n = 1$.

Let n be any positive integer, and assume the conclusion of (4.15) for every free tree on n or fewer vertices. Consider any free tree $T = (V, E)$ on $n + 1$ vertices. Let u_0 be any vertex of T. Since there are at least two vertices in the connected graph T, there is an edge e_1 in T with endpoints u_0 and u_1. Construct a walk in T beginning u_0, e_1, u_1 by next choosing an edge from u_1 other than e_1, if one exists. Continue the process, never choosing an edge twice.

Since T is acyclic, you never come to a vertex more than once. But the number of vertices is finite, so the process is forced to halt, necessarily at a vertex v of degree 1. See Figure (4.16*F*).

.(4.16*F*)

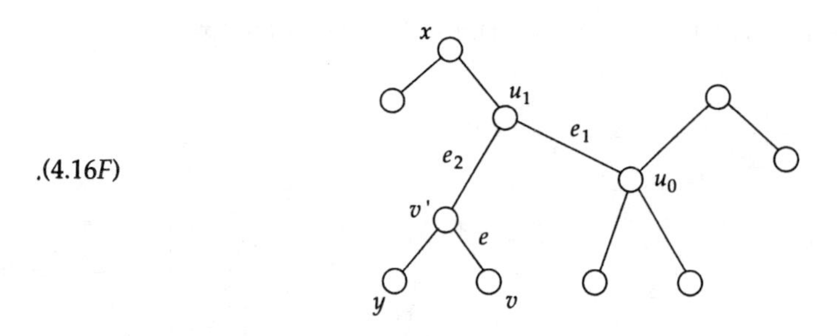

Let e be the unique edge at v, with v' as the other endpoint of e. Remove v and e from T, leaving a graph T' on n vertices. We now claim that

(4.17) T' is a free tree, i.e., T' is acyclic and connected.

That T' is acyclic is obvious, since any cycle in T' would be a cycle in T. T' is connected because for any two vertices x and y of T', there is a walk W in T joining them. Since $d(v) = 1$, if W goes through v, it does so as

(4.18) $$\ldots, v', e, v, e, v' \ldots.$$

Thus the detour through v can be eliminated, leaving a walk W' in T' joining x and y. Therefore T' is connected.

By induction, T' has exactly $n - 1$ edges, so T has exactly n. QED

Corollary (4.19) In any free tree (V, E) on n vertices,

$$\sum_{v \in V} d(v) = 2n - 2. \blacksquare$$

This result follows from (4.15) and the edge-degree lemma (2.22).

A converse of Theorem (4.15) is also true. We state it as

Theorem (4.20) If $G = (V, E)$ is any connected graph on n vertices with exactly $n - 1$ edges, then G is a free tree.

Proof (4.21) We use induction on n. In the basis case $n = 1$, and the graph is a single vertex with no edges. That is a free tree since it has no cycles. (Remember that a cycle has at least one edge, by definition. Thus the walk of length 0 from the vertex to itself is not a cycle.)

In the general case (here is the inductive step) we have $n \geq 2$ vertices and $n - 1$ edges. The edge-degree lemma (2.22) tells us that

(4.22) $$\sum_{v \in V} d(v) = 2n - 2.$$

From (4.22) we conclude that there is at least one vertex of G having degree 1. Why? Because $n \geq 2$ implies all vertices have positive degree (since G is connected). If all degrees were at least two, the n summands $d(v)$ in (4.22) would have sum at least $2n$, a contradiction. Therefore G has a vertex v of degree 1.

(4.23F)

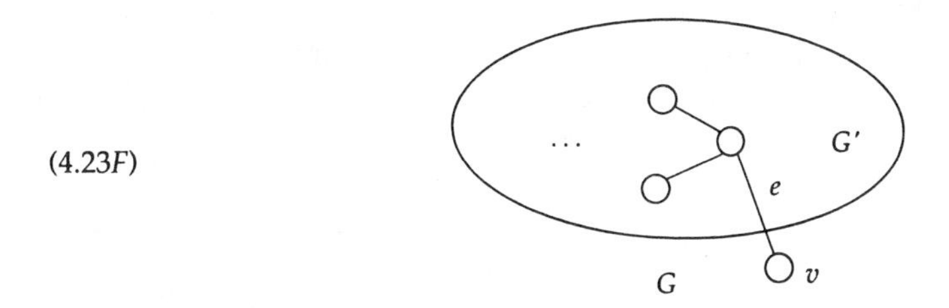

From G we remove v and the unique edge e of G with v as endpoint. There results a graph G' on $n-1$ vertices and $n-2$ edges. (See Figure (4.23F).) Because G is connected, and because the edge we took out joins G' only to v, G' is connected for the same reason T' in (4.17) is connected.

By induction, G' is a free tree. In particular, G' has no cycles.

Now look at G. Any cycle in G would have to have v as one of its vertices. But a cycle must have in it two distinct edges ending at each vertex in it. (Look at Definition (2.13) to check.) Since there is only one edge ending at v, such a cycle in G is impossible, so G too is acyclic. Therefore G is a free tree. QED

Practice (4.24P) Push the preceding argument to show that any free tree on $n \geq 2$ vertices has at least two vertices of degree 1.

As you may have begun to suspect, there are many equivalent ways to specify free trees. Berge (16.2, page 152) gives six. We state them now; some we've seen, some are new.

Theorem (4.25) *The BET Theorem (Big Equivalence on Trees).* Let $G = (V, E)$ be a graph on $n \geq 1$ vertices. The following conditions $(i), \ldots, (vi)$ are mutually equivalent. G is a free tree if and only if any one of them is true. The conditions are:

- *(i)* G is connected and acyclic;
- *(ii)* G is acyclic, and $|E| = n - 1$;
- *(iii)* G is connected and $|E| = n - 1$;
- *(iv)* G is acyclic, and if any edge is added to E, one and only one cycle is formed;
- *(v)* G is connected but becomes disconnected if any edge is removed from E;
- *(vi)* Every two vertices of G are joined by exactly one walk using no vertex or edge twice. ■

(4.26) *Comment*: We gave *(i)* as our definition of free tree. We then proved in Theorem (4.3) that *(i)* and *(vi)* are equivalent. We next proved in Theorem (4.15) that *(i)* implies *(iii)*. Finally, we showed that *(iii)* implies *(i)*, in Theorem (4.20). Berge gives an elegant proof of the BET theorem, first showing *(i)* → *(ii)*, then *(ii)* → *(iii)*, and so on, and finally he proves (*vi*) → (*i*). Thus he does it all with only six proofs, some of which are only one sentence in length. But he has more advanced machinery than we have.

We leave the missing proofs to you. There are potentially quite a few if we look at them "inefficiently." That is, we have established (*i*), (*iii*), and (*vi*) as mutually equivalent, so the following directed graph shows the implications not yet proved:

(4.27*F*)

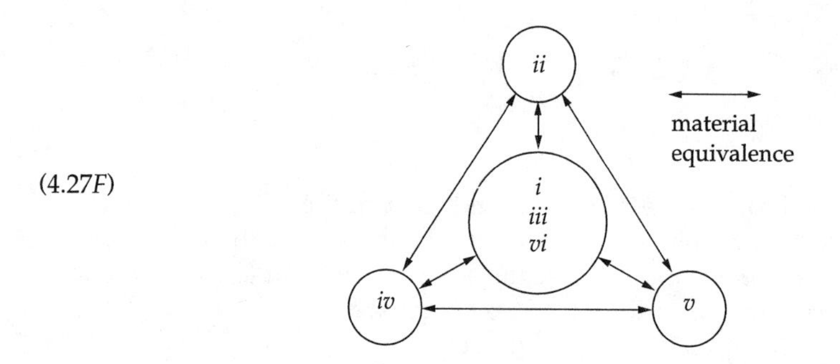

Of course, it's not necessary to prove all 12. The most efficient way to do it is to prove the four implications in a cycle through all the vertices of this graph. The transitivity of material implication will give you the other eight.

We defined *cycle* in (2.13) for undirected graphs, but the same definition is used for directed graphs. The walk must respect the directions on the edges. Besides, to some authors (e.g., Berge (16.2)) an undirected graph is by definition a symmetric directed graph, like that in (4.27*F*).

5. Subgraphs

We are overdue for the handy concept of *subgraph*.

DEFINITION (5.1) Let $G = (V, E)$ be a graph. The graph $H = (V', E')$ is called a **subgraph** of G iff

$$V' \subseteq V$$

and

$$E' \subseteq E$$

and every edge in E' has both endpoints in V'.
That is, a subgraph of G is a graph with vertices and edges belonging to G.

Examples (5.2) Here is the complete graph K_4 on four vertices (defined in Chapter 3, (12.7)) with various subgraphs drawn in bold. $V = \{a, b, c, d\}$, $E =$ the set of all 2-subsets of V.

(*i*)

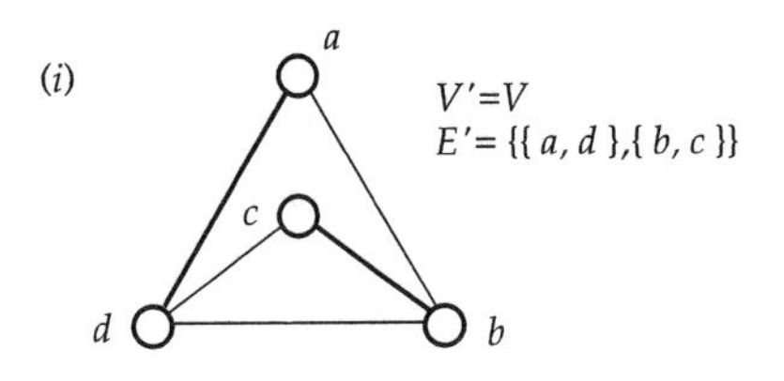

(*ii*)

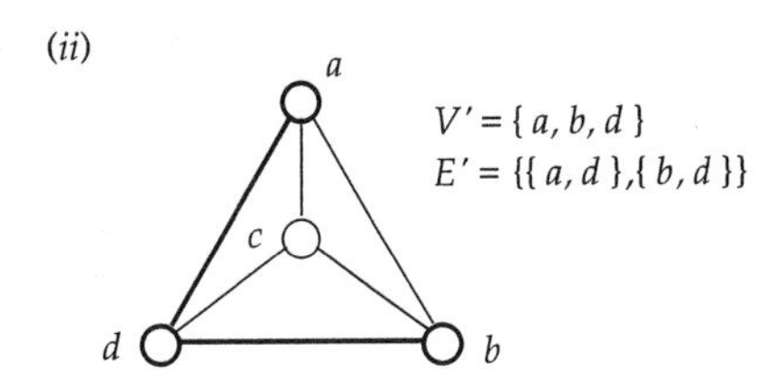

(5.2*F*)

(*iii*)

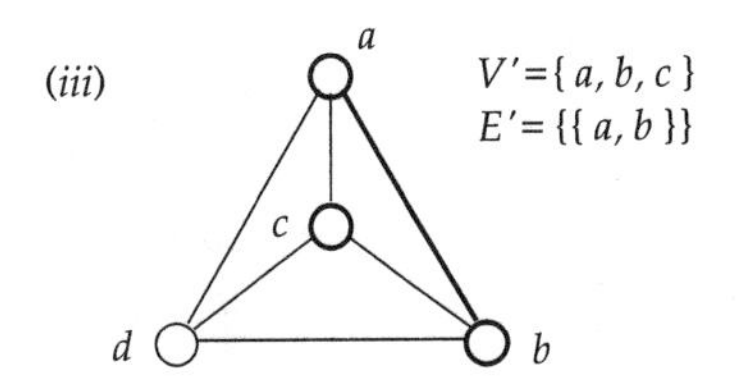

(*iv*)

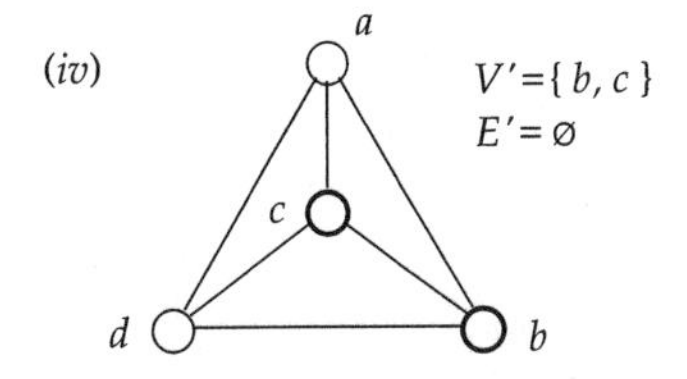

Practice (5.3P) What is the total number of subgraphs of K_4?

(5.4) *Caution:* Some authors, for example, Berge (16.2), define subgraph differently. Most use the definition (5.1), however.

(5.5) A subgraph of G having the same vertex-set as G is called a **spanning** subgraph of G. We also say it **spans** G.

Part (*i*) of (5.2*F*) is the only spanning subgraph in that set of examples.

6. Spanning Trees

We now study free trees which span a given graph.

DEFINITION (6.1) Let $G = (V, E)$ be a connected graph. A **spanning tree** of G is any free tree

$$T = (V, E')$$

having the same vertex-set as G such that

$$E' \subseteq E.$$

That is, T is a subgraph of G with the same vertex-set and is a free tree. More concisely, T is a spanning subgraph of G and T is a free tree.

Examples (6.2) None of the subgraphs in (5.2*F*) is a spanning tree. For the same K_4, however, there are many spanning trees. Three are shown with bold edges in (6.2*F*):

(6.2*F*)

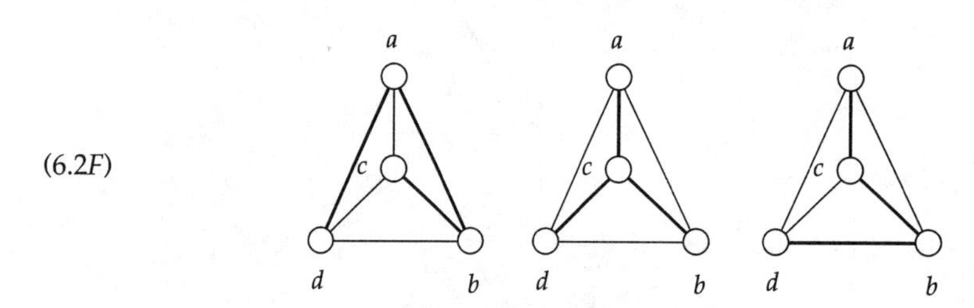

Practice *(6.3P) How many spanning trees does K_4 have? In this problem the vertices are labeled so it's not just the "shape" that matters.

Obviously a disconnected graph could not have a spanning tree, but it has a spanning forest.

(6.4) *Finding a spanning tree.* How might we find a spanning tree in a connected graph G? Answer: Look. We'll present an algorithm that does just that.

STOP. Read no further until you have thought up your own procedure for finding a spanning tree of a connected graph. Try it on a few examples to get the kinks out. Then compare it with the one given here. If yours works and is different from the two below, publish it.

The idea is simple. We start at any vertex, say u, pick any edge ending there, go along to the other endpoint v (always refusing to use loops, if present), and pick another edge ending at v. We continue the process, keeping track of the edges tried and vertices passed through. We reject any edge that takes us to a vertex we've been to already. If we use up all the edges at a vertex that way, then we go back along our most recently chosen edges to the first one we encounter with untried edges ending at it. We stop once we've reached every vertex.

Examples (6.5) (*i*) Here we start at u in (6.6*F*). We number the edges in the order in which they are chosen.

(6.6*F*)

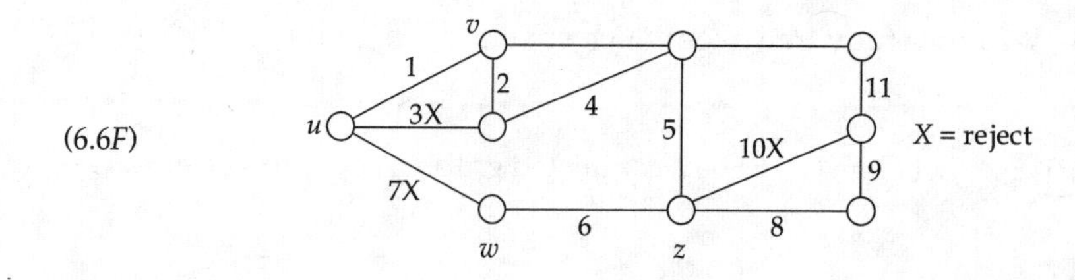

We reject edge 3 because it takes us back to u. We reject edge 7 for the same reason. We keep edge 6 in our backtracking even though there are no untried edges ending at w, and we (randomly) take edge 8 from w. Edges 8 and 9 are acceptable; edge 10 takes us back to z, so is rejected; and edge 11 takes us to the last vertex. The spanning tree is

(6.7F)

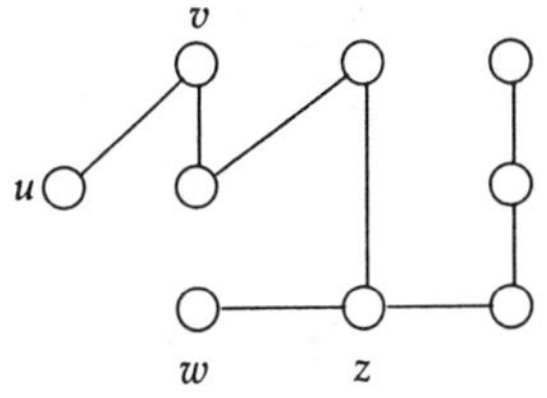

(ii) In this example we start at u (Figure 6.8F), take edge 1 to v, and then take the

(6.8F)

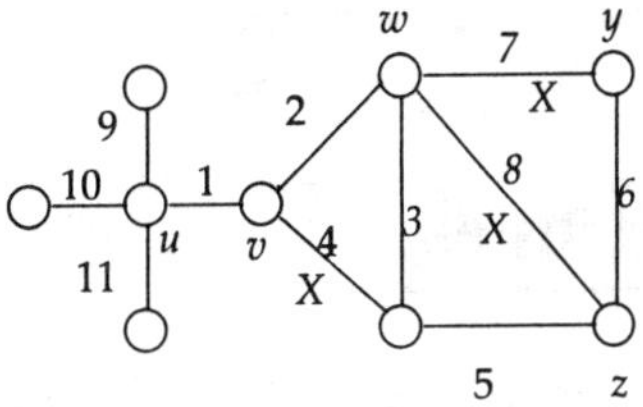

edges as numbered, rejecting 4 because it takes us back to v, and 7 and 8 because they take us back to w. After rejecting 7 we've used all the edges ending at y, so we backtrack along edge 6 to z. There we reject 8 and backtrack now all the way to u, where we finally complete the spanning tree with the choice of 9, 10, and 11. We get

(6.9F)

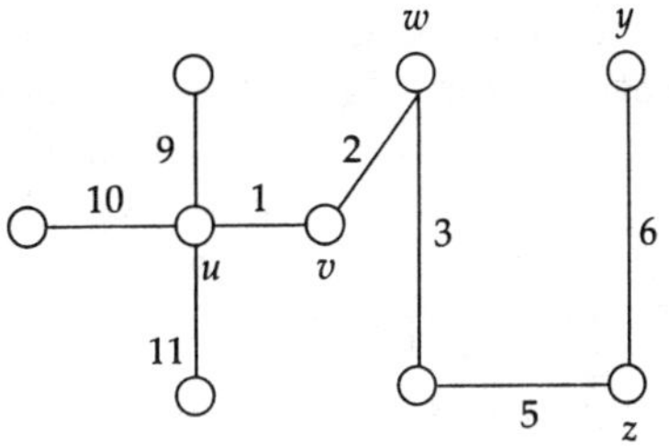

(6.10) Here is a compact algorithm slightly different from this procedure.

(6.11) **Algorithm Spantree** (Kruskal)

Input: connected graph $G = (V, E)$.

Output: A spanning tree T of G

$E_1 := E, E_2 := \varnothing$

While $E_1 \neq \varnothing$ do

  Choose $e \in E_1$

  $E_1 := E - \{e\}$

  If $(V, E_2 \cup \{e\})$ is acyclic, **then**

  $E_2 := E_2 \cup \{e\}$

$T := (V, E_2)$

End.

Practice (6.12P) How does Algorithm (6.10) differ from the procedure described in (6.4)?

(6.13) *Comment:* If the graph has many edges, it may be worthwhile to count the number of vertices touched by E_2, stopping when you reach $n = |V|$. Otherwise many steps would be required to reduce E_1 to $\varnothing$.

7. Minimum Spanning Trees

(7.1) Suppose a watering system is to be installed at a golf course. There are 18 *holes* (areas of ground, mutually disjoint), and for each two distinct holes the cost of running a water line between them is known. To save construction costs the owners decide to run the water through a "tree" of pipes. (Since the source of water is far away, there can be only one connection to it.) How can they minimize their cost of construction? What they have is a complete graph K_{18} with a cost attached to each edge. They want a spanning tree with the sum of the costs on its 17 edges a minimum. That will tell them how to lay out their water pipes.

(7.2) *The general situation.* Suppose G is a connected graph, with a cost on each edge. G may have many spanning trees. Each one has a total cost, namely, the sum of the costs of the edges of that spanning tree. Since there are only finitely many spanning trees, there is a unique minimum total cost over all spanning trees of G. Any spanning tree with total cost equal to that minimum is called a **minimum spanning tree** of G. There may be more than one such spanning tree. (If every edge has the same cost, for example, then all spanning trees are minimum.)

(7.3) *Comment.* In practice the owners would not even consider connecting two holes that were far apart. For such pairs they would omit the corresponding edge from the graph, or assign it infinite cost without bothering to estimate the true cost.

Example (7.4) Consider the following map of nine holes of a golf course. We show the costs of connection only between contiguous holes.

(7.4F)

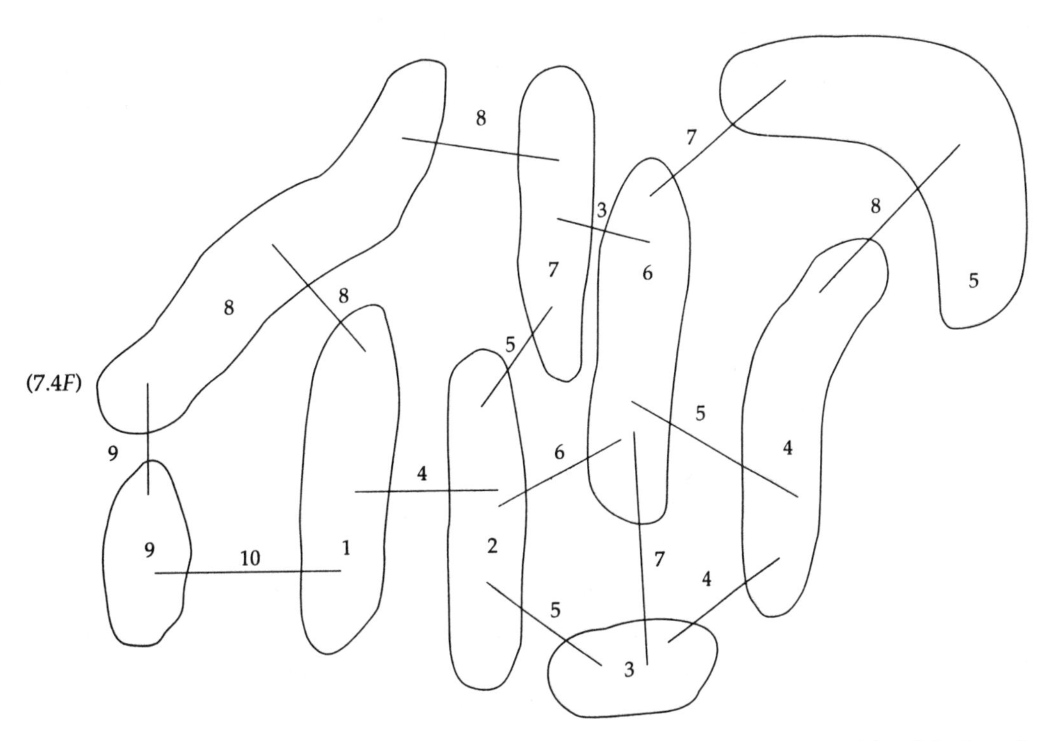

—After Mackenzie

Now again—figure out your own way of finding a spanning tree of minimum cost. STOP READING. Then compare your method with what follows.

(7.5) The costs on the edges of $G = (V, E)$ can be modeled abstractly as a function

$$f : E \to \mathbf{R}.$$

That is, for each $e \in E$, the real number $f(e)$ is the cost attached to the edge e. You see that *cost* is now abstract, too; it could be any real number we want to associate to the edge. It might stand for distance between points on the earth's surface, or the sum of the bank balances of the two people at the endpoints, or the attractive (+) or repulsive (−) force between the two particles that the endpoints represent.

Examples (7.6) (*i*) The following schema represents possible round-trip airfares in U.S. dollars:

(7.7F)

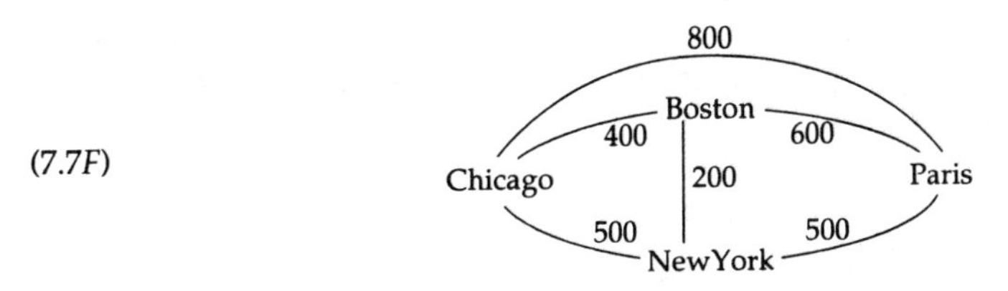

This complete graph K_4 has vertices labeled as shown and a cost on each edge.

(*ii*) There are n cities, and between each pair of them there's a cost of building a road. The graph is a complete graph K_n with a cost on each edge. A minimum spanning tree represents a road network joining the n cities at least cost.

(7.8*F*)

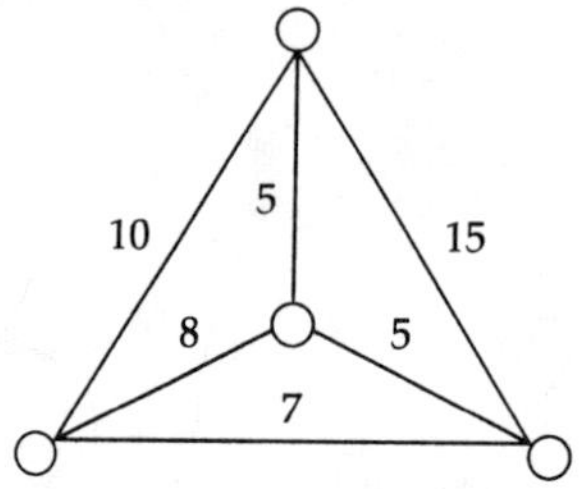

The preceding algorithm **spantree** needs only slight retrofitting to become an algorithm for finding a minimum spanning tree. All we do is have it choose the cheapest remaining edge each time.

(7.9) **Algorithm Least Spantree** (Kruskal)

Input: Any connected graph $G = (V, E)$ with a cost $f(e)$ attached to each edge $e \in E$.

Output: A minimum spanning tree T of G

$E_1 := E,\ E_2 := \varnothing$

While $E_1 \neq \varnothing$ **do**

  Choose $e \in E_1$ such that $f(e)$ is minimum

  $E_1 := E_1 - \{e\}$

  If $(V, E_2 \cup \{e\})$ is acyclic, **then** $E_2 := E_2 \cup \{e\}$.

  $T := (V, E_2)$

End

(7.10) The only change we made from **Spantree** (6.11) was to add the condition at each choice of a new edge that it be the cheapest eligible. Such a condition is called **greedy**. Many optimizing algorithms include a step like that, to pick the minimum (or maximum) from a set. The idea of such a choice could be called the *greedy tactic*, but the term is often inflated to the *greedy algorithm*.

It's less obvious that Algorithm (7.9), *Least Spantree*, is correct than it was for *Spantree* (6.11). Clearly, (7.9) produces a spanning tree, because it's the same as (6.11) except that it gives guidance on how to pick the edges that go into E_2. The trick is to prove it's minimum.

Theorem (7.11) Algorithm (7.9) *Least Spantree* is correct; it produces a minimum spanning tree of G. ∎

To prove (7.11) we'll use the concept of *free forest*, which we define as you'd expect:

DEFINITION (7.12) A **free forest** is a graph in which every component is a free tree.

Example (7.13) Here is a free forest having five free trees.

(7.13F)

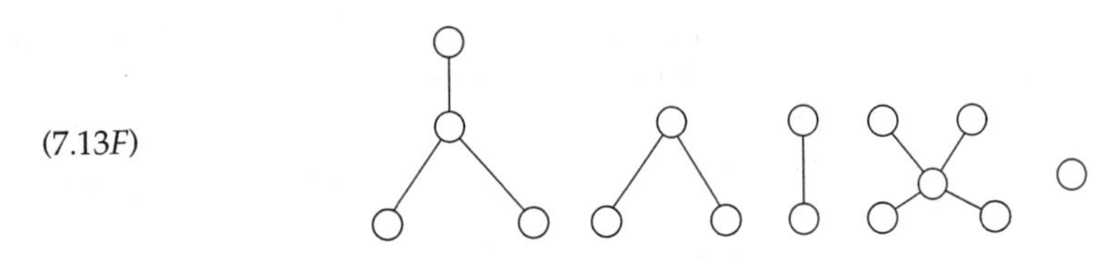

We'll also use two obvious notations: the 2-rep $[u, v]$ for the edge with endpoints u and v, and "+."

(7.14) Let $G = (V, E)$ be a graph and take $u, v \in V$. If $e := [u, v]$ is not in E, then $G + e$ denotes the graph with the new edge e added.

$$G + e := (V, E \cup \{e\}).$$

Example (7.15) We add an edge to G, and a loop to G.

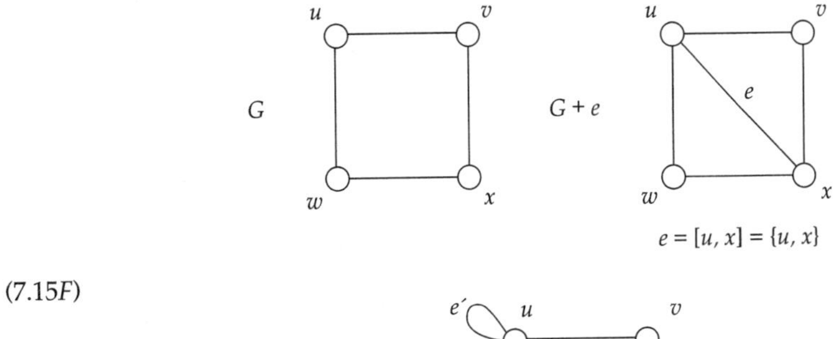

(7.15F)

Let $G = (V, E)$ be a graph with a cost function f defined on the edges of G. If $H = (W, E')$ is any graph such that $W \subseteq V$ and $E' \subseteq E$, we define $f(H)$ as the sum of the costs over all the edges of H:

$$f(H) := \sum_{e \in E'} f(e).$$

If H is a spanning tree T of G, then $f(H)$ is the cost we associate to T.

(7.16) For the proof of (7.11) we now define a **promising forest** of G as a free forest on the vertex-set of G such that there is a minimum spanning tree of G, $T_{\min} = (V, E_0)$, such that every edge in the free forest is in E_0.

Another way to say the last condition is that we may obtain $T_{\min}$ from the promising forest PF simply by adding suitable edges of G to PF. It follows from (7.16) that any promising forest is a subgraph of G.

Practice (7.17P) Prove the last sentence.

Example (7.18) Clearly, $F_0 = (V, \varnothing)$ is a promising forest. It has $|V|$ free trees each of one vertex. Hence we achieve any minimum spanning tree of G by adding its edges to F_0.

The crux of the proof of Theorem (7.11) will be the following result.

Lemma (7.19) *The Expansion Lemma*. Let G be a connected graph with a cost function defined on the edges. Let PF be any promising forest of G with at least two free trees. Partition the set of free trees of PF into two cells in any way. Let B be the set of all vertices of the trees in one cell, and $\overline{B} = V - B$ that for the other cell. Let e be any edge of minimum cost in the set of all edges with one endpoint in B and the other in $V - B$. Then PF with e added is also a promising forest.

Proof Consider the following picture. It's not really necessary for the proof, but it will help you think about the situation.

(7.20F)

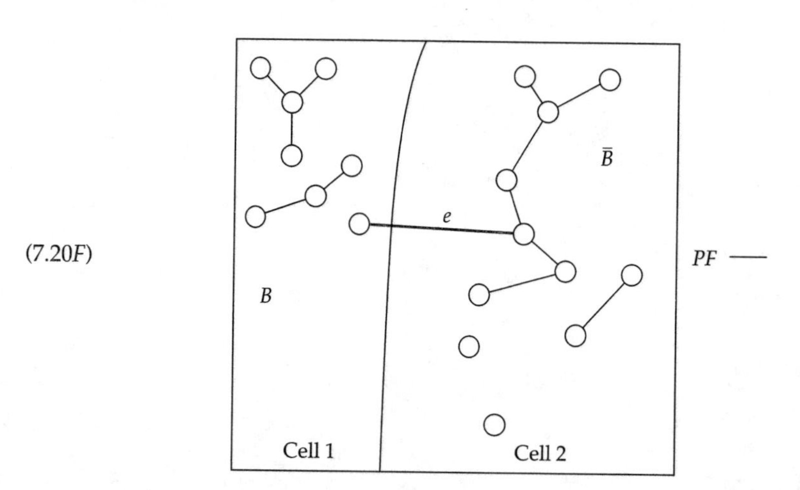

(7.21) Since PF is a promising forest of G, there is (by definition) some minimum spanning tree T of G having all the edges of PF in it.
(7.22) Adding edge e to PF makes one free tree out of two. $PF + e$ is still a forest.

Case 1. Edge e is an edge of T: Then $PF + e$ is a promising forest, since every edge is in T.

Case 2. Edge e is not an edge of T: Then by the BET Theorem (4.25*iv*), $T + e$ has exactly one cycle. See the following picture, in which the dashed black lines are the edges of T not in PF.

(7.23F)

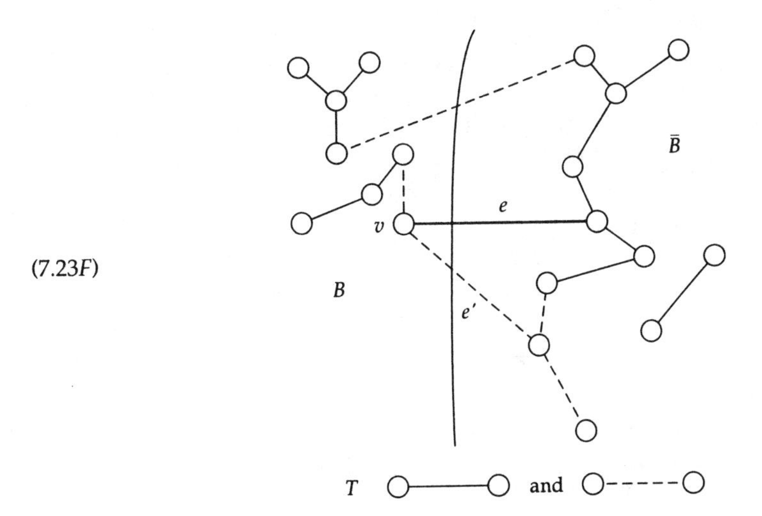

If we go around this cycle starting, say, at the vertex v of e that is in B and going along e, we go from B to $\overline{B}$. The cycle must have at least one other edge "crossing the boundary," i.e., with one end in B and one in $\overline{B}$. Otherwise we could never return to v. Call that edge e'.

By our definition of e as a least-cost edge joining B to $\overline{B}$, $f(e) \leq f(e')$.

We now define a new spanning tree T', obtained by deleting e' from T and adding e. T' is also a minimum spanning tree, because

$$f(T') = f(T) - f(e') + f(e)$$

and, therefore, since $f(e) \leq f(e')$,

$$f(T') \leq f(T).$$

But $f(T)$ is the minimum value of f over all spanning trees of G. Therefore

$$f(T') = f(T),$$

so T' is also a minimum spanning tree. Since every edge of $PF + e$ is in T', $PF + e$ is a promising forest. QED

Comment: It also follows that $f(e) = f(e')$.

Proof of (7.11) Now you easily see how the proof will go. The algorithm starts with the promising forest PF_0 of $|V|$ vertices and no edges (Example (7.18)). It picks an edge e_1 of minimum cost. If $e_1 = \{v, u\}$ we set $B :=$ any subset of V with $v \in B$ and $u \notin B$. This choice gives us a partition of PF_0 into two cells.

We set $PF_1 := PF_0 + e_1$. By the Expansion Lemma, PF_1 is a promising forest.

Practice (7.24P) Why is e_1 of minimum cost among all edges joining B and $\overline{B}$? We must show it to be so if we are to use the Expansion Lemma.

At each stage the algorithm chooses the cheapest edge, say e, that does not introduce a cycle into the promising forest PF_i chosen to date. We then imagine partitioning PF_i into two cells, one cell having one endpoint of e among its vertices, the other cell having the other endpoint of e. The edge e then satisfies the conditions of the Expansion Lemma, so $PF_i + e$ is a new promising forest PF_{i+1}.

Eventually the promising forest is a tree, hence a spanning tree, hence a minimum spanning tree. QED

8. A Second Proof of Correctness

The proof of (7.11) is cleaner if we approach it as a problem in proving a program correct. We introduce an *invariant*, i.e., a comment, into the algorithm and show that it is correct whenever we enter the *while* loop. We do that now, rewriting the algorithm with comments in braces { }.

(8.1) **Algorithm Least Spantree** (Kruskal).

Input: Any connected graph $G = (V, E)$ with a cost $f(e)$ attached to each edge $e \in E$.

Output: A minimum spanning tree T of G.

$E_1 := E,\ E_2 := \varnothing$

$\{ (V, E_2)$ is a promising forest $\}$

while $E_1 \neq \varnothing$ **do**

  Choose $e \in E_1$ such that $f(e)$ is minimum

  $E_1 := E_1 - \{e\}$

  If $(V, E_2 \cup \{e\})$ is acyclic, **then** $E_2 := E_2 \cup \{e\}$

$T := (V, E_2)$

End.

Proof that* (8.1) *is correct We prove that every time the algorithm enters the *while* loop, (V, E_2) is a promising forest. This follows from the Expansion Lemma (7.19), as before. Therefore, at the end, (V, E_2) is both a promising forest and a spanning tree, so a minimum spanning tree. QED

9. Interval Graphs

The case of the dozing dons. Three professors, unable to think of an excuse fast enough, must attend a faculty meeting. Each of them nods off for part of the meeting. Prof. $A[B,C]$ is awake until time $a[b,c]$, when $A[B,C]$ falls asleep. Once $A[B,C]$ wakes up $A[B,C]$ stays awake. For each two of the three there is a time when both are asleep. Prove that there is a moment when all three are asleep.

We'll see that this is a simple problem in interval graphs.

An *interval graph* is a graph obtained from intervals of real numbers according to the rule stated below.

(9.1) A **closed interval** of real numbers is defined to be any set of the form, for $a, b \in \mathbf{R}$,

$$\{x;\ x \in \mathbf{R},\ a \le x \le b\}. \tag{9.2}$$

For definiteness we'll speak only of closed intervals and call them simply **intervals**.

Suppose we have a finite set $\mathcal{I}$ of intervals $I_1, \ldots, I_n$. We define the interval graph of $\mathcal{I}$ as follows:

DEFINITION (9.3) The set $\mathcal{I} := \{I_1, \ldots, I_n\}$ of intervals is the vertex-set V of the associated **interval graph** G. Any two vertices I_i and I_j are joined by an edge of G iff $I_i \cap I_j \neq \varnothing$.

Example (9.4) Suppose the intervals are as drawn in (9.4F).

(9.4F)

Then the associated *interval graph* is

(9.5F)

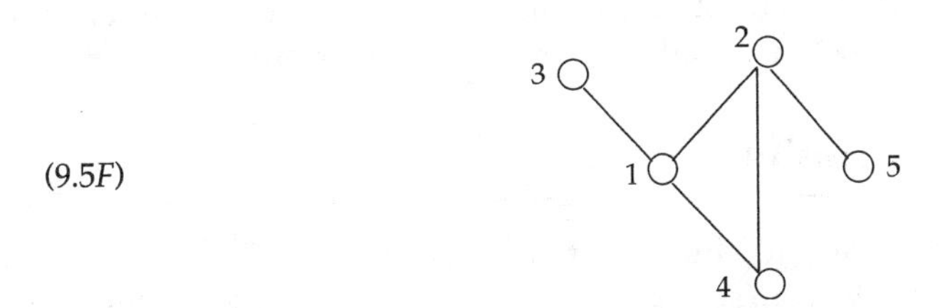

Interval graphs have practical applications, as in archeology, where artifacts can sometimes be dated to within an interval of time. The interval graph then shows at a glance which pairs of tools, say, were (or weren't) in use at the same time. (For more information, please see (16.6) and (16.7).) For us, they are a source of nice problems.

10. Directed Graphs

Up to now in this text we've used directed graphs only to discuss the relations that they are in one-to-one correspondence with. Now we look at two important applications that focus on the directed graph (sometimes called **digraph**) and not much on the associated relation.

(10.1) Informally, a directed graph is a finite, nonempty set B of **nodes** together with a set E of **arcs** (arrows), each drawn from a node to a node.

Formally, a directed graph is an ordered pair (B, E), where B is a finite, nonempty set and E is a subset of $B \times B$. The set of nodes is B and the set of arcs is E. If we want repeated arcs we make E a multiset from $B \times B$.

Subgraph is defined for digraphs with the same definition (5.1) as before. *Degree* remains an important idea but now splits in two:

DEFINITION (10.2) The **indegree** of a node is the number of arcs coming in to it. The **outdegree** of a node is the number of arcs going out from it.

Practice (10.3P) Write definitions of indegree and outdegree in terms of B and E.

(10.3*F*)

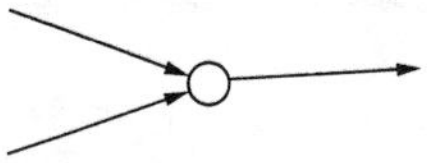

A node with indegree 2, outdegree 1.

Since every arc $(x, y) \in E$ contributes one to the outdegree of node x and one to the indegree of y, we see that

(10.4)
$$\sum_{x \in B} \text{indegree}(x) = \sum_{x \in B} \text{outdegree}(x) = |E|.$$

This result (10.4) is the directed version of the edge-degree lemma (2.22).

In fact we used the idea of indegree and outdegree in the proof of Euler's theorem (2.22). We counted $t_{\text{in}}(v)$ and $t_{\text{out}}(v)$ for a given Euler tour, which in effect assigns an arrow to each edge of an undirected graph.

11. The Shortest-Path Problem

Suppose you want to ship a crate of oranges from San Diego to Boston. You must choose the route yourself, and you want to do it at least cost. You know the costs from San Diego to several nearby places, like Needles and Azusa, and from those to the next set of choices, and so on. A partial diagram of the possibilities is shown in the graph

(11.1*F*)

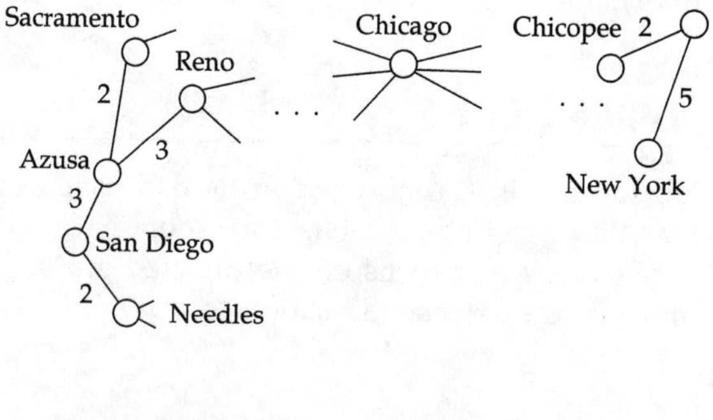

in (11.1*F*) with the costs listed on the edges. Your job is to find the route of least cost.

(11.2) The general problem: A graph, directed or undirected, has a *cost* attached to each arc or edge. These costs are real numbers that we assume to be nonnegative. For each pair (i, j) of vertices, find the cost $c(i, j)$ of the walk of least cost that joins them. In the directed case, this means a walk of least cost from node i to node j that respects the arrows. There may be a different walk from j to i, or none at all.

Example (11.3) Consider the following graphs.

(11.3*F*)

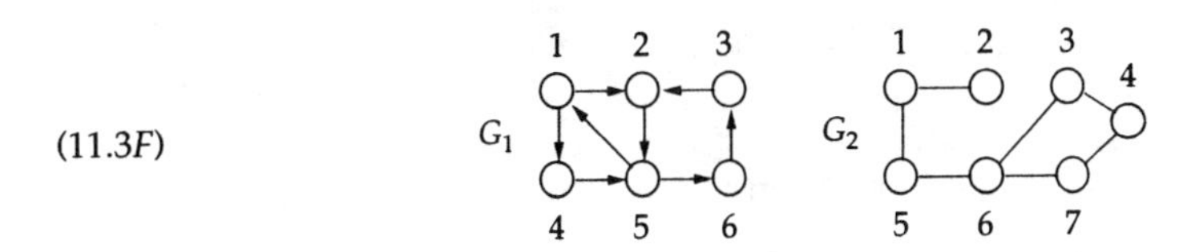

Assume all arcs or edges have cost 1. In G_1 we find

$$\begin{aligned} c(1,2) &= 1,\ c(1,4) = 1, \\ c(1,5) &= 2 \text{ via 1 to 2 to 5 or via 1 to 4 to 5;} \\ c(1,6) &= 3, \\ c(1,3) &= 4. \end{aligned}$$

In the other direction

$$\begin{aligned} c(2,1) &= 2 = c(4,1), \\ c(5,1) &= 1, \\ c(6,1) &= 4, \\ c(3,1) &= 3. \end{aligned}$$

In G_2 $c(i,j) = c(j,i)$ for all i, j because G_2 is undirected (symmetric).
Now consider

(11.4*F*)

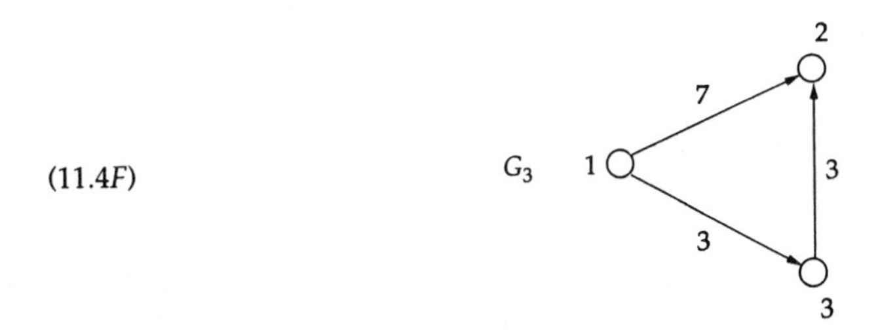

In this case $c(1,2) = 6$, but $c(2,1)$ is not defined because there is no path from 2 to 1. We represent that situation by writing $c(2,1) = \infty$.

Our term **path** here means *walk* (2.7). We use it because it's the traditional terminology for this problem.

Matrices allow us to represent the edge costs and sought-for least costs efficiently.

Example (11.5) For instance, for G_3 in (11.4F) we have

$$\begin{bmatrix} 0 & 7 & 3 \\ \infty & 0 & \infty \\ \infty & 3 & 0 \end{bmatrix} \qquad \begin{bmatrix} 0 & 6 & 3 \\ \infty & 0 & \infty \\ \infty & 3 & 0 \end{bmatrix}$$

Edge costs Least costs

The matrix of edge costs is just the relation-matrix with its 1's replaced by the costs at those edges. (In graphs the relation-matrix is usually called the **adjacency matrix.**)

Practice (11.6P) Find the matrices of edge costs and of least costs for G_2 in (11.3F).

There are several algorithms for finding the matrix $(c(i,j))$ of least costs. We present one that should remind you of another algorithm.

(11.7) **Algorithm Least Cost** (M)

Input: The $n \times n$ edge-cost matrix M of a directed graph.

Output: The matrix $(c(i,j))$, where $c(i,j) :=$ the cost of a least-cost path from i to j.

$M^{(0)} := M$

$\forall t \geq 1,\ M^{(t)} := \left(c_{ij}^{(t)}\right)$

(11.8)
$$c_{ij}^{(t)} := \min\left\{ c_{ij}^{(t-1)},\ c_{it}^{(t-1)} + c_{tj}^{(t-1)} \right\}.$$

$c(i,j) := c_{ij}^{(n)}$.

End

Yes, this is a lot like Warshall's algorithm (Chapter 12, (19.12)) for the transitive closure of a relation. In fact, this too is due to Warshall. The only difference is that we apply different functions to the entries in the matrices. In Chapter 12, (19.14), it was

$$m_{ij}^{(t)} = m_{ij}^{(t-1)} + m_{it}^{(t-1)} \cdot m_{tj}^{(t-1)},$$

where "+" was Boolean addition and "·" was multiplication. In (11.8) we have a real addition in place of "·" and $\min\{x,y\}$ in place of $x+y$.

Proof (11.9) ***that (11.7) is correct.*** $M^{(0)}$ records the costs of the edge from i to j if it exists; it has ∞ wherever no edge (i,j) is in the graph. That is, $M^{(0)}$ records the least cost of a path from i to j having no intermediate nodes.

By induction on t we see that

(11.10) $c_{ij}^{(t)}$ is the cost of the path of least cost from i to j having all intermediate nodes in $\{1,\ldots,t\}$.

The assertion (11.10) is true because of (11.8). To see why, let's adopt this temporary shorthand: Call a path a *t-path* if all its intermediate nodes are in $\{1,\ldots,t\}$. Then obviously every $(t-1)$-path is a t-path. And (11.10) says

(11.11) $\quad c_{ij}^{(t)}$ is the cost of the t-path of least cost from i to j.

Why is (11.11) true? We assume it for $t-1$. Then (11.8) defines $c_{ij}^{(t)}$ as the smaller of

$$\alpha := c_{ij}^{(t-1)}$$

and

$$\beta := c_{it}^{(t-1)} + c_{tj}^{(t-1)}.$$

Notice that β is the cost of a t-path from i to j that goes through t. See (11.12F).

(11.12F)

Therefore in all cases

(11.13) $\quad \beta \geq$ the least cost of any t-path from i to j.

Consider a least-cost t-path P from i to j.

Case 1. If P is a $(t-1)$-path, then its cost is $c_{ij}^{(t-1)}$, since it is, *a fortiori,* a least-cost $(t-1)$-path. Because of (11.13), (11.11) is true in this case.

Case 2. If P goes through t, then we have the picture of (11.12F). The first segment of P is a $(t-1)$-path from i to t. Its cost must be $c_{it}^{(t-1)}$, for otherwise it could be replaced by the minimum-cost $(t-1)$-path from i to t, which would reduce the cost of P. A similar argument exists for the second segment of P, from t to j.

Finally, we must have $\beta \leq \alpha$ in this case, for otherwise there is a $(t-1)$-path from i to j less costly than P, contradicting the minimality of P.

To sum up, $c_{ij}^{(t)}$ is defined in Case 1 to be α, and in Case 2 to be β. We proved these were the least costs in our arguments. QED

(11.14) *Comment:* An abstract explanation of how the two Warshall's algorithms can be viewed as special cases of one algorithm appears in (16.1, pages 195 ff.).

12. Complexity and Finding the Paths

(12.1) *Complexity.* At stage t in (11.7) you do one addition and one comparison in (11.8) for each pair i,j. Thus there are n^2 additions and n^2 comparisons for each $t = 1,\ldots,n$. Therefore Warshall's algorithm (11.7) has n^3 additions and n^3 comparisons in total. (That wasn't so bad, was it?)

If the graph is undirected, the same algorithm works but is wasteful by a factor of about 2, since the matrices are symmetric. In this case you may restrict (11.8) to $i \leq j$.

(12.2) *Finding the paths.* How could we find the paths of least cost? A matrix recording such information for each i and j would have paths (walks (2.7)) as entries. It ought to be possible to insert a line in algorithm (11.7) to build this matrix.

Let's define $L^{(t)}$ as the matrix of t-paths of least cost. Thus the i,j entry $\ell_{ij}^{(t)}$ of $L^{(t)}$ is a path of least cost from i to j having all intermediate nodes in $\{1,\ldots,t\}$. We'll record a path as the sequence of its nodes (since we would always use the cheapest edge if there were multiple edges).

$L^{(0)}$ is then a matrix in which the i,j entry $\ell_{ij}^{(0)}$ is the sequence i,j if $m_{ij}^{(0)} \neq \infty$ and is blank otherwise.

If we've built $L^{(0)},\ldots,L^{(t-1)}$ we build $L^{(t)}$ as follows: Look at step (11.8) in Warshall's algorithm.

- If the minimum is $c_{ij}^{(t-1)}$, then

(12.3)
$$\ell_{ij}^{(t)} := \ell_{ij}^{(t-1)}$$

because we're using the path already found.

- Otherwise

(12.4)
$$\ell_{ij}^{(t)} := \ell_{it}^{(t-1)}; \ell_{tj}^{(t-1)},$$

the catenation of the two sequences (with the duplication of t removed). So to find the paths, modify algorithm (11.7) to read as follows:

(12.5) **Algorithm Least Paths** (M)

Input: The $n \times n$ edge-cost matrix M of a directed graph.

Output: The matrices $(c(i,j))$, where $c(i,j) :=$ the cost of a least-cost path from i to j, and $L := (L_{ij})$, where $L_{ij} :=$ a path of least cost from i to j.

$$M^{(0)} := M \qquad L^{(0)} := \left(\ell_{ij}^{(0)}\right)$$

$$\ell_{ij}^{(0)} := \begin{cases} i,j & \textbf{if } m_{ij}^{(0)} \neq \infty \\ \text{blank} & \textbf{otherwise} \end{cases}$$

$$\forall t \geq 1 \quad M^{(t)} := \left(c_{ij}^{(t)}\right), \textbf{where}$$

$$c_{ij}^{(t)} := \min\left\{ c_{ij}^{(t-1)},\ c_{it}^{(t-1)} + c_{tj}^{(t-1)} \right\}$$

$$L^{(t)} := \left(\ell_{ij}^{(t)}\right), \textbf{where}$$

$$\ell_{ij}^{(t)} := \begin{cases} \ell_{ij}^{(t-1)} & \textbf{if } c_{ij}^{(t)} = c_{ij}^{(t-1)} \\ \ell_{it}^{(t-1)}; \ell_{tj}^{(t-1)} & \textbf{otherwise.} \end{cases}$$

$$(c(i,j)) := M^{(n)} \text{ and } L := L^{(n)}.$$

End

We define the semicolon ";" in this algorithm to mean "catenate and eliminate the second t."

Example (12.6) We'll run algorithm (11.7) on the graph G_1 of (11.3).

$$M^{(0)} = \begin{bmatrix} 0 & 1 & \infty & 1 & \infty & \infty \\ \infty & 0 & \infty & \infty & 1 & \infty \\ \infty & 1 & 0 & \infty & \infty & \infty \\ \infty & \infty & \infty & 0 & 1 & \infty \\ 1 & \infty & \infty & \infty & 0 & 1 \\ \infty & \infty & 1 & \infty & \infty & 0 \end{bmatrix}$$

If we look at the graph G_1 we can find a 1-path 5, 1, 2. Beyond that the job is too tedious to do by hand. I wrote an APL program for it. Here's a way to do step (11.8) easily. Define (compare Chapter 12, (19.22))

$C :=$ the $n \times n$ matrix in which each column is column t of $M^{(t-1)}$,
$R :=$ the same for rows, using row t of $M^{(t-1)}$.

Then $M^{(t)} := \min\{M^{(t-1)}, C + R\}$, where *min* is the entry-by-entry minimum of the two matrices, realizes step (11.8) without the need to loop on i and j. It is easy to set up C and R and do this *min* in APL. In fact, though I'm far from expert in APL, my program ran the first time (except for a pair of parentheses I forgot). That shows you how easy it is to do these things in APL. Here are the $M^{(t)}$'s.

$$M^{(1)} = \begin{matrix} 0 & 1 & \infty & 1 & \infty & \infty \\ \infty & 0 & \infty & \infty & 1 & \infty \\ \infty & 1 & 0 & \infty & \infty & \infty \\ \infty & \infty & \infty & 0 & 1 & \infty \\ 1 & 2 & \infty & 2 & 0 & 1 \\ \infty & \infty & 1 & \infty & \infty & 0 \end{matrix} \qquad M^{(2)} = \begin{matrix} 0 & 1 & \infty & 1 & 2 & \infty \\ \infty & 0 & \infty & \infty & 1 & \infty \\ \infty & 1 & 0 & \infty & 2 & \infty \\ \infty & \infty & \infty & 0 & 1 & \infty \\ 1 & 2 & \infty & 2 & 0 & 1 \\ \infty & \infty & 1 & \infty & \infty & 0 \end{matrix}$$

$$M^{(3)} = \begin{matrix} 0 & 1 & \infty & 1 & 2 & \infty \\ \infty & 0 & \infty & \infty & 1 & \infty \\ \infty & 1 & 0 & \infty & 2 & \infty \\ \infty & \infty & \infty & 0 & 1 & \infty \\ 1 & 2 & \infty & 2 & 0 & 1 \\ \infty & 2 & 1 & \infty & 3 & 0 \end{matrix} \qquad M^{(4)} = \begin{matrix} 0 & 1 & \infty & 1 & 2 & \infty \\ \infty & 0 & \infty & \infty & 1 & \infty \\ \infty & 1 & 0 & \infty & 2 & \infty \\ \infty & \infty & \infty & 0 & 1 & \infty \\ 1 & 2 & \infty & 2 & 0 & 1 \\ \infty & 2 & 1 & \infty & 3 & 0 \end{matrix}$$

$$M^{(5)} = \begin{matrix} 0 & 1 & \infty & 1 & 2 & 3 \\ 2 & 0 & \infty & 3 & 1 & 2 \\ 3 & 1 & 0 & 4 & 2 & 3 \\ 2 & 3 & \infty & 0 & 1 & 2 \\ 1 & 2 & \infty & 2 & 0 & 1 \\ 4 & 2 & 1 & 5 & 3 & 0 \end{matrix} \qquad M^{(6)} = \begin{matrix} 0 & 1 & 4 & 1 & 2 & 3 \\ 2 & 0 & 3 & 3 & 1 & 2 \\ 3 & 1 & 0 & 4 & 2 & 3 \\ 2 & 3 & 3 & 0 & 1 & 2 \\ 1 & 2 & 2 & 2 & 0 & 1 \\ 4 & 2 & 1 & 5 & 3 & 0 \end{matrix}$$

Notice in the graph that the only way to get to node 3 is via node 6. This fact is reflected in $M^{(5)}$, which shows no 5-paths from 1, 2, 4, or 5 to 3. Notice also, from $M^{(1)}$, there's another 1-path, 5, 1, 4.

13. Finite-State Machines

Consider now directed graphs in which each node has outdegree 2, and each outgoing arc carries a label. One arc is labeled "0" and the other, "1."

Example (13.1) Here is such a digraph M with three nodes a, b, c:

(13.1*F*)

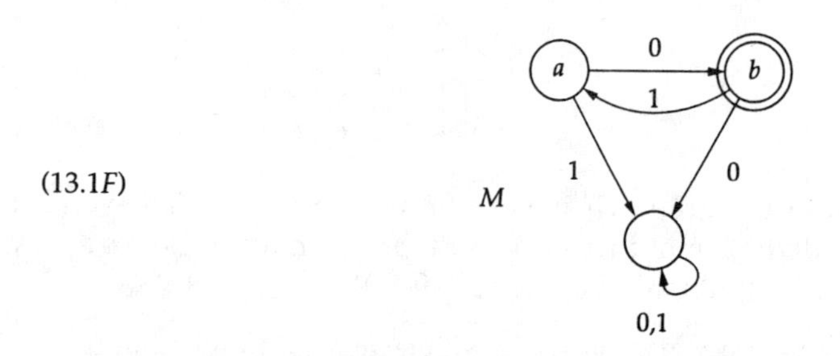

(13.2) *Comment:* This type of directed graph may be viewed as two graphs on the same vertex set. There is the "0" graph and also the "1" graph. Furthermore, in both graphs each vertex has outdegree 1. If both such arcs from x happen to go to the same vertex y, we draw just one arc from x to y and label it "0,1." If we ignore the labels, then some such graphs (as in (13.1*F*)) have repeated arcs.

(13.3) What does one do with such graphs? They represent **finite-state machines** (FSMs) or **finite-state acceptors**, which are used in compilers as lexical analyzers to check that numbers and other multicharacter symbols appear in correct form in a program for a computer. Graph M in (13.1*F*) **recognizes** or **accepts** certain strings of 0's and 1's according to the following convention: A string x is input to M. That is, the leftmost symbol x_1 in x determines the first arc in a path in M corresponding to x and *starting at an agreed-on* node, the **starting state**, here a. If x_1 is 0, the first arc is the arrow from a labeled 0. It goes to b. If x_1 is 1, the first arc in the path is the one labeled 1, from a to c. And so on until we have the whole path. We say that the input x "puts" or "leaves" M "in the state z" if z is the node at the end of the path.

Example If $x = 011$ we have the path from a to b, thence to a, and finally to c. If $x = 010$ we have the path from a to b to a to b. Thus 011 puts M in state c, and 010 puts M in state b.

We also define one or more states (nodes) as **final state(s)**, here b. The final states are usually drawn enclosed in two circles. If, and only if, the input of a string x leaves M in one of the final states, we say that M **accepts** x. The **language** L accepted by M is the set of all strings x accepted by M. Sometimes we call a string a **word**.

Example (13.4) Let's find out what language the example M in (13.1*F*) above accepts. First we try a few strings to get some ideas—we hope.

It doesn't accept 1, or any string starting with 1, because from a, the 1 arc goes to the dead end c. Similarly it doesn't accept 00 or any string starting with 00, because the first 0 takes it to b and the second 0 from b to c, the dead end.

Clearly M accepts 0, and 010, and 01010. And now we can generalize and say it accepts $0(10)^n$ for all $n \geq 0$, where $(10)^n$ denotes $1010 \cdots 10$ (n copies of "10" juxtaposed into one string)—more on that notation later. It's obvious: the first 0 takes M to state b, and every 10 after that brings it back to b.

Now we prove that the language L accepted by M is the language L' we just described: Define $L' = \{ 0(10)^n;\ n \geq 0 \}$, and show that any string x is accepted by M if and only if $x \in L'$. We have already seen that $x \in L'$ implies x is accepted by M. (Strictly speaking, a proof requires an easy induction, which we omit.) For the converse, suppose x is accepted by M. Then x must begin with a 0, as we showed above. But then $x = 0y$ for some string y that takes M from b to b. It is an easy induction to finish the proof, but this time we include it. We know that y (if nonempty) must start with 1 and cannot start with 11, or M would go to the dead end state c. So y begins with 10. Thus $y = 10y_1$, where y_1 is another string taking M from b to b. Since the same reasoning applies to y_1, we have just completed the inductive step of our proof!

That is, we are proving that the only strings taking b to b are $(10)^n$ for $n \geq 0$. Since the length of y_1 is less than that of y, the inductive assumption applies to y_1. It tells us that $y_1 = (10)^n$ for some $n \geq 0$. Therefore $y = (10)^{n+1}$. Therefore $L = L'$. QED

You may think of a finite-state machine as a kind of mechanical dictionary. If you give it a word it will tell you either "Yes, this word is in my language," or "No, it isn't one of my words." In this sense it *defines* the language that it accepts.

The whole result in (13.4), that $L = L'$, is obvious; but it's tedious to prove. The latter defect, alas, pervades the subject of formal languages, to which this section of the chapter is an introduction.

(13.5) *Formal Languages*. In this section we introduce a large subject by presenting only its most elementary part. Formal languages are a highly developed and intricate specialty. To treat them, we'd need more abstract machinery than we have developed in this book. You can, however, taste the flavor of the subject and exercise your logical skills with what is here, namely, the simplest kind of formal language, that defined by a finite-state machine.

We first need to agree on some terminology. Recall (from Chapter 1, (19.9)) that a *string* of *length n* over the set X is an element of X^n, the Cartesian product of n copies of x. If $n = 0$, then $X^0 = \{ \lambda \}$, where λ is the empty string. In writing strings, we save space and bother by omitting the parentheses and commas.

(13.6) NOTATION:
$|x|$ denotes the length of the string x.
Σ is a finite nonempty set called the **alphabet**
Σ^+ is the set of all nonempty strings over Σ
λ is the empty string, of length 0.
$\Sigma^* := \Sigma^+ \cup \{ \lambda \}$.

DEFINITION (13.7) A **(formal) language** L over Σ is defined to be any subset of Σ^*.

Examples (13.8) If $\Sigma = \{ 0, 1 \}$, then the set $L = \{ 0(10)^n;\ n \geq 0 \}$ of all strings accepted by the finite state machine M of Example (13.1) is a formal language over $\{ 0, 1 \}$. So is the set Σ^*; so is $\{ 000, 111 \}$. The set $\{ x_1x_2 \ldots x_n; \forall n \geq 1 \}$ is a language over $\{ 0, 1, \ldots, 9 \}$ if x_i is the ith decimal digit in the base 10 representation of π as an infinite decimal. Thus we're speaking of the language $\{ 1, 14, 141, 1415, 14159, 141592, \ldots \}$.

If we took Σ to be the set of all 26 letters of the Roman alphabet we could define the set of all English words† as a formal language. Since it's a finite set we could view it as the language accepted by some finite-state machine (see (13.39P)), but there's little to be gained therefrom. There is one breakthrough, though; in this language $|length| = 6$.

DEFINITION (13.9) A *finite-state machine* M for Σ is a directed graph with a finite vertex-set such that at each vertex (**node** or **state**) for each $\alpha \in \Sigma$ there is an outgoing arc labeled α. One state of M is designated as the *starting state*, and some nonempty subset of the states of M is designated as the set of *final states*.

The language accepted by M is by definition the set of all strings over Σ that are accepted by M.

(13.10) *Operations on languages.* If L_1 and L_2 are two languages over the alphabet Σ, we have the operations union and intersection on the sets L_1 and L_2. We also define the *catenation* of L_1 and L_2 as the language

$$L_1\,L_2 := \{ xy;\ x \in L_1,\ y \in L_2 \}, \tag{13.11}$$

where xy denotes the catenation of the string x and the string y.‡

DEFINITION (13.12) The **catenation** of string x and string y, denoted xy, is the string $x_1x_2 \ldots x_m\, y_1y_2 \ldots y_n$, where $x = x_1x_2 \ldots x_m$ and $y = y_1y_2 \ldots y_n$. The symbols x_i and y_j for all i, j are from the alphabet Σ.

Thus the catenation of languages is the catenation of every string in the first language with every string in the second. This operation is not in general commutative.

Remember that sometimes catenation is called *concatenation*.

Another operation performed on languages, called the **Kleene closure**, is

$$L^* := L^0 \cup L^1 \cup L^2 \cup \cdots \tag{13.13}$$

where $L^0 := \{ \lambda \}$, $L^1 := L$, and $L^n := L \cdot L^{n-1}$, the catenation of L with L^{n-1}, for all $n \geq 1$. Thus L^* is the set of all strings formed, for all $n \geq 0$, by the catenation of all choices of n strings from L. In particular, $\lambda \in L^*$.

†This definition assumes we could decide such questions as "Is 'meaningful' an English word?"

‡Catenation of strings is thus a binary, associative operation on Σ^*. The empty string λ is an *identity* element for catenation on Σ^*: For all $x \in \Sigma^*$, $\lambda x = x\lambda = x$. A set M together with a binary associative operation on M for which there is a two-sided identity element is called a *monoid*. Σ^* is a monoid *under* catenation. A set S that is closed under a binary associative operation is called a *semigroup*. Σ^+ is a semigroup under catenation. So is the set of negative integers, under addition. No course in discrete mathematics would be complete without mention of *monoid* and *semigroup*.

Notice that the alphabet Σ may itself be regarded as a language, consisting of a finite number of strings of length 1.

Examples (13.14) $\Sigma_1 = \{0\}$. $\Sigma_1^+ = \{0, 00, 000, 0000, \dots\}$ and $\Sigma_1^* = \{\lambda, 0, 00, \dots\}$.

$\Sigma_2 = \{0, 1\}$. Σ_2^+ is the set of all nonempty strings over $\{0, 1\}$.

Let L' be the language accepted by the finite-state machine of (13.1). Thus, as we saw in (13.4), $L' = \{0(10)^n; n \geq 0\}$. Then L' is the catenation

$$L' = \{0\}\{10\}^*.$$

Notice that $\{10\}^*$ is far different from $\{1, 0\}^*$.

The language

$$\{0,1\}^3 \ \{00, 11\}^*$$

is the set of all strings $x_1 x_2 \dots x_m$ in 0,1 of odd length $m \geq 3$ in which the first three symbols x_1, x_2, x_3 are unrestricted, and $\forall n \geq 4$, if n is even and less than m, then $x_n = x_{n+1}$.

There is an elaborate theory of formal languages, of which the simplest are those accepted by finite-state machines. There are many formal languages not accepted (i.e., not definable) by any finite-state machines; the following result can sometimes be used to prove such a nonexistence result for a language.

Theorem (13.15) *The Pumping Lemma.* Let M be a finite-state machine with exactly n states. Let L be the language accepted by M. Suppose some string x in L has length n or greater. Then there exist strings u, v, w (not necessarily in L) such that the following four conditions hold:

$$\begin{aligned} x &= uvw, \\ |uv| &\leq n \\ |v| &\geq 1 \\ \forall i \geq 0 \quad & uv^i w \in L. \ \blacksquare \end{aligned}$$

Note: v^i denotes the catenation of i copies of v; thus $v^2 = vv, v^3 = vvv$, and so on. The pumping lemma says that if x is any long-enough string in L, then x can be "factored" as $x = uvw$, and that x can then be "pumped up" to an arbitrarily long string *still in* L. (And we can delete v, too, making a shorter string in L; that's the case $i = 0$.)

Proof of the Pumping Lemma Presently, this result will seem as obvious as it may now seem mysterious. It's nothing but the pigeonhole principle at work, even in the identical situation where we last saw it (Chapter 12, (19.8)).

Let $x = x_1 x_2 \dots x_m$, where the x_i's are in Σ. Consider the states s_0 (the starting state), $s_1, \dots$ that M goes through as x is input:

$$s_0 \rightarrow s_1 \rightarrow \cdots \rightarrow s_n \rightarrow \cdots$$

The path continues if $|x| > n$. It goes at least as far as s_n; that is, it uses $n + 1$ nodes of M. But M has only n nodes. Therefore two of these nodes are the same: Say $s_i = s_j$ but $1 \le i < j \le n$. Now we see that the path has a circle in it, from s_i to s_i using the input symbols $x_{i+1}, \ldots, x_j$. Since $i + 1 \le j$, this part of the input string has length at least 1. We call it v, where

$$v := x_{i+1} \ldots x_j,$$

and set u and w equal to the preceding and following parts:

$$u := x_1 \ldots x_i$$
$$w := x_{j+1} \ldots x_m.$$

Since $j \le n, |uv| \le n$. Because we may travel around the circle (13.15F) at s_i as many times as we like and still get accepted when we get off and follow w to the final state, $uv^i w$ is in L for all $i \ge 0$.

(13.15F)

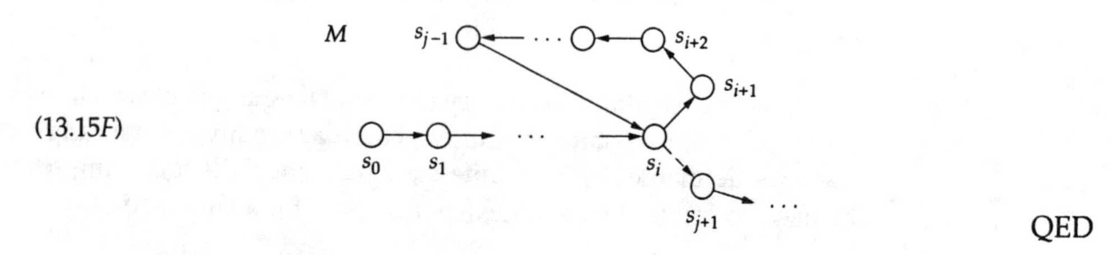

QED

As you see, the Pumping Lemma is totally obvious. All it requires is the existence of a long-enough string in the language.

Example (13.16) Consider the language $L = \{0(10)^n;\ n \ge 0\}$, which we may write as $\{0\}\{01\}^*$ in light of (13.11) and (13.13). This L is the language accepted by our three-state machine M of example (13.1). We may take $x = 010 \in L$. Then there is only one factorization of x as uvw, namely,

$$u = 0, \qquad v = 10, \qquad w = \lambda.$$

The "10" is the circle from b through a and back. The Pumping Lemma tells us that

$$0(10)^i \lambda = 0(10)^i \in L$$

for all $i \ge 0$. We knew that already, of course. No matter which x in L we choose of length ≥ 3, we must take $u = 0$ and $v = 10$ because $|uv|$ is required to be at most 3.

In the preceding example the Pumping Lemma doesn't add to our knowledge. Mostly we use it to prove that for a given language there is no finite-state machine accepting it exactly. Here are three examples of such use.

Examples (13.17) Consider the binomial coefficient $\binom{k}{2}$ for all $k \ge 1$, but express $\binom{k}{2}$ over the alphabet $\Sigma := \{0\}$ as a string of $\binom{k}{2}$ zeros. We use the special superscript notation you've seen before: if $i \geq 0, 0^i$ stands for the string of i 0's. Then

$$\binom{1}{2} = 0 \quad \text{is expressed as } \lambda \text{, the empty string}$$

$$\binom{2}{2} = 1 \quad \text{is expressed as } 0 =: 0^1$$

$$\binom{3}{2} = 3 \quad \text{is expressed as } 000 =: 0^3$$

and so on.

We define the language L_2 now as

$$L_2 := \{ 0^{\binom{k}{2}};\ k \in \mathbf{N} \}.$$

We'll prove L_2 is not the language accepted by any finite-state machine.

Suppose L_2 is the language accepted by some finite-state machine M with exactly n states. We'll derive a contradiction.

Choose k so that $\binom{k}{2} \geq n$. Then

$$x := 0^{\binom{k}{2}} \in L_2, \quad \text{and since} \quad |x| = \binom{k}{2} \geq n,$$

we may apply the pumping lemma. It tells us that there are strings $u,\ v,\ w$ over Σ such that

$$x = uvw, \quad |v| \geq 1, \quad |uv| \leq n, \tag{13.18}$$

and

$$\forall i \geq 0 \qquad uv^iw \in L_2. \tag{13.19}$$

In this problem u, v, and w must each be a string of 0's. Thus for some integers a, b, c

$$u = 0^a, \qquad v = 0^b, \qquad w = 0^c.$$

From (13.18) we know that, in particular,

$$1 \leq b \leq n.$$

Now we use (13.19). It tells us that

$$0^a\, 0^{bi}\, 0^c = 0^{a+c+bi} \tag{13.20}$$

is a word in L_2 for every integer $i \geq 0$. In particular

(13.21) we can find two arbitrarily long words in L_2 with lengths differing by b, namely, those for i and $i+1$ in (13.20), for any i.

Now we go to the definition of L_2. This is where reality comes in. Consider $\binom{k}{2}$ for any $k > n$. The words

$$0^{\binom{k}{2}} \quad \text{and} \quad 0^{\binom{k+1}{2}} \in L_2$$

differ in length by

$$\binom{k+1}{2} - \binom{k}{2} = \binom{k}{1} = k,$$

from the basic recursion of Chapter 8, (3). And if $\ell > k+1$, then the difference in length between $0^{\binom{k}{2}}$ and $0^{\binom{\ell}{2}}$ is still greater than k. In other words, any two sufficiently long words in L_2 differ in length by more than n. This result contradicts (13.21), since $b \leq n$. But (13.21) arose from the assumption that L_2 is the language accepted by M. Therefore that assumption is false, so L_2 is not the language accepted by any finite-state machine.

(13.22) Here is a more complicated example of the use of the pumping lemma to show that a language is not accepted by any finite-state machine. Let L_3 be the language over $\Sigma = \{0, 1, \ldots, 9\}$ of the "prefixes" of the decimal expansion of $\sqrt{2}$. That is, since

$$\sqrt{2} = 1.414\ldots$$

the language L_3 is

$$L_3 = \{1, 14, 141, 1414, \ldots\}.$$

For each $n \in \mathbf{N}$ L_3 has exactly one word of length n, and it is the truncation of $\sqrt{2}$ after n places.

(13.23) As usual, we assume that there is a finite-state machine M that accepts exactly the language L_3. We'll get a contradiction. Let n be the number of states of M. Take the string in L_3 of length n. Apply the pumping lemma:

$$\exists \text{ strings } u,\ v,\ w \text{ over } \Sigma \text{ such that } x = uvw, |v| \geq 1,\ |uv| \leq n, \text{ and}$$

$$\forall i \geq 0 \qquad uv^i w \in L_3. \tag{13.24}$$

As usual, it is (13.24) that upsets the applecart. In this example it says that

$$u \overbrace{v\ v\ v \ldots\ v}^{i} w \in L_3$$

for any value of i. Going back to the definition of L_3 now, we see that $uv^i \in L_3$ also. It follows that the decimal-number counterpart d_i of uv^i approaches $\sqrt{2}$ as i tends to infinity. But

$$d_i = \frac{1}{10^{|u|-1}}\left(u + \frac{v}{10^{|v|}} + \frac{v}{10^{2|v|}} + \cdots + \frac{v}{10^{i|v|}}\right). \tag{13.25}$$

With (13.25) we've shown that $\sqrt{2}$, as a repeating decimal, is rational. We know, however, that $\sqrt{2}$ is not rational. (See, e.g., Chapter 2, (12.4).) This contradiction shows that the assumption (13.23) is false.

(13.26) Our third example is still more complicated. Consider the language L' consisting of all strings over $\{0, 1\}$ that are the base 2 representations of a prime number. Thus

10	represents	2
11		3
101		5
111		7
1011		11

and so on. Thus L' is $\{10, 11, 101, 111, 1011, \ldots\}$. We use the pumping lemma to prove that L' is not the language accepted by any finite-state machine. (This example is taken from Reference (16.5), pages 57–58.)

This proof will require Fermat's Theorem.

$$\text{If } p \text{ is prime, then } \forall a \in \mathbf{Z},\ a^p \equiv a \pmod{p}$$

(see Chapter 7, (11.1); or Chapter 8, problem G10.) We also need a simple identity, namely,

(13.27) $$\forall k \geq 0 \qquad (1 + b + b^2 + \cdots + b^k)(b - 1) = b^{k+1} - 1$$

for any integer b. We noted this in Chapter 3, (16.1).

Now to the proof: Suppose there is a finite-state machine M that accepts L'. We shall derive a contradiction. Let n be the number of states in M, and choose a prime p greater than 2^n. This choice makes the string x representing p have length greater than n. (Why?) Thus we may apply the pumping lemma (13.15) to x to find strings u, v, w such that

$$\begin{aligned} x &= uvw \\ |uv| &\leq n \\ |v| &\geq 1 \\ \forall i \geq 0 &\qquad uv^iw \in L. \end{aligned}$$

We set $i = p$; thus $uv^pw \in L$. This is the binary string $uvv\ldots vw$. It is the base 2 representation of some integer q, which we now show is larger than p but divisible by p.

So we'll show that M accepts a number q that is not prime. This fact means no finite-state machine can accept exactly the set of all primes.

We work out the values of the integers p and q in terms of the integers n_u, n_v, and n_w, which have u, v, and w, respectively, as their base 2 representations.

Thus if, for example, $p = 89$, then $x = 1011001$; and if $n = 6$, then u, v, w might be $u = 10, v = 110, w = 01$. Then $n_u = 2, n_v = 6$, and $n_w = 1$. And $p = 89 = 2 \cdot 2^5 + 6 \cdot 2^2 + 1$. In general,

$$p = n_u \cdot 2^{|v|+|w|} + n_v \cdot 2^{|w|} + n_w,$$

as in the example just concluded. The integer q is expressed similarly, but we have to multiply by $2^{|v|}$ each time we repeat v. We get

(13.28) $$q = n_u \cdot 2^{|v|p+|w|} + n_v 2^{|w|}(1 + 2^{|v|} + \cdots + 2^{|v|(p-1)}) + n_w.$$

We let s denote the quantity in parentheses:

$$s = 1 + 2^{|v|} + \cdots + 2^{|v|(p-1)}.$$

Then by (13.27), with $b = 2^{|v|}$ and $k = p - 1$, we see that

(13.29) $$s(2^{|v|} - 1) = 2^{|v|p} - 1.$$

Now we use Fermat's Theorem to see that (13.29) becomes (with $a = 2^{|v|}$),

$$s(2^{|v|} - 1) \equiv 2^{|v|} - 1 \pmod{p},$$

which we rewrite as

(13.30) $$(s - 1)(2^{|v|} - 1) \equiv 0 \pmod{p}.$$

Now remember that $p > 2^n$ and $1 \le |v| \le n$. Therefore† $p > 2^{|v|} - 1 \ge 1$, so p cannot divide $2^{|v|} - 1$. By (13.30) p must then divide $s - 1$. That is,

(13.31) $$s \equiv 1 \pmod{p}.$$

Now we go back to (13.28). The key is (13.31); it tells us that

(13.32) $$\begin{aligned} q &\equiv n_u \cdot 2^{|v|p+|w|} + n_v \cdot 2^{|w|} + n_w \pmod{p} \\ &\equiv n_u \cdot 2^{|v|+|w|} + n_v \cdot 2^{|w|} + n_w \pmod{p}, \end{aligned}$$

where we used Fermat's Theorem again to say $2^{|v|p} \equiv 2^{|v|} \pmod{p}$. But the right-hand side of (13.32) equals p. Thus (13.32) says $q \equiv 0 \pmod{p}$. Thus p divides q. Finally, $q > p$ because the leading 1 in the base 2 representation uv^pw of q is at least $p-1$ places to the left of that in uvw, the base 2 representation of p. Therefore any finite-state machine with n or fewer states that accepts a prime $p > 2^n$ (as the string of 1's and 0's in the base 2 representation of p) must also accept a nonprime. QED

(13.33) *How to use the pumping lemma.* Sometimes students have trouble using the pumping lemma. The reason is that the greater number of parts in its statement may obscure the underlying simplicity of the proof by contradiction. Let's highlight the logical structure of the preceding three examples.

(*i*) We are given a language L. We want to prove that it is not the language accepted by any finite-state machine.

(*ii*) To start our proof by contradiction, we assume

(13.34) $$\exists M \quad L = L(M).$$

(The domain of M is the set of all finite-state machines.)

(*iii*) We check that L has arbitrarily long words. If so, L satisfies the hypotheses of the pumping lemma if (13.34) is true.

(*iv*) Therefore if (13.34) is true, the conclusions of the pumping lemma must hold. One of them is the proposition

(13.35) $$c: \forall i \ge 0 \quad uv^iw \in L.$$

(*v*) Now we go to the definition of L and prove

(13.36) $$\exists i \ge 0 \quad uv^iw \notin L.$$

†We chose p a little larger than the minimum value required by the pumping lemma just so we could get this inequality.

But (13.36) is the negation of c. The result is that we've shown

(13.37) $$\exists M \quad L = L(M) \rightarrow c \wedge \sim c.$$

That has the form

(13.38) $$h \rightarrow \text{false},$$

where h is the proposition (13.34)

$$h: \exists M \quad L = L(M).$$

The only way (13.38) can be true, and we showed it was true in the three examples, is for h to be false. That means

$$\forall M \quad L \neq L(M),$$

our target.

The hard part of this procedure can be Step (v), to show that for some i, uv^iw is not in L. Step (v) was tricky for the "prime problem" (13.26); we had to know to choose $i = p$, and then work hard to show $q = uv^pw$ was not prime, i.e., not in L. But that's math for you. It's like finding the connection between the inductive assumption $S(n)$ and the target $S(n+1)$; no one said it would always be easy.

Many other, in fact most, languages are not definable by finite-state machines. There is a hierarchy of ever more complicated languages and of the correspondingly more complicated machines that accept them, going up to Turing machines. We leave them to other texts.

Practice *(13.39P) Try to prove that any finite language is the language accepted by some finite-state machine. (Set a time limit for your work on this problem.)

As to graphs, they too have other uses in mathematics, computer science, social science, and engineering. We chose the present application because it is so accessible and nontechnical, and it introduces another topic in computer science. There are many other applications of graphs; see, for example, (16.7). Our final application is new and exciting.

14. Application to Computerized Tomography

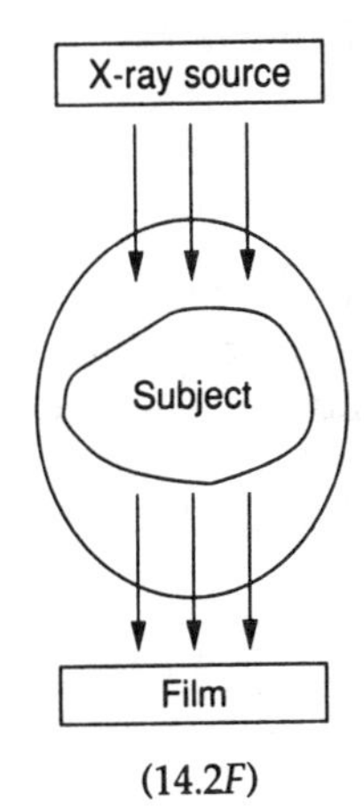

(14.2*F*)

(14.1) We now show how ideas from graph theory are used in computerized axial tomography (CAT) and magnetic resonance imaging (MRI) scanners. The CAT scanner first takes X-rays in a thin planar slice from many different angles. If the plane is that of the paper the first X-ray may be taken in the direction shown in (14.2*F*). The next one in the same plane would be taken after a slight rotation, maybe by 6°, of the source and detector. After these 50 or 60 X-ray exposures, some fancy calculus-based mathematics (incidentally, developed about a century ago, before the discovery of X-rays) is used to get the density of the tissue (bone or whatever) at each little element of volume in that slice.

Next, the equipment is moved a little (say upward off the paper in (14.2*F*)), and the process is repeated for the next slice, parallel to the first slice. It continues for as many slices as necessary, perhaps 50. The result is that the computer has data of the following type in its memory:

(14.3) Small cubes called **voxels** identified by the (x,y,z) coordinates of their centers; with each voxel is a real number, its density. (A voxel is three-dimensional; a *pixel* is two-dimensional.)

The entire space that includes the part of the patient being studied is partitioned into voxels. Figure (14.4F) shows a small space so partitioned. Each edge of each voxel is parallel to a coordinate axis. The thickness of a slice is the height of a voxel. The problem is to use the data in all these voxels to make pictures, with perspective and

(14.4F)

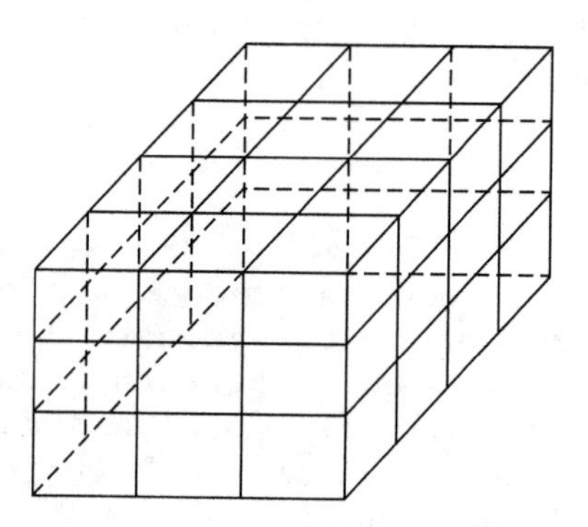

shading to simulate three dimensions. Computer graphics is advanced enough to accomplish this (another story we won't go into) if we can identify the surface of our target (say a bone or an organ). We have all these voxels—which ones are on the surface of the target? If we can only list them, the graphics package will do the rest.

This can be quite a problem, because the minimum number of voxels per slice is $64 \times 64 = 2^{12} = 4096$; the maximum is $512 \times 512 = 2^{18} \doteq 256,000$. Since there are typically 50 to 80 slices, we often must cope with more than 10^7 voxels.

(14.5) To define the surface of our target we first look at the density. We set a threshold t; all voxels with density lower than t are outside the target; those with higher density are in the target. Now

DEFINITION (14.6) The surface of our target is the set of all faces F of all voxels in the target such that some voxel adjacent to F is outside the target. Adjacent means "having a common face."

(14.7) *Representation.* In the computer the voxel is just a 4-tuple (x,y,z,d), where d is the density and the first three coordinates are the rectangular Cartesian coordinates of the center of the voxel. We scale the coordinates so that x, y, and z take integral values, say,

$$1 \le x \le b_1$$
$$1 \le y \le b_2$$
$$1 \le z \le b_3$$

for $b_1,\ b_2,\ b_3\ \in \mathbf{N}$.

The six voxels with a face in common with the voxel centered at (x,y,z) are those centered at

$$\begin{aligned} 1^- &:= (x-1,y,z), & 1^+ &:= (x+1,y,z), \\ 2^- &:= (x,y-1,z), & 2^+ &:= (x,y+1,z), \\ 3^- &:= (x,y,z-1), & 3^+ &:= (x,y,z+1), \end{aligned}$$

where we've introduced a notation to indicate which variable is changed. Clearly two voxels have a common face if and only if they differ in only one coordinate, and

(14.8F)

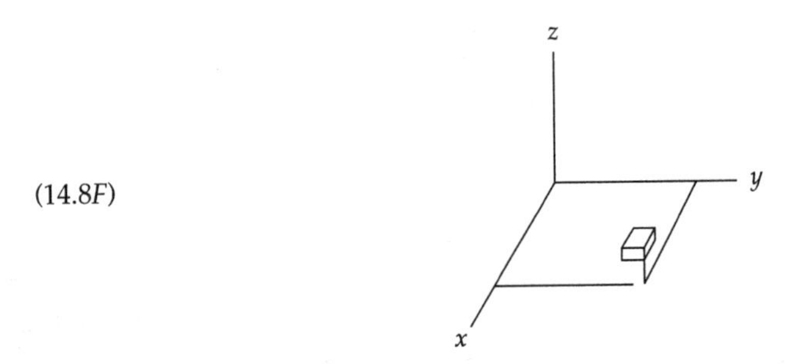

there by 1. In (14.8*F*) the voxel on top of the one shown is 3^+. The one below is 3^-. The other four are in the same horizontal plane as (x,y,z).

How does the computer *know* a face of a voxel? We tell it.

DEFINITION (14.9) A **face** is a pair of voxels that agree in two coordinates and differ by 1 in the remaining coordinate.

That defines *face* in terms of what the computer has in it.

Now in principle finding the surface of our target is simple. We have the computer look at all the faces; at each one it looks at the densities of the two voxels; if both are on the same side of the threshold, forget that face, because it's either inside or outside the target. But if one density is above the threshold and one is below, then that face is on the surface of the target.

The problem with this exhaustive procedure is that it may be too time-consuming. There are, after all, $b_1b_2b_3$ voxels. Each voxel has six faces, so there are $3b_1b_2b_3$ faces all told, since each face is on two voxels (we are inaccurate at the boundary). Since b_1 and b_2 range now from 200 to 500 and b_3 is around 60, we'd often have to look at $3 \times 60 \times 300^2$ ($\doteq 16 \times 10^6$) or $3 \times 60 \times 400^2$ ($\doteq 29 \times 10^6$) faces. For the small computers in use at hospitals that's quite a burden. Nevertheless, early scanners did just this exhaustive look at all the faces. As you can see, it's an $O(n^3)$ algorithm if $b_i = O(n)$ for $i = 1,2,3$.

(14.10) *A faster algorithm*. Here is where graph theory comes in. We can use it to define an algorithm of complexity $O(n^2)$. (At least intuitively it's of order n^2, since it's for a surface.)

We begin by assigning two arrows to each face as shown in (14.11*F*). We identify

(14.11*F*)

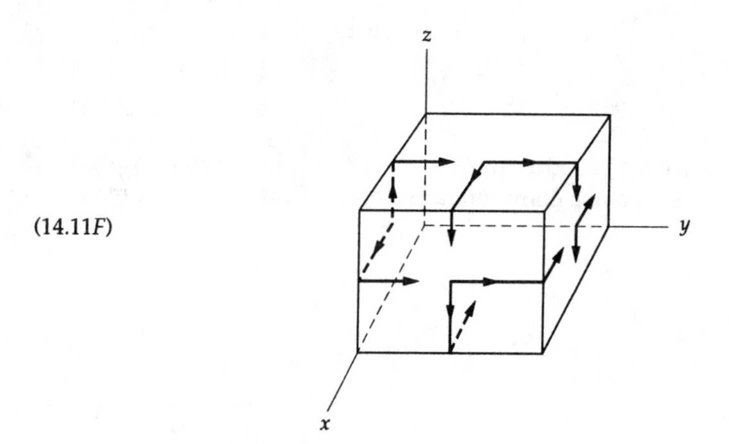

the face with its center and draw two arrows out from that center to the center of an adjoining face of the same voxel, as indicated. We replicate this set of 12 arrows on every voxel. More strictly, we *imagine* doing so. Think of it this way: pass a plane parallel to the $y - z$ plane through the center of the voxel in (14.11*F*). Choose a direction: left or right. In (14.11*F*) we chose right. Make a cycle of four arrows in the four faces through which the plane passes. Now do the same for a plane parallel to the $x - z$ plane; and finally for one parallel to the $x - y$ plane. Once you've chosen these three 4-cycles of arrows, make the same choice in every other voxel.

Notice that each of the 12 edges of a voxel is crossed by one arrow. Notice that on the six faces of a voxel you now have a digraph with indegree 2 and outdegree 2 at each node.

(14.12) *The directed graph of the surface of the target.* We now use the individual voxel-digraphs to make (or, to imagine) a big digraph for the surface we seek. The nodes are the faces on the surface. Two nodes are joined by an arc if they have a common edge. The direction of the arc is that on the two voxel-digraphs—they are always consistent. Here is a drawing (14.13*F*) that shows one possibility. The two

(14.13*F*)

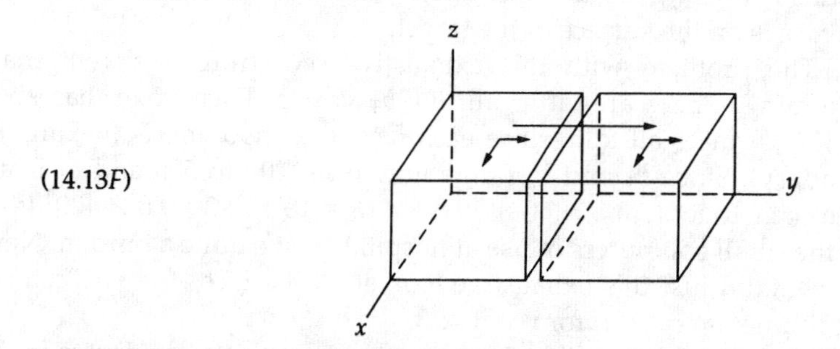

surface faces are the top faces of the voxels. The arc of the *surface digraph* goes from left to right because that's the direction of the voxel-arrows.

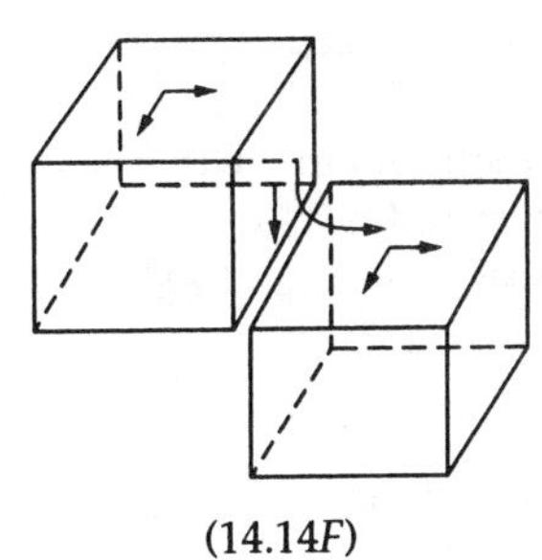

(14.14*F*)

Another possibility is shown in (14.14*F*). The y^+ face of the upper voxel and the top or z^+ face of the other voxel are on the surface. Again the arrows are compatible, so in the new graph there is an arc from the y^+ face of the one voxel to the z^+ face of the other.

The reason the voxel-arrows are always compatible is that both are in the plane bisecting the common edge of the two faces. The four voxel-arrows in that plane circle in the same direction for all voxels.

(14.15) Notice another point about the surface digraph: Before we define it, every face has indegree 2 and outdegree 2. But as we imagine forming the surface digraph by adding faces one at a time, we see that every time we do so we preserve the 2-in, 2-out property. See (14.13*F*) and (14.14*F*), which show the voxel-outarrow of one face *joining with* the voxel-inarrow of the other face to form an arc of the new digraph. (In both these figures we have omitted other voxel-arrows for clarity.)

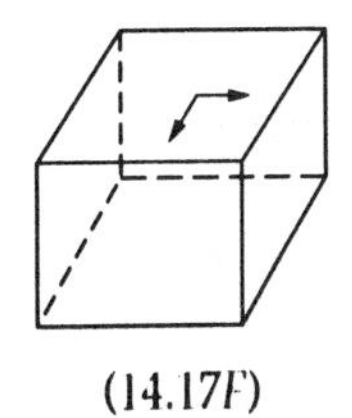

(14.17*F*)

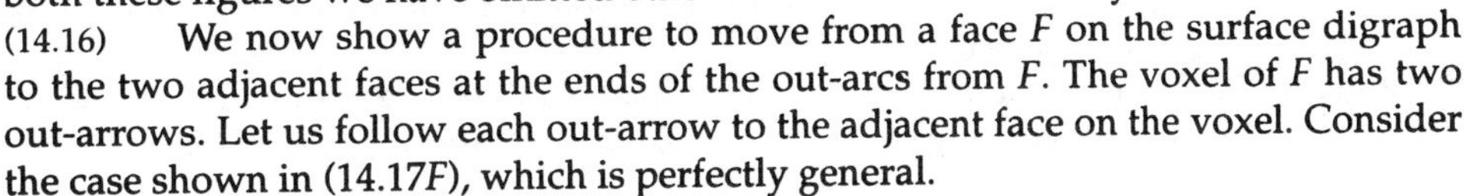

(14.16) We now show a procedure to move from a face F on the surface digraph to the two adjacent faces at the ends of the out-arcs from F. The voxel of F has two out-arrows. Let us follow each out-arrow to the adjacent face on the voxel. Consider the case shown in (14.17*F*), which is perfectly general.

Case 1. If the y^+ face of the voxel is on the surface of our target, then the voxel-arrow is the arc of the surface digraph from the z^+ face to the y^+ face.

Case 2. If the y^+ face is not on the surface, then it is inside (why?). Now the z^+ face of the adjacent voxel may be on the surface, as shown in (14.18*F*). Check to see.

(14.18*F*)

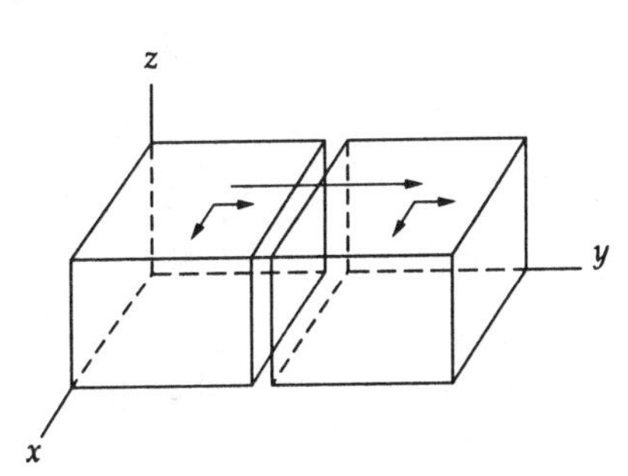

Case 2a If so, then, the original out-arrow leads to the adjacent vertex of the surface digraph.

Case 2b If not, then the face this arrow goes to must be the y^- face of the voxel above, as shown in (14.19*F*).

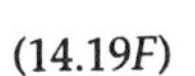

(14.19*F*)

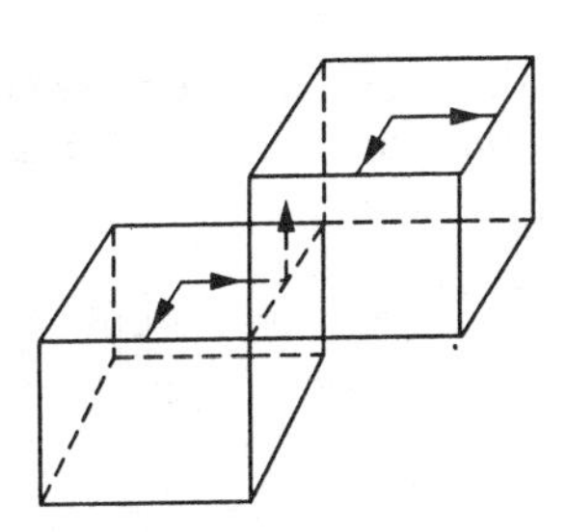

It is a matter of our definition whether we view the voxels in (14.19*F*) as connected to each other, or not.

Using the procedure (14.16) we can find all the faces in a component of the surface digraph by an $O(n^2)$ algorithm, which we now describe informally.

(14.20) We start with any face, find the two adjacent faces by (14.16), and then find the adjacent faces to those two, and so on. To avoid cycling back to a face we've already been to, we keep a check list of the faces found so far.

(14.21) Whenever we come a second time to a node, *we delete it from the check list*. That's acceptable because the surface digraph has constant indegree 2. We'll never see any node three times. And that deletion keeps the list of visited nodes from growing too large. We also keep a list of all nodes visited. At the end this list is our output of all the faces in the surface of the target.

Example (14.22) This example of the algorithm (14.20) is taken from (14.33) with permission. We have a target "organ" of three voxels, $A, B,$ and C in (14.23*F*).

(14.23*F*)

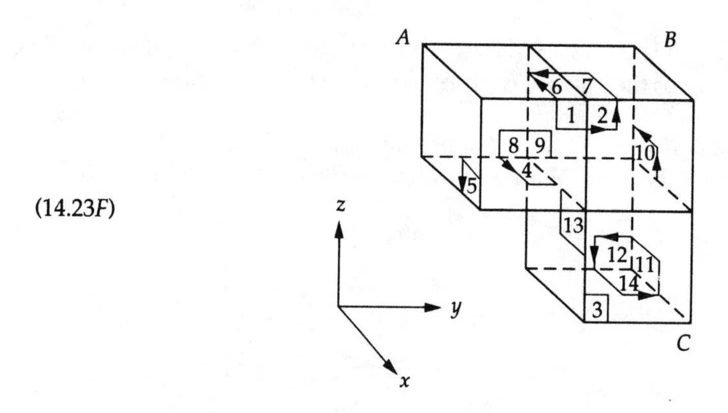

We number the faces as shown:

	z^-	z^+	y^-	y^+	x^-	x^+
voxel A	4	6	5	i	8	1
voxel B	i	7	i	10	9	2
voxel C	14	i	13	11	12	3

The i means the face is interior—we don't number those. It's necessary to specify the directions of the voxel-arrows. We have drawn some in (14.23*F*). The three 4-cycles of the faces are

z^+, y^-, z^-, y^+ (perpendicular to x-axis)

z^+, x^-, z^-, x^+ (perpendicular to y-axis)

x^+, y^+, x^-, y^- (perpendicular to z-axis).

We can use these two tables to make a table of adjacencies.

Table of Adjacencies

1 →	2, 6	6 →	5, 8	11 →	10, 12
2	7, 10	7	6, 9	12	13, 14
3	11, 2	8	4, 5	13	14, 3
4	1, 13	9	12, 8	14	11, 3
5	4, 1	10	9, 7		

This table means, for example, that there is an arrow in the face digraph from 1 to 2 and from 1 to 6. How do we know? We can look at the drawing (14.23*F*) or we can do a form of procedure (14.16):

We find 1 as the x^+-face of A in the first table. After x^+ comes z^+ in one cycle in the second table; the z^+-face of A is 6. In the last cycle y^+ comes after x^+, but the y^+-face of A is interior. It equals the y^--face of B, and after y^- comes x^+ in the last cycle. So the other arrow from 1 goes to the x^+-face of B, namely, 2.

We can use this table of adjacencies (in general we'd use subprocedure (14.16)) to find the two faces adjacent to each face as we come to it. We start with face 1.

We enter 1 twice in the check list because we will come into it on two arcs. We enter each other face once because we took one in-arc already to get to it. Thus 1 is adjacent to 2 and 6. We immediately enter "6,2" in the check list and in the output list. And so on. Here is the entire set of actions taken.

(14.24)

Node considered	*List of nodes to be considered*	*Check list*	*Output list*
	1	1,1	1
1	6,2	1,1,6,2	1,6,2
6	2,5,8	1,1,6,2,5,8	1,6,2,5,8
2	5,8,7,10	1,1,6,2,5,8,7,10	1,6,2,5,8,7,10
5	8,7,10,4	1,6,2,5,8,7,10,4	1,6,2,5,8,7,10,4
8	7,10,4	1,6,2,8,7,10	1,6,2,5,8,7,10,4
7	10,4,9	1,2,8,7,10,9	1,6,2,5,8,7,10,4,9
10	4,9	1,2,8,10	1,6,2,5,8,7,10,4,9
4	9,13	2,8,10,13	1,6,2,5,8,7,10,4,9,13
9	13,12	2,10,13,12	1,6,2,5,8,7,10,4,9,13,12
13	12,3,14	2,10,13,12,3,14	1,6,2,5,8,7,10,4,9,13,12,3,14
12	3,14	2,10,13,12,3	1,6,2,5,8,7,10,4,9,13,12,3,14
3	14,11	10,12,3,11	1,6,2,5,8,7,10,4,9,13,12,3,14,11
14	11	10,12	1,6,2,5,8,7,10,4,9,13,12,3,14,11
11	∅	∅	1,6,2,5,8,7,10,4,9,13,12,3,14,11

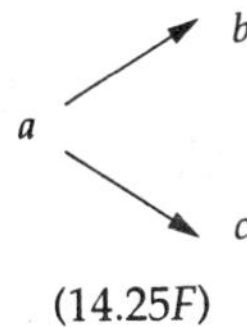

(14.25*F*)

We choose a node for consideration by picking the first item, say a, in the current list of nodes to be considered (column 2 in (14.24)). Then we update the list of nodes to be considered and the check list: We delete a from column 2 and add to both lists the two nodes a is adjacent to—*unless* one of them appears already in the check list. Say that in the digraph a is adjacent to b and c. Suppose b already appears in the check list but c does not. Then we delete b from the check list and add c at the end.

The first example of the deletion occurs when we consider node 5. Face 5 is adjacent to 4 and to 1. Since 1 is in the check list (and 4 is not) we delete 1 and append 4. Since we've already considered 1 we do not insert it into column 2, but we append 4 there also.

On the next line we consider face 8. It is adjacent to 4 and to 5, both of which are in the check list. So 4 and 5 both leave the check list, and we make no new insertions in column 2.

(14.26) Next question: How do we know this procedure gets all the faces of the surface?

The answer: Euler. Here is a simple version of Theorem (3.1) for directed graphs.

Theorem (14.27) Let D be a directed graph in which at every node v

$$\text{indegree}(v) = \text{outdegree}(v).$$

If D is also connected (as the undirected graph resulting when all arcs of D are made two-way), then there is an Euler tour of D respecting the directions on the arcs that starts at any node. ■

We leave the proof to you in the problems.

Corollary (14.28) For every ordered pair (u, v) of nodes of D there is a walk in D from u to v that respects the directions on the arcs. ■

In case you were wondering, we do not call such a walk a *diwalk*.

(14.29) We use Corollary (14.28) to see that the procedure (14.20) does produce all the faces on the component of the starting face F. The procedure has us go from each face we reach to both of its out-adjacent faces. Therefore for each node v, we try all arcs of the walk from F to v.

To prove this claim more formally, consider a walk

$$F, v_1, v_2, \ldots, v_n = v$$

from F to v. It is trivial to prove by induction that the procedure (14.20) results in listing v_1, then v_2, and so on, and eventually v_n, since it lists each of the two out-adjacent faces for each face it comes to. It may de-list a vertex, but only if it had listed it earlier. (That is, the walk may have a loop, but if so the procedure cuts it out.)

(14.30) *Complexity*. Our output is a list of the faces of the target surface. It is, therefore, of order n^2. The procedure (14.20) that constructs that output uses (14.16), which calls for up to six steps per face of the surface, and then enters items in the check list. The check list without the deletion step (14.21) would grow to the same size as the output. That's *only* another $O(n^2)$, and we know that $O(n^2) + O(n^2) = O(n^2)$—right? The algorithm must operate on the small computers attached to CAT scanners and at hospital work stations; a big increase in time and space requirements would be undesirable. That is why the deletion (14.21) is such a key idea. It cuts the expected length of the check list to a small fraction of the length of the output list. That in turn reduces the space and time needed.

(14.31) *Computer surgery*. Once a workable algorithm (14.20) exists, physicians can do *computer surgery*. If they want to see inside a vertebra, for example, they pass a plane through the set of voxels representing it, define that plane as a surface of interest, and find the surface faces. The one vertebra now has two components; they can be displayed side by side, revealing interior defects.

(14.32) To avoid technicalities we omit the representations of arcs and the formal algorithm. The essence of it is described under (14.20). If you are interested, see the original paper:

(14.33) Ehud Artzy, Gideon Frieder, and Gabor T. Herman, "The theory, design, implementation and evaluation of a three-dimensional surface detection algorithm," *Computer Graphics and Image Processing*, vol. 15 (1981), pages 1–24.

(14.34) *Postscript.* A directed graph with the property described in Corollary (14.28) is called **strongly connected.** You may remember the strongly connected components of a reflexive transitive relation (Chapter 12, (11.2)). Those are a special case of the concept for general directed graphs (relations).

(14.35) Consider the following graph:

(14.36*F*)

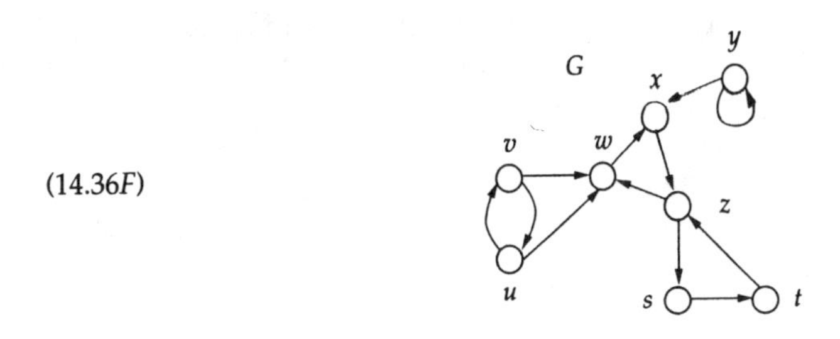

Its three strongly connected components are

(14.37*F*)

y
x
v
w
z
u
s
t

That is, a strongly connected component (SCC) of a directed graph G is a maximal strongly connected subgraph. Here *maximal* is taken in the sense of inclusion: If you added any other vertex of G to the SCC, even with all the edges of G going between it and the SCC, the result would not be strongly connected.

15. Mathematical Structures

If we were more formal, we'd define a graph with no loops or repeated edges as a pair (V, E), where V is a nonempty set and E is a collection of 2-subsets of V. If we wanted to allow loops, but not repeated edges, we'd define E as a collection of

2-reps of V. If we wanted to allow repeated edges, we'd say that E is a multiset of 2-subsets (or of 2-reps if we wanted to allow loops) of V. These definitions of graphs are examples of *mathematical structures.*

A mathematical structure is a pair (B, S), where B is a set and S, the structure set, is specified somehow as, for example,

(*i*) a subset of B
(*ii*) a set of subsets of B
(*iii*) a subset of $B \times B$.

You see that a set B with a binary relation on it comes under (*iii*). Graphs with no loops or repeated edges come under (*ii*). We'd have to expand the little list of three possible structures to allow loops or repeated edges, but perhaps you see that familiar mathematical sets with operations on them can be defined this way. For example, the integers with addition specified on them are the pair $(\mathbf{Z},+)$, where $+$ now stands for a certain subset of $\mathbf{Z} \times \mathbf{Z} \times \mathbf{Z}$, namely, the set of all triples (a, b, c) of integers such that c is the sum of a and b.

You've probably heard of Turing machines, imaginary computers that are simple to describe (it says here) but useful in studying what is or is not possible in theoretical computing. They can be specified as a grand expansion of the above structure, namely, as an ordered 7-tuple $(Q, \Sigma, \Gamma, \delta, q_0, B, F)$, where Q is a finite set of states, Γ is the alphabet of symbols to be used, $B \in \Gamma$ is the blank symbol, $\Sigma \subseteq \Gamma - \{B\}$ is the set of input symbols, and δ is the next-move function, a mapping

$$\delta : Q \times \Gamma \longrightarrow Q \times \Gamma \times \{L, R\}.$$

Here L and R stand for "move the tape of the Turing machine left (L) or right (R)." Finally, q_0 is the start state, and $F \subseteq Q$ is the set of final states. (For more about Turing machines, see (16.1) or (16.5).) I don't expect you to get anything out of this except the idea that careful study of more advanced formal languages requires more abstraction than we have employed in this book. Much the same is true for more advanced excursions into other topics of this book as well.

16. Further Reading

(16.1) Alfred V. Aho, John E. Hopcroft, and Jeffrey D. Ullman, *The Design and Analysis of Computer Algorithms*, Addison-Wesley, Reading, 1974.

(16.2) Claude Berge, *The Theory of Graphs and Its Applications*, John Wiley & Sons, New York, 1962.

(16.3) J. A. Bondy and U. S. R. Murty, *Graph Theory with Applications*, American Elsevier, New York, 1976.

(16.4) John N. Fujii, *Puzzles and Graphs*, National Council of Teachers of Mathematics, Washington, D.C., 1966.

(16.5) J. E. Hopcroft and J. D. Ullman, *Introduction to Automata Theory, Languages, and Computation*, Addison-Wesley, Reading, 1979.

(16.6) Oystein Ore, *Graphs and Their Uses*, Mathematical Association of America, Washington, D.C., 1990.

(16.7) Fred S. Roberts, *Discrete Mathematical Models*, Prentice-Hall, Englewood Cliffs, 1976.

Comment. In order of increasing difficulty, these references are ranked as follows:

Fujii, then Ore (high school to college sophomores)
[This chapter fits here.]
Roberts, Bondy and Murty, Berge (juniors to graduate students)
The other two are at a senior to graduate level and discuss topics other than graphs.

17. Problems for Chapter 14

DEFINITIONS (17.1) Let $n \in \mathbf{N}$. The *complete graph* K_n is the graph on n vertices in which every two distinct vertices are joined by exactly one edge. (There are no loops.) A *simple* graph is one with no loops or repeated edges.

The *multiset of degrees* of a graph G is an n-rep, where n is the number of vertices of G, in which the elements are the degrees of the vertices. For example, the multiset of degrees of a triangle, K_3, is [2,2,2].

Two graphs $G_1 = (V_1, E_1)$ and $G_2 = (V_2, E_2)$ are *the same* iff the vertices of G_1 can be renamed as the vertices of G_2 in such a way that the induced renaming of the edges of G_1 turns E_1 into E_2. The official term is *isomorphic*. Here is an example of the induced renaming of the edges: If $1, 2 \in V_1$ and $\{1, 2\} \in E_1$, and if we rename 1 as a and 2 as b, then edge $\{1, 2\}$ is renamed $\{a, b\}$.

Problems for Section 2

2.1 What is the smallest graph with an odd number of vertices and an odd number of edges?

2.2 Draw a graph with multiset of degrees [1,2,2,3,3,3] or explain why there is none.

2.3 How many different graphs can you draw having multiset of degrees [4,4,3,2,2,1,1]?

2.4 Let G be a graph that can be drawn on the plane or sphere without any edges crossing. Euler's formula (see Chapter 3, problem 53) says that if G is connected, then $|V| - |E| + f = 2$, where f is the number of faces. Assuming Euler's formula, prove the following generalization: If G has exactly p components, then $|V| - |E| + f = 1 + p$.

2.5 Let G be a simple graph and M its relation-matrix (also called its adjacency matrix). Show that the entries on the main diagonal in M^2 are the degrees of the vertices (if the matrix product is taken with ordinary, not Boolean, addition).

2.6 Prove that any multiset of nonnegative integers with even sum is the multiset of degrees of some graph.

2.7 Find a multiset D of positive integers with even sum such that no connected graph has D as its multiset of degrees. Explain.

2.8 (From Bondy and Murty (16.3)) Show that there is no simple graph having multiset of degrees equal to

(*i*) [2,3,3,4,5,6,7]

(*ii*) [1,3,3,4,5,6,6].

2.9 Find graphs having as their multisets of degrees those in the previous problem.

2.10[Ans] Prove that any two simple graphs on four vertices with five edges are "the same."

2.11 The simple graph in (1.1F) has five vertices and eight edges. Prove that there is only one other such simple graph. ("Other" means "not the same as.")

2.12 Let $G = (V, E)$ be a graph. Suppose V_1 is an equivalence class of the relation W defined in (2.19). Let e be any edge of G having an endpoint in V_1. Prove that both endpoints of e are in V_1. (Compare Definition (2.20).)

2.13 (Newman) There are $n \geq 2$ people at a party. Prove there are at least two people at the party each of whom has shaken hands with the same number of others present.

2.14 Suppose D is a nonempty multiset of integers with even sum. If all elements of D are two or greater, prove that some connected graph has D as its multiset of degrees.

2.15 Find two graphs that are not "the same" but have the same multiset of degrees. Try to keep the number of vertices small. Do this for two cases:

(*i*) if connectivity is not required;

(*ii*) if both graphs are connected, and simple.

In each case you will need to prove that your two examples are not "the same."

Problems for Section 3

3.1[Ans] For what values of n does the complete graph K_n have an Euler tour? Explain carefully.

3.2 A *Hamiltonian cycle* in a graph is a walk that goes through each vertex exactly once, ending at its starting vertex. (It need not use all the edges.) How many different Hamiltonian cycles does K_n have? Explain.

3.3 Let $m, n \in \mathbf{N}$. We model a rectangular grid of streets and their intersections as line-segments parallel to the coordinate axes through the points of $V = \{(x, y);\ 1 \leq x \leq m,\ 1 \leq y \leq n\}$ in the plane. Consider the graph $G = (V, E)$, where two points of V are joined by an edge iff they agree on one coordinate and differ by exactly one in the other. Find all values of m and n for which G has an Euler tour. Prove your answer.

Problems for Section 4

4.1 Is it possible for a free tree and a connected simple graph that is not a free tree to have the same multiset of degrees? Explain.

4.2 Find the minimum and maximum values of the number of vertices of degree 1 in a free tree on n vertices. Draw free trees for $n = 5$ attaining these values.

4.3 What free trees have Euler tours? Explain.

4.4 Consider a largest graph in problem 3.3 that has an Euler tour but is not a free tree. Count the total number of its Euler tours. [Hint: Make a tree showing the possibilities. Use symmetry.]

4.5 Find all free trees on six vertices. Explain how you know you have all of them, and explain how each two of yours differ.

4.6 Prove that if a free tree has a vertex of degree k, then it has at least k vertices of degree 1.

4.7 In the BET theorem (4.25) prove that (*ii*) and (*v*) are equivalent.

4.8 Complete the proof of the BET theorem. See (4.26).

***4.9** For $n \in \mathbf{N}$ let D be an n-rep from $\mathbf{N}$ such that the sum of the elements of D is $2n - 2$. Prove that there is a free tree on n vertices with D as multiset of degrees.

Problems for Section 5

5.1 How many subgraphs does K_3, the complete graph on three vertices, have? Explain.

5.2[Ans] Same question for K_n, for $n \in \mathbf{N}$.

5.3 Let S be a "star" graph. That is, $n \geq 0$, S has $n + 1$ vertices; one vertex has degree n; each other vertex has degree 1. How many subgraphs does S have? Explain.

Problems for Section 6

6.1 Suppose G_1 is a cycle, i.e., G_1 has n vertices and n edges and there is a cycle of length n in G_1. How many spanning trees does G_1 have? Explain. See (17.2F).

(17.2F)

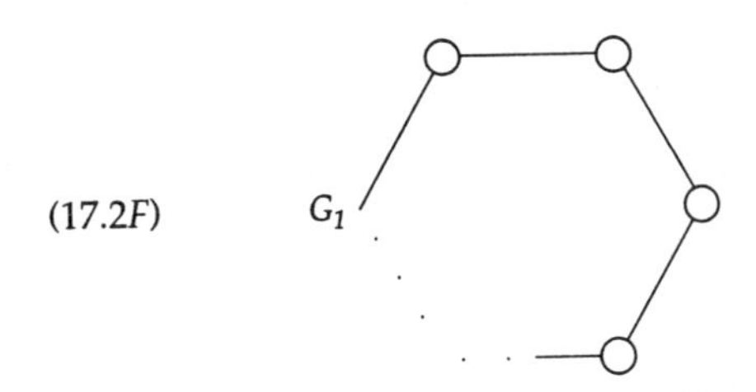

6.2[Ans] Consider the graph G_2 below. How many spanning trees does it have? Explain.

(17.3F)

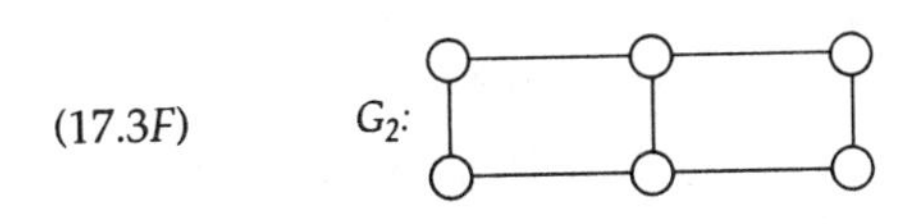

Problem for Section 7

7.1 Find a minimum spanning tree for the golf course example of (7.4*F*). Explain by stating the order in which you choose your edges. Be sure to state the total cost for your spanning tree.

Problems for Section 9

9.1 (From Ore, (16.6)) Consider the graph below. Prove it is not an interval graph.

(17.4*F*)

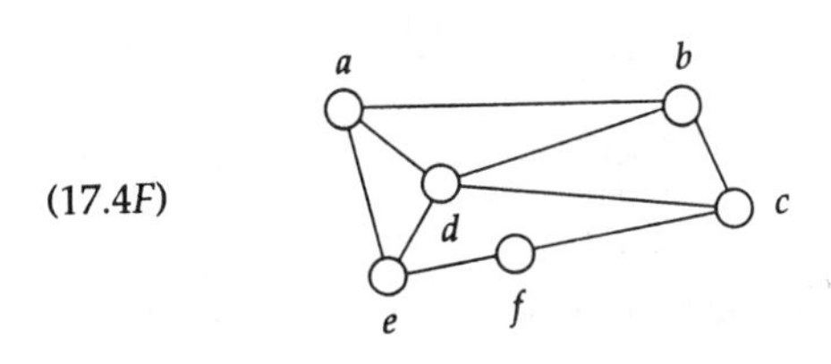

9.2 Is the graph in (17.5*F*) an interval graph? Explain.

(17.5*F*)

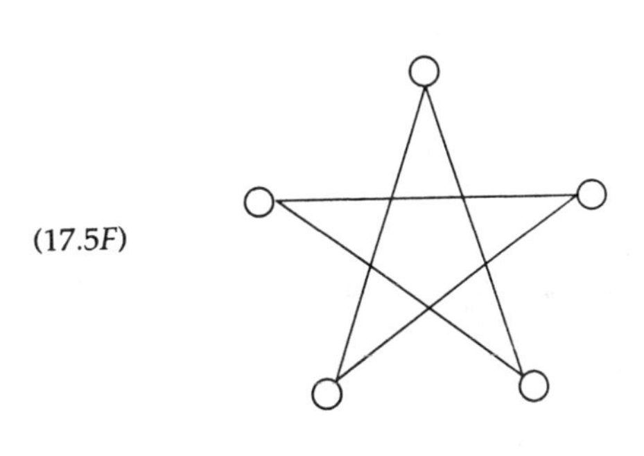

9.3 Let $n \geq 1$. Prove that if an interval graph on the vertex-set $\{ I_1, \ldots, I_n \}$ is K_n, then $I_1 \cap \cdots \cap I_n \neq \emptyset$. (The converse is obviously true.)

9.4 (*i*) Find a free tree that is not an interval graph.

(*ii*) Prove that it isn't.

(*iii*) What is the smallest number of vertices for such a free tree? Why?

9.5 Prove that for all $n \geq 1$, K_n is an interval graph.

9.6 Prove that for all $n \geq 2$, $K_n - e$, the complete graph with one edge removed, is an interval graph.

9.7 Prove that for all $n \geq 3$, $K_n - \{ e, e' \}$, a complete graph with two edges removed, is an interval graph.

9.8 Determine whether the graph in (17.6*F*) is an interval graph.

(17.6*F*)

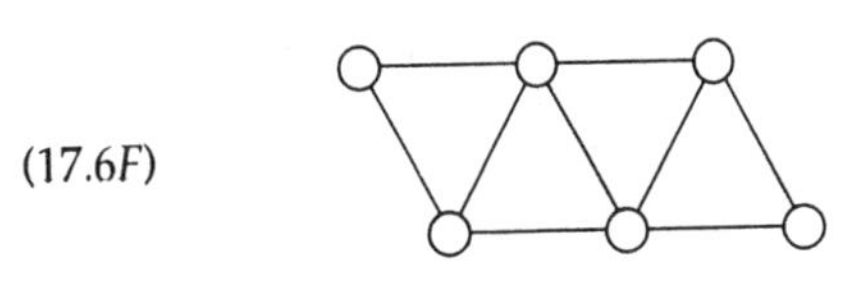

9.9 Show that the graph in (17.7*F*) is not an interval graph.

(17.7*F*)

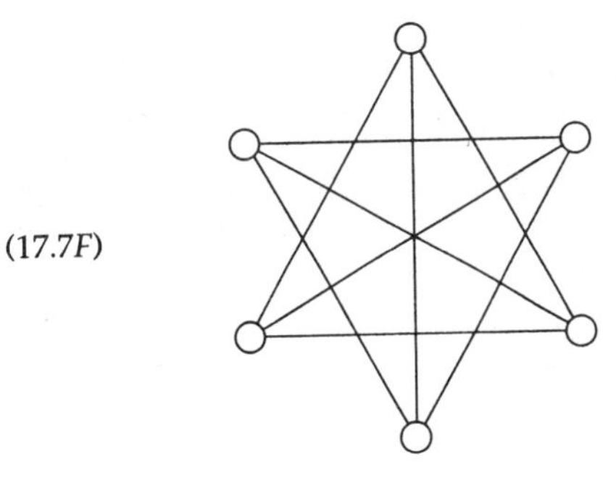

9.10 Prove that the graph drawn below is an interval graph.

(17.8*F*)

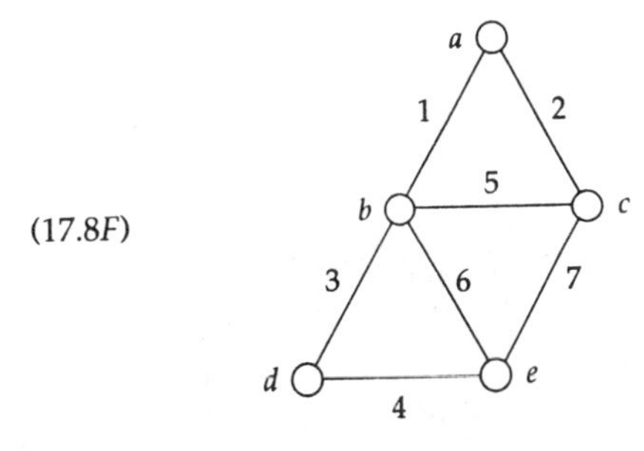

Problem for Section 11

11.1 The following matrix represents the costs of shipping oranges between five cities. Find the matrix of least costs between all pairs of cities.

0	∞	95	54	90
∞	0	92	18	56
∞	∞	0	17	11
18	29	∞	0	56
∞	∞	∞	63	0

Problems for Section 13

13.1 Prove that a language over the alphabet Σ has arbitrarily long words if and only if it is infinite.

13.2 Suppose $L \subseteq \Sigma^*$ is the language accepted by some finite-state machine. Prove that $\overline{L} := \Sigma^* - L$ is also such a language.

13.3 (*i*) Is the set of all even integers ≥ 0 recognizable by a finite-state machine? Represent them in base 2 so that $\Sigma = \{0, 1\}$. Prove your answer.

(*ii*) Represent this language in the form described in (13.10).

13.4 (*i*) Is the set of all positive integers having remainder 3 on division by 4 the language accepted by some finite-state machine? Prove your answer.

(*ii*) Represent this language in the form described in (13.10).

13.5 Find a finite-state machine that accepts the language over $\{0, 1\}$ of all strings with two consecutive 0's.

13.6 Consider the Fibonacci sequence $F_0, F_1, F_2, \ldots$ defined by the rule

$$F_0 := 0,\ F_1 := 1, \qquad \forall n \geq 0\ F_{n+2} = F_{n+1} + F_n.$$

(See Equation (1.3) and Problems 2.4 and 2.5 of Chapter 11.) Is this sequence the language accepted by some finite-state machine?

Problems for Section 14

14.1 Answer the "why" question under (14.16), Case 2.

14.2 Prove (14.27).

14.3 Prove (14.28).

General Problems

Define a *cut vertex* of a connected graph G as a vertex v such that the removal from G of v and the edges ending at v results in a disconnected graph.

G1 Prove there is no connected graph in which every vertex is a cut vertex.

G2 Consider the graph in (17.8*F*).

(*i*) How many cycles does it have? Explain.

*(*ii*) How many spanning trees? Explain.

***G3** (After J. J. Rotman) Let G be a simple, connected, undirected graph in which the total number e of edges is even. In the following special cases prove it is possible to *orient* G, meaning here to assign a direction to each edge of G so that at each vertex the total number of arrows going out is even.

(*i*) G is K_n (when e is even).

(*ii*)^{Ans} G is a free tree (when e is even).

***G4** Please see Problem 53 of Chapter 3 for the definition of planar graph and the statement of Euler's formula.

(*i*) Prove that K_5 is not planar.

(*ii*) The utilities graph $K_{3,3}$ has three houses and three utility companies as vertices. The nine edges of $K_{3,3}$ consist of an edge from each utility to each house. Prove that $K_{3,3}$ is not planar.

[Hint: Assume planarity and get a contradiction by considering flags (see Chapter 9, Section (8)) of the form (F, e), where F is a face and e is an edge that is part of the boundary of F.]

18. Answers to Practice Problems

(3.10P) This problem calls for a simple induction. We start with a vertex v of even degree. That *we can enter it* means at least one edge has v as endpoint. Since $d(v)$ is even, there are at least two such edges. Thus we may enter on one edge and exit via another edge.

There are now $d(v) - 2$ unused edges with v as

endpoint, still an even number. Again, if we can enter v on an unused edge there must be at least 2 such edges, and so forth.

(3.13P) If G has an Euler tour, we may eliminate all the loops from it to get an Euler tour of G'. Formally, suppose e_1, e_2, e_3 are edges of the Euler tour of G with e_2 as a loop $[v,v]$, and let $e_1 = [x,v]$, $e_3 = [v,y]$. Then removing the loop e_2 "works" because it starts and finishes at the same vertex:

$$\begin{array}{l} \ldots x, e_1, v, e_2, v, e_3, y, \ldots \\ \ldots x, e_1, v, e_3, y, \ldots \end{array} \tag{18.1}$$

Conversely, if G' has an Euler tour, we simply read (18.1) from bottom to top. We may insert all loops at v into the tour of G' at any visit to v.

(3.14P) It's explained already: By (3.10) we are never forced to stop at a vertex where the total number of unused edges is even. Since that's true of all vertices except v, we stop at v.

(4.8P) *(i)* If the first m edges of W' are equal to those of W, then the corresponding endpoints must be the same. So $v_1 = u_1$, $v_2 = u_2, \ldots,$ and $v_m = u_m = y$. But we have reduced our consideration to the case that no edges or vertices appear twice in a walk, so there are no more edges after f_m, because f_m takes us to the destination y. More edges would have to bring us back to y, forcing it to appear twice. Thus $n = m$ and $W' = W$.

(ii) The definition of i is that $e_{i+1} \neq f_{i+1}$, but if $i > 0$, $e_1 = f_1, \ldots, e_i = f_i$. Since there is no e_{m+1}, and since we assume $W \neq W'$, it follows from *(i)* that i exists. Therefore $i < m$.

(4.14P) I count on your intuition to help you understand the problem. It's the *shape* of the trees that's at issue, not the labels on vertices or edges. See the definition of *the same* for graphs in (16.1) for an explicit statement. There are two extremes for free trees on n vertices. One is a *star*, in which one vertex is joined to each of the others. The other is a *chain*. These two are drawn in (4.13F) for $n = 5$. The star and chain differ from each other if $n \geq 4$.

For five vertices, clearly the star is the only one if there is a vertex of degree 4. The maximum degree in the chain is 2; and clearly max $d(v) = 2$ forces us to the chain. That leaves the question: what free trees on five vertices have max $d(v) = 3$?

We draw a vertex of degree 3 in (17.2F). That accounts for four vertices.

(18.2F)

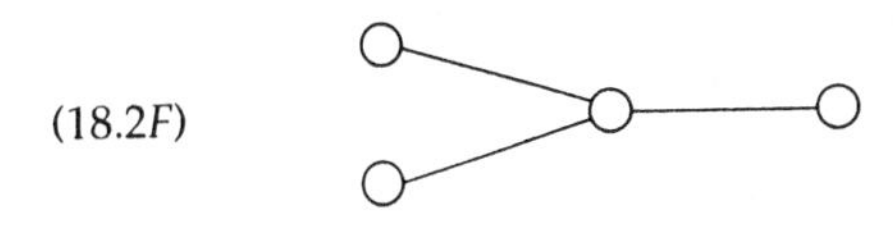

Where does the fifth vertex attach? Not to the central vertex. That would make a star. So it must be joined to one of the others, and by symmetry it doesn't matter which one.

(4.24P) The edge-degree lemma (2.22) is

$$\sum d(v) = 2|E|. \tag{18.3}$$

Our hypotheses, from Theorem (4.20), are that G is connected and that $|E| = n-1$. We showed in (4.21) that if $n \geq 2$, $d(v) = 1$ for at least one vertex v of G. If all other vertices had degree 2 or more, then (18.3) would read (if $d(v_n) = 1$)

$$1 + 2(n-1) \leq 1 + \sum_{1 \leq i < n} d(v_i) = 2n - 2.$$

That says $2n - 1 \leq 2n - 2$, a contradiction. So any free tree on $n \geq 2$ vertices has at least two vertices of degree 1.

(5.3P) K_4 has four vertices and six edges. (In this problem the labels do matter.) Since a graph has a nonempty vertex-set, we may choose any nonempty subset V_1 of V as the vertex-set of our subgraph. We are then free to choose any (or no) 2-subsets of V_1 as the edges of our subgraph. Everything depends on $|V_1|$. We display the possibilities in the table:

$\lvert V_1\rvert$	$\binom{\lvert V_1\rvert}{2} = e$	2^e	$\binom{4}{\lvert V_1\rvert}$	No. of subgraphs
1	0	1	4	4
2	1	2	6	12
3	3	8	4	32
4	6	64	1	64
			Total	112

There are $\binom{|V_1|}{2}$ edges available, from which we may choose any subset, including $\emptyset$. There are $\binom{4}{|V_1|}$ subsets of size $|V_1|$. The product is the number of subgraphs on the given number of vertices.

***(6.3P)** Now we can use some machinery. K_4 has six edges; a spanning tree has three edges. If we remove three edges from K_4 but leave the resulting subgraph connected, it will be a spanning tree by (4.22) or the BET theorem (4.25).

There are $\binom{6}{3} = 15$ 3-subsets of edges. Removing the three edges at a vertex disconnects the graph. That leaves eleven 3-subsets of edges as candidates for removal. There are two cases.

Case 1 The three edges form a triangle, say on the vertices a, b, c. The other three edges are then

$$\{a,d\}, \{b,d\}, \text{ and } \{c,d\};$$

they are the edges of a spanning tree.

Case 2 The three edges do not form a triangle. Then they touch all four vertices. Since three edges with a common vertex were ruled out already, they must be

$$\{a,b\}, \{c,d\}, \{b,c\}$$

in some labeling. The other three are

$$\{a,c\}, \{a,d\}, \{b,d\},$$

which make a spanning tree.

The answer is 11.

(6.12P) In Algorithm (6.10) we don't backtrack and we don't insist that our edges chosen to date make up a subtree.

(7.17P) Every vertex of the promising forest PF is a vertex of G. Every edge of PF is in E_0, the set of edges of a minimum spanning tree $T_{\min}$ of G. By definition of $T_{\min}$, every edge of $T_{\min}$ is an edge of G. Therefore every edge of PF is an edge of G.

(7.24P) Edge e_1 is the first one we choose. We define it as a least-cost edge of G. If $e_1 = \{u, v\}$, we *define* B as a subset of V containing u and not v. Now we see that obviously e_1 is least-cost over the subset of edges from B to $\overline{B}$, since it is least-cost over all the edges of G.

(10.3P) To define outdegree of $v \in B$ we count the number of ordered pairs in E with v as the first coordinate. Thus

$$\text{outdegree}(v) := |\{x;\ x \in B,\ (v,x) \in E\}|.$$

Similarly,

$$\text{indegree}(v) := |\{u;\ u \in B,\ (u,v) \in E\}|.$$

(11.6P) To find the edge-cost matrix we just read off the relation-matrix from G_2 in (11.3F). We then replace 0 with ∞ off the main diagonal. We write only the main diagonal and above because the matrix is symmetric.

	1	2	3	4	5	6	7
1	0	1	∞	∞	1	∞	∞
2		0	∞	∞	∞	∞	∞
3			0	1	∞	1	∞
4				0	∞	∞	1
5					0	1	∞
6						0	1
7							0

edge costs

	1	2	3	4	5	6	7
1	0	1	3	4	1	2	3
2		0	4	5	2	3	4
3			0	1	2	1	2
4				0	3	2	1
5					0	1	2
6						0	1
7							0

least costs

We can easily read off the least costs from the graph.

***(13.39P)** This problem is difficult to solve directly. If you want to understand why the stated result is true you should read about the equivalence between finite-state machines (fsm's) and **regular expressions,** in (16.5, Chapter 2), for example. Some general theorems in this subject immediately imply that if L and L' are the languages accepted by fsm's M and M', respectively, then there is an fsm with the union $L \cup L'$ as its accepted language. Since for any string s there is an fsm accepting $\{s\}$ (prove it), the result follows.

You would learn all about these ideas in a course in automata theory or formal languages.

Answers to Selected Problems

Chapter 1

12.2 First, let $X = Y$. Our target is to prove (*i*). If $X = Y$, then $X \cap \overline{Y} = \emptyset$, because $\overline{Y}$ is the set of all points of the universal set not in Y. By symmetry of argument $\overline{X} \cap Y = \emptyset$ also. Since $\emptyset \cup \emptyset = \emptyset$ (check this), we've proved (*i*).

Second, let (*i*) be true.

> Notice: For any sets $A, B, A \cup B = \emptyset$ if and only if $A = \emptyset$ and $B = \emptyset$. $\therefore$ we know that $X \cap \overline{Y} = \emptyset$ and $\overline{X} \cap Y = \emptyset$.

Our target is to prove $X = Y$. By definition (4.1) of set equality, $X = Y$ if and only if $X \subseteq Y$ and $Y \subseteq X$. So we must prove both inclusions.

To prove $X \subseteq Y$, let x be any point of X. Since $X \cap \overline{Y} = \emptyset$ we see that $x \in Y$. Why so? By definition of complement, the universal set $V = Y \cup \overline{Y}$ and $Y \cap \overline{Y} = \emptyset$. Since $x \in V, x \in Y$ or $x \in \overline{Y}$. If $x \in \overline{Y}$, then $x \in X \cap \overline{Y} = \emptyset$, an impossibility. Therefore $x \in Y$. Thus $X \subseteq Y$.
By symmetry of argument $Y \subseteq X$. Therefore $X = Y$.

13.5 (*ii*) Another "if and only if." Thus we are to prove two things: First, if $A \subseteq B$, then $A - B = \emptyset$; second, if $A - B = \emptyset$, then $A \subseteq B$. It's obvious with Venn diagrams. Here's an epsilon-argument:

For the first, if $A \subseteq B$, then every point of A is a point of B, by the definition (3.1) of inclusion. That is, there are no points of A not in B, which, by the definition (12) of set difference, tells us that $A - B = \emptyset$.

For the second: Our target is to prove $A \subseteq B$.

From our hypothesis, $A - B = \emptyset$, we know from the definition (12) of set difference that the set of all points of A that are not in B is empty. This means every point of A is in B, the definition of $A \subseteq B$.

(*vii*) Here we can get fancy. From (*ii*) we have $A \subseteq B$ if and only if $A - B = \emptyset$. From (*i*) we know that $A - B = A \cap \overline{B}$. From (*vi*) we know that $A \cap \overline{B} = \emptyset$ if and only if $\overline{B} \subseteq \overline{A}$.

13.16 You can't conclude at a glance that one of them is wrong. The reason is that the proposed cancellation law is wrong. You can easily see with Venn diagrams that $X - Y = X - Z$ if and only if $X \cap Y \cap \overline{Z} = X \cap \overline{Y} \cap Z = \emptyset$. That condition leaves much room for Y and Z to differ. For example, take $X := \{1,2,3,4\}$, $Y := \{4,5,6\}$, and $Z := \{4,7\}$. So even if $C \neq A \cap C$ you cannot conclude from that fact that the right-hand sides of (*i*) and (*ii*) are unequal.

14.7 If $x \in R \cap S$, then $x \in R \subseteq A \times A$ and $x \in S \subseteq B \times B$. It follows that $x = (a_1, a_2)$ for some $a_1, a_2 \in A$ and $x = (b_1, b_2)$ for some $b_1, b_2 \in B$. Since $(a_1, a_2) = (b_1, b_2)$, $a_1 = b_1 \in A \cap B$, and $a_2 = b_2 \in A \cap B$. $\therefore x \in (A \cap B) \times (A \cap B)$.

Chapter 2

5.1 Let p and q be the propositions "Wishes are horses," and "Beggars ride," respectively. (We relax the requirement of mathematical content.) Then the proposition in question is symbolized as $p \longrightarrow q$. It is stated with "were" and "would" because both statements are contrary to fact. It's true as an instance of $F \longrightarrow F$.

6.3 (*i*) $a = F, b = T$ makes it false.
(*iii*) $b = F = a$ makes it false for either truth value of a.

You find these either by working out the truth tables or by looking at the formulas.

7.6 (*ii*) Since logical equivalence holds iff the two formulas have the same truth table, all we need do to prove inequivalence is find a line on which the formulas have different truth-values. So $(a, b) = (T, F)$ yields F for $a \longrightarrow b$ and T for $\sim a \longrightarrow \sim b$.

8.2 *(ii)* Since $a \longrightarrow b \Leftrightarrow \sim a \vee b, (a \longrightarrow (b \longrightarrow c)) \Leftrightarrow \sim a \vee (b \longrightarrow c)$. By a De Morgan law the answer is $a \wedge b \wedge \sim c$.

9.6 For the first one, suppose g_1 and g_2 are formulas in the atoms $a_1, \ldots, a_n$ for $n \geq 0$. An n-tuple $x := (x_1, \ldots, x_n)$ over $\{T, F\}$ is in $T(g_1)$ iff $g(x_1, \ldots, x_n)$ is true. (The x_i's are T or F.) An n-tuple x is in $T(g_1) \cap T(g_2)$ iff $g_1(x)$ and $g_2(x)$ are true. But an n-tuple x is in $T(g_1 \wedge g_2)$ iff $(g_1 \wedge g_2)(x)$ is true, by definition of truth-set. From our definition of the truth table of $g_1 \wedge g_2$, we see that $(g_1 \wedge g_2)(x)$ is true iff $g_1(x)$ is true and $g_2(x)$ is true. That proves the first; I leave you the others.

9.11 *(iii)* Since the "last" connective is $\longrightarrow$, we seek the falsity-set, $\mathcal{F}$. $\sim c \longrightarrow a$ is F iff c is F and a is F. So $\mathcal{F}$ is a subset of $\{(F,T,F),(F,F,F)\} =: \mathcal{F}_1$. But $(a,c) = (F,F)$ makes $(a \wedge b) \longrightarrow c$ true for either value of b, so $\mathcal{F} = \mathcal{F}_1$.

11.2 *(ii)* These propositions have the form $(h3)$: $p \longrightarrow q$, and $(h4)$: $\sim p$. No conclusion follows, because since p is F (from $h4$), it follows that $(h3)$ is true no matter whether q is T or F.

14.4 Let f be the propositional formula of the problem. The truth table of f is

a	b	$a \longleftrightarrow b$	$a \vee b$	$(a \longleftrightarrow b) \longrightarrow (a \vee b)$
T	T	T	T	T
T	F	F	T	T
F	T	F	T	T
F	F	T	F	F

So f is logically equivalent to $a \vee b$ (which we can now see instantly by use of the shortcut). Since $a \vee b$ is a building block of CNF, the CNF of f is $a \vee b$.

19.3 The easiest solution is to list the two sets. C is defined as the set of all positive divisors of 10. Thus $C = \{1,2,5,10\}$. (That is, we've listed the elements of the truth-set of the predicate used to define C.) We obtain all the elements of D by setting $y = 0,1,2,$ and 3, producing the same elements as those of C. Thus $D = C$.

20.4 *(i)* The truth-set of $P(x) \vee Q(x)$ is $T_P \cup T_Q$. That of $P(x) \wedge Q(x)$ is $T_P \cap T_Q$. For the logical implication to hold we must have $T_P \cup T_Q \subseteq T_P \cap T_Q$. Since the reverse inclusion holds for all sets, $T_P \cup T_Q = T_P \cap T_Q$. Thus $T_P \subseteq T_P \cup T_Q \subseteq T_P \cap T_Q$, so $T_P \subseteq T_Q$. By symmetry $T_Q \subseteq T_P$. $\therefore T_P = T_Q$.

21.11 *(ii)* $\forall x,y,z \in D_x \quad P(x) \wedge P(y) \wedge P(z) \longrightarrow (x = y) \vee (x = z) \vee (y = z)$.

23.2 *(i)* To capture the sense of the quote, the answer needs

$$\forall x,y \qquad F(x) \wedge F(y) \wedge \sim A(x) \wedge A(y) \longrightarrow \sim L(x,y)$$

rather than the mechanical $\forall x,y,\ F(x) \wedge F(y) \wedge A(y) \longrightarrow \sim L(x,y)$.

23.4 *(i)* $\exists a,b \in \mathbf{N} \quad n = 2a = 7b$. (Or: $\exists a \in \mathbf{N} \quad n = 14a$.)

(ii) $n \notin \{x^2 : x \in \mathbf{N}\}$. (Or: $\forall x \in \mathbf{N} \quad n \neq x^2$).

26.1 Let B be the set of all Cobol programs, and $D[W]$ the set of all programs with "do" loops [that are well written]. Then $P1$ says $B \subseteq D$, and $P2$ says $D \subseteq W$. It follows that $B \subseteq W$, which is the conclusion (C).

Chapter 3

4. The easiest solution uses a result proved in the text. Here's how.

$$\sum(3k-2) = \left(3\sum k\right) - 2n = 3\frac{n(n+1)}{2} - 2n$$
$$= \frac{n}{2}(3(n+1) - 4),$$

which is the claimed value of the sum. A proof by induction: Let $S(n)$ be defined as the predicate

$$S(n): \sum_{1 \leq k \leq n} (3k-2) = \frac{n(3n-1)}{2}.$$

The basis case is supposed to be for $n = 0$. Any sum over the empty set is zero by convention (see Chapter 1, (10.14)), and the right-hand side equals zero if $n = 0$. For the inductive step, let n be a fixed but arbitrary integer, $n \geq 0$, and

assume $S(n)$. $\longleftarrow$ INDUCTIVE ASSUMPTION

Try to prove $S(n+1)$. $S(n+1)$ is

$$S(n+1): \sum_{1 \leq k \leq n+1} (3k-2) = \frac{(n+1)(3n+2)}{2} \longleftarrow \text{TARGET}$$

Now we seek a connection between $S(n)$ and $S(n+1)$. In this type of problem it's easy to find:

$$(\star) \qquad \sum_{1 \leq k \leq n+1} (3k-2) = 3(n+1) - 2 + \sum_{1 \leq k \leq n} (3k-2).$$

The latter sum is the left-hand side of $S(n)$. By the inductive assumption it equals $n(3n-1)/2$. Therefore

(★) becomes

$$\sum_{1\le k\le n+1} (3k-2) = 3n+1+\frac{n(3n-1)}{2}$$
$$= \frac{6n+2+3n^2-n}{2} = \frac{3n^2+5n+2}{2}$$
$$= \frac{(3n+2)(n+1)}{2},$$

and that's our target. By the first theorem on induction, $S(n)$ is true for all integers $n \ge 0$.

19. The generalization is: for $m \in \mathbf{N}$

$$(\star) \qquad \sum_{1\le k\le n} \frac{1}{(mk+1-m)(mk+1)} = \frac{n}{mn+1}.$$

The best proof uses telescoping: the summand on the left of (★) is

$$\frac{1}{m}\left[\frac{1}{m(k-1)+1} - \frac{1}{mk+1}\right].$$

The first few terms of the sum, and the last, are

$$\frac{1}{m}\left\{\left[\frac{1}{1} - \frac{1}{m+1}\right] + \left[\frac{1}{m+1} - \frac{1}{2m+1}\right]\right.$$
$$\left.+\left[\frac{1}{2m+1} - \frac{1}{3m+1}\right] + \cdots + \left[\frac{1}{(n-1)m+1} - \frac{1}{mn+1}\right]\right\}.$$

The terms cancel in pairs except for the first and last, so the sum is

$$\frac{1}{m}\left(1 - \frac{1}{mn+1}\right) = \frac{n}{mn+1}.$$

(It can also be proved by induction, but more laboriously.)

34. A gap from n to $n+1$ occurs by definition iff $S(n)$ does not imply $S(n+1)$. That can happen iff $S(n)$ is T and $S(n+1)$ is F. We're told that the only such consecutive T,F pair is at 3 and 4. This means that after 4 if there's a T, then all values after it must be T. So the distribution from place 4 onward is either all F's $(F,F,F,\ldots)$ or $F^n,T,T,T,\ldots$ ($n \ge 1$ F's for some $n \in \mathbf{N}$, followed by all T's). The values at 1 and 2 can be anything except T,F, namely, T,T; or F,F; or F,T.

48. If $n > 5$, then $n^2 \ge 5n$, so $n^2 > 5n-2$.

Chapter 4

2.1 (*iv*) R_4 is reflexive, because $\forall a \in \mathbf{N}$ $a/a = 1 = 2^0$. R_4 is not symmetric, because $(2,1) \in R_4$ (reason: $2/1 = 2^1$); but $(1,2) \notin R_4$, because $1/2 = 2^{-1}$—the exponent is negative. R_4 is transitive, because for all $a,b,c \in \mathbf{N}$ if $(a,b) \in R_4$ (thus $a/b = 2^i$ with $i \ge 0$) and if $(b,c) \in R_4$ (thus $b/c = 2^j$ with $j \ge 0$), then $a/c = (a/b)(b/c) = 2^{i+j}$; and $i+j \ge 0$.

5.3 Since, by definition, a partition of $\varnothing$ consists of cells which are nonempty subsets of $\varnothing$, there cannot be any cells in a partition of $\varnothing$. Thus any partition of $\varnothing$ must be the empty set. Let P be the empty set (of cells). Then $\bigcup_{C\in P} C = \varnothing$, by (10.13) of Chapter 1. Also, $\forall C \in P$, $C \ne \varnothing$ (since P is empty), by (25) of Chapter 2. Therefore P is a partition of $\varnothing$. There is no other partition P_1 of $\varnothing$, because P_1 would have to contain a nonempty cell.

7.5 Let D be any cell of Q. By definition of refinement, every cell C of P is either disjoint from D or is a subset of D. D is nonempty because it is a cell of a partition. Thus $\exists x \in D$. Since P is a partition, $\exists C \in P$ such that $x \in C$. Therefore $C \cap D \ne \varnothing$, so $C \subseteq D$.

The foregoing argument holds for all $x \in D$, producing a collection P_D of cells from P, all included in D and covering D (i.e., their union equals D). Thus P_D is a partition of D, because P_D consists of cells of a partition (P), which are mutually disjoint and nonempty by definition, and which cover D.

Chapter 5

3.4 $\hat{s}(\{7,11,15\}) = \{111,\ 1011,\ 1111\}$.

7.9 (*i*) P consists of cells, subsets of A that are nonempty and that cover A. First, we prove range of $\pi \subseteq P$. Let $x \in A$. Then $\pi(x) := C \in P$; C is a cell of P. Thus $\forall x \in A$, $\pi(x) \in P$. Therefore range of $\pi \subseteq P$.

Second, we prove $P \subseteq$ range of π. Let $C \in P$. Since $C \ne \varnothing$, $\exists y \in C$. By definition of π, $\pi(y) = C$.

Question: where do we use the fact about partitions that cells are mutually disjoint? Answer: it allows us to define the function π. Since each point x of A is in just one cell C of P, there is no ambiguity in defining $\pi(x)$ to be C.

9.18 The first is false. Consider $f : \{1,2\} \longrightarrow \{a,b\}$ defined by the rule $f(1) := f(2) := a$. Let $B = \{a,b\}$. Then $\widehat{f}^{-1}(B) = \{1,2\}$, and $\widehat{f} \circ \widehat{f}^{-1}(B) = \widehat{f}(\widehat{f}^{-1}(B)) = \widehat{f}(\{1,2\}) = \{a\} \ne B$. In general, if B has points not in the range of f, this equality will not hold.

The second is true. To prove this assertion of equality between sets we must prove two inclusions. First, let $x \in \widehat{f}^{-1}(B)$. Then by definition $\exists y \in B$ such that $f(x) = y$. The set on the left-hand side is by definition

$$\widehat{f}^{-1}(\widehat{f}(\widehat{f}^{-1}(B))).$$

Since $x \in \widehat{f}^{-1}(B)$, the definition of $\widehat{f}$ tells us that $f(x) \in \widehat{f}(\widehat{f}^{-1}(B))$. And now the definition of $\widehat{f}^{-1}$ tells us that $x \in \widehat{f}^{-1}(\widehat{f}(\widehat{f}^{-1}(B)))$. This proves one inclusion.

The reverse inclusion could be proved along with the first inclusion. To do it that way we'd use "if and only if" statements instead of the "if-then" statements we used above.

9.20 (*i*) By the definition of π, (range of π) $\subseteq P$. To prove the reverse inclusion, let C be any cell of P. Since C is nonempty, $\exists x \in C$. Then $\pi(x) = C$, so $C \in$ (range of π). QED

18.2 Since f is a permutation, the set $\{(x)f; x \in X\}$ is the set X. So we may describe the points of X as $(x)f$ with x running over X. So let $a, b \in X$ and suppose they are adjacent in the cycle representation of g, as $(\ldots, a, b, \ldots)$. This simply means that $(a)g = b$.

Now let's see where $(a)f$ goes under $f^{-1} \circ g \circ f$. The result is immediate:

$$\begin{aligned} ((a)f) f^{-1} \circ g \circ f &= ((a)f \circ f^{-1}) g \circ f \\ &= (a) g \circ f \qquad \text{since } f \circ f^{-1} = \text{identity} \\ &= ((a)g)f = (b)f. \end{aligned}$$

Chapter 6

1.3 For all $a \in \mathbf{Z}$, since $0 = 0 \cdot a$, we conclude that a divides 0. (The definition for "a divides x" is that there is an integer y such that $x = ay$.)

1.9 Suppose for some $n \in \mathbf{N}$ there are integers x, y, z such that $x^n + y^n = z^n$. For all divisors m of n we wish to show that there are integers X, Y, Z (depending on m) such that $X^m + Y^m = Z^m$. It's easy: $n = mk$ for some $k \in \mathbf{N}$. Therefore $x^n + y^n = z^n$ becomes $(x^k)^m + (y^k)^m = (z^k)^m$. If we could prove that there were no solutions for $n \in \{4\} \cup \{\text{odd primes}\}$, then (by the contrapositive) there would be no solutions for any integer $n > 2$.

***2.8**

$$\left\lfloor \frac{1}{m} \left\lfloor \frac{x}{n} \right\rfloor \right\rfloor = \left\lfloor \frac{x + (m-1)n}{mn} \right\rfloor,$$

which you discover by, first, using problem 2.6 and the obvious fact that

$$\forall x \in \mathbf{R}, \quad a \in \mathbf{Z} \lfloor x + a \rfloor = \lfloor x \rfloor + a.$$

Thus

$$\left\lfloor \frac{1}{m} \lfloor x/n \rfloor \right\rfloor = \left\lfloor \frac{\lfloor x/n \rfloor + (m-1)}{m} \right\rfloor,$$

and

$$\left\lfloor \frac{x}{n} \right\rfloor + m - 1 = \left\lfloor \frac{x}{n} \right\rfloor + \frac{(m-1)n}{n} = \left\lfloor \frac{x + n(m-1)}{n} \right\rfloor.$$

Now apply problem 2.5(*ii*).

***2.9** Just comments. If you had not been told the algorithm, wouldn't you try to represent m/n by first choosing the *largest* fraction of the form $1/q$ not greater than m/n? That's exactly what the algorithm does. It's curious that the world waited 3,000 years to attach someone's name to this obvious procedure. Since Fibonacci predated the printing press by more than two centuries, it couldn't have been that he was merely the first to publish it.

***2.10** It's perfectly possible to solve this by first using the division algorithm on n ($n = 5q + r$ and $0 \le r < 5$) and then breaking it into five cases, one for each possible value of r. But here's an easier way.

It's obvious that

$$\forall n \in \mathbf{Z} \text{ and } x \in \mathbf{R}, \quad n + \lfloor x \rfloor = \lfloor n + x \rfloor.$$

Thus

$$(\star) \quad n + \left\lfloor \frac{n}{5} \right\rfloor = \frac{5n}{5} + \left\lfloor \frac{n}{5} \right\rfloor = \left\lfloor \frac{5n + n}{5} \right\rfloor = \left\lfloor \frac{6n}{5} \right\rfloor.$$

To divide this by 2 and then apply $\lfloor\ \rfloor$, we use problem 2.6(*i*) to get

$$\left\lfloor \frac{3n}{5} \right\rfloor.$$

Adding n to this, as we did in $(\star)$ yields the desired right-hand side.

2.11 *(*iii*) Clearly $(\star)$ holds if $r = 0$ or 1. So from now on take $r \neq 0, 1$.

Case 1: $r < 0$. Set $x = \frac{1}{2}$. Then $(\star)$ becomes $0 = \lfloor \frac{1}{2} r \rfloor \le -1$, which is false.

Case 2: $r > 1$. We take x such that

$$\frac{1}{r} < x < 1.$$

Then $rx > 1$, so ($\star$) becomes $0 = \lfloor rx \rfloor \geq 1$— also false.

Case 3: $0 < r < 1$. Take x between 1 and 2. Then ($\star$) becomes

$$\lfloor r \lfloor x \rfloor \rfloor = \lfloor r \rfloor = 0 = \lfloor rx \rfloor.$$

If this is to hold for all such x, then rx must be less than 1. Since $x < 2$, but $x \neq 2$, we see that r must be at most $\frac{1}{2}$.

In general, let $k \in \mathbf{N}$. Assume $r < 1/k$. Take x so that

$$(\star\star) \qquad k \leq x < k + 1.$$

Then ($\star$) becomes

$$0 = \lfloor rk \rfloor = \lfloor rx \rfloor,$$

and this cannot hold for all x in the range ($\star\star$) unless $r \leq 1/(k+1)$.

In sum, once we have r in the interval $0 < r < 1$, we must step downward to $r \leq \frac{1}{2}$, or if $r < \frac{1}{2}$ to $r \leq \frac{1}{3}$, and so on. The only values available for r are the "Egyptian fractions."

4.4 Agree that λ stands for the empty string of characters and the comma in Step 2 for catenation. Then the recursive procedure is

Input: integer $b \geq 2$, integer $n \geq 1$.

Output: string $n_j n_{j-1} \ldots n_0$ over $\{0, \ldots, b-1\}$ such that $n_j \neq 0$ and

$$n = \sum_{0 \leq i \leq j} n_i b^i.$$

1. b-**ary** $0 := \lambda$
2. b-**ary** $n :=$ (b-**ary** m), n'
3. where $(m, n') :=$ **divalg**(n, b).

7.2 Let $d := \gcd(a,b)$ and $D := \gcd(ma, mb)$. Then $\exists x, y \in \mathbf{Z}$ such that

$$d = ax + by.$$

Therefore

$$md = max + mby,$$

so it follows from Corollary (7.7) that $D \mid md$. Conversely, $\exists X, Y \in \mathbf{Z}$ such that

$$D = maX + mbY.$$

Therefore

$$\frac{D}{md} = \frac{a}{d}X + \frac{b}{d}Y,$$

which is an integer since a/d and $b/d \in \mathbf{Z}$. Therefore $D = |md| = |m|d$, since D and d are both ≥ 0.

7.13 $\exists x, y \in \mathbf{Z}$ such that $1 = ax + by$. $\therefore m = max + mby$. Divide by ab. The result is

$$\frac{m}{ab} = \frac{m}{b}x + \frac{m}{a}y;$$

this is an integer because $m/a, m/b \in \mathbf{Z}$ by hypothesis.

7.15 We have $a_1 := 1, a_2 := 2$, and $\forall n \geq 1$, $a_{n+2} := a_{n+1} + a_n$. We are to prove

$$(\star) \qquad \forall n \geq 1 \quad d_n := \gcd(a_n, a_{n+1}) = 1.$$

There are several ways to do it.
One is to prove $\forall n \geq 2$, $a_{n+2}a_{n-1} - a_{n+1}a_n = (-1)^{n-1}$. Our result ($\star$) follows from this and Corollary (7.7). You aren't expected to find this proof.

Another proof: Use problem 7.16 as follows:
Use the recursion twice, getting simultaneous equations

$$a_{n+2} = 2a_n + a_{n-1}$$
$$a_{n+1} = a_n + a_{n-1}.$$

The determinant of coefficients is $2 \cdot 1 - 1 \cdot 1 = 1$, so by problem 7.16, $\gcd(a_n, a_{n-1}) = d_{n-1} = \gcd(a_{n+1}, a_{n+2}) = d_{n+1}$. Since $d_1 = d_2 = 1$, we have a proof by induction. You may not have seen this way to solve it.

Here's a proof you could find. Since $a_{n+2} - a_{n+1} = a_n$, a_n is a linear combination of a_{n+1} and a_{n+2}. $\therefore d_{n+1} \mid a_n$ by (7.7). But also $d_{n+1} \mid a_{n+1}$, by definition of d_{n+1}. $\therefore d_{n+1} \mid \gcd(a_n, a_{n+1})$ by (7.7) or (7.15). $\therefore d_{n+1} \mid d_n$. Since $d_1 = 1$, we have proved $\forall n \geq 1 \quad d_n = 1$.

8.3 (The problem assumes $b \geq 1$.) The division algorithm: If $a = bq + r$ with $0 \leq r < b$, then we set $\ell := r$ if $r \leq b/2$, and $q' := q$. If $r > b/2$, we write

$$a = bq + r = b(q+1) + r - b$$

and set $q' := q + 1$, $\ell := r - b$. Then $-b/2 < \ell \leq b/2$. The uniqueness of q' and ℓ follows from that of q and r and the fact that with q' and ℓ we can recover q and r by reversing this mapping.

The Euclidean algorithm depends on two things. One is that for all a, b with $a = bq + r$, $\gcd(a,b) = \gcd(b,r)$. Restrictions on r are irrelevant for this fact, so the same holds for $a = bq' + \ell$: $\gcd(a,b) = \gcd(b, \ell)$.

The other point is that the divisor at stage $n+1$ is always smaller in absolute value than the one at stage n. (That makes the algorithm end.) That's true here because $|\ell| \leq b/2$. At each stage the divisor is at most half the size of the prior.

G2. The hypothesis means that $n = 2^i m$ for some odd integer $m \geq 3$ and $i \geq 0$. If $i = 0$ the answer is obvious: $m = (m-1)/2 + (m+1)/2$. We can build on this solution as i increases, but only to a point. (Example: $11 = 5 + 6$, $22 = 4 + (5 + 6) + 7$, $44 = 2+3+(4+5+6+7)+8+9$. We've added numbers fore and aft of the prior sum, each pair having sum 11. If we do it once more, we run into forbidden territory, integers less than 1: $88 = 2 + (-1) + 0 + 1 + (2 + \cdots + 9) + 10 + 11 + 12 + 13$. To represent 88 "legally," we may set $88 = (3 + \cdots + 7) + 8 + (9 + \cdots + 13)$. Check that $7 + 9 = 16 = 2 \cdot 8 = 6 + 10 = 16$, and so on.) So here's a general answer. For $n = 2^i m$, set $x = (m-1)/2$. If $2^i \leq (m-1)/2$ we represent n as S_i, where

$$S_0 := x + (x+1) \quad \text{for} \quad i = 0,$$

$$S_1 := x - 1 + S_0 + x + 2 \quad \text{for} \quad i = 1,$$

$$S_2 := (x-3) + (x-2) + S_1 + (x+3) + (x+4) \quad \text{for} \quad i = 2,$$

$$\vdots$$

The number of positive integers in S_i that are less than $x + 1$ is

$$1 + 1 + 2 + \cdots + 2^{i-1} = 2^i,$$

so the sum S_i does not use zero or any negative integers if $2^i \leq (m-1)/2$.

If $2^i \geq (m-1)/2$ we represent $n = 2^i m$ as the sum of m consecutive integers, of which the middle one is 2^i. (These are m integers with mean 2^i.)

Chapter 7

1.11 If z is even, then z satisfies the first congruence $z \equiv 0 \pmod 2$.

If z is odd, then $z \equiv \pm 1 \pmod 4$. (Explain.)

The third congruence is $z \equiv 1 \pmod 4$, so we consider from now on only $z \equiv -1 \pmod 4$. But $z \equiv -1 \pmod 4$ iff

$$(\star) \qquad z \equiv 3, 7, 11, 15, 19, 23 \pmod{24}.$$

The commas stand for *or*.

The other six congruences may also be expressed mod 24, as follows:

$$z \equiv 0 \pmod 3 \text{ iff } z \equiv 0, 3, 6, 9, 12, 15, 18, 21 \pmod{24}.$$

Thus with $z \equiv 0 \pmod 3$ we dispose of the 3 and 15 in $(\star)$. Similarly $z \equiv 1 \pmod 6$ iff

$$z \equiv 1, 7, 13, 19 \pmod{24},$$

so $z \equiv 1 \pmod 6$ encompasses the 7 and 19 of $(\star)$. Now $z \equiv 3 \pmod 8$ iff

$$z \equiv 3, 11, 19 \pmod{24},$$

so this covers the 3 and the 11 of $(\star)$. Finally, $z \equiv 11 \pmod{12}$ iff

$$z \equiv 11, 23 \pmod{24},$$

so we get the last remaining number of $(\star)$.

1.21 (*i*) Any set S of m consecutive integers may be represented as $x, x+1, \ldots, x+m-1$ with $x :=$ the least element of S. We map S into $\mathbf{Z}_m = \{\lfloor 0 \rfloor_m, \lfloor 1 \rfloor_m, \ldots, \lfloor m-1 \rfloor_m\}$, the partition of $\mathbf{Z}$ induced by congruence mod m, according to the following rule f:

$$\forall s \in S, \qquad f(s) := \lfloor s \rfloor_m.$$

Each element s of S is mapped to the congruence class mod m that contains s. We now prove that f is an injection.

Let $x+i, x+j \in S$ (where $i, j \in \{0, 1, \ldots, m-1\}$). Then $f(x+i) = f(x+j)$ iff $\lfloor x+i \rfloor_m = \lfloor x+j \rfloor_m$ iff $x + i \equiv x + j \pmod m$ iff $i \equiv j \pmod m$ iff $i = j$, since no two elements of $\{0, 1, \ldots, m-1\}$ are congruent mod m. Therefore the elements of S are congruent mod m in some order to $0, 1, \ldots, m-1$, and so

$$\bigcup_{s \in S} \lfloor s \rfloor_m = \lfloor 0 \rfloor_m \cup \lfloor 1 \rfloor_m \cup \cdots \cup \lfloor m-1 \rfloor_m = \mathbf{Z}.$$

3.3 For all $x \in \mathbf{Z}$, x is congruent mod 9 to one of $0, \pm 1, \pm 2, \pm 3$, or ± 4. Therefore

$$x^2 \equiv 0, 1, 4, 9, \text{or } 16 \pmod 9,$$

which comes down to

$$x^2 \equiv 0, 1, 4, \text{ or } -2 \pmod 9.$$

3.15 Since January and October each have 31 days, it's enough to show that each starts on the same day of the week. The ith and jth days of any year fall on the same day of the week iff $j \equiv i \pmod 7$. January 1 is day 1 of the year; which is October 1? In nonleap years the first nine months have, respectively, 31, 28, 31, 30, 31, 30, 31, 31, 30 days. Let S be the sum of these nine numbers. October 1 falls on the day numbered $1 + S$. Reduced mod 7 these numbers are 3, 0, 3, 2, 3, 2, 3, 3, 2. The sum is $15 + 6 = 21 \equiv 0 \pmod 7$. So $S \equiv 0 \pmod 7$, and $1 \equiv 1 + S \pmod 7$.

11.3 It's obviously false for $n = 0$ and true for $n = 1$. Since $3^6 \equiv 4^6 \equiv 1 \pmod 7$ by Fermat's theorem, the value mod 7 of $3^n + 4^n$ repeats after $n = 5$. That is, set $f(n) := 3^n + 4^n$. Then

$$\forall n \geq 0 \quad f(n+6) \equiv f(n) \pmod 7.$$

So we need calculate only $f(0), \ldots, f(5)$.

n	0	1	2	3	4	5
$f(n)$ mod 7	2	0	3	0	1	0

Thus $n \in \lfloor 1 \rfloor_6 \cup \lfloor 3 \rfloor_6 \cup \lfloor 5 \rfloor_6$.

11.4 (Second part) Fermat's theorem may be stated as

$$\forall n \text{ if } n \text{ is prime,}$$

$$\text{then } \forall a \in \mathbf{Z}, \quad (n \nmid a \longrightarrow a^{n-1} \equiv 1 \pmod{n}).$$

Since the converse of $p \longrightarrow q$ is $\sim p \longrightarrow \sim q$, the converse of Fermat's theorem is

$$\forall n \text{ if } n \text{ is composite,}$$

$$\text{then } \exists a \in \mathbf{Z} \quad (n \nmid a \wedge a^{n-1} \not\equiv 1 \pmod{n}).$$

The reason is that $b \longrightarrow c \Longleftrightarrow \sim b \vee c$, and thus $\sim (b \longrightarrow c) \Longleftrightarrow b \wedge \sim c$. Thus the statement in the first part of problem 11.4 is the converse of Fermat's theorem. See Chapter 2, (21.11).

12.5 (*iii*) Let $n = 2^i m$ for some odd m. Then $\varphi(2n) = \varphi(2^{i+1}m) = \varphi(2^{i+1})\varphi(m) = 2^i\varphi(m)$. On the other side, $\varphi(n) = \varphi(2^i m) = \varphi(2^i)\varphi(m) = 2^{i-1}\varphi(m)$ since i is given as positive.

Notice that the result is false for all odd n.

12.8 We use the result and notation of problem 12.4.

$$(\star) \qquad \frac{n}{\varphi(n)} = \frac{1}{\prod_{p|n}(1 - 1/p)} = \frac{p_1 \cdots p_r}{(p_1 - 1)\cdots(p_r - 1)},$$

where the p_i are the distinct primes dividing n. The question is merely for which sets $\{p_1, \ldots, p_r\}$ of primes is this an integer.

If $r = 1$, then p_1 must be 2, for otherwise $p_1 - 1$ is even and cannot divide p_1. Thus for $e \geq 0$ $n = 2^e$ is one such n. If $r = 2$, then one prime is odd, and so the denominator is even. Then the other prime must be 2. If $\{p_1, p_2\} = \{2, p\}$, then $(\star)$ becomes

$$\frac{n}{\varphi(n)} = \frac{2p}{p-1}.$$

If $p - 1 = 2$, well and good. If $p - 1 > 2$, then $2p/(p-1)$ is not an integer because $p - 1$ is relatively prime to p. Thus p could be 3 but no other value, so $n = 2^e 3^f$ for $e, f \geq 1$ is another such n. There are no solutions with $r \geq 3$ because the denominator would be a multiple of 4.

G3. It is best to represent n mod 6: $n = 6k \pm 1$, since the odd integers are the set $\{6k \pm 1, 6k + 3;\ k \in \mathbf{Z}\}$. Then $n^2 = 36k^2 \pm 12k + 1 = 12k(3k \pm 1) + 1$, so $n^2 - 1 = 12k(3k \pm 1)$. Now $k(3k \pm 1)$ is always even, for if k is odd, then $3k \pm 1$ is even.

Chapter 8

2.2 Basis step: $\binom{0}{0} = 1$ as the coefficient of x^0 in $(1+x)^0 = 1$.

Inductive step: let $n \geq 0$; assume $\binom{n}{n} = 1$. Thus the coefficient of x^n in $(1+x)^n$ is 1. Since $(1+x)^{n+1} = (1+x)(1+x)^n$, we see that

$$(1+x)^{n+1} = (1+x)\left(\binom{n}{0} + \binom{n}{1}x + \cdots + \binom{n}{n-1}x^{n-1} + x^n\right) = \cdots + x^{n+1}.$$

That is, the $x \cdot x^n$ is the only term of exponent $n+1$. Thus $\binom{n+1}{n+1} = 1$.

3.5 For $r = 0$ the equation reads $\binom{n}{k} = \binom{n}{k}$. For $r = 1$ it becomes $\binom{n}{k} = \binom{n-1}{k} + \binom{n-1}{k-1}$, the basic recursion (3.2). These facts suggest a proof by induction on r, but consider instead the following: The right-hand side is the coefficient of x^k in the product of two polynomials, namely,

$$\sum_{j \geq 0} \binom{r}{j} x^j \quad \text{and} \quad \sum_{j \geq 0} \binom{n-r}{k-j} x^{k-j}.$$

That claim follows from (1.25). The first polynomial is $(1+x)^r$, and the second is $(1+x)^{n-r}$. The product is $(1+x)^n$, in which the coefficient of x^k is, by definition (2.1), $\binom{n}{k}$.

4.2

$$5 = 0 + 1 + 4 = \binom{0}{1} + \binom{2}{2} + \binom{4}{3}$$

$$10 = 0 + 0 + 10 = \binom{0}{1} + \binom{1}{2} + \binom{5}{3}$$

$$15 = 2 + 3 + 10 = \binom{2}{1} + \binom{3}{2} + \binom{5}{3}$$

6.2 There are 13 people. The choice is unrestricted. There are $\binom{13}{6}$ 6-subsets of the 13 people, so that's the answer. As we'll see, $\binom{13}{6} = 1716$.

6.11 The number of strings with no 1's is 1, because 0^n is the only such string. With one 1, it is $\binom{n}{1} = n$, because there are n places to put the 1: $10^{n-1}, 010^{n-2}, \ldots, 0^{n-1}1$ are the n strings. With two 1's it is $\binom{n}{2}$; there are that many 2-subsets of the n places.

6.17 Let Y be the set of all strings of $2m$ 0's and n 1's in which every 0 is next to another 0. Let X be the set of all strings of m 0's and n 1's. We map X into Y by the function δ, defined as follows:

$\forall x \in X$, $\delta(x)$ is the string obtained from x on replacement of each 0 in x by 00.

It is clear that $\delta(X) \subseteq Y$, and that δ is injective. Therefore the range of δ, that is, $\delta(X)$, has size $\binom{m+n}{n}$ since $|X| = \binom{m+n}{n}$. But $\delta(X) \neq Y$, because, for example, if $m = 3$ and $n = 2$, then 00010001 is in Y but not in $\delta(X)$. If $m = 1$ or 2, then $\delta(X) = Y$. Therefore

$$|Y| = \binom{m+n}{n} \quad \text{if } m \le 2$$

$$|Y| > \binom{m+n}{n} \quad \text{if } m > 2.$$

7.1 $a := \binom{11}{4} = \dfrac{11 \cdot 10 \cdot 9 \cdot 8}{4 \cdot 3 \cdot 2 \cdot 1} = 11 \cdot 10 \cdot 3 = 330.$

$$b := \binom{11}{5} = \frac{11 \cdot 10 \cdot 9 \cdot 8 \cdot 7}{5 \cdot 4 \cdot 3 \cdot 2 \cdot 1} = \frac{7}{5}a = 7 \cdot 66 = 462.$$

$$b - a = 462 - 330 = 132, \qquad \frac{b}{a} = \frac{7}{5} = 1.4.$$

By the basic recursion (3.2), $\binom{12}{6} = \binom{11}{6} + \binom{11}{5} = 2 \cdot \binom{11}{5}$ by symmetry, (5.1). So $\binom{12}{6} = 2b = 924$.

***7.9** The formula of problem 6.21(*i*) and that obtained from it on interchange of n and w are really the same. To see this, it is perhaps easiest to use problem 6.21(*i*) if $n < w$ to write

$$\binom{w}{2} = \binom{n}{2} + \binom{w-n}{2} + n(w-n).$$

This equation implies

$$\binom{n}{2} = \binom{w}{2} - \binom{w-n}{2} - n(w-n).$$

Now apply the binomial theorem to the term $\binom{w-n}{2}$ to see that the last two terms sum to $S := \frac{1}{2}(n-w)(w+n-1)$.

Finally, do the same to $\binom{n-w}{2}$ to see that $\binom{n-w}{2} + w(n-w) = S$. This result suggests a simple binomial identity, which you can easily verify:

$$\text{For all } x, y \in \mathbf{Z} \quad \binom{x}{2} + xy = -\binom{-x}{2} + (x+y)x.$$

Our problem is the case $x = n - w$, $y = w$.

8.2 The result of problem 8.1 is, for $k := 2$,

$$(i) \qquad \sum_{0 \le n \le m} \binom{n}{2} = \binom{m+1}{3}.$$

Since $\binom{n}{2} = n(n-1)\,2 = \frac{1}{2}n^2 - \frac{1}{2}n$ by the binomial theorem (7.2), we may express (*i*) as

$$(ii) \qquad \sum_{0 \le n \le m} n^2 = 2\binom{m+1}{3} + \sum_{0 \le n \le m} n.$$

We know the sum on the right, from Chapter 3; but it is also given by problem 8.1 with $k = 1$. At any rate, (*ii*) works out to

$$\sum_{0 \le n \le m} n^2 = \frac{2(m+1)(m)(m-1)}{6} + \frac{m(m+1)}{2}$$

$$= \frac{m(m+1)}{2}\left[\frac{2(m-1)}{3} + 1\right]$$

$$= \frac{m(m+1)(2m+1)}{6}.$$

9.6 These two cases ($m = n - 1$ and $m = n$) can be worked out schematically on the Pascal triangle, using only the basic recursion (3.1) and the result of problem 7.7.

***9.7** This asks us to prove that for $n \ge 4$

$$\binom{2n-4}{n-3} + \binom{2n-4}{n-1} > \frac{1}{2}\binom{2n-2}{n-2} =: \frac{1}{2}D.$$

The two terms on the left are equal, say to B, by symmetry. We are to prove that $4B > D$. We know from problem 3.3 that

$$D = A + 2B + C,$$

where $A := \binom{2n-4}{n-4}$ and $C := \binom{2n-4}{n-2}$. So all we need prove is that $A + C < 2B$. We know that $A < B < C$, so this is tougher than the previous problem.

For simplicity we set $n - 2 =: \ell$ and prove for $\ell \ge 2$

$$(\star) \qquad \binom{2\ell}{\ell-2} + \binom{2\ell}{\ell} < 2\binom{2\ell}{\ell-1}.$$

We divide by $\binom{2\ell}{\ell-1}$ and use $\binom{N}{K} \div \binom{N}{K+1} = (K+1)/(N-K)$, which follows from (7.2) or problem 7.4, to turn $(\star)$ into

$$\frac{\ell-1}{\ell+2} + \frac{\ell+1}{\ell} < 2.$$

The left-hand side is $2 - 3/(\ell+2) + 1/\ell$. It is less than 2 because

$$\frac{-3}{\ell+2} + \frac{1}{\ell} = \frac{-3\ell + \ell + 2}{\ell(\ell+2)} = \frac{2(1-\ell)}{\ell(\ell+2)}$$

is negative for $\ell \ge 2$.

G4. (*i*) $\binom{52}{13}$, which is 635,013,559,600.

(*ii*) If the hand has four aces, then it has nine other cards from the $52 - 4 = 48$ non-aces. So the answer is $\binom{48}{9} = 1{,}677{,}106{,}640$. The ratio of (*ii*) to (*i*) is the probability of being dealt a hand with four aces:

$$\binom{48}{9} \div \binom{52}{13} = \frac{13 \cdot 12 \cdot 11 \cdot 10}{52 \cdot 51 \cdot 50 \cdot 49} = 0 \cdot 002641+,$$

about 1 chance in 379.

Chapter 9

2.8 (*i*) In forming monomial terms by multiplying the factor $(x_1 + \cdots + x_r)$ by itself, we make n choices, one from each of the n factors $x_1 + \cdots + x_r$. If we are not to choose the same x_i twice, we must choose an n-subset of the r-set $\{x_1, \ldots, x_r\}$. There are $\binom{r}{n}$ of these sets, and, therefore, $\binom{r}{n}$ different monomials. (So far this count is also correct if $r < n$.) How many times is such a monomial selected?

(*ii*) *First answer:* No two of the n variables in $x_{i_1} \cdots x_{i_n}$, are the same. The first, x_{i_1}, can be chosen from any of the n factors; the second x_{i_2} from any of the $n-1$ remaining factors, and so on. Therefore each such monomial arises exactly $n!$ times.

(*iii*) *Second answer:* It is nice to derive the answer $n!$ by differentiation. If we calculate $\frac{\partial}{\partial x_{i_1}} \cdots \frac{\partial}{\partial x_{i_n}} (x_1 + \cdots + x_r)^n$ we get simply

$$n! = (\text{coefficient of } x_{i_1} \cdots x_{i_n}),$$

because any monomial other than $x_{i_1} \cdots x_{i_n}$ lacks at least one, say x_{i_j}, of the variables $x_{i_1}, \ldots, x_{i_n}$; hence it becomes zero on application of $\frac{\partial}{\partial x_{i_j}}$.

3.4 We are prescribing the function f to this extent: for each point y_i in the codomain we specify at how many points of the domain f takes the value y_i. There are to be a_1 points where y_1 is taken, and so on. There are $\binom{k}{a_1}$ choices of subsets where f takes the value y_1. There remain $k - a_1$ points not yet assigned; we may assign a_2 of them to y_2 in $\binom{k-a_1}{a_2}$ ways, and so on. The answer is

$$(\star) \quad \binom{k}{a_1}\binom{k-a_1}{a_2}\binom{k-a_1-a_2}{a_3}\cdots\binom{k-a_1-\cdots-a_{n-1}}{a_n},$$

which is the multinomial coefficient.

We justify the multiplications in (⋆). They don't fit our Cartesian product rule **C2**, since again the set of second coordinates changes with the choice of the first coordinate. We go back to basics: If $f^{-1}(y_1) \neq g^{-1}(y_1)$, then certainly $f \neq g$, so different a_1-subsets yield different functions. For each choice of $f^{-1}(y_1)$ there are

$$c := \binom{k-a_1}{a_2}$$

choices of $f^{-1}(y_2)$. Thus there are $c + c + \cdots + c$ choices of $f^{-1}(y_1)$ and $f^{-1}(y_2)$, with $\binom{k}{a_1}$ summands c. That is the product $\binom{k}{a_1}\binom{k-a_1}{a_2}$.

Similarly, we could argue like this for all the y_i's.

4.6 Call the organizations A, M, S. The numbers are consistent if we interpret the three joint memberships in two organizations as meaning not $|A \cap M|$, for example, but $|A \cap M \cap \overline{S}|$.

4.14 Suppose first that $x \in A_1 \cap \cdots \cap A_n$. Then x is counted once in each intersection of sets: in $A_1, A_2 \ldots,$ in $A_1 \cap A_2, \ldots,$ in $A_1 \cap A_2 \cap A_3, \ldots,$ and so on. There are n single intersections, $\binom{n}{2}$ double intersections, and so on. Thus x is counted this many times:

$$(\star) \qquad \binom{n}{1} - \binom{n}{2} + \binom{n}{3} - + \cdots + (-1)^{n-1}\binom{n}{n}.$$

If we subtract $1 = \binom{n}{0}$ from (⋆) we know from Chapter 8, problem 2.4, that the result is 0. ∴ the sum in (⋆) is 1.

In general, if x is in only some of the sets, say $A_1, \ldots, A_r$, we have the same proof with $n := r$, since x does not appear in any intersection of any of the sets $A_{r+1}, \ldots, A_n$. It is counted zero times in each such intersection.

5.11 If you represent a cycle with the parenthesis notation for permutations, you might think there are $n!$ cycles. But on $\{1,2,3\}$ the cycles $(1,2,3)$, $(2,3,1)$, and $(3,1,2)$ are the same because the rule is that the rightmost element is mapped to the leftmost. It's better to think of the directed graph of the permutation: it's a circle with n spots for the points. Since each point is there somewhere, normalize by putting n at the top; then there are $(n-1)!$ ways to place the remaining points.

7.1 (*i*) This is the number of 10-reps of a 4-set. The answer is

$$\binom{10+4-1}{10} = \binom{13}{10} = \binom{13}{3}$$

$$= \frac{13 \cdot 12 \cdot 11}{3 \cdot 2 \cdot 1} = 286.$$

(*ii*) If the variables must all be positive, set $w' = w - 1$, and so on, so the primed variables are nonnegative. The equation becomes $w' + x' + y' + z' = 6$. The number of solutions is the number of 6-reps of a 4-set:

$$\binom{6+4-1}{6} = \binom{9}{6} = \binom{9}{3} = \frac{9\cdot 8\cdot 7}{6} = 84.$$

7.12 Since 24 is the total number of books, it's enough to determine how many ways one person may take 12. It's the number of 12-reps of a 3-set, but with upper bounds on the variables. The problem is to find the number of solutions in integers of (*i*) and (*ii*).

(*i*) $a + b + c = 12$,
(*ii*) $0 \le a \le 7, \quad 0 \le b \le 8, \quad 0 \le c \le 9.$

We count the number without restriction and then subtract the number of solutions with variables too large. The answer to (*i*) is

$$\binom{3+12-1}{12} = \binom{14}{12} = \binom{14}{2} = 91.$$

A solution to (*i*) fails to satisfy (*ii*) if $a \ge 8$, or if $b \ge 9$, or if $c \ge 10$.

Let A, B, C be the sets of solutions to (*i*) with, respectively, $a \ge 8$, $b \ge 9$, $c \ge 10$. We seek $|A \cup B \cup C|$. By in-ex it is $|A| + |B| + |C|$, since $A \cap B = \varnothing = A \cap C = \cdots$. We calculate $|A|$ by setting $a' = a - 8$ and substituting in (*i*). The result is $a' + b + c = 4$, and the number of solutions is $\binom{4+3-1}{4} = \binom{6}{4} = \binom{6}{2} = 15$. Similarly, we find $|B| = 10$ and $|C| = 6$, so there are $15+10+6 = 31$ forbidden solutions. The answer is therefore $91 - 31 = 60$.

G3. In part (*i*) it's the number of functions from the n-set of dances to the 20-set of boys. To fill in names on the card is to make a table of the function. So the answer is 20^n. This answer assumes that the girl dances with some boy at each dance. The problem doesn't specify that condition, however. To allow for her sitting out some dances we must introduce a twenty-first "boy," representing *not dancing*. The answer then is 21^n.

In part (*ii*) the answer is the number of solutions in integers ≥ 0 of $x_1 + \cdots + x_{20} + x_{21} = n$, where x_{21} represents the phantom boy who doesn't dance. The answer is now the number of n-reps of a 21-set: $\binom{n+21-1}{n} = \binom{n+20}{20}$.

Chapter 10

2.1 (*i*) $\frac{6}{10} = 0.6$.

(*ii*) There are $\binom{10}{2} = 45$ ways to choose two letters from the word. Three of these choices yield both letters equal. Thus the probability that the letters are different is $(45 - 3) \div 45 = 42 \div 45 = 14 \div 15 = 0.933$.

(*iii*) The points of the sample space and the values of the probability function on them are

point	c	a	ℓ	u	t	o	r
point	0.2	0.2	0.2	0.1	0.1	0.1	0.1

2.9
$$\frac{112}{46{,}483} = 0.00241- .$$

In the actual game, you chose two 6-subsets, so your probability of winning was about double this figure. Still, the chance of winning (something) was less than one in 200.

4.2 (*i*) The answer is the sum of the individual probabilities, since the different outcomes are mutually exclusive. To three significant figures it is 0.0797.

(*ii*) 29 cents.

4.7 If you toss a coin $i + 1$ times, the probability that the first i tosses are heads and the next toss is tails is $\left(\frac{1}{2}\right)^{i+1}$. For $i = 0,1,\ldots,9$, if that happens you win $1 + 2 + \cdots + 2^{i-1} = 2^i - 1$ pennies.

The probability of 10 heads is $\left(\frac{1}{2}\right)^{10}$, and the winnings are $2^{10} - 1$ in that case.

You thus expect to win

$$\sum_{1\le i\le 9} \frac{2^i - 1}{2^{i+1}} + \frac{2^{10}-1}{2^{10}} = \sum_{1\le i\le 9}\left(\frac{1}{2} - \frac{1}{2^{i+1}}\right) + 1 - 2^{-10}$$
$$= \frac{11}{2} - 2^{-10} - \frac{1}{4}\left(1 + \cdots + \frac{1}{2^8}\right).$$

Since the geometric sum in parentheses is

$$\frac{1-2^{-9}}{1-2^{-1}} = 2 - 2^{-8},$$

the answer is 5.

6.1 It means $\frac{\partial y}{\partial x} = 0$ and $\frac{\partial x}{\partial y} = 0$. There may not be any probability involved. Each variable may change independently of the other.

9.2 The total number of dominoes is $\binom{7}{2} = 21$, as the number of 2-reps of a 6-set. Each domino is equally likely to be drawn.

(*i*) The probability that a particular domino d is drawn is $\frac{1}{21}$. The probability that the dice thrown match that domino is $p(d) = \frac{1}{36}\left[\frac{2}{36}\right]$ if the domino is [is not] a double. The probability of drawing domino d and matching it with a throw of the dice is $\frac{1}{21}p(d)$.

The answer is thus

$$\frac{1}{21}\sum_{d} p(d) = \frac{1}{21}.$$

The sum, taken over the 21 dominoes d, is the sum of the probability function for the dice over the whole space of the dice. Hence the value is 1.

(*ii*) We count the probability that the dice have no spot in common with the domino, and then subtract from 1. If the domino is a double, then there are five spots different, so the probability that the dice have at least one spot the same as the domino is $1 - \frac{5^2}{36} = \frac{11}{36}$. There are six double dominoes, so the probability here is $\frac{6}{21} \cdot \frac{11}{36} = \frac{11}{7 \cdot 18}$.

If the domino is not a double, then there are four spots different. The probability here is

$$\frac{15}{21}\left(1 - \frac{16}{36}\right) = \frac{5}{7} \cdot \frac{10}{18}.$$

The answer is the sum

$$\frac{1}{7 \cdot 18}(11 + 50) = \frac{61}{126} = 0.484+ .$$

(*iii*) The sample space is the Cartesian product of the 21-set D for the dominoes with the space $S \times S$, where $S = \{1, \ldots, 6\}$. The probability function takes the value $1/21 \cdot 36$ at each point $(d, (x, y))$ for $d \in D$ and $x, y \in S$.

The event for part (*i*) is the subset where the 2-rep $[x, y] = d$. The event for part (*ii*) is the subset for which $d \cap \{x, y\} \neq \varnothing$ (if we may regard d for the moment as a set).

10.4 (*i*) \$3.

(*ii*) At a \$4 fee with an honest coin the pitchman's expected profit was \$1 per play. To retain that dollar at a \$3 fee, he needs a coin with probability p of heads so low that the player's expected winning is only \$2.

The probability of one head and two tails is $3pq^2$ (where $q = 1 - p$). Of two heads and one tail it is $3p^2q$. Of three heads it is p^3. The expected winning x is

$$\begin{aligned} 1 \cdot 3pq^2 + 4 \cdot 3p^2q + 9 \cdot p^3 = x &= 3p(q^2 + 4pq + 3p^2) \\ &= 3p(q + p)(q + 3p). \end{aligned}$$

Since $p + q = 1$ it is easy to find the value of p making $x = 2$. It is $p = 0.379+$.

10.7 The answer is $0.439+$.

G4. The answer is $0.0023597-$.

Chapter 11

2.6 Let r_1 and r_2 be the roots of $x^2 - x - 1$. Then $a_n = r_1^{n+1} + r_2^{n+1}$.

*2.10 (*i*) Let $q_n := F_{n+1}F_{n-1} - F_n^2$. Then you can apply the recurrence to show that $q_n = -q_{n-1}$ for all $n \geq 1$, yielding a proof by induction.

It's possible to save that work by getting fancy, however. Notice that for all $n \geq 1$

$$(\alpha) \qquad (F_{n+1}, F_n) = (F_n, F_{n-1})\begin{pmatrix} 1 & 1 \\ 1 & 0 \end{pmatrix}.$$

This is a matrix product: The row-vectors are really 1-by-2 matrices, and the right-hand side of (α) works out to

$$(F_n + F_{n-1}, F_n).$$

Call the matrix M.

$$M := \begin{pmatrix} 1 & 1 \\ 1 & 0 \end{pmatrix}.$$

Then (α) says: $\forall n \in \mathbf{N} \ (F_{n+1}, F_n) = (F_n, F_{n-1})M$. The q_n we want to study is the dot-product of (F_{n+1}, F_n) with $(F_{n-1}, -F_n)$. That is, it is the matrix

$$(\beta) \qquad q_n = (F_{n+1}, F_n)\begin{pmatrix} F_{n-1} \\ -F_n \end{pmatrix}$$

product of the 1-by-2 matrix (F_{n+1}, F_n) with the 2-by-1 matrix ${}^t(F_{n-1}, -F_n)$, the superscript t denoting transpose.

We play the same trick as for (α) to see that

$$(\gamma) \qquad \begin{aligned} (F_{n-1}, -F_n) &= (F_{n-2}, -F_{n-1})\begin{pmatrix} 0 & -1 \\ -1 & 1 \end{pmatrix} \\ &= (F_{n-2}, -F_{n-1})N, \end{aligned}$$

where we define N to be the matrix shown in (α). If we now use (α) and (γ) to calculate q_n as expressed in (β), we get

$$q_n = (F_n, F_{n-1})M\,{}^tN\begin{pmatrix} F_{n-2} \\ -F_{n-1} \end{pmatrix}.$$

But ${}^tN = N$, and, as you can easily see,

$$MN = \begin{pmatrix} -1 & 0 \\ 0 & -1 \end{pmatrix}.$$

Therefore

$$q_n = (F_n, F_{n-1})\begin{pmatrix} -F_{n-2} \\ F_{n-1} \end{pmatrix}$$
$$= -F_n F_{n-2} + F_{n-1}^2 = -q_{n-1}.$$

Now the result is obvious by induction.

(*ii*) You may use (*i*), since it shows that a linear combination of F_n and F_{n+1} is 1. Or you may see the answer to problem 7.15 of Chapter 6.

(*iii*) Here it's straightforward to let m be any fixed positive integer and then do induction on $n \geq m$.

Another way: Press forward with the matrix M above. Applying (α) to itself, we see that

(δ)
$$(F_{n+1}, F_n) = (F_n, F_{n-1})M = (F_{n-1}, F_{n-2})M^2$$
$$= \cdots = (1,1)M^{n-1} = (1,0)M^n.$$

In particular, if $n \geq m \geq 1$, then

(ϵ) $\quad (F_n, F_{n-1}) = (F_m, F_{m-1})M^{n-m}.$

This equation motivates us to find out what the entries of M^n are. From (δ) we see that

$$(F_{n+1}, F_n) = (1,0)M^n = \text{row 1 of } M^n.$$

Therefore

$$M^n = \begin{pmatrix} F_{n+1} & F_n \\ x & y \end{pmatrix},$$

and we find x and y from (δ):

$$(F_{n+2}, F_{n+1}) = (1,1)M^n$$
$$= (1,1)\begin{pmatrix} F_{n+1} & F_n \\ x & y \end{pmatrix}$$
$$= (F_{n+1} + x, \ F_n + y),$$

from which we get x and y. Thus $\forall n \geq 1$

(ζ)
$$M^n = \begin{pmatrix} F_{n+1} & F_n \\ F_n & F_{n-1} \end{pmatrix}.$$

If we use this in (ϵ), we get the result of (*iii*).

Further comment on (*i*): If you know that the determinant of a product of matrices is the product of the individual determinants, then since $\det(M) = -1$, the result of (*i*) is immediate from (ζ).

(*iv*) Use induction on n. Prove that for all divisors $m \geq 1$ of n, F_m divides F_n. For $n \in \mathbf{N}$ assume the statement true for all $k < n$. Now if $m|n$, then $m|(n-m)$; and by assumption, $F_m|F_{n-m}$. It follows from (*iii*) that $F_m|F_n$.

(*v*) If at least one of the m and n, say m, is zero, then $\gcd(m,n) = m+n = n$ and $F_m = F_0 = 0$, so the assertion is trivial. Assume now that $1 \leq m \leq n$.

From (*iv*) we see that $F_{\gcd(m,n)}$ is a common divisor of F_m and F_n.

In the other direction it follows from (*iii*) that if d is any common divisor of F_m and F_n, then d also divides F_{n-m}, since $\gcd(F_m, F_{m-1}) = 1$. If $n - 2m \geq 0$ we may apply this argument to F_m and F_{n-m} to find that d also divides F_{n-2m}. And so on: For all j such that $n - jm \geq 0$, d divides F_{n-jm}.

Now $\gcd(m,n) = xm - kn$ (or $yn - jm$) for integers $x, y, j, k \geq 0$. Suppose the latter. Then we may apply the argument of the preceding paragraph to F_m and F_{yn} to find that d divides $F_{yn-jm} = F_{\gcd(m,n)}$. QED

*4.4 The generating function is

$$\sum_{0 \leq n} F_n x^n = \frac{x}{1-(x+x^2)} = x\sum_{0 \leq k}(x+x^2)^k;$$

we use the geometric series. If we expand $(x+x^2)^k$ by the binomial theorem and collect terms, we get the assertion.

5.3 When $m = 1$ the expression in ($\star$) is $A(z)/(z - rx)$, and that has the form (5.24) with $m_1 = 1$. If we differentiate (5.24) with respect to z, we get, for a typical term,

$$\frac{d}{dz}\frac{B_i(z)}{(z-rx)^i} = \frac{d}{dz}B_i(z)/(z-rx)^i - iB_i(z)/(z-rx)^{i+1},$$

where $1 \leq i \leq m$. If we sum over i and combine terms, we get the form (5.24) for $m+1$ instead of m. The reason is that $\frac{d}{dz}B_i(z)$ is also a rational function. (The second sentence begins the inductive step of a proof.)

G1. You could find a recurrence for a_n by using that for F_n. A second way is to express F_n as $F_n = c_1 r_1^n + c_2 r_2^n$ for appropriate c_1, c_2 and sum the resulting finite geometric series. The third way is to use a general fact about generating functions:

If $g(x) = \sum_{0 \leq n} g_n x^n$, then $\dfrac{g(x)}{1-x} = \sum_{0 \leq n}(g_0 + \cdots + g_n)x^n$.

You can prove this easily, because it depends merely

on the fact that $1/(1-x) = \sum_{0\le k} x^k$. In our case $g(x)$ is $x/(1-x-x^2)$, so the generating function $g_1(x)$ for a_n is

$$g_1(x) := \sum_{0\le n} a_n x^n = \frac{x}{(1-x)(1-x-x^2)}.$$

This is just a way to shortcut finding the recurrence satisfied by $\langle a_n \rangle$. Now we don't have to work it out from scratch; we set

$(\star)$

$$g_1(x) = \frac{x}{(1-x)(1-x-x^2)} = \frac{A}{1-x} + \frac{Bx+C}{1-x-x^2}.$$

We find A, B, and C: Set $x := 0$ to get

$$0 = A + C.$$

Multiply by $1-x$ and then set $x := 1$. This gives $A = -1$. So $C = 1$. Now multiply by $1-x-x^2$ and set $x := r_1 :=$ a root of $1-x-x^2$. This gives

$$Br_1 + C = \frac{r_1}{1-r_1}.$$

Solving yields $r_1 B = (2r_1 - 1)/(1-r_1)$; thus, if r_2 is the other root of $1-x-x^2$,

$$B = \frac{2-r_1^{-1}}{1-r_1} = \frac{2+r_2}{1-r_1}$$

since $r_1 r_2 = -1$. And now we see that $B = 1$, because $r_1 + r_2 = -1$. So we may write $(\star)$ as

$$g_1(x) = -\sum_{0\le n} x^n + \frac{x}{1-x-x^2} + \frac{1}{1-x-x^2}.$$

But we know from the text that

$$\frac{x}{1-x-x^2} = \sum_{0\le n} F_n x^n.$$

Divide by x. Since $F_0 = 0$ the result is still a power series:

$$\frac{1}{1-x-x^2} = \sum_{0\le n} F_n x^{n-1} = \sum_{0\le n} F_{n+1} x^n.$$

Put these three together to see that

$$g_1(x) = \sum_{0\le n} (-1 + F_n + F_{n+1}) x^n$$
$$= \sum_{0\le n} (F_{n+2} - 1) x^n.$$

Therefore

$$a_n := F_0 + \cdots + F_n = F_{n+2} - 1.$$

(The first thing to do with this problem is to work out the first several values of a_n. Then you could guess the result. That way you wouldn't quit too soon in using any of the foregoing methods.)

Chapter 12

4.4 (*i*) By definition of reflexivity, $\forall a \in A$ $(a,a) \in R$. By definition of converse, $R^c := \{ (y,x);\ (x,y) \in R \}$.

$$\therefore \forall a \in A, \quad (a,a) \in R^c, \quad \text{proving } (i).$$

(*ii*) We must show that for all $(x,y) \in R \circ R^c$, (y,x) is also in $R \circ R^c$.

Let $(x,y) \in R \circ R^c$. This is true iff $\exists a \in A$ such that

$$(x,a) \in R \quad \text{and} \quad (a,y) \in R^c.$$
(From the definition of $\circ$.)

From $(x,a) \in R$ follows $(a,x) \in R^c$, and from $(a,y) \in R^c$ follows $(y,a) \in R$. The composition $R \circ R^c$ thus contains (y,x), proving (*ii*). (Notice we did not use reflexivity.)

(*iii*) We suspect $R \circ R^c$ is not necessarily transitive, because it's not assumed and ought to be independent of reflexivity and symmetry. So we seek a counterexample. Set $A := \{a,b,c,d\}$, $I_A := \{(x,x); x \in A\}$, and $R := \{(b,a),(a,c),(b,d)\} \cup I_A$. This R is reflexive.

Then $R^c = \{(a,b),(c,a),(d,b)\} \cup I_A$, and $R \circ R^c = \{(a,b),(b,a),(a,c),(c,a),(b,d),(d,b)\} \cup I_A$. Both (a,b) and (b,d) are in $R \circ R^c$, but (a,d) is not there, so it's not transitive.

Therefore $R \circ R^c$ is not necessarily an equivalence relation, even when R is reflexive.

4.6 It all hinges on the intersection of the range of R with the domain of S. Set

$$I := (\text{range of } R) \cap (\text{domain of } S).$$

Then define subrelations $R' \subseteq R$ and $S' \subseteq S$ as

$$R' := \{ (a,b);\ (a,b) \in R,\ b \in I \}$$
$$S' := \{ (b,c);\ (b,c) \in S,\ b \in I \}.$$

Then domain $R \circ S =$ domain R' and range $R \circ S =$ range S'.

10.4 If A is empty then R and its complement R' are empty. Both are vacuously reflexive, symmetric, transitive, and antisymmetric.

If A is not empty, then no $(a,a) \in R'$. So R' is not reflexive. R' is symmetric, however, because $\forall (a,b) \notin R$, we have $(b,a) \notin R$. (Why? Because R is symmetric. Therefore $(b,a) \in R \longrightarrow (a,b) \in R$, the contrapositive.)

If R is transitive, R' need not be transitive. Consider the example

$$R := \{ (a,b), (b,c), (a,c), (c,a) \}.$$

Then $R' \supseteq \{ (b,a), (c,b) \}$, but $(c,a) \notin R'$.

If R is antisymmetric, then R' need not be so. We could take $R := \{ (a,a), (b,b) \}$. Then $R' = \{ (a,b), (b,a) \}$.

14.4 *i)* We must prove $L_2 \subseteq L_1$. Let $(a,b) \in L_2$. Since $a,b \in A$, and L_1 is a linear order on A, aL_1b or bL_1a. If aL_1b, we're done. If bL_1a, then $(b,a) \in L_1$ and since $L_1 \subseteq L_2$, $(b,a) \in L_2$. $\therefore a = b$ (since $(a,b) \in L_2$), and $\therefore (a,b) \in L_1$.

(ii) If $L_1 \neq L_2$ suppose (a,b) is in one of L_1, L_2 but not in the other. Then $(a,b) \notin L_1 \cap L_2$. If $L_1 \cap L_2$ were linear, then $(b,a) \in L_1 \cap L_2$. Since $(a,b) \in L_1$ (say), we get $a = b$ and $\therefore$ $(a,b) \in L_1 \cap L_2$ after all. This contradicts our hypothesis; $\therefore L_1 \cap L_2$ is not linear.

17.3 *(i)* Consider the relation-matrix, a 2×2 matrix of 0's and 1's. Since there are four entries in it, there are 2^4 different matrices, hence 16 relations on a 2-set.

(ii) Let the 2-set be $\{a,b\}$ and the relation R. If neither (a,b) nor (b,a) is in R, then R is transitive, because $R \subseteq \{ (a,a), (b,b) \}$. There are four such R.

If only one of (a,b) and (b,a) is in R, then R is transitive. (Check this.) There are eight such R, four with (a,b) and four with (b,a).

If both (a,b) and (b,a) are in R, then, as explained in Chapter 4, (2.13), transitivity forces both loops to be present. There's one such R, so there are 13 transitive relations on a 2-set.

19.2 Denote $S := R \cup \ldots \cup R^{j-1}$. We shall prove that, for all $m \geq j$, $R^m \subseteq S$. We use induction on m.

If $m = j$, that is our hypothesis: $R^j \subseteq S$. For general m, assume $R^m \subseteq S$. Consider

$$R^{m+1} := R \circ R^m.$$

Let $(a,c) \in R^{m+1}$. Thus $\exists b \in A$ such that $(a,b) \in R$ and $(b,c) \in R^m$. But since $R^m \subseteq S$, $(b,c) \in R^k$ for some $k < j$. $\therefore (a,c) =$ "$(a,b) \circ (b,c)$" $\in R \circ R^k = R^{k+1}$. Since $k+1 \leq j$, $(a,c) \in S$, either directly (if $k+1 < j$) or via $R^j \subseteq S$ (if $k+1 = j$).

QED

Chapter 13

4.4 *(i)* $\min_n = 1$. You may explain.

$*$*(ii)* (This problem gives you great guess-then-prove practice.)

Drawing shows you that $\max_n$ is 1 for $n = 1$ and 2, and 2 for $n = 3$ and 4. You can easily show that for any $n \geq 0$ there is a binary tree on n nodes with $\lceil \frac{n}{2} \rceil$ leaves, so $\max_n \geq \lceil \frac{n}{2} \rceil$. And an obvious guess is that: $\max_n = \lceil \frac{n}{2} \rceil$.

We'll use induction to prove that $\max_n \leq \lceil \frac{n}{2} \rceil$. So suppose (by way of contradiction) that n is the least integer for which $\max_n > \lceil \frac{n}{2} \rceil$. Let B be a binary tree on n nodes having more than $\lceil \frac{n}{2} \rceil$ leaves.

Remove a leaf from B, getting a binary tree B' on $n-1$ nodes having either the same number of leaves as B (an immediate contradiction), or one fewer. Let $n_0(B)$ $[n_0(B')]$ denote the number of leaves of B $[B']$.

If one fewer, then $n_0(B') \geq \lceil \frac{n}{2} \rceil$. But $n_0(B') \leq \max_{n-1}$ leaves, and by the inductive assumption, $\max_{n-1} \leq \lceil \frac{n-1}{2} \rceil$. That is,

$$\left\lceil \frac{n}{2} \right\rceil \leq n_0(B') \leq \left\lceil \frac{n-1}{2} \right\rceil,$$

which forces n to be even.

Back to B: Using Theorem (4.6), problem 4.3, we see that

$$n_0(B) = 1 + n_2(B).$$

Since n is even, we know that

$$n_2(B) = n_0(B) - 1 \geq \left\lceil \frac{n}{2} \right\rceil = \frac{n}{2}$$

and thus that

$$n \geq n_0(B) + n_2(B) > \left\lceil \frac{n}{2} \right\rceil + \left\lceil \frac{n}{2} \right\rceil = n,$$

a contradiction. Hence $\max_n \leq \lceil \frac{n}{2} \rceil$ for all n. Therefore equality holds, since we have already noted that $\max_n \geq \lceil \frac{n}{2} \rceil$.

(*iii*) One easy construction: Set $j = \lceil \frac{m}{2} \rceil$. Construct a binary tree B on m nodes with j leaves. Then run a *chain* of $n - m$ nodes upward from the root of B. That is, make each of the $n - m$ nodes have just 1 son.

***4.6** The values taken by the height function on such trees are $\lfloor \log_2 n \rfloor, \ldots, \lfloor \frac{n}{2} \rfloor$.

Denote by n_i the number of nodes with i sons, as usual. Remember that since $n_0 = \lceil \frac{n}{2} \rceil$, $n_1 = 1$ if n is even and $n_1 = 0$ if n is odd.

If n is odd, we get maximum height by putting just two nodes at each level after level 0. So the height is $(n-1)/2 = \lfloor \frac{n}{2} \rfloor$.

If n is even, there's just one node, say the root, having only one son, and the rest all have 2 or 0 sons. The subtree with x as root has an odd number of nodes, so it fits the previous case. The height is $\frac{n}{2} = \lfloor \frac{n}{2} \rfloor$.

We now prove by induction that any level between these extremes is possible. If n is even, we choose by induction a left subtree on $n-2$ nodes of height j for any j satisfying $\lfloor \log_2(n-2) \rfloor \le j \le \lfloor \frac{n-2}{2} \rfloor$. Then our tree has height $1 + j \le \lfloor \frac{n}{2} \rfloor$. If n is odd, we choose a left subtree on $n - 1$ nodes and proceed similarly.

5.4 We look at what β does. Suppose the forest F consists of trees $T_1, \ldots, T_m$, with a as the root of T_1. And suppose that the immediate subtrees of a are $T'_1, \ldots, T'_n$. Then $\beta(F)$ is a binary tree with root a; the left subtree of a is $\beta(T'_1, \ldots, T'_n)$, and the right subtree of a is $\beta(T_2, \ldots, T_m)$.

To prove β injective, let F_1 and F_2 be forests on X. Then $\beta(F_1) = \beta(F_2)$ implies the roots of the first trees are the same, say a. Now by induction on the number of nodes it follows that $F_1 = F_2$.

For surjectivity, let B be any binary tree with nodes in X. Again we use induction on the number of nodes. To define a forest F such that $\beta(F) = B$, take $r :=$ root of B, and set r as the root of a tree T_1. Now we seek a forest $F' = T'_1, \ldots, T'_n$ such that $\beta(F')$ is the left subtree of r, and F' exists by the inductive hypothesis. Similarly $\exists$ the forest F'' such that $\beta(F'')$ = the right subtree of r. We take F as the forest T_1, F'', where T_1 is the tree with root r and ordered set of immediate subtrees F'.

6.6 We use induction on the number of nodes. The basis case is obvious. In general, consider any binary tree B on $n \ge$ nodes, with left and right subtrees T_L and T_R. Since every leaf of B is a leaf of T_L or T_R, and in each of the three traversal rules we visit the left subtree before the right, and the inductive hypothesis tells us immediately that in T_L the assertion is true, also in T_R, the assertion follows for B.

***6.11** Suppose b is a descendant of a. In doing the preorder [postorder] traversal of the subtree with root a, we visit a before [after] visiting the subtrees where b is.

Conversely, if a precedes b in preorder, either a is an ancestor of b, or a and b are in the left and right subtrees, respectively, of their "lowest" common ancestor (the one of greatest level).

If also a follows b in postorder, then either a is an ancestor of b, or a and b are in the right and left subtrees, respectively, of their lowest common ancestor.

These latter possibilities are contradictory, so a is an ancestor of b.

11.3 (*i*) It's easy to argue on the binomial coefficient $\binom{2n}{n}$, but it's easier to say that the set of all binary trees on a positive even number of nodes has no fixed trees under reflection (about the vertical line through the root), since one subtree has an odd number of nodes and the other an even number.

*(*ii*) The number b_n of binary trees on n nodes is odd iff the total number of them which are their own reflection is odd. Call such a binary tree a *fixed tree*, for short.

We can easily produce all the fixed trees on n nodes (for n odd) as follows: They are a root, together with a left subtree B_L on $(n-1)/2$ nodes, and the right subtree is *defined* as the reflection of B_L.

Notice that B_L may be any tree of $(n-1)/2$ nodes. So b_n is odd iff $b_{(n-1)/2}$ is odd. The problem has become recursive: $b_1 = 1$ is odd. therefore b_3 is odd. $\therefore$ b_7 is odd, and so on. We've proved that if $n + 1 = 2^k$, then b_n is odd.

The converse is almost as obvious. If b_n is odd, then $b_{(n-1)/2}$ is odd. So we let $n + 1 = 2^k \cdot m$ for some odd $m \ge 1$. Then k successive applications of the function $f(x) := (x-1)/2$ to n yield $m-1$, an even integer; and we know b_{m-1} is even unless $m = 1$. $\therefore m = 1$.

11.8 This is another "guess-then-prove" problem. Since the height is minimum, it must be like the complete binary tree except that the leaves on the bottom row may be placed anywhere. You can work out values for h_n as follows, using direct counting:

n	1	2	3	4	5	6	7	8	9
h_n	1	2	1	4	6	4	1	8	28.

The table suggests that from $n = 2^k - 1$ to $n = 2^{k+1} - 1$ the values are the binomial coefficients $\binom{2k}{m}$ for $m = 0, 1, \dots, 2^k$.

So let $2^k - 1 \le n \le 2^{k+1} - 1$, setting $n = 2^k - 1 + m$ for $0 \le m \le 2^k$. The lowest full row of the tree has 2^{k-1} nodes. Each of these nodes may have two sons, making 2^k possible places where the m leaves may be assigned. Each m-subset of these 2^k places determines a tree on n nodes that we are trying to count. So there are $\binom{2^k}{m}$ such binary trees.

11.9 (*i*) 2^{n-1}, since after the root we have two choices for each son—will it be left or right? We never have a node with two sons, for that would lead eventually to two leaves.

*(*ii*) Using an analysis like that in the text to count binary trees, you can derive this recursion for t_n, the number of binary trees on n nodes with exactly two leaves:

$$t_n = 2t_{n-1} + (n-2)2^{n-3}.$$

You could solve this by methods of Chapter 11.

Another way is to count them directly. Since $n_0 = 1 + n_2 = 2$, there is exactly one node with two sons. Call that node z, and let it be at level k; then $0 \le k \le n-3$.

The two subtrees of z have n' and n'' nodes, respectively, and

$$(\star) \qquad n' + n'' = n - k - 1.$$

Each of these subtrees is a tree with only one leaf. Thus there are, from (*i*),

$$2^{n'-1} \cdot 2^{n''-1} = 2^{n'+n''-2} = 2^{n-k-3}$$

ways to assign these subtrees for each choice of n' and n'' satisfying ($\star$). Since n' and n'' must be positive, n' runs over $1, \dots, n-k-2$. ∴ there are

$$(n-k-2)2^{n-k-3}$$

such binary trees with z at level k.

There are also 2^k ways to choose the chain from root to z, making

$$(\star\star) \qquad (n-k-2)2^{n-3}$$

binary trees altogether with z at level k. We add these terms ($\star\star$) over $0 \le k \le n-3$ to find

$$t_n = 2^{n-3}(1 + 2 + \cdots + n - 2) = (n-2)(n-1)2^{n-4}.$$

(As a check, this t_n satisfies the recurrence and has the correct values for $n = 1$ and $n = 2$.)

Chapter 14

2.10 Any simple graph on n vertices arises from the complete graph K_n on n vertices on removal of some (or none) of its edges. K_n has $\binom{n}{2}$ edges. K_4 has six edges. It's intuitively clear that removing one edge from K_4 yields the same graph as removing another. But let's nail it down: Take the vertex set as $V = \{a, b, c, d\}$. We have K_4 on V, and we remove edge $\{a, b\}$ to get a graph G on V with five edges. We rename the vertices with an arbitrary permutation π. If we want to map G to the graph G' obtained on removal of edge $\{x, y\}$ from K_4, where $x \ne y$ are in V, we just set $\pi(a) := x$, $\pi(b) := y$ and $\pi(c)$ and $\pi(d)$ equal to the other two points of V.

3.1 If you start at v, you can go to any other vertex, and from that to any of the $n-2$ remaining, and so on. You may choose v in n ways, the next in $n-1$ ways, and so on, resulting in $n!$ Hamiltonian cycles.

If you prefer to regard two cycles as the same if they differ only in the naming of the starting vertex, then each one appears n times in the above tally, so the answer would be $(n-1)!$.

5.2 We classify the subgraphs by the size of their vertex-sets. By definition, a graph may not be empty. There are $\binom{n}{k}$ choices of vertex-set of size k, and for each such choice there are $\binom{k}{2}$ edges of K_n available for our choice as edges of the subgraph. There are $2^{\binom{k}{2}}$ different choices of edge-set. Therefore the answer is

$$\sum_{1 \le k \le n} \binom{n}{k} 2^{\binom{k}{2}}.$$

6.2 Just a hint: You know how many edges a spanning tree has. Think of the spanning tree as the result of removal of edges from G_2. The number removed is the same in each case. So, for the appropriate k, how many k-sets of edges of G_2 can be removed to give a spanning tree? Suppose you remove a k-set and the graph G' that's left is not a spanning tree? What is the most obvious property of G' in that case?

∗G3. *(ii)* Proof by induction on the number e of edges of G. If $e = 0$, the result is vacuously true. Now let G have $e \geq 2$ edges.

Note: if the orientation of the problem is possible in a graph G, then at every vertex u of degree 1 in G, the edge at u is oriented with the arrow pointing to u.

Let v be a vertex of maximum degree $k > 1$ in G. Divide G into k subgraphs $G_1, \ldots, G_k$ each with v as a vertex of degree 1 but with no other vertices in common. (For example, the *star* graph S_4, consisting of one vertex v of degree 4 and four vertices 1, 2, 3, 4 each of degree 1, splits into four free trees v o———o i, for $i = 1, 2, 3, 4$.) This division of G is possible because G is a free tree.

Case 1. One of the G_i's, say G_1, has an even total number of edges: We join $G_2, \ldots, G_k$ at v to get a subgraph G' on an even number $e' < e$ of edges. The same is true of G_1. The inductive assumption allows now an orientation for each of G' and G_1. In G_1 the outdegree of v is 0, so when we join G' and G_1 at v to recover G, we see that G is now oriented.

Case 2. For each $i = 1, \ldots, k$, G_i has an odd total number of edges: This forces k to be even.

Case 2a. $k = 2$: Since k is the maximum degree, G is a chain; and the result is obvious.

Case 2b. $k > 2$: Thus $k \geq 4$. We join G_1 and G_2 at v, and $G_3, \ldots, G_k$ at v, to get free trees G' and G'' each with an even total number of edges less than e. By induction both G' and G'' have orientations. We join G' and G'' at v and recover G, and G now has an orientation.

Index

A

B

C

D

E

F

G

H

I

J

K

L

M

N

O

Q

R

S

T

U

V

W

Z